The Olive

The Olive

Botany and Production

Edited by Andrea Fabbri

Co-editors Luciana Baldoni,
Tiziano Caruso, Franco Famiani

CABI

CABI is a trading name of CAB International

CABI
Nosworthy Way
Wallingford
Oxfordshire OX10 8DE
UK

Tel: +44 (0)1491 832111
E-mail: info@cabi.org
Website: www.cabi.org

CABI
200 Portland Street
Boston
MA 02114
USA

Tel: +1 (617)682-9015
E-mail: cabi-nao@cabi.org

A catalogue record for this book is available from the British Library, London, UK.

ISBN-13: 9781789247336 (hardback)
9781789247343 (ePDF)
9781789247350 (ePub)

DOI: 10.1079/9781789247350.0000

Commissioning Editor: Rebecca Stubbs
Editorial Assistant: Emma McCann
Production Editor: James Bishop

Typeset by Straive, Pondicherry, India
Printed and bound by CPI Group (UK) Ltd, Croydon, CR0 4YY

Contents

List of Contributors

Giovanni Enrico Agosteo, Department of Agriculture, Mediterranean University, Reggio Calabria, Italy
Luciana Baldoni, Institute of Biosciences and BioResources, National Research Council, Perugia, Italy
Giuseppe Barbera, Department of Agricultural, Food and Forest Sciences, University of Palermo, Italy
Karim Barkaoui, CIRAD, UMR ABSys, INRAE and Institut Agro, Montpellier, France
Angjelina Belaj, IFAPA – Centro Alameda del Obispo, Córdoba, Spain
Antonio Belcari, Department of Agriculture, Food, Environment and Forestry (DAGRI), University of Florence, Italy
Giora Ben-Ari, Institute of Plant Sciences, ARO, The Volcani Center, Israel
Alon Ben-Gal, Gilat Research Center, Volcani Institute, Ministry of Agriculture, Israel
Giovanni Benelli, Department of Agriculture, Food and Environment, University of Pisa, Italy
Rita Biasi, Department for Innovation in Biological, Agro-Food and Forest Systems, Tuscia University, Viterbo, Italy
Iris Biton, Institute of Plant Sciences, ARO, The Volcani Center, Israel
Konstantinos Blazakis, Department of Horticultural Genetics and Biotechnology, Mediterranean Agronomic Institute, Chania, Greece
Aureliano Bombarely, Department of Bioscience, University of Milan, Italy
Antonio Brunori, Programme for Endorsement of Forest Certification Schemes, Italy
Santa Olga Cacciola, Department of Agriculture, Food and Environment, University of Catania, Italy
Salvatore Camposeo, Dipartimento di Scienze del Suolo, della Pianta e degli Alimenti, Università Aldo Moro, Bari, Italy
Angelo Canale, Department of Agricultural, Food and Agro-Environmental Sciences, University of Pisa, Italy
Giovanni Caruso, Department of Agricultural, Food and Agro-Environmental Sciences, University of Pisa, Italy
Tiziano Caruso, Department of Agricultural, Food and Forestry Sciences, Palermo University, Italy
Sergio Castro-Garcia, Departamento de Ingeniería Rural, Universidad de Córdoba, Spain

Nicola Cinosi, Dipartimento di Scienze Agrarie, Alimentari e Ambientali, Università degli Studi di Perugia, Italy
Arnon Dag, Gilat Research Center, Volcani Institute, Ministry of Agriculture, Israel
Raul de la Rosa, Center 'Alameda del Obispo', Andalusian Institute of Agricultural and Fisheries Research and Training (IFAPA), Córdoba, Spain
Ran Erel, Gilat Research Center, Volcani Institute, Ministry of Agriculture, Israel
Andrea Fabbri, Department of Food and Drug, University of Parma, Italy
Franco Famiani, Department of Agricultural, Food and Environmental Sciences, Perugia University, Italy
Daniela Farinelli, Department of Agricultural, Food and Environmental Sciences, Perugia University, Italy
Louise Ferguson, Department of Plant Sciences, University of California at Davis, USA
James Friel, Department of Bioscience, University of Milan, Italy
Tommaso Ganino, Department of Food and Drug, University of Parma, Italy
Jesus A. Gil-Ribes, Departamento de Ingeniería Rural, Universidad de Córdoba, Spain
José Alfonso Gómez Calero, Instituto de Agricultura Sostenible, CSIC, Córdoba, Spain
Riccardo Gucci, Department of Agriculture, Food and Environment, University of Pisa, Italy
Consolación Guerrero, University of Malaga, Department of Molecular Biology and Biochemistry, Malaga, Spain
Panagiotis Kalaitzis, Department of Horticultural Genetics and Biotechnology, Mediterranean Agronomic Institute, Chania, Greece
Maurizio Lambardi, Institute of Bioeconomy, National Research Council, Sesto Fiorentino, Italy
Pierre-Eric Lauri, ABSys, CIHEAM-IAMM, CIRAD, INRAE, Montpellier, France
Lorenzo León, Center 'Alameda del Obispo', Andalusian Institute of Agricultural and Fisheries Research and Training (IFAPA), Córdoba, Spain
Riccardo Lo Bianco, Department of Agricultural, Food and Forestry Sciences, Palermo University, Italy
Enrico Maria Lodolini, Centro di Ricerca Olivicoltura, Frutticoltura e Agrumicoltura, CREA, Roma, Italy
Francisco Luque, Center for Advanced Studies in Olive Grove and Olive Oils, Genomics Unit, University of Jaén, Spain
Hanene Mairech, Department of Agricultural, Food and Environmental Sciences, Perugia University, Italy
Giulia Marino, Department of Plant Sciences, University of California at Davis, USA
Roberto Mariotti, Institute of Biosciences and Bioresources, National Research Council, Perugia, Italy
Francesco Paolo Marra, Department of Architecture, University of Palermo, Palermo, Italy
G. Medina-Alonso, Servicio Técnico de Agricultura y Desarrollo Rural, Cabildo Insular de Tenerife, Santa Cruz de Tenerife, Spain
Jose A. Mercado, Department of Botany and Plant Physiology, University of Malaga, Spain
Maurizio Micheli, Department of Agricultural, Food and Environmental Sciences, University of Perugia, Italy
Soraya Mousavi, Institute of Biosciences and Bioresources, National Research Council, Perugia, Italy
Monji Msallem, International Consultant – Olive Development, Tunis, Tunisia
Dvora Namdar, Institute of Plant Sciences, The Volcani Center, ARO, Israel
Isabel Narváez, Department of Botany and Plant Physiology, University of Malaga, Spain
Davide Neri, Dipartimento di Scienze Agrarie, Alimentari e Ambientali, Università Politecnica delle Marche, Ancona, Italy
Elena Palomo-Ríos, Department of Botany and Plant Physiology, University of Malaga, Spain

Ruggero Petacchi, Institute of Life Sciences, Sant'Anna School of Advanced Studies, Pisa, Italy
Malayka Samantha Picchi, Institute of Life Sciences, Sant'Anna University, Pisa, Italy
Pierluigi Pierantozzi, Instituto Nacional de Tecnología Agropecuaria, San Juan, Argentina
Amalia Rosa Maria Piscopo, Department of Agriculture, Mediterranean University, Reggio Calabria, Italy
Fernando Pliego-Alfaro, Department of Botany and Plant Physiology, University of Malaga, Spain
Primo Proietti, Department of Agricultural, Food and Environmental Sciences, Perugia University, Italy
Alessandro Ragazzi, Dipartimento di Scienze e Tecnologie Agrarie, Alimentari, Ambientali e Forestali, University of Florence, Italy
Pilar Rallo, Department of Agronomy, University of Seville, Spain
Hava F. Rapoport, Department of Crop Protection, Institute for Sustainable Agriculture (IAS-CSIC), Cordoba, Spain
Luca Regni, Department of Agricultural, Food and Environmental Sciences, Perugia University, Italy
Laia Ribalta, Department of Botany and Plant Physiology, University of Malaga, Spain
Adolfo Rosati, Centro di Ricerca Olivicoltura, Frutticoltura e Agrumicoltura (CREA), Spoleto, Italy
Patrizia Sacchetti, Department of Agriculture, Food, Environment and Forestry (DAGRI), University of Florence, Italy
Luca Sebastiani, Crop Science Research Center, Sant'Anna University, Pisa, Italy
Alicia Serrano, Center 'Alameda del Obispo', Andalusian Institute of Agricultural and Fisheries Research and Training (IFAPA), Córdoba, Spain
Giuseppe Surico, Dipartimento di Scienze e Tecnologie Agrarie, Alimentari, Ambientali e Forestali, University of Florence, Italy
Luca Testi, Department of Agronomy, Institute for Sustainable Agriculture (IAS-CSIC), Cordoba, Spain
Maria Cristina Valeri, Institute of Biosciences and Bioresources, National Research Council, Perugia, Italy
Andrea Vitale, Olive Technical Consultant, Pistoia, Italy
Uri Yermiyahu, Gilat Research Center, Volcani Institute, Ministry of Agriculture, Israel
Hande Yilmaz-Düzyaman, Center 'Alameda del Obispo', Andalusian Institute of Agricultural and Fisheries Research and Training (IFAPA), Córdoba, Spain
Valeria Zeni, Department of Agriculture, Food and Environment, University of Pisa, Italy
Isaac Zipori, Gilat Research Center, Volcani Institute, Ministry of Agriculture, Israel

Presentation

'... olea, quae prima omnium arborum est.' ['...the olive, which is the first among all trees.'] (Columella, *De Agricoltura*, V, 8:1)

That in the first century AD in Rome the importance and value of the olive tree and of its product, olive oil, was so much reputed, should come as no surprise. The olive had been one of the first domesticated trees in the Mediterranean basin, thousands of years before then, and its cultivation was well known, together with the properties of the oil obtained from the fruit. That it was easy to grow, and that it was always bountiful in spite of the difficult conditions in which it was often grown, was a commonplace for growers; also, because at that time it was not infested by the pests that would attack it in later centuries.

What was felt in those years was the necessity to transmit the knowledge of the crop's cultivation, in writing, and not just orally.

From Theophrastus (4th century BC) until the Roman Empire, scholars have produced treatises on cultivation of the most important crops, among which was the olive. Unfortunately, most of that production was lost, Columella's work being one of the few exceptions.

Olive and, in general, fruit-tree cultivation, lost importance in the Middle Ages and resumed agricultural relevance starting in the 15th century, for a number of reasons that are described in Chapter 1.

It was at that time that the need was felt to collect in specialized treatises the knowledge of fruit-tree cultivation. Most literature was produced for the grapevine, but olive always had a space, albeit smaller. In the 18th and 19th centuries, the moment had come for a more detailed collection of the science of the olive, and very detailed and reliable texts are produced, such as Presta's *Memoria intorno ai sessantadue saggi diversi di olio* (Napoli, 1788) and De Herrera's volume on olives in *Agricultura General* (1818, former edition 1513). The most authoritative, and modern, text of these times is probably the *Trattato Teorico-Pratico Completo sull'Olivo* (1819) by Giuseppe Tavanti, who wrote it after appointment by the Accademia Economico-Agraria dei Georgofili of Firenze. Then, many texts on olive were eventually produced by scholars, especially in the first half of the 20th century. Most are of a high level, and among them stands out the work of Morettini – *Olivicoltura* (1950), a text that was a reference for generations of scholars and olive growers.

Many textbooks of a good level have been published in Italy, Spain, France, Greece and the USA. All of them are characterized by a peculiar attention to the aspects of cultivation, of the germplasm, and of other related issues, of the specific country.

This book, besides being an update on research and technology of olive culture, has aimed at overcoming this limitation, trying to coalesce the best scientific resources in the olive research world, to produce a really international textbook, able to provide scientists, students and growers with an updated, non-parochial and easy-to-handle instrument. We hope that we have, at least in part, fulfilled this goal, and that the olive world will appreciate our efforts.

The Editors

1 A Short History of the Olive

Andrea Fabbri
Department of Food and Drug, Parma University, Italy

1.1 Introduction

The cultivated olive (*Olea europaea* L. subsp. *europaea* var. *europaea*, Family *Oleaceae*) is a tree species of the Mediterranean basin that, more than others, well represents the Mediterranean environment and climate, the latter characterized by warm-wet winters and hot-dry summers, to the point of being considered a bio-indicator (Fall *et al.*, 2002; Moriondo *et al.*, 2013) as well as one of the characteristic crops of the Mediterranean regions, where oil largely replaces butter and other animal fats, and is widely exported. The olive oil of the region represented almost 98% of all world production in 2018 (International Olive Council). Another important product is table olives, also mainly produced in the Mediterranean basin and in the Near East (96% in 2018). The importance of the olive tree in the Mediterranean area goes beyond its productive aspects, and over the millennia the tree has been endowed with symbolic, cultural and religious aspects that have characterized all societies and cultures developed around the Mediterranean Sea.

The definition of Mediterranean climate has been the object of many studies, but for botanists and agronomists there are a few spontaneous or cultivated species, which thrive from sea level up to 800–1000 m asl, all around the Mediterranean basin, that represent the said environment.

With reference to fruit crops, the most typical are fig, grape, pomegranate, olive, and date palm, which were the first to be domesticated during the so-called Neolithic Revolution, and which are, in their wild form, endemic in more or less wide areas of the region. Of the two main crops of the list, grapevine has its natural environment more shifted to the north, and can be grown only in the northernmost shores of North Africa (as the southern limit), while wild grape can be found in inland areas of central Europe. As concerns cultivation, fine wines are produced as far north as the Moselle valley and southern-central England. The olive instead is grown on all the coastal areas of the Mediterranean, with few exceptions, from Andalusia and Morocco to Egypt's Suva Oasis, from Provence to Greek and Turkish Thrace.

This suitability of the olive to define the Mediterranean has been the object of many interventions of botanists, geographers and historians. According to the work of Zohary and Spiegel-Roy (1975), the limit of the wild olive (oleaster) is in coastal areas south of 42° N. Nevertheless, the olive is cultivated, thanks to man's intervention, far north of this latitude, due to cultivation techniques, cultivar improvement and careful choice of locations. Traditionally, the

DOI: 10.1079/9781789247350.0001

northernmost latitude for its cultivation is considered to be 45° N, but only if in the proximity of seas or large lakes. Thus, for Morettini (1950), the olive can be grown in the Mediterranean between latitude 30° and 45°. He divides the area of cultivated olive in three subzones, of different suitability and productivity; while the first subzone coincides with the habitat of wild olive, the third includes all coastal areas of the northern Mediterranean, mostly Spain and Italy. With the exception of the lake districts, northern Italy is not included. A similar conclusion had been reached by Briccoli (1925). However, the cultivated olive distribution includes quite varied climates, especially as concerns the amount and distribution of rains, and Morettini's subdivision simply mirrored the distribution of cultivated olive in the mid-20th century.

It is therefore the cultivated olive that can, for many, define, from a bio-agronomical point of view, the Mediterranean world (Fig. 1.1).

> In trying to determine the extent of the Mediterranean climatic type, the distribution of the olive ... has proved more useful than purely statistical criteria. It will flourish in almost all the soils likely to be encountered in the Mediterranean ... but as regards its climatic requirements it is much more demanding. As long as there is enough sunshine in the daytime to warm it up, it will tolerate conditions in which the mean figure for the coldest month is as low as 39°F [3.9°C]; it demands a hot, dry summer sufficiently prolonged to ensure the maturation of the fruit ... [A]lthough light snowfall does no serious harm a really severe frosty period ruins the tree by bursting the tissues and splitting open the trunk and branches. Its usefulness as a Mediterranean yardstick is enhanced by taking into account automatically the important modifications introduced by altitude and the limitations imposed by infrequent but none the less critical periods of exceptional weather. The olive takes many years to mature and the farmer is not likely to plant if he knows that once every ten years or so he runs the risk of fatally severe weather. (Walker, 1960)

'The Mediterranean ends where the olive tree no longer grows.' Duhamel's (1947) sentence shows that the association of the olive with the Mediterranean basin, and with the cultures that flourished on its shores, is an old one, and is justified by the enormous importance of the olive tree and its oil for those cultures.

The cultivation of the olive tree since the first millennia of the Neolithic Revolution, in the Chalcolitic Age, has allowed the species to overcome its natural bio-climatic limits and to be grown at higher altitudes and latitudes, with the result that the distribution of the olive tree scarcely reflects that of the wild genotypes. Recently, it has been hypothesized that even northern Italy, due to an uninterrupted presence of the olive, might be considered part of the Mediterranean (Fabbri, 2017).

1.2 The Symbolism of the Olive in Religions and Legends

In many civilizations (Phoenician, Hebrew, Egyptian, Greek, Roman, Arab, etc.) and religions, the olive tree has represented a symbol of human life, and of man's relationship with the divinity.

The Jews, a people of nomadic shepherds, become sedentary when provided with olive groves, left by the land's former Canaanite occupants; as Joshua (24:13) reminds them: 'I gave you a land whereon thou hadst not labored, and cities which ye built not, and ye dwell therein; of vineyards and oliveyards which ye planted not do ye eat' (see also Deuteronomy 6:11, 8:8).

The olive is often mentioned in the Old Testament. An olive twig was brought back to Noah by a dove to demonstrate that the flood was over (Genesis 8:11). The olive is listed in Deuteronomy 8:8 as one of the seven species that are noteworthy products of the land of Israel. That it is not just an ordinary tree for the Jews is also shown in Judges 9:8–9:

> One day the trees went out to anoint a king for themselves. They said unto the olive tree, 'Reign thou over us.' But the olive tree said unto them, 'Should I leave my fatness, wherewith by me they honour

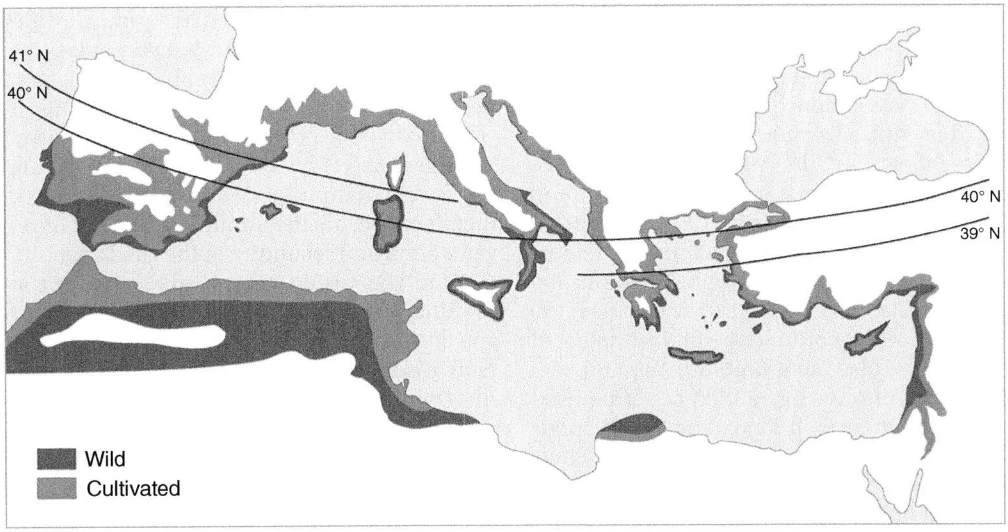

Fig. 1.1. Present-day distribution area of the wild and cultivated olive in the Mediterranean basin. (From Carriòn *et al.*, 2010)

God and man, and go to wave to and fro over the trees?'

According to the legend, in the vicinity of his own death, Adam remembered the words of the Lord, who had promised him 'the oil of mercy', namely his redemption and that of all humanity; so it was that, according to biblical tradition, he, shortly before dying, sent his son Seth to the cherub who was guarding Eden. The cherub took three seeds from the tree of good and evil and handed them over to Seth. The seeds were placed in the pit where Adam was buried. The seeds sprouted and three plants were born from the mound: one became an olive tree, another a cedar and the third became a cypress.

The magnificence of the olive tree is sung by the poets of the Old Testament. In their metaphors the olive tree symbolizes salvation and prosperity. Psalm 128 says: 'Thy wife shall be as a fruitful vine by the sides of thine house: thy children like olive plants round about thy table.' The prophet Hosea sings the strength and beauty of the olive tree: 'His branches shall spread, and his beauty shall be as the olive tree, and his smell as Lebanon' (Hosea 14:6–7).

In Jewish observance, olive oil is the only fuel allowed to be used in the seven-branched Menorah in the Mishkan service during the Exodus of the tribes of Israel from Egypt, and later in the Temple in Jerusalem. It was obtained by using only the first drop from a squeezed olive and was consecrated for use only in the temple by the priests, stored in special containers.

Another use of oil in Jewish religion is for anointing the kings of the kingdom of Israel, originating from King David. Tzidkiyahu was the last anointed king of Israel. One unusual use of olive oil in the Talmud is for bad breath, by creating a water-oil-salt mouthwash. The Talmud also states that frequent consumption of olive oil is good for one's memory.

Another culture in which the olive was particularly important was that of Greece. According to tradition, the olive was introduced into Greece from Egypt by Cecrops, the first king of Attica, founder of Athens and of the Aeropagus (Loussert and Brousse, 1978). According to Pindar, instead, the olive was brought by the Dorian hero, Heracles, from the country of the Hyperboreans to the Olympian sanctuary in Greece. The same myth was also reported in the second century AD by Pausanias in his *Description of Greece*. Another Greek myth attributes to Athena, the patron goddess of the city of Athens, the invention of olive oil and even the introduction of the olive into Attica. The god of the sea, Poseidon, competed with her to gain the sovereignty of the city of Athens and both gods vied to offer Attica the best possible gift. Poseidon offered beautiful white horses, while Athena gave an olive tree: she struck her spear into the earth and an olive tree appeared (Fig. 1.2). The council of the gods – or, according to other versions, a sentence given by Cecrops – determined that Athena was the winner because the olive not only could live for hundreds of years but it also gave edible fruits and was the source of olive oil, which could be used by men to dress their food, cure their wounds and ailments and light their dwellings (Blazquez Martinez, 1996).

The Iliad and *The Odyssey* also contain numerous references to the olive tree, the tree of Athena. Homer, in *The Iliad*, compares the fall of Euforbos, defeated by Menelaus in the battlefield, to the fall of the olive tree that 'grows handsome, agitated by all kinds of winds, covered with white flowers, that suddenly, when a hurricane comes, is from the earth uprooted and thrown to the ground'.

It was with oil that Auricle anointed the body of Ulysses, the hero. The log that blinded Polyphemus was from the olive tree, and the nuptial bed of Ulysses and Penelope was made of olive tree wood.

Fig. 1.2. The myth of Athena offering the olive to Athenians. A 100-lire coin.

The olive was regarded as a symbol of euphoria, purity, victory and honour, and olive shoots have been used to honour victory, peace and wisdom. According to Herodotus, it was also believed that the olive tree had arisen spontaneously and flourished following the fire in the Temple of Athena. The olive tree therefore became the 'luminous symbol' of regeneration and fecundity of the Greek people.

Oil was used in ceremonies and for the unction of statues of worship. Generally speaking, funerary stelae, sacred stones, were also the object of anointments, such as the Delphi oniphalos or navel of the world which was rubbed daily with perfumed oil. This use had eventually been extended to other stones, which were clearly less sacred, such as those of the crossroads.

Oil was also a common offering of which we have examples from the Mycenaean period. Among the tablets found in the palace considered of Nestor, in Pylos, almost 50 of them list the stock of oil intended for the gods and particularly for Zeus, Poseidon and a deity called Potnia, which is perhaps to be assimilated to Athena. It also happens that it is the container that is offered: oil lamps or panathenaic amphorae.

The olive tree and its oil are directly linked to the great panhellenic festivals, both religious festivals and sports competitions. The victorious athletes at Olympia received an olive wreath; in Athens, the award consisted of an amphora filled with oil. The Panathenaea was indeed the great festival dedicated to Athena. Pisistratus, since 565 BC, wanted to make it compete with the great panhellenic feasts which were held in Olympia, Delphi or Nemea. Competitions framed the party and the winning athletes were rewarded with amphorae filled with oil from the harvest of sacred olive trees which, according to tradition, were shoots of the Athena tree. The prestige of the winner was substantial: according to calculations made on the victory in the chariot race with three horses, the winner received 140 amphorae with a capacity of almost 40 liters each. At the rate of 12 drachmas per 40 liters, the gain amounts to 1680 drachmas. For comparison, the average daily wage was 1 drachma (Amouretti, 1986).

The use of olive oil was very important in funeral rites. There is the famous example in *The Iliad* of the funeral of Patroclus, who, after his death, was washed and anointed with rose oil. This treatment aims to prepare the body for a new life by preserving it from decay. In this sense it takes on a spiritual dimension and strengthens the link that unites man to the deity or to God. It is in this sense that the anointing has passed into Christianity.

The Greek symbolism of the olive passed to Roman culture, which also revered and venerated the plant. Hercules was associated with oil, according to an inscription dedicated to him by the oil traders in the Forum Boarium and dating from the third century BC. At the end of the Republic, the oil traders dedicated a temple and a statue to Hercules Olivarius in Delos. The hero's club was believed to have been made of olive wood. Even the divine twins, Romulus and Remus, according to tradition, were born under an olive tree.

The cult of the goddess Athena became in ancient Rome the cult of the goddess Minerva, and olive twigs entwined with laurel twigs were used for plaiting crowns intended as a reward for heroes. For the Romans, the olive represented a fruit crop, a religious power, and a theme of political organization; it represented the relations between the city and the countryside, between Rome and the conquered lands, where Romans regularly introduced it.

The Greco-Roman symbolism of the olive tree was incorporated in Christian traditions from the time of the persecutions until 380 AD when, by an edict of the emperor Constantine, the new religion became the official state religion. Its symbolism is evoked in the holy scriptures and represents divine blessing, a sign of salvation and the symbol of eschatological happiness. In the Bible, the olive tree is quoted no less than 70 times.

In the New Testament, the cross of Jesus Christ was made of olive wood. The very name of Jesus, 'Christos', simply means anointed.

The olive tree in Christian tradition represents many symbologies. The return of the dove with an olive twig to Noah's ark is given a double meaning: the symbol of regeneration, since after the flood's destruction the world was back flourishing; and the symbol of peace, since it marked the end of the punition and the reconciliation between God and man. Both are celebrated on Palm Sunday, when the olive represents Christ himself who, through his sacrifice, becomes an instrument of reconciliation and peace for all humanity. In this way, the olive becomes a sacred plant, and sacred also is the oil the plant's fruit produces. Olive oil is the Chrism, used in Christian liturgies: the Catholic and Orthodox Churches use olive oil for the Oil of Catechumens (used to bless those preparing for baptism) and Oil of the Sick (used to confer the sacrament of Anointing of the Sick). Olive oil mixed with a perfuming agent is consecrated by bishops as sacred Chrism, which is used to confer the sacrament of confirmation, in the rite of baptism, in the ordination of priests and bishops, in the consecration of altars and churches, and, as a Middle Ages tradition, in the anointing of monarchs at their coronation, following an ancient Jewish custom.

Olive symbolism is also present in the Gospels: Jesus was joyfully received by a crowd waving palm leaves and olive twigs; he spent his last hours before his death in the Garden of Olives.

In the Orthodox Church, olive oil is a product not consumed during Lent or penance, while Orthodox monks use it sparingly in their diet. Exceptions are feast days and Sundays.

Significantly, in the first centuries following the recognition of Christianity as the state religion, the Christian clergy conducted a crusade against the pagan cult of trees, but not against the olive.

Even before the birth of Islam, olive culture and symbolism had spread in the Near East and in the Maghreb thanks to Greek and Roman settlements and trade. Islam has the Bible as a holy book, and therefore all mention of the olive tree and of olive oil is shared by Muslims. Extensive use of olive oil and olive leaves is cited in the Bible as a natural healer.

The olive tree and olive oil are mentioned seven times in the Quran, and the olive is praised as a precious fruit. The olive tree and olive oil health benefits have been propounded in prophetic medicine. Muhammad is reported to have said: 'Take oil of olive and massage with it – it is a blessed tree'.

The cultivation of olive trees began in the Arab world by foreigners. It then became an integral part of the countryside, of the culture and of the local economy.

In the verbal tradition of the Kabylia (Algeria), the olive tree was the sacred tree *par excellence* and was at the center of numerous ceremonies. In the accounts of the wars between the Turks and the Kabylian tribes, it was cited as the 'rod which becomes green again'. For these tribes, the olive tree was an important part of daily life: women placed themselves under its protection, and asked the tree to grant them fecundity and good health (Bartolini and Petruccelli, 2002).

The olive tree was also associated with the rites of the dead and, in particular, with those of holy men. Their tombs became sanctuaries. In the Muslim religion, the olive tree is the symbol of goodness: Mohammed had faith in the horse for its speed, strength and salvation, and in the olive tree for its faithful goodness (Quran, XCV, 1).

Olives are substitutes for dates (if not available) during Ramadan fasting, and olive tree leaves are used as incense in some Muslim Mediterranean countries.

1.3 A Mediterranean Story

The natural history of olive is a long one: phylogenetic analyses (Besnard *et al.*, 2009) suggest that the most recent common ancestor of the six subspecies of *Olea europaea* was present during the late Miocene or Early Pliocene (4.0–8.3 million years ago), although fossil evidence indicates that the olive tree had its origins some 20–40 million years ago in the Oligocene, in what is now corresponding to Italy and the eastern Mediterranean basin. Miocene remains have been found near Leghorn, Italy (Therios, 2009). Pliocene fossils of *Olea* were found not far from Bologna, and near Lake Garda, Italy (Bally *et al.*, 1962). More recent archeological findings indicate the presence of the olive in Greece in the upper palaeolithic (50–60,000 years ago) (Friedrich, 1978).

The origins and natural distribution of wild olive populations have been a focus of multidisciplinary research, largely on aspects of its domestication and cultivation, and are dealt with in detail in Chapter 3.

The cultivated olive derives from the wild olive (*Olea europaea* L. subsp. *europaea* var. *sylvestris*, also called oleaster), after domestication occurred in the first millennia of the Neolithic Revolution. Where and when this took place is still the object of research. Like most fruit crops, olive utilization was, in the beginning, dependent on fruit harvest from naturally present oleasters. The earliest evidence for the use of wild olives dates to about 17,000 BC (Kislev *et al.*, 1992).

The climatic events occurred through Pleistocene (up to 11,000 years BC; Carriòn *et al.*, 2010; Kaniewski *et al.*, 2012; Besnard *et al.*, 2018) separated and diversified through natural selection the subspecies, and the oleaster was confined to the coastal areas of southern and central Mediterranean, where it can still be found, although, in most cases, with a certain difficulty due to human presence (Fig. 1.1).

A few millennia after the beginning of agriculture in various parts of the Mediterranean, an event that had originated in the Levant and spread mostly westwards at the speed of about 1 km per annum, the process of olive domestication started. It took place by the obvious technique of mass selection, and was possible thanks to olive suitability to asexual propagation, which permitted the conservation of interesting traits in the progeny, such as high fruit set, large fruit, high oil content, and resistance to biotic and abiotic stress. This biological characteristic, also present in the pomegranate, grapevine, fig and date palm, allowed these tree fruit species to be domesticated millennia before other fruit species, such as those belonging to the Rosaceae family (Zohary and Spiegel Roy, 1975).

Scholars have long debated the location and timing of domestication. De Candolle (1959) indicated Syria and Iran as the sites of first domestication. However, until the first half of the 20th century, olive was believed to have been domesticated somewhere in the Fertile Crescent, between Palestine and the Caucasus; the diffusion of the culture would then favour the dissemination of seedlings all over the Mediterranean, thus causing the presence of feral plants.

As was the case with all food crops, the production and use of olive oil anticipated olive cultivation, and consequently its domestication, with the ensuing selection of cultivars.

The use of the wild olive lasted long, in the Near East, in the early Neolithic period, as shown by the remains of olive stones found in Israel (Lipschitz, 1997). *Olea* is well recorded on the Island of Cyprus since the 9th millennium BC (Thiébault, 2003). In the central Aegean area, finds of *Olea* charcoal date olive presence to the late or final Neolithic period (4th millennium BC) (Ntinou and Tsartsidou, 2017). Earlier finds are not yet confirmed.

In the western and central Mediterranean, olive finds are abundant south of the parallel 40° N (Thiébault, 2001; Carrión *et al.*, 2013). Finds of the 6th–7th millennium BC were also in Portugal (Figueiral and Terral, 2002) and in Italy (Costantini, 1989; Nisbet, 1997).

The presence of olive in the archaeological deposits of the western Mediterranean region increases exponentially since the start of the Holocene, and studies suggest that it became more intensively used since the Neolithic (Caracuta, 2020).

Genetic investigations argue in favor of a single primary domestication event in the eastern Mediterranean basin, followed by diffusion of the first domesticated olive and diversification in the central and western Mediterranean as key processes in the olive tree history (Khadari and El Bakkali, 2018).

The findings of other studies support multiple origins of cultivars across the Mediterranean region (Diez *et al.*, 2015) but it remains unclear whether this reflects secondary diversification or multiple independent primary domestication events.

It is considered by some research that primary olive domestication, from wild progenitors, began approximately around 4000 BC in the Middle East, near the border between Turkey and Syria, close to an area inferred to be suitable for the long-term persistence of olive trees during the late Pleistocene (Kaniewski *et al.*, 2012; Besnard *et al.*, 2013; Gurbuz-Veral *et al.*, 2018; Besnard *et al.*, 2018).

Other research suggested that wild olives may have persisted over long time periods on the Mount Carmel coast (north-west Israel), where early cultivation has been reported (Kislev *et al.*, 1992), and trees from this area may have contributed to the cultivated olive gene pool (Barazani *et al.*, 2016). Based on the amounts of pollen presence, the southern Levant appears to have been the site of primary olive cultivation as early as 4500 years BC, and that a later, early/mid-4th millennium BC, cultivation process occurred in the Aegean (Crete). From these two areas of origin, the southern Levant and the Aegean, olive cultivation is believed to have spread across the Mediterranean. In Anatolia, large-scale olive horticulture was palynologically recorded by 1200 BC, in mainland Italy at 1400 BC, and in the Iberian Peninsula at mid-/late 1st millennium BC (Langgut *et al.*, 2019).

All these data might also suggest that multiple geographic origins of cultivars are possible, followed by a process of diversification and domestication in the central and west Mediterranean basin.

Most likely, domestication did not occur as a single event, but rather as a long and gradual process that might have developed in different places at different times in the wider region of the Mediterranean. Today, over 2000 cultivars are recorded (Bartolini *et al.*, 1998).

As concerns timing of domestication, i.e. the passage from collection of fruit from individual wild or selected trees to exploitation of improved cultivars for large-scale oil production, most authors agree that it should have taken place not before the 5th millennium BC.

1.4 Middle East

The Middle East was therefore the cradle of olive cultivation, and this is confirmed by archeological finds. The first references we have show that the Syrian and Palestinian region was where the production and marketing of olive oil first got underway.

The olive tree and olive oil were major components of the culture of Ancient Israel and the economy of its people. Cup-marks – small cavities chiselled on rock surfaces – found in epipalaeolithic sites (10,500–8,300 BC) may indicate a first production of olive oil by man. These were utilized to crush cereals and also to produce liquids (although not necessarily only olive oil), and they increased in numbers during the early Neolithic period (8,300–6,000 BC). In many sites, deep concave mortars have been found close to them. Both apparently served for producing oil using primitive techniques (Eitam and Heltzer, 1996).

Some of the earliest olive processing evidence has been found off the Carmel coast (near Haifa). Pits with crushed olive remains and the baskets containing them were found underwater (at about a meter depth and several meters from the shoreline). The pits date to late Neolithic/early Chalcolithic periods (as early as 4500 years BC) (Galili *et al.*, 1997). At that time, Philistines already grew olives and produced notable quantities of oil (Fig. 1.3).

Olive trees were certainly cultivated in Israel during the second half of the fourth millennium BC in small villages in the Mediterranean regions around the Golan woodlands where the first separating vessels have been found (Epstein, 1993) and in the hills of Samaria (allowing an annual consumption of at least 5 liters of oil per person) (Fig. 1.4).

Olive trees and oil production in the eastern Mediterranean can also be traced to archives of the ancient city-state of Ebla (2600–2240 BC), which were located on the outskirts of the Syrian city Aleppo. Here, some dozen documents dated 2400 BC describe lands of the king and the queen; these belonged to a library of clay tablets perfectly preserved by having been baked in the fire that destroyed the palace. Olive orchards were third in terms of crop acreage. The tablets describe how fields were demarcated by counting the number of olive trees standing in each one. One of these documents mentions the existence of three groves, two with 500 and one with 1000 olive trees. Another document talks about the different varieties of olive oil and, for the first time, mentions exports of high-quality olive oil to other kingdoms.

A later source is the frequent mentions of oil in the Tanakh.

Herodotus, in *Histories* (I, 193), writes that in Babylon the olive is not cultivated, nor is olive oil consumed, since they obtain

Fig. 1.3. Ancient oil mill, Israel.

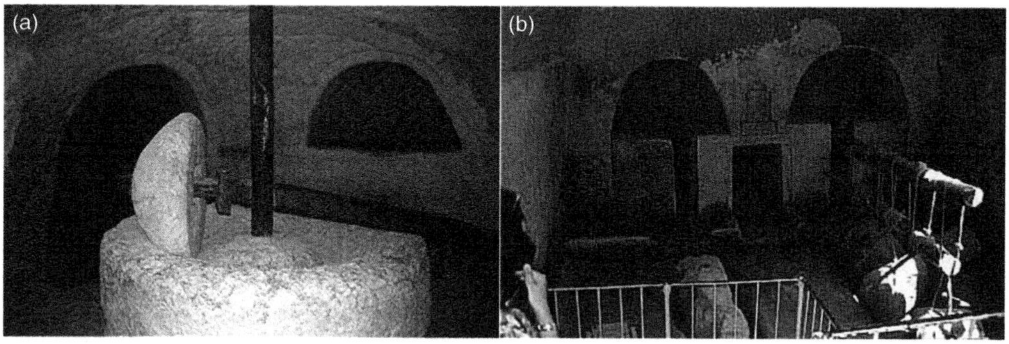

Fig. 1.4. Hypogeal oil mill, Israel. (a) mill; (b) presses.

oil from sesame seeds. However, the Babylonian Code of Hammurabi introduced olive oil into law by regulating its production and trade long before the writing of *Histories*, around 2500 BC.

About halfway through the second millennium BC, reports mentioning olive oil suddenly become much more numerous and references have been found in the city archives of Alalakh, Mari and Ugarit. In many of these documents, olive oil is mentioned as a medicine but in others, for example in the Mari documents, we are told of imports of olive oil from the city of Aleppo (close to the ancient city of Ebla). The Mari documents show that the price of olive oil was five times higher than that of wine and two-and-a-half times more expensive than seed oils (sesame or linseed) (Remesal Rodriguez, 1996).

An ancient Syrian document dating back to the start of the second millennium BC from Karum, a Syrian trading center situated in Hittite territory in central Anatolia, tells us of a local merchant's order of first-class olive oil from the capital Assur or, alternatively, Hahlum (south-east Turkey or northern Syria). Therefore, olive oil was already traded over long distances. Documents from the Ugarit city archives dating from the 13th century BC prove that olive-growing was important, although it took second place to vine-growing. Olive oil was used to pay tax to the palace and, at the same time, the palace itself paid for certain services in olive oil. However, perhaps the most important document is one which describes the trade in olive oil between Ugarit, Cyprus and Egypt.

There is a good deal of archaeological information about the development of olive cultivation in the Syrian and Palestine region at the beginning of the first millennium BC. Excavations in Israel have found the remains of many olive oil presses and have shown the techniques developed for obtaining oil, such as circular millstones and beam presses in which stones were used as counterweights (Figs 1.3 and 1.4).

Commercial mills for non-sacramental use of oil were present in the tribal confederation around 1000 BC, in an area consisting of present-day Palestine, Lebanon and Israel. Over 100 olive presses have been found in Tel Miqne (Ekron), where the Biblical Philistines also produced oil. It was oil for lighting and for balms to be exported to Egypt, to be used for embalming and other funerary rites. These presses are estimated to have had output of between 1000 and 3000 tons of olive oil per season. It can be considered one of the largest industrial complexes of antiquity (Vannucci, 2009).

In the first half of the first millennium BC, the olive spread throughout Assyria, as shown in the Nimrod-Kalkhu findings, and the olive is mentioned in the Assyrian lexicons of the time.

During the third century BC, the center of oil production moved to the Sidonian colony of Maresha, where 20 oil press caves were carved in the soft limestone around the city. The oil mills and an improved lever press with three weights (1500 kg each) indicate the 'technological revolution' of the Hellenistic period (Kloner and Sagiv, 1993).

1.5 Egypt

There is little information about olive oil in the Egypt of the Pharaohs. The olive may have reached Egypt from Syria, Anatolia, Palestine or Crete (Bartolini and Petruccelli, 2002). The oldest reference to this tree in Egypt is the import activity that took place during the 4th Dynasty, 2600 years BC, and the existence of a sacred olive tree in the city of Heliopolis, in lower Egypt, during the 5th and 6th Dynasty.

New evidence (Gerish, 2020) suggests that olive wood was present in ancient Egypt as early as 2551–2523 BC. Otherwise, the earliest known traces of olive were fruit stones found in 12th Dynasty deposits at Memphis. Even then, there are almost no other archaeological finds of olive until the 18th Dynasty (about 1552–1314 BCE).

During the New Empire, olive groves had become fairly widespread, according to an inscription from the time of Ramses II (1279–1212 BC) found in Heliopolis in the temple to the god Ra. This states that the olive groves around the city gave pure oil, the best-quality oil in Egypt, for lighting the lamps in the sacred palace. Olive oil was certainly utilized under the Ramseses: according to the 'Harris' papyrus, of the 20th Dynasty, Ramses III (1166–1154 BC) had olive trees planted around the Temple of Thebes (Meeks, 1993). In the tombs of Ramses III and Tutankhamon (1325 BC) are frescoes reproducing oil jars and olive twigs, as well as votive crowns plaited with olive leaves (Bartolini and Petruccelli, 2002; Vannucci, 2009). Sources from Greek and Roman times indicate the existence of olive orchards in Thebaid to the south of the country, in Alexandria and, above all, in the Fayum oasis. These same sources go on to say that Egyptian olive oil was of poor quality and had a strong smell, which is to be expected due to the extreme southern latitude of the oases (Remesal Rodriguez, 1996). Cultivation was probably also made in the Siwa oasis, where oil is still being produced today.

The culinary utilization of olive oil as food was, however, quite marginal, since the Egyptians had also great quantities of other oils; most references rather mention medicinal or cosmetic uses (Meeks, 1993).

1.6 Phoenicians

The expansion in the use of olive oil throughout the western Mediterranean is undoubtedly to be credited to the Phoenicians, who took it with them to northern Africa, Sicily and southern Spain starting in the 16th century BC, and to the Greeks who took it to southern Italy. Trade contacts between the two ends of the Mediterranean began to flourish around that time, and intensified with the new millennium (Hyams, 1971; Remesal Rodriguez, 1996). Carthage's trade empire was also based on olive oil, and Carthaginians are credited with the first use of grafting on seedlings for olive multiplication, thus greatly increasing the capacity for the species diffusion, if compared to the use of rooted suckers and other plant parts. The technique also enabled the use of widespread oleasters stands, by simply grafting them and exploiting the existing root system (Fiorino, 2008).

1.7 Aegean

Before reaching mainland Greece, the olive landed on the island of Crete. The earliest surviving olive oil amphorae date to 3500 BC (early Minoan times), though the production of olive oil is assumed to have started several centuries earlier (Fig. 1.5).

The cultivation of olive trees in Crete became particularly intense in the post-palatial period (starting in 1500 BC) and played an important role in the island's economy. The Minoans used olive oil in religious ceremonies. The oil became a principal product of the Minoan civilization, where it is thought to have represented wealth, and was exported to all parts of the Mediterranean, and particularly to Egypt. Olive trees, branches and leaves are commonly represented as decorations on vases, jewels and objects of daily life, apart from the famous Cnossos frescoes (Fig. 1.6).

Fig. 1.5. Large jars in Cnossos Palace, Crete.

The consumption of olive oil in Greece depended on the social class, and apparently oil as food was only consumed by the rich. More common was its use as lamp fuel, as a medicament, as a body balm and for perfumes (Bartolini and Petruccelli, 2002). It has not been proved that the Myceaneans used oil for cooking but they probably did for lighting, since lamps were common from the middle Minoan period (1700–1600 BC). However, olives were apparently included in their diets. There are numerous mentions in Pylos, Greece, and just a few in Cnossos.

In the Cnossos Palace (the mythical location of the minotaur's labyrinth), deposits of enormous amphorae (pithoi) for oil storage, up to two metres high, were excavated. Other objects relating to oil extraction were found in nearby Phaistos. Documents written in Linear B and dating from the 13th century BC indicate that on Crete, oil from both wild trees and cultivated olive trees was used. It is said that in the palace at Knossos up to 250,000 liters of olive oil could be stored, although other research has reduced this figure to one-third of that amount. It has been calculated that the stores of the palace of Malia could hold up to 23,000 liters of olive oil. Deliveries of more than 10,000 liters sent from rural areas to the palace at Knossos are recorded on tablets. It has been suggested that oil from wild olive trees was used as a base for perfume (Remesal Rodriguez, 1996).

1.8 Greece

According to tradition, the olive tree was brought from Asia Minor to Greece by Cecrops, who founded Athens in the year 1582 BC. The ancient inhabitants of Greece, who were familiar with the wild olive tree, imported cultivated varieties and techniques for oil production from the eastern Mediterranean and Crete. Olive stones and presses have been found of the same period, the second half of the 2nd millennium BC (Figs 1.7 and 1.8).

At the beginning of the Ist millennium BC, the cultivated olive had reached most shores of the Mediterranean basin, and the domestic form had substituted the wild one. The Greeks, in the oldest literature, distinguish the κότινος, the oleaster – used as for precious woodworks (famous is Ulysses's bed) – from the ελαία, the oil olive, a symbol of peace and prosperity, which gods created to enrich the poor lands of Attica. The Greek merchants and sailors would eventually spread olive cultivation, first in the northern Aegean Sea, and then along the coasts of the northern Mediterranean and to southern Italy. Thales of Miletus (7th century BC) mentions Chios and Delos as 'rich of lush olive groves'; Samos is called Ελαιοφύτος (olive grove).

Theophrastus, in *On the Causes of Plants*, reminds that the cultivated olive must be vegetatively propagated; indeed, the seeds give rise to thorny, wild-type olives, with small fruit, and will take years to bear. He reports how the bearing olive can be grafted on the wild olive. In his *Enquiry into Plants* he maintains that the olive can be propagated from a segment of the trunk, from branches, from roots or from twigs.

The olive therefore becomes a typical product of Attica towards the end of the Bronze Age: it becomes a focus of agriculture, with grapevine, and also a center of Greek culture and civilization. Homer's works are a contribution to understanding the timing of olive presence. In *The Iliad*, on events occurring at the end of the 2nd millennium BC, the olive is not mentioned, nor is it depicted on Achilles's shield, where a vineyard is represented; wine is also often mentioned. The plant and oil are both mentioned in *The Odyssey* (several centuries later) a number of times.

Fig. 1.6. Frescoes in Cnossos Palace, Crete.

Fig. 1.7. Ancient Greek oil mill.

Fig. 1.8. Bodrum, Turkey, the base of an ancient Ionian oil press.

Olive oil was a luxury for the ordinary citizen, but widely consumed by the rich. It has been calculated that gymnasium patrons consumed an average 30 liters of oil for body care, 20 liters as food, 3 liters as lubricant and lamp fuel, half a liter as a medicine, and a couple of liters for sacred rites. Among these, most important were the funerary rites and body unctions.

There was great production of perfumes and balms, a fashion imported from Egypt. Spartans criticized the habit as corrupting, but theirs was the worst crime against the olive when, during the war of Peloponnesus, they cut down the olive trees around Athens, which surrendered.

The Spartans, however, also used oil to rub themselves while exercising in the gymnasia. From its beginnings early in the 7th century BC, the decorative use of olive oil quickly spread to all of the Hellenic city states, together with naked appearances of athletes.

Another use of olive oil was as a prize for Olympiad (since 776 BC) winners. The quantities were huge. Victors were also crowned with olive leaves, according to Bally *et al.* (1962); for other scholars the leaves were from oleaster.

Olive oil from Attica was extremely important for the Athenians; according to Solon's laws (638–558 BC), olive oil was the only food product that could be exported from Athens. This demonstrates the importance of olive groves in Attica, not only as a key element of land occupation and employment but also, and above all, because of its trade value. By exporting its olive oil, Athens was able to obtain the grain it lacked (Remesal Rodriguez, 1996).

Each city produced its amphorae, with particular shape and decorations. Athens's oil was traded in an amphora called 'SOS', which guaranteed the customers quality and quantity of the content. Attica oil was considered to be the best, but oils from other locations, such as Sicion, Euboea, Samos, Cyrene, Cyprus, were also appreciated. Sophistications, however, were common even then. From the 6th century BC, the Magna Graecia colonies and Marseille started competing with the homeland for quality and quantity of oil, and ships crossed the Mediterranean loaded with oil amphorae.

1.9 Italy and the Roman World

Comprehensive studies of the presence and exploitation of olive on the Italian Peninsula show records of such presence since the early Holocene. The earliest evidence (7th millennium BC) comes from Apulia (Caracuta, 2020), although the olive exploited until the 2nd millennium BC was most likely the oleaster.

During the mid-Holocene olive appears in Liguria, and pollen records from Apulia and Tuscany show that *Olea* was present there from the 6th to the 5th millennium BC. An increase in *Olea* pollen and charcoal presence is detected in the 2nd millennium, with a peak in its last part (Mercuri *et al.*, 2013), suggesting that olive cultivation became widespread between the 2nd and 1st millennia BC. Olives were grown in several areas of southern Italy, to produce olive oil, as confirmed by the retrieval of oil residues in containers of the late Bronze Age in Sicily, Calabria and Apulia (Caracuta, 2020). Recently, residues were found in a large container discovered in a hut at the site of Castelluccio, Sicily (Fig. 1.9), which was radiocarbon-dated to the 20th century BC. The characteristics of the find reinforced the hypothesis that the olive oil that had been stored inside was locally produced, rather than imported from other areas of the Mediterranean (Tanasi *et al.*, 2018). The olive was certainly present in northern Italy at the time of the Terramare culture (2nd millennium BC) (Bartolini and Petruccelli, 2002).

Abundant evidence confirming cultivation and all transformation activities is, however, more recent, in the first centuries of the 1st millennium BC. Finds are particularly present in Etruria and Magna Graecia (southern Italy). A sudden increase of *Olea* pollen in an Apulian site of the 8th–7th century BC supports this hypothesis (Di Rita and Magri, 2009). Finds are numerous in the areas under Etruscan influence (which spread in most of central and northern Italy, and in some areas of the south), although the Etruscans are believed to have acquired the relative techniques from the Greeks of Magna Grecia, since their name for oil is clearly derived from the Greek.

At any rate, before producing it the Etruscans had imported olive oil from Attica, as shown by the many amphorae found in several settlements (Fig. 1.10). Oil was not so much used as food (use of animal fats for cooking was common) as much as a scented ointment, in the Greek fashion. Also important was its use for lighting, as suggested by the many lamps found (Fig. 1.11). The many stones found in the same sites make us think of an alimentary use of the fruit (Ampolo, 1980). More evidence comes from shipwrecks of the time.

The timing of the diffusion of olive cultivation approximately fits with the tradition; according to Pliny the Elder, olives reached Italy at the time of the reign of Lucius Tarquinius Priscus (616–579 BC), the fifth king of Rome, of Etruscan origin.

According to Cato, olives and olive oil are common in Italy in the 5th century BC, and by the 1st century, Italy was the world's first producer.

The Romans emphasized the use of oil as food, which had been neglected in former cultures. Olive oil enriches with its flavor *puls* (polenta), vegetables and cereals, which in the Republic period are central in the diet of all, also patricians.

Oil production peaked in the early part of the Roman era, mostly in Campania, Latium and Apulia, where oil facilities were connected to latifundia, great estates owned by the aristocracy.

Many improvements in oil technology were achieved in this period, as witnessed by the many agricultural treatises. The oldest is the *Liber de agricultura* or *De re rustica* by Cato the Elder (3rd century BC): 'Oil must be extracted from the olives soon after harvest, to avoid their decay'. Varro, a century later, recommends not to harvest by hitting the tree with sticks but, rather, by browsing with hands. The belief is that the olive grove is not expensive and is profitable, as asserted by Columella (1st century AD): 'Among all

Fig. 1.9. Oil jar from Castelluccio, Sicily.

Fig. 1.10. Attican amphora, from the 6th century BC, found in Vulci, Etruria.

trees the olive is the least costly, while is the first of them all'. According to the historian Pliny the Elder (*The Natural History*, Book XV), Italy had 'excellent olive oil at reasonable prices' by the 1st century AD, 'the best in the Mediterranean', he maintained. Pliny, however, also gives plenty of information about the most suitable environment: 'The olive does not grow at a distance of more than forty miles from the sea... the olive will grow neither in very cold climates, nor yet in very hot ones'. He also mentions propagation: 'It is sown in nurseries ... and if transplanted will bear fruit the following year'. Pruning is recommended as well as cultivation practices. Several cultivars are described, with their qualities and relative care. But most of all, Pliny describes fruit development and ripening:

> 'The riper the berry, the more unctuous the juice, and the less agreeable the taste.'
>
> 'The best time to gather it is when the berry is just on the point of turning black.'
>
> 'It is not with olive oil as it is with wine, for by age it acquires a bad flavour, and at the end of a year it is already old.'

Recommendations are given as to how to harvest, since beating with sticks damages the trees and next year's harvest; there is also a consideration on alternate bearing, and its causes:

> The tree is reduced to bearing fruit but once in two years, in consequence of the injury done to the buds. Not less injurious, however, are the results of waiting till the berries fall from the tree; for, by remaining on it beyond the proper time, they deprive the crop that is coming on of its due share of nutriment.

There were several oil categories: *Oleum ex albis ulivis* was the top-quality oil, obtained from unripe olives; *Oleum viride* was from olives harvested at color change, also very much esteemed; *Oleum maturum* was extracted from fully ripe and black olives of inferior quality; *Oleum caducum*, of even worse quality, was from olives collected from the ground; and finally *Oleum cibarium*, the worst quality, was from diseased and fermented olives, for slaves and other uses such as lighting (Fig. 1.12).

As the Roman legions conquered the Mediterranean territories, the population of the empire and of the urbs grew, and local production became insufficient. Other territories were used as a source of oil, and the most important were southern Spain and North Africa.

Fig. 1.11. Double oil lamp, 1st century AD.

Fig. 1.12. Oil bottle and bread loaf from Pompei, southern Italy, 79 AD.

It was Baetica, today called Andalusia, that was the first province to supply olive oil from the 2nd century onwards, and North Africa was keen competition for the Andalusian olive oil.

How much oil was consumed in Rome? According to calculations based on the amphorae accumulated in the Testaccio, an artificial hill made of broken vessels, during the empire about 320,000 amphorae reached Rome each year, that is over 22,000 tons. Accordingly, each Roman (there was 1 million of them) consumed 22 kg of oil, or a little less the 2 kg per month – not very different from today's consumption (12 kg) if we consider that in ancient Rome olive oil was utilized for many other purposes (Vannucci, 2009).

The Romans were the first to introduce advanced systems of storage and distribution, that lasted through the Republican era. The emperors instead started to rely on tributes, in oil, from subjected peoples. This allowed them to fix prices, decide modes of collection and enable stockpiling and free distribution to Roman citizens. They also encouraged private trade, giving privileges to the *negotiatores olearii*. This organization, which through the colleges of importers, and the *arca olearia* (a sort of stock exchange) had created a blooming activity, vanished with the end of the empire.

The expansion of olive cultivation in the Italian Peninsula was slow and depended to a large extent upon the social conditions brought by the Roman conquest of Italy. The most highly prized Italian oil was from Venafro in Campania, although the area where most oil was produced was still Magna Graecia.

The creation of the empire allowed the development of a more open economic system than had existed until then. The peace which Augustus established throughout the Mediterranean facilitated trade and travel. Rome enjoyed the fruits of her exploitation of the resources of all the countries she dominated. Augustus based his political power on two fundamental pillars, Rome's people and the army. He had to provide for both. Therefore, Augustus created an economic system in which each province of the empire contributed to satisfy Rome's interests and supply the army with the products it was able to produce.

Spain was one of the most important sources of food, particularly olive oil. The centers of production in Baetica were quite large, and the areas by the river Guadalquivir (as well as the tributary of the Genil) were very important.

Olive-growing had been introduced into Spain during the maritime domination of the Phoenicians, in the land of Tartessos (present-day Andalusia and southern Portugal) but did not develop to a noteworthy extent until the arrival of Scipio (212 BC) and Roman rule (45 BC). The Tartessians were important trading partners of the Phoenicians, whose presence in Iberia dated from the 11th century BC and who built a harbor of their own, Gadir (present-day Cádiz). Contacts apparently existed with Greek traders in the north-east of the peninsula. As was customary in most cultures, the Tartessians utilized olive oil as body anointment and as fuel for lamps, while animal fats were predominant in cooking. After the third Punic war (146 BC), olives occupied a large stretch of the Baetica valley and spread towards the central and Mediterranean coastal areas of the Iberian Peninsula, including Portugal. It is not clear when commercial olive groves started to be planted, but certainly an important stimulus to the industry came during the Roman rule, since Rome became an important market. At any rate, this development was initially slow, if until the time of Augustus (27 BC–14 AD) Spain imported oil from Italy (Garrido Gonzàlez, 2004).

Andalusian olive oil was distributed all over the Roman Empire – from Britain to Egypt – in amphorae, the production centers of which were located in a triangle formed by the cities of Seville, Cordova and Ecija. The identification of these centers of production has made it possible to determine that the ancient city of Axati (today Lora del Rio, Seville) was the area with the highest level of production. A host of details were noted down on the amphorae in which the oil was exported – the empty weight of the amphora, the net weight of its oil content and the name of the trader or carrier, together with a complex system of fiscal checks and the date when the transaction took place as registered by the consular authorities (Remesal Rodriguez, 1996).

The oil exports from the Baetica to Rome peaked in the first two centuries of the Christian era. Permanent witness is Mount Testaccio, formed with remnants of oil amphorae that carried the Spanish oil. Lucius Moderatus Columella, a Hispano-Roman agronomist born in Cádiz in 3 BC, describes in his treatises *De re rustica* ('On Rustic Affairs') and *De arboribus* ('About Trees') olive tree care and oil production; he even mentions the ten main varieties that were cultivated in Roman Spain, and the various flavors of their oils.

Avienus (4th century AD), in his book *Ora Marítima*, calls the river Ebro, *Oleum Flumen* ('Oil river'). Avienus indicates in this work how during the lower Roman Empire olive plantations appear on the Mediterranean coast, such as the Tarragona plantations, in the area of Alcañiz (lower Aragon).

Some Romans made great wealth by marketing olive oil. In Roman times there were various Roman traders such as M. Casius Marci, from Oducia (Lora del Río); M. Iulius Hermesianus and D. Caecilius Onesimus, from Astigi (Écija); as well as numerous merchants from Hispalis (Seville), Astigi, Carmo (Carmona) and Corduba (Córdoba).

During the empire, the barbarians were also defined by their eating and cooking habits; thus the barbarian peoples were accused of being incapable of cooking with olive oil and of being consumers of animal fats.

Indeed, the olive and oil technology had reached Gaul, through the Phocian colony of Marseille, in the same period as for Rome, or, for some, even before, in the 7th century BC. Celts then consumed oil, not as food, but for other purposes, either produced in what would become Roman Provence, or imported from Greece (Bally *et al.*, 1962) .

Nevertheless, the Romans used animal fats in certain non-culinary applications, such as the production of medicines, the preservation of food etc., as well as in culinary applications such as the feeding of soldiers and the most humble people. For this reason, Alexander Severus reduced the price of oil eight times so that it could be a basic food for the population.

The production of olive oil generated an agrarian landscape in the Roman colonies, while its production decreased in the Italian Peninsula.

The Romans were not the only ones to spread the olive tree throughout the Mediterranean areas. The Tyrians (inhabitants of the city of Tyre), whose origin was Phoenician and founders of Carthage, did so in North Africa. In 29 BC, Octavian founded the Roman colony of Julia Carthage, which became the capital of the Roman province of Africa Proconsularis, one of the most important cereal-producing areas of the empire. Its port was vital for the export of African wheat and oil to Rome. The city became the second most important in the empire with 400,000 inhabitants. Olive oil was also produced in Tripolitania.

North Africa was therefore an important source of olive oil for the empire. Production was mainly in present-day northern Tunisia and northern Morocco (Tingitania) (Fig. 1.13).

The presence of *Olea europaea* L. has been demonstrated since the Villafranchian Stage in the Sahara and since 12,000 BC in North Africa. When the Romans arrived, the Berbers already knew how to graft wild olives and the cultivation that the Carthaginians had begun was to be considerably expanded by the Romans.

The first evidence of the olive tree found in Tingitania – in a necropolis near Tangier – dates from the period of first contact between Phoenicians/Carthaginians and Numidian communities. Olive fruits were placed as an offering in a 7th century BC farmer's grave. From the start of the Roman occupation of Tingitania, the importance of olive-growing led to an impressive number of isolated rural sites round the main urban centers.

During the Republic and in the first years of the empire, Rome did not devote much attention to developing the crop, since it was worried by the long and difficult struggle against the Musulamii and other berber peoples. In Caesar's time, the

Fig. 1.13. Roman oil press in Volubilis, Morocco.

olive had not yet started to spread, even though the *Lex Manciana* had encouraged olive cultivation. The policy of the Iulio-Claudian and Flavian dynasties was surpassed by that of Antoninus and Severus, the latter being himself of African origin, effective in boosting olive cultivation. The olive became one of the peaceful methods of encouraging settlement. Olive-growing tied the inhabitants of the steppes to the land and checked nomadism up to the edge of the desert (Camps-Fabrer, 1996). The olive soon represented the main crop in large North African territories, which exported oil to Rome until at least the 5th century AD. Oil-driven prosperity, however, continued until Arabic invasion. The ensuing wealth led to the growth of local populations, and the creation of important cities, such as El Djem, today poorly inhabited.

Unlike Spanish oil, the oil from Africa had a bad reputation, according to Pliny. At any rate, if the emperor was able to impose heavy taxes in the form of olive oil from the African provinces for daily distribution in Rome, then this oil must have been up to Roman standards.

The proximity of Tingitania to southern Spain, the geographic and climatic similitude, technical similarity in the layout and construction of oil mills together with age-old influences, all meant that these two provinces shared many aspects of olive cultivation (Fig. 1.13). The oil amphorae from Baetica found in Tingitania, although not very numerous, prove that the Mauritanians were able to compare the different oils and thus try to raise the quality of their production (Ponsich, 1996).

The destiny of Roman northern Africa was parallel to the decline of the empire.

Cheap wheat, first from Sicily and then from Africa, cheap wines and olive oils imported from the provinces of Spain and Africa to cope with the increasingly scarce and expensive Italian production, combined with administrative and political corruption, are some of the many causes of the decline of the Roman Empire. Its progressive decline is accompanied by that of agriculture, so important for the Romans throughout their history, both republican and imperial.

From the 3rd century onwards, we find Italy exhausted, lacking in cultivated land –

because it was first incorporated into vast estates and then abandoned to the neglect of slaves, with swampy areas in continuous increase, due to the interruption of overly burdensome drainage works – with rampant poverty – because the peasants, who became serfs, were no longer able to cope with the exaggerated imposition of gabelles.

These situations determined two very important facts: first, the landowners, worried about the danger of rioting of the peasants, tormented by hunger and destitution, began to build defensive strongholds to control the serf population. Secondly, poverty – more and more apparent in Italy – and wealth – more and more consistent in Asia Minor – determined the shift of power to the east of the Mediterranean.

Misfortune needs companionship. Poverty is often accompanied by a degraded social condition: crimes, banditry, infanticides, diseases. For this reason, from the 3rd century AD, helped by social and ethical instability and the spread of malaria, Germans and Gauls, morally and physically advantaged, began to impose themselves more and more in Italy. Their influence, though politically important, was instead economically unimportant, because, being the barbarians too few, and too primitive in their needs and their work, they could not direct local work and production, which therefore continued as before.

One of the few changes they introduced was the barbarian use of butter – and with it of lard and other pork fats, in daily use – and of oils made from rapeseed. Butter also served as a cosmetic for their blond and red hair.

Before Pliny there was no mention of soap. Pliny himself considered it an invention of the Celts and said that they made it with tallow and used it to enhance the red color of their hair. Galen, doctor and philosopher of the 2nd century, rates as excellent the soap of the Germans.

The multiplicity of events of those times and the changes in eating habits led to a significant reduction in the protected and selected production of Italic olive trees, magnified in several classical texts. The neglected olive groves soon turned wild in many areas. As a result, the use of olive oil was further reduced, not only as food but in all its other uses. This was due to many reasons: reduction of production, changed eating habits, changed traditions and, with time, reduction of population.

Olive oil continued to be present on the tables of the rich – mainly in central and southern Italy – but a little less on the tables of the poor, with the exception of sailors, who considered it irreplaceable for frying fish, their main food.

The end of antiquity, with barbarian invasions and the Greek-Gothic wars, saw oil production become very local, serving the needs of small farmers only.

1.10 The Olive in the Middle Ages

In western Europe, with the fall of the western Roman Empire, but above all with the loss of control of sea routes, trade, and therefore the use of oil, declined fairly quickly. Until the 6th century AD, traces of the introduction of oil from North Africa to Rome are still found (Brugnoli and Varanini, 2005) but from the 7th century sea transport to the ancient capital seems to have ceased (at least in an organized form).

The fall of the Roman Empire and the consequent social, economic and demographic collapse brought about, in many places, an abandoning of arboriculture in general and of olive tree-growing in particular. In northern and central Italy, social and political instability seems to have been such that very little arboriculture was practiced, despite the favorable legal measures promulgated in the high Middle Ages by the new civil and political authorities (the Rotari edict, the first Lombard set of laws, of 643, provided measures three times harder for cutting an olive tree than for other trees). This was apparently not the case in southern Italy where a precocious rebirth of fruit tree-growing seems to have taken place.

In the same epoch, a period of instability began for North Africa, marking the general abandonment of cultivation, which only resumed with the Arab conquest around the 10th century. Even the oil of Istria, renowned for its quality, disappeared from organized trade around the 7th century, and in some marginal areas (characterized by unfavorable climatic conditions), olive cultivation was undertaken to ensure self-sufficiency at least for the needs of churches and monasteries and for industrial purposes.

The climate was also worsening. The hot phase that favored the growth of the Roman Empire had ended and a very cold period started that pushed the peoples of the north to descend towards the shores of the great sea. The newcomers, in the beginning, did not like the all-Roman preference for oil as a condiment for vegetables and polenta, and preferred meat and fat, milk and lard. Only wine prevailed over beer. With the rulers, also the culture changed. The olive tree was replaced by the oak, which turned fields, vineyards and olive groves into woods. And in the woods, under the oaks, the wild pig ate acorns and got fat, to become lard and meat. Having been replaced by several types of animal fats, olive oil became a relatively unimportant commodity as a foodstuff.

Thus, oil succumbed to lard, and if it resisted, it was only because the Church required long periods of abstinence from meat and also from animal fats. However, only the rich few and the clergy could afford fish, and oil as a condiment and to illuminate sacred places.

It may therefore appear to be a contradiction that in the high Middle Ages olive cultivation and the production of olive oil seems to have been carried on throughout the Italian Peninsula, including the Po valley and even reaching up into some of the northern Italian alpine valleys, such as the Aosta valley, the Adige river valley up to Bolzano, and the swiss Canton Valais (Grieco, 1993; Fabbri, 2017). This diffusion in northern Italy of this typically Mediterranean fruit crop was a quite new phenomenon since the Po valley was not used in Roman times to commercially produce olive oil. Such an expansion was certainly aided by an increase in average temperatures, after five centuries of low temperatures, starting in the 9th century and lasting until the 13th century.

This apparent paradox can be explained by the fact that olive oil continued to be unsubstitutable for a whole series of liturgical or para-liturgical purposes. These uses have been described above, and this need was felt by the whole of European Christianity.

The actual presence of olive trees in northern Italy during the entire medieval period must not lead us to overestimate oil production, which remained, quantitatively, very small. Surfaces, and number of trees, were usually small; the trees were in most cases in the neighborhood of churches or castles, and a few trees were usually sufficient to satisfy local needs.

Under the Norman rulers in the 11th and 12th centuries, in spite of a disastrous frost which occurred in 1009, olive became a specialized crop in central Apulia, where, with time, more than half of the land became devoted to it, and olive oil from Apulia was sold on the international market. In the following centuries, the Swabians supported olive cultivation, applying economic policies that made olive oil from Apulia a desirable item for export (Iorio, 1985).

This transformation was favored by another factor, which, conversely, marked the end of the fortunes of olive oil production in the Po valley: the so-called 'revolution of medieval transportation'. Tariffs were no longer calculated according to the weight of the goods to be transported but on their value. This new way of calculating tariffs was responsible for stimulating the circulation of an increasing amount of heavy but cheap goods (timber, wine etc.), which also included olive oil.

The 12th century saw an increase of olive groves, entrusted to peasants with unlimited contracts, in exchange for money or a share of the produce. In most cases, however, oil production was directly managed by the property (be it monastic or feudal), requiring all the produce, while other crops might be subjected to sharecropping.

The mild temperatures, therefore, favor olive planting in the north of Italy, and numerous edicts were made through the 13th and 14th centuries to support, or order, plantations; but the worsening climate and the afore-mentioned tariffs revolution made large-scale production unsustainable, while production boomed in southern Italy, namely Apulia, which became the main producer for Venice.

Venice was particularly active in oil trade, at least since the 14th century. Special officers were appointed to control both imports and exports. Custom duties were three times higher for domestic productions (on the coasts of the Adriatic) than for oil from other Mediterranean locations. Special ships were built for oil transport, the 'marciliane'. By the 16th century, Venice had become the main actor in olive oil trade in the eastern Mediterranean (Mattozzi, 1981).

With the affirmation of communal freedoms and the establishment of independent states, agriculture flourished through the following centuries. In the 16th century, the demand for olive oil intensified, with the consequent development of intense trafficking of tons of oil to and from the Mediterranean, reaching northern Europe, the East and the Black Sea. The sought-after oils came from Italy (from the ports of Genoa, Leghorn, Gaeta, Naples and Bari), from Spain (Catalonia, Andalusia), Portugal and North Africa, but not from Greece, which, indeed, in the high Middle Ages, imported oil from Apulia (Speciale Giorgi, 2003). These oils arrived in Bruges, Paris, London, Constantinople, Alexandria and beyond.

Venice and Genoa developed intense oil trade with Greece (and particularly Crete, which recovered its production), with Spain and with North Africa.

The greatest transformation occurred in Apulia, which turned into an immense grove, also thanks to massive Greek immigration. Also, Calabria, Sicily and Abruzzo increased their production for export. The Spanish domination of southern Italy further stimulated oil production, which also involved Sardinia.

Florence was developing, the first in the West, a new innovative textile activity, anticipating by a few centuries industrial development: its wool industry requires large amounts of oil. Therefore the hills around the city were planted with great numbers of olive trees, which however were not sufficient to guarantee the required amounts of oil, hence the need of imports, mainly from Calabria and Campania (Vannucci, 2009).

The vicissitudes of olive growing and oil production and consumption in Spain were somehow different. During the Visigothic kingdom of Toledo, the presence of olive groves was noted by Saint Isidore of Seville, who for the first time distinguished the types of oil obtained in Andalusia.

Since the 8th century, a new culture of olive oil was formed among the Muslims in al-Andalus, as they could not use pork fat for religious reasons. Thus, the replacement of animal fats by olive oil (of vegetable origin) as a foodstuff became widespread. As a matter of fact, vegetable oils are halal (permitted), so their consumption is accepted by the dietary rules of Islam, and this made olive oil very popular during the occupation of the Iberian Peninsula (Garcia Sanchez, 1996).

The Arabs who inhabited the peninsula for almost 800 years found very productive plantations, from the economic point of view, in the area of today's Andalusia, already established since the time of the Roman Empire, which gave a very high yield, thus promoting monoculture in these regions. The Muslims of al-Andalus distinguished between two types of oil, those extracted from the green olive (zayt al-unfāq) and the oil extracted from the ripe olive (zayt al-zaytūn).

Despite this large production, the Muslims used the oil in moderate quantities and exported most of the production (Villar Raso, 1995). The Muslim writer of the time, Al-Idrisi, mentions that the wealth of the inhabitants of Seville was largely due to the export of olive oil. Other Muslim writers of the time mention olive oil, such as Ibn Zuhr in his *Kitāb al-Aġḏiya* (Book of Food); Ibn al-Baitar describes other oils of vegetable origin such as radish oil, which resembles rancid olive oil.

During the Emirate and Caliphate periods, there was a regional specialization in olive oil. In all Andalusia, and above all around Seville, there was the largest production of oil. Andalusian oil was exported to Alexandria. The main centers of oil production were, apart from Seville, Cordoba, Jaen, Valencia, Badajoz, Coimbra, etc., Sevillian olives were said to keep for 20 years and their oil never to go rancid. In the 10th century, Al-Razi said that production was such that, if the oil were not exported, the inhabitants of Seville would not be able to store it (Comet, 1996). However, from the second half of the 11th century onwards, the Andalusian olive grove started to decline and many olive growers were ruined.

With the Castilian conquest and repopulation of Andalusia, from the second half of the 13th century onwards we only find some allusions to extensive olive cultivation. But by the end of the 14th century the olive groves in Seville had recovered (Garrido Gonzàlez, 2004).

Provence was not an olive-growing area in the Middle Ages, not until at least the 16th century. Rare mentions appear in the 11th and 12th centuries (Draguignan, Marseilles, Nice). In the 13th century toll tariffs show that olive oil was being transported. More information is available from the 14th and 15th centuries, a time of economic depression. Carpological studies shed doubt on a widespread presence of olives which are mentioned only rarely in texts before the mid-15th century (Stouff, 1988). Olive oil is present in the diet of the population, but olive cultivation remains marginal.

Another important oil-consuming country in the Middle Ages was the Ottoman Empire. The product was imported from the subjected territories, such as the Aegean islands and North Africa. Some exports were recorded from Tunisia to Marseille (Inalcik *et al.*, 1995).

1.11 Modern Period

1.11.1 Italy

During the Renaissance, southern oil became very important for lighting and industry (Fig. 1.14). Demand from all European markets was very strong. Venice dominated in Apulia, which was transformed into a large olive grove, but there were also Genoese in Calabria and as far away as the Ionian Sea, along with Tuscans, Russians, Germans, Dutch and, increasingly enterprising, the English.

In every port in the region, Bari, Brindisi, Taranto, Otranto, there was a constant traffic of ships. And Gallipoli had diplomatic legations in half of Europe and consular representatives who remained there until 1923.

Although less intensively than in Apulia, in the other regions – Calabria, Abruzzi, Campania and Sicily – olive trees were planted by monks or friars and then by feudal lords, first for their own consumption and then for export.

Throughout the 16th century, olive oil passing through Venice remained competitive in price despite periodic natural shortages and tax burdens.

In Tuscany in the 16th century, the Medici administration favored the sale of wooded hillside land to the municipalities, with the obligation to rent it at a minimum price on condition that the tenant transformed it by planting olive trees and vineyards. Large estates were also created, real farms, with their own oil mills (Fig. 1.15).

After rapid expansion, there was a lull in the 17th century. At the same time as the climate cooled down considerably, wars and the continuous upheaval of the political and economic balance also affected olive production and oil trade.

There were years of bad harvests, particularly in southern Italy, and, under Spanish rule, an increase in the tax burden. New types of fixed-term contracts were also introduced, first for three years and then, from 1670, for two. Often the olive groves were abandoned because the yield did not compensate for the expenses and work.

In this period, in a general situation of agricultural depression, which was particularly severe because it was aggravated by high taxes in the south of Italy, there was a lack of care, so important in ancient times, both in olive growing, which was associated to other crops in mixed cropping systems, but to the detriment of the more profitable cereal growing, and in oil extraction.

Fig. 1.14. Medieval illustration of Dioscorides's *Materia Medica* (Codex Vindobonensis).

However, even during the 17th century, a difficult century for the whole of Italy, Venice was always supplied with oil, which was scarce here and there (Vannucci, 2009).

In those very years, the Serenissima responded to the difficulties of, let us say, foreign supplies (especially from Apulia, where pressure from other competitors, England in particular, was beginning to be felt) with an effort to extend olive growing to the Venetian lands on the Dalmatian coast – mainly Istria – and the Ionian islands (Corfu, Zante, Cephalonia) and Candia (Mattozzi, 1981).

Until the 17th century, the Garda olive groves produced enough oil to meet the needs of Verona, Brescia, Bergamo and Mantua. There was also some left over for export via the Tyrol to Germany. This production would gradually undergo a deep crisis, due to the combination of environmental conditions and the development of transport, particularly on rail.

It is worth noting that, especially for the large productions in southern Italy, and particularly in Puglia, there were two peaks of production and interest in oil corresponding to the development of wool production, first in Florence around 1300 and then from 1600 in England and Flanders, a demand which we have seen arise from the large quantities of olive oil needed for spinning and weaving wool. All this oil came mostly from Apulia and Calabria. The port of Gallipoli, in particular, served as a collector of Salento olives, which, after being pressed in the city's 40 oil mills (hypogeal), were sent by sea to northern Europe. The residual sludge was then

Fig. 1.15. Illustrations from Mattioli's *Materia Medica,* first edited in 1544. (a) wild olive; (b) cultivated olive.

the basis for soaps that were produced in Venice, Genoa and then Marseilles.

Many writers consider the year 1709 as critical because it was sort of a turning point between old-fashioned olive growing and modern cultivation. The freezing spell in January – the 'terrible winter' described in many chronicles – virtually banished oil from the market. Prices increased dramatically and it was an occasion, as often has been the case with similar events, for a renewal of the industry: new orchards replaced the damaged or destroyed ones and today's distribution of olive cultivation began to take shape (Martìnez Ruiz, 1996).

As a matter of fact, it is the sudden and permanent frost that causes the most damage, as has been the case with all frosts up to 1985. The 1709 frost, however, was one of the worst in history and concerned the whole of southern Europe: Greece and the Balkans, all of Italy, Provence and much of France and even Spain. Olive trees in particular suffered, almost all of which dried up and were abandoned in areas already suffering from climatic difficulties, in the plains or in valleys that had a northern exposure.

In Tuscany, on the other hand, the decimation stimulated a new expansion of olive growing, but the quantity of oil production calculated at the beginning of the century at 200,000 barrels (about 66,800 hectolitres) was only reached again after the middle of the century. And since that quantity was more or less that necessary for the self-consumption of the Grand Duchy, the price of oil remained very high for all those years, driving experimentation both towards new cultivation systems and towards new oils more suitable for lighting and manufacturing use.

It was during this period that production became more specialized. Oil for food use was concentrated in the most prestigious areas, such as Tuscany, while lampante oil increasingly came from the areas of highest production in the south.

Venice lost its commercial leadership and the areas around the great lakes in the north continued to produce for local consumption. The Po valley turned, as far as possible, to butter and lard.

It is therefore interesting in this period to follow the important evolution of olive growing in Tuscany between the careful Grand Ducal government and the birth of the first European scientific institution dealing specifically with agriculture. It was the Accademia dei Georgofili that studied and proposed to the enlightened government of Leopold of Lorraine innovations in the technical, economic and social fields to improve production and also the lives of farmers, and give advice to landowners interested in developing their crops. Already in its statute, the Accademia, which brought together scholars and shrewd landowners, showed concern for the 'cultivation of olive trees and other fruits to make up for their shortage. Of the way to make the best oil and preserve it' and began to establish calls for tenders with prizes of 25 zecchini for those who presented the best written treatises on various topics. Among these was the olive tree, which was defined as 'the tree most useful to the state, decorum, wealth and amenity of the Tuscan hills'. These similarities with the classics are not at all accidental.

In 1788, an award was established for olive cultivation by the Accademia, which also commissioned in 1805 a theoretical and practical treatise on the olive. The prizes were won by Giuseppe Tavanti with a beautiful catalog of the various types of olive trees grown in Tuscany, complete with color plates, which led to the complete theoretical and practical treatise on the olive in 1819 (Fig. 1.16). All this led to a period of remarkable growth in the first half of the 19th century. It was during this period that the Italian product, both edible and lampante, spread like wildfire across Europe: France, England, Belgium, Germany and even as far as Russia. The empress Catherine received a sample from the scholar Giovanni Presta, a doctor and erudite connoisseur, many bottles in an olive wood case.

In a treatise on the olive tree and olives, written towards the end of the 18th century with loving wisdom, Presta had undertaken to uphold the importance of the quality of Apulian oil in a very competitive market in which oils from Spain, Provence, Greece, Tuscany and the Genovese area predominated, and rightly so.

In 1700, the fine oils from Provence and Languedoc, which boasted not only the accurate cold-processing method but also the prerogative of remaining unaltered for a long time, were able to reach the Indies (Speciale Giorgi, 2003).

The 19th century strengthened the trend towards improving and extending valuable olive groves. In Tuscany, under the aegis of the Georgofili, there was an important development that reached 152,000 hectares of olive groves in 1830. This was also seen in the papal states, where a notification issued by Pius VII in 1830 guaranteed a premium of one paolo (compensation for a day's work by a laborer) for each olive tree planted and tended for up to 18 months: 38,000 olive trees were planted in Umbria between 1830 and 1840.

In the second half of the 19th century, however, this growth was followed by a slowdown in development, which became a contraction at the end of the century and the beginning of the 20th century. The climate did not improve and there were recurring frosts: in 1846, 1849, 1871, 1880, 1907, until the catastrophic one in 1929. This was the period of great emigration, with a decrease in the workforce, especially in the south, and the consequent abandonment of farming.

But the great emigration did have a positive effect on oil: it spread its use to all the continents and olive cultivation spread throughout the world where it had not yet arrived. Italian emigrants taught the use of olive oil for food in the United States and then in Australia, New Zealand and South Africa.

At the turn of the 19th and 20th centuries, the most important companies producing olive oil for the domestic market and for export were established. Companies that we

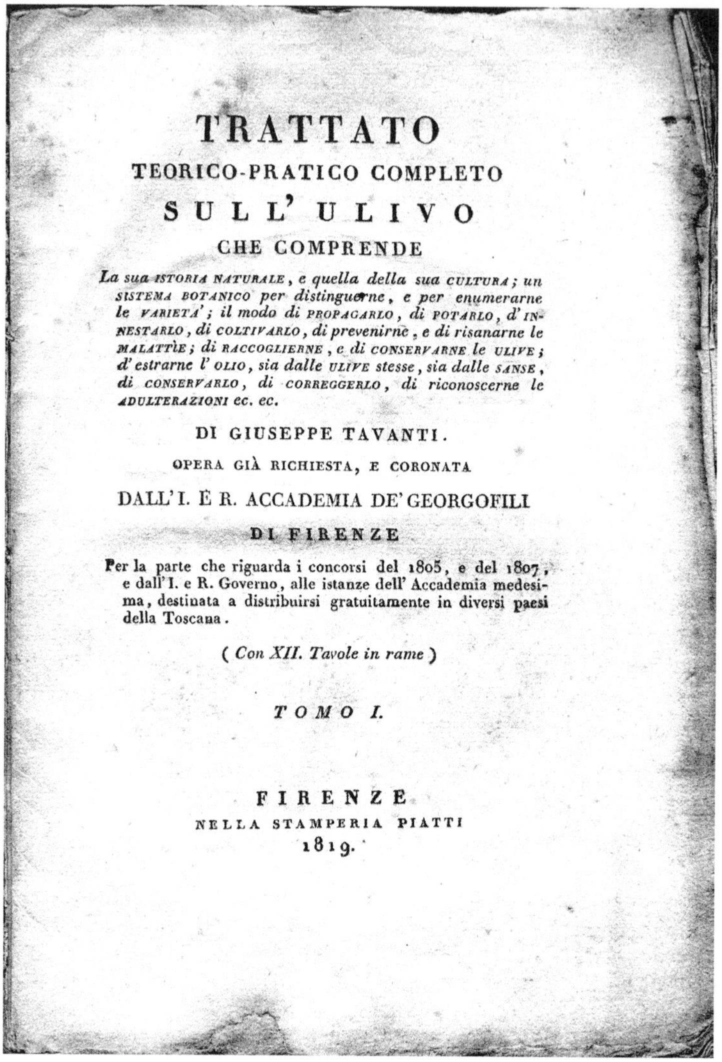

TRATTATO
TEORICO-PRATICO COMPLETO
SULL' ULIVO
CHE COMPRENDE

La sua ISTORIA NATURALE, e quella della sua CULTURA; un SISTEMA BOTANICO per distinguerne, e per enumerarne le VARIETÀ; il modo di PROPAGARLO, di POTARLO, d'INNESTARLO, di COLTIVARLO, di prevenirne, e di risanarne le MALATTÌE; di RACCOGLIERNE, e di CONSERVARNE le ULIVE; d'estrarne l'OLIO, sia dalle ULIVE stesse, sia dalle SANSE, di CONSERVARLO, di CORREGGERLO, di riconoscerne le ADULTERAZIONI ec. ec.

DI GIUSEPPE TAVANTI.

OPERA GIÀ RICHIESTA, E CORONATA

DALL'I. E R. ACCADEMIA DE' GEORGOFILI

DI FIRENZE

Per la parte che riguarda i concorsi del 1805, e del 1807, e dall'I. e R. Governo, alle istanze dell'Accademia medesima, destinata a distribuirsi gratuitamente in diversi paesi della Toscana.

(Con XII. Tavole in rame)

TOMO I.

FIRENZE
NELLA STAMPERIA PIATTI
1819.

Fig. 1.16. Tavanti's treatise.

can say would be famous when, two or three generations later, oil would lose its nature as a local product linked to the life of the self-producing farmers who kept it in the jar in the cellar, to arrive, labeled, on the shelves of the shops at the disposal of city consumers.

1.11.2 Spain

The Arabs, after their arrival on the Peninsula in 711, as great connoisseurs of the olive tree, improved the techniques of cultivation and oil extraction. In the district of Muslim Seville alone 30,000 oil mills were operating (Vannucci, 2009).

From the 11th century onwards, various schools of agronomy developed in al-Andalus, such as those of Toledo and Seville. The Andalusian agronomists studied the varieties of olive trees, the soil that was suitable for them, their fertilization, as well as the care needed to make them more productive.

In general, most of the agronomists describe the Aljarafe as an immense field, all covered with a green tapestry of olive trees. Towards the middle of the 12th century, Abu Zakariyya noted the vast expanse of olive groves surrounding Seville and the excellent quality of the oil produced in this area. Andalusian olive growing improved considerably under Muslim rule, especially in the Aljarafe region, which became a lush olive grove.

Agriculture improved its production thanks to irrigation, the Andalusians took up the Roman irrigation techniques, implanted and built watermills, irrigation ditches and canals. This made it possible to obtain a high yield from the land, to supply the large urban population and to introduce new products and horticultural practices hitherto unknown. Particular importance was given to those products whose trade was profitable, such as oil and olives, wine, oranges and vegetables and, in the dry land areas, olive oil and cereals (Garrido Gonzalez, 2004).

Between the 15th and 17th centuries, the expansion and geographical distribution of today's olive groves began to consolidate, with the greatest density of plantations in the center of Andalusia, in the provinces of Jaén, Cordoba and Seville.

Sixteenth-century inhabitants of Castile mostly consumed animal fats, especially pork lard, whereas in Aragon, Catalonia and Andalusia olive oil was basic to the local diet. These consumption patterns were closely tied up with medical and religious practices.

Nevertheless, from the 16th century onwards, two reasons favored the spread of olive growing: on the one hand, the demographic increase in Spain and, on the other, the discovery of America. But also the internal consumption, as a foodstuff as well as an industrial raw material (soap, ointments etc.) and for religious purposes, increased (Garrido Gonzalez, 2004).

Thus, the Catholic monarchs favored its cultivation, which they extended to inland areas of the peninsula, such as Aragon. In Jaén, in the region of Andùjar and Arjona, in the 40 years prior to 1517, there is evidence that more than 20,000 olive trees had been planted and the number of olive mills had increased from 7 to 22.

In Andalusia, prices shot up and between 1511 and 1559 the price of cereals increased by 209% whereas that of oil grew by 297%. Large estates were planted with olives and vines and were mostly tended by Moorish farm workers. From the descriptions of certain travelers and from the royal levies, it can be deduced that olives occupied large areas of land in Andalusia with some towns producing substantial amounts of oil.

The technological development of the industry in the following centuries was slow, albeit uninterrupted, as was oil production. The main market for Andalusian olive oil was the interior, favored by the improvement of transport from Andalusia to Madrid, with the opening of the Despefiaperros pass and the founding of La Carolina.

However, during the 17th century, perhaps due to the expulsion of the Moors from Spain between 1609 and 1614 and the great importance they had over agriculture, there was a recession in the production of oil in Spain. To this must be added the plague epidemic at the beginning of the century and the demographic contraction. This crisis, at the rural level, led to a decline in olive growing.

During the early 18th century, in spite of the growth in population and agriculture, the olive continued to be a secondary crop. Oil traders had a hard time because animal fats continued to be used for cooking, in preference to oil. Olive oil was used for religious purposes and for making soap and this may have been the reason for some new plantations (Martìnez Ruiz, 1996).

Nevertheless, in the middle of the 18th century, this crop started enjoying great importance: in some municipalities it began to spread, in others it became more established. It was from this moment onwards that the great expansion took place, culminating in the 19th century.

The central decades of the 19th century were clearly an expansive first period for the contemporary Andalusian olive culture; however, with the *fin-de-siècle* crisis between 1880 and 1896, the Andalusian olive industry suffered competition from mineral oils, animal fats and seed oils.

The formation of a new oil culture emerged in the large urban consumer markets, which were much more demanding in terms of quality, both in Andalusia and the rest of Spain, as well as abroad.

In the first third of the 20th century, the aim of Andalusian olive growers was to produce as much olive oil as possible, with less attention to quality. Cultivation was concentrated in Jaén, Córdoba and Seville. In the contemporary Andalusian olive-growing phase, from 1936 to the end of the 1950s, there was a strong intervention of the public authorities.

In the 1960s and 1970s, in a context of intense economic growth, the Andalusian olive sector experienced a negative phase, but the entry into the EEC (today the EU) in January 1986 favored the Andalusian olive industry (Garrido Gonzalez, 2004).

1.11.3 Rest of the world

The 15th century was a turning point in the history of mankind; in 1453 Constantinople fell, the last vestige of a civilization born almost through the cultivation of olives, while in 1492, in the same year, Granada, the last Muslim kingdom in Spain, fell and Columbus landed in San Salvador. The axis of Europe shifted westwards and the olive tree also sailed westwards.

Some 1520 documents in the Casa de Contratación (now at the Archive of the Indies, Seville) mention transfers of olive trees to the Americas. The first introductions were in the Antilles (Cuba), and afterwards in the continent. Mexico had regularly producing olive groves towards the end of the 16th century. From here, they expanded to Peru and then to Chile.

In Argentina, in 1562, Francisco de Aguirre planted the first olive trees from cuttings brought from Peru. As a result of multiple crossings, an indigenous olive variety called 'Arauco' was eventually established. There is a legend that says that

Fig. 1.17. Old olive tree in the Chelsea Physic Garden, London.

Fig. 1.18. Old olive tree in Itigi, Tanzania.

the viceroy of Peru, Pedro Fernández de Castro (Count of Lemos), ordered all the olive trees to be cut down because production was in competition with oils brought from the metropolis, and that an old woman kept a cutting that she hid and made it grow in other places. Nowadays, the Arauco olive tree, the only survivor of the felling ordered by the viceroy in 1770, is still cultivated; the name is from that of an Andean Indio tribe.

The olive tree reached the United States, namely California, in 1769 (Hartmann and Papaioannou, 1951), when it was introduced by Fray Junípero Serra, founder of the San Diego de Alcalá mission (today's San Diego). Years later, olive trees were planted by Franciscan fathers in the missions they established along the 600 miles of the Californian coast. Presently, an olive tree variety called 'Mission' is related to the selections of those early days.

In the next 150 years, trees were planted in several waves along with interest in olives and olive oil. Many of these older groves (80–200 years old) still exist in California. Most are in northern California. However, in the second half of the 20th century, olive growing became less profitable, and the industry had almost disappeared, at least for oil production. In the last decades, though, thanks to modern olive management, olive growing has been the object of a new wave of plantings.

In 1661, Dutch traders brought the olive tree to South Africa, where there still exists a small olive industry.

At the beginning of the 19th century, the species landed in southern Australia, where it found a favorable environment and gave rise to the phenomenon of natural spread of seedlings (feral olives), which, escaping from cultivation, repopulated wide areas of the territory with patches of olive trees, effectively following the ideal path that had been attributed to the domestic olive tree in the western Mediterranean. At the same time (1815), Franciscan monks introduced the olive in Western Australia. Olive plantings were established at a number of sites in New South Wales, South Australia, Western Australia, Victoria and, to a lesser extent, in other Australian states (Smyth 2002; Kailis and Considine, 2002). In the second half of the 20th century, interest in olive growing coincided with the post-war migration of people from European countries who were familiar with olive cultivation. Production of both olive oil and table olives in Australia is increasing, especially of quality products.

Successful attempts at introducing olive cultivation have been made in the same period in Asian countries like China, India and Japan. Although the tree can survive at more northern latitudes (Fig. 1.17), down to the equator (Fig. 1.18), its productivity is, if anything, erratic.

References

Amouretti, M.C. (1986) Le pain et l'huile dans la Grèce antique, de l'araire au moulin. *Annales littéraires del l'Université de Besançon*, 328. Paris.

Ampolo, C. (1980) Le origini di Roma e la "Cité antique". *Mélanges de l'École française de Rome* 92, 567–75.

Bally, W., Ferwerda, J.D. and Morettini, A. (1962) Tropische und subtropische Weltwirtschaftspflanzen. II. Ölpflanzen. F. Enke Verl., Stuttgart, Germany.

Barazani, O., Keren-Keiserman, A., Westberg, E., Hanin, N., Dag, A. *et al.* (2016) Genetic variation of naturally growing olive trees in Israel: from abandoned groves to feral and wild? *BMC Plant Biology* 16, 261.

Bartolini, G. and Petruccelli, R. (2002) Classification, origin, diffusion and history of the olive. FAO, Rome.

Bartolini, G., Prevost, G., Messeri, C. and Carignani, G. (1998) *Olive Germplasm: Cultivars and World-wide Collections*. FAO, Rome.

Besnard, G., Rubio de Casas, R., Christin, P.A. and Vargas, P. (2009) Phylogenetics of Olea (Oleaceae) based on plastid and nuclear ribosomal DNA sequences: tertiary climatic shifts and lineage differentiation times. *Annals of Botany* 104(1), 143–160.

Besnard, G. Khadari, B., Navascués, M., Fernandez-Mazuecos, M., El Bakkali, A. *et al.* (2013) The complex history of the olive tree: from Late Quaternary diversification of Mediterranean lineages to primary domestication in the northern Levant. *Proceedings of the Royal Society B: Biological Sciences* 280, 20122833.

Besnard, G., Teral, J.F. and Cornille, A. (2018) On the origins and domestication of the olive: a review and perspectives. *Annals of Botany* 121(3), 385–403.

Blazquez Martínez, J.M. (1996) Evolution and history: 19–20. *World Olive Encyclopedia*. International Olive Oil Council, Madrid.

Briccoli, M. (1925) Il clima dell'olivo in Italia, *Nuovi Annali di Agricoltura* V, pp. 333–372.

Brugnoli, A. and Varanini, G.M. (2005) Olivi e olio nel medioevo italiano. In: Brugnoli, A. and Varanini, G.M. (eds) Olivi e olio nel medioevo italiano. CLUEB, Bologna, pp. 3–100.

Camps-Fabrer, H. (1996) Evolution and history: 30–33. *World Olive Encyclopedia*. International Olive Oil Council, Madrid.

Caracuta, V. (2020) Olive growing in Puglia (Southeastern Italy): a review of the evidence from the Mesolithic to the Middle Ages. *Vegetation History and Archeobotany* 29, 595–620.

Carriòn, Y., Ntinou, M. and Badal, E. (2010) *Olea europaea* L. in the north Mediterranean basin during the Pleniglacial and the early-middle Holocene. *Quaternary Science Reviews* 29, 952–968.

Carriòn, Y., Ntinou, M. and Badal, E. (2013) Neolithic woodland in the north Mediterranean basin: a review on *Olea europaea* L. In: Damblon, F. (ed.) *Proceedings of the Fourth International Meeting of Anthracology*. Archaeopress, Oxford, UK, pp. 31–40.

Comet, G. (1996) The economics of olive oil: the middle ages: 50-52. *World Olive Encyclopedia*. International Olive Oil Council, Madrid.
Costantini, L. (1989) Grotta dell'Uzzo, Sicily: new evidence for the transition from Mesolithic to Neolithic subsistence in southern Europe. In: Harris, D.R. and Hillman, G.C. (eds) *Foraging and Farming: The Evolution of Plant Exploitation*. London, pp. 197–206.
De Candolle, A. (1959) [*1886*]. *Origin of Cultivated Plants*. Hafner, New York.
Di Rita, F. and Magri, D. (2009) Holocene drought, deforestation and evergreen vegetation development in the central Mediterranean: a 5500 year record from Lago Alimini Piccolo, Apulia, southeast Italy. *Holocene* 19, 295–306.
Diez, C.M., Trujillo, I., Martinez-Urdiroz, N., Barranco, D., Rallo, L., Marfil, P. and Gaut, B.S. (2015) Olive domestication and diversification in the Mediterranean Basin. *New Phytologist* 206, 436–447.
Duhamel, G. (1947) *Le Temps de la Recherche*. Hartmann, Paris.
Eitam, D. And Heltzer, M. (1996) *Olive Oil in Antiquity: Israel and Neighboring Countries from the Neolithic to the Early Arab Period*. History of the Ancient Near East / Studies – Vol. VII. Sargon, Padua, Italy.
Epstein, C. (1993) Oil production in the Golan Heights during the Chalcolithic period. *Journal of the Instiute of Archaelology of Tel Aviv University* 20, 133–146.
Fabbri A. (2017) The olive in northern Italy: a Mediterranean tale. *Rivista di Storia dell'Agricoltura* LVII(1), 25–56.
Fall, P.L., Falconer, S.E. and Lines, L. (2002) Agricultural intensification and the secondary products revolution along the Jordan Rift. *Human Ecology* 30, 445–482.
Figueiral, I. and Terral, J.F. (2002) Late Quaternary refugia of Mediterranean taxa in the Portuguese estremadura: charcoal based palaeovegetation and climatic reconstruction. *Quaternary Science Reviews* 21, 549–558.
Fiorino, P. (2008) L'evoluzione globale dell'olivicoltura. Atti Giornata di Studio 'L'olivo e il suo olio'. *I Georgofili, Quaderni* V, 23–51.
Friedrich, W.L. (1978) Fossil plants from Weichselian Interstadials, Santorini (Greece) II. In: *Thera and the Aegean World II*. London, pp. 109–128.
Galili, E., Stanley, D.J., Sharvit, J. and Weinstein-Evron, M. (1997) Evidence for earliest olive-oil production in submerged settlements off the Carmel coast, Israel. *Journal of Archaeological Science* 24, 1141–1150.
Garcia Sanchez, E. (1996) El consumo de aceite de oliva y otras grasas vegetales en Al-Andalus. *XIV Jornades d'Estudis Històrics Locals: La Mediterrània, àrea de convergència de sistemes alimentaris (segles V-XVIII) (Palma de Mallorca)* 15, 38.
Garrido Gonzàlez, L. (2004) *Historia del olivar y del aceite en Andalucìa*. Editorial Sarrià, Malaga, Spain.
Gerish, R. (2020) Egypt's oldest olive. *Aeragram* 9(2), 3.
Grieco, A.J. (1993) Olive tree cultivation and the alimentary use of olive oil in late medieval Italy (ca. 1300–1500). *Bulletin de Correspondance Hellenique* Suppl. XXVI, 297–306.
Gurbuz-Veral, M., El Bakkali, A., Essalouh, L., Tollon, C., Hakan, M. *et al.* (2018) New insights on olive domestication in Turkey. VIII International Olive Symposium, 15–20.
Hartmann, H.T. and Papaioannou, P. (1951) Olive varieties in California. *California Agricultural Experimental Station Bulletin* 720, 3–55.
Hyams, E. (1971) *Plants in the Service of Man*. Littlehampton Book Services, Littlehampton, UK.
Inalcik, H., Faroqhi, S., McGowan, B., Quataert, D. and Pamuk S. (1995) *An Economic and Social History of the Ottoman Empire, 1300–1914*. Cambridge University Press, Cambridge, UK.
Iorio, R. (1985) Olivi e olio in Terra di Bari in età normanno-sveva. In: *Quaderni Medievali* 67–102.
Kailis, S. and Considine, J. A. (2002) The olive *Olea europaea* L. in Australia: 2000 onwards. *Advances in Horticultural Science* 16 (3–4), 299–306.
Kaniewski, D., Van Campo, E., Boiy, T., Terral, J.F., Khadari, B. and Besnard, G. (2012) Primary domestication and early uses of the emblematic olive tree: palaeobotanical, historical and molecular evidences from the Middle East. *Biological Reviews* 87, 885–899.
Khadari, B. and El Bakkali, A. (2018) Primary selection and secondary diversification: two key processes in the history of olive domestication. *International Journal of Agronomy* 2018, Article ID 5607903.
Kislev, M.E., Nadel, D. and Carmi, I. (1992) Epipalaeolithic (19,000 bp) cereal and fruit diet at Ohalo II, Sea of Galilee, Israel. *Review of Palaeobotany Palynology* 73, 161–166.
Kloner, A. and Sagiv, N. (1993) The olive presses of Hellenistic Maresha, Israel. In: Amouretti, M.C. and Brun, J.P. (eds) *Oil and Wine Production in the Mediterranean Area*. Bulletin de Correspondance Hellenique, Athens and Paris, pp. 119–136.
Langgut, D., Cheddadi, R., Carrión, J.S., Cavanagh, M., Colombaroli, D. *et al.* (2019) The origin and spread of olive cultivation in the Mediterranean Basin: the fossil pollen evidence. *Holocene* 29, 902–922.

Lipschitz, N. (1997) Wood remains from two PPNB sites: Horvat Galil and Nahal Beset. *Journal of the Institute of Archaeology of Tel Aviv University* 24, 237–239.

Loussert, R. and Brousse, G. (1978) *L'Olivier*. Maisonneuve et Larose, Paris.

Martìnez Ruiz, E. (1996) The economics of olive oil: modern times: 52–53. *World Olive Encyclopedia*. International Olive Oil Council, Madrid.

Mattozzi, I. (1981) Crisi, stagnazione e mutamento nello Stato veneziano sei-settecentesco: il caso del commercio e della produzione olearia, Studi Veneziani, n.s. IV, pp. 199–276.

Meeks, D. (1993) *La production de l'huile et du vin dans l'Égypte Pharaonique*. Athens XXVI: 3–38.

Mercuri, A.M., Bandini Mazzanti, M., Florenzano, A. and Montecchi, M. (2013) *Olea, Juglans* and *Castanea*: the OJC group as pollen evidence of the development of human-induced environments in the Italian peninsula. *Quaternary International* 303, 24–42.

Morettini, A. (1950) *Olivicoltura*. Ramo Editoriale degli Agricoltori, Rome.

Moriondo, M., Trombi, G., Ferrise, R., Brandani, G., Dibari, C., Ammann, C.M. and Bindi, M. (2013) Olive trees as bioindicators of climate evolution in the Mediterranean Basin. *Global Ecology and Biogeography* 22, 818–833.

Nisbet, R. (1997) Arene Candide: charcoal remains and prehistoric woodland use. In: Maggi, R. (ed.) *Arene Candide: A Functional and Environmental Assessment of the Holocene Sequence (Excavations Bernabo' Brea-Cardini, 1940–50)*. Istituto Italiano di Paleontologia Umana, Rome, pp. 103–112.

Ntinou, M. and Tsartsidou, G. (2017) Domestic and ritual use of plants and fuels in the Neolithic Cave of Alepotrypa, southern Peloponnese, Greece: the wood charcoal and phytolith evidence. *Quaternary International* 457, 211–217.

Ponsich, M. (1996) Evolution and history: 34–36. *World Olive Encyclopedia*. International Olive Oil Council, Madrid.

Remesal Rodriguez, J. (1996) Evolution and history: 47–49. *World Olive Encyclopedia*. International Olive Oil Council, Madrid.

Smyth, J.D. (2002) Perspectives of the Australian olive industry. *Advances in Horticultural Science* 16(3–4), 280–288.

Speciale Giorgi, C. (2003) La coltivazione dell'ulivo dall'antichità ad oggi. Adda editore, Bari.

Stouff, L. (1988) L'olivier et l'huile d'olive en Provence aux derniers siècles du Moyen Age. *Provence historique t.* 38 fasc. 152, pp. 181–191.

Tanasi, D., Greco, E., Noor, R.E., Feola, S., Kumar, V., Crispino, A. and Gelis, I. (2018) 1H NMR, 1H–1H 2D TOCSY and GC-MS analyses for the identification of olive oil in Early Bronze Age pottery from Castelluccio (Noto, Italy). *Analytical Methods* 23.

Therios, I.N. (2009) *Olives*. Crop Production Science in Horticulture. CAB International, Wallingford, UK.

Thiébault, S. (2001) Anthracoanalyse des établissements néolithiques de la région liguro-provençale. *Bulletin de la Société Préhistorique Française* 98, 399–409.

Thiébault, S. (2003) Les paysages végétaux de Chypre au Néolithique: premieres données anthracologiques. *Bulletin Correspondance Hellenique* 43, 221–230.

Vannucci, S. (2009) Storia dell'olio, in 'L'ulivo'. Bayer CropScience, Milan, Italy.

Villar Raso, M. (ed.) (1995) El zoco: vida economica y artes tradicionales en Al-Andalus y Marruecos. El legado andalusí. Mini-catálogo de la exposición.

Walker, D.S. (1960) *The Mediterranean Lands*. Methuen, London.

Zohary, D. and Spiegel-Roy, P. (1975) Beginnings of fruit growing in the Old World. *Science* 187, 319–327.

2.1 Anatomy and Morphology

A. Fabbri[1] and H.F. Rapoport[2]
[1]*Department of Food and Drug, University of Parma, Italy;* [2]*Department of Crop Protection, Institute for Sustainable Agriculture (IAS-CSIC), Cordoba, Spain*

2.1.1 Description and Habit

The olive (*Olea europaea* L. subsp. *europaea*, var. *europaea*; Green, 2004) belongs to the family *Oleaceae*, a family widely distributed in temperate and tropical regions and consisting mainly of trees and shrubs but also including climbers.

The olive is an evergreen tree, with gray-green leaves and small, white flowers that produce abundant pollen, and which grows spontaneously in relatively dry, rustic conditions with a Mediterranean climate, i.e. with low rainfall concentrated in spring and fall, and dry summers. Numerous wild, cultivated and feral genotypes exist in the Mediterranean basin, where it is widely spread. The abundance of variability, due to prevalent incompatibility and to a millennia-long period following domestication has produced a range of tree architectures and fruit morphology. Environmental conditions and agricultural management practices, in particular pruning and irrigation, add greatly to both tree and fruit forms.

The cultivated olive usually grows as a tree with a large central trunk that becomes characteristically gnarled and twisted in old trees, or as a multi-trunk, bush-like structure. As a tree it can attain a height of 5–10, up to 20, meters in monumental trees (Figs 2.1.1, 2.1.2).

The tree trunk and limbs have a high capacity for forming new lateral branches from numerous meristematic zones or axillary buds formed at the time of leaf initiation, which remain latent for an indefinite number of years, and suckers sprout frequently around the collar at the trunk base. With time, the trunk base may take a conical form, also due to erosion that exposes the top part of the root system (Fig. 2.1.3). The olive tree's high branching capacity and the longevity of latent buds, which may remain dormant for a long time, have further implications for its growth habit and agricultural management. Sprouting provides healthy regrowth following pruning or grazing and has been useful for traditional vegetative propagation and for the proverbial capacity to survive, sometimes for several centuries, of individual trees; monumental trees are many in areas of traditional cultivation.

Heavy branching and the accompanying formation of numerous leaves also tend to produce a dense canopy which, for optimum fruit production, requires thinning to allow light penetration and reduce disease, particularly so in colder climates and dense planting systems.

Trunk and branches, leaves, fruits and roots all contain structural features considered xerophytic adaptations, which contribute to reduction of water loss (see Chapter 2.3). Being a woody plant, olive seedlings are

DOI: 10.1079/9781789247350.0002

Fig. 2.1.1. Huge olive tree along the Swan river, Western Australia. In spite of the size, the tree cannot be more than two centuries old.

Fig. 2.1.2. Far older, monumental olive tree, Apulia, Italy.

characterized by a long juvenile period, as much as 10–15 years under natural conditions, before the tree is capable of forming flowers. In its reproductive behavior the olive tree tends to relatively pronounced alternate bearing, in which years of heavy flowering and fruiting (on years) alternate with those with reduced crop levels (off years). While these processes can be modified by environmental conditions and agricultural management, there is a genetic component involved in the regulation of both the juvenile period length and the tendency to alternate bearing, with differences expressed among cultivars (Rapoport *et al.*, 2016).

The fruit of the olive tree is unusual in the plant kingdom in that oil, in the form of triglyceride, is present in the mesocarp, and is not destined to support embryo development after germination. Olive fruits are consumed in the wild by birds and small mammals and under domestication by humans, mainly as extracted oil but also as pickled or processed fruit.

2.1.2 Vegetative Morphology and Anatomy

2.1.2.1 Trunk and branches

Olive tree architecture, the structural pattern of trunk and lateral shoot growth, shows

Fig. 2.1.3. Suckers emerging from old, exposed roots.

high phenotypic diversity as well as ontogenetic modification as the plant matures (Rapoport *et al.*, 2016).

Young olive plants show a prevalent tendency to form a single-axis trunk. Lateral shoots sprout along the trunk from one or both of the two lateral buds present at each node. Branched and unbranched nodes are distributed vertically along the trunk in a scattered or irregular manner, although large areas of no branching are infrequent. Within this diffuse pattern there appear to be relatively equal tendencies for different dominant zones of more numerous and more vigorous branching: basitony near the trunk base, mesotony in the central region, acrotony in the upper or distal zone, or the absence of a dominant zone (Fig. 2.1.4). Branch orientation is more frequently upright (orthotropic) than horizontal (plagiotropic), and while the majority of branches are straight there may be many that bend downward due to genetic determination. Unpruned branches may also bend due to fruiting weight.

The branching system is monopodial, in which the main shoot and branch axes are maintained and their apices continue to be active, rather than being overtaken by lesser-order lateral shoots. In older trees, branching follows a proleptic pattern in which new-shoot sprouting from dormant lateral buds occurs periodically, once a year. In contrast, young trees and regrowth following heavy pruning often present syllepsis, where multiple orders of branches develop in the same growth season due to reduced bud dormancy (Rapoport *et al.*, 2016). Sylleptic branching in the olive has a significant genetic component and a strong role in early tree morphology (Ben Sadok *et al.*, 2013).

Vigor is a morphological characteristic, which refers to overall plant growth rate and size. In cultivated olive varieties, vigor is considered to be an important agronomic component which is relevant to juvenile period length and variety choice for planting density and has noted genotypic tendencies (de la Rosa *et al.*, 2006; Rallo *et al.*, 2008; Farinelli and Tombesi, 2015). Olive tree vigor may be classified qualitatively as high, medium or low, as in the official Olive Oil Council cultivar passport criteria (Barranco *et al.*, 2000), or measured as canopy volume, canopy surface parallel to the

Fig. 2.1.4. Young olive tree.

ground, or trunk diameter at a specified distance from the soil surface (Rallo *et al.*, 2005). Although trunk growth is clearly influenced by multiple factors, a molecular marker for olive trunk diameter has been recently identified at the seedling stage (Atienza *et al.*, 2014).

The trunk, always present, with few exceptions, in cultivated olive trees, has a uniform greenish-gray cortex, which, with time, tends to turn to gray, with irregular, rough and gnarled surface (Fig. 2.1.5). Trunk growth is not uniform, annual growth rings often being more than one, depending on the environment, and with different thickness in different positions of the ring. The development of branches determines further irregularity in cambial activity, with the formation of characteristic longitudinal reliefs. The trunk of old trees when sectioned transversally shows very twisted rings, made visible by the dark lines of late season wood.

The plant is rich in different orders of branches, although this depends mostly on the training system. As they age, branches tend to resemble the trunk. Branches carry shoots; as a rule, the term 'shoot' is given to vegetative formations of up to the second year since sprouting, i.e. still carrying leaves.

2.1.2.2 Shoot

The shoots are of two types, vegetative (with only leaf buds) and fertile or reproductive (with predominantly flower buds). There are intermediate shoots in this respect. Particular types of shoots are water shoots, developing on branches of different age, vertically and very vigorous; they usually develop from latent or adventitious buds. The same occurs with suckers, which originate from the collar, from 'ovules' (the name given in Italy to a type of woody burls) and other tissues, and derive either from latent buds or from adventitious meristems (sphaeroblasts). Both water shoots and suckers are very vigorous, with large leaves, often with a phyllotaxy resembling that of seedlings, as well as juvenile features such as a higher tendency to adventitious rooting.

The normally positioned shoot at the end of first year growth (September/October) is slender, lepidote (covered with scurfy scales), gray, and is made of nodes, which carry a pair of opposite, simple leaves, and internodes; one very small, lepidote-sericeous flower or leaf bud is at the axil of each leaf (Fig. 2.1.6). The internode may be 3–10 cm long, and is characterized by an elliptical (thinner shoots, 2.0–2.5 mm) or roundish to quadrangular (thicker shoots) section. The first tissue found from the outside is the epidermis, made of prismatic-rectangular cells, relatively large tangentially, thin and strongly cuticularized. Further inside is the periderm, or primary cortex, made of two to four series of dead and suberized cells (phellem or cork), the phellogen, a unicellular meristem, which can be very hard to detect because at this time it has terminated its activity, and the cork parenchyma, the phelloderm, three to six series of cells that appear polygonal in transverse section and are relatively elongated in longitudinal section (Fig. 2.1.7a, b). More internally, the cortex is separated from the phloem by a sclerenchymatous ring. This is formed by a number of fiber bundles that surround the phloem and are positioned at regular distance along the ring. Between the bundles can be present a number of sclereids, isodiametric although fairly irregular in shape; the sclereids derive from the gradual differentiation and thickening of interbundle parenchyma cells (Fabbri, 1980; Fig. 2.1.8), and at the end of the season most shoots present a continuous ring, which in a tridimensional representation is an uninterrupted sheath; the continuity of the ring is less evident in the younger parts of the shoot, i.e. the most distal and thin ones.

Fig. 2.1.5. Gnarled trunk of old olive tree, northern Cyprus.

Fig. 2.1.6. Detail of a node before sprouting.

Below the sclerenchyma is the phloem, made of up to 10–12 cell layers: its most external cell layers are made of small cells, irregularly arranged since they derive from bundles of primary phloem, which eventually constitutes a continuous tissue. Their

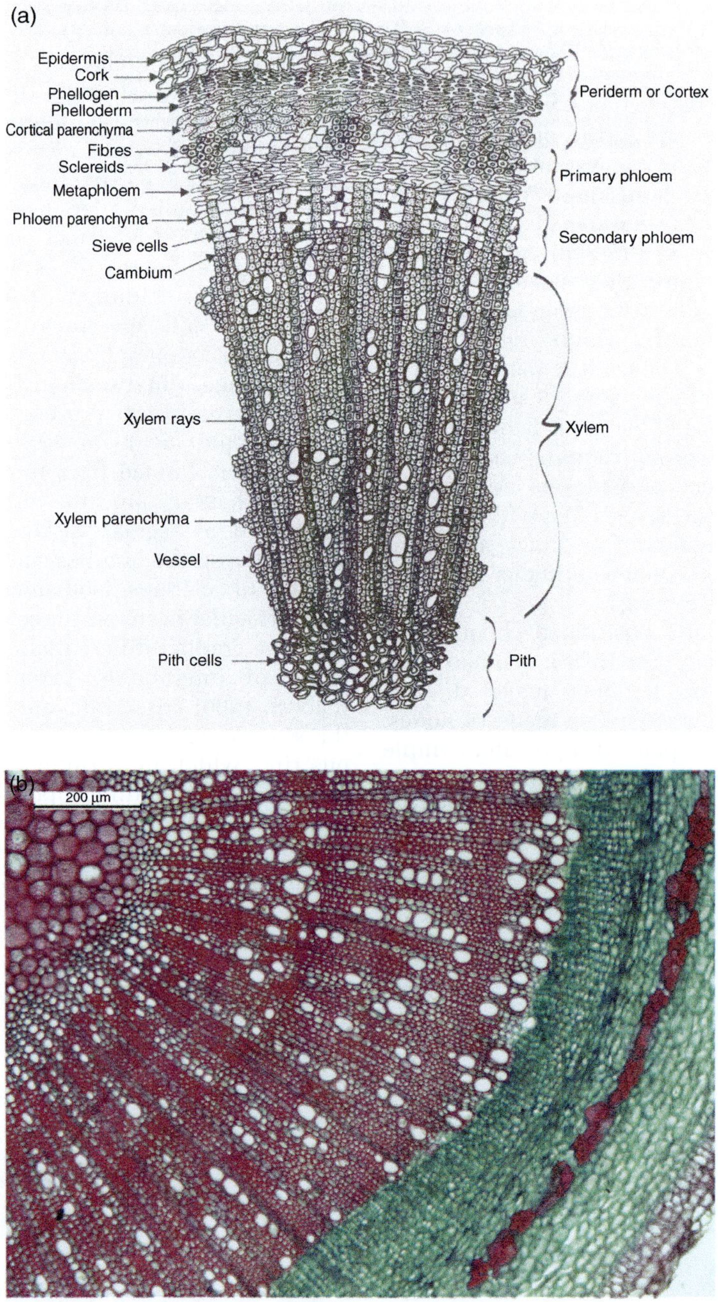

Fig. 2.1.7 a. Section of one-year shoot (from Troncoso *et al.*, 1975); **b.** Section of a one-year-old shoot at season's end. Often the section describes a succession of vegetation flushes, due to environmental conditions, mainly soil water content. Here, besides two main flushes, a minor vegetative recovery is visible on one sector of the cambium.

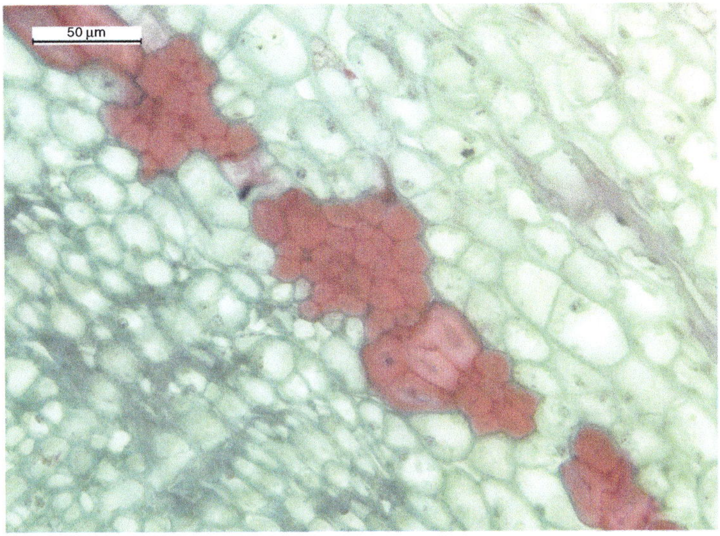

Fig. 2.1.8. Sclerenchyma ring (red), made of sclereids and fibers. Fiber bundles surround the central cylinder, and parenchyma cells between them are gradually transformed into sclereids, to form a continuous sheath at season's end. In the picture, sclereids (larger cells) are at different levels of differentiation.

size also depends on pressure exerted by the growth of inner tissues, compressing them against the sclerenchyma. More internally is the secondary phloem, characteristically made of the various cell types: sieve tubes, companion cells, parenchyma cells, phloem ray cells; these are the last-formed, and still functioning, also in view of the next growing season. Sclerenchyma and phloem (primary and secondary) constitute the pericycle (Fig. 2.1.9).

The cambium at this stage is made of only one layer of small cells, more if there still is some activity. Its cells are rich in cytoplasm, and no vacuoles (abundant in all parenchyma cells) are present.

The xylem forms a closed cylinder traversed by narrow rays. It is of the homoxylous, diffuse-porous wood type, which means that large vessels are absent, and that vessels, fibers and parenchyma cells show similar transverse size, with vessels slightly larger. Rays are frequent but thin, one to four cell rows, usually two to three (Fig. 2.1.10). The vessels are vasicentric tracheids with rare spiral thickening, 200–800 μm long (Baas *et al.*, 1988), with simple perforations (but occasional scalariform plates are also recorded); they are often in radial patterns of four to five (Metcalfe and Chalk, 1953). In a one-year-old stem, vessel diameters are typically less than 40 μm, with 90% in diameter classes around 20 μm (Lo Gullo and Salleo, 1990). In mature olive plants the vessel diameters can vary among different years and according to water availably. Tyloses are sometimes present, as are solid deposits. The xylem parenchyma, typically, is predominantly paratracheal. Fibers usually have small simple or indistinctly bordered pits. The pith is homogeneous, made of large polygonal (up to 100 μm and more) and vacuolated cells.

2.1.2.3 Leaf

The olive tree is evergreen and its leaves usually live for two to three seasons, rarely four, although most leaves abscise during the second season, especially when they are shaded. Leaf drop is determined by the activation of a pre-existing abscission layer, between the shoot and the petiole. The small, dorsiventral leaves are simple, coriaceous, with an elongated blade (3–9 cm long) with

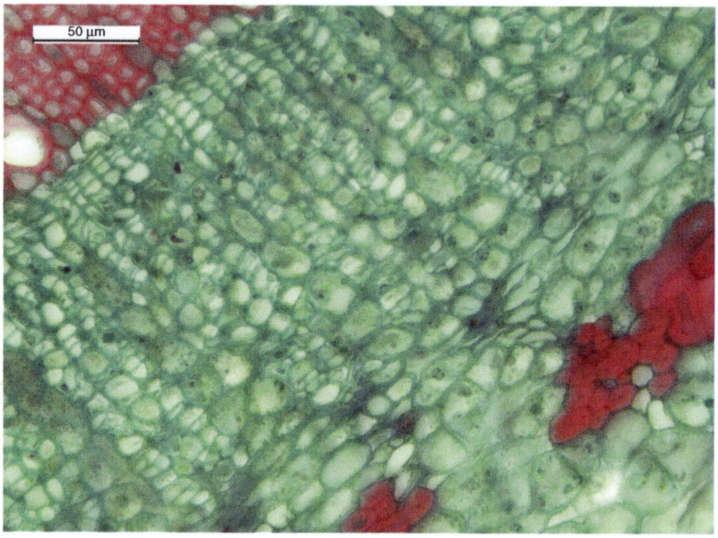

Fig. 2.1.9. Pericycle: external to xylem (*top left*) are secondary phloem, primary phloem and sclerenchyma (*bottom right*). The cambium, between xylem and phloem, is hardly detectable at this stage (autumn).

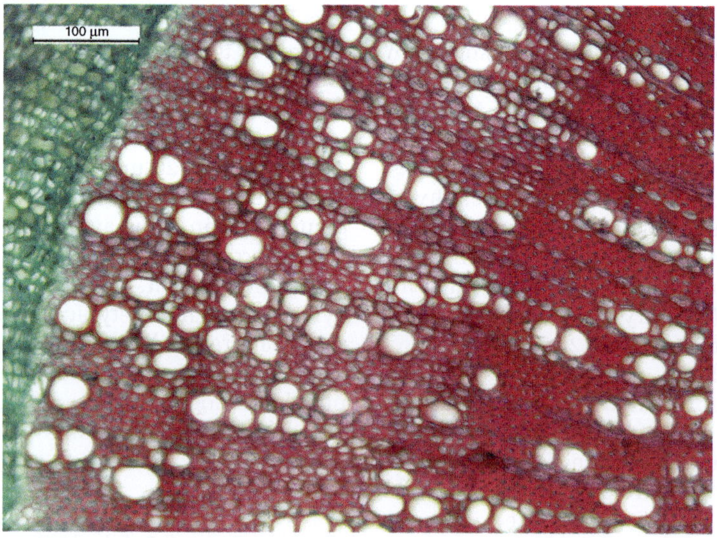

Fig. 2.1.10. Xylem. In this section (one-year shoot), different flushes of activity can be detected. A thin layer of cambium is visible.

entire (smooth) margins, and a very short petiole (approximately 0.5 cm). Width ranges between 1 and 2 cm, blade length being more than six times leaf width. The petiole, in transverse sections through the distal end, exhibits a shallow, crescentic vascular strand. Blade shape varies from the slightly wider and more symmetrical elliptical form to lanceolate or obovate; the width is often greatest at the base, but in some cultivars the largest point can be in the distal portion; the leaf is usually straight, but some cultivars can have a characteristically curved shape (Fig. 2.1.11). The tip is often pointed (mucronate). A thick, highly visible central vein divides the leaf in half along its length and protrudes along the lower surface. The leaves are positioned on the branches with decussate phyllotaxy, i.e. there are two opposite leaves (180° apart) at each node, alternating at right angles with those of the nodes above and below. Shoot growth producing new nodes and leaves may occur at any time of the year, depending on water supply, temperature and solar radiation, but usually it is only active or most active in spring and autumn (Connor and Fereres, 2005).

Olive leaf anatomy and surface characteristics are typical of drought-resistant sclerophyllous vegetation, including small size, thick cuticle, high stomatal density, high pubescence, and compact mesophyll cells (Bacelar *et al.*, 2004). The upper surface of the leaf is dark-green and glossy. The lower surface appears gray or pale in color due to the presence of numerous cutinized, multicellular trichomes in the form of umbrella-shaped peltate scales, each having a unicellular stalk and a shield with exclusively vertical divisions, appearing as transparent or sunken dots in the leaf, forming a dense protection of three to four layers against transpiration (Fig. 2.1.12a, b, c). Epidermis consists of square cells with straight or sinuous anticlinal walls; their size is apparently reduced in drought conditions (Bosabalidis and Kofidis, 2002); the upper epidermis is characteristically covered by a thick layer of cuticle (15–20 μm). The stomata, most abundant on the lower surface, but also recorded in small numbers on the upper side, usually ranunculaceous, are shielded by the mentioned peltate trichomes (Fig. 2.1.13).

Internally, the upper zone of the leaf mesophyll is composed of palisade parenchyma consisting of several layers, usually two to three, sometimes four, of compact, relatively elongated cells, rich in chloroplasts. (Fig. 2.1.14a, b and c).

The spongy mesophyll, composing the lower portion of mesophyll, is characteristically replete with relatively small intercellular spaces, and includes idioblastic sclereids (Green, 2004), also called elongated, filiform, fiber-like, T-shaped etc. sclereids or branched sclereids (Arzee, 1953), which are present throughout all mesophyll layers, a notable and widespread feature of *Oleaceae*. They can be isolated or in bunches of two to four, and are mostly present between palisade parenchyma and epidermis, and extend to spongy mesophyll. Their function is debatable, although they certainly increase leaf rigidity; they consist in pure cellulose, often lignified. They assume many shapes, often as T or Y, but can also be asymmetrically branched.

The abaxial, lower epidermis is composed by small cells like the upper epidermis, but the cuticle is thinner; stomata

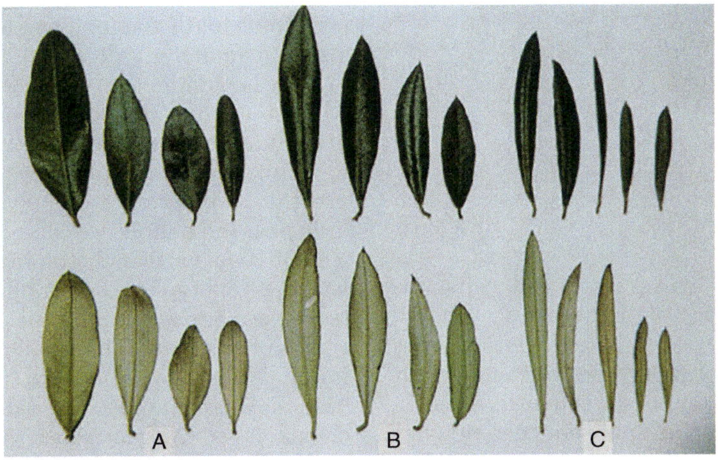

Fig. 2.1.11. Diversity of leaf shapes: wide (*left*); elliptic (*center*); narrow (*right*).

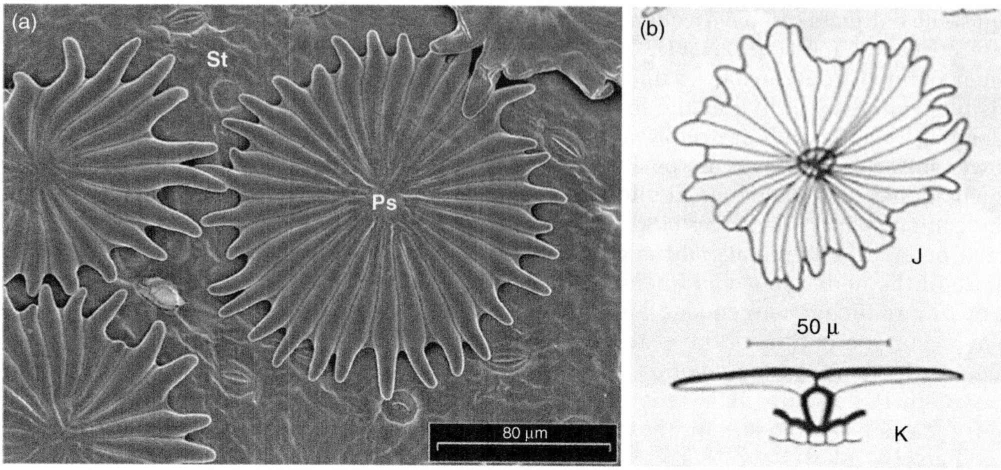

Fig. 2.1.12. Peltate hairs: (a) abaxial leaf surface surface. SEM image of young leaf (cv. 'Leccino'), lower surface. Trichomes (peltate scales (Ps)) and stomata (St). Trichomes will eventually cover the whole surface (Photo: Antonio Minnocci); (b) scheme of peltate hair structure. (From Esau, 1977)

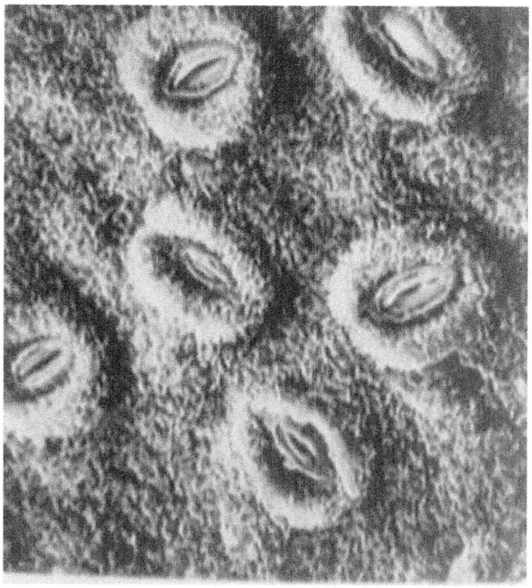

Fig. 2.1.13. Stomata on the leaf abaxial side, deprived of hairs.

have strongly cutinized guard cells. The epidermis is rich of depressions, each with a trichome. Directly below the lower-leaf epidermis, between that tissue and the spongy mesophyll, is frequently a second palisade or pseudo-palisade parenchyma, consisting of a continuous single layer of cells rich in chloroplasts, although shorter and less densely packed than the principal palisade tissue (Chartzoulakis *et al.*, 1999; Bacelar *et al.*, 2004; Moreno-Alías *et al.*, 2009). Within the leaf, where the palisade parenchyma and spongy mesophyll meet, is a dense reticulate network of secondary veins, which consist in vascular bundles, surrounded by fibers and a parenchymatic sheath rich in starch. The main bundle, that of the central vein, interrupts the palisade mesophyll, and is associated to sclerenchymatic and collenchymatic tissues (Fig. 2.1.15).

The veins, with the leaf sclerenchyma, contribute to the toughness of the olive leaf, also considered a typical structural feature of leaves adapted to dry environments. Also, small, acicular and prismatic crystals of calcium oxalate occur both in the epidermis and in the mesophyll; raphids can be present in palisade parenchyma, particularly near the petiole.

Olive leaf size and structure vary in relation to genotype and development, and in response to growth conditions. Leaves of juvenile plants tend to be smaller and have a lower length-to-width ratio, less peltate trichomes, and no second palisade parenchyma layer (Moreno-Alías *et al.*, 2009). Leaves from wild genotypes are also frequently small and more elliptical in form.

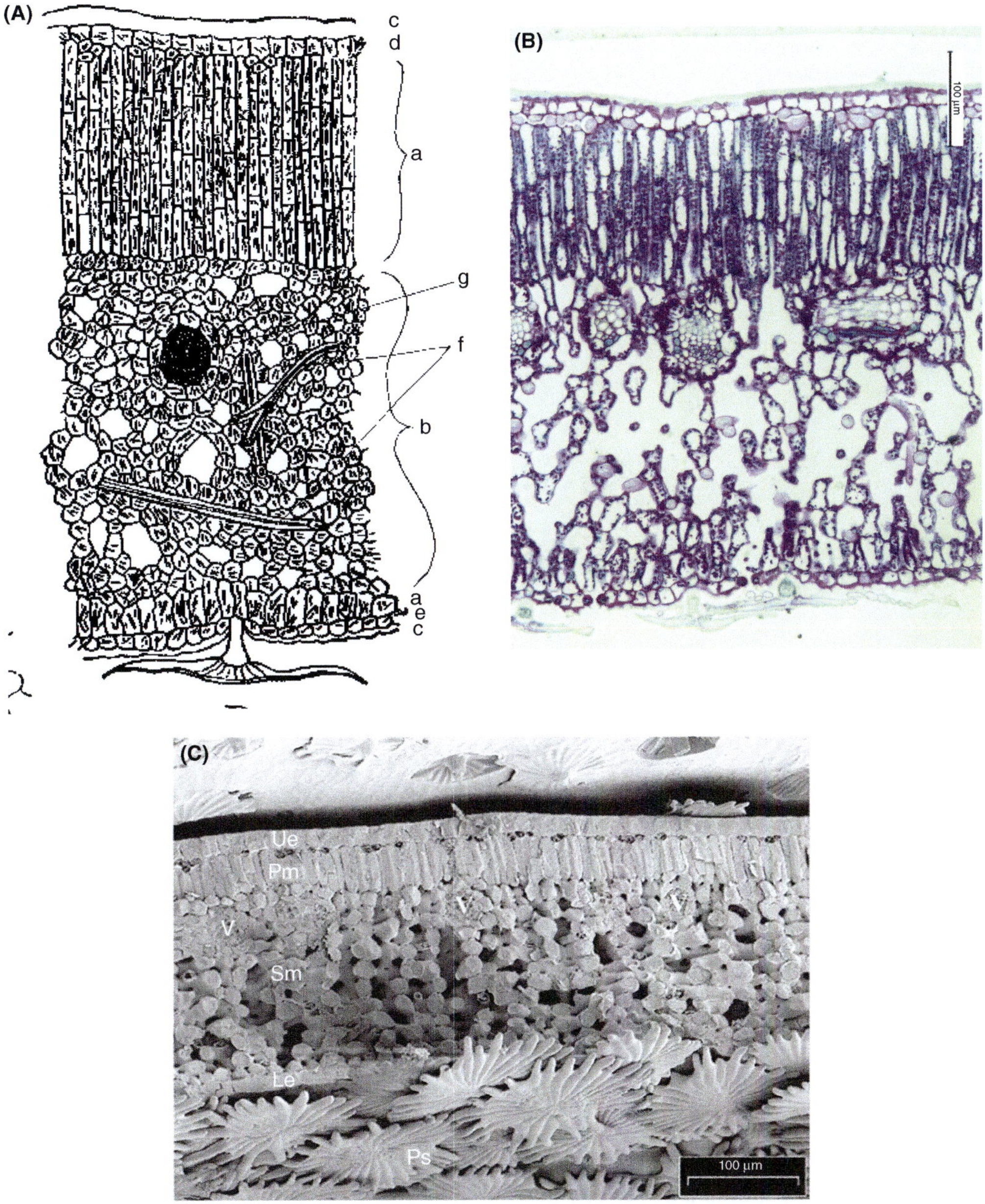

Fig. 2.1.14. Section of leaf mesophyll: **(A)** Scheme from Metcalfe and Chalk (1953); tissues: (a) palisade parenchyma; (b) spongy mesophyll; (c) cuticle; (d) upper epidermis; (e) lower epidermis; (f) sclereids; (g) cribrovascular bundle. **(B)** Section where purple dots are starch grains; **(C)** SEM image of young leaf (cv. 'Leccino'). The upper epidermis (Ue) and one layer of palisade mesophyll (Pm) are next to the upper surface; in the center is the spongy mesophyll (Sm) traversed by veins (V); and below, the lower epidermis (Le). The lower surface is covered by numerous overlapping trichomes (Ps, peltate scales). (Photo: Antonio Minnocci)

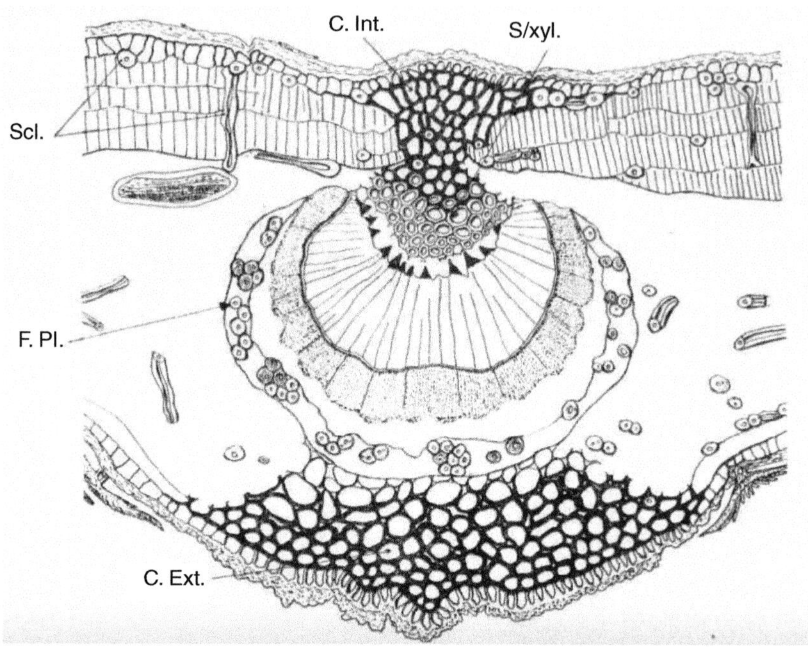

Fig. 2.1.15. Section of midrib, with collenchyma and sclerenchyma tissues. Around the central cribrovascular tissue are: external or abaxial collenchyma (C. Ext.); internal or adaxial collenchyma (C. Int.); xylem sclerenchyma (S/xyl.); periphloem fibers (F. Pl.); idioblastic sclereids (Scl.). (From Roland, 1961)

Overall palisade thickness has been correlated to resistance to *Spilocaea oleagina* (Peacock's spot) (Morettini, 1954) and the number of palisade layers to plant growth potential (Mazzamurro, 1973). The whole leaf anatomy formed during drought has also been correlated to drought resistance (Bacelar *et al.*, 2006). Under water stress, cell density increases for all leaf tissues, and the cuticle is substantially thicker (Chartzoulakis *et al.*, 1999), trichome number increases (Guerfel *et al.*, 2009; Boughalleb and Hajlaoui, 2011; Edziri *et al.*, 2018), but leaf size decreases (Ennajeh *et al.*, 2010).

Some figures on stomata, trichomes and cuticle are given by Therios (Table 2.1.1).

Table 2.1.1. Some characteristics of olive leaves. (From Therios, 2009)

	Upper surface	Lower surface
Stomates (n/mm^2)	0	470
Trichomes (n/mm^2)	18	143
Cuticle thickness (μm)	106.0	125.4

2.1.2.4 Buds

Olive buds are ordinarily either flower buds or leaf buds. Flower buds produce inflorescences while leaf buds produce shoots. In rare cases, mixed buds can be present, i.e. they produce a weak shoot carrying an inflorescence, but they can be found only on certain cultivars, and only on very vigorous trees (Fig. 2.1.16).

Another way to classify the buds is position: they can either be terminal, in which case they are always leaf buds, and lateral. Lateral buds can be flower buds or leaf buds. The prevalence of one of the two types determines the type of shoot and also has implications for pruning. Lateral buds are positioned on nodes at the axil of leaves and are therefore two per each node. In the preceding season the bud develops without sprouting, and its content is decided over the period between summer and winter. Apart from the rare occurrence of feathers, all buds develop in the year following their formation.

The buds are very small, and their content cannot be told before sprouting (Fig. 2.1.16). They do not have scales, but only a pair of smaller leaf primordia which protect the bud during the rest period, although the olive does not have a proper rest period, and sprouting of leaf buds can occur any moment the environment is favorable in terms of temperature. Flower buds also need a minimum of chilling to develop.

At any rate we conventionally distinguish resting buds from latent buds, which do not develop the next year, but stay inactive for an unpredictable number of years. These buds become numerous as a tree ages, and are responsible for the recovery after heavy frost damage or too-severe pruning. They are particularly numerous at the collar.

Another source of latent buds are accessory buds, present as a branching of the main bud, as described by Agati (1951; Fig. 2.1.17). These almost invariably become latent buds and are present in nodes as early as the second or third year of plant life. Latent buds can also originate in any tissue, apparently at random, and are also called adventitious buds. All shoots deriving from latent buds, i.e. years after their formation, are called adventitious shoots, and, accordingly, develop on branches and trunk. Sphaeroblasts are also formed on branches; these are small meristematic lumps that originate in the cortex without any connection with the central cylinder (Baldini and Scaramuzzi, 1956; Fig. 2.1.18). The connection is eventually established, when the conditions occur for an adventitious shoot to develop from them.

Fig. 2.1.16. Mixed shoot.

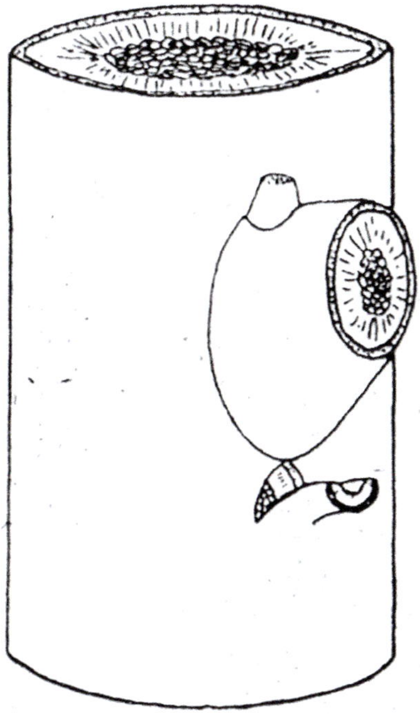

Fig. 2.1.17. Drawing of a node structure, with accessory bud. (From Agati, 1951)

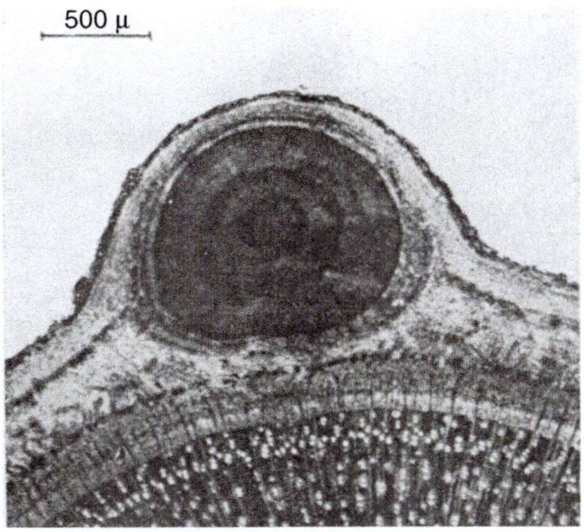

Fig. 2.1.18. Sphaeroblast. (From Baldini and Scaramuzzi, 1956)

One particular feature of the olive is the presence of vegetative ovules. These are hyperplasias that occur on the plant, at the level of the collar, i.e. at the transition between trunk and root system (Messeri, 1942), and appear as burls, lumps of tissue. After three years their size is that of a walnut, but eventually they keep growing and merge with other ovules to form larger masses that protrude from the collar, which, with time, assume a typical conical shape, particularly pronounced in very old trees, as the collar and underlying root system is gradually exposed due to erosion (Fig. 2.1.19). These formations are mainly responsible for the formation of suckers at the trunk base, and of adventitious roots when ovules develop below the soil surface. Their formation therefore appears to be an additional instrument for the plant to survive any adverse event and not a reaction to pathogens or traumas of any kind (Morettini, 1956). In the past, the ovules, separated from the collar with particular instruments, have represented an easy way to transport and clonally propagate olive cultivars.

Fig. 2.1.19. Ovules on the collar of an old olive tree.

2.1.2.5 Roots

Olive seedlings develop a single, vertical root, the taproot. In the case of transplanting (after propagation by grafting) the taproot is shortened and rapidly develops a number of secondary roots, with a mainly horizontal orientation. Olive cuttings, a major source of commercial olive trees, develop adventitious roots from shoot tissues, which grow both horizontally and towards the lower soil layers, initially with little inclination. After a few years it can be quite difficult to distinguish the propagation system from the root system's appearance, but there is no apparent influence on plant growth and productivity (Morettini, 1972; Guerriero *et al.*, 1972). For details on adventitious rooting see Chapter 4.

Olive tree root morphology depends greatly on growth conditions, including soil texture (and therefore air presence), nutrition, water availability, spacing, etc. In aerated soils, the angle of the root system is fairly narrow, and the roots can develop to a depth of 6 m and more, especially when water supply is limited (i.e. North Africa; Loussert and Brousse, 1978). Otherwise, the root system of the olive is generally considered relatively superficial, especially in the case of regular irrigation. As an average, most root systems develop in 20–70 cm depth, with exploring roots able to reach 1–1.5 m depth. Horizontally, the root system is much more extended than the canopy projection, an area two to three times larger. Again, in arid areas, the extension of the root system can be several times this value. A relationship exists between major branches and roots; this means that main roots growing less due to adverse soil conditions determine less related branch growth and unbalanced canopies.

Anatomically, the root is a typical dicotyledonous root, although the olive tree root presents attributes related to withstanding water stress, and, like the shoot, shows the capacity to rapidly modify its structure in relation to water availability and salt stress (Rapoport, 2010; Tataranni *et al.*, 2015).

In primary structure (Fig. 2.1.20), the first tissue from the outside is the epidermis, which in a subapical zone also carries root hairs. Root hairs are relatively short but numerous. More internally, there is a thick layer of parenchyma cells, the cortex. The external cell layers of the cortex, those that lie directly below the epidermis, eventually exhibit structural modifications to create a hypodermis or exodermis. The hypodermal cells undergo specialized secondary-wall differentiation, which includes thickening and deposition of suberin, acting as an additional barrier to water movement and protecting the root from water loss (Fig. 2.1.21). In fact, under water-limiting conditions, olive roots produce added hypodermal cell layers and increased suberization of hypodermal as well as endodermal cell walls (Tataranni *et al.*, 2015). The endodermis, a unicellular layer of radially suberized cells, which is a barrier to the entrance of unwanted molecules into the central cylinder, acts as a sort of border between the cortex and the central cylinder. The endodermis of olive roots is well differentiated and, furthermore, has demonstrated the capacity to undergo additional suberization in response to exposure to water (Lo Gullo *et al.*, 1998) and salinity stress (Rossi *et al.*, 2015). Salt stress determines a number of structural changes in roots, i.e. increased root diameter (which is in part caused by enlarged periderm cell diameter), and vessel modifications able to reduce overall hydraulic conductivity (Tan *et al.*, 2020). Periderm is the cortical tissue produced by roots in secondary structure.

2.1.3 Reproductive Anatomy and Morphology

2.1.3.1 Inflorescence and flower

Olive flowers are borne on inflorescences which originate as lateral buds formed in the leaf axils of the previous year's shoots or developing directly at the shoot apex. The buds are small, highly reduced shoots composed of four to five nodes, each node bearing two oppositely positioned leaf primordia (Fabbri and Alerci, 1999; de la Rosa *et al.*, 2000; Ramos *et al.*, 2018). All of the buds are structurally identical and lack any morphological differences which distinguish between a vegetative and reproductive future (Pinney and Polito, 1990; Andreini *et al.*, 2008). Subsequent bud fate depends on a complex combination of endogenous and exogenous factors, and includes the possibility of

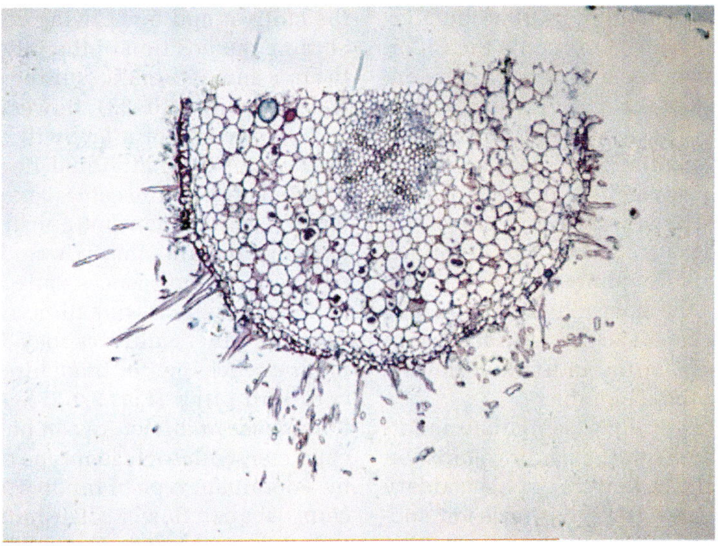

Fig. 2.1.20. Root primary structure. The cortical tissue is predominant, and root hairs are still present.

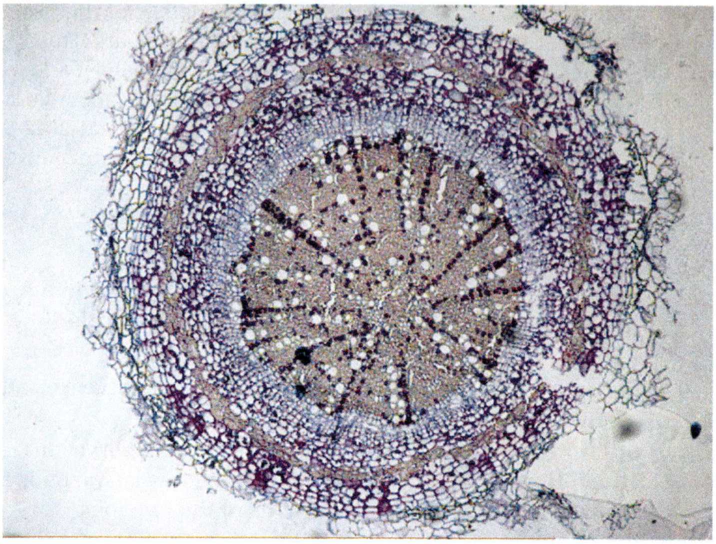

Fig. 2.1.21. Section of a root at season's end. The xylem is more homogeneous (no flushes as in shoots), the sclerenchyma ring is continuous, and the cortex is being eroded, together with the suberized hypodermis.

development of an inflorescence, growth of a shoot, continued dormancy, or senescence and abscission.

Inflorescence development occurs in the spring of the year following bud formation, when buds have received the necessary induction and initiation stimuli, with differentiation immediately following initiation. When bud differentiation is reproductive, the bud axis elongates to become the inflorescence axis; branches and flowers differentiate in the axils of the bud's leaf primordia, and leaf primordia cease growth. In contrast, when bud differentiation is vegetative, the leaf primordia expand to become leaves, while the bud axis elongates and grows as a new branch (Fig. 2.1.22, 2.1.23 a–c). Following bud break, full inflorescence development, including the elongation and branching of the inflorescence axis and the formation and development of the individual flowers requires two to three months.

The 1.0–7.0 cm-long paniculate inflorescence has a central axis or rachis on which individual flowers or secondary flower-bearing axes (rachillas) arise at successive bilaterally symmetrical branching points (Fig. 2.1.24). Individual flowers may have a short pedicel (stalk) or be sessile; that is, arising directly from the inflorescence axis. The panicle is connected to the shoot by a peduncle, and individual flowers to the inflorescence by pedicels (Fig. 2.1.25 a, b). Flowers are often arranged in triads on the panicle (Fig. 2.1.26). The number of flowers in an inflorescence is influenced by the cultivar and by growing conditions, including the position of the inflorescence on the tree and within the fruiting shoot. Of the 5–60 (usually 10–35) flowers (Morettini, 1972), only one or a few will set fruit. The pedicels of the individual flowers may be inserted both on the central axis and on secondary ones. When not apical, they are in opposed pairs at the axis nodes. In some cultivars, inflorescences only consist of a main axis, with flowers directly inserted on it, while other cultivars may instead have inflorescences with branching of second and third order (Fig. 2.1.27). Although inflorescence architecture can be a taxonomic character, with each genotype characterized by a dominant type of inflorescence, as concerns shape, flower distribution, pedicel and peduncle length, flower size and number, on the same tree their shape and size

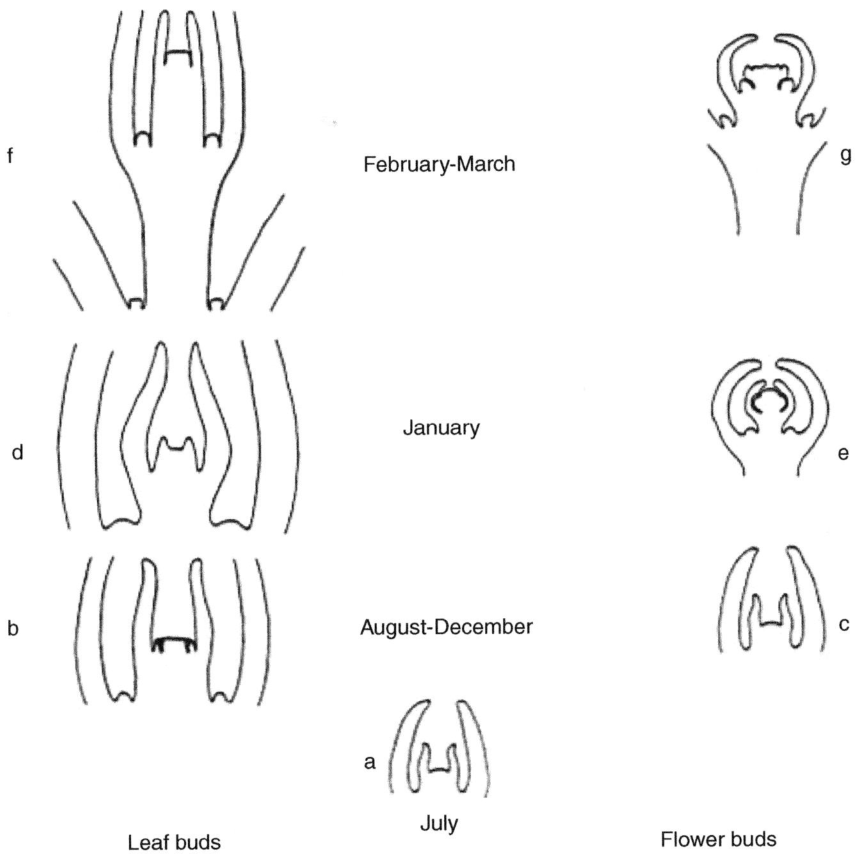

Fig. 2.1.22. Scheme of leaf and flower bud development in the summer/winter period.

can vary. The inflorescence axes are initially greenish, then turn white, and, as anthesis approaches, become yellowish-white (Morettini, 1972).

Olive flowers are small and radially symmetrical (Fig. 2.1.28). Arrangement of their component structures is hypogenous, such that the masculine and accessory flower organs are attached at the end of the pedicel in successive whorls below the ovary base (Brooks, 1948). The outermost whorl is the small green conical calyx formed by the fused sepals. Within the calyx are four white, symmetrically arranged petals, joined at their base, composing the corolla. The male reproductive system, the androecium, is composed of two stamens, each with a relatively large, plump, bright yellow anther attached to the petal bases by a short filament. The anthers contain numerous pollen grains, as is typical for anemophilous (wind-pollinated) species such as the olive tree. The bicellular pollen grains have an elliptical or rounded shape and contain three aperture regions (trizonocolpate); their size is between 15 μm and 30 μm, and their outer layer (exine) has complex ridges. These cause an external wall structure that varies according to genotype (Fig. 2.1.29).

The pollen grain has a structure that is particularly well suited to withstand even severe dehydration and can be stored in a dry place at a low temperature for a period of one year. There may be considerable differences in the shelf life of pollen between different cultivars.

Fig. 2.1.23. Sections of leaf (a) and flower (b) buds before sprouting; (c) sprouted flower bud.

The gynoecium, the female reproductive system located in the flower center, consists of a single pistil with a long, two-lobed stigma, a short white style and a small round green ovary. The stigma surface, formed by multicellular papillae, is considered as wet type because of the secretion it produces, a heterogeneous substance including carbohydrates, lipids and proteins (Serrano *et al.*, 2008). Below the papillae a funnel-shaped receptive tissue is continuous with and similar in structure to the transmitting tissue in the center of the solid style (Suarez *et al.*, 2012). In contrast to the single-locule characteristic of drupe-type fruits, the ovary has two locules or cavities, derived from a two-carpel origin. Each locule contains two anatropous ovules that have undergone a turning movement during their development, orienting the micropyle, the embryo sac's entrance-way for the pollen tube, upwards towards the style (King, 1938). Full development of the embryo sac is required for fertilization.

Macrogametophyte differentiation within the olive ovule follows a bisporic pattern such that two of the four megaspore nuclei produced during meiosis abort and two survive, generating an eight-celled embryo sac (Extremera *et al.*, 1988). The olive ovule is tenuinucellate and unitegmic, meaning that the nucellus, the ovule tissue where the embryo sac forms, is highly reduced at ovule maturity and is surrounded by only a single integument. Development of the ovules occurs during approximately three weeks prior to bloom, at which time the embryo sac is fully mature and capable of fertilization. In Tuscan hilly environments, micro- and macrosporogenesis have been found to take place in 10–15 days (Bini and Lensi, 1981). Ovule differentiation is sometimes incomplete and no embryo sac forms within the ovule, only a residual atrophied nucellus at the base of a long micropylar canal (Fig. 2.1.30). This incomplete ovule development occurs in variable numbers and is associated with both genotypic and environmental factors (Moreno-Alías *et al.*, 2012). Fertilization of the undeveloped ovules is not possible, but usually other, fully differentiated and competent ovules are also present among the four in the ovary, allowing fertilization to proceed. Although there could, theoretically, be four seeds per fruit, only one is present at fruit maturity, rarely two, and at times the only seed degenerates (stenospermocarpy), although fruit growth is nevertheless completed. That factor, along with the olive tree's tendency to produce numerous flowers, means that in normal, good-flowering years, the presence of undeveloped ovules will not likely cause a drop in fruit production (Rapoport and Rallo, 1991).

The olive tree's sexual system is therefore polygamous-monoecious, having two sexually different kinds of flowers present together within the same inflorescence:

Fig. 2.1.24. A panicle with flowers still closed, close to anthesis.

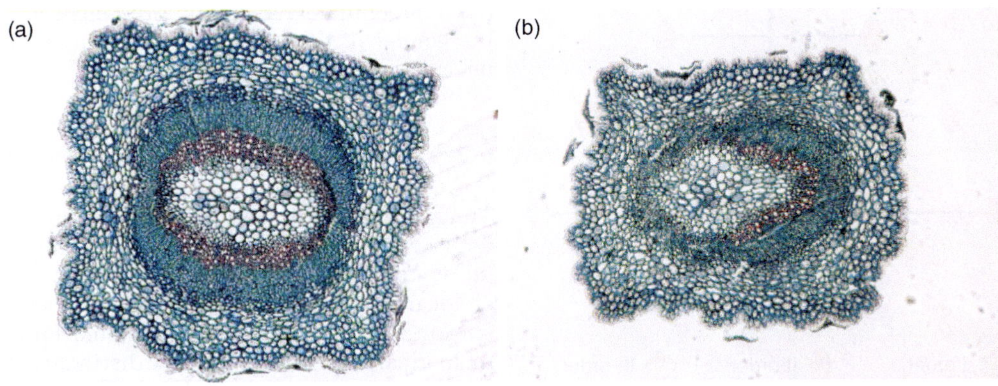

Fig. 2.1.25. Section of a peduncle (a) and of a pedicel (b). The pedicel shows tension wood, due to bending caused by fruit weight. (From Ganino *et al.*, 2011)

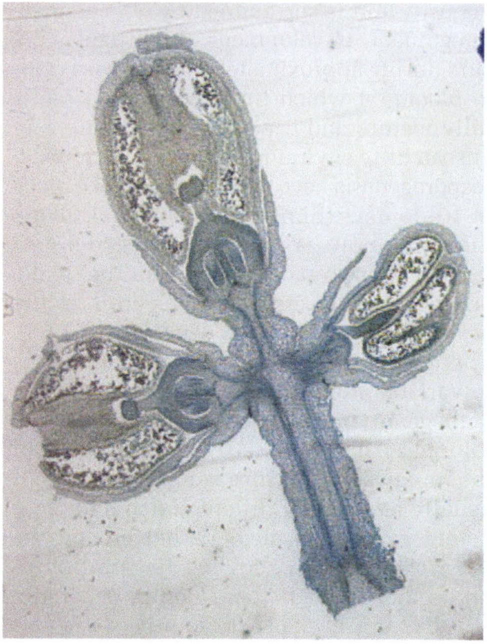

Fig. 2.1.26. Olive flower triad just prior to bloom. At this time the pollen grains and pistil are mature and ready for pollination and fertilization. Flowers on panicles are often arranged as triads in the distal parts of branching.

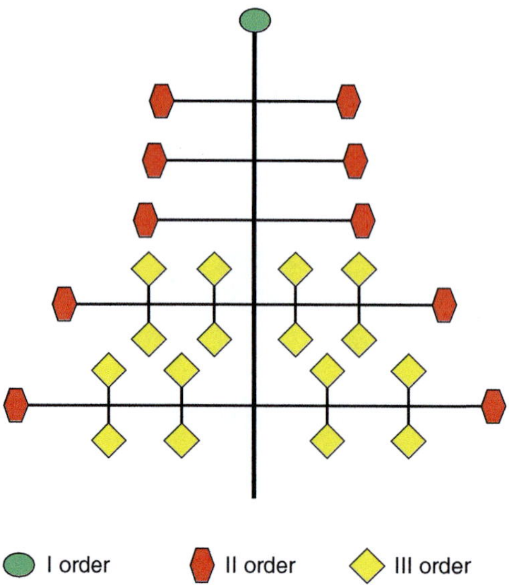

Fig. 2.1.27. Schematic drawing of a common panicle structure. (Barone and Di Marco, 2003)

(i) hermaphroditic or bisexual flowers, with both male and female reproductive organs; and (ii) staminate flowers, with only male reproductive parts. The hermaphroditic flowers are also referred to as perfect flowers and the staminate flowers as imperfect (having well-developed stamens and underdeveloped ovary) (Levin and Lavee, 2005). Staminate flower formation results from pistil abortion during development, which occurs with varying intensity (Fig. 2.1.31). Stamen development and pollen production, however, are similar and equally robust in both flower types (Cuevas and Polito, 2004). As with ovule development, the proportion of imperfect flowers varies according to genotype, environment and their interaction (Navas-López *et al.*, 2019). More rare are pistillate flowers.

Influencing environmental factors include temperature, mineral nutrition and water availability, as well as competition for assimilates among flowers when flower number is high. The critical role of substrate supply in the differentiation of the two flower types is evidenced directly by much higher starch accumulation in perfect than in staminate flowers (Reale *et al.*, 2009). Furthermore, within the inflorescence there appear to be preferred positions for perfect flower formation, also likely related to positional priorities for substrate supply (Cuevas and Polito, 2004; Seifi *et al.*, 2008). Flower-quality evaluation for the olive tree in relation to potential productivity, both traditionally and in present times, considers primarily the number of flowers per inflorescence and the gender of those flowers. No functional nectaries are present on olive flowers.

2.1.3.2 Fruit

The small green olive fruit is a drupe, an indehiscent fleshy fruit with a hard, lignified pit or stone. The three principal tissues, the endocarp (pit), mesocarp (pulp or flesh) and epi- or exocarp (skin), which together form the pericarp, each follow a very distinct and characteristic form of growth and differentiation during fruit development. The tissue of principal economic importance is the

Fig. 2.1.28. Olive flower types. In the hermaphrodite flower (a), both stamens (male) and the single pistil (female) are functional, while in the staminate flowers (b, c), pistil development is aborted and only the stamens are functional. In (c), petals and stamens have been removed showing the aborted pistil.

Fig. 2.1.29. Olive pollen grain. Two of the three aperture regions are visible (arrows). (PReBAIL (Plant Reproductive Biology and Advanced Imaging Laboratory), CSIC, Spain)

mesocarp, from which the principal amount of oil is extracted, and which is consumed in the case of table olives (Fig. 2.1.32 a, b).

At the time the flower bud opens, both the pollen grains and pistil are mature and ready for pollination and fertilization (Fig. 2.1.26). Following successful pollination and pollen tube growth, one of the four ovules, from this time called the functional ovule, is fertilized, and fruit growth is set in

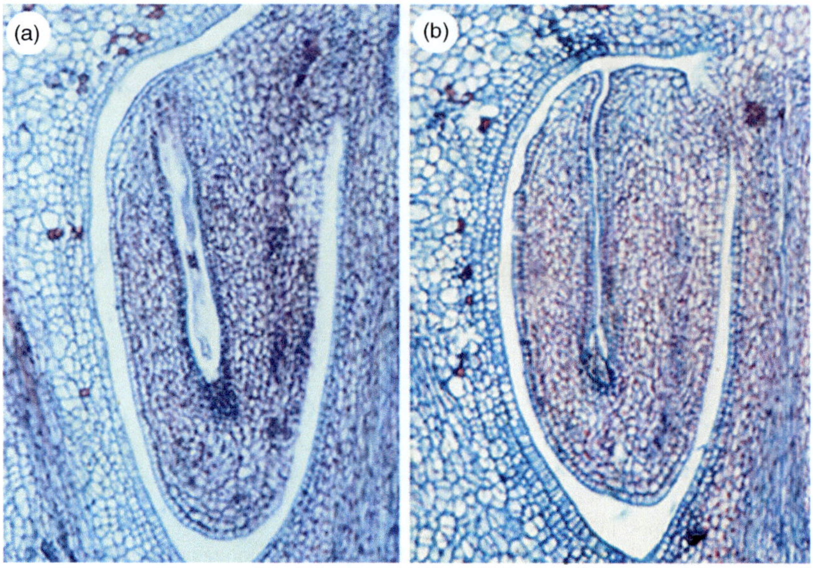

Fig. 2.1.30. Mature ovules with (a) fully and (b) partially developed embryo sacs. Fertilization requires a fully developed embryo sac (a). In the partially developed embryo sacs, frequent in the olive tree, differentiation of the embryo sac is incomplete, resulting in a residual atrophied nucellus at the base of a long micropylar canal (b).

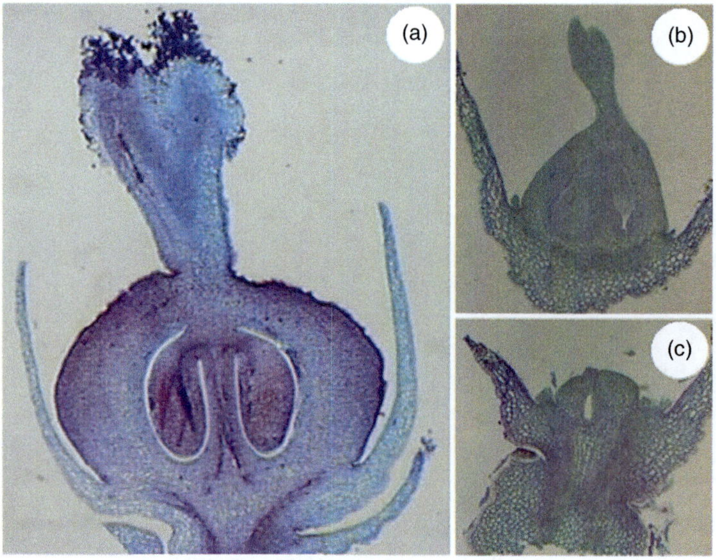

Fig. 2.1.31. Olive pistil structure: (a) fully developed pistil with stigma, style and ovary; (b, c) pistils of staminate flowers showing different degrees of abortive development.

motion (Fig. 2.1.33b). At bloom the ovary mesocarp and endocarp are composed of small, isodiametric parenchymatic cells, and can be distinguished from each other by the ring of vascular bundles which separates them (Fig. 2.1.33a; see also Fig. 2.1.32a). Studies of the olive ovary cells and tissues have shown that ovary size and cell number

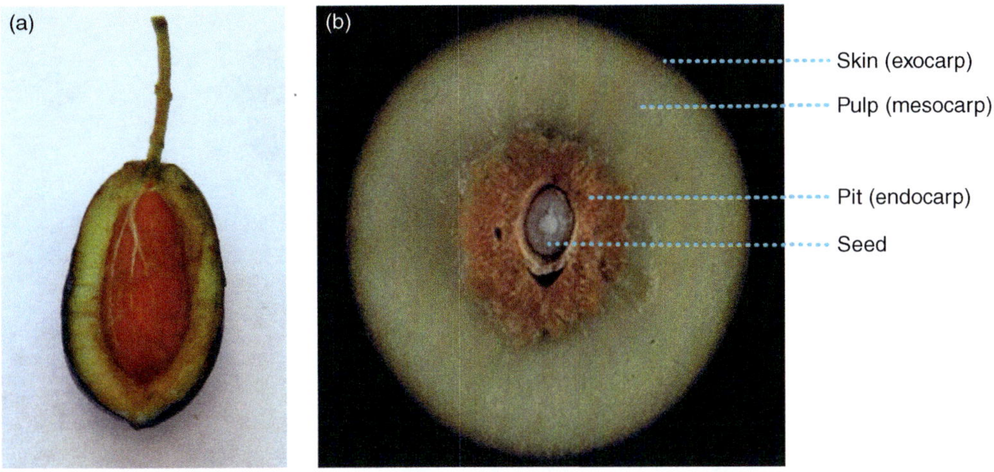

Fig. 2.1.32. Olive fruit cut to show the component structures: (a) longitudinal section: peduncle (inflorescence axis), a thin exocarp, a thick mesocarp and at the center the endocarp; (b) transversal section, with the same tissues, a protective skin (exocarp), a fleshy pulp (mesocarp) and a woody pit (endocarp). The seed grows within the pit, and contains the embryo (bright, solid white) and the endosperm (cloudy white in color).

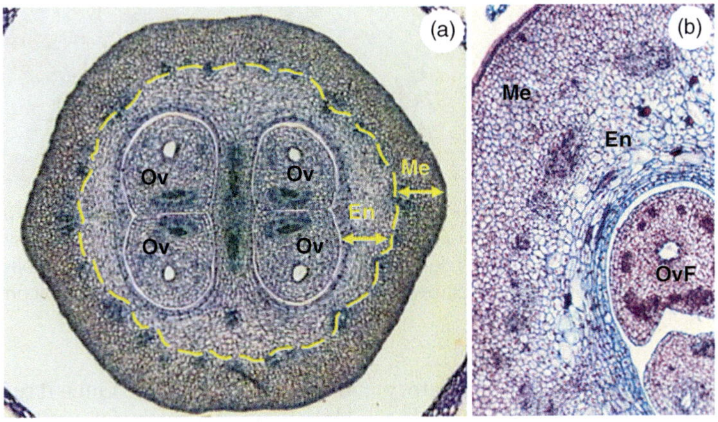

Fig. 2.1.33. Olive ovary equatorial section at anthesis: (a) the mesocarp (Me) and endocarp (En) tissues are separated from each other by a ring of vascular bundles (dotted line). There are two locule cavities, each containing two ovules (Ov) at anthesis; (b) following successful pollination and fertilization of one ovule, now called the functional ovule (OvF), fruit development begins.

at bloom are highly correlated with cultivar mature fruit size (Rosati *et al.*, 2012), that differences in ovary size among cultivars are mainly due to cell number in both the endocarp and mesocarp (Rosati *et al.*, 2011), and that cell number is more important in determining fruit growth and sink strength (Rosati *et al.*, 2020).

The endocarp is the fruit pit, a hard, woody structure surrounding the seed in the fruit center (Fig. 2.1.34a). During fruit growth the endocarp cells undergo the process of sclerification, in which stone cells form by the deposition of a thick secondary wall rich in lignin (Fig. 2.1.34b). As a consequence of sclerification, the endocarp hardens, producing the agriculturally recognized phase of pit hardening, considered a useful indicator for timing agricultural management practices such as irrigation

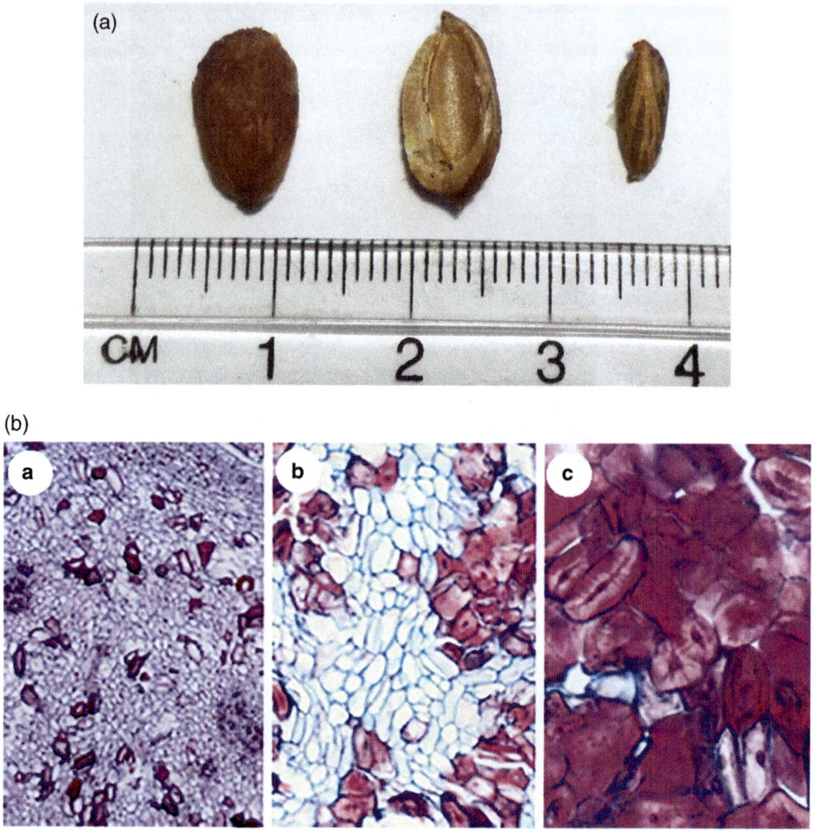

Fig. 2.1.34. (a) kernel, whole (*left*), and sectioned (*center*); seed (*right*). (b) endocarp sclerification, in which a lignin-rich secondary cell wall forms, producing a hard, woody structure. Early in fruit growth, scattered endocarp cells undergo sclerification – **a.** As development continues, greater numbers of cells initiate the process, while wall-thickening increases for those that have started earlier – **b**, eventually converting all of the cells into fully sclerified stone cells – **c.**

strategies or the application of regulatory products. Physical measurement of the pressure required to break the pit has shown that pit-hardening takes place during approximately two months, much longer than traditionally interpreted from simple evaluation by hand-cutting (Rapoport *et al.*, 2013).

The endocarp represents 20–40% of total fruit fresh weight and is made up of large cells with thick cell walls that harden with the deposition of lignin. The endocarp is light-brown in color and has different shapes, sizes and structures that are used for variety identification (Fig. 2.1.32a).

Endocarp expansive growth (Fig. 2.1.34b) occurs during the first two months of fruit development. Once it reaches its final size, there is a further period of growth with weight gain, due to lignin deposition while the sclerification process finalizes. Endocarp sclerification begins gradually while the endocarp tissue is still growing and intensifies over time. The individual stone cells appear, and their number and proportion within the endocarp tissue increase. Meanwhile, those cells that have started sclerification continue to undergo lignin deposition and cell-wall thickening, and physical hardening intensifies as there are more stone cells and their degree of sclerification intensifies (Hammami *et al.*, 2013). Once a cell begins sclerification, cell-wall structure changes and neither cell division nor expansion can take place, so that when the majority of endocarp cells have initiated sclerification, endocarp expansion slows and eventually stops. Afterwards, hardening continues, as does cell sclerification, and the associated endocarp dry-weight increases (Rapoport *et al.*, 2013). Under water limitation, endocarp expansion is reduced and sclerification, although its onset is hastened, occurs at a much slower rate.

Exterior to the endocarp is the mesocarp, or fruit pulp, the edible part of the olive fruit, and the site for oil metabolism and storage. The mesocarp is a fleshy tissue composed of parenchyma cells, similar to many other fleshy fruits (Fig. 2.1.35a, b). These cells maintain the capacity for division and expansion, producing expansive growth throughout the major part of fruit development. Oil storage takes place in the cytoplasm of the mesocarp cells, initially in the form of small oil bodies or droplets, which then fuse and may eventually fill the entire cell.

The mesocarp is the most developed part of the fruit, accounting for about 60–70% of total weight, and consists of cells rich in protoplasm and with a large nucleus. The vacuole contains sugars, weak acids, tannins, pigments and, among these, chlorophyll at the beginning of fruit development, which, later, collects oil droplets that are synthesized in the plastids and in the mitochondria of the cytoplasm. The cell wall is rigid and, together with the cell constituents, contributes to the consistency of the pulp.

Mesocarp expansive growth is driven by the high capacity of its parenchyma cells for cell division and expansion. The highest rate and amount of mesocarp cell division in the olive fruit occurs in the period immediately following fertilization and fruit set, although some cell division continues at a reduced rate throughout fruit development (Hammami *et al.*, 2011). Cell expansion is also highly active in early fruit development but is masked by the high rate of cell division (Rapoport *et al.*, 2017). Cultivar differences in fruit size depend principally on mesocarp cell number, not cell size, and average mesocarp cell size is similar among cultivars and the seasonal pattern of cell-division intensity is consistent for olive cultivars with a wide range of fruit size (Hammami *et al.*, 2011). Lower fruit mesocarp growth under reduced water status is principally due to smaller cell size, although with severe water status reduction cell number is also affected (Gucci *et al.*, 2009). Irradiance level differences, explored at different positions in the canopy, appear to affect fruit size via cell size (Trentacoste *et al.*, 2016).

The epicarp or exocarp, the fruit skin, constitutes the external fruit tissue. The

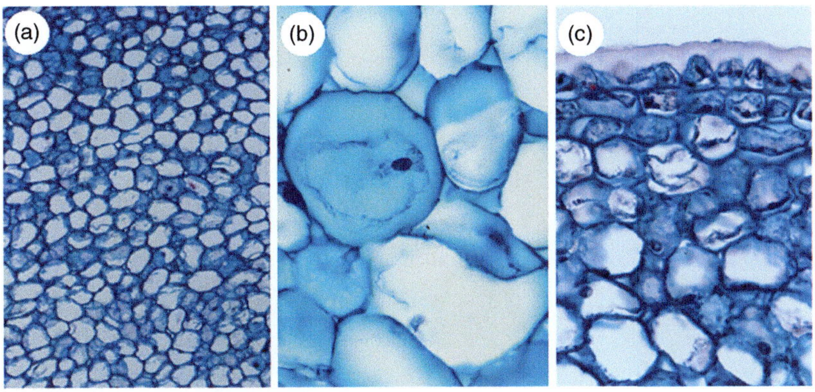

Fig. 2.1.35. Mesocarp (a, b) and exocarp (c). The mesocarp (pulp) is formed by relatively isodiametric parenchyma cells which have simple cell walls and are capable of cell division and expansion: (a) mesocarp of young fruit; (b) mesocarp of mature fruit, shown at higher magnification; (c) exocarp (fruit skin) is composed of the epidermis with a thick cuticle (upper part of figure) and a number of subepidermal layers that are similar to and continuous with the mesocarp.

exocarp is composed of the epidermis, the external layer of cells with its thick protective cuticle in addition to a series of subepidermal cell layers. Recent studies of cell dimensions (Hammami and Rapoport, 2012) indicate that approximately five layers of subepidermal cells, although continuous with and similar in structure to the mesocarp cells, show structural and developmental patterns more unified with the epidermis than the mesocarp (Fig. 2.1.35c). Scattered across the exocarp are lenticels, small patches derived from the ovary stomata (Fig. 2.1.36a), which become atrophied during fruit growth. The lenticels are most visible during mature green or reddish stages of fruit growth as light-colored specks, and their size and frequency are characters used in cultivar identification (Barranco *et al.*, 2000) (Fig. 2.1.36 b, c). Under environmental conditions of water shortage, the fruit increases the thickness of cuticle, cutin and wax layers to prevent water and nutrient loss. Under conditions of high water availability, the cuticle thins out.

The fruits of cultivated olive varieties are 1–4 cm long, 0.5–2 cm in diameter and can weigh as much as 10 g fresh weight, 2.5 g

Fig. 2.1.36. Lenticels. The surface of the olive fruit is covered with lenticels, semi-porous zones in the epidermis, which arise from the stomates on the ovary surface (a) and are visible as white spots during some ripening stages (b, c).

Fig. 2.1.37. Fruits of some popular olive cultivars. (Photo: Banco Mundial de Germoplasma del Olivo (BMGO) del IFAPA, Centro 'Alameda del Obsipo', Córdoba, Andalucia, Spain)

dry weight. Fruit shape ranges from spherical to ellipsoidal or elongated, with varying degrees of asymmetry and the apex either rounded or pointed (Fig. 2.1.37). While there is a tendency for the cultivars with smaller fruits to be destined for oil production and those with larger fruits for table use, many are of double aptitude and size is not a strict determinant of use. Olive fruit parameters appreciated for their agronomic values are the pulp-to-pit ratio and the oil content.

Fig. 2.1.38. Indices for external (1–7) and internal (4–7) color change during olive fruit maturation. (Photo: IFAPA (Instituto de Investigación y Formación Agraría y Pesquera), Junta de Andalucía, Spain)

Pulp-to-pit ratio is understandably of particular interest for table-olive production, and values among cultivars can range from 3 to 8 on a fresh-weight basis, and 1 to 3 on a dry-weight basis. Oil content is generally indicated as percentage of total fruit dry weight. In seven years of observations of 61 Spanish cultivars, oil was found to range 30–60%, with the cultivar determining 50% of the variation (Rallo *et al.*, 2005).

During maturation, the olive fruit goes through a characteristic series of external color changes or veraison. The green fruit first becomes lighter, and then progressively darkens, acquiring a reddish blush, which spreads over the fruit, turns purple and, finally, black. Within the black fruit, following veraison, the pulp darkens progressively from the exterior to the interior. These changes are one of the factors used by growers to determine harvest time (Fig. 2.1.38).

2.1.3.3 Seed and embryo

The olive seed is contained and protected within the fruit woody endocarp. Generally,

Fig. 2.1.39. Seed and embryo: (a) seed within the endocarp; (b) mature, healthy seeds; (c) degenerating or aborting seeds from mature fruit (this occurs frequently in the olive, at different times throughout fruit development); (d) the embryo with large flat cotyledons and a short radicle.

only one seed is formed, although sometimes there are two or, very rarely, three seeds, a characteristic which seems to be associated with the cultivar. The seedcoat, derived from the single ovule integument, is thin, tough and permeated by numerous vascular ridges. Between seedcoat and embryo is a layer of endosperm, rich in starch (King, 1938).

The embryo has two long, flat, spatula-shaped cotyledons, and fills most of the seed volume (Figs 2.1.39 and 2.1.40). A short radicle, located at the lowest end of the embryonic axis, will give rise to the root system. Between the cotyledons is a small plumule, from which will develop the future epigeic system, the plant parts that will be exposed to the open atmosphere. The embryo starts growing within three to four weeks after bloom and reaches the globular stage at week 6; after eight weeks the cotyledons appear to be well developed. The embryo is usually completely formed after five months from full bloom. No further morphological or anatomical changes appear to occur in the embryo, although dormancy is imposed on the seed late in the season. Seed growth means a gradual embryo enlargement, which, at the end, occupies most of the space inside the endocarp, at the expense of the endosperm.

The endosperm, rich in starch, proteins and oil (King 1938; Alché *et al*., 2006) is located between the embryo and the seedcoat.

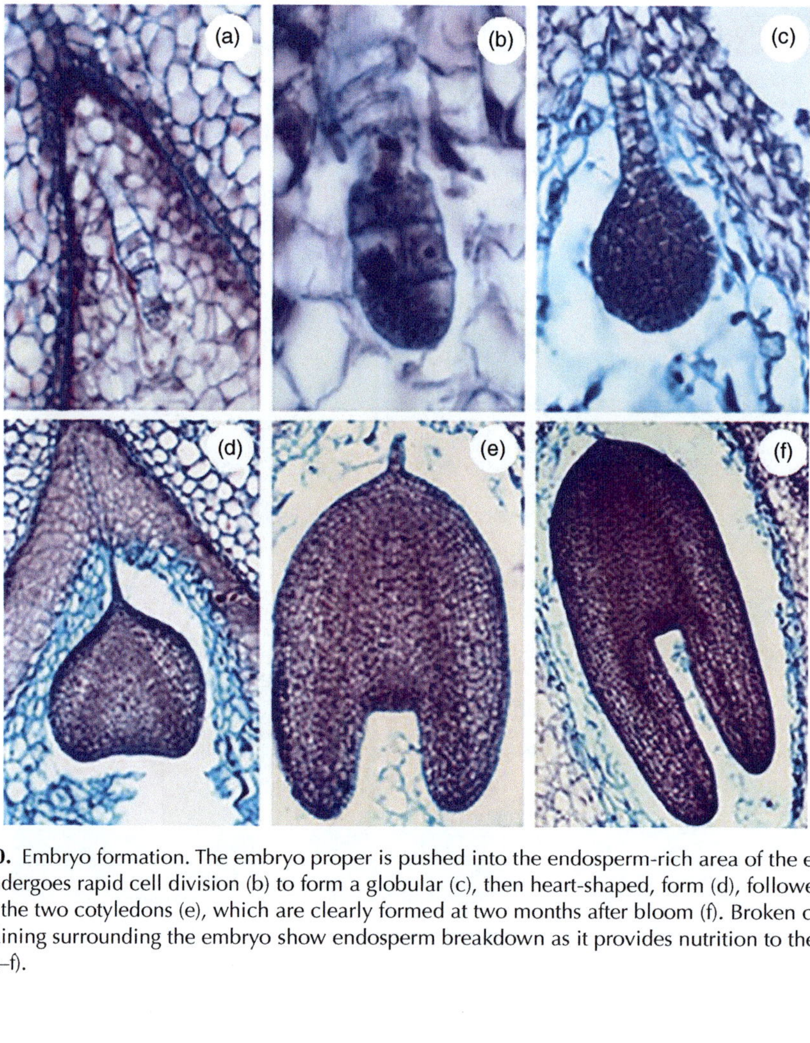

Fig. 2.1.40. Embryo formation. The embryo proper is pushed into the endosperm-rich area of the embryo sac (a), undergoes rapid cell division (b) to form a globular (c), then heart-shaped, form (d), followed by the growth of the two cotyledons (e), which are clearly formed at two months after bloom (f). Broken cells and altered staining surrounding the embryo show endosperm breakdown as it provides nutrition to the growing embryo (c–f).

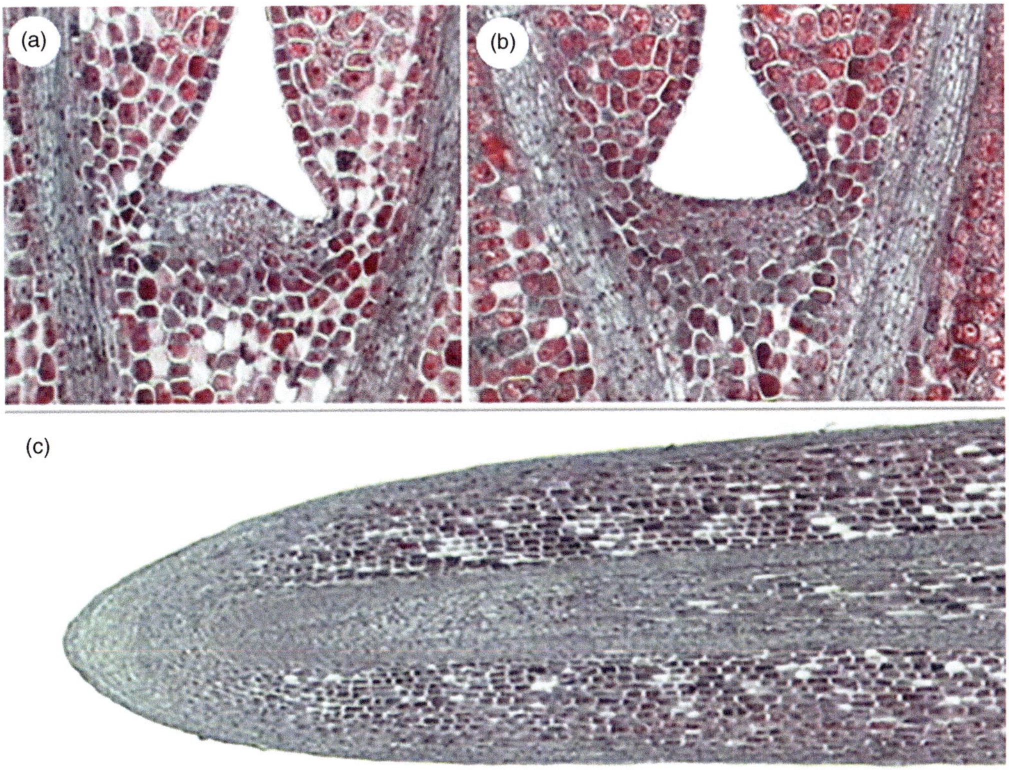

Fig. 2.1.41. Tissue differentiation in the mature embryo. The shoot apical region is slightly domed (a) or completely flat (b). The radicle, the embryonic root, contains a well-developed apex covered by a clearly distinguished root cap (c). Appreciable provascular strands (internal elongated gray zones) and stored reserves (red) are observed.

Immediately following successful pollination and fertilization, the multicellular endosperm is formed and commences rapid growth, accompanied by notable expansion of the ('functional') ovule and differentiation of ovule vascular tissues (Fig. 2.1.33b). Embryo differentiation and growth, however, is delayed by approximately 3-4 weeks, while the endosperm increases substantially. The embryo proper is then pushed by proembryo elongation into the endosperm, by now plentiful and occupying much of the embryo sac. Embryo growth and differentiation follow the classical sequence from multicellular embryo proper to globular and then heart-shaped, and the two cotyledons are clearly formed at two months after bloom (Fig. 2.1.40). At four to five months the mature olive embryo contains a well-differentiated root apex at the tip of a short axis composed of hypocotyl and radicle, the embryonic root; appreciable provascular strands and storage substances are observed in the axis and cotyledons. Between the cotyledons, the shoot apex is flat, or sometimes slightly domed, with a subtending zone of intense metabolic activity but no other visible differentiation or plumule formation (Germanà *et al.*, 2014; Fig. 2.1.41, 2.1.42),

As the fruit matures, physiological dormancy is imposed on the seed, with factors apparently located in both the endosperm and the embryo itself (Germanà *et al.*, 2014). Thus, olive seed germination procedures require removing the mechanical barrier of the hardened pit, and cold stratification to remove physiological dormancy. Among tested genotypes, optimum germination

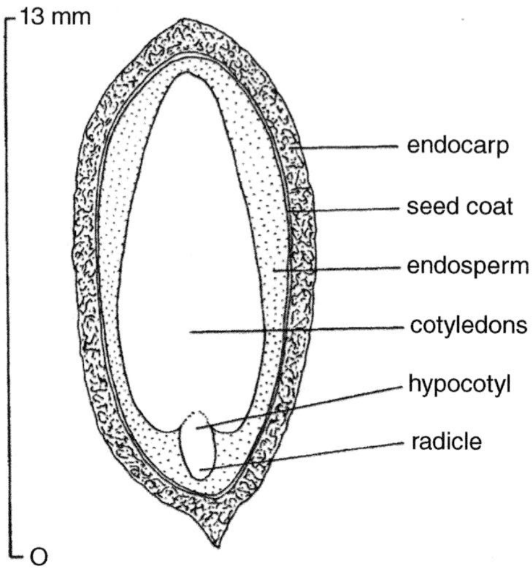

Fig. 2.1.42. Schematic representation of the olive seed. (From Krugman, 1974)

was found to be consistent with stratification at 14°C for 30 days, although germination rate was highly related to genotype (Morales-Sillero *et al.*, 2012).

At maturity, numerous olive fruits contain empty pits with no seed present or else with seeds at different stages of abortion or degeneration (parthenocarpy and stenospermocarpy). It appears that the initiation of the embryo is required for fruit development to begin, but that, subsequently, seed abortion may occur though fruit growth is not deterred. This behavior is evidenced by the consistent observation of growing fertilized ovules in young fruits, the occurrence of aborted embryos at two months, and by the enlargement of one of the ovary locules in the same manner as when a seed is present; that is, the locule expands to accommodate the growing fertilized ovule (Rapoport and Moreno-Alías, 2017). Aborted or degenerating seeds are also observed at different times throughout fruit growth, and flawed seed quality may also be evidenced by seedling production of 40–60% of sown seeds, over up to three years of permanence in seedbeds, in major Italian nurseries (Fabbri *et al.*, 2004). In some Spanish cultivars, the occurrence of nucellar polyembryony has been reported.

Fruit shape and size, and pit size and surface morphology vary greatly among cultivars, and are the most reliable morphological features to distinguish between them.

References

Agati, G. (1951) Contributo allo studio delle gemme dell'olivo. *Annali della Sperimentazione Agraria*, Roma, 847–858.

Alché, J.D., Jimenez-Lopez, J.C., Wang, W., Castro-Lopez, A.J. and Rodriguez-Garcia, M.I. (2006) Biochemical characterization and cellular localization of 11S type storage proteins in olive (*Olea europaea* L.) seeds. *Journal of Agricultural and Food Chemistry* 54, 5562–5570.

Andreini, L., Bartolini, S., Guivarc'h, A., Chriqui, D. and Vitagliano, C. (2008) Histological and immunohistochemical studies on flower induction in the olive tree (*Olea europaea* L.). *Plant Biology* 10, 588–595.

Arzee, T. (1953) Morphology and ontogeny of foliar sclereids in Olea europaea. I. Distribution and structure. *American Journal of Botany* 40, 680–687.

Atienza, S.G., de la Rosa, R., León, L. Martín, A. and Belaj, A. (2014) Identification of QTL for agronomic traits of importance for olive breeding. *Molecular Breeding* 34, 725–737.

Baas, P., Esser, P.M., van der Westen, M.E.T. and Zandee, M. (1988) Wood anatomy of the *Oleaceae*. *IAWA Bulletin* 9(2), 103–182.

Bacelar, E.A., Correia, C.M., Moutinho-Pereira, J.M., Goncalves, B.C., Lopes, J.I. and Torres-Pereira, J.M.G. (2004) Sclerophylly and leaf anatomical traits of five field-grown olive cultivars growing under drought conditions. *Tree Physiology* 24, 233–239.

Bacelar, E.A., Santos, D.L., Moutinho-Pereira, J.M., Gonçalves, B.C., Ferreira, H.F. and Correia, C.M. (2006) Immediate responses and adaptative strategies of three olive cultivars under contrasting water availability regimes: changes on structure and chemical composition of foliage and oxidative damage. *Plant Science* 170(3), 596–605.

Baldini, E. And Scaramuzzi, F. (1956) Indagini e osservazioni sugli sferoblasti dell'olivo. *Annali della Sperimentazione Agraria*, Roma.

Barone, E. and Di Marco, L. (2003) Morfologia e ciclo di sviluppo. In: Fiorino, P. (ed.) *Olea*. Edagricole, Bologna, Italy, 13–33.
Barranco, D., Cimato, A., Fiorino, P., Rallo, L., Touzani, A. *et al.* (2000) *World Catalogue of Olive Varieties*. International Olive Oil Council, Madrid.
Ben Sadok, I., Celton, J.M., Essalouh, L., El Aabidine, A.Z., Garcia, G. *et al.* (2013) QTL mapping of flowering and fruiting traits in Olive. *PLOS ONE* 8 (5), e62831.
Bini, G. and Lensi, M (1981) Osservazioni su alcuni aspetti dell'ontogenesi fiorale nell'olivo. *Rivista dell'Ortoflorofrutticoltura Italiana* 65(5), 371–380.
Bosabalidis, A. and Kofidis, G. (2002) Comparative effects of drought stress on leaf anatomy of two olive cultivars. *Plant Science* 163, 375–379.
Boughalleb, F. and Hajlaoui, H. (2011) Physiological and anatomical changes induced by drought in two olive cultivars (cv. Zalamti and Chemlali). *Acta Physiologiae Plantarum* 33, 53–65.
Brooks, R.M. (1948) Seasonal incidence of perfect and staminate olive flowers. *Proceedings of the American Society of Horticultural Science* 52, 213–218.
Chartzoulakis, K., Patakas, A. and Bosabalidis, A. (1999) Changes in water relations, photosynthesis and leaf anatomy induced by intermittent drought in two olive cultivars. *Environmental Experimental Botany* 42, 113–120.
Connor, D.J. and Fereres, E. (2005) The physiology of adaptation and yield expression in olive. In: Janick, J. (ed.) *Horticultural Reviews*. Wiley, Portland, Oregon, pp. 155–229.
Cuevas, J. and Polito, V.S. (2004) The role of staminate flowers in the breeding system of *Olea europaea* (Oleaceae): an andromonoecious, wind-pollinated taxon. *Annals of Botany* 93, 547–553.
de la Rosa, R., Rallo, L. and Rapoport, H.F. (2000) Olive floral bud growth and starch content during winter rest and spring budbreak. *Hortscience* 35, 1223–1227.
de la Rosa, R., Kiran, A.I., Barranco, D. and Leòn, L. (2006) Seedling vigor as a preselection criterion for short juvenile period in olive breeding. *Australian Journal of Agricultural Research* 57, 477–481.
Edziri, H., Jaziri, R., Aissaoui, F., Verschaeve, L., Flamini, G. *et al.* (2018) Impact of water deficit on physiological parameters, bioactive content and antioxidant activity of three olive cultivars. *South African Journal of Botany* 118, 268–273.
Ennajeh, M., Vadel, A.M., Cochard, H. and Khemira, H. (2010) Comparative impacts of water stress on the leaf anatomy of a drought-resistant and a drought-sensitive olive cultivar. *Journal of Horticultural Science & Biotechnology* 85(4), 289–294.
Esau, K. (1977) *Anatomy of Seed Plants*, 2nd edn. John Wiley & Sons, Chichester, UK.
Extremera, G., Rapoport, H.F and Rallo, L. (1988) Caracterización del desarrollo normal del saco embrionario en olivo (*Olea europaea* L.). *Anales del Jardin Botánico de Madrid* 45, 197–211.
Fabbri, A. (1980) Influenza di alcuni caratteri anatomici sulla radicazione di talee di olivo cv. 'Frangivento'. *Rivista della Ortoflorofrutticoltura Italiana* 64(4), 325–335.
Fabbri, A. and Alerci, L. (1999) Reproductive and vegetative bud differentiation in *Olea europaea*, L. *Journal of Horticultural Science and Biotechnology* 74, 522–527.
Fabbri, A., Bartolini, G., Lambardi, M. and Kailis, S.G. (2004) *Olive Propagation Manual*. CSIRO Publishing, Canberra.
Farinelli, D. and Tombesi, S. (2015) Performance and oil quality of 'Arbequina' and four Italian olive cultivars under super high density hedgerow planting system cultivated in central Italy. *Scientia Horticulturae* 192, 97–107.
Ganino, T., Rapoport, H.F. and Fabbri, A. (2011) Anatomy of the olive inflorescence axis at flowering and fruiting. *Scientia Horticulturae* 129, 213–219.
Germanà, M.A., Chiancone, B., Hammami, S.B. and Rapoport, H.F. (2014) Olive embryo in vitro germination potential: role of explant configuration and embryo structure among cultivars. *Plant Cell, Tissue and Organ Culture* 118, 409–417.
Green, P.S. (2004) Oleaceae. *The Families and Genera of Vascular Plants*, Vol. VII. Springer.
Gucci, R., Lodolini, E.M. and Rapoport, H.F. (2009) Water deficit induced changes in mesocarp cellular processes and the relationship between mesocarp and endocarp during olive fruit development. *Tree Physiology* 29, 1575–1585.
Guerfel, M., Baccouri, O., Boujnah, D., Chaıbi, W. and Zarrouk, M. (2009) Impacts of water stress on gas exchange, water relations, chlorophyll content and leaf structure in the two main Tunisian olive (*Olea europaea* L.) cultivars. *Scientia Horticulturae* 119(3), 257–263.
Guerriero, R., Scaramuzzi, F., Crescimanno, F.G. and Sottile I. (1972) Ricerche comparative tra olivi innestati ed autoradicati. *Tecnica Agricola* 4.

Hammami, S.B.M. and Rapoport, H.F. (2012) Quantitative analysis of cell organization in the external region of the olive fruit. *International Journal of Plant Sciences* 173, 1–13.
Hammami, S.B.M., Manrique, T. and Rapoport, H.F. (2011) Cultivar-based fruit size in olive depends on different tissue and cellular processes throughout growth. *Scientia Horticulturae* 130, 445–451.
Hammami, S.B.M., Costagli, G. and Rapoport, H.F. (2013) Cell and tissue dynamics of olive endocarp sclerification vary according to water availability. *Physiologia Plantarum* 149, 571–582.
King, J.R. (1938) Morphological development of the fruit of the olive. *Hilgardia* 11, 437–458.
Krugman, S.L. (1974) *Olea europaea* L., Olive. In: Schopmeyer, C.S. (ed.) *Seeds of Woody Plants in the United States*. Agriculture Handbook 450, USDA, Washington, pp. 558–559.
Levin, A.G. and Lavee S. (2005) The influence of girdling on flower type, number, inflorescence density, fruit set, and yields in three different olive cultivars ('Barnea', 'Picual', and 'Souri'). *Australian Journal of Agricultural Research* 56, 827–831.
Lo Gullo, M.A. and Salleo, S. (1990) Wood anatomy of some trees with diffuse- and ring-porous wood: some functional interpretations. *Giornale Botanico Italiano* 124, 601–614.
Lo Gullo, M.A., Nardini, A., Salleo, S. and Tyree, M.T. (1998) Changes in root hydraulic conductance (Kr) of *Olea oleaster* seedlings following drought stress and irrigation. *New Phytologist* 140, 25–31.
Loussert, R. and Brousse, G. (1978) *L'olivier*. Paris.
Mazzamurro, M. (1973) Rapporto fra struttura fogliare e potenziale accrescimento nell'olivo. *Rivista di Ortoflorofrutticoltura Italiana* 5, 369–373.
Messeri, A. (1942) Osservazioni anatomiche sui primi stadi di sviluppo degli 'ovuli' in giovani piante di olivo. *Atti Convegno di Studi Olivicoli*, Florence, Italy, pp. 181–186.
Metcalfe, C.R. and Chalk, L. (1953) *Anatomy of the Dicotyledons*. Clarendon Press, Oxford, UK.
Morales-Sillero, A., Paz Suárez, M., Jiménez, M.R., Casanova, L., Ordovás, J. and Rallo, P. (2012) Olive seed germination and initial seedling vigor as influenced by stratification treatment and the female parent. *HortScience* 47, 1672–1678.
Moreno-Alías, I., Leon, L., de la Rosa, R. and Rapoport, H.F. (2009) Morphological and anatomical evaluation of adult and juvenile leaves of olive plants. *Trees – Structure and Function* 23, 181–187.
Moreno-Alias, I., Rapoport, H.F and Martins, P.C. (2012) Morphological limitations in floral development among olive tree cultivars. *Acta Horticulturae* 932, 23–27.
Morettini, A. (1956) *Alcuni aspetti della biologia dell'olivo e loro applicazioni nel perfezionamento della tecnica colturale*. Florence.
Morettini, A. (1972) *Olivicoltura*. Ramo Editoriale degli Agricoltori, Roma.
Morettini, I. (1954) Ricerche sull'anatomia delle foglie delle più note varietà di olivo toscane in relazione alla loro resistenza al Cicloconium oleaginum. *Notiziario sulle Malattie delle Piante* 28 (n.s. 7).
Navas-Lopez, J.F., León, L., Rapoport, H.F., Moreno-Alías, I., Lorite, I.J. and de la Rosa, R. (2019) Genotype, environment and their interaction effects on olive tree flowering phenology and flower quality. *Euphytica* 215, 1–13.
Pinney, K. and Polito, V.S. (1990) Flower initiation in 'Manzanillo' olive. *Acta Horticulturae* 286, 203–206.
Rallo, L., Barranco, D., Caballero, J.M., Del Rio, C., Martìn, A., Tous, J. and Trujillo, I. (2005) *Variedades de Olivo en España*. Junta de Andalucía/Ediciones Mundi Prensa.
Rallo, P., Jiménez, R., Ordovás, J. and Suarez, M.P. (2008) Possible early selection of short juvenile period olive plants based on seedling traits. *Australian Journal of Agricultural Research* 59, 933–940.
Ramos, A., Rapoport, H.F., Cabello, D. and Rallo, L. (2018) Chilling accumulation, dormancy release temperature, and the role of leaves in olive reproductive budburst: evaluation using shoot explants. *Scientia Horticulturae* 231, 241–252.
Rapoport, H.F. (2010) Botany and morphology. In: Barranco, D., Fernández-Escobar, R. and Rallo, L. (eds) *Olive Growing*. Rural Industries Research and Development Corporation, Australia, pp. 33–57.
Rapoport, H.F. and Moreno-Alías, I. (2017) Botánica y morfología. In: Barranco, D., Fernández Escobar, R. and Rallo, L. (eds) *El Cultivo del Olivo*. Mundi Prensa, Madrid/Junta de Andalucía, España, pp. 35–64.
Rapoport, H.F. and Rallo, L. (1991) Postanthesis flower and fruit abscission in 'Manzanillo' olive. *Journal of the American Society for Horticultural Science* 116, 720–723.
Rapoport, H.F., Pérez-López, D., Hammami, S.B.M., Agüera, J. and Moriana, A. (2013) Fruit pit hardening: physical measurement during olive fruit growth. *Annals of Applied Biology* 163, 200–208.
Rapoport, H.F., Fabbri, A. and Sebastiani, L. (2016) Olive biology. In: Rugini, E., Baldoni, L., Muleo, R. and Sebastiani, L. (eds) *The Olive Tree Genome*. SpringerLink.

Rapoport, H.F., Hammami, S.B.M., Rosati, A. and Gucci, R. (2017) Advances in olive fruit cell and tissue development. *Acta Horticulturae* 1177, 209–214.
Reale, L., Sgromo, C., Ederli, L., Pasqualini, S., Orlandi, F. *et al.* (2009) Morphological and cytological development and starch accumulation in hermaphrodite and staminate flowers of olive (*Olea europaea* L.). *Sexual Plant Reproduction* 22, 109–119.
Roland, J.C. (1961) Compléments relatifs à la connaissance des stéréomes. I. Étude des tissus de soutien de la feuille de l'Olivier: *Olea europea* L. (Oléacées). *Bulletin de la Société Botanique de France* 108(9), 393–413.
Rosati, A., Caporali, S., Hammami, S.B.M., Moreno-Alias, I., Paoletti, A. and Rapoport, H.F. (2011) Differences in ovary size among olive (*Olea europaea* L.) cultivars are mainly related to cell number, not to cell size. *Scientia Horticulturae* 130, 185–190.
Rosati, A., Caporali, S., Hammami, S.B., Moreno-Alías, I., Paoletti, A. and Rapoport, H.F. (2012) Tissue size and cell number in the olive (*Olea europaea*) ovary determine tissue growth and partitioning in the fruit. *Functional Plant Biology* 39, 580–587.
Rosati, A., Caporali, S., Hammami, S.B.M., Moreno-Alias, I., Paoletti, A. and Rapoport, H.F. (2020) Fruit growth and sink strength in olive (*Olea europaea*) are related to cell number, not to tissue size. *Functional Plant Biology* 47, 1098–1104.
Rossi, L., Francini, A., Minnocci, A. and Sebastiani, L. (2015) Salt stress modifies apoplastic barriers in olive (*Olea europaea* L.): a comparison between a salt-tolerant and a salt-sensitive cultivar. *Scientia Horticulturae* 192, 38–46.
Seifi, E., Guerin, J., Kaiser, B. and Sedgley, M. (2008) Inflorescence architecture of olive. *Scientia Horticulturae* 116, 273–279.
Serrano, I., Suarez, C., Olmedilla, A., Rapoport, H.F. and Rodriguez-Garcia, M.I. (2008) Structural organization and cytochemical features of the pistil in Olive (*Olea europaea* L.) cv. Picual at anthesis. *Sexual Plant Reproduction* 21, 99–111.
Suarez, C., Castro, A.J., Rapoport, H.F. and Rodríguez-García, M.I. (2012) Morphological, histological and ultrastructural changes associated with olive pistil pre- to post-anthesis events. *Sexual Plant Reproduction* 24, 133–146.
Tan, J., Ben-Gal, A., Shtein, I., Bustan, A., Dag, A. and Erel, R. (2020) Root structural plasticity enhances salt tolerance in mature olives. *Environmental and Experimental Botany* 179, 104224.
Tataranni, G., Santarcangelo, M., Sofo, A., Xyloyannis, C., Tyerman, S.D. and Dichio, B. (2015) Correlations between morpho-anatomical changes and radial hydraulic conductivity in roots of olive trees under water deficit and rewatering. *Tree Physiology* 35, 1356–1365.
Therios, I. (2009) *Olives*. CAB International, Wallingford, UK.
Trentacoste, E.R., Gómez-del-Campo, M. and Rapoport, H.F. (2016) Olive fruit growth, tissue development and composition as affected by irradiance received in different hedgerow positions and orientations. *Scientia Horticulturae* 198, 284–293.
Troncoso, A., Valderrey, L., Prieto, J. and Liñan, J. (1975) Algunas observaciones sobre la capacidad de enraizamiento de variedades de *Olea europaea* L. bajo tecnicas de nebulizacion. I. *Anales de Edafologia y Agrobiologia*, XXXIV (7–8), 461–471.

2.2 Vegetative and Reproductive Physiology

A. Fabbri[1] and R. Gucci[2]

[1]*Department of Food and Drug, University of Parma, Italy;* [2]*Department of Agriculture, Food and Environment, University of Pisa, Italy*

2.2.1 Life Cycle

Olive trees are long-lived. There are records of trees several centuries old and, in some instances, their age has been claimed to exceed 1000 years. In cold climates, the longevity of trees is much shorter due to freezing temperatures but, because of the high capacity to regenerate the canopy from suckers, even under such conditions plants can live long and end up being formed by an ancient root system and a young canopy formed by multiple stems. The longevity and resilience of olive trees after stress events make them unique markers of the Mediterranean landscape.

In commercial olive growing, the timeframe over which an economic analysis should be projected is about 40 years. This means that business plans for new plantations should consider 40 years for the profitability of the investment. However, old trees can be very productive despite their age. Under many circumstances they continue to produce high yields once they are successfully rejuvenated by pruning. The main constraint to rejuvenating old trees is not their future recovery or productivity, but rather the high cost for pruning large canopies and the number of unproductive years after rejuvenation. Hence, rejuvenation of old trees, albeit technically feasible, is often uneconomical.

The olive tree goes through various stages of development during its life cycle. During the juvenile stage, the reproductive activity is absent. This is followed by a phase of increasing productivity, during which the vegetative development of the plant remains high while productivity progressively increases, and a phase of maturity during which it reaches the maximum productive potential. As the tree grows, the canopy expands laterally and takes on a typical globular shape, while juvenile plants tend to be more elongated in height. The juvenile and mature growth phases are often distinguishable by morphological characteristics, which in *Olea europaea* can include leaf shape and size, phyllotaxy and the ability to form adventitious roots and/or buds. Juvenile leaves are smaller, thicker and rounder than mature leaves, and are usually of a more intense green color. The juvenile plant usually produces several vigorous upright shoots. The juvenile shoot can be identified by morphological characters, including growth rate, thorns in place of terminal buds and different internode length. Because of the high growth rate of these shoots, the points of insertion of the opposite leaves of each node are a few millimeters apart. Juvenile shoots

DOI: 10.1079/9781789247350.0003

often bear three leaves per node rather than the typical two of mature shoots.

Juvenility is the first phase after seed germination that a plant experiences during its life cycle. The main feature of juvenility is that the plant is unable to flower even if environmental conditions are suitable for it. In olive seedlings the juvenile phase can last several years. Seedlings usually require up to ten years before they start flowering. However, under highly inductive conditions (growth concentrated on a few apices, high fertilization and irrigation regimes, sunlight supplemented with artificial light during winter), the first flowering can occur at the end of the third year after seed germination. Refinements of this protocol have been reported by Rallo (1999), who induced 5–40% of plants (depending on the cultivar) to flower 30 months after germination.

The trees propagated from mature tissue are much more precocious than seedlings. If olive trees are correctly managed in the field they will flower and begin production in the third or fourth year after planting. Some cultivars (e.g. 'Arbequina', 'Arbosana') are able to flower and fruit in the second year after planting. However, climatic conditions or irrational practices (e.g. excessive pruning) that stimulate vegetative growth rather than reproductive activity can delay the onset of flowering.

Phase change describes the period during which a plant undergoes the transition from juvenility to maturity. This transition is a gradual and continuous process. The term maturation describes the transition from juvenility to maturity.

If environmental conditions are appropriate, mature plants are able to flower. The foliage of mature plants shows the typical characteristics of the cultivar, which are instead not evident during juvenility. The canopy of the mature plant is broader than that of the juvenile plant.

Aging involves the loss of vigor. Old trees have a higher ratio of wood to foliage than young trees. Old trees maintain the ability to fruit and flower under proper environmental conditions, but their vegetative potential is less than in mature trees. Their potential reproductive activity is high but, because of little vegetative activity and growth, old trees generally yield less than mature ones.

2.2.2 Growth Habit

If left unpruned, the olive plant tends to grow as a shrub rather than a tree, with the canopy often formed by multiple stems originating directly from the crown. Even in a seedling, the central stem does not grow vigorously and is progressively overgrown by lateral shoots. The overall result is a bushy appearance of the canopy. The taproot is replaced by lateral roots often originating from burr-knots (ovuli); hypertrophic tissue of high metabolic and meristematic activity is present a few centimeters below the soil line. Burr-knots can produce a whole tree if removed from the mother plant. The number of suckers originating from a stump after the canopy is cut off is extraordinarily high. Over 300 new suckers can emerge from 30-year-old stumps of about 1.5 m circumference. The self-thinning of newly formed suckers is also exceptional; their number decreases by about tenfold by the end of the second growing season after the removal of the canopy.

The shape of the olive plant is the result of both the basitonic growth habit of shoots and the tendency to produce many stems from adventitious buds. A basitonic gradient consists in lateral shoots growing in the proximal part of the main axis more vigorously than those growing at the distal end. The profile of a vertical shoot growing undisturbed with a basitonic gradient is an asymmetrical diamond. Shoots emerging from the abaxial (lower) side of an inclined branch or axis are less vigorous than those growing on the adaxial (upper) side. In basitonic species, the response to inclination further emphasizes the tendency of basal shoots to prevail over apical ones. The dominance of the main axis is progressively weakened from the base to the terminal end by competition with lateral shoots. Basal shoots inserted on branches inclined at over 45° from the vertical axis tend to grow upright and become vigorous and are usually termed watersprouts or watershoots.

The basitonic habit and the response to inclination of olive trees have important practical implications. First, unlike peach, cherry and other tree-fruit species in which the supremacy of the central leader is more easily maintained, the central axis of olive plants does not grow as vigorously as do lateral shoots. In practice, it is more difficult and requires more work to train an olive plant to a central leader than to train other species. Repeated thinning of the uppermost lateral shoots and elimination of vigorous laterals inserted in the lower part of the stem are needed to maintain the central leader in most olive cultivars. This holds true for both the single vertical axis of central leader systems and the primary branches of the polyconic vase (see Chapter 6.4).

It should also be noted that olive is an evergreen species that retains leaves all year round. Leaves persist for two to three years. However, if trees are shaded, affected by diseases or subject to abiotic stress, leaves last less than two years. The energetic cost to form a new leaf is more expensive in olive trees than in deciduous fruit trees. This is partially due to the evergreen character of the leaf and the abundance of complex compounds (waxes, celluloses, carbohydrates, lignins, phenolic substances and oils) in the cuticle, cell wall and cytoplasm of cells. Most of these compounds serve mainly for protection against biotic and abiotic stress. The investment is partially compensated for by the longevity of the olive leaf and by the greater annual photosynthetic gain (Meriño, 1987).

The presence of leaves during winter months makes the plant susceptible to freezing-temperature damage. On the other hand, leaves are able to photosynthesize during mild winter days contributing to the carbon gain of the plant. The small size and rigidity of the olive leaf allow efficient exchange of heat when the temperature is high in the summer. The presence of trichomes seems to protect stomata and mesophyll from UV radiation, which is high in Mediterranean climates.

2.2.3 Vegetative Growth and Phenology

The annual development of vegetative organs is under environmental control. Temperature is the main factor driving growth of plant organs and the developmental process can be simulated by appropriate mathematical models. In general, temperatures above 9°C are effective in determining growth of olive buds, shoots and leaves. Simple models calculating the sum of mean daily (growing degree days (GDD)) or hourly (growing degree hours (GDH)) temperatures above the baseline temperature (9°C) can be used to predict the development of vegetative organs of olive trees. Temperatures above 35°C are considered ineffective in determining growth. As the plant develops, it goes through different phenological stages. The main phenological phases are: vegetative stasis, budding, bud growth, inflorescence elongation, flowering (anthesis), fruit set, fruit growth and veraison (Fig. 2.2.1). Vegetative stasis can be defined as a suspension or slowdown in the growth of the vegetative organs. During the budding and sprout growth phase, the apical and lateral buds swell, and the emission of new leaves, nodes and internodes begins. The inflorescence elongation phase consists in the formation of inflorescences from the flower buds while the phase between the opening of the flower buds and the fall of the stamens and petals represents flowering.

Starting from flowering, and for the following 20–30 days, flower fertilization and fruit set take place. Subsequently, there is a phase of intense cell division and distension which leads to a rapid growth of the fruit for about 20 days. This is followed by the hardening phase of the stone, characterized by a slowdown in the growth in volume of the fruit and by the lignification of the cells of the endocarp, and the phase of accumulation of oil during which the growth is linked to cellular expansion and to the accumulation of oil in the vacuoles. The last phase is that of fruit ripening, during which the set of processes that determine the morphological and biochemical changes in the fruit take place. In this phase,

Fig. 2.2.1. (a) fully elongated inflorescences; (b) anthesis; (c) fruit set; (d) pit hardening; (e) green fruits; (f) veraison; (g) fruits fully colored.

the accumulation of pigments continues, leading to the coloring of the exocarp and mesocarp typical of each variety. The fruit development process, from fruit set to ripening, follows a double sigmoid curve and lasts longer than six months (see also Rallo *et al.*, 1994).

To standardize the progression of phenological stages and make them comparable across locations and years, the

BBCH (Biologische Bundensanstalt Bundensortenamt and Chemical Industry) scale has been proposed as a reference method (Sanz-Cortes *et al.*, 2005). Identifying the main phenological stages is important for orchard management as it allows monitoring of the progression of development of the different organs during the annual cycle. For instance, pit hardening, which can be simply assessed by knife-cutting the developing fruit, is the stage when irrigation can be restricted without major consequences for yield. This concept is the basis of deficit irrigation to save water while maintaining production (see Chapter 6.2). Another example is given by symptoms of fruit color change that indicate the approaching of harvest period and can be easily assessed visually.

2.2.4 Reproductive Biology

In trees that are no longer juvenile, or that have overcome the unproductive period after asexual propagation, flowering takes place. The entire reproductive cycle from flower bud induction to fruit ripening takes longer than 12 months.

2.2.4.1 Flower bud induction

At the level of the individual shoot, the ability of a meristem to become mature is strictly related to the cultivar. It has been shown that the aging time can vary among cultivars. 'Koroneiki' has a very short aging time, and inflorescences develop from the lateral buds formed during the second year of growth of the apical meristem. For instance, the cultivar 'Frantoio' has a long cycle, with lateral buds developing inflorescences only on those extensions of the vegetation that have exceeded the third year of growth under proper environmental conditions (Fiorino *et al.*, 2020).

The formation of flower-bearing buds is a process requiring the transition of the meristematic apex of the bud, undifferentiated in its early stages of growth, to a structure carrying flowers. Unlike deciduous species, where bud differentiation starts immediately after induction in the summer, in olive trees, bud meristems remain exposed to the environment for a long period, during which their structure changes little or not at all. Anatomical investigations have shown that flower buds seem to be meristems the plant somehow ignores during the summer, while the available resources are diverted to leaf buds (Fabbri and Alerci, 1999). It appears that the plant, therefore, separates, at a quite early stage, the existing buds into two groups, which, with few exceptions, will only be able to produce one kind of structure. The process is commonly divided into two stages: during induction, the bud undergoes a series of conditionings, both internal and external to the plant, following which, in the meristematic apex, such biochemical modifications occur as to 'commit' it to the formation of reproductive structures. This is also termed 'irreversibility stage' to indicate a condition of the bud, in which this has only left one ontogenic possibility, i.e. of developing into a fruit bud; otherwise it will not undergo any further development.

When the irreversibility stage is overcome, the bud is to be considered 'induced' to flower, and the next phase begins, termed 'differentiation'. The exact time of flower induction is difficult to determine, mostly because the modifications involved are of a physiological nature, and therefore non-detectable at the microscopic level. In addition, the duration of this phase does not depend so much on the length of the process in the individual bud; it takes place in different moments in the various parts of the canopy, with a graduality that is present in all cultivars, although to different extents.

The criteria the plant follows for the choice of bud destiny are not completely understood, although factors such as bud and shoot position, light radiation, fruit load, water balance and nutrient availability are likely to be determinant.

The abundant literature on flower bud induction indicates that an important influence is exerted by factors acting in the October–December period. On the other hand, reports (Fernandez-Escobar *et al.*, 1992) of very early flower induction activity have apparently contrasted this view, often based on complete plant shading at different times (Fabbri *et al.*, 1997). The results of Fabbri and Alerci (1999) are consistent with both sets of data. It might therefore be postulated that a flowering potential is established in the summer by the plant, and that to which extent such potential will be expressed in the next spring is something the plant will decide in the course of the following autumn.

This would agree with a two-step explanation of olive flower bud induction and differentiation (Lavee, 1996). The first set of stimuli occurring in the spring would select a number of likely flower buds, which do not differentiate into leaf buds, and suspend their growth. These 'candidate' flower buds will need further favorable conditions for a second triggering of their developmental pathway, to be enabled to form flower structures. In both stages the number of buds to be stimulated will depend on environmental conditions, nutrition, hormonal balance and other factors, all strictly related and dependent on the plant vegetative and reproductive history.

Recently, a possible role was identified in the interpretation of environmental signals and endogenous factors affecting flower induction in olive trees for genes encoding the FT protein involved in flowering processes. In particular, low temperatures affect the subsequent processes of induction and differentiation of the lateral buds in potted plants of cv. 'Barnea', reevaluating the importance of meeting cold requirements as a necessary step to allow the evolution of the lateral buds toward inflorescence formation (Haberman *et al.*, 2017). The authors underline, however, that environmental mechanisms leading to flower induction can only be activated when the plant has passed the juvenile stage during which there is a form of internal inhibition that would block the subsequent flowering evolution of the buds.

Inhibition of flower development in olive trees seems to be mainly determined by nutritional factors. Accurate measurements comparing the behavior of apex elongation in plants in the on year and in the off year showed that flowering in the following year was reduced, in plants grown in on years, mainly by a decrease in apex elongation, and therefore in the number of nodes laid and inflorescences formed (Fiorino *et al.*, 2020). However, a more detailed analysis of the data showed that the relative percentage of buds evolved to flower was also reduced, with a concomitant twofold effect of a reduction in the number of sites potentially able to bear inflorescences and a lower percentage of the number of sites that evolved; this inhibition effect, in addition to being nutritional, can also be attributed to hormonal inhibition caused by the olive tree fruitlets whose embryos, at a stage prior to hardening of the kernel, were able to inhibit the flowering of newly formed buds or buds in formation in the upper part of the branch, similar to what happens in other species (apple). This behavior is called alternate bearing (Stutte and Martin, 1986a).

Light is an important factor in flower induction: shading reduces flower bud differentiation (Tombesi and Cartechini, 1986). Shading following flower bud differentiation does not affect flowering; however, it can cause morphological sterility or abortion of the ovary. Shading or leaf removal may postpone flowering.

Light intensity and quality are modified as light passes through the top of olive trees and varies according to the shape of the trees and the density of their leaves. The optimal leaf area index (LAI) value is 2.5. The maximum net assimilation of CO_2 under optimum conditions is not particularly high (see Chapter 2.3).

Several climatic factors are apparently critical for olive performance, with respect to flowering, fruit growth and olive oil production and quality. Some winter chilling is required for floral initiation, but this requirement varies quantitatively between varieties (Hartmann and Porlingis, 1957; Hackett and Hartmann, 1963, 1964, 1967; Navarro *et al.*, 1990; Pinney and Polito, 1990; Ferrara *et al.*, 1991; Rallo and Martin, 1991). Varieties originating from Italy, Spain, northern Greece and California require more winter chilling than those grown in southern Greece (Crete), Israel or Egypt

(Martin *et al.*, 1994), where the chilling periods range from 86 to 1400 h (Hartmann and Whisler, 1975). Temperature fluctuation affects flowering: fluctuation between 2°C and 15°C for 80 days produced adequate flowering, while fluctuation between 7°C and 18°C produced a smaller number of inflorescences.

A chilling period is required prior to flowering. Afterwards, the determination of heat requirements in the first developing phases of plants has been expressed as temperature accumulation by means of parameters such as heat units, which are usually expressed as GDD. Their determination is useful for forecasting when flowering will occur. The most suitable threshold temperatures were carried out in a range of 7–15°C. The causes responsible for varying threshold temperatures recorded in areas of the same latitude include different climatic conditions and the various cultivars present within olive groves at the same latitude (Therios, 2009).

Due to the late blooming of olives, flower damage due to late spring frosts is rare. On the contrary, hot and dry winds during flowering reduce fruit set. Strong winds in June can increase the percentage of fruit drop.

The effect of temperature on the percentage of perfect flowers is not clear. Therefore, the level of temperature during the first stages of flower bud differentiation affects, at least in some cultivars, the percentage of flower buds and number of flowers per inflorescence, with no effect either on pistil development or the percentage of perfect flowers. In some olive cultivars, such as 'Arbequina', flowering and fruiting can be achieved in the absence of chilling temperatures ($\leq$7.2°C) (Malik and Bradford, 2005, 2006). The absence of flowering in subtropical climates is not due to lack of chilling but rather to high temperatures during winter days. Hence, temperature during the day > 23°C could produce adverse effects on flowering.

The levels of ABA, IAA and GA_3 in leaves, nodes and fruits during the induction, initiation and differentiation periods in on years were lower than those in off years. However, GA_3 levels during these periods were higher in the on years. The fact that GA_3 decreased and GA_4 levels increased during the induction and initiation periods in the off years suggests that these affect flower bud formation. Zeatin increase during the induction period in the off year suggests that an increase in cytokinin during the induction period, possibly, has a positive effect on floral formation (Ulger *et al.*, 1999, 2004).

The ABA concentration changes are thought to be due to the shedding of bracts and imperfect flowers during the period of inflorescence development prior to full bloom.

Exogenous application of GA_3, asparagine and glutamine affected the induction of flowering buds (Proietti and Tombesi, 1996).

Differences in concentration of any of the sugars, with the exception of fructose, were not significant in the on and off years (Ulger *et al.*, 2004). The highest macro- and microelement concentrations were found to be calcium and iron, respectively. The results suggest that carbohydrates and mineral nutrients may not have a direct effect on flower initiation (Stutte and Martin, 1986b; Ulger *et al.*, 2004).

Oleuropein levels decreased sharply during the transition from vegetative to flower buds (Malik and Bradford, 2006), while levels increased rapidly with the expansion of fertilized pistils and then declined quickly with fruit maturation. Exogenous application of oleuropein inhibited flowering in *Kalanchoe blossfeldiana* (Bongi, 1986). Therefore, oleuropein could play a role in flowering. Drastic reduction in oleuropein levels was observed in floral buds, as compared with vegetative buds. After floral bud differentiation, oleuropein levels progressively increased from the flowering to fruiting stage and then declined as the fruit started to mature, and were negligible in the fully mature black fruits (Amiot *et al.*, 1989).

The date of harvesting affects significantly the percentage of flower bud differentiation. Late harvesting depletes the olive tree and results in a lower percentage of flower bud differentiation (Therios, 2009). Fruit thinning instead increases the percentage of flowering bud differentiation and flowering.

The percentage of flower bud differentiation also decreases after defoliation due to diseases such as peacock spot (*Venturia oleaginea*), nutrient and water stress.

2.2.4.2 Bud differentiation

A bud is considered 'differentiated' when visible or measurable changes occur in its structure, other than those leading to vegetative growth; anatomical differentiation, i.e. the appearance of structures that indicate that morphologically an inflorescence is being formed, has been dated from November to February, the very variable data depending on the area where the research was carried out and, probably, on the cultivar.

Thus, research carried out in Italy and Spain has detected the first symptoms of anatomical differentiation at the end of winter and just before vegetative recovery (January–February; Fabbri and Alerci, 1999; de la Rosa *et al.*, 2000), while in California these processes appeared before winter (November) (Pinney and Polito, 1990).

By utilizing other biochemical methods, the time of differentiation has been further anticipated, even to the month of July, i.e. just after two or three months after bud formation, in relation to the appearance of protocytokinins (Fiorino *et al.*, 2020). As a rule, however, most floral differentiation occurs during the 40–60 days before anthesis; therefore the process is completed as the inflorescence emerges and develops (Fabbri and Benelli, 2000).

The morphological differentiation leading to the formation of an inflorescence consists in the formation of particular structures of the meristematic apex; Italian researchers saw in the so-called 'orceolo' shape (that of a small bowl) of the bud apex the first sign of flower differentiation. Troncoso de Arce (1966) ascribed the earliest sign of differentiation to the formation, in the apical meristem, of three layers of cells with large nuclei, from which the axis of the inflorescence would eventually develop, producing branching with similar apices.

The first signs of differentiation in the individual flowers are a broadening of the bud apex towards a more or less flattened conical mass, with the four sepal primordia appearing as minute protuberances on the sides of the apex itself (Fig. 2.2.2). At the axils of the leaflets enclosing the apex, more meristems (decussate couples) appear and develop, forming a branching that will form the inflorescence, which will comprise a large number of flowers. Meantime, in each individual flower primordium, the various floral organs (petals, stamens, ovary) are formed centripetally. While the terminal flower develops, the inflorescence develops several nodes and branching, with lateral flowers developing with varied degrees of delay if compared to the terminal flower.

2.2.4.3 Formation of the inflorescence and flower

Inflorescence development begins in early spring, roughly one month after the onset of flower differentiation, usually starting on the south side of the tree (in the northern hemisphere). Development is gradual, and the time between inflorescence emission and anthesis is usually around four weeks (up to six or more in warmer climates) in separate flushes. As a rule, the earlier the emission of inflorescences the higher the expected production, as fruit set may take place in less-dry conditions. However, environmental events may markedly alter the forecast. Flowers are usually borne on one-year-old shoots. Only occasionally, inflorescences develop on two- or three-year-old branches.

The newly formed inflorescence has a compact appearance, with a first internode (10–12 mm) developing quickly, and soon distancing from the central axis the meristems due to produce the inflorescence structure. Full bloom is achieved within 8–10 weeks from bud sprouting.

Three types of flowers are present each season: 'perfect flowers', containing stamens and pistils; 'staminate flowers', containing

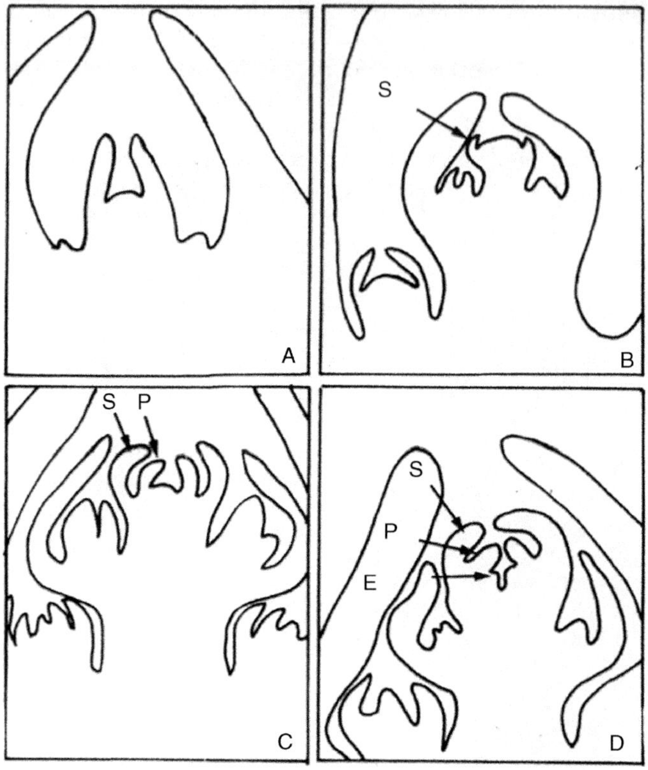

Fig. 2.2.2. Flower differentiation of an olive bud. A – leaf bud; B – elongation of the inflorescence axis and appearance of the terminal flower with sepal primordia (S); C – well-developed sepal (S) and petal (P) primordia; D – sepal and petal primordia at a more advanced stage, with the appearing of stamens primordia (E). The formation of carpels primordia completes floral differentiation. (From Loussert and Brousse, 1978, redrawn from Hartmann, 1951)

aborted pistils and functional stamens; and 'pistillate flowers', in which the stamens are degenerated. 'Ovary abortion' refers to absence of ovary or to small imperfect, non-persistent ovary. All olive trees display ovary abortion, although to different extents, depending on cultivar, environment, year, inflorescence and type of shoot. Its incidence is variable, from less than 10% to 70% or more. In spite of that, production is usually not depressed as normal harvests require no more than 4% of fruit set, usually much less. One hundred percent abortion cannot exist in a commercial cultivar, with the exception of 'Swan Hill', an ornamental cultivar, selected by Hartmann in Australia (Hartmann, 1967), which displays this character, advantageous for olive trees utilized in urban forestry or as ornamentals.

Ovary abortion seems to be caused mainly by competition among flowers, namely among ovaries, i.e. the future fruits, for the tree's resources, which are insufficient to develop all the flowers, which are in an excessive number if compared to the plant's productive potential. The competition arises very early and also influences fruit set. Conditions that accentuate competition between flowers/fruits or that diminish resources generally lead to an increase in abortion and a decrease in fruit set. These conditions include nitrogen deficiency, leaf diseases, a low leaf-to-flower ratio, water shortages, insufficient light, unfavorable climatic conditions, abundant production in the previous year, very abundant flowering and the location of inflorescences in less favorable areas of the canopy.

Abortion varies greatly with cultivar (Table 2.2.1). Indeed, ovary abortion tends to be higher in large-fruited cultivars, which have larger flowers and ovaries. This implies a higher energy cost for flower formation and thus a higher competition for resources, and consequently a higher incidence of abortion (Rosati *et al.*, 2009, 2010, 2011b). Recent work has shown that this genetic component can also be traced, at least in part, to competition for resources, degeneration of a part of the flower, originally hermaphroditic, but which retains only one of the two functions when blooming. Staminiferous flowers derive from the abortion of the ovary and/or the drying out of the pistil, phenomena that make fertilization impossible in any case, while the anthers regularly develop for pollen production; all cultivars have a more or less variable percentage of staminiferous flowers, a genetically controlled character, even if strongly influenced by nutritional factors. Pistillate flowers, on the other hand, are characterized by the abortion of the male part, with the anthers withering before the flower opens, or by the inability of the anthers to produce pollen, or by the inability of the anthers to produce viable pollen (cv. 'Obliça', Croatia). The three cases are variously mixed and probably more widespread than has really been studied; classic pistillate cultivars are the Italian 'Cerasuola', the Algerian 'Chemlal' and the French 'Lucque', but Spanish and Croatian cultivars are also mentioned in the literature. The phenomenon is under strict genetic control, and pistillate varieties depend for their production on the presence of suitable pollinators.

Table 2.2.1. Percentages of ovary abortion in several olive cultivars in the Chianti region. (Magherini. 1971)

Cultivar	1961	1962	1963	1964	4 year average
Frantoio	8.2	9.4	6.2	6.6	7.4
Moraiolo	29.3	18	7.2	12.95	16.08
Pendolino	26.2	22.5	31.4	18.45	23.4
Ascolana Tenera	49.7	40.4	56.1	49.65	49.1
Uovo di Piccione	57.7	46.4	56.4	46.45	50.66
Morellona di Grecia	32.6	32	45.1	32.95	35.12
Grossa di Spagna	34.8	49.6	46.8	86.6	60.88
Sant'Agostino	36.8	45.4	45.6	64.85	51.5
Gordal Sevillana	49.1	56.6	59.2	69.15	60.64
Averages	36.03	35.57	39.33	43.07	

The percentage of perfect flowers on a tree depends on: (i) cultivar; (ii) shoot vigor; and (iii) environmental conditions, such as the lack of soil moisture during spring, when the inflorescences develop. Furthermore, the leaf:bud ratio affects the percentage of perfect flowers. When the leaf:bud ratio is increased by disbudding, the production of perfect flowers is promoted.

The presence of imperfect or staminate flowers (see Fig. 2.1.28b) grants certain advantages to olive trees, since they can enhance pollen donation to fertilize other flowers. Hence, perfect or pistillate flowers can receive enough pollen. Therefore, a correlation exists between the quantity of pollen released and olive productivity. Consequently, staminate flowers are not a non-functional byproduct; they instead perform a vital role for olive trees (Cuevas and Polito, 2004). The dry weight of staminate flowers is 9% less than that of hermaphrodite flowers.

The formation of staminate flowers is due to arrested pistil growth as a consequence of water stress or N deficiency during the period of flowering bud differentiation. A small percentage of setting (1–2%) is, however, sufficient for a good crop.

The distribution of different flower types within the inflorescence (Lavee *et al.*, 1996) is correlated with the time of anthesis and the percentage of imperfect flowers. Most hermaphrodite flowers are located at the apex and on primary pedicels, while flowers on secondary pedicels are mainly staminate and flower later.

2.2.4.4 Flowering time

A flower is fully opened when both anthers and petals are separated. During the hottest part of the day, anther dehiscence takes place and an abundant amount of pollen is shed.

Olive full bloom is recorded in Mediterranean areas from April to the first days of June. The flowering period varies between cultivars (Fig. 2.2.3) and this should be taken into account when selecting the most appropriate pollinators. The blooming period

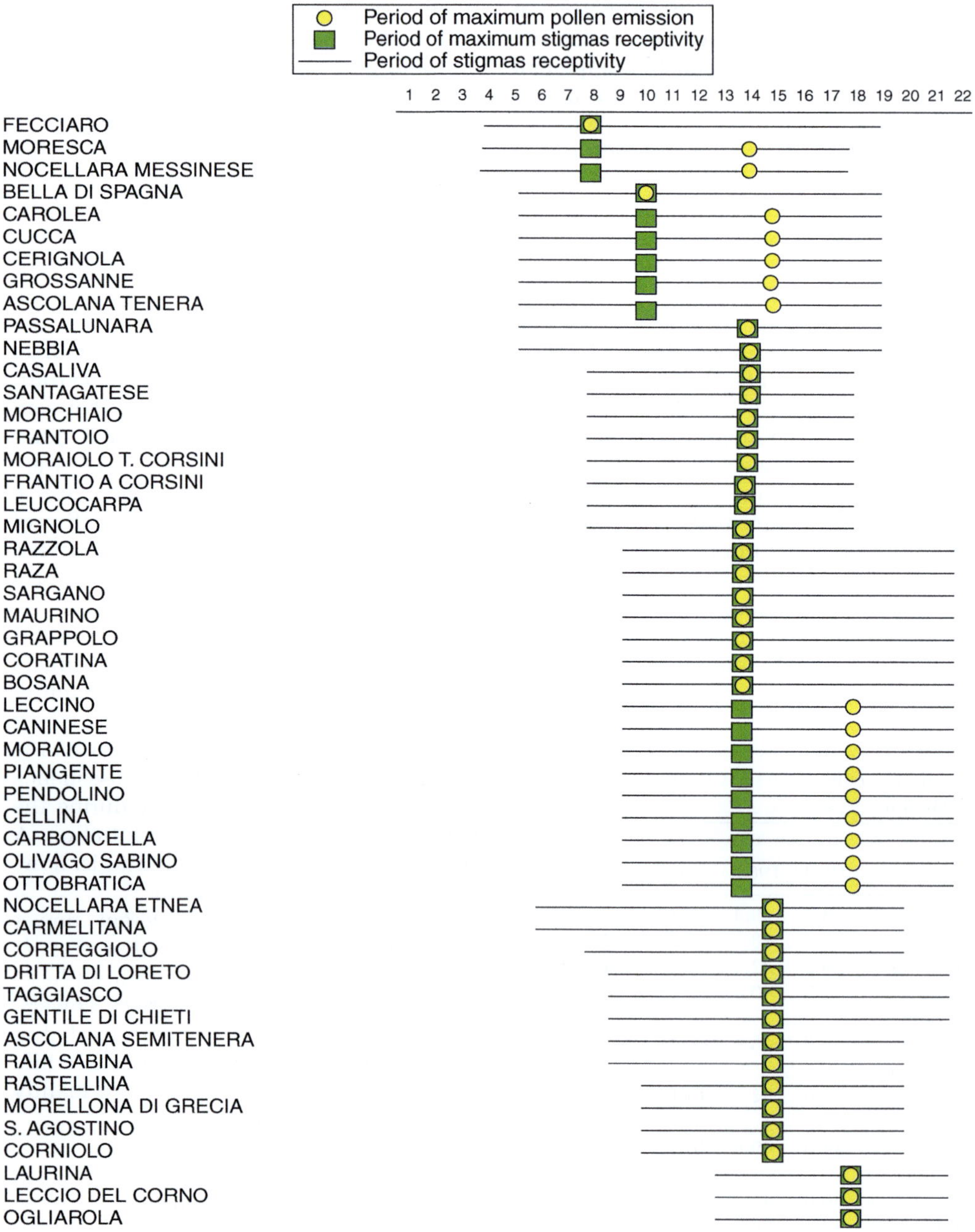

Fig. 2.2.3. Relative phenology of flowering in Italian olive cultivars. Data are average of two years, in Perugia (hilly) environment. (From Tombesi, 2003)

lasts five to six days in any particular cultivar, up to two weeks on the individual tree, and a little more when average temperature is lower than usual.

The flowering time can differ between years and locations, mainly due to the effect of the temperatures in late winter and early spring. As this effect differs among cultivars, the flowering sequence also varies with time.

The flowering time can vary in the same environment between the different cultivars by up to three to four weeks; references from large collections, where numerous varieties from different locations can be compared, are reliable in this respect. However, the flowering dates recorded in different years have shown that the variations in time fluctuations between years for the same cultivar can be greater than the fluctuations between cultivars in the same year.

The factors that determine this variability in flowering time for the same cultivar are not completely known, and even in this case, recent studies have shown that the most influential factor is the sum of temperatures above 8.5°C from January onwards. Again, from the literature, it appears that the succession between the different cultivars should remain unchanged, although exceptions to this rule are reported. Information limited to a few cultivars would suggest that the flowering time intervals between the different cultivars are amplified as the latitude decreases. It is well known that, in the hottest olive-growing areas in Italy, the combination of 'Pendolino', 'Frantoio' and 'Leccino', which is typical of the inland areas of Tuscany and widely used to cover the pollen requirements of the three varieties, which flower almost simultaneously, can no longer be used, since 'Pendolino' tends to flower much earlier than 'Leccino', which is generally later than 'Frantoio'. At the basis of this difficulty in establishing a hierarchy in flowering dates lies the reduced knowledge of the thermal requirements of the flowering buds after their formation, especially on the basis of the length and intensity of the cold period which must be overcome in any case, but certainly with a duration and temperature very different from those of the fruit species of the temperate zones to which reference is generally made.

The main phenological stages from inflorescence development to full bloom are given in Fig. 2.2.1.

2.2.4.5 Pollination

The olive tree is an anemophilous species (pollen is transported by wind) and tends to be allogamous, i.e. it prefers or requires cross-pollination between cultivars. These characteristics of the species condition the presence and arrangement of pollinators within the specialized crop. The olive tree flower is proterogynous; i.e. in the perfect flower, at the moment of opening, the ovary, style and stigma complex is already receptive, while the anthers have yet to complete the pollen-formation processes. For some authors, the rule is the opposite (Fig. 2.2.4). This delay in pollen availability, even for a potentially self-fertile cultivar, must be assessed in intensive plantations for the convenience of having a supply of pollen in the environment from the moment the first flowers of the main cultivars open. However, the planting of more cultivars in the same plot is an ordinary practice, although it may represent a problem in very high density orchards.

Pollen is released one or two days after flower opening. When it is mature, the two anthers open longitudinally, releasing it into the environment. This occurs during the hottest and driest hours of the day. In humid or poorly ventilated environments, the pollen grains tend to remain somewhat agglutinated, falling in small clusters on the leaves below or as far as the ground, but in suitable conditions they can be transported quite easily by the wind to a distance of one to two kilometers.

The amount of pollen is a function of the cultivar, e.g. 'Leccino' and 'Frantoio' (two oil cultivars very common in central Italy) produce small amounts of pollen; larger quantities are produced by 'Ascolana', 'Manzanilla' and 'Pendolino'. More important

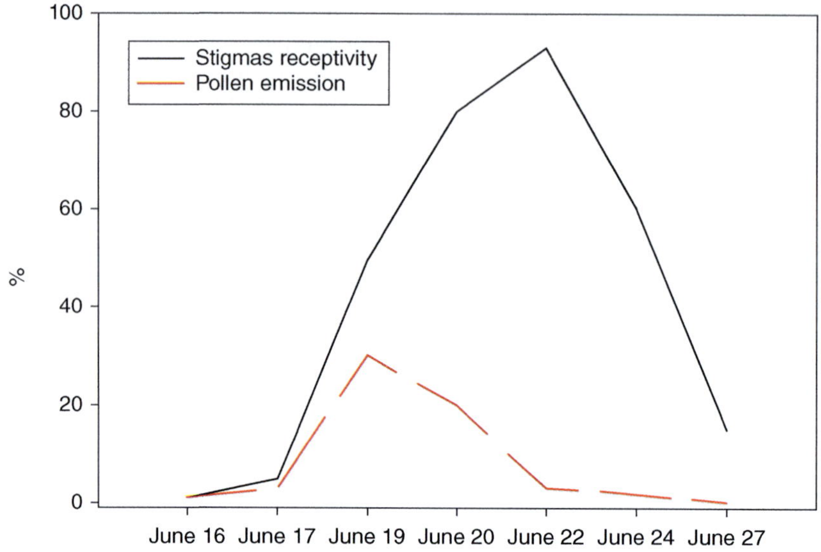

Fig. 2.2.4. Course of stigmas receptivity and pollen emission in 'Frantoio' inflorescences in Perugia. (From Tombesi, 2003)

is pollen's ability to germinate: this characteristic appears to fluctuate (*in vitro*) between 12% and 60% (Zito and Spina, 1956; Fernandez-Escobar *et al.*, 1983).

Pollination in olives occurs through either self-pollination – as in the case with 'Koroneìkì' – or cross-pollination (Cuevas and Rallo, 1990; Cuevas *et al.*, 2001), as in 'Kalamon', and the cultivars are characterized as either self-fruitful (Androulakis and Loupassaki, 1990) or non-self-fruitful (Diaz *et al.*, 2006). In the case of self-pollination, the pollen of a flower pollinates the pistils of flowers of the same tree. The pollen is transferred by wind, gravity or insects. In cross-pollinated cultivars, the pollen is transferred by wind. Olive pollen is poor in sugars, nectaries are rudimentary, and therefore bees seldom visit olive flowers.

Pollination is influenced by several factors, the most important being temperature, which has the effect of enhancing tube growth (Galàn *et al.*, 2001), although when it is too high the stigma may get dry. For anther dehiscence, the optimum is 30°C or a little less with 50% RH (relative humidity); a good value of RH also enhances pollen germination. Very low temperatures reduce the speed of pollen tube growth, while high temperatures, besides drying out the stigmas, do not allow pollen germination. Wind and rain can also have a relevant effect. Rain is always negative; indeed, it may determine pollen grain plasmolysis, dilute stigma secretions and hinder pollen transport. Wind, which is fundamental for this anemophilous species (Griggs *et al.*, 1975), when too strong, may transport masses of pollen away from the grove. Although olive pollen can be found as far as 12 km from the originary tree (Morettini, 1972), the effective range is considered to be up to 30 m. However, dry and hot winds destroy pollen grains and nullify pollen germination and pollen tube growth. When pollinizers are planted, their distance from the main cultivar should not be more than 30 m, in order to attain an adequate fruit-setting percentage.

Among nutritional factors, good nitrogen fertilization has been proven to elongate the effective pollination period, and therefore to improve fruit set.

For efficient fertilization, the ovules should develop completely. However, of the four ovules contained in an embryo sac, only one is fertilized and the other three degenerate and shrink. In olives, degeneration of ovules can depend on various stress conditions, such as water stress and heat stress. Also, the type of cultivar affects embryo sac degeneration.

When olive trees are not well supplied with N, ovule viability declines rapidly. Nitrogen concentrations in leaves of <1.3% markedly reduce ovule longevity. The size of the ovary is affected by water and N availability during the period of flower bud differentiation. Subsequently, through the process of cell division, expansion and/or differentiation, ovarian size increases 50-fold (Rapoport *et al.*, 2004) to produce the olive drupe.

Fertilization does not always follow pollination; in some varieties, as a result of pollination, there is a form of stimulatory parthenocarpy that leads to the formation of roundish fruitlets deriving from a mere enlargement of the ovary, therefore of small size, with a hollow stone inside; these fruits are parthenocarpic (lacking seed), pseudodrupes known as shotberries (Fig. 2.2.5). The reason why they are produced is unknown. Shotberries mature earlier than normal fruits.

Differences between various olive cultivars were observed regarding the onset and the start and length of the flowering period. These differences are strictly correlated to differences in micro- and megagametogenesis and are cultivar-specific characteristics.

Factors responsible for inadequate fertilization and subsequent fruitlet drop include: (i) incompatible pollinating cultivars; (ii) lack of pollinator trees; (iii) pest and disease problems; (iv) strong winds; (v) water stress and nutrient stress; and (vi) unfavorable climatic conditions (Cuevas *et al.*, 1995). Usually, one fertilized fruit is maintained per inflorescence (Fiorino *et al.*, 2020); more drupes per inflorescence are more common in small-fruited cultivars.

Fig. 2.2.5. Olive inflorescence with many shotberries.

2.2.4.6 Sterility

Sterility may be due to factors different to those affecting ovary abortion, such as anomalies during meiosis producing imperfect gametophytes (cytological sterility), quite rare in olive, or incompatibility (factorial sterility). Self-sterility occurs with pistilliferous flowers (e.g. cv. 'Ogliarola messinese'), in which the functional female part is obliged to receive pollen from another cultivar; another form of physiological incompatibility is characteristic of cultivars that are unable to produce viable pollen, such as 'Obliça'. Incompatibility occurs when a perfect pollen grain fails to germinate on the stigma, or germinates, but its tube growth is insufficient, or too slow-growing, for fertilization; this incompatibility may be between two cultivars (inter- or cross-incompatibility) or, as is often the case with olive, when a cultivar is genetically programmed not to be fertilized by its own pollen (self-incompatibility).

Most olive cultivars (especially the Italian ones) are self-incompatible, a feature that has been the main reason for the large genetic variability typical of the species. This aspect has been extensively studied due to its recognized importance in the establishment of olive orchards, to assure inter-compatibility among cultivars and, hence, productivity, and care must be taken to establish orchards with more than one cultivar. It is therefore important to know which cultivars are best suited to fertilize the one you want to be the principal one. In any case, growing three or four cultivars in the same plot guarantees good set, even if the cultivar you are interested in is considered self-fertile, as is the case for

'Frantoio'. Notwithstanding, contradictory results have often been reported as to the behavior of some of the main important cultivars (e.g. 'Arbequina', 'Manzanilla de Sevilla', 'Picual' and the already mentioned 'Frantoio'), according to different studies, either 'self-incompatible' or 'self-compatible' (Diaz *et al.*, 2006). Other cultivars are definitely self-incompatible, such as 'Leccino' and 'Picholine de Languedoc', and require the presence of pollinators. More investigation seems necessary on this aspect, particularly to ascertain the genetic control of the 'self-compatibility' mechanism in olive. This aspect is also important when breeding programs are designed.

No evidence was instead found to indicate incompatibility between different cultivars, the only factor limiting fertilization being the contemporaneity between the opening of the receptive flower and the availability of pollen. Farinelli *et al.* (2006) have proposed the most appropriate pollinators for 21 olive cultivars (Table 2.2.2).

The particular aspects of self-incompatibility have to be taken into account, especially when operating in superdense orchards, which reduce wind effects, and in climates characterized by high temperatures during the pollination period, as an inhibition effect on pollen tube growth related to high temperatures is known in the case of self-fertilization.

2.2.4.7 Fruit set

In the absence of sterility barriers, the pollen germinates on the stigma, and a recognition takes place between ovary tissues and those of the pollen, which allows the development through the style of a pollen tube, which, in a few days, through the style central tissue, reaches the ovary, to the membrane of the embryo sac. Between the

Table 2.2.2. The best pollinators for a number of olive cultivars, after a four-year trial in a hilly area of central Italy. The indicated pollinators increased fruit set if compared to free pollination and self-pollination. (From Farinelli *et al.*, 2006)

Cultivar	Suitable pollinators
Arbequina	Carolea, Kalamon
Ascolana Semitenera	Carolea, Kalamon, Itrana, Picholine
Ascolana Tenera	Maurino
Bella di Spagna	Nocellara Etnea
Carolea	Ascolana Semitenera, Bella di Spagna, Leccino, Maurino,Moresca
Giarraffa	Ascolana Tenera, Leccino, Nocellara Etnea
Gordal Sevillana	Ascolana Tenera, Manzanilla
Grossa di Spagna	Giarraffa, Itrana, Santa Caterina
Itrana	Ascolana Tenera, Carolea, Manzanilla
Kalamon	Arbequina, Ascolana Semitenera, Carolea, Giarraffa, Koroneiki, Leccino, Manzanilla, Maurino
Koroneiki	Kalamon
Leccino	Arbequina, Ascolana Tenera, Carolea, Kalamon, Manzanilla, Pendolino, Taggiasca
Manzanilla	Gordal S., Leccino, Maurino, Nocellara Etnea, Picholine , S. Caterina
Maurino	Ascolana Tenera, Leccino, Manzanilla, Picholine, Picual
Moresca	Ascolana Semitenera, Carolea, Leccino, Picholine, Picholine Marocaine, Sorani
Nocellara Etnea	Bella di Spagna, Giarraffa, Leccino
Pendolino	Ascolana Tenera
Picholine	Ascolana Tenera, Ascolana Semitenera, Maurino, Manzanilla, Moresca
Picholine Marocaine	Moresca, Sorani
Picual	Leccino, Maurino
Santa Caterina	None
Sorani	Gordal S., Moresca, Picholine Marocaine
Taggiasca	Ascolana Tenera, Carolea, Leccino

tissues of the two membranes, a second recognition occurs that determines their lysis and allows the penetration of the male gametes and the ensuing fertilization ('double fertilization'): one sperm cell fuses with the egg, producing the zygote, which, after division, produces the embryo, which, with the accessory tissues, is destined to become the olive seed. Cold weather reduces the speed of pollen tube growth; in this case, therefore, it reaches the embryo sac late, when the latter can be deteriorated. After pollination and fertilization, the ovary turns into a fruit. Each ovary has two cavities, the carpels, each one containing two ovules (in two separate locules), i.e. four ovules per ovary. Only one developed ovule is required for seed formation; occasionally, there may be none (parthenocarpy or stenospermocarpy) or two. The presence of well-developed embryo sacs in the ovules and ovule longevity are determining factors in their capacity for fertilization.

Many factors determine fruit set, such as: (i) inflorescence load and distribution (Lavee *et al.*, 1999); (ii) size and number of florets per inflorescence; (iii) percentage of hermaphrodite flowers; (iv) ovarian size; and (v) physiological condition of the ovules and their longevity.

After fertilization, petals and stamens are abscised and the ovary starts to grow and enlarge (Rapoport and Rallo, 1991), gradually taking the fruit shape.

Within a month from full bloom, only 7–8% of flowers are still on the tree as fruitlets; of them, no more than 25% is present by late summer. The average fruit set is therefore no more than 2%, although yearly fluctuations can be wide (even 0.5% in 'on' years). In 'off' years, with limited flowering, flower quality is good and fruit set percentage the highest (Cuevas *et al.*, 1994; Martins *et al.*, 2006). Of the 5–60 flowers present on the inflorescence, only one is present at harvest; however, in some cultivars, two or more fruits can be found, even in off years. The presence of numerous drupes on the same inflorescence is quite constant in several small-fruited cultivars; in France these cultivars are named 'bouteilleau'.

Fruit drop is utilized by the plant to adapt production to its elaborating surface; other factors may influence fruit drop, such as: nutritional and water deficiencies, weather conditions during bloom, sterility, lack of pollinators, and pests. Fruitlets compete for survival, starting about 10–15 days after full bloom. This phenomenon seems to be caused mainly by competition between flowers, i.e. between ovaries, the future fruits, for the tree's resources, which are insufficient to develop all the flowers, which, as mentioned, are excessive compared to the plant's productive possibilities. This competition arises very early and influences both the abortion of the ovary and, subsequently, the setting of the fruit. Conditions that accentuate competition between the flowers/fruits or that diminish resources generally lead to an increase in abortion and a decrease in fruit set. These conditions include nitrogen deficiency, leaf diseases, a low leaf-to-flower ratio, water shortages, insufficient light, unfavorable climatic conditions, abundant production in the previous year, very abundant flowering and the location of inflorescences in less favorable areas of the canopy.

Fruit set is completely inhibited at a constant temperature of 30°C, which significantly reduces pollen germination but does not prevent pollen tube growth. The most favorable temperatures for flowering, pollination and fertilization are between 18°C and 25°C, when faster tube growth and higher fruit set were both observed (Porlingis and Voyiatzis, 1999). Most fruits are formed at the periphery of the tree, and very few fruits set in the interior of the olive canopy.

Fruit set is the point of transition from flower to fruit and begins with the formation and first division of the zygote. The phenomenon is marked by a rapid petal drop, thanks to the formation of a zone of abscission located between petals and the base of the calyx, and by a rapid increase in the size of the ovary, due to a resumption of cell divisions in its various parts.

2.2.4.8 Pollination and fruit set in non Mediterranean areas

Olive cultivation is expanding not only in areas with a Mediterranean climate with mild, rainy mid-seasons and long, warm, dry summers but also in other places with very different environmental conditions, e.g. Argentina, Mexico (Malik and Bradford, 2006) and other areas of the southern hemisphere. In most cases, the closeness to the equator has an impact on both temperature regimes and photoperiod. The trees may show problems with fruit set and yield, with plant growth and metabolism being affected by the very different environmental conditions found in those regions. Thus, for instance, average winter temperature can be too high for flower induction and differentiation; lack of excursion in photoperiod may lead to continuous vegetation and flowering, making necessary to harvest several times each year, with the obvious increase of costs.

High temperatures during olive flowering increase pollen incompatibility. Pollen tubes of the same cultivar frequently become blocked between the stigma and embryo sac in this condition. Furthermore, under hot, arid conditions, the distance for effective pollination is reduced and is less than 30 m. Increasing the number of pollinator trees in a grove could solve some of these problems, although it introduces a problem with monocultivar orchards, as is fashionable in new plantings. Cross-pollination may improve fertilization and fruit set under hot environments, as well as in years of poor flower quality. The need for cross-pollination exists when temperatures during olive bloom exceed 30°C (Rallo, 1997). Another problem in warm, dry areas is a very short vernalization period (Ayerza and Sibbett, 2001). This increases the incompatibility of self-pollination for many cultivars, including 'Manzanillo', but it does not affect open pollination (Lavee *et al.*, 2002).

The problem of fertility when temperatures never go below 13–14°C is more difficult to overcome. Rather than resorting to chemical treatments, research should single out the best cultivars for any individual environment, exploiting the wide genetic pool of this species (Bartolini and Fabbri, 1994).

2.2.4.9 Fruit growth and seed development

The olive fruit is a drupe, which means it is made of two main parts: pericarp and seed. The pericarp is made of (i) the skin (exocarp), free of hairs and with stomata; (ii) the flesh (mesocarp), the tissue containing oil; and (iii) the pit (endocarp), a lignified shell enclosing the seed. The 'true seed' (see Fig. 2.1.42 in Chapter 2.1) consists in a seedcoat and a thick endosperm that ensheath a large embryo made of flat cotyledons and of a short radicle and plumule. As a rule, there is only one seed, rarely two. The occurrence of nucellar polyembryony in some Spanish cultivars has been reported. Fruit shape and size, pit size and surface morphology vary greatly among cultivars, and are the most reliable morphological features to distinguish between cultivars.

Fruit size is related to that of the flower and, in particular, of the ovary (Figs 2.2.6, 2.2.7). It depends on the number of its cells, as well as on the intercellular spaces, while cell size tends to be similar. In general, larger fruits result from larger ovaries already at flowering. Fruit size varies greatly between cultivars, with differences of five to six times or more. The dry weight of the ripe olive is, on average, 2000 times that of the ovary (the actual ratio can vary from 1000 to 4000).

In the ovary, the cells are more numerous and smaller in the tissues that will give rise to the mesocarp than in those that will originate the endocarp, regardless of cultivar. It can be hypothesized that the genetic difference in fruit size between olive cultivars depends on a different intensity of cell division in the ovary before flowering, leading to ovaries (and therefore fruits) with different numbers of cells and therefore proportionally different sizes. Recent studies have confirmed that it is the number of cells in the ovary that determines the potential size of the fruit and not the mass of the ovary (Rosati *et al.*, 2011a, 2012).

The final size of the fruit, in addition to genetic control, also depends on environmental and plant conditions, which allow, or not, the genetic potential to unfold.

Fruit growth is represented by the double sigmoid curve, typical of drupes (Fig. 2.2.8). Five stages can be separated.

Stage 1: Fertilization and cytokinesis. From flowering to 20–30 days after fertilization of the flowers and fruit set. In this stage

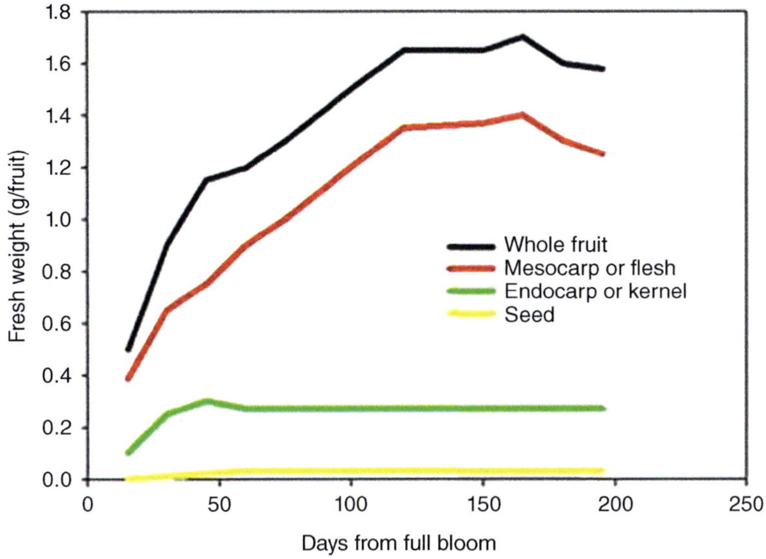

Fig. 2.2.6. Growth of the different parts of the olive fruit as fresh weight. (From Tombesi *et al.*, 2020)

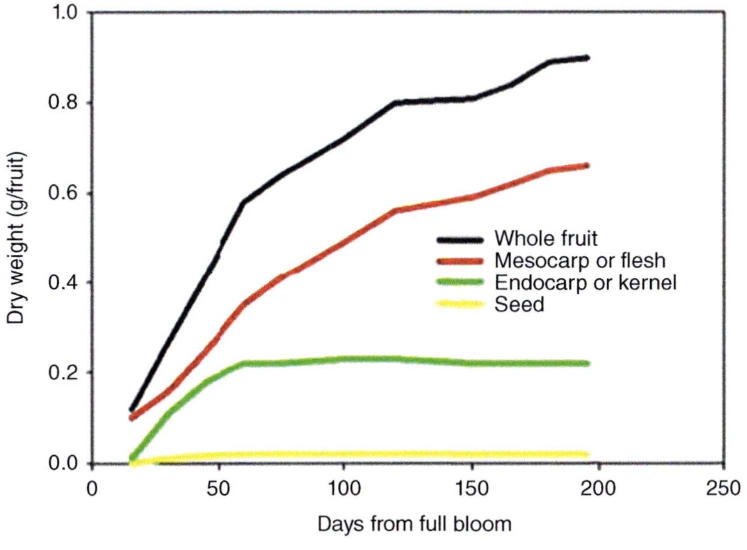

Fig. 2.2.7. Growth of the different parts of the olive fruit as dry weight. (From Tombesi *et al.*, 2020)

there is an intense cell multiplication with the growth of the embryo, the fruits are visible already 10–15 days after pollination with the enlargement of the ovary. During this period, many fruits drop due to competition for nutrients; therefore favorable conditions of availability of assimilates favor their permanence and cell division (cytokinesis), which is the prerequisite for reaching the final fruit size. This last aspect is particularly influenced by ovary size, which influences fruit growth. The process of cell division is a fundamental step because the number of cells determines the final size and the fruit's ability to attract carbohydrates. However, an excess of fruit set can

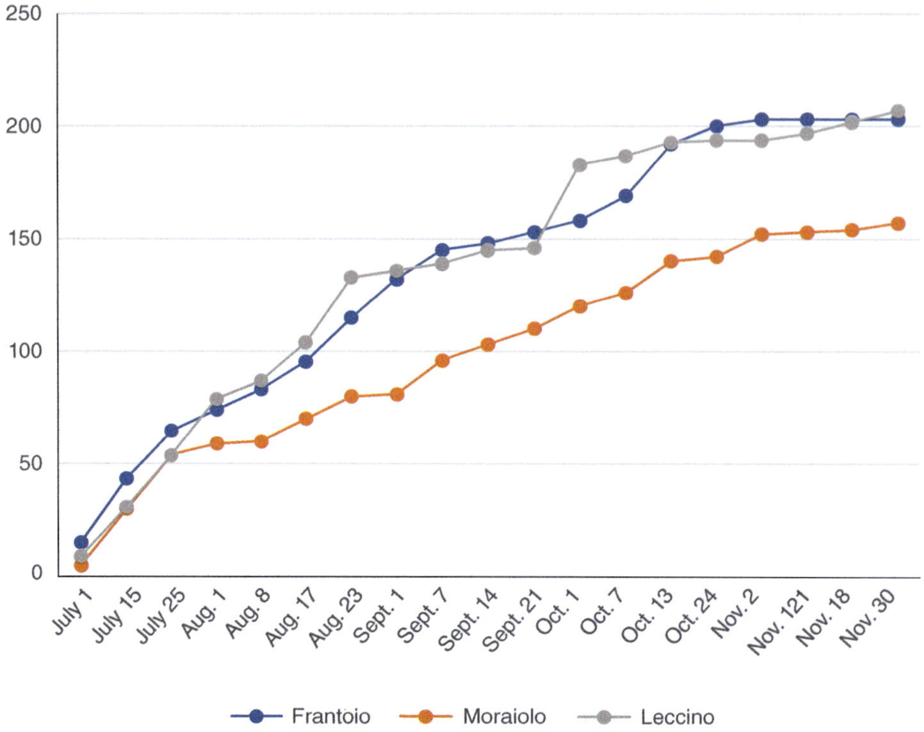

Fig. 2.2.8. Growth of 100 drupes (grams) of three Tuscan cultivars. (From Morettini, 1972)

increase fruit load and reduce their size. Therefore, in table olives, where size is an important commercial factor, it is necessary that fruit set is not excessive (this is especially true for dual-purpose varieties) and, if necessary, thinning may be carried out to achieve high-quality sizes.

Stage 2: Development of the embryo and endocarp. A period of 20 days characterized by rapid growth of the fruit due to intense cell division and distension, mainly affecting the seed and endocarp.

Stage 3: Kernel hardening. It occurs in a period of about 20 days, during which the growth in volume is markedly reduced, endocarp cells stop dividing and become lignified. The duration can vary by a few weeks depending on fruit load or the occurrence of prolonged drought stress. Moreover, it should be noted that under conditions of good soil water availability in summer or in irrigated olive groves, the reduction of fruit growth in Stage 3 is modest or even absent.

Stage 4: Development of the mesocarp and oil accumulation. This takes place over a long period of about 60 days and represents the second largest period of fruit growth, due to the development of the mesocarp, mainly by expansion of the existing cells, and the intense accumulation of oil. This stage continues into autumn when the olives begin to change color from deep green to yellow-green and then darken. Stress can have an effect by reducing oil formation and fruit growth.

Stage 5: Fruit ripening. This is the final stage during which the accumulation of pigments continues until they take on black-purple or intermediate colors. The accumulation of oil continues, although at a lower rate.

This growth model is less evident if growth is represented by weight increases, rather than diameter or volume. The curve can even turn into a single sigmoid if dry weight is taken into account, thus showing that Stage 2 is the one requiring most dry weight accumulation.

In the course of fruit formation, and tightly connected to it, the vital ovule develops to form the seed. The embryo makes up for most of seed volume.

Usually pollination and fecundation are essential for fruit set and early seed development. The presence of a vital seed in a growing drupe is not essential, as many fruits have their embryos aborted. As a consequence, many apparently normal fruits have no seed (stenospermocarpy). The fruit can also develop without the presence of a fertilized ovule (parthenocarpy), but in this case the fruit remains distinctly smaller.

During ripening, the cell walls become thinner and there is a gradual separation of the cells due to the solubilization of pectins and hemicelluloses and to the softening of the pulp. The pulp contains water, oil, carbohydrates, minerals, vitamins and proteins. Unlike the drupes of other species, olive fruits are rich in oil and low in soluble carbohydrates. A specific feature of olive fruits is their high phenol content, including oleoeuropein, which gives the typical bitter taste to the fresh fruit.

The seed, included and protected by the endocarp, represents 1–2% of the weight of the fruit and has an oil content of 20–30%, with a different composition to that of the pulp.

The fruits accumulate dry matter at a constant rate during fruit growth, but each component shows a different pattern (Fig. 2.2.9). The endocarp grows rapidly only during the first 45 days. The mesocarp, on the other hand, develops at a regular rate during the entire fruit-growth period. The seed shows an early increase in weight. Oil synthesis, considered in milligrams of oil per day and per fruit, is very intense between 60 and 120 days after full flowering (Fig. 2.2.10). During the specific growth periods, the limitation and availability of resources may markedly influence fruit development and components.

Fruit ripening in certain cultivars is relatively concentrated in time (a few days); in others it is gradual, over one or two weeks, or more.

As a rule the drupe at harvest is separated from the pedicel thanks to the formation of an abscission zone, already defined in the first months of fruit growth, but it

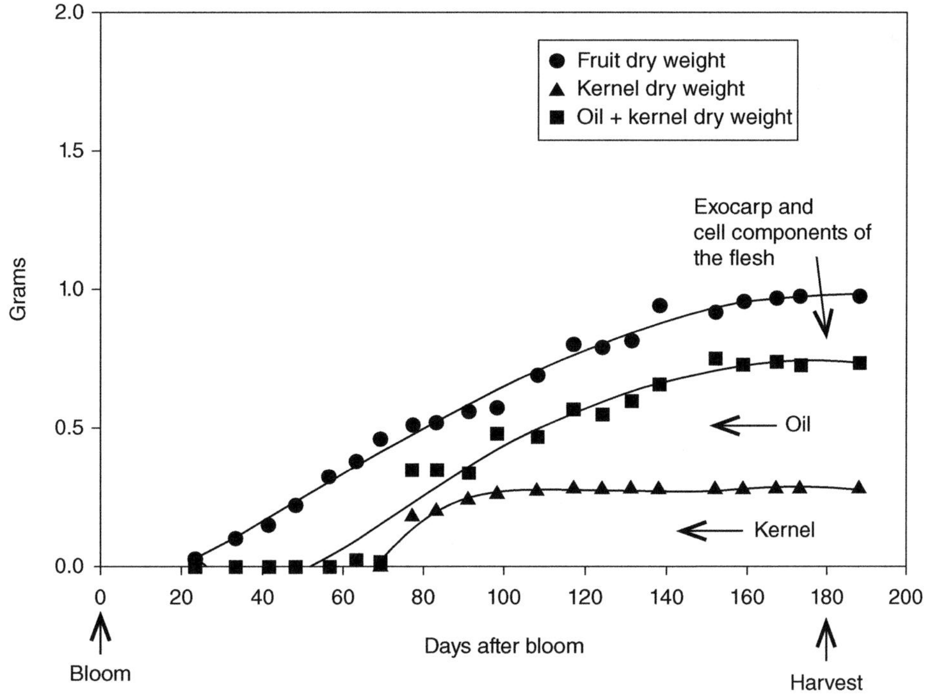

Fig. 2.2.9. Time course of the different fruit components, as dry weight. (From Tombesi, 2003)

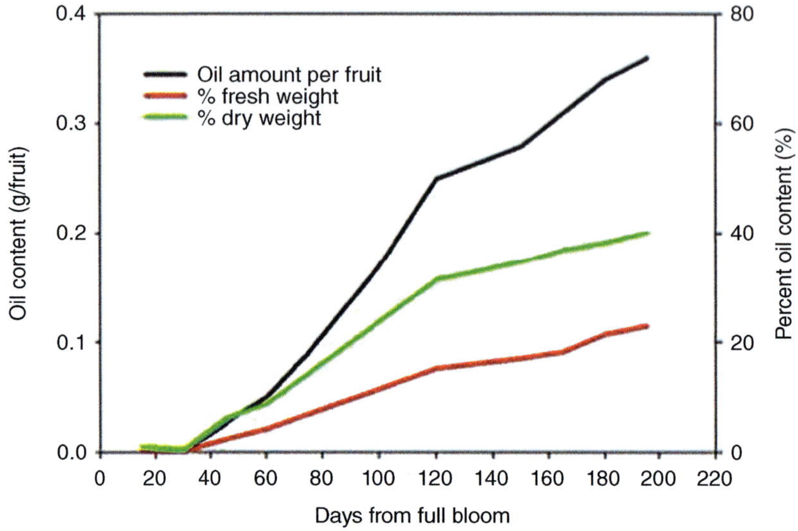

Fig. 2.2.10. Course of oil accumulation in the mesocarp. (From Tombesi *et al.*, 2020)

may retain it, or even retain the whole axis of the inflorescence. Natural abscission, however, can be distributed over months, and harvesting through the winter was customary in certain areas when farm hands were scarcely available (Fig. 2.2.11).

2.2.4.10 Chemical composition of the fruit

The fruit of the olive tree, with a fresh weight ranging from less than 1 g to more than 10 g, when fully ripe consists of the mesocarp, which contains about 60% water, 30% oil, 4% sugars, 3% protein and the rest is mainly fiber and ashes. The endocarp contains 10% water, 30% cellulose, 50% other carbohydrates and about 1% oil. The seed has 30% water, 27% oil, 27% carbohydrates and 10% protein.

Fatty substances

Lipids account for 10–30% of the fresh weight of the fruit and are concentrated in the mesocarp, up to 98% of the total. They are almost absent in the endocarp, while the seeds contain 20–28% fats. Lipids are divided into: (i) neutral lipids; (ii) polar lipids; and (iii) free fatty acids. At maturity, the neutral lipids account for more than 98% of the total and are mostly triglycerides; only 1.1% are diglycerides. The accumulation of triglycerides starts at the end of July and, for about five weeks, the accumulation is slow. During the next two months there is intense synthesis and for another five weeks accumulation is moderate, then the oil content remains constant or undergoes a small reduction just before fruit drop (Fig. 2.2.12). During fruit growth, the amount of all fatty acids increases. The most obvious increase is in oleic acid, which accounts for 70–80% at the end of ripening, followed by palmitic acid 10–15%, linoleic acid 5–10%, and stearic acid 2–3%. Polar fats include phospholipids and galactolipids, the former containing oleic acid, the latter linoleic acid. They are membrane lipids and specific constituents of the chloroplasts and plastids of the fruit, and increase as the fruit grows. Free fatty acids, in high quantity in the young fruit (2%), are metabolized during olive growth and decrease to 0.15% at maturity. Some of them are used for the synthesis of triglycerides. The accumulation of oil that occurs from pit hardening until harvest can be early or late depending on the variety and climate; in cold climates, the oleic acid content is higher than in warm climates. The formation of oil in the fruit, according to the review

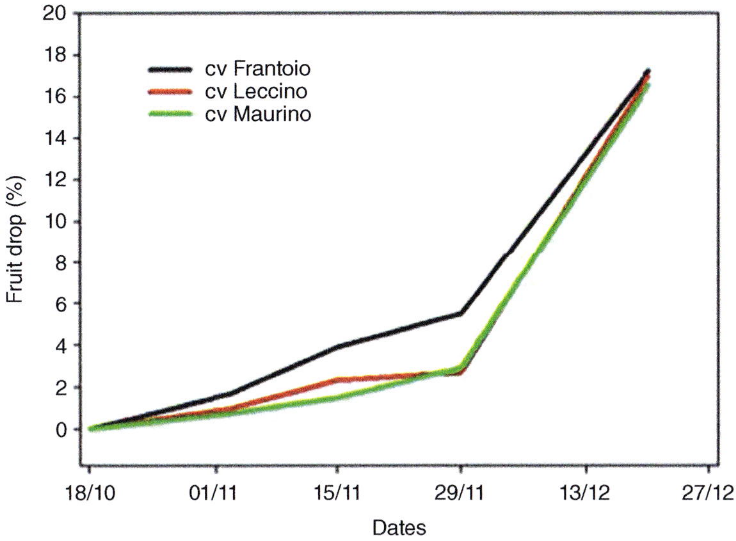

Fig. 2.2.11. Course of natural fruit drop in the ripening period in three Tuscan cultivars. (From Tombesi *et al.*, 2020)

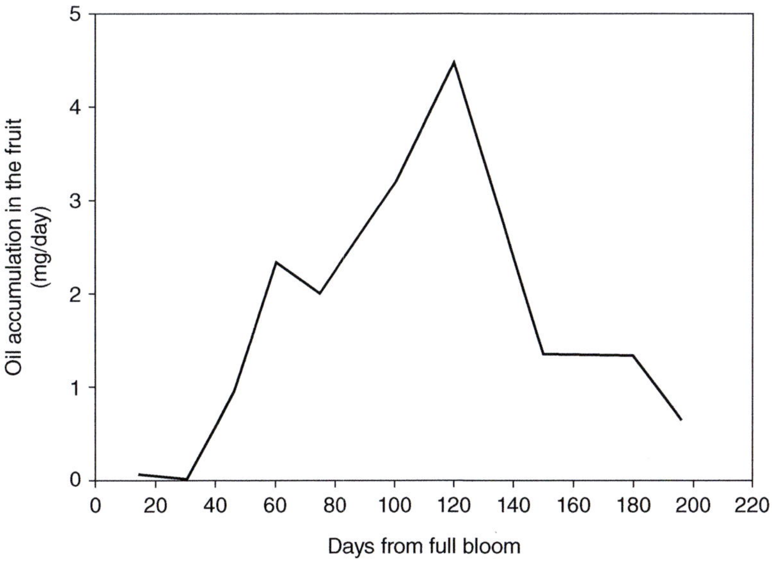

Fig. 2.2.12. Rate of oil accumulation in the mesocarp. (From Tombesi *et al.*, 2020)

by Mazliak (1970) and more recent studies (Sanchez and Harwood, 2002), begins with the generation of Acetyl Coenzyme A. This compound is derived from the glycolysis of carbohydrates accumulated in the fruit and supplied as assimilation compounds by translocation from the leaves. Acetyl Coenzyme A is transformed into Malonyl Coenzyme A and, with the intermediation of Acyl-Transport Proteins and the enzyme complex of Fatty Acid Synthases, fatty acids are formed, which, in combination with glycerol, form the oil particles. The factors influencing this fundamental process are not yet clear. They can only be indicated in the translocation of assimilates and the availability of Acetyl Coenzyme A Carboxylase and ATP. Most are under genetic control and some can be influenced by cultivation practices and environmental conditions.

Carbohydrates

In olives, there are soluble sugars and complex sugars. The main soluble sugars are glucose, fructose and mannitol. Levels vary from 0.5% to 5% depending on the variety. Some research shows higher amounts of glucose in comparison with fructose and mannitol; other work supports the prevalence of mannitol. The latter compound has been indicated as a derivative of soluble sugars used for translocation. Soluble sugars decrease during fruit development with the start of oil synthesis, so there are twice as many in green olives as in olives at an advanced stage of ripening. They are more abundant in olives from rainfed groves than in irrigated ones. The complex sugars are polymers of the simple sugars and make up cellulose, hemicelluloses or pectins, and lignin, the latter concentrated in the pit. Hemicelluloses and cellulose are structural components of the cell wall and are involved in the connection with adjacent cells. The consistency of the mesocarp is due to the layers of pectins bound by esters to each other and to the cell wall; the higher their number, the higher the number of protopectin layers and the higher the consistency of the mesocarp. In the intercellular spaces, the polygalacturon chains are bound by esters to the cell wall polysaccharides and then to each other to form a rigid three-dimensional reticulum.

When the pectins are hydrolysed or broken, the pulp loses its structure and softens. Among the sugar derivatives, the most important are an ester of sucrose with oleoeuropic acid (a monoterpenoid monocarboxylic acid) and oleuropein, which is a complex ß-glucoside; oleuropein, the most abundant phenolic compound in olive mesocarp, accumulates during fruit growth and is slowly converted to elenolic acid glycoside and dimethyloleuropein as ripening progresses.

Phenols

Phenols, also called polyphenols, are secondary compounds in the tissues, with antioxidant activity and protection from external stress. Among the phenols in olives we find oleoeuropein, hydroxytyrosol, caffeic acid and tyrosol. They are present in the pulp in quantities of 1–2%, prevalently as oleoeuropein. The latter accumulates during growth and is converted into enolic acid glycoside and dimethyloleuropein during ripening. There are also other compounds: verbascoside, luteolin 7-O-glucoside and rutin.

Proteins

There are low levels of insoluble proteins in the mesocarp – 1.5% – made up with numerous amino acids.

Pigments

In the pulp we find chlorophyll a and b (green), carotenoids and hydrocarbons (yellow), anthocyanins (pink, purple). The green and yellow pigments are soluble in oil, while the anthocyanins are soluble in water. In the early stages of fruit development, chlorophyll is the most important pigment, then it decreases as beta-carotene and anthocyanins increase. Anthocyanins, of which cyanidin is the most common, give the black-purple color and develop initially in the skin. They may expand into the flesh and finally into the stone.

2.2.5 Balance Between Vegetative and Reproductive Activities

In the olive tree, achieving and maintaining a balance between vegetative and reproductive activity is essential for obtaining high yields. Errors in pruning or cultivation can delay the establishment of this balance. Similarly, if external factors upset this balance in the adult tree, it will mainly develop vegetative structures with significant losses in yield.

The concept of balance between functions can be extended to plant parts and individual organs. The root:shoot ratio (R/S), which is usually expressed as the dry matter of the root system in relation to that of the aerial part, and the ratio between the biomass of young wood (which bears leaves and fruits) compared to that of old wood (transport and support function), are useful indicators of the relationships between structural and functional parts. A similar index between structure and function can be calculated for the root system, with the fine roots present above all in the periphery and the large roots that mainly perform functions of support, transport and storage. In any case, it should be noted that trees with a similar R/S ratio can be in different physiological conditions, so the simple measurement of biomass is not sufficient to assess the balance of functions within the tree. Typical indicators of an unbalanced R/S ratio are the vigorous shoots and branches, long internodes, large leaves, abundant suckers and pollards.

In perennial species, the balance between vegetative and reproductive activity is further complicated by residual effects of cultivation practices and environmental conditions. Previous events during the life of the tree determine effects that become progressively more evident as the plant ages. The fact that in the olive tree the flowering cycle embraces, from the induction of the flower buds to flowering, two growing seasons and can partially overlap with the growth of the branch and the development of the fruit explains, within certain limits, the inverse relationship between fruit load in one year and flowering in the following year (Fig. 2.2.13).

Factors altering the source–sink relationships of the plant are reflected in shoot growth patterns and the status of root carbohydrate reserves. The most active sink in the fruiting shoot or branch is the fruit, which reduces considerably terminal elongation and dry matter in the leaf or shoot (Rallo and Suarez, 1989).

2.2.5.1 Alternate bearing

One of the practical consequences of the balance between vegetative and reproductive

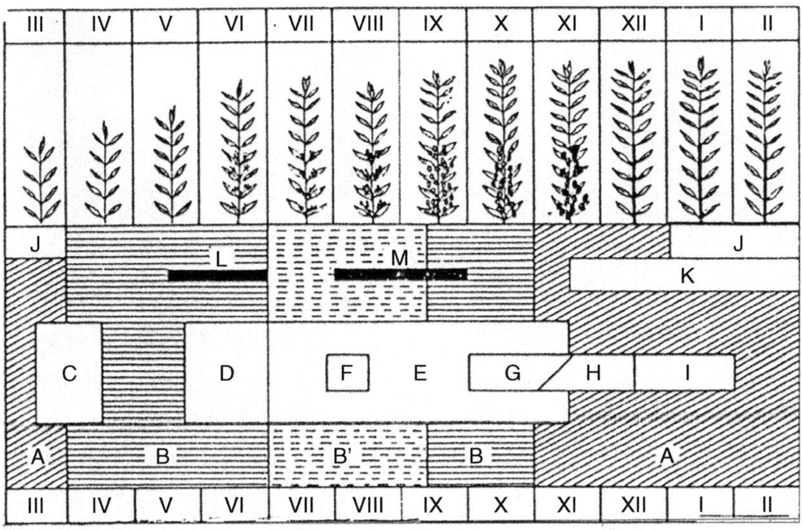

Fig. 2.2.13. Olive phenological stages. A, rest period; B, period of active vegetative growth; B′, period of reduced vegetative growth; C, bud differentiation; D, bloom – fruit set; E, fruit growth; F, pit hardening; G, veraison; H, ripening; I, vernalization; J, pruning; K, harvest; L, critical period for nitrogen nutrition; M, critical period for water supply. (Modified from Pansiot and Rebour, 1960)

activity is alternate bearing. Olive trees tend to produce at alternate years, so that after a year of high yields (on) it is likely that a low (off) year will follow. Although this behavior cannot be explained only by the vegetative/reproductive balance (the phenomenon is cultivar-specific), this ratio is certainly important in mitigating the amplitude of alternate bearing. In other words, a well-balanced olive tree will more likely maintain yields stable over the years.

In olive trees, we observe two growth flushes, the first one in spring and the second in autumn (October), under adequate soil moisture and temperature. Alternate bearing is influenced by various factors such as water stress, soil temperature, climatic factors and the cultivar. Significant genetic variation on alternate bearing is detected among cultivars. The main causal factor of alternate bearing is the inhibition of flower bud induction by the seed of growing fruits, and this is due to competition for nutrients in the period in which both fruits and buds develop (Fig. 2.2.13).

Research has collected abundant data on the endogenous events concomitant with alternate bearing. Endogenous plant hormones play a significant role in alternate bearing (Lavee, 1989; Ulger *et al.*, 1999). The total extractable protein content in the leaves of off trees is significantly lower than in the leaves of on trees (Lavee and Avidan, 1994). This was observed in the cultivars 'Koroneiki', 'Uovo di piccione', 'Manzanillo' and 'Barnea'. The least difference was recorded in the leaves of 'Koroneiki', which demonstrates the least alternating capacity. Chlorogenic acid is associated with alternate bearing and levels are high in the on years and low in the off years. Endogenous control of alternate bearing includes the possible involvement of phenolic acids, with the end product of caffeoylquinic acid (Lavee and Avidan, 1982).

The two main factors that can reduce the intensity of alternate bearing are pruning and harvest time. Furthermore, the irrigation schedule may reduce alternate bearing. Mineral nutrition is also a determining factor. Nutrient deficiency, especially of N during the period of flower bud differentiation, affects vegetative vigor, the size of fruits and the flowering of the following spring (Fernandez-Escobar and Marin, 1999; Fernandez-Escobar *et al.*, 2004).

References

Amiot, M.J., Fleuriet, A. and Macheix J.J. (1989) Accumulation of oleuropein derivatives during olive maturation. *Phytochemistry* 28(1), 67–69.

Androulakis, I.I. and Loupassaki, M.H. (1990) Studies on the self-fertility of some olive cultivars in the area of Crete. *Acta Horticulturae* 286, 159–162.

Ayerza, R. and Sibbett, G.S. (2001) Thermal adaptability of olive (*Olea europaea* L.) to the Arid Chaco of Argentina. *Agriculture, Ecosystems and Environment* 84, 277–285.

Bartolini, G. and Fabbri, A. (1994) The olive tree in a monsoon climate: first results in India on the performance of 12 cultivars. Second International Symposium on Olive Growing, Jerusalem, 6–10/9/1993. *Acta Horticulturae* 356, 119–122.

Bongi, G. (1986) Oleuropein: an *Olea europaea* secoiridioid active on growth regulation. *Acta Horticulturae* 179, 245–249.

Cuevas, J. and Polito, V.S. (2004) The role of staminate flowers in the breeding system of *Olea europaea* (Oleaceae): an andromonoecious, wind-pollinated taxon. *Annals of Botany* 93, 547–553.

Cuevas, J. and Rallo, L. (1990) Response to cross pollination in olive trees with different levels of flowering. *Acta Horticulturae* 286, 179–182.

Cuevas, J., Rallo, L. and Rapoport, H.F. (1994) Crop load effects on floral quality in olive. *Scientia Horticulturae* 59, 123–130.

Cuevas, J., Rallo, L. and Rapoport, H.F. (1995) Relationship among reproductive processes and fruitlets abscission in Arbequina olive. *Advances in Horticultural Science* 9, 92–96.

Cuevas, J., Diaz-Hermoso, A.J., Galiàn, D., Hucso, J.J., Pinillos, V., Prieto, M., Sola, D. and Polito, V.S. (2001) Response to cross-pollination and choice of pollinisers for the olive cultivars (*Olea europaea* L.) 'Manzanilla de Sevilla', 'Hojiblanca' and 'Picual'. *Olivae* 85, 26–35.

de la Rosa, R., Rallo, L. and Rappoport, H.F. (2000) Olive floral bud growth and starch content during winter rest and spring budbreak. *HortScience* 35, 1223–1227.

Diaz, A., Martin, A., Rallo, P., Barranco, D. and de La Rosa, R. (2006) Self-incompatibility of 'Arbequina' and 'Picual' olive assessed by SSR Markers. *Journal of the American Society Horticultural Science* 131(2), 250–255.

Fabbri, A. and Alerci, L. (1999) Reproductive and vegetative bud differentiation in *Olea europaea* L. *Journal of Horticultural Science & Biotechnology* 74, 522–527.

Fabbri, A. and Benelli, C. (2000) Flower bud induction and differentiation in olive. *Journal of Horticultural Science and Biotechnology* 75(2), 131–141.

Fabbri, A., Tamburino, A. and Cicala, A. (1997) Preliminary observations on flower induction in 'Nocellara Etnea' olive trees in Sicily. 9th Consultation, FAO Inter-Regional Cooperative Research Network, Hammamet, Tunisia, 20–23 September 1995. *Olea* 23, 36.

Farinelli, D., Boco, M. and Tombesi, A. (2006) Results of four years of observations on self-sterility behaviour of several olive cultivars and significance of cross-pollination. *Proceedings of the 2nd International Seminar Olivebioteq, 5–10 November, Marsala-Mazara del Vallo, Italy* 1, 275–282.

Fernandez-Escobar, R. and Marin, L. (1999) Nitrogen fertilization in olive orchards. *Acta Horticulturae* 474, 333–335.

Fernandez-Escobar, R., Gomez-Valledor, G. and Rallo, L. (1983) Influence of pistil extract and temperature on in vitro pollen germination and pollen tube growth of olive cultivars. *Journal Horticultural Science* 58(2), 219–227.

Fernandez-Escobar, R., Benlloch, M., Navarro, C. and Martin. G.C. (1992) The time of floral induction in the olive. *Journal of the American Society for Horticultural Science* 117, 304–307.

Fernandez-Escobar, R., Moreno, R. and Sanchez-Zamora, M.A. (2004) Nitrogen dynamics in the olive bearing shoot. *HortScience* 39(6), 1406–1411.

Ferrara, E., Lorusso, G. and Lampanelli, E. (1991) A study of floral biology and the technological features of seven olive cultivars of different origins. *Acta Horticulturae* 474, 279–283.

Fiorino, P., Marone, E., Rosati, A., Caporali, S. and Paoletti, A. (2020) *Il fiore e la biologia fiorale*. Accademia Nazionale dell'Olivo e dell'Olio, Spoleto, Italy.

Galàn, C., Garcia-Mozo, H., Cariñanos, P., Alcàzar, P. and Dominguez-Vilches, E. (2001) The role of temperature in the onset of the of the *Olea europaea* L. pollen season in southwestern Spain. *International Journal of Biometeorology* 45, 8–12.

Griggs, W.H., Hartmann, H.T., Bradley, M.V., Iwakiri, B.T. and Whisler, J.E. (1975) Olive pollination in California. *California Agriculture Experimental Station Bulletin* 869, 49.

Haberman, A., Bakhshian, O., Cerezo-Medina, S., Paltiel, J., Adler, C. *et al.* (2017) A possible role for flowering locus T-encoding genes in interpreting environmental and internal cues affecting olive (*Olea europaea* L.) flower induction. *Plant, Cell & Environment* 40(8), 1263–1280.

Hackett, W.P. and Hartmann, H.T. (1963) Morphological development of buds of olive as related to low temperature requirement for inflorescence formation. *Botanical Gazette* 124, 383–387.

Hackett, W.P. and Hartmann, H.T. (1964) Inflorescence formation in olive as influenced by low temperature, photoperiod, and leaf area. *Botanical Gazette* 125, 65–72.

Hackett, W.P. and Hartmann, H.T. (1967) The influence of temperature on floral initiation in the olive. *Physiologia Plantarum* 20, 430–436.

Hartmann, H.T. (1951) Time of floral differentiation of the olive in California. *Botanical Gazette* 112, 323–327.

Hartmann, H.T. (1967) 'Swan Hill' a new ornamental fruitless olive for California. *California Agriculture* 21, 4–5.

Hartmann, H.T. and Porlingis, I.C. (1957) Effect of different amounts of winter chilling on fruitfulness of several olive varieties. *Botanical Gazette* 119, 102–104.

Hartmann, H.T. and Whisler, J.E. (1975) Flower production in olive as influenced by various chilling temperature regimes. *Journal of the American Society for Horticultural Science* 100, 67–74.

Lavee S (1989) Involvement of plant growth regulators and endogenous growth substances in the control of alternate bearing. *Acta Horticulturae* 239, 311–322.

Lavee, S. (1996) Olive biology and physiology. In: *World Olive Encyclopedia* (COI edn). Barcelona, Spain.

Lavee, S. and Avidan, N. (1982) The involvement of phenolic substances in controlling alternate bearing in the olive (*Olea europaea* L.). In: *XXI International Horticultural Congress 29 August–4 September*. Hamburg, Germany. Abst. 1371.

Lavee, S. and Avidan N. (1994) Protein content and composition of leaves and shoot bark in relation to alternate bearing of olive trees (*Olea europaea* L.). *Acta Horticulturae* 356, 143–147.
Lavee, S., Rallo, L., Rapoport, H.F. and Troncoso A. (1996) The floral biology of the olive: effect of flower number, type and distribution on fruit set. *Scientia Horticulturae* 66, 149–158.
Lavee, S., Rallo, L., Rapoport, H.F. and Troncoso A. (1999) The floral biology of the olive. II. The effect of inflorescence load and distribution per shoot on fruit set and load. *Scientia Horticulturae* 82, 181–192.
Lavee, S., Taryan, J., Levin, J. and Haskal, A. (2002) The significance of cross-pollination for various olive cultivars under irrigated intensive growing conditions. *Olivae* 91, 25–36.
Loussert, R. and Brousse, G. (1978) *L'olivier*. Paris.
Magherini, R. (1971) Osservazioni sull'aborto dell'ovario nell'olivo. *Agricoltura Italiana* 71(5), 291–301.
Malik, N.S.A. and Bradford, J.M. (2005) Is chilling a prerequisite for flowering and fruiting in 'Arbequina' olives? *International Journal of Fruit Science* 5, 29–39.
Malik, N.S.A. and Bradford, J.M. (2006) Flowering and fruiting in 'Arbequina' olives in subtropical climates where olives normally remain vegetative. *International Journal of Fruit Science* 5(4), 47–56.
Martin, G.C., Ferguson, L. and Polito, V.S. (1994) Flowering, pollination, fruiting: alternate bearing and abscission. In: Ferguson, L., Sibbett, G.S. and Martin G.C. (eds) *Olive Production Manual*. University of California, Division of Agriculture and Natural Resources, Berkeley, California. Publication 3353, 117–128.
Martins, P.C., Cordeiro, A.M. and Rapoport, H.F. (2006) Flower quality in orchards of olive, *Olea europaea* L., cv. 'Morisca'. *Advances in Horticultural Science* 20(4), 262–266.
Mazliak, P. (1970) Lipids. In: Hulme, A.C. (ed.) *The Biochemistry of Fruits and Their Products*. Academic Press, London, pp. 209–238.
Meriño, J. (1987) The costs of growing and maintaining leaves of Mediterranean plants. In: Tenhuned, J.D, Catarino, F.M., Lange, O.L. and Oechel, W.V. (eds) *Plant Response to Stress: Functional Analysis in Mediterranean Ecosystems*. NATO ISI Series, Vol. G15. Springer-Verlag, Berlin, pp. 553–564.
Morettini, A. (1972) *Olivicoltura*. Ramo Editoriale degli Agricoltori, Rome.
Navarro, C., Fernandez-Escobar, R. and Benlloch, M. (1990) Flower bud induction in 'Manzanillo' olive. *Acta Horticulturae* 286, 195–198.
Pansiot, F.P. and Rebour, H. (1960) *Amélioration de la culture de l'olivier*. FAO, Rome.
Pinney, K. and Polito, V.S. (1990) Flower initiation in 'Manzanillo' olive. *Acta Horticulturae* 286, 203–205.
Porlingis, I.C. and Voyiatzis, D.G. (1999) Paclobutrazol decreases the harmful effect of high temperatures on fruit set in olive trees. *Acta Horticulturae* 474, 241–244.
Proietti, P. and Tombesi, A. (1996) Effects of gibberellic acid, asparagines and glutamine on flower bud induction in olive. *Journal of Horticultural Science and Biotechnology* 71(3), 383–388.
Rallo, L. (1997) Fructificaciòn y producciòn. In: Barranco D., Fernandez- Escobar, R. and Rallo, L. (eds). *El cultivo del olivo*. Junta de Andalucia, Andalucia, Spain, pp. 107–136.
Rallo, L. (1999) Miglioramento delle risorse genetiche. In: *Proceedings of the International Seminar on Scientific Innovation and Their Application in Olive Growing and Elaiotechnique*. 10–12 March. Florence, Italy, pp. 1–28.
Rallo, L. and Martin, G.C. (1991) The role of chilling in releasing olive floral buds. *Journal of the American Society for Horticultural Science* 116(6), 1058–1062.
Rallo, L. and Suarez, M.P. (1989) Seasonal distribution of dry matter within the olive fruit-bearing limb. *Advances in Horticultural Science* 3, 55–59.
Rallo, L., Torreño P., Vargas, A. and Alvarado, J. (1994) Dormancy and alternate bearing in olive. *Acta Horticulturae* 356.
Rapoport, H.F. and Rallo, L. (1991) Fruit-set and enlargement in fertilized and unfertilized olive ovaries. *HortScience* 26(7), 896–898.
Rapoport, H.F, Manrique, T. and Gucci, R. (2004) Cell division and expansion in the olive fruit. *Acta Horticulturae* 636, 461–465.
Rosati, A., Zipanćič, M., Caporali, S. and Padula, G. (2009) Fruit weight is related to ovary weight in olive (*Olea europaea* L.). *Scientia Horticulturae* 122(3), 399–403.
Rosati, A., Zipanćič, M., Caporali, S. and Paoletti, A. (2010) Fruit set is inversely related to flower and fruit weight in olive (*Olea europaea* L.). *Scientia Horticulturae* 126(2), 200–204.
Rosati, A., Caporali, S., Paoletti, A. and Famiani, F. (2011a) Pistil abortion is related to ovary mass in olive (*Olea europaea* L.). *Scientia Horticulturae* 127(4), 515–519.
Rosati, A., Caporali, S., Hammami, S.B.M., Moreno-Alias, I., Paoletti, A. and Rapoport H.F. (2011b) Differences in ovary size among olive (*Olea europaea* L.) cultivars are mainly related to cell number, not to cell size. *Scientia Horticulturae* 130(1), 185–190.

Rosati, A., Caporali, S., Hammami, S.B.M., Moreno-Alias, I., Paoletti, A. and Rapoport H.F. (2012) Tissue size and cell number in the olive (*Olea europaea*) ovary determine tissue growth and partitioning in the fruit. *Functional Plant Biology* 39(7), 580–587.

Sanchez, J. and Harwood, J.L. (2002) Biosynthesis of triacylglycerols and volatiles in olives. *European Journal of Lipid Science and Technology* 104, 564–573.

Sanz-Cortés, F., Martinez-Calvo, J., Badenes, M.L., Bleiholder, H., Hack, H., Llacer, G. and Meier, U. (2005) Phenological growth stages of olive trees (*Olea europaea*). *Annals of Applied Biology* 140(2), 151–157.

Stutte, G.W. and Martin G.C. (1986a) Effect of killing the seed on return bloom of olive. *Scientia Horticulturae* 29(1–2), 107–113.

Stutte, G.W. and Martin G.C. (1986b) Effect of light intensity and carbohydrate reserves on flowering in olive. *Journal of the American Society for Horticultural Science* 111(1), 27–31.

Therios, I. (2009) *Olives*. CAB International, Wallingford, UK.

Tombesi, A. (2003) Biologia fiorale e di fruttificazione. In: Fiorino, P. (ed.) *Olea: trattato di olivicoltura*. Edagricole, Bologna, Italy, pp. 35–55.

Tombesi, A. and Cartechini, A. (1986) L'effetto dell'ombreggiamento della chioma sulla differeziazione delle gemme a fiore dell'olivo. *Rivista della Ortoflorofrutticoltura Italiana* 80, 277–285.

Tombesi, A., Tombesi, S. and Gucci R. (2020) *Sviluppo del frutto*. Collana divulgativa, vol. 2. Accademia Nazionale dell'Olivo e dell'Olio, Spoleto, Italy.

Troncoso de Arce, A. (1966) Alcune osservazioni sullo sviluppo delle gemme dell'olivo (*Olea europaea*). *Frutticoltura* 28, 439–447.

Ulger, S., Baktir, I. and Kaynak, L. (1999) Determination of the effects of endogenous plant hormones on alternate bearing and flower bud formation in olives. *Turkish Journal of Agriculture and Forestry* 23(7), 619–623.

Ulger, S., Sonmez, S., Karkacier, M., Ertoy, N., Akdesir, O. and Aksu, M. (2004) Determination of endogenous hormones, sugars and mineral nutrition levels during the induction, initiation and differentiation stage and their effects on flower formation in olive. *Plant Growth Regulation* 42, 89–95.

Zito, F. and Spina, P. (1956) Come germina il polline dell'olivo. *Italia Agricola* 93(5), 413–425.

2.3 Ecophysiology

R. Gucci[1] and L. Sebastiani[2]
[1]*Department of Agriculture, Food and Environment, University of Pisa, Italy;* [2]*Crop Science Research Center, Sant'Anna University, Pisa, Italy*

2.3.1 Photosynthesis and Respiration

Olive leaves assimilate carbon dioxide converting radiant energy into chemical bonds by means of pigments and enzymes. The photosynthetic process occurs via a C_3 mechanism, whereby the first stable product of the carboxylation reaction is 3-phosphoglyceric acid, a compound with three carbon atoms. Carbohydrates and oxygen are the main end products of the photosynthetic reaction. Carbohydrates that are not used for dark respiration either accumulate in the leaf cells or are transported to other tissues and organs, where they are utilized or stored; oxygen diffuses back into the atmosphere. Photorespiratory rates are high; about 10% of fixed carbon is used in the glycolate pathway, which occurs in the peroxisomes (Bongi and Palliotti, 1994).

The compact mesophyll of olive leaves implies a small volume of intercellular spaces in both the palisade and the spongy tissue, which determines high resistances to gas diffusion inside the leaf (Bongi *et al.*, 1987a). Further physical limitations to gas exchange are due to the thickness of the cuticle and cell wall of both epidermal and mesophyll cells, and to the presence of small stomata only on the abaxial (lower) side of the mature leaf. Stomatal density, which varies widely depending on cultivar and growing conditions (Table 2.3.1), affects neither photosythetic rate (*A*) nor stomatal conductance (Bongi *et al.*, 1987b; Hagidimitriou and Pontikis, 2005; Tognetti *et al.*, 2007). Values of stomatal density are comparable to those reported for many other fruit crops. Thus, the transport pathway is subjected to diffusional and non-diffusional limitations to CO_2. Diffusional limitations are regulated by boundary layer conductance (g_b), stomatal conductance (g_s) and mesophyll conductance (g_m), while non-diffusional limitations involve biochemical processes responsible for carboxylation itself and photosynthetic efficiency including photoinhibition.

Photosynthetic rates in olive leaves are not particularly high. Besides high resistance to gas diffusion, the relatively low rate of *A* is also attributed to the inherently low content in leaf proteins and enzymes, and particularly that of ribulose-1,5-disphosphate carboxylase/oxygenase, the key enzyme for assimilation. Maximum *A* values can reach 22–25 μmol CO_2 m^{-2} s^{-1} in mature leaves of field-grown trees under optimal conditions of temperature, water availability and mineral nutrition. More typically, *A* ranges 10–18 μmol CO_2 m^{-2} s^{-1} depending on cultivar and

DOI: 10.1079/9781789247350.0004

Table 2.3.1. Maximum photosynthetic rates reported for different olive cultivars.

Cultivar	Maximum *A* (μmol m^{-2} s^{-1})	Stomatal density (no. mm^{-2})	Author
Koroneiki	21	399	Hagimiditriou and Pontikis (2005)
Konservolia	16	450	Hagimiditriou and Pontikis (2005)
Megaritiki	18	458	Hagimiditriou and Pontikis (2005)
Linolia	13	507	Hagimiditriou and Pontikis (2005)
Kalamon	14	550	Hagimiditriou and Pontikis (2005)
Frantoio	14		Gucci *et al.* (1997)
Frantoio	25	385	Tognetti *et al.* (2002)
Leccino	25	380	Tognetti *et al.* (2002)
Dolce Agogia	5*	≅300	Bongi *et al.* (1987a)
Leccino	5*	≅300	Bongi *et al.* (1987a)
Coratina	5*	≅250	Bongi *et al.* (1987a)
Manzanilla	5*	≅220	Bongi *et al.* (1987a)
Arbequina		451	Bacelar *et al.* (2004)
Blanqueta		420	Bacelar *et al.* (2004)
Cobrancosa		455	Bacelar *et al.* (2004)
Manzanilla		442	Bacelar *et al.* (2004)
Negrinha		537	Bacelar *et al.* (2004)
Frantoio	≅19		Tognetti *et al.* (2005)
Leccino	15		Tognetti *et al.* (2005)
Frantoio	≅16		Proietti and Famiani (2002)
Maurino	≅15		Proietti and Famiani (2002)
Leccino	≅14		Proietti and Famiani (2002)

*Assimilation determined on plants grown in growth chambers

growing conditions (Table 2.3.1). Values below that range are frequently measured in leaves of plants grown in containers or controlled environments, or when environmental conditions inhibit photosynthesis. Under non-limiting conditions of soil humidity, the seasonal course of *A* in one-year-old leaves shows high rates in late spring and early fall and a decrease in summer, presumably due to the effect of high temperatures. Because of their evergreen character, olive leaves photosynthesize throughout the annual cycle.

Photosynthetic rates of current-year leaves also increase rapidly in spring, decrease at mid-summer to reach maximum values in September–October and then decrease again (Hagimiditriou and Pontikis, 2005; Proietti and Famiani, 2002). In winter, *A* is often limited by low temperatures (Gucci *et al.*, 1997), but when environmental conditions are favorable, it can be relatively high and contribute substantially to the daily carbon budget.

Leaf age affects carbon assimilation. Very young leaves (less than 8 d after completion of full expansion) or still-expanding leaves have lower *A* values than older leaves, especially in spring and early summer. Photosynthetic rates of fully expanded, current-year leaves are slightly higher than those of leaves older than 1.5 years (Bongi *et al.*, 1987b; Gucci *et al.*, 1997; Proietti and Famiani, 2002). Differences in *A* may be evident in the first two to three months after completion of expansion, but they gradually decrease as the growing season progresses, and disappear by the end of the summer. One-year-old leaves show higher *A* than current-year ones early in the season while the reverse is true from summer onwards (Fig. 2.3.1). In their study, Hagimiditriou and Pontikis (2005) showed that most differences in *A* between one-year-old and current-year leaves could be accounted for by the higher stomatal conductance (g_s) and chlorophyll concentration of older leaves, the latter being more than double in spring and about 30% higher in autumn.

The diurnal course of *A* in field-grown trees closely follows changes in photosynthetically active radiation and stomatal conductance. The tight relationship between light-saturated *A* and g_s in field-grown trees is

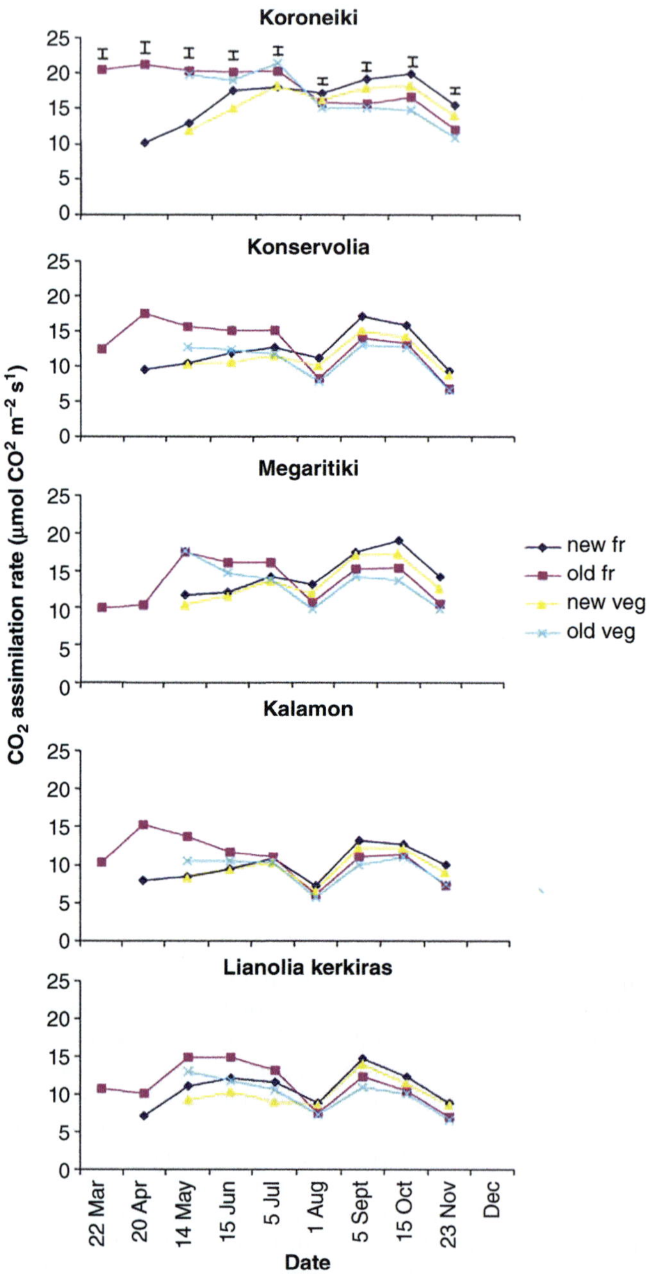

Fig. 2.3.1. Seasonal changes in CO_2 assimilation rate of five Greek olive cultivars in different leaf ages; current-season vegetative (new veg) and fruiting (new fr), and one-year-old vegetative (old veg) and fruiting (old fr) shoots. Differences in *A* were determined by analysis of variance (n=12). Vertical bars represent LSD0.05 values within each sampling date. (From Hagimiditriou and Pontikis, 2005)

reported in Fig. 2.3.2. Assimilation increases rapidly after sunrise, reaches maximum values in the morning, and usually declines in the afternoon. Typical diurnal courses of assimilation and stomatal conductance in plants grown in containers under different irrigation regimes are illustrated in Fig. 2.3.3. The shape of the diurnal profile may change

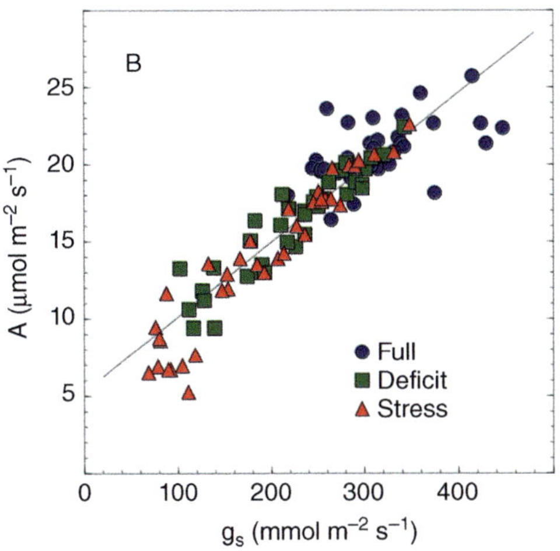

Fig. 2.3.2. The relationship between assimilation rate (*A*) and stomatal conductance (g_s) in leaves of olive trees (cv. 'Frantoio') grown in the field and subjected to different irrigation regimes at Venturina in 2009. Symbols are means of two measurements per tree. Regression equation: y=5.03+0.05x, R^2=0.85***.

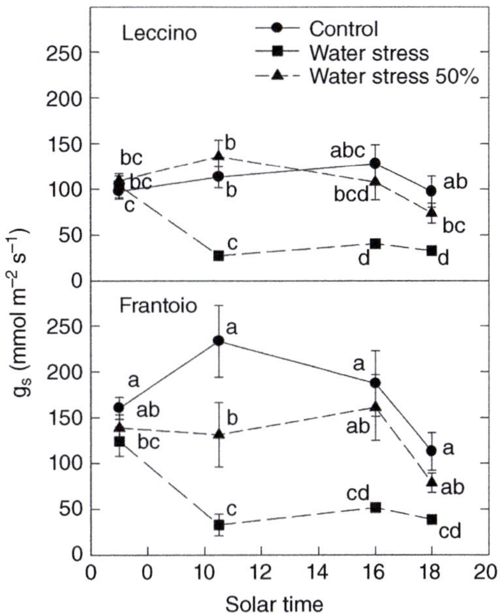

Fig. 2.3.3. Diurnal course of carbon assimilation (*A*) in fully expanded leaves of two olive cultivars ('Frantoio' and 'Leccino') grown in containers. Well-irrigated plants received 100% of their weekly water requirement, severely stressed plants did not receive any water for ten days, 50%-deficit irrigated plants received only half the water of well-irrigated plants. Different letters indicate significant differences within each time of measurements at $P < 0.05$ (n = 4). (From Gucci *et al.*, 2002)

depending on available soil moisture. A mid-day depression of photosynthesis has often been measured in olive and other species growing under Mediterranean climate conditions due to high vapour pressure deficit between the leaf and the atmosphere (VPD), occurring during the hottest part of the day, which causes partial stomatal closure (Moriana *et al.*, 2002). The degree and duration of the mid-day depression varies with environmental conditions and tree water status. As water deficit develops, *A* declines; in severely stressed plants the peak of *A* is small and occurs at early morning hours, after which assimilation does not change (Angelopoulos *et al.*, 1996). Transient conditions of water deficit are not unusual in the canopy of even well-irrigated trees during the central part of the day in the summer and under these circumstances the mid-day depression of *A* may become evident.

Varietal differences in *A* have often been reported (Bongi *et al.*, 1987a; Chartzoulakis *et al.*, 2002; Hagimiditriou and Pontikis, 2005). In one study, comparing five Greek cultivars, maximum *A* rates in the highly phosynthesizing cultivar 'Koroneiki' were almost double those of 'Kalamon' and 'Lianolia Kerkyras' and these differences were consistently maintained throughout the growing season (Fig. 2.3.2; Table 2.3.1). Large differences in *A* of well-irrigated plants were also reported between 'Kerkira' and five other Greek cultivars (Loreto *et al.*, 2003), but not in rainfed trees of three Italian cultivars (Proietti and Famiani, 2002).

Olive fruits exhibit photosynthetic activity from the early stages of their development until fruit color change, when chlorophyll degradation becomes substantial and photosynthetic activity is almost completely suppressed. Photosynthetic rate is well correlated with chlorophyll content throughout fruit development. Chlorophyll concentration is high compared with fruits of other species and declines according to a parabolic curve from approximately 0.8 to 0.1 g kg^{-1} fruit d.w. (equivalent to 0.25–0.15 g chlorophyll m^{-2} fruit surface area) from 25 to 150 days after full bloom (Proietti *et al.*, 1999). Fruit photosynthetic rate, expressed on a dry-weight basis, is relatively high during the first three weeks after fruit set (Proietti *et al.*, 1999) when fruit stomata open and close actively, the surface-to-volume ratio is high, and chlorophyll is abundant. As the fruit develops, CO_2 diffusive resistances along the pathway to the carboxylation sites become high and the stomata gradually lose their functionality for gas exchange. Moreover, stomata present on the fruit surface have a lower density frequency than on the leaf abaxial surface and, since their number is set at anthesis, their frequency decreases as the fruit expands. Fruit stomatal conductance reaches very low values by 60 days AFB, when the stomata appear to have lost their typical shape and are covered by wax. Olive fruit photosynthesis at 120 days AFB is only about 10% of the initial maximum rate (Proietti *et al.*, 1999), a decline similar to that measured in apple and grape from after anthesis until the onset of ripening (Blanke and Lenz, 1989). Photosynthetic characteristics of bark tissue have been reported by Filippou *et al.* (2007).

2.3.1.1 Responses of leaf and canopy photosynthesis to environmental conditions

Environmental factors limiting photosynthesis are light, temperature, CO_2 concentration, soil moisture, VPD, nitrogen and chlorophyll content, salinity, and pollutants.

Light is the major driving force for photosynthesis. The relationship between *A* and photosynthetic photon flux (PPF) is illustrated in Fig. 2.3.4. At low light levels, *A* increases almost linearly and the slope of that line estimates maximum apparent quantum yield (ϕ). Under optimal conditions of carbon assimilation, ϕ can reach 0.05 mol CO_2 mol^{-1} photon, but, more typically, ϕ values for mature leaves of field-grown trees are lower, 0.02–0.03 mol CO_2 mol^{-1} photon, due to photochemical limitations and losses in the amount of incoming irradiance actually reaching the chloroplasts. The light compensation point of olive mature leaves is about 20–60 μmol CO_2 m^{-2} s^{-1} PPF. Net photosynthesis of potted plants becomes saturated at PPF values

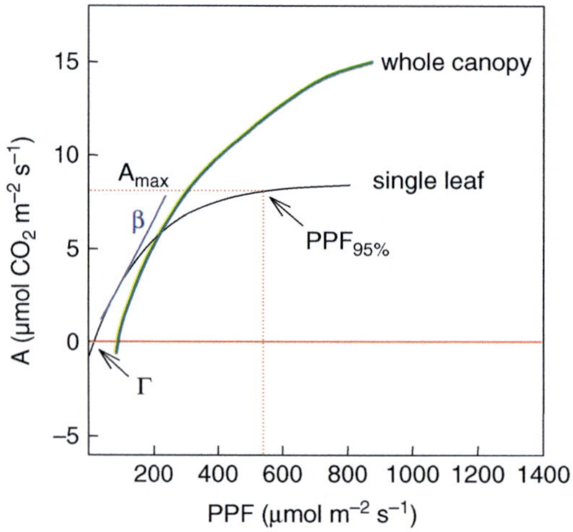

Fig. 2.3.4. The relationship between net photosynthetic rate (*A*) and incident photosynthetically active radiation (PPF) for a single leaf of young olive plants (cv. 'Leccino') grown in containers. The dotted line indicates the radiation level at which photosynthetic saturation occurs for single leaves. The CO_2 compensation point (Γ), which is the light level at which photosynthesis equals respiration, and the apparent quantum yield (β) estimated as the slope of the linear part of the curve for single leaves are also indicated. (From Gucci and Cantini, 2000)

higher than 500–800 μmol CO_2 m^{-2} s^{-1} μmol photon m^{-2} s^{-1} depending on cultivar and experimental conditions (Bongi *et al.*, 1987a). In field-grown, mature trees, Proietti and Famiani (2002) reported a saturation threshold of 1000–1200 μmol photon m^{-2} s^{-1}. When incident PPF exceeds that threshold, *A* does not increase any further because biochemical limitations occur. As in other fruit tree species, the light saturation threshold and light compensation point are smaller for leaves that developed under shade rather than full sunlight.

In areas where olive trees are grown, incident PPF can exceed 1600 μmol photon m^{-2} s^{-1} under sunny conditions. Excessive light may be harmful for the photosynthetic apparatus.

Photoinhibitory effects include a decrease in photochemical efficiency of PSII, chlorophyll over-excitation and consequent induction of reactive oxygen species (ROS). The result is a decrease in photosynthetic efficiency due to the use of reducing power (NADPH) and CO_2. To dissipate the excess energy, several protective mechanisms have been identified such as quenching heat (non-photochemical quenching), moving electrons (water-water cycle, cyclic transport around PS I, glutathione-ascorbate cycle, and photorespiration) and scavenging the produced reactive oxygen species (ROS), some of which are active in olive leaves too (Bacelar *et al.*, 2007).

Photosynthesis at the whole-canopy level is not saturated within the range of PPF that normally saturate single leaves, due to reciprocal leaf shading in the canopy of adult trees (Fig. 2.3.4). In dense canopies, most leaves are shaded and photosynthesize at much lower rates than their maximum potential. As a result, the *A* of the canopy continues to increase as PPF increases and does not reach saturation even at the high light levels of sunny days. Photosynthetic photon flux in areas where olive trees are grown is usually sufficient to fully exploit the photosynthetic potential of olive leaves, but the high leaf-area density of adult canopies can dramatically decrease the amount of light that penetrates and, thereafter, not only ϕ values for the whole canopy are smaller than for individual leaves but also inner or poorly exposed parts of the canopy may receive only 10–30% of full sunlight depressing *A* as well as reproductive processes.

The photosynthetic response to temperature shows a parabolic curve with maximum values measured over the 25–28°C range (Fig. 2.3.5). Bongi *et al.* (1987a) reported that cultivars native of relatively hot climates had maximum values of *A* at temperatures of about 2°C higher than cultivars native of cooler environments. Net assimilation rate decreases sharply above 35°C because of the marked increase in dark respiration, and below 5–10°C. Such temperatures occur frequently in winter and summer under Mediterranean climate conditions, with consequent decreases in the amount of carbohydrates produced by the evergreen leaves. On the contrary, Centritto *et al.* (2005) reported no difference in *A* between 20°C and 40°C in field-grown trees subjected to partial root drying (PRD) irrigation regimes in an arid region.

Inhibitory effects of limiting temperatures often last after the period of exposure and the duration of this lag phase depends on both the absolute temperature values and prevailing conditions during the recovery phase. Photoinhibition consequent to exposure to extreme temperatures depends on the radiation level to which the leaf was exposed later. Bongi and Long (1987) estimated a 10% and 75% reduction in ϕ values of adult leaves exposed to 5°C for 12 h followed by exposure to low (μmol photon m^{-2} s^{-1}) or high irradiance (> 650 μmol photon m^{-2} s^{-1}), respectively. The recovery of *A* to values similar to those prior to temperature stress occurred only if the period of chilling at high PPF was less than 8 h. Photoinhibition damage increases further if the chilling periods are repeated in cycles and becomes even more evident if the whole olive plant rather than the single leaf is chilled. Short periods (12 h overnight) of sub-lethal freezing temperatures (from –1°C through –4°C) determined drastic decreases in leaf *A* of two cultivars ('Frantoio' and 'Leccino'). Measurements were made at saturating PPF (> 1000 μmol m^{-2} s^{-1}) in the

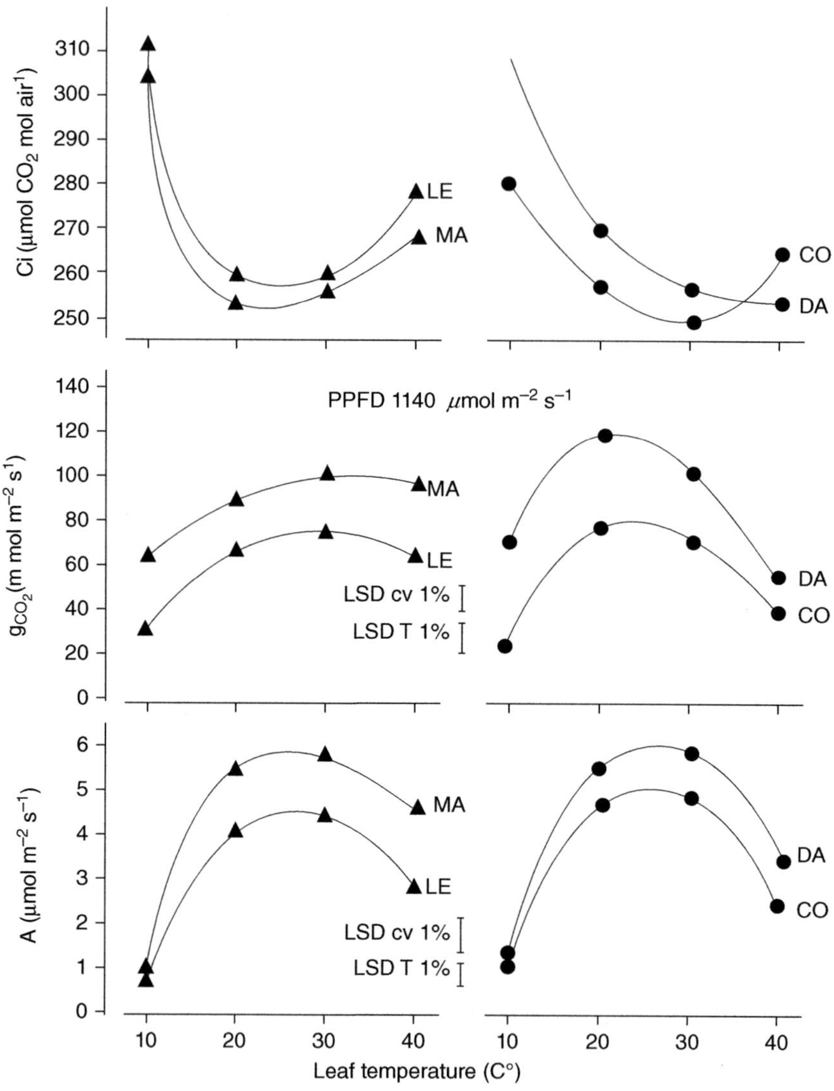

Fig. 2.3.5. Relationship between photosynthetic rate (*A*), stomatal conductance to CO_2 (g_s), intercellular CO_2 concentration (*Ci*) and leaf temperature for four olive cultivars: 'Manzanilla' (Ma) and 'Leccino' (Le) (*left*); 'Dolce Agogia' (DA) and 'Coratina' (Co) (*right*). LSD between cultivars and temperatures were calculated with ten replications per point. PPF was 1140 µmol photon m^{-2} s^{-1}, MFD was 50% of saturated water vapor for each temperature. (From Bongi *et al.*, 1987b)

afternoon following exposure to freezing temperatures and *A* was decreased in both cultivars to about one-third of rates measured in control plants kept in a cold glasshouse overnight (Sebastiani *et al.*, 2002).

Carbon dioxide penetrates by diffusion through the stomata and is incorporated into carbon products via a carboxylation reaction. The relationship between *A* and CO_2 follows a pattern similar to that described for light. The initial part of the *A* response to internal CO_2 partial pressure (C_i) is almost linear, but as Ci increases the curve becomes asymptotic. The slope of the linear part (usually between 0 and 200 µmol m^{-2} s^{-1}) indicates the carboxylation efficiency (k_{CO2}).

Values of k_{CO2} for the olive leaf are typical of C_3 species. A-C_i curves are useful to partition stomatal from non-stomatal (mesophyll) limitations of A in response to stress conditions. For instance, a decrease in A at constant C_i is an indication of mesophyll limitations, which include both biochemical constraints and resistances internal to the leaf along the pathway of CO_2 diffusion to the chloroplast. Low CO_2 concentrations make stomata open and eliminate stomatal limitations.

Limitations to CO_2 diffusion towards the carboxylation sites in the chloroplast can be partially overcome by CO_2 enrichment of the atmosphere around the leaf. Under these conditions, limitations due to CO_2 diffusion are of little significance compared with enzymatic limitations such as the carboxylation reaction and reactions leading to the regeneration of substrates for carboxylation and pentose-phosphate pathway. However, heterogeneous stomatal closure, which may occur in olive leaves, may lead to errors in interpretation of *A-Ci* curves. An additional problem may be that the *A-Ci* curve varies with VPD, as it has been shown comparing curves derived at 3.2 or 1.0 kPa (Loreto and Sharkey, 1990). To overcome this problem it is suggested to associate *A-Ci* gas exchange curves with chlorophyll fluorescence measurements, and remove biochemical limitations using low O_2.

Experiments conducted on olive plants grown in a CO_2-enriched atmosphere under field conditions (FACE facility) showed that long exposure to high CO_2 concentrations increased A and decreased g_s, determining an increase in water-use efficiency by the plant (Tognetti *et al.*, 2002). Carbon dioxide enrichment also determined a decrease in stomatal density, but did not affect the C_i/C_a ratio. This ratio is usually 0.75–0.8 in species with C_3 photosynthetic mechanism, but tends to increase when non-stomatal limitations to photosynthesis prevail over stomatal ones.

2.3.1.2 Soil humidity

Photosynthesis occurs at the expense of water loss. Carbon is assimilated concomitantly with transpiration through stomata, so that for each unit of carbon acquired, about 1000 times as much water is lost. As stomata open, both photosynthesis and transpiration increase. However, beyond a certain degree of stomatal opening, photosynthesis does not increase any further because biochemical limitations overcome the physical limitations to gas diffusion via stomata whereas transpiration continues to increase. As a result, the highest efficiency for plant assimilation is reached when stomata are not fully open because the unit of carbon gained costs less water (Fig. 2.3.6).

Low water availability decreases photosynthesis; however, the tolerance of the photosynthetic apparatus to water deficit is high in olive trees. In a detailed study on potted plants of cv. 'Coratina', Angelopoulos *et al.* (1996) reported that maximum A of plants experiencing pre-dawn Ψ_w −1.2, −4.5 and −5.7 MPa was 98%, 61% and 28% that of fully irrigated plants (pre-dawn Ψ_w of −0.45 MPa), respectively (Fig. 2.3.7). Relatively high photosynthetic rates (6.5 μmol m^{-2} s^{-1}) were still measured in plants with a leaf pre-dawn Ψ_w of about −4 MPa (Angelopoulos *et al.*, 1996).

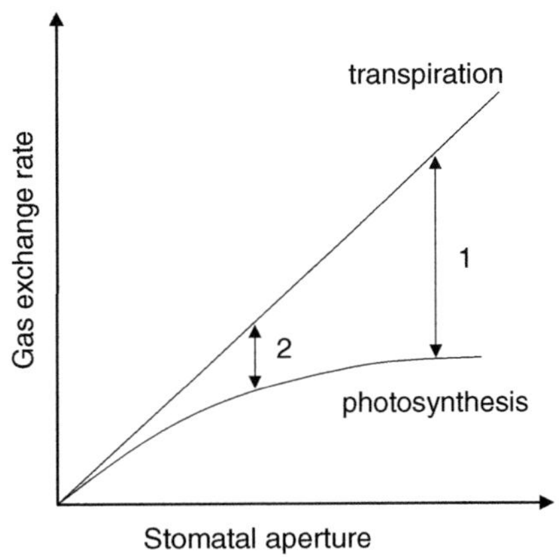

Fig. 2.3.6. The compromise between transpiration and photosynthesis expressed as a function of stomatal opening in a hypothetical leaf. Biochemical limitations prevent further increases of net photosynthetic rate when stomata are fully open, whereas transpiration continues to increase (1). When stomata are fully open, more carbon is accumulated per unit of water vapor transpired (2). (From Gucci and Cantini, 2000)

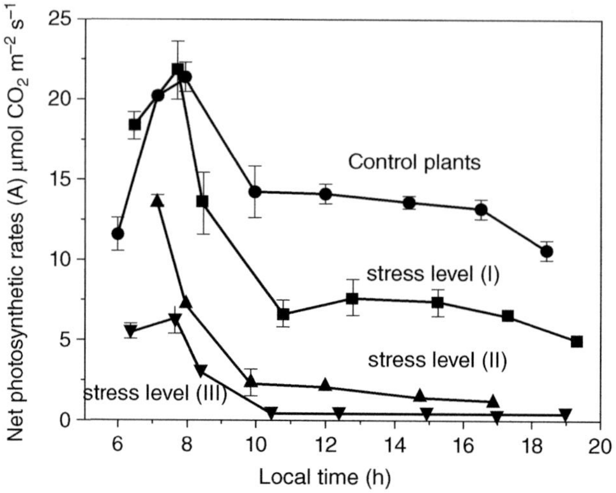

Fig. 2.3.7. Diurnal course of net photosynthetic rates in control plants and in different stress-level plants. (From Angelopoulos *et al.*, 1996)

In field-grown trees subjected to various irrigation regimes, *A* reached 50% and 25% of maximum values at pre-dawn Ψ_w of about −1.75 and −4.5 MPa, respectively (Gucci and Caruso, 2011), confirming the high tolerance of the photosynthetic apparatus of the olive tree to water deficit.

During initial or intermediate stages of water deficit, the inhibition of assimilation is mainly due to the decrease of stomatal conductance, whereas the damage to photosystem II is minor. Olive trees show a more marked capability than other fruit tree species to recover once water stress is relieved. At greater level of stress, photoinhibitory effects increase with consequent less reductions in the capacity to repair damage during the recovery period. Under these circumstances, high temperatures (> 35°C) and high irradiance (PPF > 1300 µmol m^{-2} s^{-1}) increase the probability of photoinhibitory damage (Bongi and Long, 1987). Evidence for energy-expensive repair mechanisms of damaged PSII from long-term photo-oxidative damage (Angelopoulos *et al.*, 1996), water availability enhancement of antioxidant protection, and decrease of oxidative damage (Bacelar *et al.*, 2007) have been reported. The maintenance of a positive daily net budget of assimilation even during periods of prolonged drought is an important adaptive trait, as it allows the olive plant to produce energy and materials to sustain growth and developmental processes during the dry season.

On a diurnal basis the decrease in *A* in young olive plants grown in containers at saturating PPF is proportional to the degree of water deficit (Angelopoulos *et al.*, 1996). At extreme levels of stress the diurnal pattern becomes virtually flat with little photosynthetic activity only in the early morning hours (Fig. 2.3.7). In well irrigated plants, g_s reaches maximum values in the morning and the course rresembles a bell-shaped curve. Under severe deficit maximum values are less than half of the control plants and the peak is shifted to early morning hours. An intermediate condition of water deficit, achieved by irrigating plants with only 50% of their daily water need determines g_s values intermediate between the previous two curves. Other studies correlated the A_n with the relative water content (RWC) and the Ψ_w of the leaf. Jorba *et al.* (1985) measured a reduction of A_n by 85% when RWC declined from 96% to 65%, while Larcher *et al.* (1981) registered a significant reduction in A_n when the ψ_{leaf} was lower than −1.3 MPa and a 50% reduction in A_n when the ψ_{leaf} was lower than −2.2 MPa. Regarding genetic variability to drought,

Tugendhaft *et al.* (2016) found that 'Barnea' cultivar that has relatively high g_s and *A* also showed a lower drought tolerance when compared to 'Picual' and 'Souri' cultivars, suggesting a trade-off between higher yields and drought tolerance. In general, the preservation of *A* enables olive plants to produce photoassimilates under drought that can sustain root growth, increase the root-to-crown ratio and help olive to explore new soil zones where water is still available. Conditions of moderate water deficit also improve the pulp-to-pit ratio of olive fruits (Gucci *et al.*, 2009).

The gradient in vapor pressure deficit between the leaf and the atmosphere is the major driving force for transpiration through stomata and also a major factor regulating stomatal aperture and g_s (Gucci *et al.*, 1996). Stomata open and close rapidly (within 10 min) in response to changing VPD (Bongi *et al.*, 1987a) so that changes in g_s are driven by changes in VPD. The response of leaf conductance to increasing VPD for olive plants experiencing different levels of stress is illustrated in Fig. 2.3.8. Leaf conductance decreases linearly as VPD increases, but for severely-stressed plants ($\Psi_w < -4$ MPa) conductance is virtually zero and non responsive to VPD (Moriana *et al.*, 2002). A sharp decline in mid-day leaf conductance was reported between 1.5 and 3 kPa atmospheric VPD in field-grown trees of cv. 'Manzanillo' growing in an arid region in winter (Rousseaux *et al.*, 2008). Loreto and Sharkey (1990) reported that when VPD was less than 3.0 kPa *A* of mature leaves was unaffected by humidity except as a result of stomatal closure and that stomatal conductance was about 30% less at high than at low VPD (1 kPa). In young leaves high VPD determines a greater reduction in *A* at the leaf tip than at the base, but this gradient, attributed to patchy stomatal closure induced by low air humidity, is not apparent in mature leaves (Loreto and Sharkey, 1990). Finally, it should be noted that in all cited studies the response curves of *A* to VPD were obtained by correlating values measured at different times of day, when changes in VPD were the result of concomitant changes in both air temperature and humidity. This is not the best way to determine this relationship because adjustments in VPD should be obtained by modifying independently the dew point (humidity) of air entering the cuvette by heating or cooling the water bath for humidity control.

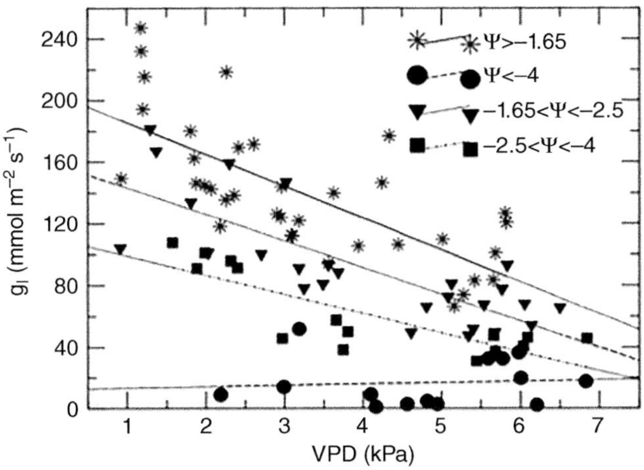

Fig. 2.3.8. Relationship between mid-day leaf conductance (g_l) and the vapour pressure deficit (VPD) for four different levels of water stress. Each point is the mean of 12 observations at various levels of water status as follows: $\psi_w > -1.65$ MPa; y=206.4 –20.7x; R^2=0.5***; n=36; RMSE=30.7; $-1.65 < \psi_w < -2.5$ MPa; y=160.6 –17.3x; R^2=0.6***; n=26; RMSE=23.7; $-2.5 < \psi_w < -4$ MPa; y=111.5 –12.4x; R^2=0.66***; n=15; RMSE=15.9; $\psi_w < -4$ MPa; ns; n=14. (From Moriana *et al.*, 2002)

2.3.1.3 Nitrogen and chlorophyll concentration

Photosynthesis includes a number of biochemical reactions strictly dependent on enzymes. The chlorophyll molecule, which is synthesized in the presence of light, also contains nitrogen. Hence, it is not surprising that there is a close relationship between assimilation and availability of nitrogen. Leaf N content and light-response dependence have been indicated as the most critical factors to simulate canopy assimilation. Changes in leaf N and chlorophyll concentrations occurring seasonally or in response to cultural practices such as irrigation (Gucci *et al.*, 2010), fertilization or the cropping condition can explain, partially or entirely, differences in photosynthesis reported in the literature. In organs like fruits, where the main source of CO_2 for assimilation originates from respiratory processes via an internal recycling mechanism, the investment in N is less than in leaves and other organs that photosynthesize through atmospheric CO_2.

2.3.1.4 Endogenous factors

The sink demand for assimilates has been reported to affect photosynthesis under certain conditions. The activity of carbon sinks varies seasonally and in relation to organ development so that the same organ can be a net importer or exporter of carbon depending on its stage of development. Leaves are initially relying on imported carbon for their expansion but later become major photosynthetic organs. The activities of sinks and sources of C are usually balanced and the concept is expressed as source:sink ratio. The leaf:fruit ratio is often used to describe the source:sink ratio of trees, leaves and fruit being major sources and sinks, respectively. The implied physiological mechanism for regulation of A by source–sink relationship is feedback inhibition due to accumulation of carbon products when the production of assimilates largely exceeds utilization. In apple and orange, the fruiting condition of the tree (crop load, removal of fruits) determines transient changes in leaf A (Gucci *et al.*, 1995; Syvertsen *et al.*, 2003), but in olive trees the presence of fruits does not seem to affect leaf A (Hagimiditriou and Pontikis, 2005).

2.3.1.5 Respiration

Stomata are virtually closed (g_s < 10 mmol CO_2 m^{-2} s^{-1}) at night. In the absence of light, leaves respire rather than photosynthesize. Rates of dark respiration (R_d) in mature leaves of well-watered plants typically range from −1.0 to −3.0 μmol CO_2 m^{-2} s^{-1} (Aganchich *et al.*, 2009; Centritto *et al.*, 2005; Gucci *et al.*, 1998; Proietti and Famiani, 2002), similar to those measured in other tree crops. Temperature and tree water status have a strong influence on R_d.

In olive leaves, differences between daylight net photosynthesis and night respiration are relatively small, which implies that only a limited availability of assimilated carbon can be used for organ and tissue growth even under conditions of optimum photosynthetic activity. Over the 24 h diurnal cycle, leaf assimilation was estimated equal to 0.375 mol CO_2 m^{-2} leaf area in potted plants (light period of 14.5 h) on a summer day under clear-sky conditions, whereas leaf respiration during the night amounted to 0.104 mol CO_2 m^{-2} leaf area. The net balance of carbon available per day resulted, therefore, in 0.271 mol CO_2 m^{-2}, corresponding to 0.0662 mol glucose m^{-2} (Gucci *et al.*, 2002). These estimates were made using values of the daily requirement for maintenance respiration of adult leaves (0.0136 g glucose g^{-1} d.w.) and of the energetic cost for growth (1.66 g glucose g^{-1} d.w.) reported by Meriño (1987). Contribution by other green parts is minor and can be assumed to be negligible.

Fruit respiration occurs actively throughout development and CO_2 efflux is always greater than uptake. Fruit growth is essentially a heterotrophic process and carbohydrates imported via the phloem are needed to sustain it. Carbohydrate respiration generates CO_2, which is partially re-fixed by fruit photosynthesis.

Olive fruit R_d is high after fruit set, decreases to about 70% 45 days after full bloom (AFB) and reaches very low values afterwards. Fruit exposure to full sunlight reduced CO_2 efflux by about 40% at initial stages of fruit development (20 d AFB) and by 75% after two months due to fruit photosynthesis. By 130–150 d AFB, the effect of light on CO_2 efflux had become negligible. The contribution to photosynthesis to the overall carbon budget of olive fruits is minor and fruits should be regarded as net importers (Proietti *et al.*, 1999). However, defoliation or fruit-shading experiments started at 12 weeks AFB of fruit development hint at an important contribution of fruit light environment to oil synthesis, whereas imported carbon appears more important for the attainment of final fruit size (Sanchez and Harwood, 2002).

Dark respiration rates of olive stems range 5–10 μmol CO_2 m^{-2} s^{-1}. Rates are relatively stable across age classes (two- through 30-year-old) with the exception of one-year-old stems that had higher rates than the others. Little information exists on respiration rates of olive roots. Soil CO_2 efflux rates of 0.05–0.15 mg CO_2 m^{-2} s^{-1} have been measured, using portable chambers, on summer days in an irrigated, young orchard, kept weed-free by herbicide applications. Soil moisture conditions (zone wetted by drippers vs. alley) determined large differences in CO_2 efflux, which showed little diurnal variation or temperature dependence (Testi *et al.*, 2008). Soil efflux rates are only indicative of root respiration and the results are site-specific as they depend on soil physical and chemical properties, moisture and orchard management.

CARBON FLUXES AND SOIL CARBON STORAGE IN THE OLIVE ORCHARD

The contribution of agriculture to annual emissions of greenhouse gases is 10–12% and the carbon footprint of individual crops may soon be considered by growers and administrators in the making of decisions and policies. Plants assimilate carbon via the photosynthetic process and contribute to carbon emissions by respiration. The gain between assimilated and respired carbon is the net primary production (NPP). To calculate the carbon budget and determine whether orchards contribute to alleviate the global problem of carbon emissions, soil carbon pools and fluxes are to be measured in addition to photosynthetic uptake and autotrophic respiration of tree organs. This is not an easy task because of the perennial lifespan of trees and the high variability of soil respiration (Rs) components, such as root respiration, microbial respiration, decomposition of plant litter, and root exudation, as they depend on soil type, temperature, humidity, tree age and cultural practices (Scandellari *et al.*, 2016).

Carbon fluxes in and out of olive orchards have been determined in only a few studies and the information available is site-specific. In southern Italy it has been estimated, by destructive samplings, that irrigated olive orchards fix above-ground 2.74 and 9.54 t CO_2 ha^{-1} per year depending on tree age (Sofo *et al.*, 2005). The potential amount of humus produced annually out of senescent leaves and pruned material was 0.45 and 1.69 t ha^{-1} in the young and mature orchard, respectively (Table 2.3.2). When the CO_2 released into the atmosphere by mineralization of soil organic matter was taken into account emissions were higher than assimilation for the young orchard, whereas CO_2 fixed by the mature orchard was three times greater than emissions (Sofo *et al.*, 2005). Testi *et al.* (2008) used eddy covariance techniques to measure daily net ecosystem exchange flux (NEE) and night-time ecosystem respiration (R_{ECO}) in an irrigated orchard (408 trees ha^{-1}) in Andalusia. NEE was virtually zero when LAI was 0.4% and 11% of incoming daily radiation was intercepted, whereas it was about 10 mg CO_2 m^{-2} per day at an LAI of 1.9 and 50% intercepted radiation. R_{ECO} ranged from 0.05 to 0.15 mg CO_2 m^{-2} s^{-1}, the lower limit being close to eterotrophic R_s for that site (Testi *et al.*, 2008). About 50% of the calculated NEE of 2.8 kg CO_2 m^{-2} per year was recovered as above ground dry matter (Villalobos *et al.*, 2006), whereas the remaining 50% was attributed to root processes (growth, turnover and exudation) and litter production (Testi *et al.*, 2008). Because of the discontinuous planting system, root density is not uniform in orchards and lowest values of Rs are measured in the alley. Maximum values of Rs occur in mid-spring and autumn, minima in the summer under rainfed conditions (Bertolla *et al.*, 2014).

Table 2.3.2. Above-ground biomass production, annual CO_2 fixed and humus produced in a young (less than ten years old cv. 'Coratina') and a mature (older than 50 cv. 'Maiatica') olive orchard planted at a 6 x 3 m and 8 x 8 m distance, respectively. (From Sofo *et al.*, 2005)

Orchard	Organ	Dry matter(t ha^{-1} year^{-1})	Fixed CO_2(t ha^{-1} year^{-1})	Isohumic coefficient	Humus(t ha^{-1} year^{-1})
Young	Senescent leaves	0.5	0.91	0.2	0.1
	Pruned material	1	1.83	0.35	0.35
	Total	1.5	2.74		0.45
Mature	Senescent leaves	0.5	1.67	0.2	0.18
	Pruned material	1	7.87	0.35	1.51
	Total	1.5	9.54		1.69

The soil is potentially a large sink for carbon storage. In natural ecosystems the carbon stored in dry matter eventually returns to the soil pool due to leaf and fruit drop and death of trees; in turn, soil organic carbon goes back to the atmosphere by decomposition. Heterotrophic respiration represents a substantial CO_2 flux out of the soil.

Although highly variable, estimates vary between 15% and 55% of total flux in forests and from 40% and 95% in non-forest ecosystems. Agricultural development has drastically depleted organic matter and carbon in the soil and the decline has been particularly sharp in the last 50 years. Roots of higher plants are responsible for 25–33% of the respiration occurring in the soil, rhizodeposition accounts for 2–30% of total dry-matter production in young plants (Weil and Brady, 2017). Soil respiration is the second-largest source of CO_2 into the atmosphere after the oceans (Schlesinger and Andrews, 2000). From one-third to one-fifth of the pruned material left to decompose onto the soil remains in the soil (Sofo *et al.*, 2005). Since mineralization is strongly dependent on organic matter content and soil type, more information under different growing conditions is also needed. Root decomposition increases rapidly in a young olive tree. In young olive trees, 73% of CO_2 was fixed in the above-ground biomass and the remaining CO_2 in root biomass over a five-year period (Sofo *et al.*, 2005). The value of fixed CO_2 in leaves of young olive plants decreased from 17.9 in year 1 to 9.0% in year 5; in contrast the value of fixed CO_2 in woody parts increased *from* 64.9 to 71.9% in year 5. Olive fruits immobilized from 0.18 to 6.7 t ha^{-1} in years 1 and 5, respectively. Soil humidity, organic matter and cover crops markedly influence soil respiration. Soil respiration rates of the orchard with a grass cover were higher than those of the periodically tilled soil and totaled 11.8 and 7.8 t ha^{-1} of annual emissions out of the soil. The estimated annual amount of respired carbon for the tilled orchard was double that of non-irrigated olive orchards in southern Spain (Almagro *et al.*, 2010), but similar to that reported for a 30-year-old vineyard in California (Carlisle *et al.*, 2006). The presence of grass roots contributes to CO_2 fluxes because of their respiration and decomposition as well as a high content of organic matter and total N in the soil makes conditions more favourable for microbial activity and eterotrophic respiration.

2.3.2 Carbon Partitioning and Translocation

2.3.2.1 Carbon partitioning

Carbohydrates are structural compounds, the main source of energy for plant metabolism and indispensable precursors for biosynthesis of other compounds, including lipids, in the plant cell. Carbohydrates are compounds where most energy is stored. When broken down by respiration, carbohydrates supply carbon units and energy needed for plant metabolism.

Many soluble carbohydrates are present in olive leaves. Glucose, mannitol and sucrose are the three main non-structural carbohydrates in the olive leaf (Table 2.3.3). The glucose content is over 50% of total soluble carbohydrates, whereas that in mannitol and sucrose is about 30% and 8%, respectively (Fig. 2.3.9). Although conclusive evidence is still lacking, it is likely that the high concentrations of glucose and mannitol in the olive leaf contribute to maintain low osmotic potentials, which may turn useful to maintain turgor and water-potential gradient between the canopy and the root under drought or salinity stress. Total soluble carbohydrate content ranges 1000–2000 µg cm^{-2} leaf area (200–400 µmol g^{-1} f.w.) under optimal conditions of carbon assimilation. Leaf starch content is usually comprised between 200 µg cm^{-2} and 800 µg cm^{-2} leaf area.

Table 2.3.3. Major carbohydrates found in different tissues of olive plants. (From Flora and Madore, 1993; Gucci *et al.*, 1998; Priestley, 1977; Tattini *et al.* 1996)

Leaf mesophyll	Fruit	Phloem sap
Glucose	Sucrose	Stachyose
Mannitol	Mannitol	Raffinose
Sucrose	Glucose	Sucrose
Fructose	Fructose	Mannitol
Galactose	Galactose	Verbascose
Galactinol		Glucose
myo-inositol		Fructose
Raffinose		*myo*-inositol
Stachyose		
Verbascose		

The diurnal course of non-structural carbohydrate shows that the content of single compounds is relatively stable except that of mannitol, which tends to increase from the beginning of the light period (Fig. 2.3.10). However, this increase is not evident in diurnal courses measured during periods of the growing season when light and temperature are suboptimal for photosynthesis. Despite the wide diurnal differences in photosynthetic rates (Fig. 2.3.10), the contents of soluble carbohydrates and starch were fairly stable during the day, which seems to indicate a strong tie between availability and utilization of carbohydrates on a diurnal basis. The seasonal course of leaf mannitol of field-grown trees shows an increase (from about 200 to 300 µmol g^{-1} d.w.) from October through February and a decrease (returns to 200) in early spring in Sicily (Oddo *et al.*, 2002).

The mannitol content seems to increase with leaf age, whereas fructose, glucose and sucrose contents are unaffected by leaf age. On a seasonal basis the leaf mannitol content is higher in winter than during the growing season in leaves (less than one year old), whereas the opposite trend occurs for glucose (Priestley, 1977).

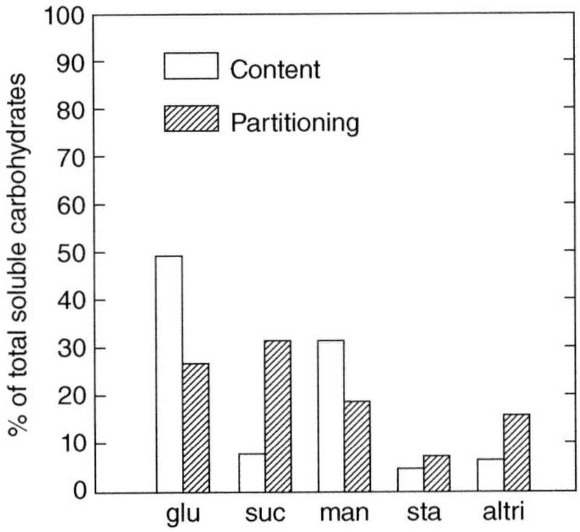

Fig. 2.3.9. Concentration and partitioning of major soluble carbohydrates, expressed as percentage of total, in fully expanded leaves of young olive plants. (Gucci, 2003) glu, glucose; suc, sucrose; man, mannitol; sta, stachyose + raffinose. Others include fructose, myo-inositol, galactose and galactinol

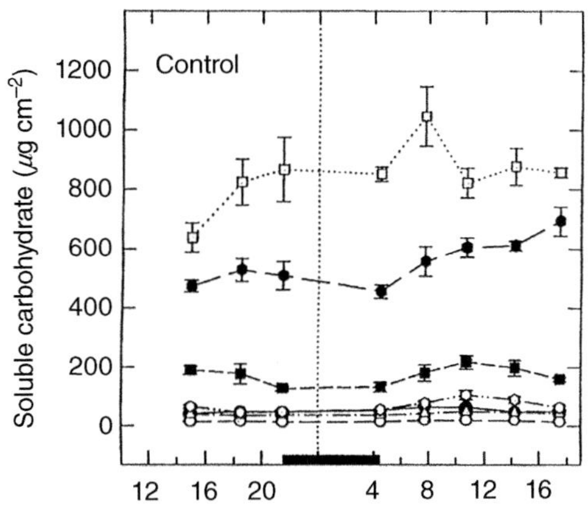

Fig. 2.3.10. Diurnal pattern of soluble carbohydrate pools in fully expanded leaves of well-watered olive plants. The black rectangle represents the night period. Symbols are means + standard errors (not shown if smaller than symbol) of four replicate plants. (Modified from Gucci *et al.*, 1998)

Carbohydrate partitioning is more complex in olive than in other tree crops (apricot, cherry, apple, peach, privet), in which carbon is partitioned only between sucrose and a sugar-alcohol (mannitol or sorbitol) (Loescher and Everard, 1996; Stoop *et al.*, 1996). In fully expanded olive leaves, about 90% of fixed carbon is partitioned to soluble carbohydrates, while the remaining 10% is recovered mainly as starch. Glucose, mannitol, fructose and sucrose are primary products of photosynthesis. In pulse-chase experiments using $^{14}CO_2$, these carbohydrates and the precursors for the raffinose oligosaccharides (galactinol and sucrose) became labeled within 30 seconds of chase (Flora and Madore, 1993; Gucci *et al.*, 1998), whereas it took about 10 minutes before stachyose, raffinose and verbascose were appreciably labeled (Flora and Madore, 1993; Gucci *et al.*, 1998). This delay in labeling has been interpreted as evidence that the oligosaccharides of the raffinose series are probably synthesized in the intermediary cells or in a compartment close to the phloem, where the precursors are transported after being synthesized in the mesophyll (Flora and Madore, 1993).

Photosynthetic carbon is partitioned into glucose and sucrose in similar proportions and both compounds account for over 58% of total soluble carbohydrates. Glucose accounts for about 28% of label recovered in soluble carbohydrates under non-stressed conditions and this large partitioning is consistent with the large glucose pool in the leaf mesophyll. About 20–30% of carbon is recovered as mannitol and less than 10% in raffinose, stachyose and verbascose, under optimal and stable conditions of assimilation and partitioning (Flora and Madore, 1993; Gucci *et al.* 1998). The remaining percentage of carbohydrates is partitioned into fructose, myo-inositol, galactose and galactinol.

Carbon partitioning is altered under stress conditions. In plants exposed to water deficit or high salinity in the soil, there is an increase in partitioning towards mannitol and a decrease into glucose and sucrose, whereas the other leaf-soluble carbohydrates remain unaffected (Gucci *et al.*, 1998). Mannitol seems to play an important role in stress resistance because of its action as a compatible solute in the cytoplasm (Loescher and Everard, 1996). An increase in leaf mannitol content was measured in plants subjected to 100 mM NaCl stress for five weeks, with changes in mannitol concentrations more rapid than those of glucose. A greater rate of decrease of mannitol concentration was observed when salt stress was relieved by watering olive plants with non-saline solutions (Tattini *et al.*, 1996) (see also response to salt or drought).

2.3.2.2 Translocation

Stachyose, raffinose, verbascose, sucrose and mannitol are the main carbohydrates recovered from the olive phloem sap. In addition to sucrose and mannitol, often found as translocated sugars in plants because highly soluble and chemically inert, carbohydrates of the raffinose series (raffinose, stachyose, verbascose) and their precursors (galactose, galactinol, myo-inositol) are also found in olive phloem. Stachyose and raffinose are likely synthesized in the intermediary cells associated to the phloem, as it occurs in species that mainly translocate these carbohydrates (Flora and Madore, 1993). Stachyose is the most abundant carbohydrate (about 50% of the label recovered in phloem exudate of adult leaves, followed by sucrose, mannitol and raffinose) (Flora and Madore, 1993). However, the amount of mannitol, early product of photosynthesis, present in the phloem sap is less than 5% of labeled carbon recovered in the exudate.

Photoassimilates exported out of the leaf and their transport in the phloem stream are rather slow (3–6 mm min^{-1}) compared with rates measured in herbaceous species. Diurnal changes in transport speed and the fraction of carbon exported out of fully expanded leaves have also been measured using ^{11}C-techniques (Gucci and Minchin, 2002). The short-lived isotope ^{11}C has a half-life of 20.4 minutes and so it allows *in vivo* monitoring of newly assimilated carbon. Higher carbon flows were measured in the afternoon than in the morning and flows varied in intensity and direction according to the leaf position from the apex (Gucci and Minchin, 2002). In particular, the pattern of export out of the leaf depends on the leaf position along the olive vegetative shoot. There was no evidence of short-term export (within 120 min) when the youngest fully expanded leaf was ^{11}C-labeled. Labeling the second-youngest leaf resulted in a strong tracer pulse in the portion of the stem immediately proximal to the petiole of the labeled leaf, but no activity a further 5 cm below; only a very small amount of radioactivity was detected in the acropetal direction (Table 2.3.4).

Flows of newly assimilated carbon showed typical diurnal courses, with an increase in the exported radioactivity in the afternoon (12 h after the beginning of the photoperiod). It has been hypothesized that the long time lag before newly fixed carbon is exported out of the fully expanded leaf is probably due to the different compartmentalization and retention of the various carbohydrates present (Gucci and Minchin, 2002). For instance, the retention time of mannitol in leaf mesophyll cells after its biosynthesis is longer than that of other carbohydrates. As a result, mannitol appears in the phloem sap later than raffinose and stachyose, which are, instead, readily exported in the phloem as soon as they are synthesized in the intermediary cells. Flows of newly assimilated carbon showed typical diurnal courses, with an increase in the exported radioactivity in the afternoon (Table 2.3.4). Differences in export were not caused by changes in carbon assimilation since photosynthetic uptake remained relatively stable during the experiments (Gucci and Minchin, 2002).

Table 2.3.4. Newly assimilated carbon exported basipetally out of the second- or fourth-youngest fully expanded leaf in the vegetative shoot of olive. Leaves were labeled 4 h (a.m.) and 12 h (p.m.) after the beginning of the photoperiod. Export is expressed as a fraction of tracer in the load zone 100 minutes after labeling. Transport was calculated as the speed between the two detectors placed basipetally from the load zone. (From Gucci and Minchin, 2002)

	Export		Transport speed (mm/min)	
Leaf position	a.m.	p.m.	a.m.	p.m.
4th youngest	0.12	0.14	6.7	5.0
4th youngest	0.05	0.06	3.0	2.7
2nd youngest	0.04	0.05		

Transport systems for mannitol have been characterized. Mannitol transcripts increase during olive fruit maturation (Conde *et al.*, 2007). Unlike other species, the polyol/H+ symport has low selectivity for mannitol, as it also accepts dulcitol, sorbitol and xylitol, which act as competitive inhibitors for mannitol. In olive, the uptake of monosaccharides and polyols appears mediated by two different transport systems, whereas glucose uptake and mannitol uptake do not inhibit each other (Conde *et al.*, 2008).

In fully expanded olive leaves, photosynthetic carbon is partitioned almost equally into glucose and sucrose (together amounting to over 58% of total soluble carbohydrates), whereas only about 20% is recovered as mannitol under optimal and stable conditions of assimilation and partitioning. Partitioning of carbon products is altered under conditions of water deficit or excess salinity with an increase towards mannitol and a decrease in glucose and sucrose while other soluble fractions remain unaffected (Gucci *et al.*, 1998). Mannitol seems to play an important role in stress resistance because it works as a compatible solute in the cytoplasm (Conde *et al.*, 2007).

Oil accumulation is inversely proportional to soluble carbohydrates concentration in the fruit. The latter decreases sharply during ripening. Glucose, fructose, sucrose and mannitol are the main carbohydrates found in fruit mesocarp. Carbohydrates also supply the carbon skeleton for biosynthetic pathways of shikimic acid and mevalonic acid, which are the precursors of several phenolic compounds and secoiridoids, respectively.

Little is known about phloem unloading into the olive fruit. Based on studies in other fruit systems, such as tomato and grape, it can be hypothesized that both symplastic and apoplastic routes are active. The monosaccharide transporter OeMST2 (*Olea europaea* Monosaccharide Transporter 2) is highly expressed during fruit ripening, which implies that apoplastic transport may be prevalent in ripe fruits. This transporter, in fact, carries monosaccharides (fructose, galactose and glucose) derived from the breakdown of phloem transported carbohydrates into the mesocarp parenchymatic cells to supply energy for metabolic functions and materials for oil biosynthesis (Conde *et al.* 2007).

2.3.2.3 Accumulation of carbohydrate reserves in different organs

Carbohydrates are ubiquitous in the tree tissues and their composition varies in different organs. Table 2.3.3 lists the major soluble carbohydrate fractions in various plant organs of olive trees.

Carbohydrates that have a hydroxyl group replaced by an aldo or keto group are commonly called sugar alcohols. Sugar alcohols are common in higher plants. Alditols include sorbitol, mannitol, galactinol, galactitil and others; myo-inositol is the precursor for cyclitols like pinitol and ononitol. Sorbitol is present in many *Rosaceae* species of economic interest, such as apple, pear, peach, cherry, plum, apricot. Mannitol is present in celery (the most commonly used species for studies of phloem transport of mannitol) and in many species of the *Oleaceae* family, including *Olea, Phillyrea, Fraxinus* and *Ligustrum* (Flora and Madore, 1993; Oddo *et al.*, 2002).

Mannitol is a primary photosynthetic product synthesized from triose –P via mannitol 6-P and mannitol 1-P after two reactions catalyzed by mannose-6-phospate reductase and mannitol 1-phosphatase, respectively (Loescher *et al.*, 1992). Mannose-6-phospate reductase, a cytoplasmic enzyme, catalyzes the key step that uses reducing power, that is NADPH is oxidized to NADP (a NADPH-dependent reductase is also involved in sorbitol biosynthesis). Mannitol is translocated in the phloem of olive as well as in celery, ash and privet. Mannitol has been localized in vacuole, stroma and cytosol of celery mesophyll cells (Nadwodnik and Lohaus, 2008) with higher symplastic concentration than apoplastic. Myo-inositol concentration is high in the stroma (celery). Mannitol catabolism occurs by oxidation to mannose by 1-oxidoreductase and that has been fully characterized in celery sink tissue (Stoop and Pharr, 1993). Mannitol has multiple roles in the olive tree. Storage compounds in rings contribute to osmotic regulation and play a protective role under stress conditions.

There are not many studies on carbohydrate composition in olive organs other than the leaf. Priestley (1977) determined the seasonal course of carbohydrates in different organs of potted plants (cv. 'Ascolana') and measured high amounts of mannitol and glucose in current-year shoots growth. The glucose content was less in the two-year-old wood. There were not appreciable seasonal changes of glucose, mannitol, sucrose and fructose in the root, whereas a larger content of mannitol was present in wood tissue during winter months. Recent studies confirmed the abundance of mannitol, glucose, fructose and sucrose, both in stems and roots of olive plants grown under either field or controlled environment conditions. The dry-weight content of soluble carbohydrates in shoots and leaves is greater than in the roots, whereas the starch content is greater in the roots and shoots than in the leaves.

The main non-structural carbohydrates in fruits are glucose, fructose and mannitol (in order of abundance). The content in soluble carbohydrates tends to decrease during ripening when the oil content increases (Wodner *et al.*, 1988). Despite the importance of carbohydrates on fruit development, oil synthesis and accumulation, there is still little information available on carbohydrate metabolism in olive fruit tissue.

2.3.2.4 Dry-matter distribution and harvest index

In fruit trees, the relationship between the amount of intercepted light and fruit production is not a simple one. In fact, carbon not only has to be assimilated through photosynthesis but also has to be transported to the reproductive sinks. The ratio between the dry matter accumulated into the usable part of the plant product (fruit for fruit trees) and total dry matter is defined as the harvest index. The importance of the harvest index, which represents a simple way to measure dry-matter allocation toward the economic objective of cultivation, can be fully understood when considering that most of the increase in yield of wheat cultivars achieved in the last century has been obtained by breeding for higher dry-matter allocation to the seed rather than for carbon assimilation in itself (Gifford *et al.*, 1984). In fruit trees, the harvest index is quite small in comparison with that of more productive annual crops or perennial crops used for wood or fiber.

There are periods when fruits and vegetative growth actively compete for assimilates, so that dry-matter distribution in the shoot and flower induction are influenced by the presence, number and degree of development of reproductive organs. At the whole-plant level, competition between vegetative meristems and fruits is of the utmost importance for dry-matter distribution and stability of fruit production considering the susceptibility of the olive tree to alternate bearing.

Dry-matter distribution also depends on other factors such as plant age, amount of intercepted PPF, planting density and, if any, conditions of environmental stress and soil properties. In one-year-old plants, 42% of canopy dry matter is allocated to the leaves and the remaining 58% to shoots and stem. In ten-year-old plants, 9.5%, 81.0% and 8.5% of canopy dry matter were distributed to leaves, wood and fruit, respectively (unpublished results). The root-to-canopy ratio decreases with plant age and increases under conditions of water deficit in the soil (Celano *et al.*, 1999). In four-year-old plants (cv. 'Coratina'), the harvest index was about 16% and 13% in irrigated and non-irrigated plants, respectively (Celano *et al.*, 1999). Dry-matter partitioning coefficients between leaf and wood and between canopy and root, measured on young, non-fruiting olive plants over two growing seasons, are reported in Table 2.3.5. At lower planting densities, the partitioning coefficient towards the leaf increases, whereas that towards the wood decreases. The partitioning coefficient towards the root decreases from the first to the second growing season (Table 2.3.5). The estimated cost of dry-matter production was 0.669 g g^{-1} dry weight in young plants, a high value compared with that of other species, due to high content in lignin and structural materials in olive tissues (Mariscal *et al.*, 2000).

Table 2.3.5. Partitioning coefficients between leaf, shoot and root of olive trees (cv. 'Picual') planted at either a 0.5 or 2 trees m^{-2} density. Partitioning coefficients were calculated from destructive samplings over two growing seasons. (Modified from Mariscal *et al.*, 2000)

Planting density (trees m^{-2})	Organ	Partitioning coefficient
0.5	Leaf	0.37
0.5	Wood	0.60
2.0	Leaf	0.28
2.0	Wood	0.69
0.5	Root	0.26

2.3.3 Water Relations

Since the 1980s, the water relations of the olive trees have been extensively studied to elucidate the mechanisms whereby this species is able to withstand the long periods of drought typical of Mediterranean summers. The marked tolerance to soil water deficit is the result of a series of adaptive traits, which include anatomical, morphological and physiological adaptations to environmental conditions. At the cellular level, water makes up the largest proportion of the cellular volume and cell expansion largely depends on water uptake into the vacuole. At the whole-plant level, almost 97% of the water entering the plant is lost into the atmosphere, fundamentally through the transpiration process; 2% is necessary for the volume increase or cell expansion and only 1% is consumed in the metabolic process (i.e. photosynthesis). Insufficient water availability, referred to as water deficit, is a common phenomenon in natural and agricultural habitats and it is caused by intermittent, or even continuous, precipitation. Drought describes the condition that arises during a period of insufficient precipitation and results in plant water deficit. Water scarcity in the Mediterranean climate exposes yearly olive plants to stress during the active period of the growth cycle. Olive trees evolved several adaptive traits and acclimation strategies to cope with drought. Nevertheless, water deficit easily becomes one of the main factors limiting growth and productivity.

2.3.3.1 Water potential

Most of the water is absorbed by roots from soil. The water potential (Ψ_w) of the soil solution approaches zero only when the soil water content is at saturation (all soil spaces are filled with water). The saturation state is temporary for well-drained soils, as the excess water quickly drains out of the larger pores under the influence of gravity and is replaced by air. Soil drying gradually reduces the soil Ψ_w making water less available for roots up to Ψ_w values that could induce a permanent wilting of the plant. At this point, the soil contains only non-available water for root absorption; plants become permanently wilted and cannot be revived even if placed in a water-saturated atmosphere. For many species this point corresponds to a soil Ψ_w below –1.5 MPa.

The transport of sap in the xylem depends on the characteristics of the conducting tissue and the hydraulic architecture of the canopy. Olive has porous diffuse wood with several xylem vessels (up to 100 mm^{-2}), little accessory parenchyma and abundance of fibers. Xylem vessels are rather short and seldom wider than 40 µm in diameter. In *Olea oleaster*, over 90% of vessels have diameters less than 20 µm. The few vessels with wider diameter, which transport up to 30% of xylem sap in the olive tree, are the first ones to cavitate, so that the loss of hydraulic functioning of the system is not proportional to the number of cavitated vessels (Salleo and Nardini, 1999). These features limit the stem hydraulic conductivity, which is directly proportional to the fourth power of vessel diameters based on Poiseuille's Law.

Stem conductivity, expressed as a function of leaf area supplied, changes according to the order of insertion and stem diameter. The LSC is greater in stems of order 1 and smaller in those of order 2 and 3. LSC usually increases in stems with a large diameter (Thompson *et al.*, 1983). However, the lesser efficiency in sap transport is compensated for by fewer risks of cavitative events, since vessels of small diameter are less prone to embolism. The narrowest vessels are usually at the petiole–node junctions, which are more resistant to cavitation than vessels in nodes and internodes (Salleo and Nardini, 1999).

The leaf water potential (ψ_w) undergoes large changes both on a diurnal and a seasonal basis. Unlike isohydric species (e.g. kiwi fruit, maize, poplar), in which stomatal closure reduces differences in ψ_w between droughted and non-stressed plants during the hottest part of the day, the diurnal course of ψ_w in the olive tree reflects both tissue-water status and evapotranspirative demand by the environment (Fig. 2.3.11). The diurnal marked decrease of ψ_w is the result of tissue dehydration (the leaf is capable of losing up to 40% of its relative water content during a day of high evaporative demand) and the rigidity of cell walls. As a consequence, the ψ_w reaches values close to the turgor loss point during the hottest time of the day even when soil moisture is adequate. The turgor loss point for the olive leaf has been estimated at –3.0–3.5 MPa of ψ_w and 75–80% relative water content (Lo

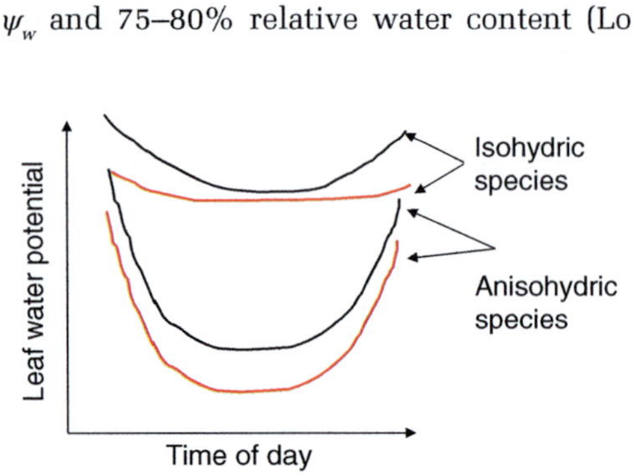

Fig. 2.3.11. Diurnal course of leaf water potential in isohydric (i.e. kiwi) and anisohydric species (i.e. olive). Key: well watered plants (black), water deficit plants (red)

Gullo and Salleo, 1988). Therefore, there are wide differences between the optimal water content and that of zero turgor and the free energy in plant tissue is low. Differences in ψ_w between irrigated and non-irrigated plants (at initial levels of stress) are not always evident and often become apparent only in the afternoon when tissues are unable to release water to maintain the transpiration stream. The lowest values in ψ_w under severe deficit conditions are –8–10 MPa. Interestingly, even at these extreme levels of stress the olive tree maintains the capacity to rehydrate and recover once stress conditions are relieved. A recent study surveyed results from different olive-growing areas to establish the following stem-water potential non-stressed baselines: between –2 and –3 MPa (mild water stress from fruit set to pit hardening); between –3 and –4 MPa (pit hardening); and between –2 and –3 MPa during the period of rapid oil accumulation (Shackel *et al.*, 2021).

The diurnal and seasonal courses of ψ_w in olive trees are indicative of a non-conservative strategy for transpiration (Lo Gullo and Salleo, 1988). This is confirmed by the relatively slow stomatal closure after the onset of water deficit and as stress develops and progresses. The diurnal course of g_s depends on soil water availability even at initial stages of water stress (Fig. 2.3.11). The relationship between g_s and ψ_w for olive trees has a stable negative slope over a wide interval of ψ_w. As ψ_w decreases, stomatal closure is less ready than that of other woody species grown for fruit production such as kiwi fruit and apricot (2.3.12). It should be noted that the range of variation of ψ_w before the plant undergoes irreversible damage is limited for kiwi fruit, intermediate for apricot and wide for olive (Fig. 2.3.12).

An important mechanism to maintain cell turgor is the decrease in tissue osmotic potential ($\psi\pi$) (Lo Gullo and Salleo, 1988). Leaves can lose, daily, up to a quarter of their relative water content under conditions of water deficit, with an increase of concentration of cell sap (Salleo and Nardini, 1999). In addition, under conditions of scarce soil moisture, solutes are accumulated and this can bring about a decrease in $\psi\pi$ of over

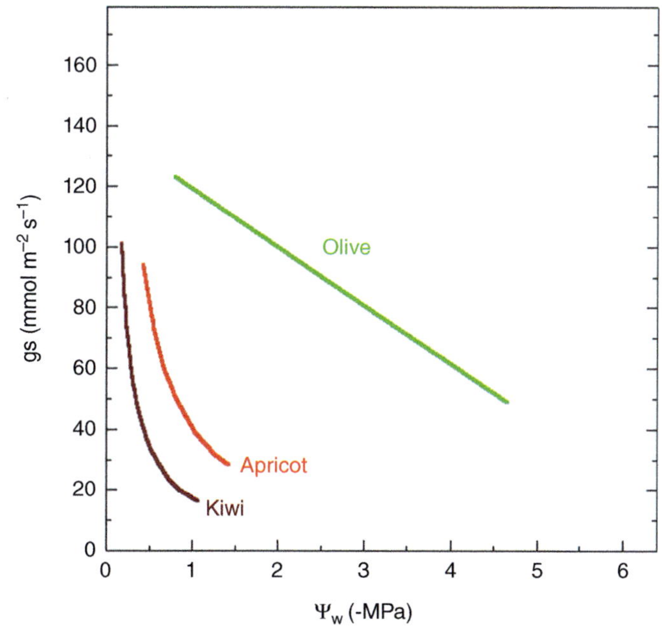

Fig. 2.3.12. The relationship between stomatal conductance (g_s) and pre-dawn leaf water potential (ψ_w) in young kiwi (brown), apricot (red) and olive (green) plants. The relationships have been derived from experimental data.

1.5 MPa. The decrease in ψ_w and $\psi\pi$ involves the maintenance of a high-water-potential gradient between the canopy and the root which is the driving force to extract water at soil water potential as low as a –2.5 MPa, a value much lower than the wilting point for most herbaceous and woody crops.

2.3.3.2 Water uptake by roots

The root system is the plant organ that perceives variation in soil water content and its structure, and distribution in soil is important to understand how plants behave under water-limiting conditions. Water uptake mostly depends on the expansion and depth of the root system. Unfortunately, there are few data available on olive root distribution and activity under different levels of water content in soil. In general, there is a consensus on the fact that the root system of olive trees is well adapted to explore the topsoil layers and take advantage from the scarce and intermittent rainfall events typical of the Mediterranean climate (Searles *et al.*, 2009).

The root system of olive trees is not particularly deep and it is usually comprised in the upper 1 m layer of soil. The root system expands almost horizontally even beyond the canopy projection, but maximum root density is usually found within 1 m of the trunk and 0.4 m depth in trees grown under rainfed conditions. As a result, the high resistance to drought is not necessarily due to a deep root system, which grows deep and explores and extracts water from the deep layers of soil. On the contrary, the sensitivity of olive roots to low O_2 concentrations forces the root system towards growing in superficial layers of soil, and in this way the root system is capable of intercepting soil moisture due to intermittent and scarce rains. Soil texture and fertility, planting density and the irrigation system influence the distribution and functionality of the root system. With localized irrigation, root density is high close to the drippers. Only a few roots can explore the soil to 2–3 m depth. The presence of active roots was also found in soil volumes close to the trunk that could increase the water-absorbing efficiency from rainfall running down the stem. The root apparatus showed also a good phenotypic plasticity to water availability in soil. Root distribution and root activity measurement under dry land conditions or under drip irrigation showed that adult trees adapted their rooting system, by concentrating the roots within the wet soil zones near the drippers. The highest root activity and densities occur in wet soil zones, down to a depth of 0.6 m with roots < 0.5 mm in diameter being the most abundant. Similarly, fine roots (< 1 mm) were positively influenced by the proximity of drippers, around which they were more numerous, with most fine roots in the top 20–30 cm of the soil profile within the crown projection. Less affected by the position of drippers were the coarse roots (1–5 mm). These roots were found at 40–50 cm soil depth and between 50 cm and 80 cm from the trunk. Since all these studies were conducted in low-rainfall areas, this could explain the few fine roots present outside the wetted zone. Rhizotron observations done by Polverigiani *et al.* (2012), in drip-irrigated trees from November to August in central Italy, showed that root extension activity was present in November to mid-March (winter) with a peak of very fine (< 0.2 mm) roots growth in February. Fine roots (> 0.2 mm) growth peaked in April–May, and from mid-June to late August root growth almost ceased.

Most roots are less than 5 mm in diameter and, hence, specialized in water and nutrients (Fernandez and Moreno, 1999). The roots of plants grown under rainfed conditions grow especially in spring, whereas under irrigated conditions growth continues throughout the summer. From the anatomical point of view, the roots of rainfed plants have the transition zone to secondary growth closer to the root apex than irrigated plants (Fernandez *et al.*, 1994).

It has been shown that roots can change the xylem sap flow depending on soil moisture conditions (Fernandez and Moreno, 1999); for example, the root sap flow in non-irrigated plants is greater in deep roots than in superficial ones, but the latter ones are able to transport more sap as soon as they are irrigated. Sap flow occurs in the outermost layers of the xylem in the roots (and in the stem, too) in irrigated plants and in the innermost layers for non-irrigated plants (Fernandez and Moreno, 1999). Olive roots show a marked capability to rehydrate and recover hydraulic conductivity after long periods of drought.

In some tree species with different resistance to water deficit, it has been shown that root hydraulic conductance (L_p) is similar when soil moisture is optimal, but it can be very different under stressed conditions (Rieger, 1995). The decrease in L_p in non-irrigated olive plants is compensated for by osmotic adjustment in the leaf, which allows, in theory, maintainence of a higher turgor pressure even when stomata are open (Rieger, 1995). In *Olea oleaster* seedlings, hydraulic conductance varies depending on the season and in August it decreases by 40% of the values measured in May, then remains stable between August and November when soil moisture returns to optimal levels (Salleo and Nardini, 1999). The irreversible decrease in hydraulic conductance when the water deficit is severe has been attributed to the suberification of cell walls of the cells in the inner cortex, which drastically decreases the radial flow of water and forces the plant to form new roots starting from the end of winter (Salleo and Nardini, 1999).

2.3.3.3 Stomatal regulation

Stomatal characteristics are one of the main factors affecting gas exchange. The olive is hypostomatous with stomatal density ranging from 200 to 700 stomata mm^{-2} depending on the cultivar, leaf position and growing conditions (Bongi *et al.*, 1987b; Fernandez and Moreno, 1999). Stomatal conductance of the olive leaf, which is similar to that of other fruit tree species when grown under conditions of ample water availability, remains relatively high even when water deficit is severe.

Stomata open and close depending on many factors. Stomata are closed in the dark and open as light increases. CO_2 enrichment tends to make stomata close, whereas stomata open when CO_2 concentrations are less than the atmospheric concentration. The degree of stomatal opening depends on soil water availability and vapour pressure between the leaf and the atmosphere (VPD), that is the vapour pressure gradient which accounts for differences in temperature and relative humidity between the leaf and the atmosphere.

A high VPD reduces g_s and increases heterogeneity in stomatal opening (Bongi and Palliotti, 1994). This heterogeneity was found in young leaves depending on the duration of the exposure period to high VPD (Loreto and Sharkey, 1990). Besides affecting stomatal response, VPD is the driving force for the transpiration process (E) out of the leaf, according to the equation:

$$E = g_l \times VPD \qquad [1]$$

where g_l is leaf conductance to water vapour. In general, stomatal opening is slower than closure, but stomatal response to light is more rapid than that to either CO_2 or VPD.

The hairiness of the abaxial surface of the olive leaf determines a high boundary layer. The leaf boundary layer is a thin zone immediately in contact with the leaf surface where the friction between air and surface decreases the speed of the airflow over the surface itself. The wind reduced the boundary layer resistance and, hence, usually evaporation increases proportionally to wind velocity itself. The removal of trichomes involves the increase in g_s and of transpiration rate (*E*). In windy areas the olive canopy may develop asymmetrically with the downwind side growing much more than the upwind side. In extreme cases, canopies grow flag-shaped.

At the whole-canopy level, the boundary layer resistance is very low so that gas exchanges markedly depend on stomatal conductance. This means that leaf temperature is similar to that of the air surrounding it under most circumstances and that the tree leaf surface is coupled to the surrounding environmental conditions. Therefore, canopy conductance is the main endogenous factor determining tree transpiration and crop water use.

2.3.3.4 Transpiration and water use efficiency

The driving force for the water lost by transpiration is the vapor pressure gradient that is determined by the water concentration differences between the atmosphere and the leaf stomatal cavity. In atmospheric conditions of very low relative humidity, a large vapor pressure gradient can lead to a water deficit in the plant, even if the soil water content is well above the permanent wilting point. More generally, the plant water status depends on the soil water content, the root capacity to absorb water, and the hydraulic conductivity of root and shoot tissues (Fig. 2.3.13). Even under optimal conditions the RWC – the actual plant water content as a percentage of the maximum water content at full turgor – of the tissues during the day is lower than 100%. Only at night when the evaporative demand from the leaves is low can the rehydration process take place and the RWC can approach 100%. The *Olea europaea* L. is commonly considered a species with low water needs. In the literature we find average values of transpiration rates ranging from 2.6 to 8.0 x 10^{-5} L m^{-2} s^{-1} = 0.000026-0.00008 L m^{-2} s^{-1}. Fruit transpiration is small and equal to 11% of water consumed by a tree with 31.40 m^2 leaf area and 22 kg of fresh crop load at harvest, assuming an average g_s of 120 mmol m^{-2} s^{-1} (Bongi and Palliotti, 1994). Transpiration rates of olive trees are similar to those of apricot trees, and just slightly less than the grapevine under irrigated conditions and high evapotranspirative demand. Olive transpiration is less than that of the peach tree and equal to only 34% of that of the apple tree under similar environmental conditions. The water consumption of the olive tree has been estimated equal to 790 g m^{-2} leaf area per day. Fernandez and Moreno (1999) report water consumption rates of container-grown young olive trees between 1.0 m^{-2} and 1.7 l m^{-2} per day. These values were confirmed (1.2 l m^{-2} leaf area per day) in studies on field-grown, irrigated trees using non-destructive techniques of measurement. These estimated values are indicative as they depend on environmental conditions, measurement techniques, stage of

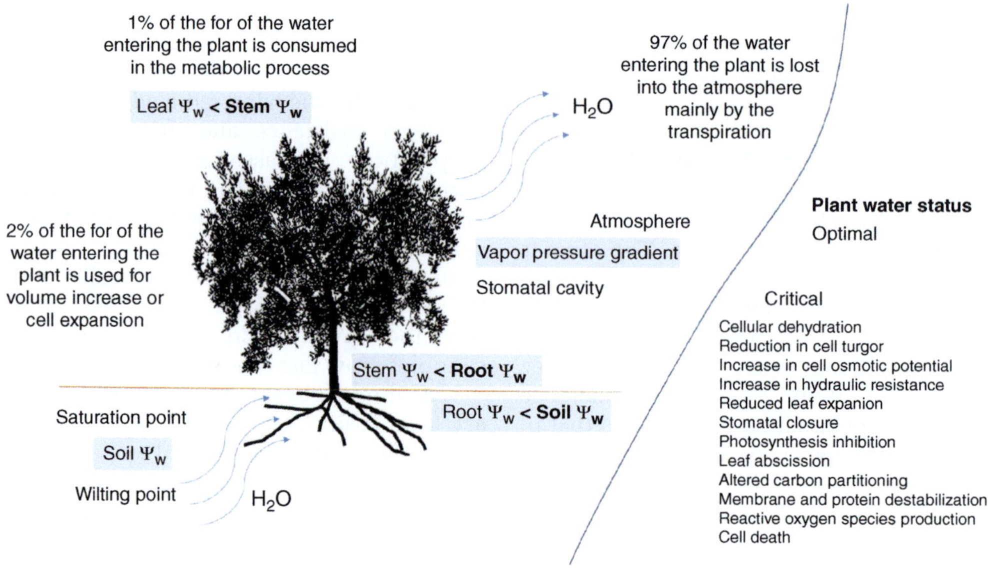

Fig. 2.3.13. Main driving forces for water flow from the soil through the plant to the atmosphere (*left*) and physiological and biochemical changes caused by critical plant water status.

development and canopy density. On a typical summer day, transpiration rates tend to exceed water uptake by the roots during the morning independently of soil water potential. As a result, tree tissues release water to maintain and supply transpiration. Leaves can contribute up to 60% of their water content and attribute to the high capacitance of tissues the capability to maintain transpiration when the evapotranspirative demand by the environment is high. In the afternoon, tissues start recovering their water content since water uptake prevails over transpiration and this opposite trend continues during the night.

Transpiration increases linearly as the stomata open (Fig. 2.3.11). On the other hand, photosynthesis increases up to a certain degree of stomatal opening and then reaches a plateau because of biochemical, rather than stomatal, limitations. As a result, we can identify an optimal range of stomatal opening for water use efficiency that is the quantity of assimilated carbon per unit of water transpired. Water use efficiency estimates reported in the literature are quite variable, but the reported values of 5.5 and 9.6 g CO_2 kg^{-1} H_2O are quite high for C_3 species. As far as dry matter is concerned, Bongi and Palliotti (1994) report that the olive tree is able to produce 3.17 g of fruit dry matter per kg of transpired water, and that value is higher than those measured in *Citrus* (2.46 g) or *Prunus* (1.78 g).

References

Aganchich B., Wahbi, S., Loreto, F. and Centritto, M. (2009) Partial root zone drying: regulation of photosynthetic limitations and antioxidant enzymatic activities in young olive (*Olea europaea*) saplings. *Tree Physiology* 29, 685–696.

Almagro, M., López, J., Boix-Fayos, C., Albaladejo, J. and Martínez-Mena, M. (2010) Belowground carbon allocation patterns in a dry Mediterranean ecosystem: a comparison of two models. *Soil Biology and Biochemistry* 42, 1549–1557. http://dx.doi.org/10.1016/j.soilbio.2010.05.031.

Angelopoulos, K., Dichio, B. and Xiloyannis, C. (1996) Inhibition of photosynthesis in olive trees (*Olea europaea*) during water stress and rewatering. *Journal of Experimental Botany* 47, 1093–1100.

Bacelar, E.A., Correia, C.M., Moutinho-Pereira, J.M., Goncalves, B.C., Lopes, J.I. and Torres-Pereira, J.M.G. (2004) Sclerophylly and leaf anatomical traits of five field-grown olive cultivars growing under drought conditions. *Tree Physiology* 24, 233–239.

Bacelar, E.A., Moutinho-Pereira, J.M., Goncalves, B.C., Ferreira, H.F. and Correia, C.M. (2007) Changes in growth, gas exchange, xylem hydraulic properties and water use efficiency of three olive cultivars under contrasting water availability regimes. *Environmental Experimental Botany* 60, 183–192.

Bertolla, C., Caruso, G. and Gucci, R. (2014) Seasonal changes in soil respiration rates in olive orchards. *Acta Horticulturae* 1057(1), 275–280.

Blanke, M.M. and Lenz, F. (1989) Fruit photosynthesis. *Plant, Cell and Environment* 12, 31–46.

Bongi, G. and Long, S.P. (1987) Light dependent damage to photosynthesis in olive leaves during chilling and high temperature stress. *Plant, Cell and Environment* 10, 241–249.

Bongi, G. and Palliotti, A. (1994) Olive. In: Schaffer, B. and Andersen, P.C. (eds) *Handbook of Environmental Physiology of Fruit Crops*. CRC Press, Boca Raton, Florida, pp. 165–187.

Bongi, G., Mencuccini, M. and Fontanazza, G. (1987a) Photosynthesis of olive leaves: effects of light flux density, leaf age, temperature, peltates, and H_2O vapour pressure deficit on gas exchange. *Journal of the American Society for Horticultural Science* 112, 143–148.

Bongi, G., Soldatini, G.F. and Hubick, K.T. (1987b) Mechanism of photosynthesis in olive tree (*Olea europaea* L.). *Photosynthetica* 21, 572–578.

Carlisle, E.A., Steenwerth, K.L. and Smart, D.R. (2006) Effects of land use on soil respiration: conversion of oak woodlands to vineyards. *Journal Environmental Quality* 35, 1396–1404.

Celano, G., Dichio, B., Montanaro, G., Nuzzo, V., Palese, A.M. and Xiloyannis, C. (1999) Distribution of dry matter and amount of mineral elements in irrigated and non-irrigated olive trees. *Acta Horticulturae* 474, 381–384.

Centritto, M., Wahbi, S., Serraj, R. and Chaves, M.M. (2005) Effects of partial root drying (PRD) on adult olive tree (*Olea europaea*) in field conditions under arid climate. II. Photosynthetic responses. *Agriculture, Ecosystems and Environment* 106, 303–311.

Chartzoulakis, K., Loupassaki, M., Bertaki, M. and Androulakis, I. (2002) Effects of NaCl salinity on growth, ion content, and CO2 assimilation rate of six olive cultivars. *Scientia Horticulturae* 96, 235–247.

Conde, C., Silva, P., Agasse, A., Lemoine, R., Delrot, S., Tavares, R. and Gerós, H. (2007) Utilization and transport of mannitol in *Olea europaea* and implications for salt stress tolerance. *Plant and Cell Physiology* 48(1), 42–53.

Conde C., Delrot, S. and Geros, H. (2008) Physiological, biochemical and molecular changes occurring during olive development and ripening. *Journal of Plant Physiology* 165, 1545–1562.

Fernandez, J.E. and Moreno, F. (1999) Water use by the olive tree. *Journal of Crop Production* 2, 101–162.

Fernandez, J.E., Moreno, F., Martin-Aranda, J. and Rapoport, H.F. (1994) Anatomical response of olive roots to dry and irrigated soils. *Advances in Horticultural Science* 8, 141–144.

Filippou, M., Fasseas, C. and Karabourniotis, G. (2007) Photosynthetic characteristics of olive tree (*Olea europaea*) bark. *Tree Physiology* 27, 977–984.

Flora L.J. and Madore, M.A. (1993) Stachyose and mannitol transport in olive (*Olea europaea* L.). *Planta* 189, 484–490.

Gifford, R.M., Thorne, J.H., Hitz, W.D. and Giaquinta, R.T. (1984) Crop productivity and photoassimilate partitioning. *Science* 225, 801–807.

Gucci, R. (2003) Ecofisiologia. In: Fiorino, P. (ed.) *Olea – Trattato di Olivicoltura*. Edagricole. Edizioni Agricole de Il Sole 24 ORE Edagricole Srl pp 77–89.

Gucci, R. and Cantini, C. (2000) Pruning and training systems for modern olive growing. CSIRO Publishing, Collingwood, Australia. ISBN 0643-06443-5

Gucci, R. and Caruso, G. (2011) Environmental stresses and sustainable olive growing. XXVIII International Horticultural Congress, Lisboa, Portugal, 22–27 August 2010. *Acta Horticulturae* (ISHS) 924, 19–30.

Gucci, R. and Minchin, P.E.H. (2012) Newly-assimilated carbon in the vegetative shoot of olive. *Acta Horticulturae* 586(586), 461–463.

Gucci, R., Corelli Grappadelli, L., Tustin, S. and Ravaglia, G. (1995) The effect of defruiting at different stages of fruit development on leaf photosynthesis of 'Golden Delicious' apple. *Tree Physiology* 15, 35–40.

Gucci, R., Massai, R., Xiloyannis, C. and Flore, J.A. (1996) The effect of drought and vapour pressure deficit on gas exchange of young kiwifruit (*Actinidia deliciosa* var. *deliciosa*) vines. *Annals of Botany* 77, 605–613.

Gucci, R., Lombardini, L. and Tattini, M. (1997) Analysis of leaf water relations of two olive (*Olea europaea*) cultivars differing in tolerance to salinity. *Tree Physiology* 17, 13–21.

Gucci, R., Moing, A., Gravano, E. and Gaudillere, J.P. (1998) Partitioning of photosynthetic carbohydrates in leaves of salt-stressed olive plants. *Australian Journal of Plant Physiology* 25(5), 571–579.

Gucci, R., Grimelli, A., Costagli, G., Tognetti, R., Minnocci, A. and Vitagliano, C. (2002) Stomatal characteristics of two olive cultivars 'Frantoio' and 'Leccino'. *Acta Horticulturae* 586(2), 541–544.

Gucci, R., Lodolini, E.M. and Rapoport, H.F. (2009) Water deficit-induced changes in mesocarp cellular processes and the relationship between mesocarp and endocarp during olive fruit development. *Tree Physiology* 29, 1575–1585.

Gucci, R., Caruso, G. and Sebastiani, L. (2010) Seasonal changes in leaf nitrogen of olive trees grown under different irrigation regimes and crop level. *Journal of Plant Nutrition* 33, 1849–1859.

Hagimiditriou, M. and Pontikis, C.A. (2005) Seasonal changes in CO_2 assimilation in leaves of five major Greek olive cultivars. *Scientia Horticulturae* 104, 11–24.

Jorba, J., Tapia, L., and Sant, D. (1985) Photosynthesis, leaf water potential, and stomatal conduct ance in *Olea europaea* under wet and drought conditions. *Acta Horticulturae* 171, 237–246.

Larcher, W., de Moraes, J.A.P.V. and Bauer, H. (1981) Adaptive responses of leaf water potential, CO2-gas exchange and water use efficiency of *Olea Europaea* during drying and rewatering. In: Margaris, N.S. and Mooney, H.A. (eds) *Components of Productivity of Mediterranean-climate Regions: Basic and Applied Aspects*. Tasks for Vegetation Science, vol. 4. Springer, Dordrecht, The Netherlands. https://doi.org/10.1007/978-94-009-8683-1_12

Lo Gullo, M.A. and Salleo, S. (1988) Different strategies of drought resistance in three Mediterranean sclerophyllous trees growing in the same environmental conditions. *New Phytologist* 108, 267–276.

Loescher, W.H. and Everard, J.D. (1996) Sugar alcohol metabolism in sinks and sources. In: Zamski, E., and Schaffer, A.A. (eds) *Photoassimilate Distribution in Plants and Crops: Source-Sink Relationships*. Marcel Dekker Inc., New York, pp. 185–207.

Loescher W.H., Tyson, R.H., Everard, J.D., Redgwell, R.J. and Bieleski, R.L. (1992) Mannitol synthesis in higher plants: evidence for the role and characterization of a NADPH-dependent mannose-6-phosphate reductase. *Plant Physiology* 98, 1396–1402.

Loreto, F. and Sharkey, T.D. (1990) Low humidity can cause uneven photosynthesis in olive (*Olea europaea* L.) leaves. *Tree Physiology* 6, 409–415.

Loreto, F., Centritto, M. and Chartzoulakis, K. (2003) Photosynthetic limitations in olive cultivars with different sensitivity in salt stress. *Plant, Cell & Environment* 26, 595–601.

Mariscal, M.J., Orgaz, F. and Villalobos, F.J. (2000) Radiation-use efficiency and dry matter partitioning of a young olive (*Olea europaea*) orchard. *Tree Physiology* 20, 65–72.

Meriño, J. (1987) The costs of growing and maintaining leaves of Mediterranean plants. In: Tenhunen, J.D.. Catarino, F.M., Lange, O.L. and Oechel, W.C. (eds) *Plant Response to Stress – Functional Analysis in Mediterranean Ecosystems*. NATO ASI Series, Springer-Verlag, Heidelberg, Germany, pp. 553–564.

Moriana A., Villalobos, F. and Fereres, E. (2002) Stomatal and photosynthetic response of olive (*Olea europaea* L.) leaves to water deficits. *Plant, Cell & Environment* 25, 395–405.

Nadwodnik, J. and Lohaus, G. (2008) Subcellular concentrations of sugar alcohols and sugars in relation to phloem translocation in *Plantago major, Plantago maritima, Prunus persica, and Apium graveolens*. *Planta* 227, 1079–1089.

Oddo, E., Saiano, F., Alonzo, G. and Bellini, E. (2002) An investigation of the seasonal pattern of mannitol content in deciduous and evergreen species of the Oleaceae growing in Northern Sicily. *Annals of Botany* 90, 239–243.

Polverigiani, S., McCormack, M.L., Muller, C. and Eissenstat, D.M. (2012) Growth and physiology of olive pioneer and fibrous roots exposed to soil moisture deficits. *Tree Physiology* 31(11), 1228–1237.

Priestley, C.A. (1977) The annual turnover of resources in young olive trees. *Journal of Horticultural Science* 52, 105–112.

Proietti, P. and Famiani, F. (2002) Diurnal and seasonal changes in photosynthetic characteristics in different olive (*Olea europaea*) cultivars. *Photosynthetica* 40, 171–176.

Proietti, P., Famiani, F. and Tombesi, A. (1999) Gas exchange in olive fruit. *Photosynthetica* 36, 423–432.

Rieger, M. (1995) Offsetting effects of reduced root hydraulic conductivity and osmotic adjustment following drought. *Tree Physiology* 15, 379–385.

Rousseaux, M.C., Benedetti, J.P. and Searles, P.S. (2008) Leaf-level responses of olive trees (*Olea europaea*) to the suspension of irrigation during the winter in an arid region of Argentina. *Scientia Horticulturae* 115, 135–141.

Salleo, S. and Nardini, A. (1999) Ecofisiologia di *Olea oleaster* Hoffmgg. et Link: verso un modello predittivo dell'adattamento all'aridità. *Olivo e Olio* 2(4), 70–79.

Sanchez, J. and Harwood, J.L. (2002) Biosynthesis of triacylglycerols and volatiles in olives. *European Journal Lipid Science and Technology* 104, 564–573.

Scandellari, F., Caruso, G., Liguori, G., Meggio, F., Palese, A.M. *et al.* (2016) A survey of carbon sequestration potential of orchards and vineyards in Italy. *European Journal of Horticultural Science* 81(2), 106–114.

Schlesinger, W.H. and Andrews, J.A. (2000) Soil respiration and the global carbon cycle. *Biogeochemistry* 48, 7–20.

Searles, P., Saravia, D.A. and Rousseaux, C.A. (2009) Root length density and soil water distribution in drip-irrigated olive orchards in Argentina under arid conditions. *Crop and Pasture Science* 60(3), 280–288.

Sebastiani, L., Scebba, F., Grimelli, A. and Gucci, R. (2002) L'effetto di brevi periodi di congelamento sulla resistenza al freddo di giovani piante di olivo. *Atti del Convegno Internazionale di Olivicoltura*. Spoleto, 22–23 April, pp. 381–386.

Shackel, K., Moriana, A., Marino, G., Corell, M., Pérez-López, D. *et al.* (2021) Establishing a reference baseline for midday stem water potential in olive and its use for plant-based irrigation management. *Frontiers in Plant Science* 12, 791711. DOI: 10.3389/fpls.2021.791711

Sofo, A., Nuzzo, V., Palese, A.M., Xiloyannis, C., Celano, G., Zukowsky, P. and Dichio, B. (2005) Net CO_2 storage in Mediterranean olive and peach orchards. *Scientia Horticulturae* 107, 17–24.

Stoop, J.M.H. and Pharr, D.M. (1993) Effect of different carbon sources on relative growth rate, internal carbohydrates, and mannitol 1-oxido-reductase activity in celery suspension cultures. *Plant Physiology* 103, 1001–1008.

Stoop, J.M.H., Williamson, J.D. and Pharr, D.M. (1996) Mannitol metabolism in plants: a method for coping with stress. *Trends in Plant Science* 1, 139–144.

Syvertsen, J.P., Goñi, C. and Otero, A. (2003) Fruit load and canopy shading affect leaf characteristics and net gas exchange of 'Spring' navel orange trees. *Tree Physiology* 23, 899–906.

Tattini, M., Gucci, R., Romani, A., Baldi, A. and Everard, J.D. (1996) Changes in non-structural carbohydrates in olive (*Olea europaea*) leaves during root zone salinity stress. *Physiologia Plantarum* 97, 117–124.

Testi, L., Orgaz, F. and Villalobos, F.J. (2008) Carbon exchange and water use efficiency of a growing, irrigated orchard. *Environmental and Experimental Botany* 63, 168–177.

Thompson, R.G., Tyree, M.T., Lo Gullo, M.A. and Salleo, S. (1983) The water relations of young olive trees in a Mediterranean winter: measurements of evaporation from leaves and water conduction in wood. *Annals of Botany* 52, 399–406.

Tognetti, R., d'Andria, R., Morelli, G. and Alvino, A. (2005) The effect of deficit irrigation in seasonal variations of plant water use in *Olea europaea* L. *Plant and Soil* 273, 139–155.

Tognetti, R., d'Andria, R., Sacchi, R., Lavini, A., Morelli, G. and Alvino, A. (2007) Deficit irrigation affects seasonal changes in leaf physiology and oil quality of *Olea europaea* (cultivars Frantoio and Leccino). *Annals of Applied Botany* 150, 169–186.

Tognetti, R., Sebastiani, L., Minnocci, A., Vitagliano, C. and Raschi A. (2002) Foliar responses of olive trees (*Olea Europaea* L.) under field exposure to elevated CO_2 concentration. *Acta Horticulturae* 586, 449–452.

Tugendhaft, I., Eppel, A., Kerem, Z., Barazani, O., Ben-Gal, A., Kadereit, J. and Dag, A. (2016) Drought tolerance of three olive cultivars alternatively selected for rain fed or intensive cultivation. *Scientia Horticulturae* 199, 158–162.

Villalobos, F.J., Testi, L., Hidalgo, J., Pastor, M. and Orgaz, F. (2006) Modelling potential growth and yield of olive (*Olea europaea* L.) canopies. *European Journal of Agronomy* 13, 1155–1163.

Weil, R.R. and Brady, N.C. (2017) *The Nature and Properties of Soils*, 15th edn. Pearson Press, Upper Saddle River, New Jersey.

Wodner, M., Lavee, S. and Epstein, E. (1988) Identification and seasonal changes of glucose, fructose and mannitol in relation to oil accumulation during fruit development in *Olea europaea* (L.). *Scientia Horticulturae* 36, 47–54.

3.1 Olive Genome Structure, Evolution and Domestication

A. Bombarely[1], J. Friel[1] and F. Luque[2]
[1]*Department of Bioscience, Università degli Studi di Milano, Milan, Italy;*
[2]*Center for Advanced Studies in Olive Grove and Olive Oils, Genomics Unit, University of Jaén, Jaén, Spain.*

3.1.1 The Ploidy and Size of the Nuclear Olive Genome

The size of the nuclear genome of the cultivated olive trees has not yet been well determined, even if genomes of numerous genotypes have been assembled and published (Cruz *et al.*, 2016; Unver *et al.*, 2017; Jimenez-Ruiz *et al.*, 2020; Rao *et al.*, 2021; Vatansever *et al.*, 2022). Our estimation of the size of the haploid genome of the 'Picual' variety using Kmer analysis ranged between 1.63 Gb and 1.81 Gb (Jiménez-Ruiz *et al.*, 2020). The haploid genome size of other varieties has been estimated using different methods and ranged from 1.48 Gb for the 'Cobrançosa' variety (Loureiro *et al.*, 2007) to 2.21 Gb for the 'Leccino' variety (Rugini *et al.*, 1996; de la Rosa *et al.*, 2003). These discrepancies in size are probably due to differences in the methodologies used. However, the possibility of real differences in the nuclear genome size between varieties cannot be discarded, because, as we will see below, the olive genome is somewhat unstable, with chromosome rearrangements in the form of duplications, deletions and translocations. Indeed, a number of major alterations are known to have occurred during the evolution of the *O. europaea* L. genome, notably chromosomal rearrangements, have been observed within chromosome pairs I, V and VII (Minelli *et al.*, 2000). To have a more comprehensive view of the cultivated olive genome, more varieties should be sequenced and their genomes assembled. Indeed, this goal likely will be addressed in the coming years.

The cultivated and wild olive trees are considered varieties of the same subspecies, *O. europaea* L. subsp. *europaea*, var. *europaea* and var. *sylvestris*, respectively. In addition to this, *O. europaea* L. is known to consist as a complex of six subspecies and at least one hybrid; *O. europaea* L. subsp. *europaea*; subsp. *laperrinei*; subsp. *guanchica*; subsp. *cerasiformis*; subsp. *maroccana*; subsp. *cuspidata*; with possible hybrids between subsp. *europaea* and subsp. *cuspidata* (Besnard *et al.*, 2009). Most of these subspecies are diploid, except for the subsp. *cerasiformis*, which is a tetraploid, and the subsp. *maroccana*, which is a hexaploid (Besnard *et al.*, 2008). The *O. europaea* L. has a large diploid chromosome number (2n=46) (Minelli *et al.*, 2000), resulting from ancient whole-genome duplications (WGD), which will be described later. Chromosomal rearrangements and the complex genomic history of the olive account for some of the difficulties encountered by different groups attempting to assemble high-quality olive genomes.

DOI: 10.1079/9781789247350.0005

3.1.2 The Olive Genome and the Number of Genes

The first olive genome sequenced (Oe6) was the variety 'Farga' (Cruz *et al.*, 2016). This genome was improved four years later to the new version Oe9 (Julca *et al.*, 2020). This latest version anchors the assembly to chromosomes using the previously published genetic map (Unver *et al.*, 2017). A putative wild genome (Oe451) was assembled by Unver *et al.* (2017) in '*sylvestris*' background, where the authors also anchored their scaffolds to chromosomes. However, some evidence suggests that it is not a sample of *sylvestris* but a feral form of a cultivated variety (Julca *et al.*, 2020); though it is important to note that to distinguish between real wild olives and feral forms is problematic. Indeed, there is evidence that not all apparent wild individuals are true wild trees (Belaj *et al.*, 2007). In both genome assemblies the gene number was not widely different, 56,349 coding sequences (cds) in Oe6 and 50,684 in Oe451, and in both cases the size of the assembled genome was lower than 1.4 Gb (Table 3.1.1).

The third assembled genome available corresponds to the 'Picual' variety and was obtained using PacBio long-sequencing technology (Jiménez-Ruiz *et al.*, 2020). The assembly of the *Olea europaea* 'Picual' genome (Oleur061 assembly) was based on the integration of two sequencing technologies, Illumina short reads and PacBio long reads. Illumina libraries were sequenced delivering 623.28 Gb of paired-end (with an insert size of ~300 bp) and 363.59 Gb of three mate-pair (with an insert size of ~5 Kb, ~8Kb and ~10 Kb). The 'Picual' genome size was estimated using the Li *et al.* (2010) approach, and ranged from 1.63 Gb to 1.81 Gb. Additionally, 75 PacBio Smart Cells were sequenced from the same samples, yielding 68.30 Gb of long reads with an average size of 12.06 Kb. The PacBio long reads were the basis for the genome assembly of the Oleur061 genome. This technology allowed detection of highly similar duplicated sequences that would otherwise have been wrongly collapsed using just short reads. In addition to the PacBio long reads, the tool Pseudohaploid was applied and no haploid contig redundancy was

Table 3.1.1. Olea genome assemblies statistics.

Assembly stats	Oe451 GCA_002742605.1 (Unver *et al.*, 2017)	Oe6 GCA_900603015.1 (Cruz *et al.*, 2016)	Oe9 GCA_902713445.1 (Julca *et al.*, 2020)	Oleur0.6.1 (Jiménez-Ruiz *et al.*, 2020)	(Rao *et al.*, 2021)
Assembly size (Gb)	1.14	1.32	1.31	1.68	1.10**
Scaffolds	41256	11038	9753	9174	
Longest sequence (bp)	46026434	2580346	36438822	4142698	68066766**
Shortest sequence (bp)	452	500	500	1017	
Average sequence length (bp)	27688.5	119464.7	135002.2	183123.7	
N90, number of sequences	3410	3099	2019	4503	
L90 (bp)	22596	110965	116062	86918	279924**
N50, number of sequences	23	901	162	1145	
L50 (bp)	12567911	443100	734380	410451	42601851
% BUSCO complete*	85.9	96.6	96.6	96.6	94.2**
% BUSCO duplicated*	13.4	51.5	23.6	23.6	20.4**
% BUSCO fragmented*	5.9	1.9	2	1.7	2.4**
% BUSCO missing*	8.2	1.5	1.4	1.5	3.4**
Variety/wild plant	*sylvestris*	FARGA	FARGA	PICUAL	ARBEQUINA

*Results obtained using BUSCO v.5 with the eudicots_odb10 data set.
**Reported stats, data was not available for analysis at the time of writing.

detected in the Oleur061 assembly. The final assembly was 1.68 Gb, larger than the two previous olive genomes, and contained a higher number of gene models, 78,079 (Jimenez-Ruiz *et al.*, 2020).

In 2021, the olive genome of cv. 'Arbequina' was published, using a combination of Oxford Nanopore and HiC, delivering 23 pseudomolecules with an assembly size of 1.1 Gb, accounting for the 82% of the estimated genome size. The genome was annotated with 53,318 coding-protein gene models (Rao *et al.*, 2021). A summary of each assembly including size and completeness can be viewed in Table 3.1.1. The completeness of each of the assemblies was recently contextualized for variety genotyping using whole-genome sequencing (WGS) and genotyping-by-sequencing (GBS) approaches pointing to the Oleur061 as the most complete assembly although lack of anchoring to chromosomes may limit some of its applications (Friel *et al.*, 2021).

More recently, the cv. 'Ayavalik' was sequenced with Illumina short reads, assembled with CLC Genomics Workbench and re-scaffolded into chromosomes using RaGOO (Vatansveer *et al.*, 2022). Although no standardized QC values were supplied, this new pseudo-assembly was used to successfully study the variation of the functional gene space across the different varieties with assembled genomes.

A full understanding of the olive genome involves the interpretation of the genomes in the context of their genetic diversity. This requires that the genome of many olive cultivars should be assembled, due to the great diversity among varieties, and between cultivated and wild olives, in terms of genome size and number of duplicated genes. In fact, our data indicate that the Picual genome (Oleur061 assembly) could be larger and contain a high number of genes than a supposed wild olive genome (Oe451 assembly). Illumina reads of four re-sequenced wild genomes, two of subsp. *europaea* and two of subsp. *guanchica* (Wild 5, Wild 6, Wild 7 and Wild 9), were mapped to the Oleur061 genome uncovering 175.4 ± 25.3 Mb and 8941 ± 805 genes in 'Picual' that were not present in the wild genome (Jimenez-Ruiz *et al.*, 2020). This point will be clarified once there are more wild and cultivated genomes assembled using long inserts from PacBio or Oxford Nanopore technologies. At present, the whole-genome sequencing data for 50 more genomes, 40 of cultivated olive varieties and 10 of putative wild trees, are available (https://genomaolivar.dipujaen.es/db/index.php) (Table 3.1.2). The genome resequencing approach was also used to identify the origin of the monumental Vouves olive tree stating that non-grafted part of the tree is one of the oldest domesticated living trees sequenced to date (Bombarely *et al.*, 2021).

3.1.3 The Evolution of the Olive Genome

Plants and vertebrate animals have by far the largest genomes compared to all other organisms. Plants tend to have exceedingly complex genomes; this is in part due to high levels of chromosomal plasticity and their ability to tolerate aberrant chromosome numbers and large chromosome rearrangements. Polyploidy is also a frequent and recurring source of diversity amongst angiosperms, with several ancestral WGD events identified in all major clades (Soltis *et al.*, 2009; Van de Peer *et al.*, 2009). Although both cultivated and wild olive trees are mostly diploid, they have a complex evolutionary history.

What we know about the olive genome evolution is that it seems to have undergone at least two ancestral WGD events. Distribution of synonymous substitutions per synonymous site for duplicated genes and duplicates retained only in collinear regions showed two clear peaks of duplicates in the *sylvestris* background (Oe451) (Unver *et al.*, 2017) and in 'Picual' (Oleur061) genomes (Jiménez-Ruiz *et al.*, 2020; Fig. 3.1.1). Modern *Oleaceae* species, such as *Olea* and *Fraxinus*, probably share a common ancestral allotetraploid hybrid ancestor that resulted from the hybridization of an ancestral *Fontanesia* related species with 12 pairs of chromosomes and an ancestral *Jasminum-Forsythia* species with 11 pairs of chromosomes (Taylor, 1945). This event was previously dated between 34

Table 3.1.2. *Olea europaea* re-sequenced genotypes. (From World Olive Germplasm Collection of IFAPA, Córdoba, Spain)

Varieties-accessions	Origin
Abbadi Abou Gabra-842	Syria
Abou Satl Mohazam	Syria
Abou Kanani	Syria
Arbequina	Spain
Barnea	Israel
Barri	Syria
Chemlal de Kabylie	Algeria
Dokkar	Turkey
Forastera de Tortosa	Spain
Frantoio	Italy
Grappolo	Italy
Hojiblanca	Spain
Jabali	Syria
Kalamon	Greece
Klon-14-1812	Albania
Koroneiki	Greece
Leccino	Italy
Lechín de Sevilla	Spain
Llumeta	Spain
Maarri	Syria
Manzanilla de Sevilla	Spain
Manzanillera de Huercal Overa	Spain
Mari	Syria
Mastoidis	Greece
Mavreya	Greece
Majhol-1013	Syria
Majhol-152	Syria
Megaritiki	Greece
Menya	Spain
Morrut	Spain
Myrtolia	Greece
Ocal	Spain
Picudo	Spain
Piñonera	Spain
Royal	Spain
Shengeh	Iran
Temprano	Spain
Uslu	Turkey
Verdial de Velez-Malaga-1	Spain
Zarza	Spain
Wild trees	
Wild 1 (W1R198) (subsp. *europaea*)	Croatia
Wild 2 (W2R74) (subsp. *europaea*)	Spain
Wild 3 (W3R78) (subsp. *europaea*)	Spain (Menorca island)
Wild 4 (W4R183) (subsp. *europaea*)	Spain
Wild 5 (W5R121) (subsp. *guanchica*)	Spain (Gran Canaria island)
Wild 6 (W6R1048) (subsp. *guanchica*)	Spain (Tenerife island)
Wild 7 (W7R224) (subsp. *europaea*)	Morocco
Wild 8 (W8R225) (subsp. *europaea*)	Morocco
Wild 9 (W9R302) (subsp. *europaea*)	Albania
Wild 11 (W11R37) (subsp. *europaea*)	Spain

and 65 million years ago (mya). The recent genomic analysis has revealed a more probable peak for a WGD event at 59 (between 57 and 63 mya) and 62 mya for the wild/feral 'sylverstris' and 'Picual' backgrounds, respectively. Both estimations are quite similar and indicate that this tetraploid appeared very close to the cataclysm at the end of the Cretaceous period. The polyploidy events seem to be favored by global extinction events and can be linked with periods of extreme environmental changes or cataclysms (Van de Peer *et al.*, 2017). In fact, at that time, independent WGD events have likely occurred in other plant groups such as *Solanales* (Vanneste *et al.*, 2014), a close relative of *Oleaceae*.

Published *O. europaea* genomes, as well as that of another *Oleaceae* species, *Fraxinus excelsior* (Sollars *et al.*, 2017), supported the hypothesis of a latter duplication event, with a peak around 0.25 in the estimation synonymous substitution rate (Ks) for gene pairs. Again, the estimation of the WGD event was very close in the analysis of both the wild/feral 'sylvestris' and 'Picual' genomes (28 and 25 mya, respectively). This is in line with the estimated age of speciation of the *Olea-Fraxinus* ancestor.

In addition to these two ancestral WGD events towards the edge of the Cretaceous period, the 'Picual' variety appears to have undergone a very recent partial genome duplication (PGD) event (Jiménez-Ruiz *et al.*, 2020). The absence of this observation in the cv. 'Farga' (Oe6/OLEA9) and 'sylvestris' (Oe4561) genomes might be due to the lack of long-read sequencing used in these assemblies. As the PGD is so recent, most duplicated coding sequences will likely still be so

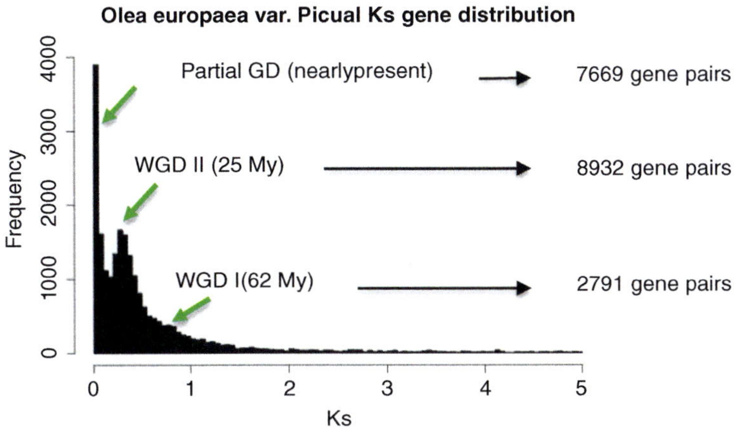

Fig. 3.1.1. Gene duplication analysis of 'Picual' genome. Gene duplication analysis showing putative partial or whole-genome duplication events. Ks, synonymous substitution rate

similar that they are undistinguishable from allelic sequences, resulting in them being collapsed during the assembly of these two previous genomes. The paralog genes, even those that have high sequence similarity, may be regulated in a very different manner, including silencing some of them (unpublished results; Faize *et al.*, 2020). This very recent PGD appears as duplications and chromosomal rearrangements throughout the genome. A source for this chromosomal instability may be an extensive transposon element (TE) activity detected in wild and cultivated olive trees. In fact, LTR retrotransposons have impacted the olive genome to a higher extent than in other angiosperms (Barghini *et al.*, 2015). In some cultivated varieties, such as 'Picual', TE activity was especially high. The analysis of the TE activity revealed a very recent activation, coinciding with the end of the last Ice Age, around 10,000 years ago, and the beginning of olive domestication (Jimenez-Ruiz *et al.*, 2020; Fig. 3.1.2). The high transposon activity may have provided a source of variability and new mutations in nearby genes (Kidwell and Lisch, 1997) that may have been positively selected during olive domestication. We found eight genes surrounded by these dynamic TEs that were associated with reproduction, photosynthesis, seed development and oil production (Jiménez-Ruiz *et al.*, 2020). Another set of genes nearby these TEs were annotated to abiotic and biotic stress resistance. Potentially, these genes have provided farmers with high phenotypic diversity during early domestication.

3.1.4 The Domestication Process and the Origin of Varieties

The domestication of the olive tree is thought to have begun around 6000 years ago in the Levant area of the Mediterranean basin (Kaniewski *et al.*, 2012; Besnard *et al.*, 2013). They also have a long juvenile phase of around 15–20 years (García-López *et al.*, 2014), making sexual reproduction a very slow process, which makes it difficult to obtain new varieties. This produces a net result of thousands of years of domestication appearing to be very recent on a generational timescale relative to annual plants (Diez *et al.*, 2015). Despite the difficulties, there are hundreds of cultivated olive tree varieties. These varieties are vegetatively propagated to avoid loss of their desired traits, as descendants segregate in an extremely wide range of phenotypes, many of them becoming more susceptive to disease, low- or no-fruit production and are no longer desirable for agriculture. Thus, the use of vegetative propagation has been necessary for maintenance and spread of superior individuals while at the same time decreasing the

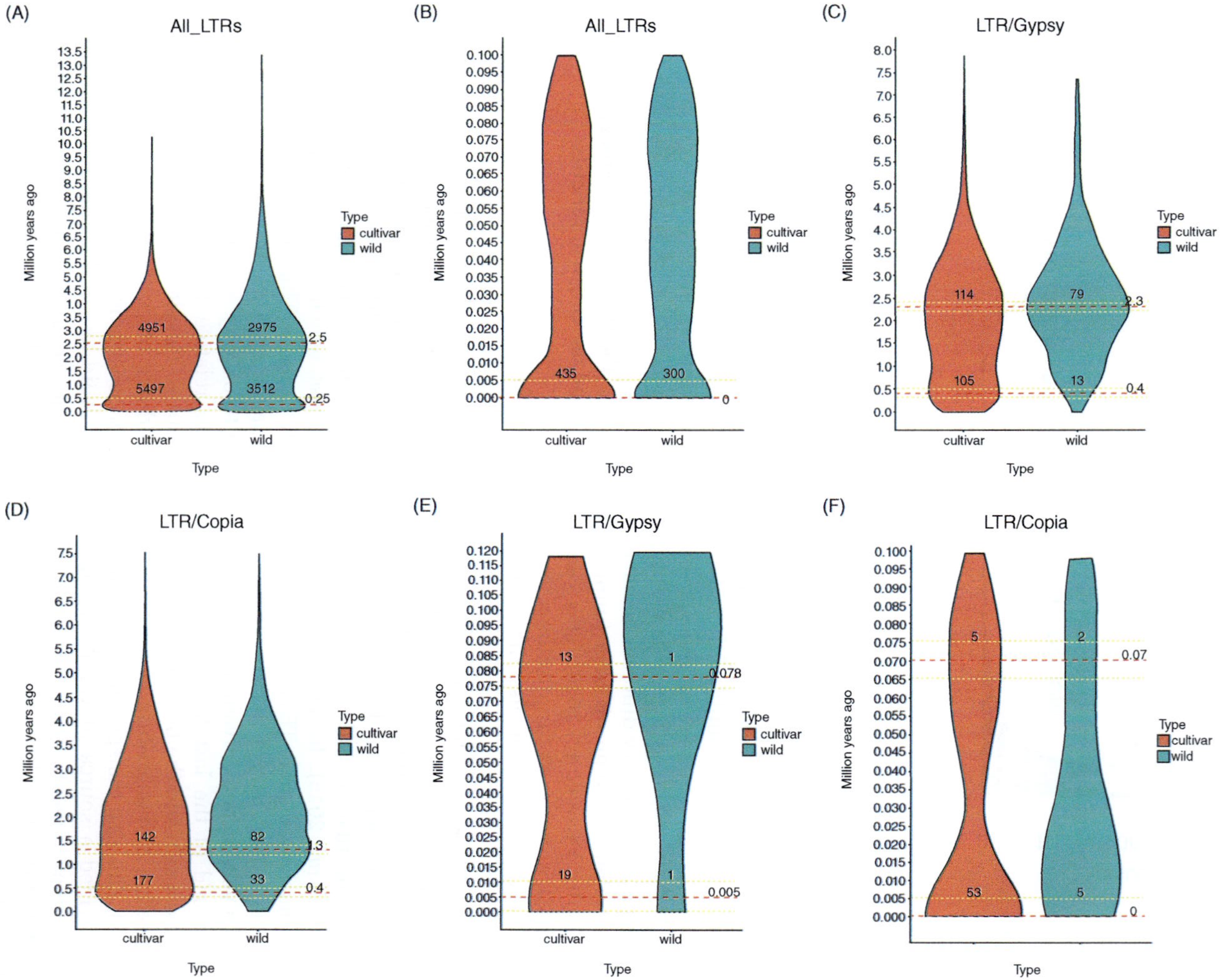

Fig. 3.1.2. Long terminal repeat (LTR) expansion analysis in 'Picual' variety and wild olive trees. All LTRs expansion in the last 13 my (A) and in detail of the last 100,000 years (B). LTR/'Gypsy' expansion in the last 8 my (C) and the last 100,000 years (E). LTR/ 'Copia' expansion in the last 7.5 my (D) and the last 100,000 years (F).

number of sexual cycles (Mckey and Elias, 2010; Miller and Gross, 2011).

The domestication process produced a bottleneck in the initial stages and a loss of genetic variability (Diez *et al.*, 2015; Jimenez-Ruiz *et al.*, 2020; Julca *et al.*, 2020). However, cultivated varieties have recovered a great deal of genetic variability. Although TEs activity may produce genetic variability, they cannot produce the recovery of such sequence variability that we have observed between varieties. The most likely explanation is that high levels of admixture and introgression of wild sequences has produced, in a few generations of sexual reproduction, the recovery of large amounts of the genetic variability. In fact, there is evidence of such genetic admixture and introgression with wild olives (Diez *et al.*, 2015; Gros-Balthazard *et al.*, 2019; Julca *et al.*, 2020).

There is still an open debate around the question of whether olive domestication occurred as a single event or if there were two, or indeed multiple, domestication events. While transcriptomic analysis and plastid markers seem to fit better with a single domestication event and subsequent admixture with wild olives (Gros-Balthazard *et al.*, 2019; Julca *et al.*, 2020), the occurrence of such large numbers of nuclear variants are not so clearly explained and may fit more with the two domestication events hypothesis (Jiménez-Ruiz *et al.*, 2020; Julca *et al.*, 2020). Phylogenetic trees, with robust bootstrapping values, based on randomly selected nuclear markers, consistently show two clearly separated branches in cultivated olives (Fig. 3.1.3). This is compatible with two major domestication events, at it should be understood that the domesticated olives derivate from two different genetic pools. This is consistent with the described primary origin from the Levant (Besnard *et al.*, 2018). However, it might be also explained by a primary domestication event and a

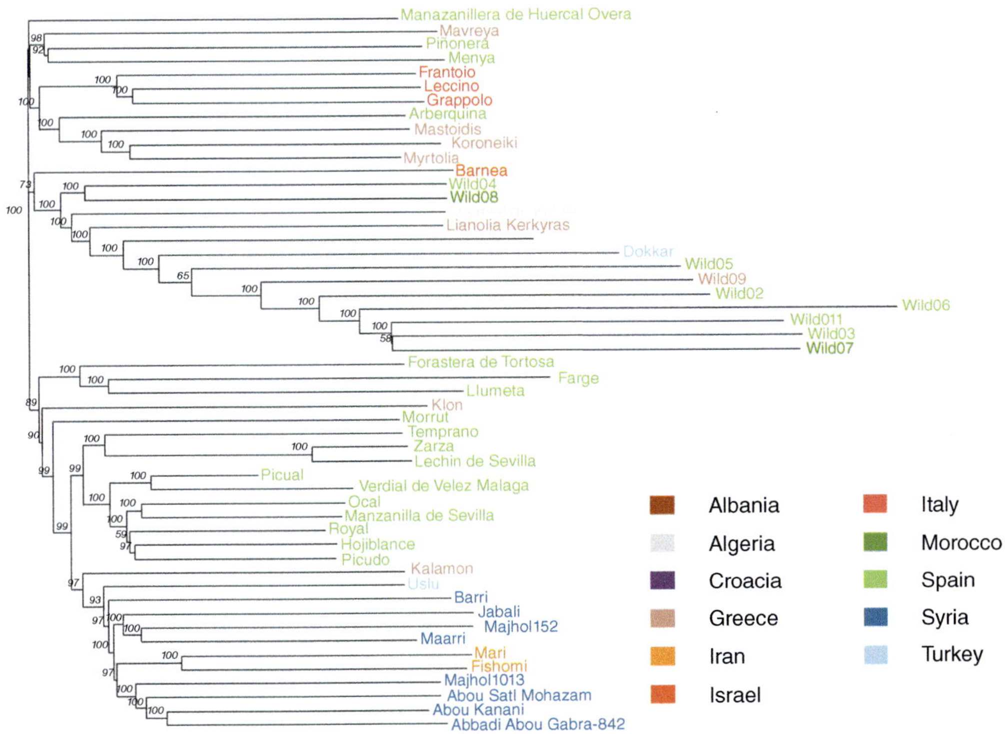

Fig. 3.1.3. Rooted by outgroup Maximum Likelihood phylogenetic tree (hierarchical view) of wild and cultivated olives. The tree was made with PhyML using 100,000 random SNPs. Bootstrap values represent the percentage of 100 replicates.

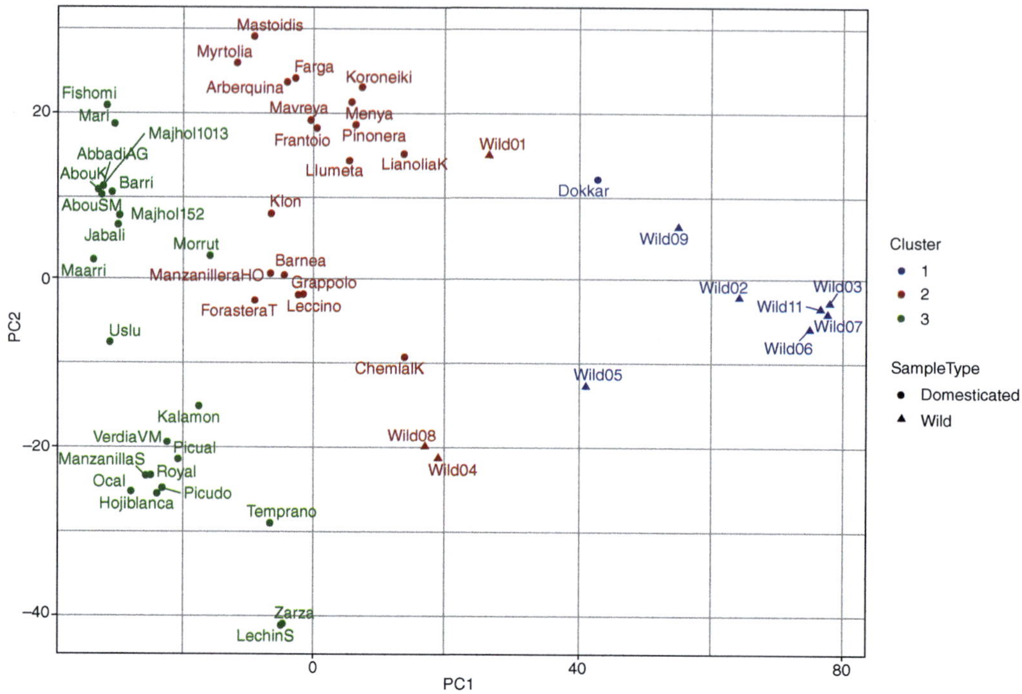

Fig. 3.1.4. Principal component analysis (PCA) of sequenced genomes of cultivated and wild trees.

subsequent split into two large groups. As observed in Fig. 3.1.3, one main branch is comprised of varieties from the north Mediterranean side, with the exception of southern Spain. These Spanish varieties form part of another branch with the south Mediterranean varieties, Fig. 3.1.4. This is in agreement with the Diez *et al.* (2015) proposal of two alternative domestication scenarios in which there may have been two main domestications events, but this still does not exclude the possibility of a single event, followed by later diversification into two main groups after repetitive admixture with the wild olives during geographical migration. The complex origins make it difficult to clarify the history of olive domestication and migration.

References

Barghini, E., Natali, L., Giordani, T., Cossu, R.M., Scalabrin, S. *et al.* (2015) LTR retrotransposon dynamics in the evolution of the olive (*Olea europaea*) genome. *DNA Research* 22(1), 91–100.

Belaj, A., Muñoz-Diez, C., Baldoni, L., Porceddu, A., Barranco, D. and Satovic, Z. (2007) Genetic diversity and population structure of wild olives from the North-Western Mediterranean assessed by SSR markers. *Annals of Botany* 100(3), 449–458.

Besnard, G., Garcia-Verdugo, C., Rubio De Casas, R., Treier, U.A., Galland, N. and Vargas, P. (2008) Polyploidy in the olive complex (*Olea europaea*): evidence from flow cytometry and nuclear microsatellite analyses. *Annals of Botany* 101, 25–30.

Besnard, G., Rubio de Casas, R., Christin, P.A. and Vargas, P. (2009) Phylogenetics of Olea (Oleaceae) based on plastid and nuclear ribosomal DNA sequences: tertiary climatic shifts and lineage differentiation times. *Annals of Botany* 104(1), 143–160.

Besnard, G., Khadari, B., Navascués, M., Fernández-Mazuecos, M., El Bakkali, A., *et al.* (2013) The complex history of the olive tree: from Late Quaternary diversification of Mediterranean lineages

to primary domestication in the northern Levant. *Proceedings of the Royal Society B: Biological Sciences* 280(1756), 20122833.
Besnard, G., Terral, J.-F. and Cornille, A. (2018) On the origins and domestication of the olive: a review and perspectives. *Annals of Botany* 121(3), 385–403.
Bombarely, A., Doulis, A.G., Lambrou, K.K., Zioutis, C., Margaritis, E. and Koubouris, G. (2021) Elucidation of the origin of the Monumental Olive Tree of Vouves in Crete, Greece. *Plants* 10, 2374.
Cruz, F., Julca, I., Gómez-Garrido, J., Loska, D., Marcet-Houben, M., *et al.* (2016). Genome sequence of the olive tree, *Olea europaea. GigaScience* 5, 29.
de la Rosa, R., Angiolillo, A., Guerrero, C., Pellegrini, M., Rallo, L. *et al.* (2003) A first linkage map of olive (*Olea europaea* L.) cultivars using RAPD, AFLP, RFLP and SSR markers. *Theoretical and Applied Genetics* 106, 1273–1282.
Diez, C.M., Trujillo, I., Martinez-Urdiroz, N., Barranco, D., Rallo, L. *et al.* (2015) Olive domestication and diversification in the Mediterranean Basin. *New Phytologist* 206(1), 436–447.
Faize, M., Fumanal, B., Luque, F., Ramírez-Tejero, J.A., Zou, Z., *et al.* (2020) Genome wild analysis and molecular understanding of the Aquaporin diversity in olive trees (*Olea europaea* L.). *International Journal of Molecular Sciences* 2020(21), 4183.
Friel, J., Bombarely, A., Fornell, C.D., Luque, F. and Fernández-Ocaña, A.M. (2021) Comparative analysis of genotyping by sequencing and whole-genome sequencing methods in diversity studies of *Olea europaea* L. *Plants* 10(11), 2514.
García-López, M.C., Vidoy, I., Jiménez-Ruiz, J., Muñoz-Mérida, A., Fernández-Ocaña, A., de la Rosa, R. *et al.* (2014) Genetic changes involved in the juvenile-to-adult transition in the shoot apex of *Olea europaea* L. occur years before the first flowering. *Tree Genetics & Genomes* 10(3), 585–603.
Gros-Balthazard, M., Besnard, G., Sarah, G., Holtz, Y., Leclercq, J. *et al.* (2019) Evolutionary transcriptomics reveals the origins of olives and the genomic changes associated with their domestication. *Plant Journal* 100(1), 143–157.
Jiménez-Ruiz, J., Ramírez-Tejero, J.A., Fernández-Pozo, N., Leyva-Pérez, M., Yan, H., *et al.* (2020) Transposon activation is a major driver in the genome evolution of cultivated olive trees (*Olea europaea* L.). *Plant Genome* 2020, e20010.
Julca, I., Marcet-Houben, M., Cruz, F., Gómez-Garrido, J., Gaut B.S. *et al.* (2020) Genomic evidence for recurrent genetic admixture during the domestication of Mediterranean olive trees (*Olea europaea* L.). *BMC Biology* 18, 148.
Kaniewski, D., Van Campo, E., Boiy, T., Khadari, B. and Besnard, G. (2012) Primary domestication and early uses of the emblematic olive tree: palaeobotanical, historical and molecular evidence from the Middle East. *Biological Reviews* 87(4), 885–899.
Kidwell, M.G. and Lisch, D. (1997) Transposable elements as sources of variation in animals and plants. *Proceedings of the National Academy of Sciences USA* 94, 7704–7711.
Li, R., Fan, W., Tian, G., Zhu, H., He, L. *et al.* (2010) The sequence and de novo assembly of the giant panda genome. *Nature* 463, 311–317.
Loureiro, J., Rodriguez, E., Costa, A. and Santos, C. (2007) Nuclear DNA content estimations in wild olive (*Olea europaea* L. ssp. *europaea* var. *sylvestris* Brot.) and Portuguese cultivars of *O. europaea* using flow cytometry. *Genetic Resources and Crop Evolution* 54(1), 21–25.
Mckey, D. and Elias, M. (2010) The evolutionary ecology of clonally propagated domesticated plants. *New Phytologist* 186(2), 318–332.
Miller, A.J. and Gross, B.L. (2011) From forest to field: perennial fruit crop domestication. *American Journal of Botany* 98(9), 1389–1414.
Minelli, S., Maggini, F., Gelati, M.T., Angiolillo, A. and Cionini, P.G. (2000) The chromosome complement of *Olea europaea* L.: characterization by differential staining of the chromatin and in-situ hybridization of highly repeated DNA sequences. *Chromosome Research* 8, 615–619.
Rao, G., Zhang, J., Liu, X., Lin, C., Xin, H., Xue, L. and Wang, C. (2021) De novo assembly of a new *Olea europaea* genome accession using nanopore sequencing. *Horticulture Research* 8(1), 1–12.
Rugini, E., Pannelli, G., Ceccarelli, M. and Muganu, M. (1996) Isolation of triploid and tetraploid olive (*Olea europaea* L.) plants from mixoploid cv. 'Frantoio' and 'Leccino' mutants by in vivo and in vitro selection. *Plant Breeding* 115, 23–27.
Sollars, E.S.A., Harper, A.L., Kelly, L.J., Sambles, C.M., Ramirez-Gonzalez, R.H. *et al.* (2017) Genome sequence and genetic diversity of European ash trees. *Nature* 541, 212–216.
Soltis, D.E., Albert, V.A., Leebens-Mack, J., Bell, C.D., Paterson, A.H. *et al.* (2009) Polyploidy and angiosperm diversification. *American Journal of Botany* 96(1), 336–348.

Taylor, H. (1945) Cyto-taxonomy and phylogeny of the Oleaceae. *Brittonia* 5, 337–367.

Unver, T., Wu, Z., Sterck, L., Turktas, M., Lohaus, R. *et al.* (2017) Genome of wild olive and the evolution of oil biosynthesis. *Proceedings of the National Academy of Sciences USA* 114, E9413–E9422.

Van de Peer, Y., Fawcett, J.A., Proost, S., Sterck, L. and Vandepoele, K. (2009) The flowering world: a tale of duplications. *Trends in Plant Science* 14(12), 680–688.

Van de Peer, Y., Mizrachi, E. and Marchal, K. (2017) The evolutionary significance of polyploidy. *Nature Reviews Genetics* 18, 411–424.

Vanneste, K., Baele, G., Maere, S. and Van de Peer, Y. (2014) Analysis of 41 plant genomes supports a wave of successful genome duplications in association with the Cretaceous-Paleogene boundary. *Genome Research* 24(8), 1334–1347.

Vatansever, R., Hernandez, P., Escalante, F.J., Dorado, G. and Unver, T. (2022) Genome-wide exploration of oil biosynthesis genes in cultivated olive tree varieties (*Olea europaea*): insights into regulation of oil biosynthesis. *Functional & Integrative Genomics* 22(2), 171–178.

3.2 Cultivated and Wild Olive Germplasm: Distribution, Management, Exploitation

L. Baldoni[1], S. Mousavi[1], R. Mariotti[1], C. Guerrero[2], A. Belaj[3] and M.C. Valeri[1]
[1]*Institute of Biosciences and Bioresources, National Research Council, Perugia, Italy;*
[2]*University of Malaga, Department of Molecular Biology and Biochemistry, Malaga, Spain;* [3]*IFAPA – Centro Alameda del Obispo, Córdoba, Spain*

3.2.1 Introduction

Olive may count on a very rich genetic patrimony, represented by the cultivated varieties (*O. europaea*, subsp. *europaea*, var. *europaea*), the wild relatives (*O. europaea*, subsp. *europaea*, var. *sylvestris*) and the related subspecies *cuspidata*, *cerasiformis*, *guanchica*, *laperrinei* and *maroccana* (Green, 2002; Habib, *et al.*, 2021; Kassa *et al.*, 2019; Table 3.2.1).

The Mediterranean basin contains the almost entire genetic patrimony of olive cultivars and wild natural populations, while the related *Olea europaea* subspecies are distributed beyond the Mediterranean area, in the subtropical regions of the Asian, African and Australian continents, in a spontaneous or naturalized form (Besnard *et al.*, 2008; Brito *et al.*, 2008; Mousavi *et al.*, 2017; Besnard *et al.*, 2018; Belaj *et al.*, 2022).

The analysis of ploidy level on samples of the different subspecies revealed that most of them are diploid, as all cultivated and wild olives, also sharing the same chromosome number (2n=2x=46) (Jiménez-Ruiz *et al.*, 2020). The subspecies *cerasiformis* and *maroccana* were reported as tetraploid and hexaploid, respectively (Brito *et al.*, 2008), and the coexistence of two ploidy types (diploid and triploid) has been shown only for subsp. *laperrinei* (Besnard and Baali-Cherif, 2009; García-Verdugo *et al.*, 2009; Besnard and Rubio de Casas, 2016).

The longevity of the species has allowed the survival of centennial cultivated and wild olive trees, which represent an important living reservoir of variability, carrying a climate-resilience gene pool (Petruccelli *et al.*, 2014; Lazovic *et al.*, 2016; Anestiadou *et al.*, 2017; Ninot *et al.*, 2018; Rotondi *et al.*, 2018; Yazbeck *et al.*, 2019; Miazzi *et al.*, 2020; Valeri *et al.*, 2022; Fig. 3.2.1).

3.2.2 Diversity and Relationships among Olive Varieties

The olive germplasm still under cultivation in traditional olive groves is represented by local and minor cultivars, seedlings derived by the cross among cultivars or with wild olives and the occurrence of ecotypes of uncertain origin, pollinators and ancient trees. Thus, most of the large and partially unexplored olive genetic diversity is still available on-farm. All locally grown varieties, empirically selected by farmers and naturally tested by the environment, represent a high-value source of variability for the sustainability of olive

DOI: 10.1079/9781789247350.0006

Table 3.2.1. *Olea europaea* species, subspecies and botanical varieties and their natural and human-driven geographical distribution.

Species	Sub-species	Botanical variety	Geographical distribution
Olea europaea	*europaea*	*europaea* – cultivated olives	Mediterranean area, with recent expansion to Argentina, Chile, Peru, Uruguay, California, Australia, South Africa, China, India, and other small areas of America, Asia and Africa
		sylvestris – wild olives	Mediterranean area, islands and inland isolated areas, Atlantic coasts of Portugal, Spain and Morocco
	cuspidata		Eastern Africa, southern Asia (Saudi Arabia, Yemen, Oman, Iran, Pakistan, Afghanistan, India, Nepal, China), Australia and New Zealand (invasive African olives)
	cerasiformis		Madeira islands (tetraploid)
	guanchica		Canary Islands
	laperrinei		Algeria, Hoggar mountains
	maroccana		Morocco, western High Atlas (hexaploid)

cultivation in a context of climate change and reduction of water, energy and chemical resources.

The reasons leading to the high genetic variability of cultivated olive are mainly related to its allogamy and prevalent self-incompatibility, to the long tradition of cultivation, the survival of ancient varieties, the lack of turnover with new genotypes and the great tree longevity (Kabassi *et al.*, 2021; Schicchi *et al.*, 2021).

The genetic and phenotypic variability of local varieties should be preserved and properly evaluated in a perspective of olive orchard diversification, variability conservation and for breeding purposes.

The varietal assortment is represented by more than 1200 cataloged cultivars (Bartolini, 2008) and an uncertain number of other genotypes, including pollinators, local ecotypes and centennial trees, often represented by few or single trees (Dominguez-Garcia *et al.*, 2012; Cicatelli *et al.*, 2013; Hosseini-Mazinani *et al.*, 2014; Linos *et al.*, 2014; Lazovic *et al.*, 2016; Sakar *et al.*, 2016; Belaj *et al.*, 2018; D'Agostino *et al.*, 2018; Hmmam *et al.*, 2018; Ninot *et al.*, 2018; Gago *et al.*, 2019; Mousavi *et al.*, 2019c; Salazar-García *et al.*, 2019; Sion *et al.*, 2019; Mariotti *et al.*, 2020; Valeri *et al.*, 2022). These numbers are certainly destined to increase, as prospecting efforts will be extended to more regions and deepened at farm level, and genotyping technologies will become more precise, fast and reliable.

The cultivated germplasm is distributed in about 54 countries, but over two-thirds of the olive cultivars have been cataloged from southern European countries: 538 in Italy, 272 in Spain, 88 in France and 52 in Greece (Bartolini *et al.*, 2008). In addition, most of the genetic resources are still preserved on-farm, in traditional olive-producing countries and outside Europe, such as Syria, Turkey, Israel, Lebanon, Jordan, Egypt, Libya, Tunisia, Algeria, Morocco, Iran, Pakistan, Azerbaijan, USA and China (Belaj *et al.*, 2003; Koehmstedt *et al.*, 2010; El Bakkali *et al.*, 2013; Abdessemed *et al.*, 2015; Beghè *et al.*, 2015; Zhan *et al.*, 2015; Sakar *et al.*, 2016; Sorkheh and Khaleghi, 2016; Cheng *et al.*, 2017; Mousavi *et al.*, 2017; Veloso *et al.*, 2018; Khadari *et al.*, 2019; Naz *et al.*, 2019; Atrouz *et al.*, 2021; Islam *et al.*, 2021; Omri *et al.*, 2021; Topi *et al.*, 2021; Duran *et al.*, 2022).

Despite all efforts to collect, preserve and identify olive varieties, the consistency of the cultivated olive heritage still remains controversial and some data outline the existence of more than 2600 different olive varieties (FAO, 2010; Belaj *et al.*, 2016). The large number of cultivars, their local distribution, the complex process of domestication and hybridization

Fig. 3.2.1. Examples of ancient olive trees: (a, b) in Greece (Crete); (c) in Israel; (d, e) in Italy (Sardinia and Umbria); (f) in Spain (Valencia). (Photos: Luciana Baldoni)

with the wild forms and their continuous geographical redistribution over time, represent some of the factors that make it difficult to identify all the olive varieties and to estimate correctly the amount of resources (D'Agostino *et al.*, 2018; Julca *et al.*, 2020; Belaj *et al.*, 2022).

While in traditional olive groves most diverse varieties are still widely represented, in the new intensive cultivation systems, only a limited number of varieties, well suited to this type of plantation, have been widely spread over many olive-producing areas.

3.2.3 The Origin of the Olive Cultivars

The olive was fully domesticated during the early Bronze Age, about 5300–4000 years BP, presumably from the initial cultivation of wild olives and oil extraction in the eastern

Mediterranean (Kaniewski *et al.*, 2012; Gros-Balthazard *et al.*, 2019).

After this primary domestication event, first cultivars were likely brought towards the west of the Mediterranean, by Phoenicians and Minoans first, then by Greeks and Romans (Kostelenos and Kiritsakis, 2017).

The varieties brought along the shores of the Mediterranean, in turn crossed with each other or with the wild olives, representing local natural populations interfertile with cultivated olives (Besnard *et al.*, 2013, 2018). Through the analysis of maternally inherited chloroplast genomes, it was possible to establish that wild olives are clearly differentiated along the Mediterranean into three distinct lineages: E1 from the east, E2 from the center and E3 only present in the west Mediterranean (Besnard *et al.*, 2013) and, within each lineage, into many different chlorotypes. Although most cultivars (89.9%) carry the eastern lineage E1 (Besnard *et al.*, 2013), the signature of cultivar-wild crosses was detected in some widely cultivated varieties, carrying the lineages E2 (5.4%) or E3 (4.7%) (Besnard *et al.*, 2013; Besnard and Rubio de Casas, 2016) (Table 3.2.2).

A discussion is still underway on whether the exchanges between cultivars and wild olives in the center and west Mediterranean may represent secondary domestications or only new diversification events (Gros-Balthazard *et al.*, 2019; Julca *et al.*, 2020).

Furthermore, other cases of cultivars derived from wild pollination may remain hidden due to the maternal inheritance of chloroplasts, because they would carry the E1 lineage despite half of the nuclear genome could have derived from wild lineages (Mariotti *et al.*, 2010; Valeri *et al.*, 2022).

3.2.4 The Clonal Propagation of Olive Cultivars

Most of the varieties currently under cultivation represent ancient genotypes, 'landraces' empirically selected by growers and well-adapted to local conditions. They have not undergone any genetic improvement, nor genetic transformation.

All olive varieties are clonally propagated, either by cutting or grafting, allowing for the accurate reproduction of the best-performing genotypes. Some of them may have originated thousands of years ago and millions of clones may have been obtained since that time, making possible the development of somatic mutations within supposedly identical plants. Recent studies in several plant species have shown that evolution in clonal populations is more dynamic than usually thought, due to somatic mutations and epigenetic changes as sources of heritable variation (Schoen and Schultz, 2019). Somatic mutations, in fact, can produce genetic changes that contribute to adaptive evolution (Orr *et al.*, 2020). Mutations may

Table 3.2.2. Some important varieties carrying lineages/chlorotypes typical of the wild olive populations of the east (E1), center (E2) and west (E3) Mediterranean. (From Besnard *et al.*, 2013)

Cultivar	Country of main diffusion	Lineage/ chlorotype
Amygdalolia, Bella di Spagna[1], Biancolilla, Carolea, Chetoui, Galega Vulgar, Gordal Sevillana[1], Moresca, Zaity	Greece, Italy, Tunisia, Portugal, Syria, Spain	E1/E1.2
Blanqueta, Souri	Spain, Syria	E1/E1.3
Sivigliana da Olio	Italy	E2/E2.1
Kerkiras (Lianolia)	Greece	E2/E2.2
Lechin de Sevilla	Spain	E2/E2.3
Farga, Kalokerida[2], Olivière[2]	Spain, Greece, France	E3/E3.1
Cerasuola	Italy, Algeria	E3/E3.2

[1,2]Cultivars identical to each other (El Bakkali *et al.*, 2019; Belaj *et al.*, 2022)

accumulate over time, different branches may inherit different mutations, and growing plants can become genetic mosaics (Zahradníková *et al.*, 2020). The presence of over-millennial trees and phenotypical differences observed among trees of the same variety indirectly suggest the existence of intra-cultivar genetic diversity within some olive genotypes (Caruso *et al.*, 2014; Bradaï *et al.*, 2019; Jiménez-Ruiz *et al.*, 2020).

3.2.5 Geographical Distribution of Olive Cultivar Variability

Olive cultivars can be classified following their present geographical distribution within and outside the Mediterranean area. A list of the main olive cultivars is reported in Table 3.2.3, along with the country/region of diffusion, synonyms and main product destination (oil, table olives, double-purpose). The number of cultivars reported for each country only gives an indication of the main grown varieties. For example, until recent years, only one variety was reported for Morocco ('Picholine Marocaine'), because it is by far the most widespread in the country, but, more recently, many other local varieties have been recorded (El Bakkali *et al.*, 2019). On the other hand, Italy was represented by 578 cultivars (likely including cases of synonymy), as the varietal heritage of this country has been very well preserved and there are still hundreds of varieties under cultivation, equally contributing to production (Caruso *et al.*, 2014; Hmmam *et al.*, 2018; Rotondi *et al.*, 2018; Mousavi *et al.*, 2019c). In Spain, the main olive-producing country, the intensification and mechanization of olive-growing systems has triggered the reduction of cultivars in new plantations, favoring the use of only a handful of cultivars able to fulfill the requirements for a sustainable modern olive industry. However, recent ongoing prospecting studies have identified many additional and unique Spanish local cultivars that have been preserved on-farm and are being collected and stored in germplasm collections (Belaj *et al.*, 2022; WOGB of Córdoba database).

3.2.6 Phenotypical Variability among Olive Cultivars

To the large varietal heritage of the olive corresponds a great variability in terms of plant morphology, agronomical performance, environmental adaptation and oil and table olive characteristics (Fig. 3.2.2).

3.2.6.1 Plant morphology

The most distinctive morphological characteristics of each variety are related to the fruit (size, shape, color, pulp:stone ratio, ripening period etc.) (Figs 3.2.2–3.2.4); the endocarp (size, shape, surface roughness etc.); the leaves (size, shape etc.); and the flowers (number of flowers/cluster, cluster size) (Barranco *et al.*, 2000; Trujillo *et al.*, 2014).

3.2.6.2 Agronomical performance

Agronomical performance includes:

(a) plant architecture: vigor, habit, canopy density and branching index (Gómez-Gálvez *et al.*, 2021; Paoletti *et al.*, 2021) (Fig. 3.2.5); and
(b) productivity: flowering time, self-fertility, male sterility, incompatibility group, average production, early bearing, alternate fruiting, fruit drop, fruit oil content, fruit removal force and harvesting time (Abou-Saaid *et al.*, 2022; Barranco *et al.*, 2000; Bartolini *et al.*, 2008; Mousavi *et al.*, 2019a, 2022; Belaj *et al.*, 2020; Medina-Alonso *et al.*, 2020; Gómez-Galvez *et al.*, 2021; Mariotti *et al.*, 2021).

3.2.6.3 Environmental adaptation

Environmental adaptation includes all aspects related to the cultivar/environment interaction, including abiotic and biotic factors.

Table 3.2.3. List of main olive cultivars, their area of origin/main current diffusion (country, region), cases of synonymy and product destination (oil, table olives, double-purpose).

Cultivar name	Country	Region	Synonyms	Purpose
KALINJOT	Albania (2)			oil
KOKERRMADHI iBERATIT			Konservolia, Sant'Agostino, Tekirdag Cizmelik	table
AZERADJ	Algeria (3)	Alger		oil
CHEMLAL DE KABILYE		Kabilye	Chemlal	oil
SIGOISE		Northern Algeria	Picholine Marocaine	oil
ARAUCO	Argentina		Azapa	double purpose
AZAPA	Chile		Arauco	double purpose
LASTOVKA	Croatia (3)	Dalmatia		oil
OBLICA		Dalmatia		oil
CRNICA		Istria	Buga, Cnjaka	oil
LADOELIA	Cyprus (2)		Ladoelia, Lagoydhera, Baladi (group)	double purpose
KATO DRYS			Klirou, Ervihoy_1, Flasou	double purpose
TOFFAHI	Egypt (3)			table olives
AGGEZI SHAMI			Agazy, Aggezi Shami, Aghizi Shami	table olives
MARAKI				oil
BOUTEILLAN	France (8)			oil
GROSSANE				double purpose
LUCQUES				table olives
OLIVIERE			Kalokerida	oil
PICHOLINE DU LANGUEDOC			Abou Chawkeh/Choki, Balah, Jlot	double purpose
SALONENQUE				double purpose
TANCHE				double purpose
VERDALE				oil
ADRAMITINI	Greece (11)	Aegean islands	Ayvalik, Beyaz Yaglik, Edrremit Yaglik	oil
AMIGDALOLIA		Attiki		double purpose
CHALKIDIKIS		Chalkidiki		double purpose
KALAMON		Peloponnese	Kalamata, Chondrolia	table olives
LIANOLIA KERKIRAS		Ionian islands	Ulliri i Holle i Himares	oil
KONSERVOLIA		Epirus, Thessaly	Sant'Agostino, Kokkermadh i Beratit, Tekirdag Cizmelik	table olives
KORONEIKI		Crete	Girit zeytini	oil
MASTOIDIS		Crete		oil
MAVRELYA		Athens	Mavrolia	table olives
MEGARITIKI		Attiki, Beotia	Megaron	double purpose
MYRTOLIA			Koutsourelia	oil
MARI	Iran (4)			double purpose
ROWGHANI				oil
SHENGEH				double purpose
ZARD				double purpose
BARNEA	Israel (3)			double purpose
KADESH				table olives
NABALI			Baladi	double purpose

Continued

Table 3.2.3. Continued.

Cultivar name	Country	Region	Synonyms	Purpose
DRITTA	Italy (57)	Abruzzo	San Felice, Ogliarola del Vulture	oil
GENTILE DI CHIETI			Gentile, Nostrana, Olivastro	oil
INTOSSO			Indosso, Indossa, Grossa	double purpose
BELLA DI CERIGNOLA		Apulia	Grossa di Spagna, Dužica, Esek Zeytini (Odemis), Gaydoyrelia	table olives
CELLINA DI NARDÒ			Leccese, Oliva di Lecce, Cellina	oil
CORATINA			San Pedro , Cima di Corato, Coratese, Racioppa di Corato	oil
DOLCE D'ANDRIA			Pasola di Andria	table olives
NOCIARA			Fra' Michele	oil
OGLIAROLA BARESE			Frantoio	oil
OGLIAROLA LECCESE			Mignola, Ogliarola Salentina	oil
PERANZANA			Bosana, Cerasa di Montenero	oil
SANT'AGOSTINO			Konservolia, Kokkermadh i Beratit, Tekirdag Cizmelik	table olives
MAJATICA		Basilicata		double purpose
CIMA DI MELFI			Donna Francesca, Nostrale di Melfi	oil
OGLIAROLA DEL BRADANO				oil
CAROLEA		Calabria	Olivo dolce/Dolce di Cerchiara	double purpose
CASSANESE			Grossa di Cassano	double purpose
DOLCE DI ROSSANO			Dolce, Rossanese, Olivella	double purpose
OTTOBRATICA			Ottobrarico, Perciasacchi	oil
SINOPOLESE			Scialorea, Coccitano	oil
CAIAZZANA		Campania	Carica, Oliva a Sauce	oil
ORTICE				double purpose
PISCIOTTANA			Olivo dell'Ascea	oil
RAVECE			Olivin, Olivon	oil
NOSTRANA DI BRISIGHELLA		Emilia Romagna	Nostrana, Nostrale di Brisighella	oil
CAPOLGA			Capolga dell'Emilia-Romagna	oil
BIANCHERA		Friuli	Istarska Bjelica, Plementa Bjelica, Istrska Belica	oil
CANINO		Latium	Caninese, Olivo di Canino	oil
CARBONCELLA			Moraiolo, Petit Ribier, Morcaio, Filare, Moraiolo, Morcaio, Tondello	oil
RAJA SABINA			Frantoio	oil

Continued

Table 3.2.3. Continued.

Cultivar name	Country	Region	Synonyms	Purpose
ROSCIOLA			Rossello	oil
SIROLE			Sierola	oil
ITRANA			Tonda di Villacidro, Corsicana da Mensa, Nera di Oliena, Paschixedda, Pulazeqin, Terza Grande, Terza Piccola	double purpose
TAGGIASCA		Liguria	Frantoio	double purpose
CASALIVA		Lombardy-Veneto	Frantoio	oil
ASCOLANA TENERA		Marche	Asolano, Oliva Dolce, Oliva Grossa	table olives
PIANTONE DI MOGLIANO			Piantone di Macerata	oil
RAGGIOLA			Frantoio	oil
CERASA DI MONTENERO		Molise	Peranzana, Bosana	oil
GENTILE DI LARINO			Gentile, Oliva Nera	oil
OLIVA NERA DI COLLETORTO			Gentile di Colletorto, Gentile Nera di Colletorto, Nera di Colletorto	oil
BOSANA		Sardinia	Peranzana, Cerasa di Montenero	oil
NERA DI GON-NOS			Tonda di Cagliari, Manna, Majorca, Confetto, Olia Nuxi, Sivigliana da Mensa, Nocellara Messinese	oil
SEMIDANA			Cariasina	oil
BIANCOLILLA		Sicily	Abunara, Buscionetto	oil
CALATINA			Galatina, Montagna Calatina	oil
CERASUOLA			Zarrazi Douirat, Zarrazi Injassi, Zarrazi, Zarzis, Zarrazyet Sned	oil
NOCELLARA DEL BELICE			Beldi, Fakhari, Meski Zarzis	double purpose
NOCELLARA ETNEA			Nuciddara	double purpose
TONDA IBLEA			Cetrara, Citrala	double purpose
LECCINO		Tuscany	Gremignolo, Kallmet Kokerrvogel, Manzanilla Chilena, Palmarola	double purpose
LECCIO DEL CORNO			Grappolo	oil
MAURINO			Ginestrino, Boise	oil
OLIVASTRA SEGGIANESE			Olivastra, Olivastra di Montalcino	oil

Continued

Table 3.2.3. Continued.

Cultivar name	Country	Region	Synonyms	Purpose
FRANTOIO		Umbria	Augellina, Taggiasca, Casaliva, Ogliarola Barese, Correggiolo di Pallese, Raggiola, Raja Sabina, Corsicana da Olio, Larcianese, Maelia, Puntino, Ghermana, Calletier, Oblonga, Razzo, Razzola, San Lazzero	oil
MORAIOLO			Carboncella, Morcaio, Filare, Morcaio. Petit Ribier, Tondello	oil
DOLCE AGOGIA			Dolce, Agogia	oil
NOSTRALE DI RIGALI			Nostrale	oil
NABALI	Jordan (2)		Baladi, Sourani	double purpose
RASEI			Nabali Muhasan	double purpose
AYROUNI	Lebanon (3)			oil
BALADI			Nabali Baladi (group, see Belaj *et al.*, 2022)	double purpose
SOURI				oil
ŽUTICA	Montenegro (1)		Bjelica, Maiatica di Ferrandina, Ulliri i Bardhe i Tiranes, Ulliri i Bardhe Kruje	oil
PICHOLINE MAROCAINE	Morocco (1)		Sigoise, Cañivano Blanco, Alameño de Marchena, Hamrani, Haouzia, Menara, Misión de San Vicente, Mission Nieland, Zeitun lkhl	double purpose
BLANQUETA	Portugal (4)		Blanqueta de Alicante	oil
COBRANCOSA				oil
CORDOVIL DE SERPA				double purpose
GALEGA VULGAR			Galego	oil
ISTARSKA BJELICA	Slovenia (1)		Bianchera	oil
ALOREÑA				table
ARBEQUINA	Spain (11)	The most wide-spread cultivar for superintensive orchards, origin in Catalonia		oil
ARBOSANA		Spread in superintensive orchards in all olive-growing countries, origin in Catalonia		oil
CASTELLANA			Racimo	oil
CHANGLOT REAL			Torcio de cabra	oil
CORNICABRA				double purpose

Continued

Table 3.2.3. Continued.

Cultivar name	Country	Region	Synonyms	Purpose
EMPELTRE			Grosal de Albocácer	double purpose
FARGA		Catalonia, Valencia, Aragon (north-east Spain)	Solimar	oil
SEVILLENCA				oil
GORDAL SEVILLANA		Castilla la Mancha/ Andalusia/Ex-tremadura or central-south and south-west Spain	Giarraffa, Santa Caterina	table olives
HOJIBLANCA			Verdial de Montefrio	double purpose
LECHIN DE SEVILLA			Mawi Stanboli, Nevadillo de San Vicente, Nevadillo Valle las Palmeras S. George Greys, Souri, Zarza	oil
MANZANILLA CACEREÑA			Azeitera, Negrinha	double purpose
MANZANILLA DE SEVILLA			Redondil, Chesna, Dulzal, Stanbouli, Manzanilla de Valle de las Palmeras	double purpose
MORISCA			Santulhana	oil
PICUAL			Olivo Macho de Santiste-ban Pto, Picual de Hoja Clara, Picual de Hoja Oscura, Marteño	oil
PICUDO			Picudo de Fruto Rojo	oil/double purpose
VERDIAL DE BADAJOZ			Cordobés de la Aliseda, Cordovil de Castelo Branco, Corneja, Enagua de Arenas, Llorón de Iznalloz	double purpose
VERDIAL DE HUEVAR			Ayrouni	oil/double purpose
DOEBLI	Syria (6)	Latakia, Tartus		double purpose
SORANI		Idlib, As-Suwaida		double purpose
SOURI		Latakia	Baladi (Group of synony-mies)	oil
ZAITY		Aleppo	Halhali (Hatay), Kalembezi, Kilis Yaglik, Sayfi-2325, Tesbih Celebi, Yaglik Sari Zeytin	oil
SAFRAWI		Aleppo, Hama	Throumbolia, Sant'Agatese, Grossolana, Cirujal, Sefraoui, Antawi, Baladi Remadieh, Celebi, Cirujal, Dal, Dan, Dilmit, Erkence, Hurma Kaba, Kanabisi, Marksi, Sari Habesi, Shami, Yag Zeytini	double purpose
KAISSY		Halab, Idlib	Khalkhali-789 (SYR) Shami-953 (SYR)	double purpose

Continued

Table 3.2.3. Continued.

Cultivar name	Country	Region	Synonyms	Purpose
CHEMLALI	Tunisia (6)	Sfax	Chemlal de Sfax, Chemlal du Nord, Zalmati Chemlali Ouled Youssef Mellahi Kotti, Zalmati Zarzis	oil
CHETOUI		Bizerte	Blanquette de Gastu, Blanquette de Guelma, Chaïbi Antha, Chetoui Tunis, Chetous, Sayali T-1677	oil
GERBOUI		Jendouba	Marsaline	double purpose
MESKI		Beja		table olives
OUESLATI		Kairouan	Ouslati	oil
ZALMATI		Medenine	Chamlal de Sfax, Chemlali	oil
AYVALIK	Turkey (9)	Northern Turkey	Adramitini, Edremit Yaglik, Edrremit Yaglik	oil
BUYUK TOPAK ULAK		Aegean, Mediterranean	Cilli	table olives
DOMAT		Aegean		table olives
GEMLIK		Antalya, Bilecik	Erdek Yaglik, Kfarzeta, Samsun Ufak Tuzlamalik, Sinop 2, Siyah Salamuralik (Tekirdag), Sukare	table olives
IZMIR SOUFRALIK		Aegean		table olives
KAN CELEBI		South-eastern Anatolia	Hamza celebi	table olives
MEMECIK		Southern Turkey	Tasarasi (Aydin), Yerli Yaglik, Asi Yeli, Tasarasi (Kusadasi)	double purpose
USLU		Aegean		table olives
YUN CELEBI		Eastern & southern Anatolia	Belluti, Hirhali Celebi	oil
MISSION	USA (1)	California		double purpose
LEUCOCARPA	Different countries	Different regions	Numerous cases of homonymy, known also as Leucolea. In Malta, cv. Bajda produces white table olives	double purpose
ZARZA	Italy-Spain		Mora, Lechin de Sevilla, Zarzamora	without a clear cultivation purpose

(1) The number in parentheses close to the country name indicates the number of cultivars reported in the table for that country.

Resistance to biotic stresses

Main biotic stresses include the following: the insect olive fruit fly (*Bactrocera oleae*) (Grasso *et al.*, 2017; Quesada-Moraga *et al.*, 2018), the diseases caused by fungi, such as Verticillium wilt (*Verticillium dahliae*) (Ramírez-Tejero *et al.*, 2021) and olive peacock spot (*Spilocaea oleagina*); or by bacteria, such as olive knot (*Pseudomonas savastanoi*), anthracnose (*Colletotrichum* spp. fungal complex) (Moral *et al.*, 2017) and olive quick decline syndrome (*Xylella fastidiosa* subsp. *pauca*) (Petit *et al.*,

Fig. 3.2.2. Olive fruits of mixed varieties showing different size, color and shape. (Photo: Luciana Baldoni)

2021; Surano *et al.*, 2022). Cultivars respond to insect and pathogen attack in different ways, virtually from total resistance to total susceptibility. V if varieties totally resistant to the olive fly, *X. fastidiosa* or Verticillium wilt have not yet been observed.

Tolerance to environmental stresses

The olive varieties are strongly geo-climatically structured, on the basis of local environmental conditions and the ability of each variety to resist extreme temperatures, low water availability, or anomalous chemical and physical soil composition. They can escape from environmental stress damages on the basis of their physiology and phenology.

Cold can damage buds, young shoots, flowers or, in the most serious cases, branches or the entire aerial part of the plant, and each variety can show tolerance, or escape from one or more stresses, or differently recover from damage (Mougiou *et al.*, 2020; Petruccelli *et al.*, 2022).

In arid environments, varieties able to withstand low water availability and high temperatures have been selected, but drought and heat tolerance depends on the frequency, timing and intensity of the events (Dias *et al.*, 2018; Razouk *et al.*, 2022).

Also, plant response to salinity or anomalous soils is very different among varieties, and few of them are able to live in salty soils (Rahemi *et al.*, 2017; Mousavi *et al.*, 2019b; Regni *et al.*, 2019; Tan *et al.*, 2020) or produce on stony, poor, clayey or low soil depth (Fernández-Escobar, 2019).

3.2.6.4 Fruit and olive oil characteristics

Fruits and oil derived from each cultivar show peculiar fatty acid composition, mainly concerning oleic and linoleic acids, and different composition in secondary metabolites that contribute to defining the health and organoleptic hallmarks of extra-virgin olive oils: phenols (oleuropein, oleocanthal, hydroxytyrosol, tyrosol, demethyloleuropein, verbascoside, oleacein, ligstroside and lignans), sterols, tocopherols, triterpenes and squalene. All these compounds are under a strict genetic control; thus, each cultivar has its own specific set-up, even if plant x environment interaction may differently shape the metabolic profile of cultivars according to the site of cultivation. For this reason, data derived from many cultivars grown together in separate and well-defined environments, as is the case for germplasm collections, can provide useful information on this aspect (Beltran *et al.*, 2016; Kyçyk *et al.*, 2016; Miho *et al.*, 2018; El Riachy *et al.*, 2019; Jiménez-Carvelo *et al.*, 2019; Peréz *et al.*, 2019; Mousavi *et al.*, 2019a, 2022; Mikrou *et al.*, 2020; Hernández *et al.*, 2021).

3.2.6.5 Product destination

Olive varieties may also be grouped according to the product destination: oil, table olives or both, based on various criteria, such as fruit size, oil content, pulp properties and fruit composition at pre- and post-processing.

Cultivars for olive oil production

The varieties intended for oil production must have a high oil yield, excellent oil characteristics, tree architecture suitable for intensive cultivation and the use of mechanical machineries for pruning and harvesting (Rallo, 2014; Biton *et al.*, 2022).

Fig. 3.2.3. Fruiting shoots of olive cultivars showing different fruit color, size, shape, lenticels, peduncle length, ripeness and distribution along the branch. (a) cultivars from Italy; (b) cultivars from Spain and Portugal; (c) cultivars from other countries. (Plants of the CNR Olive Cultivar Collection, Lugnano in Teverina, Italy; photos: Luciana Baldoni and Maurizio Cenci)

Fig. 3.2.3. Continued.

Fig. 3.2.4. Fruiting shoots of some olive cultivars: (a) 'Sigoise' (Algeria); (b) 'Toffahi' (Egypt); (c) 'Picholine' (France); (d,e,f) 'Chalkidikis', 'Konservolia', 'Mastoidis' (Greece); (g–j) 'Bella di Cerignola', 'Frantoio', 'Nocellara Messinese', 'Nociara' (Italy); (k) 'Picholine Marocaine' (Morocco); (l) 'Cobrancosa' (Portugal); (m–r) 'Arbequina', 'Gordal Sevillana', 'Lechin de Sevilla', 'Manzanilla de Sevilla', 'Picual' (Spain); (s) 'Souri' (Lebanon); (t) 'Doebli' (Syria); (u) 'Zarza'. (Plants of the Olive Cultivar Collection of INRA, Marrakech, Morocco; photos: Luciana Baldoni and Claudio Ranocchia)

Fig. 3.2.5. Tree architecture of different olive cultivars showing different habit, canopy density, vigor, tree shape and bearing earliness. (a, b) upward and weeping tree habit; (c, d) low and high canopy density; (e, f) low and high tree vigor; (g, h) columnar and umbrella canopy shape; and (i, j) early- and late-bearing trees. (Plants of the CNR Olive Cultivar Collection, Lugnano in Teverina, Italy; photos: Luciana Baldoni)

There is great interest in product quality: oil fatty acid composition, oil content in total and specific secondary metabolites, oil sensorial profile, oil shelf life, oil use as healthy food, gourmet food or flavoring dressing.

Furthermore, the remodeling of the entire olive grove management should rely on cultivars with low requirements in water, nutrients, pesticides and energy resources. These climate-ready cultivars should make it possible to improve soil conservation, reducing mechanical and chemical interventions and moving towards environmentally friendly practices, with beneficial effects on the environment, farm welfare and product quality. They should also carry the potential for producing bioactive compounds with high nutritional and industrial value.

Table olives characteristics

Cultivars intended for table olive production differ from those for oil, having bigger fruit size, high pulp:stone ratio and good texture. Meanwhile, sugars and fat content mainly depend on the ripening stage at harvest (Evangelou *et al.*, 2018; Conte *et al.*, 2020). Other characteristics to take under consideration to evaluate table olive varieties include fruit removal force, fruit microflora composition, low susceptibility to the olive fruit fly and to fruit pathogens, harvest efficiency and fruit quality (Ben Ayed *et al.*, 2017; Jiménez *et al.*, 2017; Rallo *et al.*, 2018).

It would be necessary to provide indications on the behavior of varieties for each of the listed traits, but it is still very difficult, because research carried out so far concerns small groups of varieties in specific environments and conditions, while only a direct comparison among many varieties in highly contrasting environments and conditions should provide valid information. There are still few studies conducted in this area (Mousavi *et al.*, 2019a; Anguita-Maeso *et al.*, 2021), but this gap should soon be filled, and very interesting data should become available on the agronomical profile of each variety and its response and plasticity to environmental stimuli.

3.2.7 Wild Olives and Other Related Forms

Cultivated olives represent only a subset of the entire genetic variation of the *O. europaea* subsp. *europaea* genetic resources. The wild relatives (also known as oleasters) (Fig. 3.2.6), contain traits that could be introgressed into cultivated olives, either by conventional breeding, genome editing or gene transfer. At present, genuinely wild olive populations are limited to untouched areas or dispersed as scattered trees in rural and semi-natural habitats along the shores of the Mediterranean, while most other

Fig. 3.2.6. Wild olive populations naturally occurring along the Mediterranean shores: (a) forest of ancient wild olives in Sardinia (Italy); (b) wild olive plants showing the typical juvenile habitus (Crete, Greece). (Photos: Luciana Baldoni)

wild-looking olives may include feral forms, i.e. seedlings derived from varietal inter-crosses or from the cross between cultivars and true wilds (Baldoni *et al.*, 2006; Falek *et al.*, 2021; Valeri *et al.*, 2022). However, in order to maximize the exploitation of germplasm resources and introgression useful variability from wilds into cultivars, it is crucial to conserve and evaluate their genetic diversity.

Wild olives have colonized diverse environments characterized by semi-arid climatic conditions with different altitudes, vegetative communities and soils, including those with extreme levels of drought, low temperatures and salinity (Hamid *et al.*, 2022). Various studies revealed the presence of genuine wild olive forests, with high levels of diversity (Belaj *et al.*, 2007; Erre *et al.*, 2010; Besnard *et al.*, 2018; Kyriakopoulou and Kalogianni, 2020).

Wild olives may represent a useful source of variability to introduce into the cultivated olive traits, such as biotic (Colella *et al.*, 2008; Trapero *et al.*, 2015; Arias-Calderón *et al.*, 2015) and/or abiotic stress resistance (Murillo *et al.*, 2005; Aranda *et al.*, 2011), as well as to improve the olive oil health value and taste (Hannachi *et al.*, 2008; Baccouri *et al.*, 2011: León *et al.*, 2018).

3.2.8 Olive Germplasm Conservation

3.2.8.1 Threats to olive germplasm and strategies to deal with them

Despite the great genetic richness, olive cultivation is seriously threatened by various phytosanitary emergencies, among others, the *Xylella fastidiosa* epidemic, which in ten years has decimated millions of trees in Salento (Puglia, Italy), currently expanding in other olive-growing areas. One of the solutions to this threat could be represented by the selection of resistant genotypes, likely derived from related wild forms, as occurred in other *Xylella* host plant species such as grape (Agüero *et al.*, 2022; Morales-Cruz *et al.*, 2022).

In the near future, the olive germplasm could be subjected to drastic reductions due to the diffusion of intensive plantings, where only a few cultivars are able to satisfy the needs of these new cultivation systems, thus drastically reducing the cultivated varietal platform.

The globalization of the olive oil, table olives and plant market, inside and outside the borders of the Mediterranean area, could further reduce the diversification of cultivars in modern olive groves.

All the aforementioned factors may lead to a real erosion of the cultivated olive germplasm through progressive abandonment, substitution and, finally, loss of autochthonous local cultivars (Belaj *et al.*, 2016).

3.2.8.2 The need for conservation and resource evaluation

The exploration, collection, conservation, characterization and evaluation of olive genetic resources are necessary steps against the risk of genetic erosion and towards their efficient use in olive growing and breeding.

This represents valuable insurance against potential risks due to, for example, pest and disease attacks, catastrophic events or political restrictions, which are unlikely to occur simultaneously in different locations.

The conservation and further evaluation of the world olive diversity in different environments may also provide useful information for the scientific community and olive-growing industry.

3.2.9 Conservation, Management and Characterization of the Olive Germplasm

Despite the great wealth of genetic resources of the olive, the number of varieties that significantly contribute to the production of olive oil and table olives remains reduced to a few dozen, due to a strong specialization of the crop and the increasing diffusion of intensive plantings for which only few suitable cultivars may be used (Rallo *et al.*, 2013; Yılmaz-Düzyaman *et al.*, 2022). For this reason, the genetic and phenotypic variability of olive cultivars should be preserved and suitably evaluated with a view to conservation and genetic improvement.

3.2.9.1 *In situ*/on-farm conservation

The olive varieties are usually conserved *in situ*, at farm level, where traditional groves contain the intact cultivar patrimony. Different forms of olive trees can coexist in these orchards: cultivars, pollinators, unknown local cultivars, wild forms and ecotypes. Other cultivars still survive also in abandoned or reforested fields.

3.2.9.2 *Ex situ* conservation

In order to guarantee the preservation of as many olive genotypes as possible, and protect them from the risk of extinction due to abandonment, crop reconversion, phytosanitary emergencies, climate constraints, urban sprawl etc., the most efficient strategy is the establishment and management of public collections or common gardens, where clonal plants of cultivars may be stocked and evaluated (Fig. 3.2.7).

Over the past 50 years, olive cultivars have been maintained in *ex situ* field collections, which are aimed at acquiring, maintaining, documenting, assessing and making accessible the genetic diversity of the crop and avoiding or minimizing the loss of total variation or the risk of cultivar extinction (El Bakkali *et al.*, 2019; Belaj *et al.*, 2022).

The creation of international collections established in different agro-ecological conditions, aims at protecting the olive genetic patrimony of all olive-growing countries.

Currently, olive germplasm collections have been developed by approximately 100 institutions (Bartolini, 2008). At present, four world olive germplasm banks (WOGB) have been established under the IOC coordination: in Cordoba (Spain) (Belaj *et al.*, 2022), Marrakech (Morocco) (El Bakkali *et al.*, 2019), Izmir (Turkey) (Kaya *et al.*, 2019) and, very recently, San Juan (Argentina). Many other international, national and regional collections have been established in all main olive-producing countries.

Fig. 3.2.7. International olive cultivar collections established in different countries: Italy (CNR-3APTA, Lugnano in Teverina, Terni), Spain (IFAPA, Cordoba) and Morocco (INRA, Marrakech) (Photos: Luciana Baldoni)

Main activities carried out in these collections include preservation, identification, comparison and bio-agronomical evaluation of the established cultivars. They also allow for the exchange of information among the scientific and technical communities, and represent the source of material for breeding programs (Belaj *et al.*, 2018; El Bakkali *et al.*, 2019; Mousavi *et al.*, 2019a).

The availability of large sets of well characterized and highly different cultivars is critical to increase the ability to face new production challenges (De Gennaro *et al.*, 2012; Larbi *et al.*, 2015) and future climatic constraints (Moriondo *et al.*, 2013; Alfieri *et al.*, 2019; Mousavi *et al.*, 2019a; Lorite *et al.*, 2022), preserving unique genetic traits currently available and offering different metabolic and sensory profiles of extra-virgin olive oils.

Olive cultivars show high levels of polymorphism, genetic diversity values and allelic variations at regional or local level, as well as at national and Mediterranean scale. Thus, in each olive-growing country breeders may rely on the autochthonous olive genetic resources for the design of breeding programs, since there is sufficient genetic variability among native cultivars (Belaj *et al.*, 2016).

Knowledge of the genetic similarity among varieties may facilitate the efficient sampling and utilization of germplasm resources by identifying unique or very distinctive gene pools, overrepresentations, or gaps from certain areas. Recent studies, mainly based on large sets of SNP markers or on genomic sequence data, are revealing strict relationships among some important cultivars at regional level (Khadari *et al.*, 2019; Friel *et al.*, 2021).

On the varieties of the collections data on main phenotypic data and bio-agronomic behavior, are currently gathered with particular reference to:

- early bearing, productivity, fruit characteristics;
- metabolic and sensory profile of fruit and/or oil;
- incompatibility group, cold requirement, flowering and fruit harvesting time;
- tree architecture in terms of vigor, habit, crown density, branching index;
- resistance to olive fruit fly and pathogens, with particular reference to olive peacock spot, olive knot and Verticillium wilt.

References

Abdessemed, S., Muzzalupo, I. and Benbouza, H. (2015) Assessment of genetic diversity among Algerian olive (*Olea europaea* L.) cultivars using SSR marker. *Scientia Horticulturae* 192, 10–20.

Abou-Saaid, O., El Yaacoubi, A., Moukhli, A., El Bakkali, A., Oulbi, S. *et al.* (2022) Statistical approach to assess chill and heat requirements of olive tree based on flowering date and temperatures data: towards selection of adapted cultivars to global warming. *Agronomy* 12(12), 2975.

Agüero, C.B., Riaz, S., Tenscher, A.C., Bistué, C. and Walker, M.A. (2022) Molecular and functional characterization of two RGA type genes in the PdR1b locus for Pierce's disease resistance in *Vitis arizonica/candicans*. *Plant Cell, Tissue and Organ Culture* 151, 497–510.

Alfieri, S.M., Riccardi, M., Menenti, M., Basile, A., Bonfante, A. and De Lorenzi, F. (2019) Adaptability of global olive cultivars to water availability under future Mediterranean climate. *Mitigation and Adaptation Strategies for Global Change* 24(3), 435–466.

Anestiadou, K., Nikoloudakis, N., Hagidimitriou, M. and Katsiotis, A. (2017) Monumental olive trees of Cyprus contributed to the establishment of the contemporary olive germplasm. *PLOS ONE* 12(11), e0187697.

Anguita-Maeso, M., Haro, C., Montes-Borrego, M., de la Fuente, L., Navas-Cortés, J.A. and Landa, B.B. (2021) Metabolomic, ionomic and microbial characterization of olive xylem sap reveals differences according to plant age and genotype. *Agronomy* 11(6), 1179.

Aranda, S., Montes-Borrego, M., Jiménez-Díaz, R.M. and Landa, B.B. (2011) Microbial communities associated with the root system of wild olives (*Olea europaea* L. subsp. europaea var. sylvestris) are good reservoirs of bacteria with antagonistic potential against *Verticillium dahliae*. *Plant and Soil* 343, 329–345.

Arias-Calderón, R., León, L., Bejarano-Alcázar, J., Belaj, A., de la Rosa, R. and Rodríguez-Jurado, D. (2015) Resistance to *Verticillium* wilt in olive progenies from open-pollination. *Scientia Horticulturae* 185, 34–42.

Atrouz, K., Bousba, R., Marra, F.P., Marchese, A., Conforti, F.L. *et al.* (2021) Algerian olive germplasm and its relationships with the central-western Mediterranean varieties contributes to clarify cultivated olive diversification. *Plants* 10, 678.

Baccouri, B., Guerfel, M., Zarrouk, W., Taamalli, W., Daoud, D. and Zarrouk, M. (2011) Wild olive (*Olea europaea* L.) selection for quality oil production. *Journal of Food Biochemistry* 35, 161–176.

Baldoni, L., Tosti, N., Ricciolini, C., Belaj, A., Arcioni, S. *et al.* (2006) Genetic structure of wild and cultivated olives in the Central Mediterranean Basin. *Annals of Botany* 98(5), 935–942.

Barranco, D., Cimato, A., Fiorino, P., Rallo, L., Touzani, A., Castañeda, C. *et al.* (2000) *World Catalogue of Olive Varieties*. International Olive Oil Council, Madrid.

Bartolini, G. (2008) Olive germplasm (*Olea europaea* L.): cultivars, synonyms, cultivation area, collections, descriptors. Available at: http://www.oleadb.it/ (accessed 1 May 2023).

Beghè, D., Garcìa Molano, J.F., Fabbri, A. and Ganino, T. (2015) Olive biodiversity in Colombia: a molecular study of local germplasm. *Scientia Horticulturae* 189, 122–131.

Belaj, A., Muñoz-Diez, C., Baldoni, L., Porceddu, A., Barranco, D. and Satovic, Z. (2007) Genetic diversity and population structure of wild olives from the north-western Mediterranean assessed by SSR markers. *Annals of Botany* 100, 449–458.

Belaj A., Gurbuz Veral M., Sikaoui H., Moukhli A., Khadari B. *et al.* (2016) Olive genetic resources. In: Rugini, E., Baldoni, L., Muleo, R. and Sebastiani, L. (eds) *The Olive Tree Genome*. Springer, Cham, Switzerland, pp. 27–35.

Belaj, A., Caballero, J. M., Barranco, D., Rallo, L., and Trujillo, I. (2003). Genetic characterization and identification of new accessions from Syria in an olive germplasm bank by means of RAPD markers. *Euphytica* 134, 261–268.

Belaj, A., de la Rosa, R., Lorite, I.J., Mariotti, R., Cultrera N.G.M. *et al.* (2018) Usefulness of a new large set of high throughput EST-SNP markers as a tool for olive germplasm collection management. *Frontiers in Plant Science* 9, 1320–1332.

Belaj, A., de la Rosa, R., León, L., Gabaldón-Leal, C., Santos, C. *et al.* (2020) Phenological diversity in a World Olive Germplasm Bank: potential use for breeding and climate change studies. *Spanish Journal of Agricultural Research* 18(1), e0701.

Belaj, A., Ninot, A., Gómez-Gálvez, F.J., El Riachy, M., Gurbuz-Veral, M. *et al.* (2022) Utility of EST-SNP markers for improving management and use of olive genetic resources: a case study at the Worldwide Olive Germplasm Bank of Córdoba. *Plants* 11(7), 921.

Beltran, G., Bucheli, M.E., Aguilera, M.P., Belaj, A. and Jimenez, A. (2016) Squalene in virgin olive oil: screening of variability in olive cultivars. *European Journal of Lipid Science and Technology* 118(8), 1250–1253.

Ben Ayed, R., Ennouri, K., Ben Amar, F., Moreau, F., Triki, M.A. and Rebai, A. (2017) Bayesian and phylogenic approaches for studying relationships among table olive cultivars. *Biochemical Genetics* 55, 300–313.

Besnard, G. and Baali-Cherif, D. (2009) Coexistence of diploids and triploids in a Saharan relict olive: evidence from nuclear microsatellite and flow cytometry analyses. *Comptes Rendus Biologies* 332, 1115–1120.

Besnard, G. and Rubio de Casas, R. (2016) Single vs multiple independent olive domestications: the jury is (still) out. *New Phytologist* 209(2), 466–470.

Besnard, G., Garcia-Verdugo, C., Rubio de Casas, R., Treier, U.A., Galland, N. and Vargas, P. (2008) Polyploidy in the olive complex (*Olea europaea* L.): evidence from flow cytometry and nuclear microsatellite analyses. *Annals of Botany* 101, 25–30.

Besnard, G., Khadari, B., Navascués, M., Fernández-Mazuecos, M., El Bakkali, A. *et al.* (2013) The complex history of the olive tree: from Late Quaternary diversification of Mediterranean lineages to primary domestication in the northern Levant. *Proceedings of the Royal Society B* 280(1756), 2012–2833.

Besnard, G., Terral, J.F. and Cornille, A. (2018) On the origins and domestication of the olive: a review and perspectives. *Annals of Botany* 121(3), 385–403.

Biton, I., Namdar, D., Mani, Y. and Ben-Ari, G. (2022) Table olives: toward mechanical harvesting. In: Yonar, T. (ed.) *Olive Cultivation*. IntechOpen.

Bradaï, F., Sánchez-Romero, C. and Martín, C. (2019) Somaclonal variation in olive (*Olea europaea* L.) plants regenerated via somatic embryogenesis: influence of genotype and culture age on genetic stability. *Scientia Horticulturae* 251, 260–266.

Brito, G., Loureiro, J., Lopes, T., Rodriguez, E. and Santos, C. (2008) Genetic characterization of olive trees from Madeira Archipelago using flow cytometry and microsatellite markers. *Genetic Resources and Crop Evolution* 55, 657–664.

Caruso, T., Marra, F.P., Costa, F., Campisi, G., Macaluso, L. and Marchese, A. (2014) Genetic diversity and clonal variation within the main Sicilian olive cultivars based on morphological traits and microsatellite markers. *Scientia Horticulturae* 180, 130–138.

Cheng, Z., Zhan, M., Yang, Z., Zumstein, K., Chen, H. and Huang, Q. (2017) The major qualitative characteristics of olive (*Olea europaea* L.) cultivated in southwest China. *Frontiers in Plant Science* 8, 559.

Cicatelli, A., Fortunati, T., De Feis, I. and Castiglione, S. (2013) Oil composition and genetic biodiversity of ancient and new olive (*Olea europaea* L.) varieties and accessions of southern Italy. *Plant Science* 210, 82–92.

Colella, C., Miacola, C., Amenduni, M., D´Amico, M., Bubici, G. and Cirulli, M. (2008) Sources of Verticillium wilt resistance in wild olive germplasm from the Mediterranean region. *Plant Pathology* 57, 533–539.

Conte, P., Fadda, C., Del Caro, A., Urgeghe, P.P. and Piga, A. (2020) Table olives: an overview on effects of processing on nutritional and sensory quality. *Foods* 9(4), 514.

D'Agostino, N., Taranto, F., Camposeo, S., Mangini, G., Fanelli, V. *et al.* (2018) GBS-derived SNP catalogue unveiled wide genetic variability and geographical relationships of Italian olive cultivars. *Scientific Reports* 8, 15877.

De Gennaro, B., Notarnicola, B., Roselli, L. and Tassielli, G. (2012) Innovative olive-growing models: an environmental and economic assessment. *Journal of Cleaner Production* 28, 70–80.

Dias, M.C., Correia, S., Serôdio, J., Silva, A.M.S., Freitas, H. and Santos, C. (2018) Chlorophyll fluorescence and oxidative stress endpoints to discriminate olive cultivars tolerance to drought and heat episodes. *Scientia Horticulturae* 231, 31–35.

Dominguez-Garcia, M.C., Laib, M., de la Rosa, R. and Belaj, A. (2012) Characterisation and identification of olive cultivars from North-eastern Algeria using molecular markers. *The Journal of Horticultural Science and Biotechnology* 87(2), 95–100.

Duran, S.T., Aghayeva, S., Akparov, Z., Mammadov, A., Asgarova, R., Uslu, O.Y. *et al.* (2022) Genetic variation and relationships between Azerbaijani and Turkish olive genetic resources. *Molecular Biology Reports* 49, 5209–5217.

El Bakkali, A., Haouane, H., Hadiddou, A., Oukabli, A., Santoni, S. *et al.* (2013) Genetic diversity of on-farm selected olive trees in Moroccan traditional olive orchards. *Plant Genetic Resources* 11(2), 97–105.

El Bakkali, A., Essalouh, L., Tollon, C., Rivallan, R., Mournet, P. *et al.* (2019) Characterization of Worlwide Olive Germplasm Banks of Marrakech (Morocco) and Córdoba (Spain): towards management and use of olive germplasm in breeding programs. *PLOS One* 14(10), e0223716.

El Riachy, M., Hamade, A., Ayoub, R., Dandachi, F. and Chalak, L. (2019) Oil content, fatty acid and phenolic profiles of some olive varieties growing in Lebanon. *Frontiers in Nutrition* 6, 94.

Erre, P., Chessa, I., Muñoz-Diez, C., Belaj, A., Rallo, L. and Trujillo, I. (2010) Genetic diversity and relationships between wild and cultivated olives (*Olea europaea* L.) in Sardinia as assessed by SSR markers. *Genetic Resources and Crop Evolution* (57), 41–54.

Evangelou, E., Kiritsakis, K., Sakellaropoulos, N. and Kiritsakis, A. (2018) Table olives production, postharvest processing and nutritional qualities. In: Siddiq, M. and Uebersax, M.A. (eds) *Handbook of Vegetables and Vegetable Processing*, vol. II, pp. 727–744. John Wiley & Sons, Chichester, UK.

Falek, W., Sion, S., Montemurro, C., Mascio, I., Gadaleta, S. *et al.* (2021) Molecular diversity and ecogeographic distribution of Algerian wild olives (*Olea europaea* subsp. *europaea* var. *sylvestris*). *Scientia Agricola* 79.

FAO (2010) *The Second Report on the State of the World's Plant Genetic Resources for Food and Agriculture.* Food and Agricultural Organization, Rome.

Fernández-Escobar, R. (2019) Olive nutritional status and tolerance to biotic and abiotic stresses. *Frontiers in Plant Science* 10, 1151.

Friel, J., Bombarely, A., Fornell, C.D., Luque, F. and Fernández-Ocaña, A.M. (2021) Comparative analysis of genotyping by sequencing and whole-genome sequencing methods in diversity studies of *Olea europaea* L. *Plants* 10(11), 2514.

Gago, P., Santiago, J.L., Boso, S. and Martínez, M.C. (2019) The forgotten, ancient olive trees of the Spanish northwest: a first molecular and botanical analysis. *Spanish Journal of Agricultural Research* 17(2), e0702–e0702.

García-Verdugo, C., Fay, M.F., Granado-Yela, C., Rubio de Casas, R., Balaguer, L. *et al.* (2009) Genetic diversity and differentiation processes in the ploidy series of *Olea europaea* L.: a multiscale approach from subspecies to insular populations. *Molecular Ecology* 18(3), 454–467.

Gómez-Gálvez, F.J., Pérez-Mohedano, D., de la Rosa-Navarro, R. and Belaj, A. (2021) High-throughput analysis of the canopy traits in the worldwide olive germplasm bank of Córdoba using very high-resolution imagery acquired from unmanned aerial vehicle (UAV) *Scientia Horticulturae* 278, 109851.

Grasso, F., Coppola, M., Carbone, F., Baldoni, L., Alagna, F., Perrotta, G. *et al.* (2017) The transcriptional response to the olive fruit fly (*Bactrocera oleae*) reveals extended differences between tolerant and susceptible olive (*Olea europaea* L.) varieties. *PLOS ONE* 12(8), e0183050.

Green, P.S. (2002) A revision of *Olea* L. (Oleaceae). *Kew Bulletin* 57, 91–140.

Gros-Balthazard, M., Besnard, G., Sarah, G., Holtz, Y., Leclercq, J. *et al.* (2019) Evolutionary transcriptomics reveals the origins of olives and the genomic changes associated with their domestication. *The Plant Journal* 100(1), 143–157.

Habib, N.A., Müller, M., Gailing, O., Patzelt, A., Al Issai, G. *et al.* (2021) Genetic diversity and differentiation of *Olea europaea* subsp. *cuspidata* (Wall. and G. Don) Cif. in the Hajar Mountains of Oman. *Genetic Resources and Crop Evolution* 68(3), 865–883.

Hamid, A.R., Hag Husein, H. and Bäumler, R. (2022) Characteristics of some wild olive phenotypes (oleaster) selected from the western mountains of Syria. *Sustainability* 14(9), 5151.

Hannachi, H., Breton, C., Msallem, M., Ben El Hadj, S., El Gazzah, M. and Bervillé, A. (2008) Differences between native and introduced olive cultivars as revealed by morphology of drupes, oil composition and SSR polymorphisms: a case study in Tunisia. *Scientia Horticulturae* 116(3), 280–290.

Hernández, M.L., Sicardo, M.D., Belaj, A. and Martínez-Rivas, J.M. (2021) The oleic/linoleic acid ratio in olive (*Olea europaea* L.) fruit mesocarp is mainly controlled by OeFAD2-2 and OeFAD2-5 genes together with the different specificity of extraplastidial acyltransferase enzymes. *Frontiers in Plant Science* 12, 653997.

Hmmam, I., Mariotti, R., Ruperti, B., Cultrera, N., Baldoni, L. and Barcaccia, G. (2018) Venetian olive (*Olea europaea*) germplasm: disclosing the genetic identity of locally grown cultivars suited for typical extra virgin oil productions. *Genetic Resources and Crop Evolution* 65(6), 1733–1750.

Hosseini-Mazinani, M., Mariotti, R., Torkzaban, B., Sheikh-Hassani, M., Ataei, S. *et al.* (2014). High genetic diversity detected in olives beyond the boundaries of the Mediterranean Sea. *PLOS ONE* 9(4), e93146.

Islam, A.F., Sanders, D., Mishra, A.K. and Joshi, V. (2021) Genetic diversity and population structure analysis of the USDA olive germplasm using genotyping-by-sequencing (GBS). *Genes* 12(12), 2007.

Jiménez, M.R., Casanova, L., Suárez, M.P., Rallo, P. and Morales-Sillero, A. (2017) Internal fruit damage in table olive cultivars under superhigh-density hedgerows. *Postharvest Biology and Technology* 132, 130–137.

Jiménez-Carvelo, A.M., Cruz, C.M., Olivieri, A.C., González-Casado, A. and Cuadros-Rodríguez, L. (2019) Classification of olive oils according to their cultivars based on second-order data using LC-DAD. *Talanta* 195, 69–76.

Jiménez-Ruiz, J., Ramírez-Tejero, J.A., Fernández-Pozo, N., Leyva-Pérez, M.D.L.O., Yan, H. *et al.* (2020) Transposon activation is a major driver in the genome evolution of cultivated olive trees (*Olea europaea* L.) *The Plant Genome* 13(1), e20010.

Julca, I., Marcet-Houben, M., Cruz, F., Gómez-Garrido, J., Gaut, B.S. *et al.* (2020) Genomic evidence for recurrent genetic admixture during the domestication of Mediterranean olive trees (*Olea europaea* L.). *BMC Biology* 18(1), 1–25.

Kabassi, K., Martinis, A. and Minotou, C. (2021) Estimating the value of monumental olive trees: designing a tool using multi-criteria decision making. *Trees, Forests and People* 6, 100151.

Kaniewski, D., Van Campo, E., Boiy, T., Terral, J.F., Khadari, B. and Besnard, G. (2012) Primary domestication and early uses of the emblematic olive tree: palaeobotanical, historical and molecular evidence from the Middle East. *Biological Reviews* 87(4), 885–899.

Kassa, A., Konrad, H. and Geburek, T. (2019) Molecular diversity and gene flow within and among different subspecies of the wild olive (*Olea europaea* L.): a review. *Flora* 250, 18–26.

Kaya, H.B., Akdemir, D., Lozano, R., Cetin, O., Sozer Kaya, H. *et al.* (2019) Genome wide association study of 5 agronomic traits in olive (*Olea europaea* L.). *Scientific Reports* 9(1), 1–14.

Khadari, B., El Bakkali, A., Essalouh, L., Tollon, C., Pinatel, C. and Besnard, G. (2019) Cultivated olive diversification at local and regional scales: evidence from the genetic characterization of French genetic resources. *Frontiers in Plant Science* 10, 1593.

Koehmstedt, A.M., Aradhya, M.K., Soleri, D., Smith, J.L. and Polito, V.S. (2010) Molecular characterization of genetic diversity, structure and differentiation in the olive (*Olea europaea* L.) germplasm collection of the United States Department of Agriculture. *Genetic Resources and Crop Evolution* 58, 519–531.

Kostelenos, G. and Kiritsakis, A. (2017) Olive tree history and evolution. In: Shahidi, F. and Kiritsakis, A. (eds) *Olives and Olive Oil as Functional Foods: Bioactivity, Chemistry and Processing*. John Wiley & Sons Ltd, Chichester, UK, pp. 1–12.

Kyçyk, O., Aguilera, M.P., Gaforio, J.J., Jiménez, A. and Beltrán, G. (2016) Sterol composition of virgin olive oil of forty-three olive cultivars from the World Collection Olive Germplasm Bank of Cordoba. *Journal of the Science of Food and Agriculture* 96(12), 4143–4150.
Kyriakopoulou, C.I. and Kalogianni, D.P. (2020) Genetic identification of the wild form of olive (*Olea europaea* var. *sylvestris*) using allele-specific real-time PCR. *Foods* 9(4), 467.
Larbi, A., Vázquez, S., El-Jendoubi, H., Msallem, M., Abadía, J. *et al.* (2015) Canopy light heterogeneity drives leaf anatomical, eco-physiological and photosynthetic changes in olive trees grown in a high-density plantation. *Photosynthesis Research* 123(2), 141–155.
Lazovic, B., Adakalić M., Pucci, C., Perovic, T., Bandelj, D. *et al.* (2016) Characterizing ancient and local olive germplasm from Montenegro. *Scientia Horticulturae* 209, 117–123.
León, L., De la Rosa, R., Velasco, L. and Belaj, A. (2018). Using wild olives in breeding programs: Implications on oil quality composition. *Frontiers in Plant Science* 9, 232.
Linos, A., Nikoloudakis, N., Katsiotis, A. and Hagidimitriou, M. (2014) Genetic structure of the Greek olive germplasm revealed by RAPD, ISSR and SSR markers. *Scientia Horticulturae* 175, 33–43.
Lorite, I.J., Cabezas, J.M., Ruiz-Ramos, M., de la Rosa, R., Soriano, M.A. *et al.* (2022). Enhancing the sustainability of Mediterranean olive groves through adaptation measures to climate change using modelling and response surfaces. *Agricultural and Forest Meteorology* 313, 108742.
Mariotti, R., Cultrera, N.G., Díez, C.M., Baldoni, L. and Rubini, A. (2010) Identification of new polymorphic regions and differentiation of cultivated olives (*Olea europaea* L.) through plastome sequence comparison. *BMC Plant Biology* 10(1), 1–13.
Mariotti, R., Belaj, A., de la Rosa, R., Leòn, L., Brizioli, F. *et al.* (2020) EST–SNP study of *Olea europaea* L. uncovers functional polymorphisms between cultivated and wild olives. *Genes* 11(8), 916.
Mariotti, R., Pandolfi, S., De Cauwer, I., Saumitou-Laprade, P., Vernet, P. *et al.* (2021) Diallelic self-incompatibility is the main determinant of fertilization patterns in olive orchards. *Evolutionary Applications* 14(4), 983–995.
Medina-Alonso, M.G., Navas, J.F., Cabezas, J.M., Weiland, C.M., Ríos-Mesa, D. *et al.* (2020) Differences on flowering phenology under Mediterranean and subtropical environments for two representative olive cultivars. *Environmental and Experimental Botany* 180, 104239.
Miazzi, M.M., Di Rienzo, V., Mascio, I., Montemurro, C., Sion, S. *et al.* (2020) Re.Ger.O.P.: an integrated project for the recovery of ancient and rare olive germplasm. *Frontiers in Plant Science* 11, 73.
Miho, H., Díez, C.M., Mena-Bravo, A., De Medina, V.S., Moral, J. *et al.* (2018) Cultivar influence on variability in olive oil phenolic profiles determined through an extensive germplasm survey. *Food Chemistry* 266, 192–199.
Mikrou, T., Pantelidou, E., Parasyri, N., Papaioannou, A., Kapsokefalou, M., Gardeli, C. and Mallouchos, A. (2020) Varietal and geographical discrimination of Greek monovarietal extra virgin olive oils based on squalene, tocopherol and fatty acid composition. *Molecules* 25(17), 3818.
Moral, J., Xaviér, C.J., Viruega, J.R., Roca, L.F., Caballero, J. and Trapero, A. (2017) Variability in susceptibility to anthracnose in the world collection of olive cultivars of Cordoba (Spain). *Frontiers in Plant Science* 8, 1892.
Morales-Cruz, A., Aguirre-Liguori, J.A., Massonet, M., Zaccheo, M., Minio, A. *et al.* (2022) Multigenic resistance to *Xylella fastidiosa* in wild grapes (*Vitis* sps.) and its implications within a changing climate. *BioRxiv.*
Moriondo, M., Trombi, G., Ferrise, R., Brandani, G., Dibari, C. *et al.* (2013) Olive trees as bio-indicators of climate evolution in the Mediterranean Basin. *Global Ecology and Biogeography* 22(7), 818–833.
Mougiou, N., Baalbaki, B., Doupis, G., Kavroulakis, N., Poulios, S., Vlachonasios, K.E. and Koubouris, G.C. (2020) The effect of low temperature on physiological, biochemical and flowering functions of olive tree in relation to genotype. *Sustainability* 12(23), 10065.
Mousavi, S., Mariotti, R., Bagnoli, F., Costantini, L. Cultrera, N.G. *et al.* (2017) The eastern part of the Fertile Crescent concealed an unexpected route of olive (*Olea europaea* L.) differentiation. *Annals of Botany* 119(8), 1305–1318.
Mousavi, S., de la Rosa, R., Moukhli, A., El Riachy, M., Mariotti, R. *et al.* (2019a) Plasticity of fruit and oil traits in olive among different environments. *Scientific Reports* 9(1), 1–13.
Mousavi, S., Regni, L., Bocchini, M., Mariotti, R., Cultrera, N.G. *et al.* (2019b) Physiological, epigenetic and genetic regulation in some olive cultivars under salt stress. *Scientific Reports* 9(1), 1–17.
Mousavi, S., Stanzione, V., Mencuccini, M., Baldoni, L., Bufacchi, M. and Mariotti, R. (2019c) Biochemical and molecular profiling of unknown olive genotypes from central Italy: determination of major and minor components. *European Food Research and Technology* 245(1), 83–94.

Mousavi, S., Stanzione, V., Mariotti, R., Mastio, V., Azariadis, A. *et al.* (2022) Bioactive compound profiling of olive fruit: the contribution of genotype. *Antioxidants* 11(4), 672.
Murillo, J.M., Madejón, E., Madejón, P. and Cabrera, F. (2005) The response of wild olive to the addition of a fulvic acid-rich amendment to soils polluted by trace elements (SW Spain). *Journal of Arid Environments* 63(1), 284–303.
Naz, S., Khan, M.R., Awan, A.A., Hussain, M. and Ali, S. (2019). Diversity and divergence in cultivated and wild olive germplasm collected from northern Pakistan. *International Journal Agricultural Biology* 22, 1109–1115.
Ninot, A., Howad, W., Aranzana, M.J., Senar, R., Romero, A. *et al.* (2018) Survey of over 4,500 monumental olive trees preserved on-farm in the northeast Iberian Peninsula, their genotyping and characterization. *Scientia Horticulturae* 231, 253–264.
Omri, A., Abdelhamid, S., Ayadi, M., Araouki, A., Gharsallaoui, M. *et al.* (2021) The investigation of minor and rare Tunisian olive cultivars to enrich and diversify the olive genetic resources of the country. *Journal of Food Composition and Analysis*, 95, 103657.
Orr, A.J., Padovan, A., Kainer, D., Külheim, C., Bromham, L. *et al.* (2020) A phylogenomic approach reveals a low somatic mutation rate in a long-lived plant. *Proceedings of the Royal Society B* 287(1922), 20192364.
Paoletti, A., Rosati, A. and Famiani, F. (2021) Effects of cultivar, fruit presence and tree age on whole-plant dry matter partitioning in young olive trees. *Heliyon* 7(5), e06949.
Pérez, A.G., León, L., Pascual, M., de la Rosa, R., Belaj, A. and Sanz, C. (2019) Analysis of olive (*Olea europaea* L.) genetic resources in relation to the content of vitamin E in virgin olive oil. *Antioxidants* 8(8), 242.
Petit, G., Bleve, G., Gallo, A., Mita, G., Montanaro, G. *et al.* (2021) Susceptibility to *Xylella fastidiosa* and functional xylem anatomy in *Olea europaea*: revisiting a tale of plant–pathogen interaction. *AoB Plants* 13(4), plab027.
Petruccelli, R., Giordano, C., Salvatici, M.C., Capozzoli, L., Ciaccheri, L. *et al.* (2014) Observation of eight ancient olive trees (*Olea europaea* L.) growing in the Garden of Gethsemane. *Comptes Rendus Biologies* 337(5), 311–317.
Petruccelli, R., Bartolini, G., Ganino, T., Zelasco, S., Lombardo, L. *et al.* (2022) Cold stress, freezing adaptation, varietal susceptibility of *Olea europaea* L.: a review. *Plants* 11(10), 1367.
Quesada-Moraga, E., Santiago-Álvarez, C., Cubero-González, S., Casado-Mármol, G., Ariza-Fernández, A. and Yousef, M. (2018) Field evaluation of the susceptibility of mill and table olive varieties to egg-laying of olive fly. *Journal of Applied Entomology* 142(8), 765–774.
Rahemi, M., Karimi, S., Sedaghat, S. and Rostami, A.A. (2017) Physiological responses of olive cultivars to salinity stress. *Advances in Horticultural Science* 31(1), 53–60.
Rallo, L., Barranco, D., Castro-García, S., Connor, D. J., Gómez del Campo, M., and Rallo, P. (2013). High-density olive plantations. *Horticultural Reviews*, 41, 303–384.
Rallo, L. (2014) Breeding oil and table olives for mechanical harvesting in Spain. *HortTechnology* 24(3), 295–300.
Rallo, P., Morales-Sillero, A., Brenes, M., del Rocio Jiménez, M., Sánchez, A.H. *et al.* (2018). Elaboration of table olives: assessment of new olive genotypes. *European Journal of Lipid Science and Technology* 120(6), 1800008.
Ramírez-Tejero, J.A., Jiménez-Ruiz, J., Serrano, A., Belaj, A., León, L. *et al.* (2021) Verticillium wilt resistant and susceptible olive cultivars express a very different basal set of genes in roots. *BMC Genomics* 22(1), 1–16.
Razouk, R., Hssaini, L., Alghoum, M., Adiba, A. and Hamdani, A. (2022) Phenotyping olive cultivars for drought tolerance using leaf macro-characteristics. *Horticulturae* 8(10), 939.
Regni, L., Del Pino, A.M., Mousavi, S., Palmerini, C.A., Baldoni, L. *et al.* (2019) Behavior of four olive cultivars during salt stress. *Frontiers in Plant Science* 10, 867.
Rotondi, A., Ganino, T., Beghè, D., Di Virgilio, N., Morrone, L. *et al.* (2018). Genetic and landscape characterization of ancient autochthonous olive trees in northern Italy. *Plant Biosystems* 152(5), 1067–1074.
Sakar, E., Unver, H., Bakir, M., Ulas, M. and Sakar, Z.M. (2016) Genetic relationships among olive (*Olea europaea* L.) cultivars native to Turkey. *Biochemical Genetics* 54(4), 348–359.
Salazar-García, D.C., Malheiro, R., Pereira, J.A. and Lopéz-Cortés, I. (2019) Unexplored olive cultivars from the Valencian Community (Spain): some chemical characteristics as a valorization strategy. *European Food Research and Technology* 245(2), 325–334.
Schicchi, R., Speciale, C., Amato, F., Bazan, G., Di Noto, G. *et al.* (2021) The monumental olive trees as biocultural heritage of Mediterranean landscapes: the case study of Sicily. *Sustainability* 13(12), 6767.
Schoen, D.J., and Schultz, S.T. (2019). Somatic mutation and evolution in plants. *Annual Review of Ecology, Evolution, and Systematics* 50, 49–73.

Sion, S., Taranto, F., Montemurro, C., Mangini, G., Camposeo, S. *et al.* (2019) Genetic characterization of Apulian olive germplasm as potential source in new breeding programs. *Plants* 8, 268–281.
Sorkheh, K. and Khaleghi, E. (2016) Molecular characterization of genetic variability and structure of olive (*Olea europaea* L.) germplasm collection analyzed by agro-morphological traits and microsatellite markers. *Turkish Journal of Agriculture and Forestry* 40(4), 583–596.
Surano, A., Abou Kubaa, R., Nigro, F., Altamura, G., Losciale, P., Saponari, M. and Saldarelli, P. (2022) Susceptible and resistant olive cultivars show differential physiological response to *Xylella fastidiosa* infections. *Frontiers in Plant Science* 13, 96893.
Tan, J., Ben-Gal, A., Shtein, I., Bustan, A., Dag, A. and Erel, R. (2020) Root structural plasticity enhances salt tolerance in mature olives. *Environmental and Experimental Botany* 179, 104224.
Topi, D., Guclu, G., Kelebek, H. and Selli, S. (2021) Olive oil production in Albania, chemical characterization and authenticity. In: Akram, M. (ed.) *Olive Oil: New Perspectives and Applications*. IntechOpen.
Trapero, C., Rallo, L., Lopez-Escudero, F.J., Barranco, D. and Díez, C.M. (2015) Variability and selection of *Verticillium* wilt resistant genotypes in cultivated olive and in the *Olea* genus. *Plant Pathology* 64, 890–900.
Trujillo, I., Ojeda, M.A., Urdiroz, N.M., Potter, D., Barranco, D., Rallo, L. and Diez, C.M. (2014) Identification of the Worldwide Olive Germplasm Bank of Córdoba (Spain) using SSR and morphological markers. *Tree Genetics & Genomes* 10, 141–155.
Valeri, M.C., Mifsud, D., Sammut, C., Pandolfi, S., Lilli, E. *et al.* (2022) Exploring olive genetic diversity in the Maltese islands. *Sustainability* 14(17), 10684.
Veloso, M.M., Simões-Costa, M.C., Carneiro, L.C., Guimarães, J.B., Mateus, C., Fevereiro, P. and Pinto-Ricardo, C. (2018) Olive tree (*Olea europaea* L.) diversity in traditional small farms of Ficalho, Portugal. *Diversity* 10(1), 5.
Yazbeck, E.B., Abi Rizk, G., Battistini, A., Hassoun, G., Geagea, L. *et al.* (2019) Morphological and biochemical characterization of ancient olive trees in Bshaaleh Lebanon. *International Journal of Current Research* 11(03), 2337–2345.
Yılmaz-Düzyaman, H., de la Rosa, R., and León, L. (2022). Seedling selection in olive breeding progenies. *Plants,* 11(9), 1195.
Zahradníková, E., Ficek, A., Brejová, B., Vinař, T. and Mičieta, K. (2020) Mosaicism in old trees and its patterns. *Trees* 34(2), 357–370.
Zhan, M.M., Cheng, Z.Z., Su, G.C., Wang, A.Y., Chen, H.P. *et al.* (2015) Genetic relationships analysis of olive cultivars grown in China. *Genetics and Molecular Research* 14(2), 5958–5969.

3.3 Genotyping Technologies for Olive Cultivar Characterization and Development of Functional Markers

S. Mousavi[1], R. Mariotti[1], A. Belaj[2], C. Guerrero[3], M.C. Valeri[1] and L. Baldoni[1]
[1]*National Research Council, Institute of Biosciences and Bioresources, Perugia, Italy;* [2]*IFAPA – Centro Alameda del Obispo, Córdoba, Spain;* [3]*University of Malaga, Department of Molecular Biology and Biochemistry, Malaga, Spain*

3.3.1 Introduction

To a very rich and complex olive germplasm corresponds great difficulty in identifying and characterizing each single genotype.

For the correct identification of olive varieties, it is necessary to take into consideration some main aspects:

- the availability of reliable, precise and shared molecular markers;
- the possibility of accessing safe, credited and validated sources of plant material, currently represented by the institutional public varietal collections, where each variety has a known origin and a reliable and validated consensus molecular profile;
- the access to databases including the molecular information on varieties, as well as any other morphological and bio-agronomic data.

For these reasons, many olive cultivar collections have been established at regional, national and international level, accessions have been genotyped and molecular profiles have been made available to the scientific and stakeholder community (Trujillo *et al.*, 2014; Ben Ayed *et al.*, 2016; Mousavi *et al.*, 2017a, b; El Bakkali *et al.*, 2019). To identify samples of unknown varieties, now it is possible to genotype them and compare their profiles with those present in the databases and check for any matches (Belaj *et al.*, 2022).

3.3.2 Genotyping for Cultivar Identification

Since the early 1990s, various studies have been launched for the identification of olive varieties, using different types of molecular markers. These studies have made it possible to draw an overall picture of the richness of the varietal heritage. However, the dominant nature of some of the used markers, the lack of knowledge of DNA sequence and the absence of reference molecular profiles for the main varieties have, in some cases, contributed to further confusion, especially as regards intra-cultivar variability, sustained by some and denied by other authors.

For these reasons, we found it more useful to report only the markers that have demonstrated the greatest reliability and polymorphism. We then reported those most used today both for identification and

DOI: 10.1079/9781789247350.0007

discrimination work, for the study of the distribution of variability, the domestication process and for other purposes of great application relevance.

3.3.2.1 Simple sequence repeats – SSRs

SSRs represent the most popular markers for olive genotyping, due to the high polymorphism, particular abundance and fast transferability (Baldoni *et al.*, 2009; Belaj *et al.*, 2012; Díez *et al.*, 2015; Mousavi *et al.*, 2017a, b). They have been used to characterize most of the olive germplasm collections (Haouane *et al.*, 2011; Trujillo *et al.*, 2014; El Bakkali *et al.* 2019).

Other than for cultivar fingerprinting, SSRs have been used for many other purposes, including construction of linkage maps, traceability of olive oil cultivar composition, as well as for phylogenetic and domestication studies and population genomics (Dominguez-Garcia *et al.*, 2012; Kaniewski *et al.*, 2012; Díez *et al.,* 2015; Besnard *et al.*, 2018; Gomes *et al.*, 2018; Crawford *et al.*, 2019; Montemurro *et al.*, 2019; Mousavi *et al.*, 2019a; Mariotti *et al.*, 2020a).

About 100 SSR markers have been developed to date in olive, but only a small percentage of them, after a deep evaluation, have been selected and extensively used for cultivar identification and discrimination (Baldoni *et al.*, 2009). El Bakkali *et al.* (2019) have used 20 SSRs to study the molecular identity of the cultivars in common between the World Olive Germplasm Banks of Marrakech (Morocco) and Cordoba (Spain).

Most of the SSR markers used for fingerprinting contain dinucleotide repeats, which represent very short motifs that make it difficult to discriminate among alleles that differ only for two bases. To overcome this problem, it should be preferable to use polynucleotide SSRs, differing for 3, 4, 5, 6 or more bases. Recent advances in olive transcriptomics and genome sequencing have enabled access to sequence repositories, such as ESTs and untranscribed sequences, as a source of long-core repeat SSRs (Muñoz-Mérida *et al.*, 2013; Mariotti *et al.*, 2016; Arbeiter *et al.*, 2017) (Fig. 3.3.1). Even if long repeat SSRs are easier to analyze, their variation has proved to be lower compared to dinucleotide markers, likely because they are mainly developed from expressed gene sequences, and their application

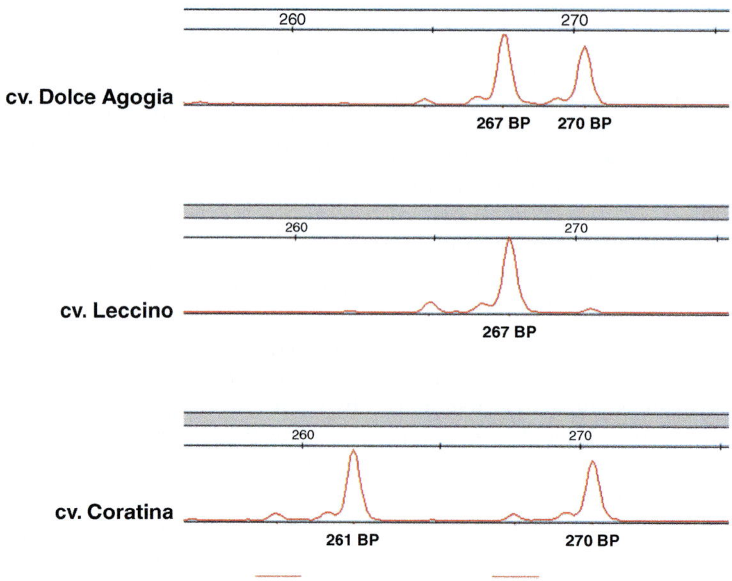

Fig. 3.3.1. SSR trinucleotide profiles for three cultivars.

for cultivar discrimination still remains rarely utilized (de la Rosa et al., 2013; Mousavi *et al.*, 2017a; Li *et al.*, 2020).

Numerous identification and characterization projects have been undertaken in all countries and at regional level by the use of SSR markers, allowing a detailed analysis of the large variability of olive cultivars (Cantini *et al.*, 2008; Corrado *et al.*, 2011; Abdessemed *et al.*, 2015; Belaj *et al.*, 2010; Beghè *et al.*, 2011; Marra *et al.*, 2013; Hosseini-Mazinani *et al.*, 2014; i Martí *et al.*, 2015; Lazović *et al.*, 2016; Hmmam *et al.*, 2018; Ninot *et al.*, 2018; Avramidou *et al.*, 2020; Li *et al.*, 2020; Gómez-Rodríguez *et al.*, 2021; Saddoud Debbabi *et al.*, 2021; Yadav *et al.*, 2021).

SSR markers are successfully applied for paternity tests in order to identify cultivars, progenies or embryo parents (Seifi *et al.*, 2012; Dridi *et al.*, 2018; Mariotti *et al.*, 2020a; Vuletin Selak *et al.*, 2021).

Few efforts have been devoted to the identification of nursery plant material for true-to-type stock certification (Rallo *et al.*, 2018), while markers are being developed to detect the presence of dangerous pathogens for propagated plants, such as *Xylella fastidiosa* (Loconsole *et al.*, 2021), or *Verticillium dahliae* (Altaae, 2019).

3.3.2.2 Single nucleotide polymorphisms – SNPs

SNPs are point mutations on single nucleotides, generally biallelic and very frequent along the genomic sequence, particularly in olive where SNP frequencies of 1/53 bp or 1/19.5 bp were detected within gene sequences, respectively shown by Cultrera *et al.* (2019) and Salimonti *et al.* (2020). They are randomly distributed along the entire genome, are less mutable than SSRs and readily assayed using high-throughput genotyping protocols and automated data analysis.

Based on transcriptome and genome sequence data made available for many cultivars, wide sets of SNP markers are being identified and used for genotyping purposes (Kaya *et al.*, 2013; Biton *et al.*, 2015; D'Agostino *et al.*, 2018; Belaj *et al.*, 2018; Mariotti *et al.*, 2020b).

Sequences related to expressed tags developed from transcript collections contributed to evaluate gene variations among different olive cultivars, highlighting possible functional effects of these polymorphisms. Thanks to a wide set of EST–SNPs, the study of functional polymorphisms in olive is finally having a boost.

Among the strategies for SNP analysis, genotyping-by-sequencing (GBS) technology is a method based on next-generation sequencing, that enables multi-sample, high-throughput parallel sequencing.

SNP genotyping has been successfully applied for olive cultivar discrimination (Hakim *et al.*, 2010; Zhu *et al.*, 2019; Friel *et al.*, 2021; Islam *et al.*, 2021). Kaya *et al.* (2013) and Besnard and El Bakkali *et al.* (2014) analyzed 2987 SNPs and characterized high levels of genetic variation in Turkish olive genotypes. Belaj *et al.* (2018) reported the clear cut-off between inter- and intra-cultivar variation in olive by using 1043 EST-SNPs on a representative number of accessions from 20 olive-growing countries (Fig. 3.3.2). Using 22,088 GBS SNP

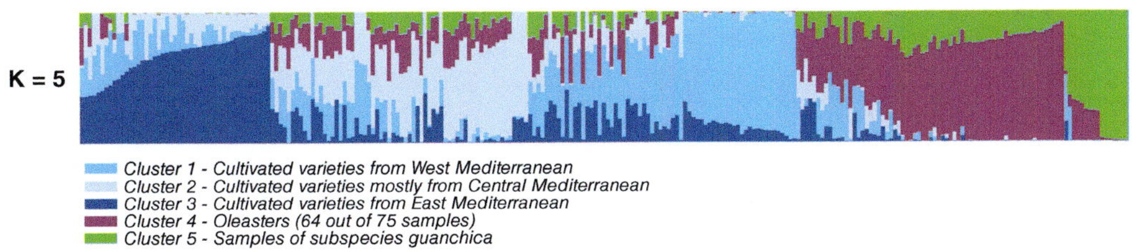

Fig. 3.3.2. Population structure estimated by the Bayesian clustering based on the genetic diversity of 260 *Olea europaea* accessions including cultivated and wild olives and a related subspecies analyzed by 1040 SNP markers. (From Mariotti *et al.*, 2020b)

data, a wide set of Italian cultivars was analyzed, detecting unexpected molecular diversity (D'Agostino *et al.*, 2018).

Studies have been carried out also to develop SNP markers able to distinguish between cultivated and wild forms of olive (Kaya *et al.*, 2013; Besnard and El Bakkali, 2014; Kassa *et al.*, 2019; Kyriakopoulou and Kalogianni, 2020; Mariotti *et al.*, 2020b) (Fig. 3.3.3).

3.3.3 Functional Markers

In order to develop new functional markers useful for discriminating among plant functions rather than among genotypes, SNP markers have been largely used for genetic mapping and for genome-wide association studies (GWAS), the most powerful strategies to detect marker-phenotype linkage, that is the association between a molecular

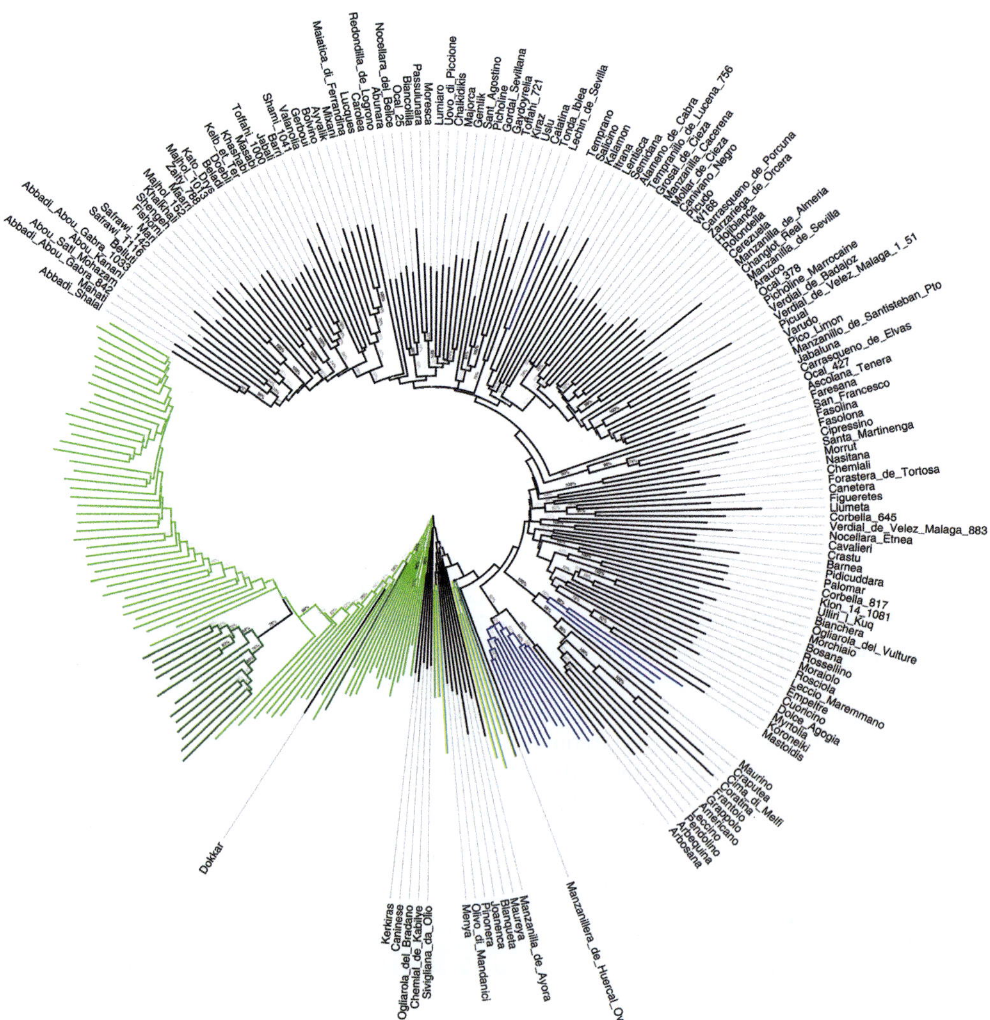

Fig. 3.3.3. Neighbor joining dendrogram of wild olives, cultivated varieties and cross-bred genotypes. Black branches: olive cultivars; green branches: wild olives; light-green: wild olives from Canary Islands; blue branches: crossbred progenies. Bootstrap values are reported as percentage at each branch node. (From Belaj *et al.*, 2018)

marker and an agronomic or qualitative character of the tree. The so-called functional markers developed by these approaches allow screening of many individuals (varieties, cross-progenies etc.) for a certain character by doing a simple molecular analysis rather than complex and long *in vivo* evaluations (Kaya *et al.*, 2019; Moret *et al.*, 2023).

3.3.3.1 Transcriptomics

The most widely used approach to capture genes putatively involved in specific characters (biosynthetic metabolic pathways, resistance to pathogens and pests, tolerance to environmental stress etc.) refers to the identification of gene transcripts differentially (over- or under-) expressed under the conditions of interest (i.e. flowering, oil accumulation, plant response to pathogen attack or to drought stress etc.).

Studies have focussed on many aspects of olive plant development, production and stress response. Transcriptomics has allowed the identification differentially expressed sequences involved in tolerance to *Verticillium dahliae* (Leyva-Pérez *et al.*, 2018), or in response to the olive fruit fly *Bactrocera oleae* attack (Grasso *et al.*, 2017), or under salt stress defence and cold acclimation (Bazakos *et al.*, 2015; Guerra *et al.*, 2015).

Muñoz-Merida *et al.* (2013) got information about genes determining oil content and composition in fruit mesocarp of cvs. 'Picual' and 'Arbequina', and numerous transcripts differentially transcribed during fruit fatty acid synthesis were identified by Niu *et al.* (2022).

Transcripts putatively involved in phenolic synthesis have been identified by comparative large-scale transcriptome analysis performed on fruits of high- and low-phenolic cultivars at different developmental stages (Alagna *et al.*, 2009; Guodong *et al.*, 2019), as well as those controlling flavonoid and anthocyanin metabolism in cultivars producing white and black drupes (Iaria *et al.*, 2016).

Transcriptome analysis revealed insights into hormonal and vesicle trafficking regulation among fruit tissues (Briegas *et al.*, 2020) and changes in gene expression were observed in drupes of cv. 'Carolea' at different maturation stages and cultivation areas (Bruno *et al.*, 2019).

After the release of wide transcriptome data on pollen and pistil at different developmental stages (Carmona *et al.*, 2015), transcripts potentially involved in flower development, pollen–pistil interaction or ovary abortion were investigated by comparing different cultivars (Alagna *et al.*, 2016a).

Finally, transcriptome analysis has allowed the identification miRNAs (non-coding RNAs that play important roles in regulating gene expression) involved in alternate bearing (Yanik *et al.*, 2013) or in fruit ripening (Carbone *et al.*, 2019).

3.3.3.2 Genetic mapping

The generation of high-resolution linkage maps represents a powerful way to identify markers linked to characters of interest, a prerequisite for gene positional identification. For example, a cross-progeny segregating for incompatibility, putatively controlled by a single locus, has been used for a deep SNP genotyping in order to identify SNP markers tightly linked to this trait (Mariotti *et al.*, 2020c). If the character of interest is controlled by multiple genes, as is the case for most agronomic traits, it is possible at least to identify major quantitative trait loci (QTL).

Linkage analysis can be performed on biparental population families derived from the cross, backcross or selfing of homozygous or heterozygous lines. In olive, the use of these progenies is hindered by the long generation time and the lack of homozygous genotypes; thus, only full-sib F1 families derived from intervarietal crosses of highly heterozygous genotypes have been used for olive mapping, and linkage analysis has been separately conducted for each parent.

Dense maps have been produced using AFLP, DArT, SSR and SNP markers on cross-progenies of different cultivars and with the aim to identify markers for specific traits (de la Rosa *et al.*, 2003; Sadok *et al.*, 2013; Atienza *et al.*, 2014; Biton *et al.*, 2015; Baldoni *et al.*, 2016; Ipek *et al.*, 2016, 2017; Kaya *et al.*, 2016; Marchese *et al.*, 2016; Unver *et al.*, 2017; Mariotti *et al.*, 2020c) (Fig. 3.3.4).

3.3.3.3. Genome-wide association studies

GWAS, based on the SNP analysis on unrelated individuals from natural or cultivated populations, requires whole-genome scans with large sets of markers directly applied on phenotypically characterized genotypes, to detect linkage between markers and traits. It mostly provides a high resolution and does not need segregating populations. The development of this kind of marker requires a high level of genome sequence information, it is therefore not surprising if only a few GWA studies were reported in olive before the release of sound and informative genome sequences of numerous varieties (Khadari *et al.*, 2014; Kaya *et al.*, 2019). Another limitation for the application of GWAS refers to the need of precise, deep and climate-referenced phenotyping of as many cultivars as possible (Moret *et al.*, 2023; Mousavi *et al.*, 2019b).

3.3.3.4 Epigenetic studies

In addition to genetic diversity in olive, the first studies on the epigenetic changes were also recently carried out, considering the increasing knowledge on how epigenetics may affect the way genes work. The methylation status of the olive genome during

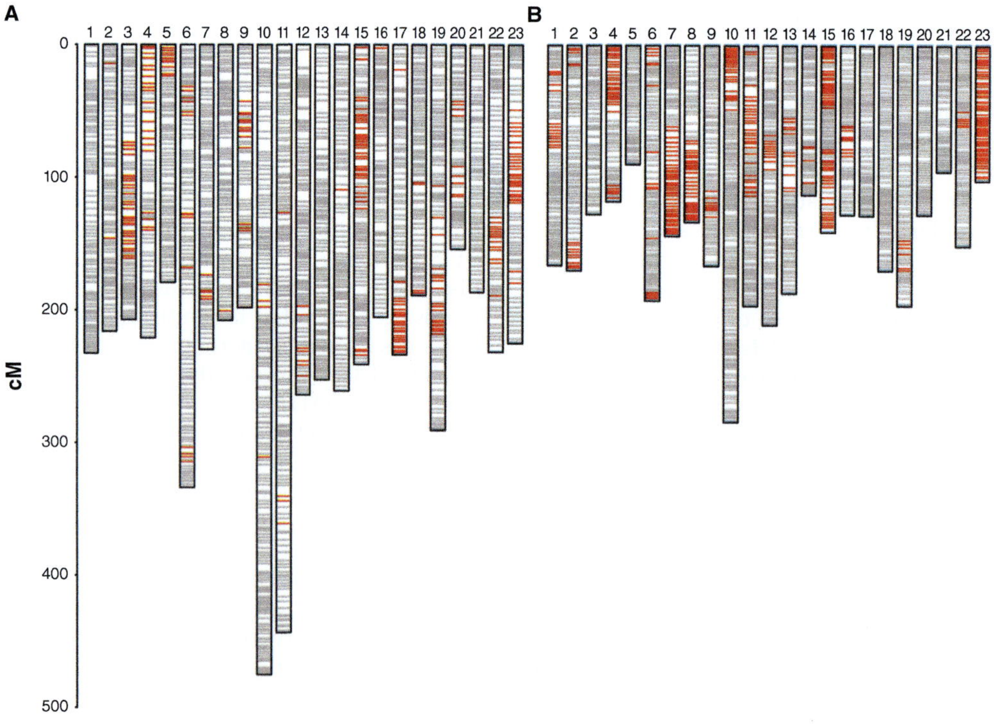

Fig. 3.3.4. Parental genetic maps: diagrams show linkage groups (LGs) in the progeny parental cultivars 'Leccino' (panel A) and 'Dolce Agogia' (panel B). Markers are plotted in gray. Markers with distorted segregation (a = 0.05) are plotted in red. LGs are not oriented with respect to the conventional orientation of chromosomes in the 'sylvestris' olive genome assembly. (From Mariotti *et al.*, 2020c)

fruit ripening has been revealed through methylated DNA immunoprecipitation sequencing (MeDIP-seq) approach, profiling leaves and fruits during fruit development in wild and cultivated olives (Badad *et al.*, 2021) and differentially methylated gene sequences under salt stress were identified distinguishing tolerant and susceptible cultivars (Mousavi *et al.*, 2019c).

3.3.3.5 Genes and gene polymorphisms involved in plant performance

Massive EST and genomic sequence data have made available data on annotated gene sequences, their organization and distribution on the genome and their allele differences, representing unprecedented information that should be fully exploited in the coming years.

Numerous works have focussed on the evaluation of gene-sequence polymorphisms, in order to detect potential markers linked to agronomical traits of interest, for germplasm screening and for breeding purposes.

Great advancements have been reached on the identification of genes involved in the biosynthetic pathways leading to the most important secondary metabolites of olive, such as phenolics, sterols, triterpenes and tocopherols (Alagna *et al.*, 2016b; Koudounas *et al.*, 2017; Georgiadou *et al.*, 2019; Inês *et al.*, 2019; Volk *et al.*, 2019; García-Vico *et al.*, 2021; Rodríguez-López *et al.*, 2021; Wang *et al.*, 2022).

Detailed analyses on the sequence of genes involved in lipid synthesis and fatty acid desaturation have allowed the identification of SNPs and indels (insertions/deletions) putatively controlling oil content and fatty acid composition in olive fruits (Cultrera *et al.*, 2019; Hernández *et al.*, 2019, 2020; Contreras *et al.*, 2020; Vatansever *et al.*, 2022).

Candidate genes putatively involved in the juvenile-to-adult transition in olive cultivars have been identified (Fernández-Ocaña *et al.*, 2010; Jiménez-Ruiz *et al.*, 2015). Nucleotide diversity observed in candidate genes for Verticillium wilt resistance in olive has been reported as responsible for plant response to pathogen attack (Cardoni *et al.*, 2022; Serrano *et al.*, 2020). Differentially methylated and regulated genes in tolerant and susceptible olive cultivars under salt stress have been identified (Mousavi *et al.*, 2019c; Skodra *et al.*, 2021).

3.3.4 DNA Tracking of Olive Products, Olive Oil and Table Olives

An important application of molecular analysis refers to the evaluation of food products derived from the olive crop: olive oil and table olives, that represent main targets of food fraud in Europe, including mislabeling of origin and composition. Oil and table olives contain traces of DNA that can be amplified and analyzed (Baldoni *et al.*, 2013). It represents *a posteriori* analysis, an alternative to documentary traceability, also applicable to products on the shelf, able to detect the raw materials from which they were obtained. But, while the markers used for tree-derived samples can be directly applied to olives, in the case of oil, the analysis of molecular markers is complicated by the small quantity of DNA, by its degradation level and, above all, by the fact that oils are generally extracted not from a single variety but from blends of several cultivars, in unpredictable proportions. For this reason, the application of standard markers on oils has proved difficult to apply to commercial oils. Many efforts have been dedicated to improve methods of DNA extraction from olive oil (Raieta *et al.*, 2015; Piarulli *et al.*, 2019), as well as to identify suitable and reliable markers to distinguish different cultivars in a blended oil and thus revealing its varietal composition (Pereira *et al.*, 2018; Chedid *et al.*, 2020; Ben Ayed *et al.*, 2022).

Other works have demonstrated the possibility to identify the cultivars of table olives (Crawford *et al.*, 2020; Sion *et al.*, 2021), also through the identification of the microbial flora in processed fruits (Benítez-Cabello *et al.*, 2020).

For DNA barcoding, chloroplast markers represent an ideal system and some chloroplast regions have been used for DNA food tracking and forensic studies on olive cultivars and olive oil composition (Uncu *et al.*, 2017; Partovi *et al.*, 2021), but the low variability of plastid markers has made this method poorly applied for food analysis.

3.3.5 Population Genomics and Olive Domestication

Through genome sequence data on a number of cultivated and wild olives, it became possible to establish genomic evidence for recurrent genetic admixture during the domestication of Mediterranean olive trees (Julca *et al.*, 2020), and evolutionary transcriptomics has revealed the genomic changes associated with olive domestication (Gros-Balthazard *et al.*, 2019).

Cytoplasm markers, represented by the polymorphisms of chloroplast and mitochondrial genomes, have been used in olive for different purposes. Since these organelles pass to offspring from the female parent also in olive, as in many other plant species (Besnard *et al.*, 2000), they are able to specifically trace the maternal lineage, allowing for reconstructing the routes of olive diversification along time and space.

Chloroplast markers thus represent a valuable tool in population genomic analysis, phylogenesis, domestication and cultivar-wild discrimination (Mariotti *et al.*, 2010; Besnard and de Casas 2016; Van de Paer *et al.*, 2018). A strong genetic differentiation for chloroplast markers was observed between eastern and western Mediterranean olives (Besnard *et al.*, 2011, 2018), and phylogenetic studies in cultivated and wild olives showed three sublineages, E1, E2 and E3, and about 40 chlorotypes (Besnard, 2008; Besnard *et al.*, 2007). Meanwhile it is still quite difficult to distinguish between cultivated varieties and wild olives based on nuclear markers (Mariotti *et al.*, 2020b), chlorotypes are strongly geographically structured in the wild olive populations, while most of the varieties belong to a single chlorotype (E1.1), typical of the eastern Mediterranean, and only a small percentage of them show polymorphisms typical of the wild populations of central and western Mediterranean (Mariotti *et al.*, 2010; Besnard *et al.*, 2013, 2018).

The mitochondrial genome of olive has been recently sequenced, but the evaluation of its polymorphisms among cultivars and other olive taxa is still ongoing (Van de Paer *et al.*, 2018).

References

Abdessemed, S., Muzzalupo, I. and Benbouza, H. (2015) Assessment of genetic diversity among Algerian olive (*Olea europaea* L.) cultivars using SSR marker. *Scientia Horticulturae* 192, 10–20.

Alagna, F., D'Agostino, N., Torchia, L., Servili, M., Rao, R. *et al.* (2009) Comparative 454 pyrosequencing of transcripts from two olive genotypes during fruit development. *BMC Genomics* 10(1), 1–15.

Alagna, F., Cirilli, M., Galla, G., Carbone, F., Daddiego, L. *et al.* (2016a) Transcript analysis and regulative events during flower development in olive (*Olea europaea* L.). *PLOS ONE* 11(4), e0152943.

Alagna, F., Geu-Flores, F., Kries, H., Panara, F., Baldoni, L. *et al.* (2016b) Identification and characterization of the iridoid synthase involved in oleuropein biosynthesis in olive (*Olea europaea*) fruits. *Journal of Biological Chemistry* 291(11), 5542–5554.

Altaae, H.H. (2019) Using nested PCR to detect the non-defoliating pathotype of *Verticillium dahliae* on olive orchard. *Crop Research* 54(5–6), 139–142.

Arbeiter, A.B., Hladnik, M., Jakše, J. and Bandelj, D. (2017) Identification and validation of novel EST-SSR markers in olives. *Scientia Agricola* 74, 215–225.

Atienza, S.G., de la Rosa, R., León, L., Martín, A. and Belaj, A. (2014) Identification of QTL for agronomic traits of importance for olive breeding. *Molecular Breeding* 34(2), 725–737.

Avramidou, E.V., Koubouris, G.C., Petrakis, P.V., Lambrou, K.K., Metzidakis, I.T. and Doulis, A.G. (2020) Classification binary trees with SSR allelic sizes: combining regression trees with genetic molecular data in order to characterize genetic diversity between cultivars of *Olea europaea* L. *Agronomy* 10(11), 1662.

Badad, O., Lakhssassi, N., Zaid, N., El Baze, A., Zaid, Y. *et al.* (2021) Genome wide MeDIP-Seq profiling of wild and cultivated olives trees suggests DNA methylation fingerprint on the sensory quality of olive oil. *Plants* 10(7), 1405.

Baldoni, L., Cultrera, N.G.M., Mariotti, R., Ricciolini, C., Arcioni, S. *et al.* (2009) A consensus list of microsatellite markers for olive genotyping. *Molecular Breeding* 24, 213–231.

Baldoni, L., Cultrera, N.G.M. and Mariotti, R. (2013) DNA tracking for the authentication of edible plant oils. In: Sforza, S. (ed.) *Food Authentication Using Bio-organic Molecules. DEStech Publications* Inc, pp. 221–258.

Baldoni, L., Khadari, B. and de la Rosa, R. (2016) Genetic mapping and detection of quantitative trait loci. In: Rugini, E., Baldoni, L., Muleo, R. and Sebastiani, L. (eds) *The Olive Tree Genome*. Springer, Cham, Switzerland, pp. 65–74.

Bazakos, C., Manioudaki, M.E., Sarropoulou, E., Spano, T. and Kalaitzis, P. (2015) 454 pyrosequencing of olive (*Olea europaea* L.) transcriptome in response to salinity. *PLOS ONE* 10(11), e0143000.

Beghè, D., Ferrarini, A., Ganino, T. and Fabbri, A. (2011) Molecular characterization and identification of a group of local *Olea europaea* L. varieties. *Tree Genetics & Genomes* 7(6), 1185–1198.

Belaj, A., Muñoz-Diez, C., Baldoni, L., Satovic, Z. and Barranco, D. (2010) Genetic diversity and relationships of wild and cultivated olives at regional level in Spain. *Scientia Horticulturae* 124(3), 323–330.

Belaj, A., Dominguez-García, M., Atienza, S.G., Urdíroz, N.M., de la Rosa, R. *et al.* (2012) Developing a core collection of olive (*Olea europaea* L.) based on molecular markers (DArTs, SSRs, SNPs) and agronomic traits. *Tree Genetics & Genomes* 8, 365–378.

Belaj, A., de la Rosa, R., Lorite, I.J., Mariotti, R., Cultrera, N.G. *et al.* (2018) Usefulness of a new large set of high throughput EST-SNP markers as a tool for olive germplasm collection management. *Frontiers in Plant Science* 9.

Belaj, A., Ninot, A., Gómez-Gálvez, F.J., El Riachy, M., Gurbuz-Veral, M. *et al.* (2022) Utility of EST-SNP markers for improving management and use of olive genetic resources: a case study at the Worldwide Olive Germplasm Bank of Córdoba. *Plants* 11(7), 921.

Ben Ayed, R., Ben Hassen, H., Ennouri, K., Ben Marzoug, R. and Rebai, A. (2016) OGDD (Olive Genetic Diversity Database): a microsatellite markers' genotypes database of worldwide olive trees for cultivar identification and virgin olive oil traceability. *Database (Oxford)*, 2016.

Ben Ayed, R., Hanana, M., Ercisli, S., Karunakaran, R., Rebai, A. and Moreau, F. (2022) Integration of innovative technologies in the agri-food sector: the fundamentals and practical case of DNA-based traceability of olives from fruit to oil. *Plants* 11(9), 1230.

Benítez-Cabello, A., Romero-Gil, V., Medina-Pradas, E., Garrido-Fernández, A. and Arroyo-López, F.N. (2020) Exploring bacteria diversity in commercialized table olive biofilms by metataxonomic and compositional data analysis. *Scientific Reports* 10(1), 1–13.

Besnard, G. (2008) Chloroplast DNA variations in Mediterranean olive. *The Journal of Horticultural Science and Biotechnology* 83(1), 51–54.

Besnard, G. and de Casas, R.R. (2016) Single vs multiple independent olive domestications: the jury is (still) out. *New Phytologist* 209(2), 466–470.

Besnard, G. and El Bakkali, A. (2014) Sequence analysis of single-copy genes in two wild olive subspecies: nucleotide diversity and potential use for testing admixture. *Genome* 57(3), 145–153.

Besnard, G., Khadari, B., Villemur, P. and Bervillé, A. (2000) Cytoplasmic male sterility in the olive (*Olea europaea* L.). *Theoretical and Applied Genetics* 100(7), 1018–1024.

Besnard, G., Rubio de Casas, R. and Vargas, P. (2007) Plastid and nuclear DNA polymorphism reveals historical processes of isolation and reticulation in the olive tree complex (Olea europaea). *Journal of Biogeography* 34(4), 736–752.

Besnard, G., Hernández, P., Khadari, B., Dorado, G. and Savolainen, V. (2011) Genomic profiling of plastid DNA variation in the Mediterranean olive tree. *BMC Plant Biology* 11(1), 1–12.

Besnard, G., El Bakkali, A., Haouane, H., Baali-Cherif, D., Moukhli, A. and Khadari, B. (2013) Population genetics of Mediterranean and Saharan olives: geographic patterns of differentiation and evidence for early generations of admixture. *Annals of Botany* 112(7), 1293–1302.

Besnard, G., Terral, J.F. and Cornille, A. (2018) On the origins and domestication of the olive: a review and perspectives. *Annals of Botany* 121(3), 385–403.

Biton, I., Doron-Faigenboim, A., Jamwal, M., Mani, Y. and Eshed, R. (2015) Development of a large set of SNP markers for assessing phylogenetic relationships between the olive cultivars composing the Israeli olive germplasm collection. *Molecular Breeding* 35(4), 1–14.

Briegas, B., Corbacho, J., Parra-Lobato, M.C., Paredes, M.A., Labrador, J. *et al.* (2020) Transcriptome and hormone analyses revealed insights into hormonal and vesicle trafficking regulation among *Olea europaea* fruit tissues in late development. *International Journal of Molecular Sciences* 21(14), 4819.

Bruno, L., Picardi, E., Pacenza, M., Chiappetta, A., Muto, A. *et al.* (2019) Changes in gene expression and metabolic profile of drupes of *Olea europaea* L. cv. Carolea in relation to maturation stage and cultivation area. *BMC Plant Biology* 19(1), 1–17.

Cantini, C., Cimato, A., Autino, A., Redi, A. and Cresti, M. (2008) Assessment of the Tuscan olive germplasm by microsatellite markers reveals genetic identities and different discrimination capacity among and within cultivars. *Journal of the American Society for Horticultural Science* 133(4), 598–604.

Carbone, F., Bruno, L., Perrotta, G., Bitonti, M.B., Muzzalupo, I. and Chiappetta, A. (2019) Identification of miRNAs involved in fruit ripening by deep sequencing of *Olea europaea* L. transcriptome. *PLOS ONE* 14(8), e0221460.

Cardoni, M., Cabanás, C.G.L., Valverde-Corredor, A., Villar, R. and Mercado-Blanco, J. (2022) Unveiling differences in root defense mechanisms between tolerant and susceptible olive cultivars to *Verticillium dahliae*. *Frontiers in Plant Science*, 13.

Carmona, R., Zafra, A., Seoane, P., Castro, A.J., Guerrero-Fernández, D. *et al.* (2015) ReprOlive: a database with linked data for the olive tree (*Olea europaea* L.) reproductive transcriptome. *Frontiers in Plant Science* 6, 625.

Chedid, E., Rizou, M. and Kalaitzis, P. (2020) Application of high resolution melting combined with DNA-based markers for quantitative analysis of olive oil authenticity and adulteration. *Food Chemistry* 6, 100082.

Contreras, C., Mariotti, R., Mousavi, S., Baldoni, L., Guerrero, C., Roka, L. *et al.* (2020) Characterization and validation of olive FAD and SAD gene families: expression analysis in different tissues and during fruit development. *Molecular Biology Reports* 47, 4345–4355.

Corrado, G., Imperato, A., La Mura, M., Perri, E. and Rao, R. (2011) Genetic diversity among olive varieties of Southern Italy and the traceability of olive oil using SSR markers. *The Journal of Horticultural Science and Biotechnology* 86(5), 461–466.

Crawford, L.M., Carrasquilla-Garcia, N., Cook, D. and Wang, S.C. (2019) Analysis of microsatellites (SSRs) in processed olives as a means of cultivar traceability and authentication. *Journal of Agricultural and Food Chemistry* 68(4), 1110–1117.

Crawford, L.M., Janovick, J.L., Carrasquilla-Garcia, N., Hatzakis, E. and Wang, S.C. (2020) Comparison of DNA analysis, targeted metabolite profiling, and non-targeted NMR fingerprinting for differentiating cultivars of processed olives. *Food Control* 114, 107264.

Cultrera, N.G., Sarri, V., Lucentini, L., Ceccarelli, M., Alagna, F. *et al.* (2019) High levels of variation within gene sequences of *Olea europaea* L. *Frontiers in Plant Science* 9, 1932.

D'Agostino, N., Taranto, F., Camposeo, S., Mangini, G., Fanelli, V. *et al.* (2018) GBS-derived SNP catalogue unveiled wide genetic variability and geographical relationships of Italian olive cultivars. *Scientific Reports* 8, 15877.

de la Rosa, R., Angiolillo, A., Guerrero, C., Pellegrini, M., Rallo, L., Besnard, G. and Baldoni, L. (2003) A first linkage map of olive (*Olea europaea* L.) cultivars using RAPD, AFLP, RFLP and SSR markers. *Theoretical and Applied Genetics* 106(7), 1273–1282.

de la Rosa, R., Belaj, A., Muñoz-Mérida, A., Trelles, O., Ortíz-Martín, I. *et al.* (2013) Development of EST-derived SSR markers with long-core repeat in olive and their use for paternity testing. *Journal of American Society for Horticultural Science* 138, 290–296.

Díez, C.M., Trujillo, I., Martinez-Urdiroz, N., Barranco, D., Rallo, L. *et al.* (2015) Olive domestication and diversification in the Mediterranean Basin. *New Phytologist* 206(1), 436–447.

Dominguez-Garcia, M.D.C., Belaj, A., de la Rosa, R., Satovic, Z. and Heller-Uszynska, K. (2012) Development of DArT markers in olive (*Olea europaea* L.) and usefulness in variability studies and genome mapping. *Scientia Horticulturae* 136, 50–60.

Dridi, J., Fendri, M., Breton, C.M. and Msallem, M. (2018) Characterization of olive progenies derived from a Tunisian breeding program by morphological traits and SSR markers. *Scientia Horticulturae* 236, 127–136.

El Bakkali, A., Essalouh, L., Tollon, C., Rivallan, R., Mournet, P. *et al.* (2019) Characterization of Worlwide Olive Germplasm Banks of Marrakech (Morocco) and Córdoba (Spain): towards management and use of olive germplasm in breeding programs. *PLOS ONE* 14 (10).

Fernández-Ocaña, A., García-López, M.C., Jiménez-Ruiz, J., Saniger, L., Macías, D. *et al.* (2010) Identification of a gene involved in the juvenile-to-adult transition (JAT) in cultivated olive trees. *Tree Genetics and Genomes* 6(6), 891–903.

Friel, J., Bombarely, A., Fornell, C.D., Luque, F. and Fernández-Ocaña, A.M. (2021) Comparative analysis of genotyping by sequencing and whole-genome sequencing methods in diversity studies of *Olea europaea* L. *Plants* 10(11), 2514.
García-Vico, L., Sánchez, R., Fernández, G., Sanz, C. and Pérez, A.G. (2021) Study of the olive β-glucosidase gene family putatively involved in the synthesis of phenolic compounds of virgin olive oil. *Journal of the Science of Food and Agriculture* 101(13), 5409–5418.
Georgiadou, E.C., Koubouris, G., Goulas, V., Sergentani, C., Nikoloudakis, N. *et al.* (2019) Genotype-dependent regulation of vitamin E biosynthesis in olive fruits as revealed through metabolic and transcriptional profiles. *Plant Biology* 21(4), 604–614.
Gomes, S., Breia, R., Carvalho, T., Carnide, V. and Martins-Lopes, P. (2018) Microsatellite high-resolution melting (SSR-HRM) to track olive genotypes: from field to olive oil. *Journal of Food Science* 83(10), 2415–2423.
Gómez-Rodríguez, M.V., Beuzon, C., González-Plaza, J.J. and Fernández-Ocaña, A.M. (2021) Identification of an olive (*Olea europaea* L.) core collection with a new set of SSR markers. *Genetic Resources and Crop Evolution* 68(1), 117–133.
Grasso, F., Coppola, M., Carbone, F., Baldoni, L., Alagna, F. *et al.* (2017) The transcriptional response to the olive fruit fly (*Bactrocera oleae*) reveals extended differences between tolerant and susceptible olive (*Olea europaea* L.) varieties. *PLOS ONE* 12(8), e0183050.
Gros-Balthazard, M., Besnard, G., Sarah, G., Holtz, Y., Leclercq, J. *et al.* (2019) Evolutionary transcriptomics reveals the origins of olives and the genomic changes associated with their domestication. *Plant Journal* 100, 143–157.
Guerra, D., Lamontanara, A., Bagnaresi, P., Orrù, L. and Rizza, F. *et al.* (2015) Transcriptome changes associated with cold acclimation in leaves of olive tree (*Olea europaea* L.). *Tree Genetics and Genomes* 11(6), 1–24.
Guodong, R., Jianguo, Z., Xiaoxia, L. and Ying, L. (2019) Identification of putative genes for polyphenol biosynthesis in olive fruits and leaves using full-length transcriptome sequencing. *Food Chemistry* 300, 125246.
Hakim, I.R., Kammoun, N.G., Makhloufi, E. and Rebaï, A. (2010) Discovery and potential of SNP markers in characterization of Tunisian olive germplasm. *Diversity* 2(1), 17–27.
Haouane, H., El Bakkali, A., Moukhli, A., Tollon, C., Santoni, S. *et al.* (2011) Genetic structure and core collection of the World Olive Germplasm Bank of Marrakech: towards the optimised management and use of Mediterranean olive genetic resources. *Genetica* 139, 1083–1094.
Hernández, M.L., Sicardo, M.D., Alfonso, M. and Martínez-Rivas, J.M. (2019) Transcriptional regulation of stearoyl-acyl carrier protein desaturase genes in response to abiotic stresses leads to changes in the unsaturated fatty acids composition of olive mesocarp. *Frontiers in Plant Science* 10, 251.
Hernández, M.L., Sicardo, M.D., Arjona, P.M. and Martínez-Rivas, J.M. (2020) Specialized functions of olive FAD2 gene family members related to fruit development and the abiotic stress response. *Plant and Cell Physiology* 61(2), 427–441.
Hmmam, I., Mariotti, R., Ruperti, B., Cultrera, N., Baldoni, L. and Barcaccia, G. (2018) Venetian olive (*Olea europaea*) germplasm: disclosing the genetic identity of locally grown cultivars suited for typical extra virgin oil productions. *Genetic Resources and Crop Evolution* 65(6), 1733–1750.
Hosseini-Mazinani, M., Mariotti, R., Torkzaban, B., Sheikh-Hassani, M., Ataei, S. *et al.* (2014) High genetic diversity detected in olives beyond the boundaries of the Mediterranean Sea. *PLOS ONE* 9, e93146.
i Martí, A.F., i Forcada, C.F. and Rubio-Cabetas, M.J. (2015) Genetic relationships and population structure of local olive tree accessions from Northeastern Spain revealed by SSR markers. *Acta Physiologiae Plantarum* 37(1), 1–12.
Iaria, D.L., Chiappetta, A. and Muzzalupo, I. (2016) A de novo transcriptomic approach to identify flavonoids and anthocyanins 'switch-off' in olive (*Olea europaea* L.) drupes at different stages of maturation. *Frontiers in Plant Science* 6, 1246.
Inês, C., Corbacho, J., Paredes, M.A., Labrador, J., Cordeiro, A.M. and Gomez-Jimenez, M.C. (2019) Regulation of sterol content and biosynthetic gene expression during flower opening and early fruit development in olive. *Physiologia Plantarum* 167(4), 526–539.
İpek, A., Yılmaz, K., Sıkıcı, P., Tangu, N.A., Öz, A.T. *et al.* (2016) SNP discovery by GBS in olive and the construction of a high-density genetic linkage map. *Biochemical Genetics* 54, 313–325.
İpek, A., İpek, M., Ercişli, S. and Tangu, N.A. (2017) Transcriptome-based SNP discovery by GBS and the construction of a genetic map for olive. *Functional and Integrative Genomics* 17(5), 493–501.
Islam, A.F., Sanders, D., Mishra, A.K. and Joshi, V. (2021) Genetic diversity and population structure analysis of the USDA olive germplasm using genotyping-by-sequencing (GBS). *Genes* 12(12), 2007.

Jiménez-Ruiz, J., García-López, M.C., Vidoy, I., de la O Leyva-Pérez, M., Fernández-Ocaña, A. *et al.* (2015) Transcriptional analysis of adult cutting and juvenile seedling olive roots. *Tree Genetics and Genomes* 11(4), 1–12.

Julca, I., Marcet-Houben, M., Cruz, F., Gómez-Garrido, J., Gaut, B.S. *et al.* (2020) Genomic evidence for recurrent genetic admixture during the domestication of Mediterranean olive trees (*Olea europaea* L.). *BMC Biology* 18(1), 1–25.

Kaniewski, D., Van Campo, E., Boiy, T., Terral, J.F., Khadari, B. *et al.* (2012) Primary domestication and early uses of the emblematic olive tree: palaeobotanical, historical and molecular evidence from the Middle East. *Biological Reviews* 87, 885–899.

Kassa, A., Konrad, H. and Geburek, T. (2019) Molecular diversity and gene flow within and among different subspecies of the wild olive (*Olea europaea* L.): a review. *Flora* 250, 18–26.

Kaya, H.B., Cetin, O., Kaya, H., Sahin, M. and Sefer, F. *et al.* (2013) SNP discovery by Illumina-based transcriptome sequencing of the olive and the genetic characterization of Turkish olive genotypes revealed by AFLP, SSR and SNP markers. *PLOS ONE* 8(9), e73674.

Kaya, H.B., Cetin, O., Kaya, H.S., Sahin, M., Sefer, F. and Tanyolac, B. (2016) Association mapping in Turkish olive cultivars revealed significant markers related to some important agronomic traits. *Biochemical Genetics* 54(4), 506–533.

Kaya, H.B., Akdemir, D., Lozano, R., Cetin, O., Kaya, H.S. *et al.* (2019) Genome wide association study of 5 agronomic traits in olive (*Olea europaea* L.). *Scientific Reports* 9(1), 1–14.

Khadari, B., El Bakkali, A., Zine El Aabidine, A., Essalouh, L., Contreras S. *et al.* (2014) How can we efficiently characterize genes of agronomic interest in olive: towards the genetic association studies? *Acta Horticulturae* 1057, 551–558.

Koudounas, K., Thomopoulou, M., Michaelidis, C., Zevgiti, E., Papakostas, G. *et al.* (2017) The C-domain of oleuropein β-glucosidase assists in protein folding and sequesters the enzyme in nucleus. *Plant Physiology* 174(3), 1371–1383.

Kyriakopoulou, C.I. and Kalogianni, D.P. (2020) Genetic identification of the wild form of olive (*Olea europaea* var. *sylvestris*) using allele-specific real-time PCR. *Foods* 9, 467.

Lazović, B., Adakalić, M., Pucci, C., Perović, T., Bandelj, D. *et al.* (2016) Characterizing ancient and local olive germplasm from Montenegro. *Scientia Horticulturae* 209, 117–123.

Leyva-Pérez, M.D.L.O., Jiménez-Ruiz, J., Gómez-Lama Cabanás, C., Valverde-Corredor, A., Barroso, J.B. *et al.* (2018) Tolerance of olive (*Olea europaea*) cv. Frantoio to *Verticillium dahliae* relies on both basal and pathogen-induced differential transcriptomic responses. *New Phytologist* 217(2), 671–686.

Li, D., Long, C., Pang, X., Ning, D., Wu, T. *et al.* (2020) The newly developed genomic-SSR markers uncover the genetic characteristics and relationships of olive accessions. *PeerJ* 8, e8573.

Loconsole, G., Zicca, S., Manco, L., El Hatib, O., Altamura, G., Potere, O. and Saponari, M. (2021) Diagnostic procedures to detect *Xylella fastidiosa* in nursery stocks and consignments of plants for planting. *Agriculture* 11(10), 922.

Marchese, A., Marra, F.P., Caruso, T., Mhelembe, K., Costa, F. *et al.* (2016) The first high-density sequence characterized SNP-based linkage map of olive (*Olea europaea* L. subsp. *europaea*) developed using genotyping by sequencing. *Australian Journal of Crop Science* 10(6), 857.

Mariotti, R., Cultrera, N.G., Díez, C.M., Baldoni, L. and Rubini, A. (2010) Identification of new polymorphic regions and differentiation of cultivated olives (*Olea europaea* L.) through plastome sequence comparison. *BMC Plant Biology* 10(1), 1–13.

Mariotti, R., Cultrera, N.G.M., Mousavi, S., Baglivo, F., Rossi, M. *et al.* (2016) Development, evaluation, and validation of new EST-SSR markers in olive (*Olea europaea* L.). *Tree Genetics and Genomes* 12, 120.

Mariotti, R., Pandolfi, S., De Cauwer, I., Saumitou-Laprade, P. and Vernet, P. (2020a) Diallelic self-incompatibility is the main determinant of fertilization patterns in olive orchards. *Evolutionary Applications* 14(4), 983–995.

Mariotti, R., Belaj, A., de la Rosa, R., Leòn, L., Brizioli, F., Baldoni, L. and Mousavi, S. (2020b) EST–SNP study of *Olea europaea* L. uncovers functional polymorphisms between cultivated and wild olives. *Genes* 11(8), 916.

Mariotti, R., Fornasiero, A., Mousavi, S., Cultrera, N.G., Brizioli, F. *et al.* (2020c) Genetic mapping of the incompatibility locus in olive and development of a linked Sequence-Tagged Site marker. *Frontiers in Plant Science* 10, 1760.

Marra, F.P., Caruso, T., Costa, F., Di Vaio, C., Mafrica, R. and Marchese, A. (2013) Genetic relationships, structure and parentage simulation among the olive tree (*Olea europaea* L. subsp. *europaea*) cultivated in Southern Italy revealed by SSR markers. *Tree Genetics and Genomes* 9(4), 961–973.

Montemurro, C., Dambruoso, G., Bottalico, G. and Sabetta, W. (2019) Self-incompatibility assessment of some Italian olive genotypes (*Olea europaea* L.) and cross-derived seedling selection by SSR markers on seed endosperms. *Frontiers in Plant Science* 10, 451.

Moret, M., Ramírez-Tejero, J.A., Serrano, A., Ramírez-Yera, E., Cueva-López, M.D. *et al.* (2023) Identification of genetic markers and genes putatively involved in determining olive fruit weight. *Plants* 12(1), 155.

Mousavi, S., Mariotti, R., Regni, L., Nasini, L., Bufacchi, M. *et al.* (2017a) The first molecular identification of an olive collection applying standard simple sequence repeats and novel expressed sequence tag markers. *Frontiers in Plant Science* 8, 1283.

Mousavi, S., Mariotti, R., Bagnoli, F., Costantini, L., Cultrera, N.G. *et al.* (2017b) The eastern part of the Fertile Crescent concealed an unexpected route of olive (*Olea europaea* L.) differentiation. *Annals of Botany* 119(8), 1305–1318.

Mousavi, S., Stanzione, V., Mencuccini, M., Baldoni, L., Bufacchi, M. and Mariotti, R. (2019a) Biochemical and molecular profiling of unknown olive genotypes from central Italy: determination of major and minor components. *European Food Research and Technology* 245(1), 83–94.

Mousavi, S., De la Rosa, R., Moukhli, A., El Riachy, M., Mariotti, R. *et al.* (2019b) Plasticity of fruit and oil traits in olive among different environments. *Scientific Reports* 9(1), 1–13.

Mousavi, S., Regni, L., Bocchini, M., Mariotti, R., Cultrera, N.G. *et al.* (2019c) Physiological, epigenetic and genetic regulation in some olive cultivars under salt stress. *Scientific Reports* 9(1), 1–17.

Muñoz-Mérida, A., González-Plaza, J.J., Canada, A., Blanco, A.M., García-López, M.D.C. *et al.* (2013) De novo assembly and functional annotation of the olive (*Olea europaea*) transcriptome. *DNA Research* 20(1), 93–108.

Ninot, A., Howad, W., Aranzana, M.J., Senar, R., Romero, A. *et al.* (2018) Survey of over 4,500 monumental olive trees preserved on-farm in the northeast Iberian Peninsula, their genotyping and characterization. *Scientia Horticulturae* 231, 253–264.

Niu, E., Hu, W., Ding, J., Wang, W., Romero, A., Shen, G. and Zhu, S. (2022) GC-MS/LC-MS and transcriptome analyses revealed the metabolisms of fatty acid and flavonoid in olive fruits (*Olea europaea* L.). *Scientia Horticulturae* 299, 111017.

Partovi, R., Iranbakhsh, A. and Ebadi, M. (2021) The use of DNA barcoding to avoid adulteration in olive plant leaf products. *Asian Journal of Forestry* 5(1).

Pereira, L., Gomes, S., Barrias, S., Fernandes, J.R. and Martins-Lopes, P. (2018) Applying high-resolution melting (HRM) technology to olive oil and wine authenticity. *Food Research International* 103, 170–181.

Piarulli, L., Savoia, M.A., Taranto, F., D'Agostino, N., Sardaro, R. *et al.* (2019) A robust DNA isolation protocol from filtered commercial olive oil for PCR-based fingerprinting. *Foods* 8(10), 462.

Raieta, K., Muccillo, L. and Colantuoni, V. (2015) A novel reliable method of DNA extraction from olive oil suitable for molecular traceability. *Food Chemistry* 172, 596–602.

Rallo, L., Barranco, D., Díez, C.M., Rallo, P., Suárez, M.P. *et al.* (2018) Strategies for olive (*Olea europaea* L.) breeding: cultivated genetic resources and crossbreeding. *Advances in Plant Breeding Strategies* 3, 535–600.

Rodríguez-López, C.E., Hong, B., Paetz, C., Nakamura, Y., Koudounas, K. *et al.* (2021) Two bi-functional cytochrome P450 CYP72 enzymes from olive (*Olea europaea*) catalyze the oxidative C-C bond cleavage in the biosynthesis of secoxy-iridoids–flavor and quality determinants in olive oil. *New Phytologist* 229(4), 2288–2301.

Saddoud Debbabi, O., Rahmani Mnasri, S., Ben Amar, F., Ben Naceur, M.B., Montemurro, C. and Miazzi, M.M. (2021) Applications of microsatellite markers for the characterization of olive genetic resources of Tunisia. *Genes* 12(2), 286.

Sadok, I.B., Celton, J.M., Essalouh, L., El Aabidine, A.Z., Garcia, G. *et al.* (2013) QTL mapping of flowering and fruiting traits in olive. *PLOS ONE* 8(5), e62831.

Salimonti, A., Carbone, F., Romano, E., Pellegrino, M., Benincasa, C. *et al.* (2020) Association study of the 5′ UTR intron of the *FAD2-2* gene with oleic and linoleic acid content in *Olea europaea* L. *Frontiers in Plant Science* 11, 66.

Seifi, E., Guerin, J., Kaiser, B. and Sedgley, M. (2012) Sexual compatibility of the olive cultivar 'Kalamata'assessed by paternity analysis. *Spanish Journal of Agricultural Research* 10(3), 731–740. https://doi.org/10.5424/sjar/2012103-501-11

Serrano, A., León, L., Belaj, A. and Román, B. (2020) Nucleotide diversity analysis of candidate genes for Verticillium wilt resistance in olive. *Scientia Horticulturae* 274, 109653.

Sion, S., Savoia, M.A., Gadaleta, S., Piarulli, L., Mascio, I. *et al.* (2021) How to choose a good marker to analyze the olive germplasm (*Olea europaea* L.) and derived products. *Genes* 12(10), 1474.

Skodra, C., Michailidis, M., Dasenaki, M., Ganopoulos, I., Thomaidis, N.S., Tanou, G. and Molassiotis, A. (2021) Unraveling salt-responsive tissue-specific metabolic pathways in olive tree. *Physiologia Plantarum* 173(4), 1643–1656.

Trujillo, I., Ojeda, M.A., Urdiroz, N.M., Potter, D., Barranco, D. *et al.* (2014) Identification of the Worldwide Olive Germplasm Bank of Córdoba (Spain) using SSR and morphological markers. *Tree Genetics and Genomes* 10, 141–155.

Uncu, A.T., Uncu, A.O., Frary, A. and Doganlar, S. (2017) Barcode DNA length polymorphisms vs fatty acid profiling for adulteration detection in olive oil. *Food Chemistry* 221, 1026–1033.

Unver, T., Wu, Z., Sterck, L., Turktas, M., Lohaus, R., Li, Z. *et al.* (2017) Genome of wild olive and the evolution of oil biosynthesis. *Proceedings of the National Academy of Sciences* 114(44), E9413–E9422.

Van de Paer, C., Bouchez, O. and Besnard, G. (2018) Prospects on the evolutionary mitogenomics of plants: a case study on the olive family (*Oleaceae*). *Molecular Ecology Resources*, 18(3), 407–423.

Vatansever, R., Hernandez, P., Escalante, F.J., Dorado, G. and Unver, T. (2022) Genome-wide exploration of oil biosynthesis genes in cultivated olive tree varieties (*Olea europaea*): insights into regulation of oil biosynthesis. *Functional and Integrative Genomics* 22(2), 171–178.

Volk, J., Sarafeddinov, A., Unver, T., Marx, S., Tretzel, J. *et al.* (2019) Two novel methylesterases from *Olea europaea* contribute to the catabolism of oleoside-type secoiridoid esters. *Planta* 250(6), 2083–2097.

Vuletin Selak, G., Baruca Arbeiter, A., Cuevas, J., Perica, S., Pujic, P., Raboteg Božiković, M. and Bandelj, D. (2021) Seed paternity analysis using SSR markers to assess successful pollen donors in mixed olive orchards. *Plants* 10(11), 2356.

Wang, C., Xue, L., Cui, Q., Liu, Q., Zhang, J. and Rao, G. (2022) Genome-wide identification of the cytochrome P450 superfamily in *Olea europaea* helps elucidate the synthesis pathway of oleuropein to improve the quality of olive oil. *Scientia Horticulturae* 304, 111291.

Yadav, S., Carvalho, J., Trujillo, I. and Prado, M. (2021) Microsatellite markers in olives (*Olea europaea* L.): utility in the cataloging of germplasm, food authenticity and traceability studies. *Foods* 10(8), 1907.

Yanik, H., Turktas, M., Dundar, E., Hernandez, P., Dorado, G. and Unver, T. (2013) Genome-wide identification of alternate bearing-associated microRNAs (miRNAs) in olive (*Olea europaea* L.). *BMC Plant Biology* 13(1), 1–22.

Zhu, S., Niu, E., Shi, A. and Mou, B. (2019) Genetic diversity analysis of olive germplasm (*Olea europaea* L.) with genotyping-by-sequencing technology. *Frontiers in Genetics* 10, 10.

3.4 Morphological Phenotyping

K.N. Blazakis and P. Kalaitzis
Mediterranean Agronomic Institute of Chania (M.A.I.Ch.), Department of Horticultural Genetics and Biotechnology, Chania, Crete-Greece

3.4.1 Introduction

Olive cultivars, selected over thousands of years of cultivation, represent a valuable source of genetic variability while a significant effort is still required for complete and detailed classification, cataloging and authentication. Olive germplasm collections represent a valuable tool for conservation and comparative phenotypic characterization. For this reason, various efforts have been initiated, in terms of European research projects (e.g. Before, Gen4olive etc.), supported by the International Olive Council, to perform thorough phenotypic evaluation (Rallo *et al.*, 2018). The ability to accurately phenotype the olive germplasm is important in order to select resilient-to-climate-change cultivars and breed for better adapted ones (Beiki *et al.*, 2012; Moriondo *et al.*, 2013; Gabaldon-Leal *et al.*, 2017; De Ollas *et al.*, 2019; Mousavi *et al.*, 2019).

Previously, most of the research effort was concentrated on molecular genotyping to assess genetic diversity and ability to discriminate olive cultivars using various DNA-based methodologies (Owen *et al.*, 2005; D'Imperio *et al.*, 2011; Mousavi *et al.*, 2014; Muleo *et al.*, 2016; Sebastiani and Busconi, 2017; El Bakkali *et al.*, 2019; Zhu *et al.*, 2019; Gomez-Rodriguez *et al.*, 2020; Belaj *et al.*, 2022). Complementary to this approach is the characterization of the phenotypic diversity of olive germplasm, in order to make available sound morphological and bio-agronomical data useful for the identification of the traits under selection and build solid tools to speed up the breeding programs. The olive phenotypic diversity is easily detectable in olive flowers, pollen, tree vigor and canopy density but it is mostly remarkable in fruit, leaf and endocarp shape and size. Therefore, morphological analysis might represent an efficient tool for the characterization and discrimination of cultivars and the establishment of phenological relationships (Ganino *et al.*, 2006; Diez *et al.*, 2012; Al-Ruqaie *et al.*, 2016; Belaj *et al.*, 2016).

The overall objective of this chapter is to review widely used efforts for phenomic characterization of olive cultivars and report additional potential methodologies that could be used for their morphological analysis. The main focus is concentrated on methodologies for morphological analysis of leaves, fruits and endocarps, rather than those used for the phenotypic description of trees, inflorescences, pollen or roots. Advantages and drawbacks, as well as their future possible upgrades, are discussed.

DOI: 10.1079/9781789247350.0008

3.4.2 How to Collect Plant Material for Morphological Analyses

The number of fruits, leaves and endocarps is of vital importance for their morphological analysis. At least 30 samples for each organ should be considered, in order to capture as much variability as possible. The plant material should be collected from healthy trees, representative of the bearing, vigor and fruit production of the variety to be characterized. Fruit samples should be collected at a predetermined ripening stage or at different ripening stages, depending on the purpose of the work. Fruits should be collected from external, randomly chosen branches of the canopy that bear fruits, approximately at 1.5 m from the ground, harvesting drupes only at the same ripening stage from the middle part of stems, avoiding large, small or irregularly shaped fruits. Fruit load of trees should be taken into account, and only trees with regular fruit load should be considered. After image acquisition, the endocarps should be extracted from the same olive fruits and, after careful pulp removal, washed with bleach, rinsed and dried. After this treatment the stones should be ready for imaging and phenotypic observations (UPOV (International Union for the Protection of New Varieties of Plants, 2011).

Fully developed leaves should be collected from the most representative, one-year-old shoots on the south-facing side of the tree at shoulder level, avoiding apical, basal and atypical ones.

In order to keep as much information as possible, the morphological analysis of fruits and endocarps is usually performed on two positions; position A refers to fruit or endocarp side displaying the greatest asymmetry, whereas position B is reached by turning them 90° from position A (UPOV and IOC).

3.4.3 Morphological Analysis for Cultivar Discrimination

Morphological markers still represent an efficient approach for the description and discrimination of olive cultivars, despite numerous limitations, such as variability due to environmental conditions, age of trees, cultivation systems and phenological stage of plants (Trujillo *et al.*, 2014; Blazakis *et al.*, 2017).

Most of the studies on morphological analysis of leaves, fruits and endocarps are based on the scheme adopted by UPOV and focus on the morphological characteristics of these plant organs (Trujillo *et al.*, 2014; UPOV, 2011).

The defined descriptors have been used widely to discriminate among olive cultivars (Belaj *et al.*, 2002; Hannachi *et al.*, 2008; Haouane *et al.*, 2011; Trujillo *et al.*, 2014; Beyaz *et al.* 2017; Blazakis *et al.* 2017; Rallo *et al.*, 2018; Koubouris *et al.* 2019).

Quantitative characteristics cover the full range of variation from one extreme to the other and usually refer to length, width, ratios and discrete size measurements. UPOV employs a one-dimensional, continuous or discrete, linear scale for scoring the range of quantitative measurements. On the other hand, qualitative characteristics refer to the need to identify and describe visual features, even where these might be contained within a wide range and are mostly discontinuous. For example, the endocarp shape can be classified into ovate, oblong, elliptic, circular and obovate; the endocarp symmetry could be scored as symmetric, weakly asymmetric and strongly asymmetric, etc. In Table 3.4.1, the most commonly used descriptors for olive are reported as indicated in the UPOV guidelines. For further details on the morphological description of olive fruits, leaves and endocarps, refer to UPOV (2011).

3.4.4 Review on Image Analysis Computerized Methodologies and Softwares

New digital imaging, computing techniques and image analysis represent promising tools for phenotypic characterization of biological objects, and nowadays are widely used in agricultural research. Image analysis offers many advantages, because it is an objective, fast, even, repeatable and detailed measurement. However, manual measurements, among others, depend on the operator's skills, the reliability of the instruments used for the measurements and the seasonality of the samples. Image analysis allows automated measurements of size, shape, color, texture etc., as well as quality evaluation and visualization. Computerized algorithms could serve as useful research tools in the direction of morphological analysis for improving the visual appearance of an image to the human eye or for the measurement of various characteristics.

A comprehensive review on the recent scientific software for plant image analysis is available at https://www.quantitative-plant.org/ (accessed 3/5/23) in which computer-based tools for quantitative measurements related to fruits, leaves, shoots and seeds for different crop species are described. In general, the morphological analysis of a plant organ by using image analysis and data elaboration proceeds as follows: image acquisition of the target object, pre-processing of the image to facilitate further processing, selection of the area of interest and, finally, acquisition of morphological features. To achieve a close comprehension of the morphological analysis of standalone software-based methodologies for different species, the most appropriate software is presented, which might be used for olive fruit, leaf and endocarp description (Table 3.4.2).

3.4.5 Morphological Analysis of Olive Fruits, Leaves and Endocarps by Using Image-analysis Tools

Numerous applications for image analysis of olive fruits, stones and leaves have been recently developed. Although some of these applications have been developed for different purposes (i.e. identification of olive cultivar, assessment of external fruit damage etc.), all of them can provide quantitative calculations of morphological characteristics. The first approach for the identification of olive cultivars was developed by Mancuso and Nicese (1999) based on different leaf

Table 3.4.1. Morphological descriptors proposed by UPOV.

Plant organ	Morphological characteristic
Fruit	length (QN), width (QN), weight (QN), shape (PQ), ratio: length/width (QN), symmetry (QN), shape of apex (PQ), presence of fruit nipple (QN), shape of base (QN), bloom of surface (QN)
Leaf blade	length (QN), width (QN), ratio: length/width (QN), intensity of green color of upper side (QN), curvature of longitudinal axis (QN), twisting (QN)
Endocarp	shape (PQ), length (QN), width (QN), ratio: length/width (QN), weight (QN), symmetry (QN), symmetry (QN), number of grooves on basal end (QN), distribution of grooves on basal end (QN), shape of apex (PQ), presence of mucron (QL), shape of base (QN), rugosity (QN)

QN, quantitative characteristic; QL, qualitative characteristic; PQ, pseudo-qualitative characteristic

Table 3.4.2. Main softwares for morphological analysis of plant samples.

Software name	Plant organ	Morphological descriptors	Publication
Tomato Analyser	Fruit	Shape descriptors	Gonzalo *et al.* (2009)
LAMINA	Leaf	Shape descriptors	Tanabata *et al.* (2012)
Leaf Analyser	Leaf	Shape descriptors	Weight *et al.* (2008)
SmartGrain	Seed	Shape descriptors	Tanabata *et al.* (2012)
ImageJ	Fruit, leaf, seed	Shape descriptors	Abramoff *et al.* (2004)
LeafJ	Leaf	Shape descriptors	Maloof *et al.* (2013)
GrainScan	Seed	Shape descriptors	Whan *et al.* (2014)
PlantCV	Leaf	Shape descriptors	Gehan *et al.* (2017)

morphological parameters, such as area, perimeter, roundness and length. Bari *et al.* (2003) and Neto *et al.* (2006) used purely geometrical and mathematical tools, such as fractal geometry, moments and elliptic fourier analysis, in order to extract conventional size descriptors related to olive endocarp morphology, including length, width, eccentricity and others. Riquelme *et al.* (2008) developed a methodology to classify defective olive fruits according to their external appearance and also to provide some basic morphological features only for the fruits such as length and perimeter.

Vanloot *et al.* (2014) provided size and shape features for olive endocarp, such as area, perimeter, length, width, circularity and elongation, among other measurements, through an identification workflow by using artificial vision. In a similar manner, Beyaz and Öztürk (2016) and Beyaz *et al.* (2017) developed a new image analyzer for olive cultivar determination based on the measurements of fruit and endocarp length and width. Biometric features of endocarp, including morphometric, colorimetric and textural characters are determined by image-analysis techniques to investigate relationships among different olive cultivars (Piras *et al.*, 2016). Ponce *et al.* (2018) presented a methodology for the estimation of olive fruit mass and size by using image analysis for the major and minor axis length of olive fruits. Finally, Martinez *et al.* (2018) provided measurements for endocarp shape, length, width, symmetry, surface texture, fibrovascular furrow distribution, and apex and base shape by a computer vision approach.

3.4.5.1 OliveID

Despite the wide use of morphological analysis for fruits, leaves and endocarps for cultivar identification, there is a lack of integrated, automated methodologies to further assist the development of this field. Towards this direction, Blazakis *et al.* (2017) developed an integrated image-analysis tool called OliveID, based on an automated methodology to describe the morphologies of these olive organs. The methodology quantifies many features of these organs based on strictly mathematical morphological parameters and provides accurate numerical measurements regarding the olive morphology attributes in a robust and efficient way.

OliveID is a semi-automatic methodology of estimating various morphological traits in a quantitative manner based on image-analysis tools. By applying intuitive mathematical descriptors for shape analysis, the methodology determines quantitative and qualitative characters of these organs such as size, shape, symmetry, contour roughness and presence of additional structures (nipple, petiole etc.). The methodology consists mostly of three main steps: raw image acquisition, image segmentation and morphological analysis (Fig. 3.4.1). Here we

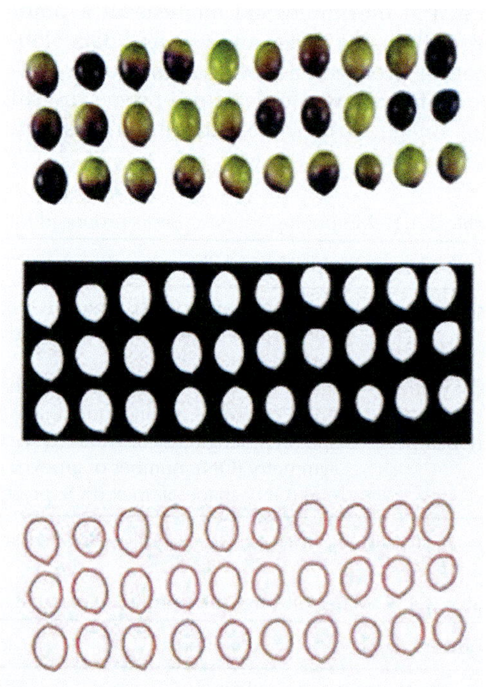

Fig. 3.4.1. Pipeline of Blazakis *et al.* (2017) methodology: (*top*) raw imaging data from cv. 'Megaritiki'; (*center*) segmented imaging data: a binary (black-and-white) image after the use of color thresholding in ImageJ; and (*bottom*) shape boundaries: representation of each shape by a discrete sequence with all its boundary points (red line).

describe briefly the three steps, referring to Blazakis *et al.* (2017) and the references therein for further details.

3.4.5.2 Image data acquisition

The imaging data of fruits and endocarps are produced by using an imaging stand. In order to avoid shadows around the objects, a thin (2 mm) glass lifted 10 cm above a light-blue background must be used to place at least 30 samples for each organ (fruits or stones) and for each cultivar; whereas the camera is installed on top of them and set to obtain high-resolution images with the focus on the outline of the images. All samples are lit with a set of lamps with a correlated color temperature (CCT) value of 5400 K, supposed to provide the same or similar illumination as natural daylight. The imaging data of leaves are generated by using a scanner at a resolution at least 600 dpi.

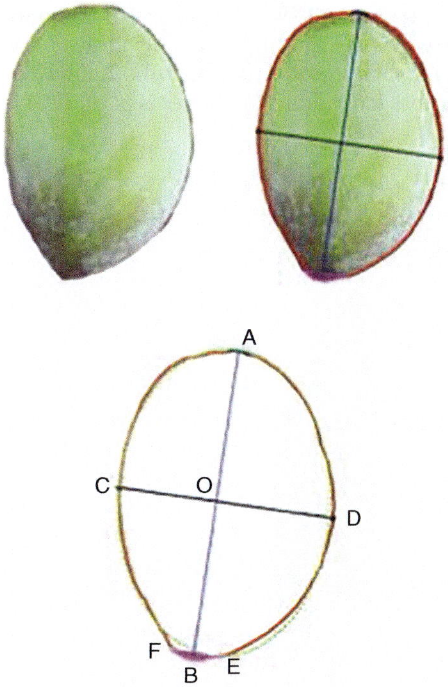

Fig. 3.4.2. Morphological characteristics of the fruit: (*top left*) raw image data; (*top right*) raw image data with measurements; (*below*) morphological measurements: contour fruit boundary (red line); height A–B; maximum transverse diameter C–D; position of the maximum transverse diameter segment O–B; best fit ellipse (green curve); fruit nipple: pink curve (arc (EBF).

3.4.5.3 Image segmentation

Segmentation is the process in which the objects are separated from the background in an image. Segmentation is one of the most important steps, from which a binary image will be derived and then the representation of each shape will be produced. The binarized image has only two values, either black or white. The black values correspond to the background, and the white ones to the describing objects. Different image-analysis softwares can be used for the segmentation. OliveID uses a manual adjustment of the image's values related to 'Hue', 'Saturation' and 'Brightness', in ImageJ, which is one of the most commonly used open-source scientific image-analysis software; brightness and other parameters may be set.

3.4.5.4 Morphological analysis

The methodology has been implemented in MATLAB (The Mathworks Inc., Natick, MA, USA) by using the Image Processing Toolbox. A set of state-of-the-art automatic algorithms for object contour extraction from imaging data (fruits, endocarps and leaves) has been developed that is able to identify various geometrical characteristics assigned to different morphological traits. The final outcome of the algorithm is the representation of each shape by a discrete sequence with all its boundary points. This sequence represents each shape and can be considered mathematically as a closed polygonal line where vertices are the boundary points and its (undirected) edges are defined by subsequent vertices, connecting the last with the first vertex. This mathematical representation of the imaging data enables analysis of the morphology of the olives, endocarps and leaves of each cultivar, quantitatively and qualitatively. In Table 3.4.3, the most representative morphological descriptors calculated in Blazakis *et al.* (2017) are reported.

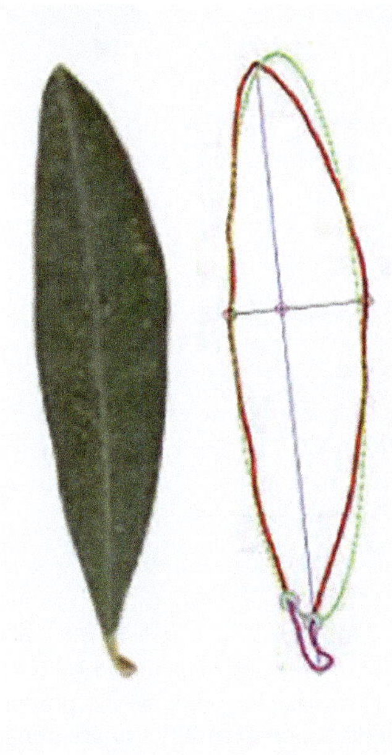

Fig. 3.4.3. Morphological characters of a leaf: (*left*) raw image data; (*right*) morphological measurements: leaf contour (red line); tip curve (green line); blade height A–B (blue line); width C–D (black line); petiole (pink curve).

3.4.6 Applications of Olive Morphological Analysis

Nowadays, there are available methodologies for olive cultivar identification based mainly on chemical and genetic fingerprinting techniques. However, these methodologies require a high level of specialization, specific and expensive infrastructure and laboratory work. The concept of cultivar identification through phenotypic characterization has been primarily used to evaluate the biodiversity of olive's genetic resources and its varietal heritage (Beiki *et al.*, 2012; Trujillo *et al.*, 2014). Different accessions in olive germplasm banks were characterized for efficient management of genetic resources and for the study and analysis of their taxonomic relationships (Beiki *et al.*, 2012). There is currently a growing interest amongst the olive sector in quality olive plants in the context of a certification program that ensures both varietal authenticity and an ideal sanitary state (Reboredo-Rodriguez *et al.*, 2018; Campos *et al.*, 2019). The use of plants with authenticated varietal origin and free of pathogens is of vital importance for successful breeding programs (Owen *et al.*, 2005; Campos *et al.*, 2019). The study of fruit shape might assist towards yield improvement and quality of olive oil by using morphological analysis and cultivar identification (Williams *et al.*, 2013). Moreover, cultivar identification is an important component to overcome issues

Table 3.4.3. List of the most representative morphological descriptors calculated in Blazakis *et al.* (2017).

Parameter	Description
Area (cm^2)	Area inside the polygonal line-contour
Perimeter (cm)	Total length of the polygonal line-contour
Hght (cm)	Height: length between the top-most and bottom-most point of each contour
MaxTrDiam (cm)	Longest segment perpendicular to the height
MinCntTr (cm)	Minimum distance between the height and the contour
VerSym	Vertical symmetry index
TrSym	Transversal symmetry index
ShIdx	Shape index
ApCur	Mean curvature of the apex
NippleIdx	Index for the presence, or not, of a nipple in a fruit
BladeHgt (cm)	Length of the leaf blade
TipCur	Mean curvature of the leaf tip curve

concerning varietal names such as synonyms, which is currently a common obstacle related to olive research (Ganino *et al.*, 2006). Finally, the authentication of olive cultivars has been used in the agri-food sector in order to ensure that the commercial label matches the varietal origin of olive oil (Trujillo *et al.*, 2014). Discrimination of table olives, sorting olive batches for the milling process to obtain monovarietal olive oil, post-harvest classification of olives according to different features such as fruit surface, size and condition, is attractive for food industry (Diaz *et al.*, 2000; Riquelme *et al.*, 2008; Puerto *et al.*, 2015; Ponce *et al.*, 2018, 2019; Gomes *et al.*, 2020). Therefore, feasible and modern image-based methodologies could improve the identification of the olive cultivars in a fast and accurate manner, highly valuable for any kind of stakeholders involved such as farmers, authorities and/or researchers.

The effectiveness and accuracy of image-based methodologies for morphological characterization of different crop species is usually linked to the cultivars classification (Pandolfi *et al.*, 2009; Fuentes *et al.*, 2018; Ishikawa *et al.*, 2018; Waldchen *et al.*, 2018; Yang *et al.*, 2019; Dheer and Singh, 2019). Traditionally, a rapid cultivar identification is a manual process through visual inspection based on organ characteristics that might be considered as discriminatory keys for each cultivar (Waldchen and Mader, 2018). Morphological characteristics of fruits, leaves, endocarps or other organs are commonly used for a quick cultivar identification based on their appearance, but visual observations require experience and sometimes appear to be very subjective, inconsistent and inaccurate (Grinblat *et al.*, 2016; Backhaus *et al.*, 2010; Waldchen and Mader, 2018; Fuentes *et al.*, 2018). Automated methodologies for cultivar identification based only on shape morphologies are still at early, preliminary stages in olive (Ishikawa *et al.*, 2018).

A high degree of automation in cultivar identification in different crop species using morphological characteristics is a challenging task and the olive sector is no exception (Ponce *et al.*, 2019). To this end, the identification of different morphological features from olive organs by imaged-based installations results in a plethora of available numerical data corresponding to specific traits. Therefore, multivariate data analysis and classical statistical methods, such as one-way analysis of variance (ANOVA), cluster analysis and principal component analysis, could evaluate the main features for cultivar classification and identification. In addition, artificial neural networks have been commonly used mostly for the olive cultivar identification for cases where the statistical standards prove to be insufficient. An initial attempt has been made by using the parameters related to leaf morphologies in an artificial neural network methodology in order to distinguish different olive cultivars (Mancuso and Nicese, 1999). Similarly, different artificial neural network models have been used from a commercial software in order to assess their application in the identification of a small set of Spanish and Turkish cultivars, by utilizing the color properties, and the length and width of olive fruits and stones (Beyaz and Öztürk, 2016; Beyaz *et al.*, 2017).

Furthermore, more attention was focussed on the use of discriminant analysis and linear regression from quantitative data of morphological traits (Diaz *et al.*, 2000; Bari *et al.*, 2003; Diaz *et al.*, 2004; D'Imperio *et al.*, 2011). Likewise, the classification of a small set of Spanish cultivars was performed by partial least-squares discriminant analysis (PLS-DA) based on morphological characteristics and texture features of olive endocarps (Martinez *et al.*, 2018). Partial least-squares discriminant analysis is a versatile algorithm that can be used for discriminative variable selection and it has been used as an additional methodology to principal component analysis (Vanloot *et al.*, 2014). However, there is a tendency for the use of automatic machine-learning algorithms for classification and other applications (Fuentes *et al.*, 2018; Ishikawa *et al.*, 2018). More recently, Koubouris *et al.* (2019) developed a statistical methodology based on classification binary trees for the discrimination of different endocarps morphologies. In addition, Ponce *et al.* (2019) used a powerful image-processing, artificial intelligence approach by designing a procedure based on convolutional neural networks (CNN) and image processing to classify images of olive fruits.

3.4.7 Conclusions

Morphological analysis is of great importance for the characterization of olive germplasm collection and the identification of olive cultivars. Morphological affinity may serve as a basis for giving insights about the evolutionary drivers and adaptive strategies of particular olive traits. Nowadays, international initiatives are established to identify and authenticate olive cultivars using a common protocol for molecular and morphological studies, in order to serve as control samples. Farmers and plant breeders have been selecting the best olive genotypes based upon their phenotype. The traditional phenotyping methods of olive cultivars deal mostly with the morphological analysis of fruits, leaves and endocarps by using a visual grading and scoring. This methodology remains labor-dependent and time-intensive phenotyping. However, there is a research interest for the development of low-cost, automated and semi-automated methodologies for data acquisition and morphological analysis, which could create useful, rapid and reliable image-based tools to describe and identify olive cultivars.

There is significant progress in molecular profiling of different crop species including olive. However, the progress of morphological analysis of olive cultivars was not adequate to optimally associate genotypes with phenotypes of olive fruits, leaves and endocarps. Different studies were performed attempting to associate genomics with phenomics data. These studies resulted in more accurate conclusions and have overcome issues regarding genetic diversity and variability in olive germplasm collections. Therefore, data extracted from the morphological analysis of olive fruits, leaves and endocarps, along with available genomic data, could lead to the dissection of the genetic inference of complex traits.

Olive morphological analysis encounters certain limitations such as the generation and handling of data and processing costs. Moreover, complex handling of scientific algorithms limits practical applications by biologists. The collection, storage and retrieval of vast amounts of data from computerized methodologies is still a challenge. Integration of data from different techniques poses an additional challenge to utilize such data generated and therefore attention is required, as well as serious efforts to harmonize these data for efficient exploitation. All the techniques and tools which have been described refer to the two-dimensional morphological analysis of olive cultivars. A future direction is the use of a three-dimensional setting. Although the three-dimensional settings are not at all trivial, they can become a less biased methodology for olive morphological characterization since the whole morphology of the plant organ will be considered.

Acknowledgements

This research has been partially co-financed by the European Union and Greek national funds through the Operational Program Competitiveness, Entrepreneurship and Innovation, under the call RESEARCH - CREATE - INNOVATE (project code: T2EDK-01315: ELIADA-Characterization and valorization of olive genetic resources aiming at climate change adaptation and production of high nutritional and organoleptic value products of olive oil - MIS 5074539 and project code: T2EDK-02637: ELAION-Innovative Methodological Tools for Traceability, Certification and Authenticity Assessment of Olive Oil and Olives - MIS 5075094). Moreover, this research has been partially funded by the European Union under the ENI CBC Mediterranean Sea Basin Program, within the Cross Border Living Laboratories for Agroforestry – LIVINGAGRO project: Grant Contract identification number [38/1315 29/08/2019].

References

Abramoff, M.D., Magalhaes, P.J. and Ram, S.J. (2004) Image processing with ImageJ. *Biophotonics International* 11(7), 36–42.

Al-Ruqaie, I., Al-Khalifah, N. and Shanavaskhan, A. (2016) Morphological cladistic analysis of eight popular olive (*Olea europaea* L.) cultivars grown in Saudi Arabia using Numerical Taxonomic System for personal computer to detect phyletic relationship and their proximate fruit composition. *Saudi Journal of Biological Sciences* 23(1), 115–121.

Backhaus, A., Kuwabara, A., Bauch, M., Monk, N., Sanguinetti, G. and Fleming, A. (2010) Leafprocessor: a new leaf phenotyping tool using contour bending energy and shape cluster analysis. *New Phytologist* 187(1), 251–261.

Bari, A., Martin, A., Boulouha, B., Gonzalez-Andujar, J.L., Barranco, D., Ayad, G. and Padulosi, S. (2003) Use of fractals and moments to describe olive cultivars. *The Journal of Agricultural Science* 141(1), 63.

Beiki, A.H., Saboor, S. and Ebrahimi, M. (2012) A new avenue for classification and prediction of olive cultivars using supervised and unsupervised algorithms. *PLOS ONE* 7(9), e44164.

Belaj, A., Satovic, Z., Rallo, L. and Trujillo, I. (2002) Genetic diversity and relationships in olive (*Olea europaea* L.) germplasm collections as determined by randomly amplified polymorphic DNA. *Theoretical and Applied Genetics* 105(4), 638–644.

Belaj, A., Veral, M.G., Sikaoui, H., Moukhli, A., Khadari, B. *et al.* (2016) Olive genetic resources. In: Rugini, E., Baldoni, L., Muleo, R. and Sebastiani, L. (eds) *The Olive Tree Genome*. Springer, Cham, Switzerland, pp. 27–54.

Belaj, A., Ninot, A., Gómez-Gálvez, F.J., El Riachy, M., Gurbuz-Veral, M. *et al.* (2022) Utility of EST-SNP markers for improving management and use of olive genetic resources: a case study at the Worldwide Olive Germplasm Bank of Córdoba. *Plants* 11(7), 921.

Beyaz, A. and Öztürk, R.(2016) Identification of olive cultivars using image processing techniques. *Turkish Journal of Agriculture and Forestry* 40(5), 671–683.

Beyaz, A., Özkaya, M.T. and İçen, D. (2017) Identification of some Spanish olive cultivars using image processing techniques. *Scientia Horticulturae* 225, 286–292.

Blazakis, K.N., Kosma, M., Kostelenos, G., Baldoni, L., Bufacchi, M. and Kalaitzis, P. (2017) Description of olive morphological parameters by using open access software. *Plant Methods* 13(1), 111.

Campos, M.D., Zellama, M.S., Varanda, C., Materatski, P., Peixe, A. *et al.* (2019) Establishment of a sensitive qpcr methodology for detection of the olive-infecting viruses in Portuguese and Tunisian orchards. *Frontiers in Plant Science* 10, 694.

De Ollas, C., Morillon, R., Fotopoulos, V., Puertolas, J., Ollitrault, P. *et al.* (2019) Facing climate change: biotechnology of iconic Mediterranean woody crops. *Frontiers in Plant Science* 10, 427.

Dheer, P. and Singh, R. (2019) Identification of Indian rice varieties using machine learning classifiers. *Plant Archives* 19(1), 55–158.

Diaz, R., Faus, G., Blasco, M., Blasco, J. and Molto, E. (2000) The application of a fast algorithm for the classification of olives by machine vision. *Food Research International* 33(3–4), 305–309.

Diaz, R., Gil, L., Serrano, C., Blasco, M., Molto, E. and Blasco, J. (2004) Comparison of three algorithms in the classification of table olives by means of computer vision. *Journal of Food Engineering* 61(1), 101–107.

Diez, C.M., Imperato, A., Rallo, L., Barranco, D. and Trujillo, I. (2012) Worldwide core collection of olive cultivars based on simple sequence repeat and morphological markers. *Crop Science* 52(1), 211–221.

D'Imperio, M., Viscosi, V., Scarano, M.-T., D'Andrea, M., Zullo, B.A. and Pilla, F. (2011) Integration between molecular and morphological markers for the exploitation of olive germplasm (*Olea europaea*). *Scientia Horticulturae* 130(1), 229–240.

El Bakkali, A., Essalouh, L., Tollon, C., Rivallan, R., Mournet, P. *et al.* (2019) Characterization of Worldwide Olive Germplasm Banks of Marrakech (Morocco) and Córdoba (Spain): towards management and use of olive germplasm in breeding programs. *PLOS ONE* 14(10), e0223716.

Fuentes, S., Hernandez-Montes, E., Escalona, J., Bota, J., Viejo, C.G. *et al.* (2018) Automated grapevine cultivar classification based on machine learning using leaf morpho-colorimetry, fractal dimension and near-infrared spectroscopy parameters. *Computers and Electronics in Agriculture* 151, 311–318.

Gabaldon-Leal, C., Ruiz-Ramos, M., de la Rosa, R., Leon, L., Belaj, A. *et al.* (2017) Impact of changes in mean and extreme temperatures caused by climate change on olive flowering in southern Spain. *International Journal of Climatology* 37, 940–957.

Ganino, T., Bartolini, G. and Fabbri, A. (2006) The classification of olive germplasm. *The Journal of Horticultural Science and Biotechnology* 81(3), 319–334.

Gehan, M.A., Fahlgren, N., Abbasi, A., Berry, J.C., Callen, S.T. *et al.* (2017) PlantCV v2: image analysis software for high-throughput plant phenotyping. *PeerJ* 5, e4088.

Gomes, L., Nobre, T., Sousa, A., Rei, F. and Guiomar, N.R.G.N. (2020) Hyperspectral reflectance as a basis to discriminate olive varieties: a tool for sustainable crop management. *Sustainability* 12(7), 3059.

Gomez-Rodriguez, M.V., Beuzon, C., Gonzalez-Plaza, J.J. and Fernandez-Ocaña, A.M. (2020) Identification of an olive (*Olea europaea* L.) core collection with a new set of SSR markers. *Genetic Resources and Crop Evolution* 1–17.

Gonzalo, M.J., Brewer, M.T., Anderson, C., Sullivan, D., Gray, S. and Van der Knaap, E. (2009) Tomato fruit shape analysis using morphometric and morphology attributes implemented in Tomato Analyzer software program. *Journal of the American Society for Horticultural Science* 134(1), 77–87.

Grinblat, G.L., Uzal, L.C., Larese, M.G. and Granitto, P.M. (2016) Deep learning for plant identification using vein morphological patterns. *Computers and Electronics in Agriculture* 127, 418–424.

Hannachi, H., Breton, C., Msallem, M., El Hadj, S.B., El Gazzah, M. and Bervillé, A. (2008) Differences between native and introduced olive cultivars as revealed by morphology of drupes, oil composition and SSR polymorphisms: a case study in Tunisia. *Scientia Horticulturae* 116(3), 280–290.

Haouane, H., El Bakkali, A., Moukhli, A., Tollon, C., Santoni, S. *et al.* (2011) Genetic structure and core collection of the world olive germplasm bank of Marrakech: towards the optimised management and use of Mediterranean olive genetic resources. *Genetica* 139(9), 1083–1094.

Ishikawa, T., Hayashi, A., Nagamatsu, S., Kyutoku, Y., Dan, I., Wada, T. *et al.* (2018) Classification of strawberry fruit shape by machine learning. *International Archives of the Photogrammetry, Remote Sensing & Spatial Information Sciences* 42(2).

Koubouris, G., Avramidou, E., Metzidakis, I., Petrakis, P., Sergentani, C. and Doulis, A. (2019) Phylogenetic and evolutionary applications of analyzing endocarp morphological characters by classification binary tree and leaves by SSR markers for the characterization of olive germplasm. *Tree Genetics & Genomes* 15(2), 26.

Maloof, J.N., Nozue, K., Mumbach, M.R. and Palmer, C.M. (2013) LeafJ: an ImageJ plugin for semi-automated leaf shape measurement. *Journal of Visualized Experiments: JoVE* (71).

Mancuso, S. and Nicese, F. (1999) Identifying olive (*Olea europaea*) cultivars using artificial neural networks. *Journal of the American Society for Horticultural Science* 124(5), 527–531.

Martinez, S.S., Gila, D.M., Beyaz, A., Ortega, J.G. and Garcia, J.G. (2018) A computer vision approach based on endocarp features for the identification of olive cultivars. *Computers and Electronics in Agriculture* 154, 341–346.

Moriondo, M., Trombi, G., Ferrise, R., Brandani, G., Dibari, C. *et al.* (2013) Olive trees as bio-indicators of climate evolution in the Mediterranean basin. *Global Ecology and Biogeography* 22(7), 818–833.

Mousavi, S., Mazinani, M.H., Arzani, K., Ydollahi, A., Pandolfi, S. *et al.* (2014) Molecular and morphological characterization of Golestan (Iran) olive ecotypes provides evidence for the presence of promising genotypes. *Genetic Resources and Crop Evolution* 61(4), 775–785.

Mousavi, S., de la Rosa, R., Moukhli, A., El Riachy, M., Mariotti, R. *et al.* (2019) Plasticity of fruit and oil traits in olive among different environments. *Scientific Reports* 9(1), 1–13.

Muleo, R., Morgante, M., Cattonaro, F., Scalabrin, S., Cavallini, A. *et al.* (2016) Genome sequencing, transcriptomics, and proteomics. In: Rugini, E., Baldoni, L., Muleo, R. and Sebastiani, L. (eds) *The Olive Tree Genome*. Springer, Cham, Switzerland, pp. 141–161.

Neto, J.C., Meyer, G.E., Jones, D.D. and Samal, A.K. (2006) Plant species identification using Elliptic fourier leaf shape analysis. *Computers and Electronics in Agriculture* 50(2), 121–134.

Owen, C.A., Bita, E.-C., Banilas, G., Hajjar, S.E., Sellianakis, V. *et al.* (2005) AFLP reveals structural details of genetic diversity within cultivated olive germplasm from the eastern Mediterranean. *Theoretical and Applied Genetics* 110(7), 1169–1176.

Pandolfi, C., Messina, G., Mugnai, S., Azzarello, E., Masi, E., Dixon, K. and Mancuso, S. (2009) Discrimination and identification of morphotypes of *Banksia integrifolia* (Proteaceae) by an artificial neural network (ANN), based on morphological and fractal parameters of leaves and flowers. *Taxon* 58(3), 925–933.

Piras, F., Grillo, O., Venora, G., Lovicu, G., Campus, M. and Bacchetta, G. (2016) Effectiveness of a computer vision technique in the characterization of wild and farmed olives. *Computers and Electronics in Agriculture* 122, 86–93.

Ponce, J.M., Aquino, A., Millán, B. and Andujar, J.M. (2018) Olive-fruit mass and size estimation using image analysis and feature modeling. *Sensors* 18(9), 2930.
Ponce, J.M., Aquino, A. and Andujar, J.M. (2019) Olive-fruit variety classification by means of image processing and convolutional neural networks. *IEEE Access* 7, 147629–147641.
Puerto, D.A., Gila, D.M.M., Garcia, J.G. and Ortega, J.G. (2015) Sorting olive batches for the milling process using image processing. *Sensors* 15(7), 15738–15754.
Rallo, L., Barranco, D., Diez, C.M., Rallo, P., Suarez, M.P. *et al.* (2018) Strategies for olive (*Olea europaea* L.) breeding: cultivated genetic resources and crossbreeding. In: Al-Khayri, J.M., Jain, S.M. and Johnson, D.V. (eds) *Advances in Plant Breeding Strategies: Fruits*. Springer, Cham, Switzerland, pp. 535–600.
Reboredo-Rodriguez, P., Gonzalez-Barreiro, C., Cancho-Grande, B., Simal-Gandara, J. and Trujillo, I. (2018) Genotypic and phenotypic identification of olive cultivars from north-western Spain and characterization of their extra virgin olive oils in terms of fatty acid composition and minor compounds. *Scientia Horticulturae* 232, 269–279.
Riquelme, M., Barreiro, P., Ruiz-Altisent, M. and Valero, C. (2008) Olive classification according to external damage using image analysis. *Journal of Food Engineering* 87(3), 371–379.
Sebastiani, L. and Busconi, M. (2017) Recent developments in olive (*Olea europaea* L.) genetics and genomics: applications in taxonomy, varietal identification, traceability and breeding. *Plant Cell Reports* 36(9), 1345–1360.
Tanabata, T., Shibaya, T., Hori, K., Ebana, K. and Yano, M. (2012) SmartGrain: high throughput phenotyping software for measuring seed shape through image analysis. *Plant Physiology* 160(4), 1871–1880.
Trujillo, I., Ojeda, M.A., Urdiroz, N.M., Potter, D., Barranco, D., Rallo, L. and Diez, C.M. (2014) Identification of the Worldwide Olive Germplasm Bank of Cordoba (Spain) using SSR and morphological markers. *Tree Genetics & Genomes* 10(1), 141–155.
UPOV (2011) Descriptors for Olive (*Olea europaea* L.). International Union for the Protection of New Varieties of Plants, Geneva, Switzerland, p. 56.
Vanloot, P., Bertrand, D., Pinatel, C., Artaud, J. and Dupuy, N. (2014) Artificial vision and chemometrics analyses of olive stones for varietal identification of five French cultivars. *Computers and Electronics in Agriculture* 102, 98–105.
Waldchen, J. and Mader, P. (2018) Plant species identification using computer vision techniques: a systematic literature review. *Archives of Computational Methods in Engineering* 25(2), 507–543.
Waldchen, J., Rzanny, M., Seeland, M. and Mader, P. (2018) Automated plant species identification: trends and future directions. *PLOS Computational Biology* 14(4), e1005993.
Weight, C., Parnham, D. and Waites, R. (2008) LeafAnalyser: a computational method for rapid and large-scale analyses of leaf shape variation. *The Plant Journal* 53(3), 578–586.
Whan, A.P., Smith, A.B., Cavanagh, C.R., Ral, J.-P.F., Shaw, L.M. *et al.* (2014) Grainscan: a low cost, fast method for grain size and colour measurements. *Plant Methods* 10(1), 1–10.
Williams, K., Munkvold, J. and Sorrells, M. (2013) Comparison of digital image analysis using elliptic fourier descriptors and major dimensions to phenotype seed shape in hexaploid wheat (*Triticum aestivum* L.). *Euphytica* 190(1), 99–116.
Yang, X., Zhang, R., Zhai, Z., Pang, Y. and Jin, Z. (2019) Machine learning for cultivar classification of apricots (*Prunus armeniaca* L.) based on shape features. *Scientia Horticulturae* 256, 108524.
Zhu, S., Niu, E., Shi, A. and Mou, B. (2019) Genetic diversity analysis of olive germplasm (*Olea europaea* L.) with genotyping-by-sequencing technology. *Frontiers in Genetics* 10, 755.

3.5 Developing Tools for Genomics-assisted Breeding

D. Namdar, I. Biton and G. Ben-Ari
Institute of Plant Sciences, ARO, The Volcani Center, Rishon LeZion 7528809, Israel

3.5.1 Introduction

Cross-breeding between cultivars or inter-cross between cultivated, wild olives or other closely related taxa with specific characters, still represents the main roads to obtain and select new olive cultivars and rootstocks (León *et al.*, 2018; Rapoport *et al.*, 2022; Serrano-García *et al.*, 2022).

However, phenotype-based selection is not sufficiently effective for developing new genotypes because it takes a very long time for olive plants to express their characters and because of demand for complex and tedious phenotyping campaigns (Ben-Ari *et al.*, 2021; Caruso *et al.*, 2021; Torres-Sánchez *et al.*, 2022).

Unfortunately, the genetic information on trait heredity and identification of specific genotype-phenotype associations is still missing in olive, and a shift from empirical phenotype-based breeding to informed genotype-based selection is still far from being applicable (Baldoni *et al.*, 2016; Rallo *et al.*, 2016; Purugganan and Jackson, 2021), especially when characters under selection are multigenic (Mousavi *et al.*, 2022; Vatansever *et al.*, 2022).

Breeding of fruit trees is hampered by their long generation time, large plant size, long juvenile phase, and the need to wait for plant maturity achievement, to assess the marketable product (fruit and oil) (Iwata *et al.*, 2016).

New genomic and metabolomic tools could instead be extremely useful to guide in the choice of the parents to be used to make new crosses, to evaluate cross progenies and for the early selection of new genotypes. These will enable to study those traits not easily detectable by traditional phenotyping, such as disease resistance, and will aid in avoiding environmental effects on fruit quality and oil yield during the selection procedure.

3.5.1.1 Choice of cross parents

To enable suitable selection of cross parents for achieving well-defined goals, the establishment of comprehensive genomic and phenotypic databases is required. In this respect, only a few outdated and incomplete databases are currently available on the web for olive germplasm. A good collection of information retrieved from the scientific literature (http://www.oleadb.it/olivodb.html) provides data on areas of diffusion, synonyms and main agronomical/morphological characters. Important information on SSR markers, tree and oil characteristics of about 200 worldwide cultivars can

DOI: 10.1079/9781789247350.0009

be visualized at http://www.bioinfo-cbs.org/ogdd/) (Ben Ayed *et al.*, 2016). The network of germplasm banks promoted by the International Olive Council is the most extensive ongoing worldwide initiative to date and a database is currently under construction as an output of the international effort (https://www.internationaloliveoil.org/the-ioc-network-of-germplasm-banks/).

The clear identification of the olive cultivars is a prerequisite for any breeding initiative. In this respect, the extensive use of molecular markers, such as simple sequence repeat (SSRs) and single nucleotide polymorphism (SNPs), has represented an important step for the identification of varieties potentially serving as cross-genitors. The analysis of large germplasm collections has enabled establishment of genetic diversity patterns, varietal relationships, synonymy cases and previous identification errors within each germplasm collection (El Bakkali *et al.*, 2019; Belaj *et al.*, 2018).

3.5.2 Genomics-assisted Breeding of Other Crops

Genomics-assisted breeding has proved efficient and rewarding for many other fruit crops. The development of high-throughput genotyping platforms of high quality, flexibility and affordable genotyping costs is critical for genomics-assisted breeding (Guo *et al.*, 2019; Zahid *et al.*, 2022). Various molecular techniques have been used to improve plant yield, quality and disease resistance (Padikasan *et al.*, 2018), and marker-assisted selection has proved to be a useful technique for the selection of desired characteristics in breeding.

In fruit tree species, many works have focussed on the identification of gene determinants or functional markers linked to the main traits of agronomical, commercial and ecological interest, leading to very important outputs. In apple, genotyping and disease phenotyping of different varieties led to the identification of two quantitative trait loci (QTL) linked to scab resistance, and candidate genes possibly controlling the resistance trait are now under evaluation (Mir *et al.*, 2022). Gene and promoter sequence variations directly involved on specific traits were also identified, as in the case of the promoter of an R2R3–MYB transcription factor determining fruit malate content (Jia *et al.*, 2021). In sweet orange, numerous fruit-related genes have been identified, controlling pigmentation, carotenoid concentration, firmness or juiciness (Lu *et al.*, 2021a; Zhang *et al.*, 2022). In peach, gene alleles controlling anthocyanin deficiency related to flower and fruit color were identified (Lu *et al.*, 2021 b), as well as those regulating fruit acidity (Wang *et al.*, 2021) or tree architecture (Dutt *et al.*, 2022).

For the four main crops forming the base of human nutrition – rice, maize, soybean and wheat – a host of genomic tools have been developed, mainly based on genome-wide association studies (GWAS). For rice, a second-generation SNP array was developed and many SNP markers associated with preferred traits are available (Morales *et al.*, 2020). A similar platform was developed in maize, where two genotype panels were developed, with markers evenly distributed across the genome, proving the reliability of SNP markers and GBTS (genotyping by target sequencing) genotyping platforms, facilitating molecular breeding and enabling germplasm evaluation, genetic mapping, marker-assisted selection and plant variety protection (Guo *et al.*, 2019). In wheat, the use of GWAS showed promise for targeted genotyping and marker-assisted selection for genetic improvement (Li *et al.*, 2018, 2019; Sun *et al.*, 2020). QTL mapping was also used to dissect the genetic basis of complex traits and haplotype-based GWAS in soybean, enabling the assessment of the genetic determinants of agronomic traits such as seed weight, plant height and seed yield (Contreras-Soto *et al.*, 2017).

3.5.3 Toward Genomics-assisted Olive Breeding

GWAS and genomic selection (GS) are two novel approaches especially useful in fruit tree breeding (Iwata *et al.*, 2016) because, based on the association between the polymorphisms of the genomic sequence and the characters to be selected, they make it possible to simplify and speed up the breeding programs, which can no longer be based on the phenotype analysis, i.e. the characteristics of the plant, but on the markers related to those characteristics. So, for example, to select new varieties with a higher fruit oil content, it will no longer be necessary to grow all individuals derived from the crosses and await their development and entry into production in order to analyze the fruits, but it will be possible to analyze the newly germinated seedlings with a simple DNA test and verify whether or not the polymorphisms that determine the high-oil phenotype are present, keeping and continuing the selection on the positive plants and discarding all others. Only a few works have started to apply this powerful approach in olive breeding.

In a pioneering work, by using SNP analysis on a set of 119 cultivars maintained in the Israeli germplasm collection, Biton *et al.* (2015) showed that cultivars mainly grouped on the basis of their function (oil, table or double-purpose), rather than on their geographic origin. First QTLs (genome regions controlling very important polygenic traits under selection, such as productivity, fruit characters, pathogen resistance etc.) have been identified for pulp:stone ratio, fruit weight and trunk diameter (Atienza *et al.*, 2014). Almost 25,000 SNPs have been used to perform a GWAS on a panel of 183 accessions, allowing for the identification of hits and mapping of 19 candidate genes related to leaf length, fruit weight, stone weight and fruit flesh-to-pit ratio (Kaya *et al.*, 2019). In order to identify functional allele variants of the *FAD2-2* gene, mainly responsible for the linoleic acid content in the olive fruit mesocarp, a study on the sequence variability and its role in the gene expression regulation was conducted, highlighting new structural variants putatively involved in the regulation mechanisms (Salimonti *et al.*, 2020). A deep sequence analysis of genes involved in fatty acid (*OeACP1, OeACP2*), triterpene synthesis (*OeLUS*) and sucrose transport (*OeSUT1*) in 90 cultivars revealed a wide range of polymorphisms and unexpected allele frequencies and genotype combinations and allowed identification of regions representing blocks of recombination putatively containing functional markers (Cultrera *et al.*, 2019). Nucleotide diversity in the coding regions of some genes involved in different resistance mechanisms against the infection of the *Verticillium dahliae* phytopathogen, analyzed on different genotypes from cultivated and related *Olea europaea* taxa, revealed a total of 92 functional SNPs and two indels. Only a few interesting associations between sequence polymorphisms and Verticillium wilt resistance were observed, potentially applicable as functional SNPs (Serrano *et al.*, 2020). Two SNPs have been identified, able to predict the group of incompatibility (G1 or G2) of any olive cultivar, enabling an early and rapid marker-assisted selection (Mariotti *et al.*, 2020). This represented the first application of functional markers in olive.

DNA methylation has also been observed to play a key role in modulating the biosynthesis of volatile aromatic compounds (Badad *et al.*, 2021). This evidence has made it possible to make known the role of epigenetic variability in the olive tree and how it can contribute to the rapid adaptation to different environments and to new scenarios evoked by ongoing climate change.

Genomics-assisted breeding has the potential of overcoming the barriers hindering conventional fruit tree breeding. However, genotypic x environmental interaction must be taken under consideration, thus enabling a ranking of genotypes valid in varying environmental conditions (Pérez *et al.*, 2018; Mousavi *et al.*, 2019).

3.5.4 Conclusions and Future Prospects

Developing collaboration between research teams in conducting association-mapping studies at contrasting environmental sites using worldwide core collections should also be encouraged.

Current breeding programs are based on crosses between cultivars, involving very long screening processes of progeny performance, carried out over several selection cycles. This process takes decades and demands large plots of land, high costs and specialized workers. In order to speed up this process, traditional breeding programs should be assisted by efficient and reliable genomic tools through a genomics-assisted breeding (GAB) approach (Fig. 3.5.1).

In order to do so, large numbers of DNA markers linked to genes involved in various traits regarded as agriculturally and economically important should be identified.

With the publication of comprehensive genome sequences of different olive varieties, availability of large and well-characterized

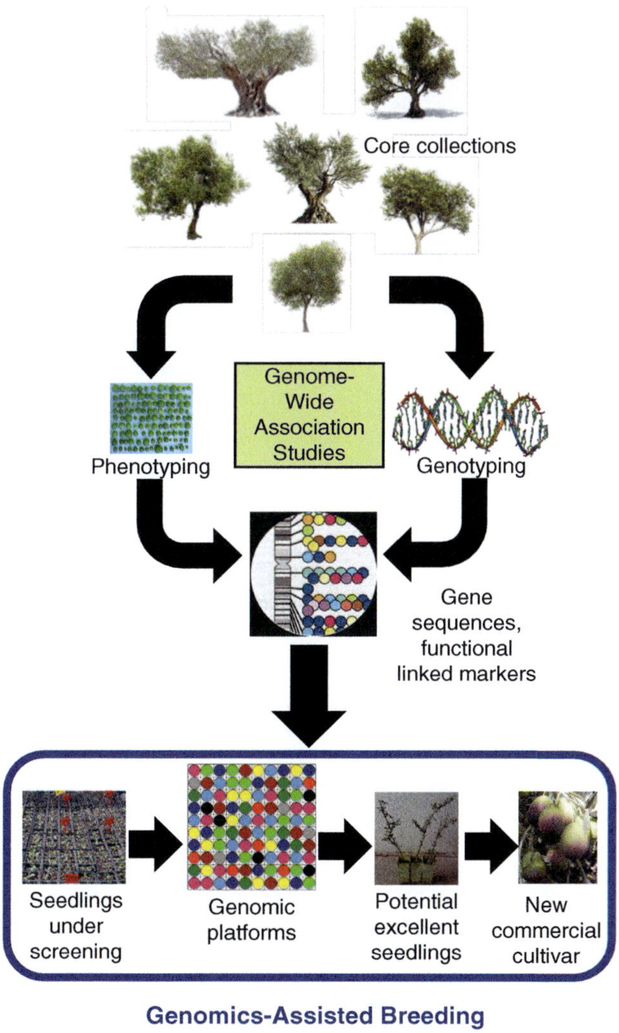

Fig. 3.5.1. A general scheme of a genome-directed breeding program for olive: germplasm collections used as plant material for genome-wide association studies, which enable the identification of many markers linked to agriculturally and economically important traits. These markers are then used for genomics-assisted breeding by fast screening of DNA extracted from young seedlings. In the final stage, field phenotype evaluation is performed on genomics pre-selected seedlings whose genotype indicates high potential for desired traits.

olive germplasm collections, and new genetic technologies, this goal is achievable.

But GAB, other than on very efficient genomic platforms, is also based on deep and precise data on plant phenotyping that, with the progress of sequencing technologies, is becoming the main bottleneck for identifying the genomic regions of interest and then characterizing genes, alleles and nucleotides responsible for the traits. International, national and regional germplasm collections, after proving that the material has been identified and authenticated by comparison with validated samples of different collections, should serve as the source of plant material required for large genome-wide association studies, including systematic phenotyping over multiple seasons and environments. The GWAS should focus on important traits, related to vegetative and reproductive aspects, oil content and composition, disease resistance and environmental stress tolerance. Tens of thousands of SNPs should be identified and genotyped for all cultivars included in the germplasm collections, using GBS or other equivalent newly developed methods, to create mega-databases. Markers linked with important traits should then be identified and then used for screening cross-progenies obtained by breeding programs practiced worldwide.

Only those seedlings with promising genotype should be included in further selection steps, grown and tested for their performance, and surplus seedlings should be abandoned. This process should prevent growing many unproductive seedlings, saving costs, speeding up the breeding process and enabling the screening of many more progenies. The selected progenies should be grown until their mature phase in several locations representing different agricultural and climatic conditions. Genomics-assisted breeding should also speed up the phenotyping procedures, reducing the number of selection cycles to one or two. This will further shorten the timespan from seeds to newly registered commercial cultivars.

References

Atienza, S.G., de la Rosa, R., León, L., Martín, A. and Belaj, A. (2014) Identification of QTL for agronomic traits of importance for olive breeding. *Molecular Breeding* 34(2), 725–737.

Badad, O., Lakhssassi, N., Zaid, N., El Baze, A., Zaid, Y., Meksem, J. and Meksem, K. (2021) Genome wide MeDIP-Seq profiling of wild and cultivated olives trees suggests DNA methylation fingerprint on the sensory quality of olive oil. *Plants* 10(7), 1405.

Baldoni, L., Khadari, B. and Rosa, R.D.L. (2016) Genetic mapping and detection of Quantitative Trait Loci. In: Rugini, E., Baldoni, L., Muleo, R. and Sebastiani, L. (eds) *The Olive Tree Genome*. Springer, Cham, Switzerland, pp. 65–74.

Belaj, A., de la Rosa, R., Lorite, I.J., Mariotti, R., Cultrera, N.G. *et al.* (2018) Usefulness of a new large set of high throughput EST-SNP markers as a tool for olive germplasm collection management. *Frontiers in Plant Science* 9, 1320.

Ben Ayed, R., Ben Hassen, H., Ennouri, K., Ben Marzoug, R. and Rebai, A. (2016) OGDD (Olive Genetic Diversity Database): a microsatellite markers genotypes database of worldwide olive trees for cultivar identification and virgin olive oil traceability. *Database*, 2016.

Ben-Ari, G., Biton, I., Many, Y., Namdar, D. and Samach, A. (2021) Elevated temperatures negatively affect olive productive cycle and oil quality. *Agronomy* 11(8), 1492.

Biton, I., Doron-Faigenboim, A., Jamwal, M., Mani, Y., Eshed, R. *et al.* (2015) Development of a large set of SNP markers for assessing phylogenetic relationships between the olive cultivars composing the Israeli olive germplasm collection. *Molecular Breeding* 35, 107.

Caruso, G., Palai, G., Marra, F.P. and Caruso, T. (2021) High-resolution UAV imagery for field olive (*Olea europaea* L.) phenotyping. *Horticulturae* 7(8), 258.

Contreras-Soto, R.I., Mora, F., de Oliveira, M.A.R., Higashi, W., Scapim, C.A. *et al.* (2017) A genome-wide association study for agronomic traits in soybean using SNP markers and SNP-based haplotype analysis. *PLOS ONE* 12, e0171105.

Cultrera, N.G., Sarri, V., Lucentini, L., Ceccarelli, M., Alagna, F., Mariotti, R. and Baldoni, L. (2019) High levels of variation within gene sequences of *Olea europaea* L. *Frontiers in Plant Science* 9, 1932.
Dutt, M., Mahmoud, L.M., Nehela, Y., Grosser, J.W. and Killiny, N. (2022) The *Citrus sinensis* Tiller Angle Control 1 (*CsTAC1*) gene regulates tree architecture in sweet oranges by modulating the endogenous hormone content. *Plant Science*, 111401.
El Bakkali, A., Essalouh, L., Tollon, C., Rivallan, R., Mournet, P. *et al.* (2019) Characterization of Worldwide Olive Germplasm Banks of Marrakech (Morocco) and Córdoba (Spain): towards management and use of olive germplasm in breeding programs. *PLOS ONE* 14(10), e0223716.
Guo, Z., Wang, H., Tao, J., Ren, Y., Xu, C. *et al.* (2019) Development of multiple SNP marker panels affordable to breeders through genotyping by target sequencing (GBTS) in maize. *Molecular Breeding* 39, 1–12.
Iwata, H., Minamikawa, M.F., Kajiya-Kanegae, H., Ishimori, M. and Hayashi, T. (2016) Genomics-assisted breeding in fruit trees. *Breeding Science* 66, 100–115.
Jia, D., Wu, P., Shen, F., Li, W., Zheng, X., Wang, Y. and Han, Z. (2021) Genetic variation in the promoter of an R2R3–MYB transcription factor determines fruit malate content in apple (*Malus domestica* Borkh.). *Plant Physiology* 186(1), 549–568.
Kaya, H.B., Akdemir, D., Lozano, R., Cetin, O., Sozer Kaya, H., Sahin, M. and Jannink, J.L. (2019) Genome wide association study of 5 agronomic traits in olive (*Olea europaea* L.). *Scientific Reports* 9(1), 1–14.
León, L., de la Rosa, R., Velasco, L. and Belaj, A. (2018) Using wild olives in breeding programs: implications on oil quality composition. *Frontiers in Plant Science* 9, 232.
Li, F., Wen, W., He, Z., Liu, J., Jin, H. *et al.* (2018) Genome-wide linkage mapping of yield-related traits in three Chinese bread wheat populations using high-density SNP markers. *Theoretical and Applied Genetics* 131, 1903–1924.
Li, H., Zhou, Y., Xin, W., Wei, Y., Zhang, J. *et al.* (2019) Wheat breeding in northern China: achievements and technical advances. *The Crop Journal* 7, 718–729.
Lu, S., Ye, J., Zhu, K., Zhang, Y., Zhang, M., Xu, Q. and Deng, X. (2021a) A fruit ripening-associated transcription factor CsMADS5 positively regulates carotenoid biosynthesis in citrus. *Journal of Experimental Botany* 72(8), 3028–3043.
Lu, Z., Cao, H., Pan, L., Niu, L., Wei, B., Cui, G. and Wang, Z. (2021b) Two loss-of-function alleles of the glutathione S-transferase (GST) gene cause anthocyanin deficiency in flower and fruit skin of peach (*Prunus persica*). *The Plant Journal* 107(5), 1320–1331.
Mariotti, R., Fornasiero, A., Mousavi, S., Cultrera, N.G., Brizioli, F. *et al.* (2020) Genetic mapping of the incompatibility locus in olive and development of a linked Sequence-Tagged Site marker. *Frontiers in Plant Science* 10, 1760.
Mir, S., Sakina, A., Masoodi, K.Z., Bhat, K.M., Padder, B.A., Murtaza, I. and Shikari, A.B. (2022) Mapping of quantitative trait loci for scab resistance in apple (*Malus* x *domestica*) variety, Shireen. *Molecular Biology Reports*, 1–12.
Morales, K.Y., Singh, N., Perez, F.A., Ignacio, J.C., Thapa, R., *et al.* (2020) An improved 7K SNP array, the C7AIR, provides a wealth of validated SNP markers for rice breeding and genetics studies. *PLOS ONE* 15, e0232479.
Mousavi, S., de la Rosa, R., Moukhli, A., El Riachy, M., Mariotti, R., Torres, M. and Leon, L. (2019) Plasticity of fruit and oil traits in olive among different environments. *Scientific Reports* 9(1), 1–13.
Mousavi, S., Stanzione, V., Mariotti, R., Mastio, V., Azariadis, A., Passeri, V. and Bufacchi, M. (2022) Bioactive compound profiling of olive fruit: the contribution of genotype. *Antioxidants* 11(4), 672.
Padikasan, I.A., Chinnannan, K., Kumar, S. and Subramaniyan, G. (2018) Agricultural biotechnology: engineering plants for improved productivity and quality. *Omics Technologies and Bio-Engineering* 2, 87–104.
Pérez, A.G., León, L., Sanz, C. and de la Rosa, R. (2018) Fruit phenolic profiling: a new selection criterion in olive breeding program. *Frontiers in Plant Science* 9, 241.
Purugganan, M.D. and Jackson, S.A. (2021) Advancing crop genomics from lab to field. *Nature Genetics* 53(5), 595–601.
Rallo, L., Caruso, T., Díez, C.M. and Campisi, G. (2016) Olive growing in a time of change: from empiricism to genomics. In: Rugini, E., Baldoni, L., Muleo, R. and Sebastiani, L. (eds) *The Olive Tree Genome.* Springer, Cham, Switzerland, pp. 55–64.
Rapoport, H.F., Moreno-Alías, I., de la Rosa-Peinazo, M.Á., Frija, A., de la Rosa, R. and León, L. (2022) Floral quality characterization in olive progenies from reciprocal crosses. *Plants* 11(10), 1285.
Salimonti, A., Carbone, F., Romano, E., Pellegrino, M., Benincasa, C., Micali, S. and Zelasco, S. (2020) Association study of the 5′ UTR intron of the FAD2-2 gene with oleic and linoleic acid content in *Olea europaea* L. *Frontiers in Plant Science* 11, 66.

Serrano, A., León, L., Belaj, A. and Román, B. (2020) Nucleotide diversity analysis of candidate genes for Verticillium wilt resistance in olive. *Scientia Horticulturae* 274, 109653.
Serrano-García, I., Olmo-García, L., Polo-Megías, D., Serrano, A., León, L., de la Rosa, R. and Carrasco-Pancorbo, A. (2022) Fruit phenolic and triterpenic composition of progenies of *Olea europaea* subsp. *cuspidata*, an interesting phytochemical source to be included in olive breeding programs. *Plants* 11(14), 1791.
Sun, C., Dong, Z., Zhao, L., Ren, Y., Zhang, N. *et al.* (2020) The Wheat 660K SNP array demonstrates great potential for marker-assisted selection in polyploid wheat. *Plant Biotechnology Journal* 18, 1354–1360.
Torres-Sánchez, J., Dd la Rosa, R., León, L., Jiménez-Brenes, F.M., Kharrat, A. and López-Granados, F. (2022) Quantification of dwarfing effect of different rootstocks in 'Picual' olive cultivar using UAV-photogrammetry. *Precision Agriculture* 23(1), 178–193.
Vatansever, R., Hernandez, P., Escalante, F.J., Dorado, G. and Unver, T. (2022) Genome-wide exploration of oil biosynthesis genes in cultivated olive tree varieties (*Olea europaea*): insights into regulation of oil biosynthesis. *Functional and Integrative Genomics* 22(2), 171–178.
Wang, L., Jiang, X., Zhao, L., Wang, F., Liu, Y., Zhou, H. and Han, Y. (2021) A candidate PpRPH gene of the D locus controlling fruit acidity in peach. *Plant Molecular Biology* 105(3), 321–332.
Zahid, G., Aka Kaçar, Y., Dönmez, D., Küden, A. and Giordani, T. (2022) Perspectives and recent progress of genome-wide association studies (GWAS) in fruits. *Molecular Biology Reports*, 1–12.
Zhang, M., Wang, J., Liu, R., Liu, H., Yang, H., Zhu, Z. and Cheng, Y. (2022) CsMYB96 confers resistance to water loss in citrus fruit by simultaneous regulation of water transport and wax biosynthesis. *Journal of Experimental Botany* 73(3), 953–966.

3.6 Breeding Strategies and Achievements for New Olive Oil Varieties

L. León[1], A. Serrano[1], G. Medina-Alonso[2], H. Yilmaz-Düzyaman[1] and R. de la Rosa[1]
[1]Andalusian Institute of Agricultural and Fisheries Research and Training (IFAPA), Center 'Alameda del Obispo', Córdoba, Spain; [2]Servicio Técnico de Agricultura y Desarrollo Rural, Cabildo Insular de Tenerife, Santa Cruz de Tenerife, Spain

3.6.1 Introduction

Breeding programs are underway in several olive-producing countries, based on crosses between cultivars with known merit, to obtain progenies and select new genotypes that combine the good qualities of the parents (Bellini *et al.*, 2008). Main objectives of olive breeding for olive oil production include early bearing, high productivity, high oil content, good oil quality, suitability to different growing systems and resistance to main pests and diseases (Fernandez-Escobar *et al.*, 2013). In fruit tree breeding, the selection process is usually arranged in several steps, and the number of selected genotypes is reduced with each step, while the number of replicates per genotype and characters under evaluation increases, until the final release of new cultivars. This process can be highly variable, according to the aims of the breeding program, the size of progenies and the financial and human resources available to manage the trials, so that many different possibilities of resource allocation have been reported for different crops (Badenes and Byrne, 2012). Due to the juvenile phase and unproductive period of olive, the whole selection process may take up to 20 years.

3.6.2 Breeding Steps

3.6.2.1 Inter-varietal crossing

The adequate choice of parents for crossing represents the first step of the breeding process. In each country, the most common cultivars have been usually selected as parents for their breeding programs. Thus, cultivars such as 'Frantoio' in Italy, 'Chemlali de Sfax' in Tunisia, 'Memecik' in Turkey, 'Galega' in Portugal and 'Picual' in Spain have been used for breeding purposes (Fontanazza *et al.*, 1998; León *et al.*, 2007; Dabbou *et al.*, 2012; Dıraman *et al.*, 2020; Sales *et al.*, 2020).

It should be considered that olive breeding could also benefit from the complex genetic resources available for the species (Lavee, 2013), which include the wild populations (*O. europaea* subsp. *europaea* var. *sylvestris*) and other related subspecies (*cuspidata*, *laperrinei*, *maroccana*, *guanchica* and *cerasiformis*) endemic in non-Mediterranean areas, which also may hold traits of interest that could be introgressed in the cultivated varieties (Cáceres *et al.*, 2015; Li *et al.*, 2021). The potential use of olive wild genetic resources as a source of genetic variability has been suggested for important

DOI: 10.1079/9781789247350.0010

agronomic traits, particularly regarding resistance to specific adverse biotic and abiotic environmental conditions (Lavee and Zohary, 2011). However, it is noteworthy that wild olives are characterized by very negative characters, such as low fruit size and oil content (Belaj *et al.*, 2011). These undesirable traits may be transmitted to their progenies, implying that additional backcross generations would be necessary, further slowing down the selection process (Klepo *et al.*, 2014). These difficulties have limited the potential use of wild relatives as parents in olive-breeding programs up to now.

Compatibility relationships among cultivars should be carefully considered for optimal design of the breeding crosses. Paternity testing by DNA microsatellite markers has been designed and successfully applied for this purpose (de la Rosa *et al.*, 2004). The diallelic self-incompatibility system has been recently uncovered for olive (Mariotti *et al.*, 2021). The system involves only two groups, i.e. cultivars belonging to one group are incompatible among each other but compatible with all cultivars of the other group. This will help to select best compatible parental combinations to design future crosses.

3.6.2.2 Juvenility of seedlings and first selection of cross-progenies

After seed germination, numerous growing methods may be applied to reduce the juvenile period (Lavee *et al.*, 1996; Santos-Antunes *et al.*, 2005). Forcing seedling growth in the greenhouse, selective pruning and intensive field growing may allow, together with adequate parent selection, a consistent decrease of the juvenile period and, thus, facilitate the breeding process (Fig. 3.6.1).

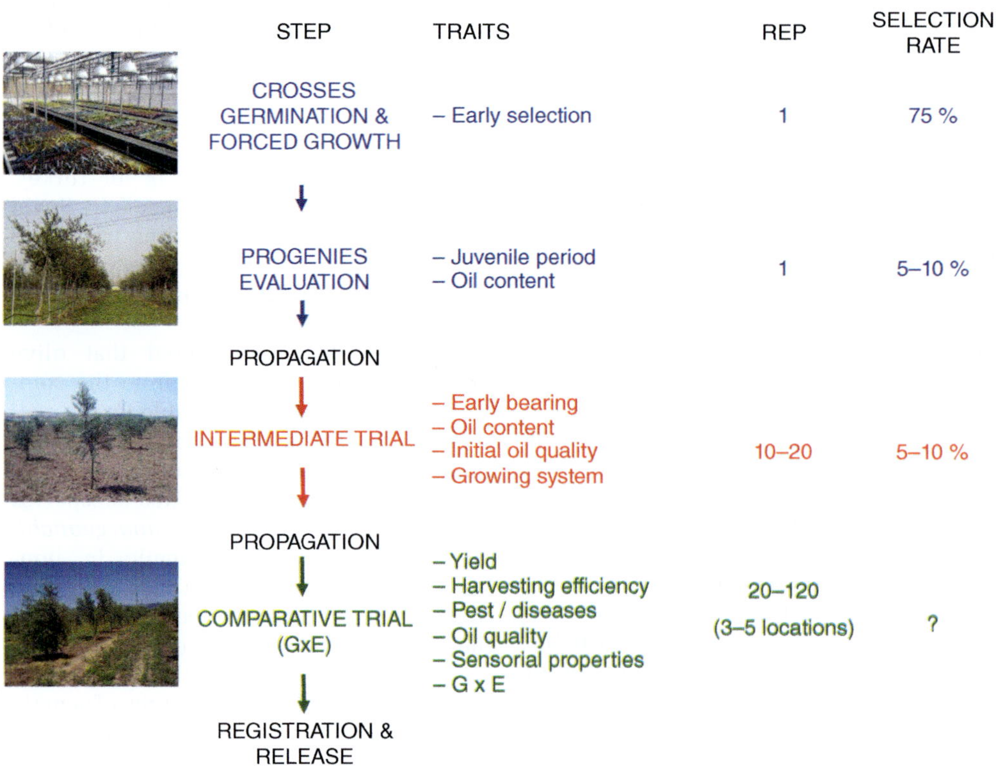

Fig. 3.6.1. Olive-breeding scheme followed at IFAPA Alameda del Obispo in Córdoba, Spain. REP: number of replicates (plants) per genotype in the trials.

Early selection criteria have been also established, based on the seedling vigor, intended as plant height or stem diameter (Pritsa *et al.*, 2003; de La Rosa *et al.*, 2006), or more complex qualitative descriptors of the whole-plant architecture (Hammami *et al.*, 2012, 2021; Yılmaz-Düzyaman *et al.*, 2022).

The application of these techniques and criteria may allow a consistent reduction of the plant growth period, thus reducing the time needed for selection, and removing one of the main hindrances to olive breeding.

3.6.2.3 Further steps of selection and genotype x environment (G x E) evaluation

After a first round of selection on the initial seedling populations, two more screening steps are recommended to fulfill the whole cycle of selection: an intermediate step of selection and a final set of comparative trials in different environments (León *et al.*, 2015).

The evaluation of a certain number of tree replicates per genotype at the intermediate step allows the evaluation for some traits that cannot be evaluated on single trees. For instance, evaluation for yield, plant vigor or growth habit can be initiated at this stage. Afterwards, a final set of comparative trials should be established for testing the best selections, previously identified at the intermediate step, under different environmental and growing conditions.

It should be noted that most interesting breeding traits are modulated by complex biological processes, regulated by a high number of genes and/or highly influenced by the environment. It is necessary to determine the heritability of these traits by analyzing the contribution of genetic and environmental effect on the observed phenotypic variance (Hansche *et al.*, 1986). The estimation of the environmental variance includes that among years, locations and trees of each genotype. In general, heritability studies in olive are still quite limited and there is an urgent need for increasing knowledge on this topic. Anyhow, the influence of genetic and environmental factors on several traits has been investigated in the framework of some olive-breeding programs.

Field evaluation of new olive selections from crossbred progenies initiated in 1971 in three locations of central and southern Italy allowed evaluation of the influence of genotype and environment on productivity, fruit characteristics and oil-quality components (Padula *et al.*, 2008; Ripa *et al.*, 2008). Among the evaluated traits, fruit oil content, pulp:pit ratio and fatty acid composition resulted mainly under genetic control. Navas-Lopez *et al.* (2019a, b, 2020) evaluated the genotype and environment interaction (G x E) for several traits, such as flowering phenology and flower quality, patterns of oil accumulation and oil quality components, on breeding selections and commercial cultivars in a multi-environment trial established in various locations of Andalusia, Spain, under different agro-climatic conditions.

A high genetic influence was observed for flowering quality parameters and maximum oil content. Significant genetic effects were also reported for several extra virgin olive oils (EVOO) components, such as tocopherol, squalene and sterol content, while the location effect was higher for total phenols and volatile contents (Serrano *et al.*, 2020, 2021a). Interestingly, these differences were not clearly translated into differences of sensory properties of EVOO.

All these works showed some significant environmental and G x E effects, allowing identification of some stable genotypes showing optimal performance under different environmental conditions, as well as some others particularly adapted to specific environments. According to these results, extensive multi-environment trials are of paramount importance in the final selection process in olive breeding, in order to select the best genotypes adapted to different environments.

3.6.3 Molecular Tools for Assisting Selection

The scarce knowledge on the genetic control of different agronomic traits still limits the progress of olive-breeding programs compared to other fruit tree species (Lavee, 2013). Several attempts to implement genomics-assisted breeding (GAB) approaches in olive have been based on the development of genetic linkage maps and localization of quantitative trait loci (QTLs) associated to agronomical traits of interest, as previously widely reported in other fruit tree species. This represents a promising strategy for improving breeding efficiency, guiding in the choice of parents and allowing selection of interesting genotypes directly from seedling populations (Collard *et al.*, 2005). Thus, first attempts for QTL mapping related to flowering, fruiting, plant vigor, fruit and oil traits (fatty acid composition) have been reported (Ben Sadok *et al.*, 2013a, b; Atienza *et al.*, 2014; Hernández *et al.*, 2017). However, the low saturation of the genetic maps and the extension of QTL regions have made it difficult to identify polymorphisms directly linked to the traits under selection.

Transcriptomic approaches have been also widely used in the last years to reveal candidate genes associated with important agronomic traits. Among others, first characters where genes putatively responsible for their regulation were identified include cold acclimation (Leyva-Pérez *et al.*, 2015), fruit abscission (Gil-Amado and Gómez-Jiménez, 2013), plant architecture (González-Plaza *et al.*, 2016), early plant development, juvenile period length (Jiménez-Ruiz *et al.*, 2018), oil quality components (Alagna *et al.*, 2012) and response to biotic stresses, such as olive fruit fly *Bactrocera oleae* (Grasso *et al.*, 2017) and *V. dahliae* (Jiménez-Ruiz *et al.*, 2019). However, deciphering the whole picture involved in these processes is still far from accomplished.

Recently, the *Olea europaea* genome has been sequenced for several cultivars and wild olive genotypes (Unver *et al.*, 2017; Julca *et al.*, 2020; Jiménez-Ruiz *et al.*, 2020). The availability of these whole-genome sequences paves the way towards the development of new genomic tools that have just started to produce interesting results for characters such as aquaporin diversity in olive trees (Faize *et al.*, 2020).

Genome-wide association studies (GWAS) represent another very promising alternative strategy with high potential compared to QTL mapping to identify gene determinants linked to traits under selection. It is based on a dense SNP genotyping of diverse populations with low levels of linkage disequilibrium, which allow narrowing of the genomic regions to search for potential candidate genes and causal polymorphisms (Korte and Farlow, 2013). GWAS of several agronomic traits in olive from genotyping-by-sequencing approaches have already been reported or are currently underway (Kaya *et al.*, 2019; Islam *et al.*, 2021).

3.6.4 Objectives and Results of Breeding

3.6.4.1 Earliness of bearing, productivity and oil content

A short unproductive period currently represents a fundamental trait for new cultivars suited to modern intensive systems. Initial selection of seedlings for a short juvenile period has been demonstrated to be correlated to a shorter unproductive period of the subsequent new cultivars (León *et al.*, 2007), underlying the need to include short juvenile period as a primary selection criterion in olive-breeding programs.

High productivity is also considered a main criterion for new cultivars. Recent works for trait prioritization of potential olive-breeding objectives showed that high productivity was the main objective for olive researchers, olive producers and orchard managers (León *et al.*, 2021). However, productivity is a complex trait subjected to high environmental influence and, therefore, accurate selection for this trait is difficult. Results from comparative trials involving olive cultivars and breeding selections report high experimental errors on the determination of productivity traits (Astorga and Mora, 2005; León *et al.*, 2007), underlying the difficulties in selection for this trait.

A wide variability for fruit traits has been reported on the evaluation of olive progenies from different origin (Bellini, 1992; Avidan *et al.*, 2012; Arias-Calderon *et al.*, 2014). As for other fruit traits, wide variability for oil content has been reported for most breeding progenies (León *et al.*, 2004a; Lavee and Avidan, 2011). Extreme phenotypes with oil content higher than that shown by parental trees have been obtained and selected. Thus, increased oil content is considered as one of the main advantages in new cultivars (Lavee *et al.*, 1986; León *et al.*, 2007).

3.6.4.2 Oil quality

EVOO fatty acid composition, mainly represented by monounsaturated fatty acids, as well as myriad minor components, are mainly responsible for health properties, particularly regarding cardiovascular, neurodegenerative, inflammatory diseases and cancer and a general increase in life expectancy for EVOO consumers (Gouvinhas *et al.*, 2014; Diez-Bello *et al.*, 2019; Gutiérrez-Miranda *et al.*, 2020).

As for fruit traits, a wide genetic variability has also been observed in progenies for chemical composition of EVOO, such as for fatty acid composition, triacylglycerols, phenolic compounds, squalene and phytosterols (Dabbou *et al.*, 2012; El Riachy *et al.*, 2012; de la Rosa *et al.*, 2016; Dıraman *et al.*, 2020). Transgressive segregation is also common, as for oil content. In fact, wider variability was observed in a single cross than in a cultivar sample set (Pérez *et al.*, 2014, 2019). The potential usefulness of wild genetic resources regarding oil composition has also been widely reported (Baccouri *et al.*, 2008; León *et al.*, 2018; Elgadi *et al.*, 2021). Therefore, the possibilities offered by breeding to improve oil quality composition from new cross combinations should be significantly implemented in the future.

The sensorial profile of new bred genotypes has been studied to a lesser extent. Only a few works reported the organoleptic properties of bred materials, mainly based on the main sensorial attributes, such as bitter, pungent and fruity taste (Serrano *et al.*, 2020; Dıraman *et al.*, 2020). A complex world of sensations is yet to be explored.

All these works would eventually allow the selection of promising new genotypes with improved EVOO composition and organoleptic properties (Baccouri *et al.*, 2007; Manai *et al.*, 2008; León *et al.*, 2011; El Riachy *et al.*, 2013; de la Rosa *et al.*, 2013; Velasco *et al.*, 2015).

An additional outcome of breeding works related to improved oil quality has been the contribution on the development of easier and faster phenotyping procedures able to reduce the time and effort for advancing the selection process. Assessment of olive fruit traits by near infrared spectroscopy (NIR) allowed a rapid, simultaneous and non-destructive analysis of several fruit and oil traits with non-destructive tests, low cost per sample, portable devices allowing even direct evaluation in the field (León *et al.*, 2004b; Gracia and León, 2011; Bellincontro *et al.*, 2012; Cayuela and García, 2017). Similarly, adaptation of protocols for chemical analysis indicates that selection for some oil-quality components, such as fatty acids, tocopherols, phytosterols and squalene, could be efficiently carried out by direct analysis of fruit flesh, avoiding the need for previous oil extraction, and thus allowing the analysis of a large number of genotypes at the initial stage of selection (Velasco *et al.*, 2014; Perez *et al.*, 2018).

3.6.4.3 Suitability to different growing systems

Changes in socio-economic conditions have promoted a deep transformation in olive growing over the last 50 years (Fig. 3.6.2). Since then, a process of olive-growing intensification is underway, from traditional manual harvesting and non-irrigated cultivation to new orchards with increasing planting density, mechanization and irrigation (Fernandez-Escobar *et al.*, 2013). The emergence and spread of high-density hedgerow olive orchards from the mid-1990s is a paradigmatic exemplification of this process (www.olint.es; accessed 5/5/23).

Fig. 3.6.2. Traditional, intensive and high-density hedgerow olive orchards.

The features of cultivars suited to intensive olive orchards are very different from those required for traditional cultivation. This is especially true for the hedgerow growing system, where only cultivars with low vigor, flexible branches and high canopy density are suitable. In fact, until very recently, for most of the new olive orchards planted with this growing system, only two cultivars were used, 'Arbequina' and 'Arbosana' (Rallo *et al.*, 2014).

For this reason, plant architecture is being extensively studied, to understand plant growth, branching pattern and yield (Rosati *et al.*, 2013, 2021; Carella *et al.*, 2022). The evaluation of tree vigor and architectural parameters (quantitative and qualitative) has been incorporated into breeding programs due to its relationship with both plant architectural traits and duration of juvenile period (Hammami *et al.*, 2012, 2021; Ben Sadok *et al.*, 2015). Assessment of olive tree crown parameters from high-resolution airborne Unmanned Aerial Vehicle (UAV) imagery has also been explored recently, as a potential useful tool for application in breeding (Díaz-Varela *et al.*, 2015; Torres-Sánchez *et al.*, 2022). This line of work will likely receive increasing attention in the future, as the spreading of high hedgerow orchards will continue. There is an urgent need for widening the offer of hedgerow-adapted cultivars.

3.6.4.4 Disease resistance

Breeding for disease resistance has been attempted for the main olive diseases. Among them, Verticillium wilt of olive (VWO) represents a major limitation for olive productivity in many producing areas (Montes-Osuna and Mercado-Blanco, 2020). Therefore, this disease received main attention in breeding works, due to the difficulties in eradicating its causal agent, *Verticillium dahliae*, in infested soils and the scarce number of traditional cultivars showing a significant level of resistance to this disease (Erten *et al.*, 2010; Arias-Calderon *et al.*, 2015b; Trapero *et al.*, 2015).

Several schemes have been proposed for olive-breeding programs aiming at obtaining resistant cultivars to VWO, depending on the step of the selection process at which the disease resistance assessment is carried out. Screening for resistance at the initial seedling stage could imply high risk of selecting genotypes that subsequently will probably show poor agronomic performance. Alternatively, later screening at the intermediate step of the selection process would increase the chances of eventually selecting genotypes that combine both, high disease resistance and appropriate agronomic characters (Arias-Calderón *et al.*, 2015b). Additionally, screening at the intermediate stage may also allow the availability of multiple plant replicates that will provide material for their vegetative propagation, instead of single seedlings used at the initial step. In this way, appropriate experimental designs can be used for testing, which guarantee more consistent evaluation of plant reaction to disease. Besides, different inoculation methods, type of inoculum, environments, age and type of plants have been used to screen olive for resistance to VWO (Lopez-Escudero *et al.*, 2007; Cirulli *et al.*, 2008; Trapero *et al.*, 2013). In any case, final testing under natural field conditions is always necessary to confirm previous results under controlled conditions (Fig. 3.6.3). This is particularly important for complex diseases such as VWO, where some discrepancies have been reported between the level of resistance obtained in field vs. controlled conditions (Trapero *et al.*, 2013; García-Ruiz *et al.*, 2014). Testing artificially infested plants with *V. dahliae* in microplots under natural environmental conditions has been recently proposed as an intermediate alternative, producing homogeneous results (Serrano *et al.*, 2021b).

High variability for resistance to *V. dahliae* infection was demonstrated in progenies from open-pollinated and crossbred progenies. Higher levels of resistance to VWO were observed in progenies derived from resistant parents, although some resistant genotypes were also obtained from VWO-susceptible parents but with a good agronomic behavior (Arias-Calderon *et al.*, 2015a, b; Trapero *et al.*, 2015). Several works for testing new genotypes from second generation of crosses between F1-resistant parents are currently ongoing.

Wild olives have also been used as parents in breeding programs for VWO resistance and resistant genotypes were detected among progenies; however, to also obtain an adequate agronomic performance,

Fig. 3.6.3. Screening for Verticillium wilt resistance in field trials.

additional backcross generations are required to obtain genotypes with good agronomic behavior (Colella *et al.*, 2008; Arias-Calderon *et al.*, 2015c; Trapero et al., 2015).

As an alternative approach, the selection of resistant rootstocks has been suggested, potentially useful for a soilborne disease such as VWO. Introgressing resistance characters from wild olives and related subspecies could be particularly useful for this purpose. However, contradictory results have been obtained regarding VWO resistance conferred by rootstocks in olive, and several examples of uncertain results have been reported for susceptible cultivars grafted onto resistant 'Frantoio' or wild-olive rootstocks (Porras Soriano *et al.*, 2003; Bubici and Cirulli, 2011; Jiménez-Fernández *et al.*, 2016). But, more importantly, a long-term evaluation in infested fields including different rootstock/scion combinations only revealed a delay in development of symptoms (Valverde *et al.*, 2021).

Breeding attempts for olive diseases other than VWO have been only scattered. A program to induce resistance to peacock spot caused by *Spilocaea oleagina* was carried out in Israel starting from a single tree with low infection in a highly infected area (Lavee *et al.*, 1999). In this work, progenies of this tree obtained from controlled self-pollination were submitted to several selection steps, initially both under controlled and field conditions, to the final selection of the new VWO-resistant cultivar 'Maalot'.

In other cases, the evaluation of new materials to other olive diseases have been reported, as a key point to consider prior to their release to the nursery industry. Thus, the resistance response of 15 new selections and their genitors was evaluated for the three main aerial diseases caused by *Spilocaea oleagina*, *Colletotrichum acutatum* and *Pseudocercospora cladosporioides* in Spain (Moral *et al.*, 2015). Moreover, high incidence of some diseases due to pathogens generally characterized as weakly virulent in olive, such as fruit rot caused by *Alternaria alternata*, has been described when testing newly bred cultivars (Moral *et al.*, 2008).

3.6.5 Achievements: Released Cultivars by Breeding Works

The need for high-performing cultivars prompted the development of olive-breeding programs in recent years. Breeding offers the opportunity to develop new cultivars taking into account the needs of modern olive growing. From the agronomical point of view, earliness of bearing, high productivity and oil content, disease resistance and suitability to mechanical harvesting are mandatory requirements for plant products deriving from any olive-breeding program. Production of high-quality EVOO is also considered an objective of paramount importance. Interest in olive breeding has increased in recent years, particularly due to the development of high-density growing systems, which demand suitable cultivars. A summary of the main new cultivars released in recent years is presented in Table 3.6.1.

Some other interesting selections from these breeding programs are currently under examination. Besides, in different countries, other breeding programs are carried out, with different scopes and based on different genetic backgrounds. Most of the obtained progenies are currently under evaluation, as in the case of the Egyptian program (Mikhail, 2015; Osman, 2015). In Tunisa, there is the breeding program aimed at improving the oil quality of the local cultivar 'Chemlali Sfax', mainly characterized

Table 3.6.1. Granted plant breeders' rights at the Community Plant Variety Office from cross-breeding works. (From https://cpvo.europa.eu/en)

Denomination	Origin*	Year	Institution	Country	Reference
Barnea	Unknown 1967	1992	Agricultural Research Organization	Israel	Lavee *et al.* (1986)
Maalot	Chemlali s.p. 1979	2001	Agricultural Research Organization	Israel	Lavee *et al.* (1999)
Fs-17	Frantoio o.p. 1980	1993	National Research Council	Italy	Fontanazza *et al.* (1998)
Don Carlo	Frantoio o.p. 1980	2002	National Research Council	Italy	–
Giulia	Moraiolo o.p. 1980	2009	National Research Council	Italy	–
Masepo	Manzanillo s.p. 1990	2007	Agricultural Research Organization	Israel	Lavee *et al.* (2004)
Sepoka	Kadesh s.p. 1990	2007	Agricultural Research Organization	Israel	Lavee *et al.* (2004)
Kadeshon	Kadesh s.p. 1990	2011	Agricultural Research Organization	Israel	Lavee *et al.* (2004)
Sikitita	Picual x Arbequina 1991	2009	Córdoba University/Andalusian Institute of Agricultural and Fisheries Research and Training, Córdoba	Spain	Rallo *et al.* (2008)
Vas One	Unknown 1988	2015	Vivai Attilio Sonnoli S.A.S.	Italy	–
Askal	Barnea x Manzanillo 1990	2016	Agricultural Research Organization	Israel	Lavee *et al.* (2003)
Bitsyoac	Arbequina x Arbosana 1988	2019	Agromillora Iberia, S.L.	Spain	–
Lecciana	Arbosana x Leccino 1998	2019	Agromillora Iberia, S.L.	Spain	Camposeo *et al.* (2021)

o.p, open pollination; s.p, self-pollination

by low oleic acid content. Leaf, fruit and stone morphological evaluation allowed selection of five new genotypes showing a significant increase in fruit size, new morphological states for fruit and stone, and a better fatty acid composition of oil (Fathi *et al.*, 2019; Guellaoui *et al.*, 2019; Amar, *et al.*, 2021; Ayadi *et al.*, 2022).

References

Alagna, F., Mariotti, R., Panara, F., Caporali, S., Urbani, S. *et al.* (2012) Olive phenolic compounds: metabolic and transcriptional profiling during fruit development. *BMC Plant Biology* 12, 162.

Amar, F.B., Guellaoui, I., Ayadi, M., Elloumi, O., Triki, M.A. and Boubaker, M. (2021) 'Zeitoun Ennour': a new olive (*Olea europaea* L.) cultivar in Tunisia with high oil quality. *Genetic Resources* 2(4), 1–6.

Arias-Calderon, R., León, L., Bejarano-Alcazar, J., Belaj, A., de la Rosa, R. and Rodríguez-Jurado, D.(2015a) Resistance to Verticillium wilt in olive progenies from open-pollination. *Scientia Horticulturae* 185, 34–42.

Arias-Calderon, R., Rodríguez-Jurado, D., Bejarano-Alcazar, J., Belaj, A. and León, L. (2015b) Evaluation of Verticillium wilt resistance in selections from olive breeding crosses. *Euphytica* 206, 619–629.

Arias-Calderon, R., Rodríguez-Jurado, D., León, L., Bejarano-Alcazar, J., de la Rosa, R. and Belaj, A. (2015c) Pre-breeding for resistance to Verticillium wilt in olive: fishing in the wild relative gene pool. *Crop Protection* 75, 25–33.

Arias-Calderon, R., Rouiss, H., Rodríguez-Jurado, D., de la Rosa, R. and León, L. (2014) Variability and heritability of fruit characters in olive progenies from open-pollination. *Scientia Horticulturae* 169, 94–98.

Astorga, M. and Mora, F. (2005) Componentes de varianza e interacciòn variedad – sitio del vigor, producciòn y productividad de *Olea europea* en Chile. *Cerne* 11, 25–33.

Atienza, S.G., de la Rosa, R., León, L., Martín, A. and Belaj, A. (2014) Identification of QTL for agronomic traits of importance for olive breeding. *Molecular Breeding* 34, 725–737.

Avidan, B., Meni, Y. and Lavee, S. (2012) Inheritance potential of fruit quality traits within olive (*Olea europaea* L.) cultivars. *Acta Horticulturae* 949, 105–112.

Ayadi, M. and Amar, F.B. (2022) Fatty acid composition in olive (*Olea europaea*. L) oil of progenies obtained from Tunisian cross breeding program. *Journal of Arid Arboriculture and Olive Growing* 1(1), 11–23.

Baccouri, B., Ben Temime, S., Taamalli, W., Daoud, D., M'Sallem, M. and Zarrouk, M. (2007) Analytical characteristics of virgin olive oils from two new varieties obtained by controlled crossing on Meski variety. *Journal of Food Lipids* 14, 19–34.

Baccouri, B., Zarrouk, W, Baccouri, O., Guerfel, M., Nouairi, I. and Krichene, D. *et al.* (2008) Composition, quality and oxidative stability of virgin olive oils from some selected wild olives (*Olea europaea* L. subsp *oleaster*). *Grasas y Aceites* 59, 346–351.

Badenes, M.L. and Byrne, D.H. (2012) *Fruit Breeding*. Springer Science and Business Media.

Belaj, A., León, L., Satovic, Z. and de la Rosa, R. (2011) Variability of wild olives (*Olea europaea* subsp. *europaea* var. *sylvestris*) analyzed by agro-morphological traits and SSR markers. *Scientia Horticulturae* 129, 561–569.

Bellincontro, A., Taticchi, A., Servili, M., Esposto, S., Farinelli, D. and Mencarelli, F. (2012) Feasible application of a portable NIR-AOTF tool for on-field prediction of phenolic compounds during the ripening of olives for oil production. *Journal of Agricultural and Food Chemistry* 60, 2665–2673.

Bellini, E. (1992) Behaviour of some genetic characters in olive seedlings obtained by cross-breeding. *Acta Horticulturae* 317, 197–208.

Bellini, E., Rosati, A. and Giordani, E. (2008) Genetic improvement of olive from clonal selection to cross-breeding programs. *Advances in Horticultural Science* 22(2), 73–86.

Ben Sadok, I., Celton, J.-M., Essalouh, L., El Aabidine, A.Z., Garcia, G. *et al.* (2013a) QTL mapping of flowering and fruiting traits in olive. *PLOS ONE* 8, e62831.

Ben Sadok, I., Moutier, N., Garcia, G., Dosba, F., Grati-Kamoun, N. *et al.* (2013b) Genetic determinism of the vegetative and reproductive traits in an F1 olive tree progeny. *Tree Genetics and Genomes* 9(1), 205–221.

Ben Sadok, I., Martinez, S., Moutier, N., Garcia, G., León, L. *et al.* (2015) Plasticity in vegetative growth over contrasted growing sites of an F1 olive tree progeny during its juvenile phase. *PLOS ONE* 10, 19.

Bubici, G. and Cirulli, M. (2011) Control of Verticillium wilt of olive by resistant rootstocks. *Plant and Soil* 352, 363–376.

Cáceres, M.E., Ceccarelli, M., Pupilli, F., Sarri, V. and Mencuccini, M. (2015) Obtainment of inter-subspecific hybrids in olive (*Olea europaea* L.). *Euphytica* 201, 307–319.
Camposeo, S., Vivaldi, G.A., Montemurro, C., Fanelli, V. and Cunill Canal, M. (2021) Lecciana, a new low-vigour olive cultivar suitable for super high density orchards and for nutraceutical EVOO production. *Agronomy* 11(11), 2154.
Carella, A., Massenti, R., Milazzo, G., Caruso, T. and Lo Bianco, R. (2022) Fruiting, morphology, and architecture of 'Arbequina'and 'Calatina' olive branches. *Horticulturae* 8(2), 109.
Cayuela, J.A, and García, J.F. (2017) Sorting olive oil based on alpha-tocopherol and total tocopherol content using near-infra-red spectroscopy (NIRS) analysis. *Journal of Food Engineering* 202, 79–88.
Cirulli, M., Colella, C., D'Arnico, M., Amenduni, M. and Bubici, G. (2008) Comparison of screening methods for the evaluation of olive resistance to *Verticillium dahliae* kleb. *Journal of Plant Pathology* 90, 7–14.
Colella, C., Miacola, C., Amenduni, M., D'Amico, M., Bubici, G. and Cirulli, M. (2008) Sources of verticillium wilt resistance in wild olive germplasm from the Mediterranean region. *Plant Pathology* 57, 533–539.
Collard, B.C.Y., Jahufer, M.Z.Z., Brouwer, J.B. and Pang, E.C.K. (2005) An introduction to markers, quantitative trait loci (QTL) mapping and marker-assisted selection for crop improvement: the basic concepts. *Euphytica* 142, 169–196.
Dabbou, S., Chaieb, I., Rjiba, I., Issaoui, M., Echbili, A. *et al.* (2012) Multivariate data analysis of fatty acid content in the classification of olive oils developed through controlled crossbreeding. *Journal of the American Oil Chemists Society* 89, 667–674.
de la Rosa, R., James, C.M. and Tobutt, K.R. (2004) Using microsatellites for paternity testing in olive progenies. *HortScience* 39, 351–354.
de la Rosa, R., Kiran, A.I.I., Barranco, D. and León, L. (2006) Seedling vigour as a preselection criterion for short juvenile period in olive breeding. *Australian Journal of Agricultural Research* 57, 477–481.
de la Rosa, R.R., Talhaoui, N., Rouis, H., Velasco, L and León, L. (2013) Fruit characteristics and fatty acid composition in advanced olive breeding selections along the ripening period. *Food Research International* 54, 1890–1896.
de la Rosa, R., Arias-Calderón, R., Velasco, L. and León, L. (2016) Early selection for oil quality components in olive breeding progenies. *European Journal of Lipid Science and Technology* 118, 1160–1167.
Díaz-Varela, R.A., de la Rosa, R., León, L. and Zarco-Tejada, P.J. (2015) High-resolution airborne UAV imagery to assess olive tree crown parameters using 3D photo reconstruction: application in breeding trials. *Remote Sensing* 7(4), 4213–4232.
Díez-Bello, R., Jardin, I., Lopez, J.J., El Haouari, M., Ortega-Vidal, J. *et al.* (2019) (–)-Oleocanthal inhibits proliferation and migration by modulating Ca2+ entry through TRPC6 in breast cancer cells. *Biochimica et Biophysica Acta (BBA)-Molecular Cell Research* 1866, 474–485.
Dıraman, H., Karaman, H.T., Sefer, F., Ersoy, N. and Arsel, A.H. (2020) Using nutritional lipid and sensorial profiles for the characterization of Turkish olive (Memecik x Gemlik cv.) hybrids obtained from controlled crossbreeding. *Journal of the American Oil Chemists' Society* 97, 943–954.
El Riachy, M., Priego-Capote, F., León, L., Dolores, M., de Castro, L. and Rallo, L. (2012) Virgin olive oil phenolic profile and variability in progenies from olive crosses. *Journal of the Science of Food and Agriculture* 92, 2524–2533.
El Riachy, M., Priego-Capote, F., Rallo, L., Dolores, M. de Castro, L. and León, L. (2013) Phenolic composition of virgin olive oils in cultivars for narrow hedgerow olive orchards. *European Journal of Lipid Science and Technology* 115, 800–810.
Elgadi, S., Ouhammou, A., Zine, H., Maata, N., Babahmad, R.A. and El Antari, A. (2021) Comparative oil composition study of the endemic Moroccan olive (*Olea europaea* subsp. *maroccana*) and wild olive (var. *sylvestris*) in Central West Morocco. *Journal of Food Quality* 2021, 1–10.
Erten, L., Yildiz, M. and Yıldız, M. (2010) Screening for resistance of Turkish olive cultivars and clonal rootstocks to Verticillium wilt. *Phytoparasitica* 39, 83–92.
Faize, M., Fumanal, B., Luque, F., Ramírez-Tejero, J.A., Zou, Z. *et al.* (2020) Genome wild analysis and molecular understanding of the aquaporin diversity in olive trees (*Olea europaea* L.). *International Journal of Molecular Sciences* 21, 4183.
Fathi, B.E.N., Guellaoui, I., Triki, M.A., Ayadi, M. and Boubaker, M. (2019) Zeitoun Ennwader: a new olive oil cultivar in Tunisia with high oil quality and low alternate bearing. *Revue Marocaine des Sciences Agronomiques et Vétérinaires* 7(4), 522–525.
Fernandez-Escobar, R., de la Rosa, R., León, L., Gomez, A, Testi, F. *et al.* (2013) Evolution and sustainability of the olive production systems. In: Arcas, N., Arroyo López, F.N., Caballero, J., D'Andria, R., Fernández, M. *et al. Options Méditerranéennes Séries A: Mediterranean Seminars* 106, 11–42.

Fontanazza, G, Bartolozzi, F. and Vergati, G. (1998) Fs-17. *Rivista di Frutticoltura e di Ortofloricoltura* 5, 61.

García-Ruiz, G.M., Trapero, C. and López-Escudero, F.J. (2014) Shortening the period for assessing the resistance of olive to Verticillium wilt using continuous lighting. *Hortscience* 49(9), 1171–1175.

Gil-Amado, J.A. and Gomez-Jimenez, M.C. (2013) Transcriptome analysis of mature fruit abscission control in olive. *Plant and Cell Physiology* 54, 244–269.

González-Plaza, J., Ortiz-Martín, I., Muñoz-Mérida, A., García-López, C. and Sánchez-Sevilla, J.F. (2016) Transcriptomic analysis using olive varieties and breeding progenies identifies candidate genes involved in plant architecture. *Frontiers in Plant Science* 7, 240.

Gouvinhas, I., Machado, J., Gomes, S., Lopes, J., Martins-Lopes, P. and Barros, A.I.R.N.A (2014) Phenolic composition and antioxidant activity of monovarietal and commercial Portuguese olive oils. *Journal of the American Oil Chemists' Society*, 91, 1197–1203.

Gracia, A. and León, L. (2011) Non-destructive assessment of olive fruit ripening by portable near infrared spectroscopy. *Grasas y Aceites* 62, 268–274.

Grasso, F., Coppola, M., Carbone, F., Baldoni, L., Alagna, F. *et al.* (2017) The transcriptional response to the olive fruit fly (*Bactrocera oleae*) reveals extended differences between tolerant and susceptible olive (*Olea europaea* L.) varieties. *PLOS ONE* 12, e0183050.

Guellaoui, I., Amar, F.B., Ayadi, M. and Boubaker, M. (2019) Chemometric classification of new *Olea europaea* L. cultivars developed through a crossbreeding program in Tunisia. *Journal of Scientific Agriculture* 3, 22–27.

Gutiérrez-Miranda, B., Gallardo, I., Melliou, E., Cabero, I., Álvarez, Y. *et al.* (2020) Oleacein attenuates the pathogenesis of experimental autoimmune encephalomyelitis through both antioxidant and anti-inflammatory effects. *Antioxidants* 9, 1161.

Hammami, S.B., de la Rosa, R., Sghaier-Hammami, B., León, L. and Rapoport, H.F. (2012) Reliable and relevant qualitative descriptors for evaluating complex architectural traits in olive progenies. *Scientia Horticulturae* 143, 157–166.

Hammami, S.B., León, L., Rapoport, H.F. and de la Rosa, R. (2021) A new approach for early selection of short juvenile period in olive progenies. *Scientia Horticulturae* 281, 10993.

Hansche, P.E., Beres, V. and Brooks, R.M. (1986) Heritability and genetic correlation in the sweet cherry. *Journal of the American Society for Horticultural Science* 88, 173–183.

Hernandez, M.L., Belaj, A., Sicardo, M.D.D., León, L., de la Rosa, R. Martín, A. *et al.* (2017) Mapping quantitative trait loci controlling fatty acid composition in olive. *Euphytica* 213, 12.

Islam, A.S.M., Sanders, D., Mishra, A.K. and Joshi, V. (2021) Genetic diversity and population structure analysis of the USDA olive germplasm using genotyping-by-sequencing (GBS). *Genes* 12, 2007.

Jiménez-Fernández, D., Trapero-Casas, J.L., Landa, B.B., Navas-Cortés, J.A., Bubici, G., Cirulli, M. and Jiménez-Díaz, R.M. (2016) Characterization of resistance against the olive-defoliating *Verticillium dahliae* pathotype in selected clones of wild olive. *Plant Pathology* 65, 1279–1291.

Jiménez-Ruiz, J., de la O Leyva-Pérez, M., Vidoy-Mercado, I., Barceló, A. and Luque, F. (2018) Transcriptomic time-series analysis of early development in olive from germinated embryos to juvenile tree. *BMC Genomics* 19, 824.

Jiménez-Ruiz, J., Leyva-Pérez, M. de la O, Gómez-Lama Cabanás, C., Barosso, J.B., Luque, F. and Mercado-Blanco, J. (2019) The transcriptome of *Verticillium dahliae* responds differentially depending on the disease susceptibility level of the olive (*Olea europaea* L.) cultivar. *Genes* 10, 251.

Jiménez-Ruiz, J., Ramírez-Tejero, J.A., Fernández-Pozo, N., de la O Leyva-Pérez, M., Yan, H. (2020) Transposon activation is a major driver in the genome evolution of cultivated olive trees (*Olea europaea* L.). *The Plant Genome* e20010, 1–19.

Julca, I., Marcet-Houben, M., Cruz, F., Gómez-Garrido, J., Gaut, B.S. *et al.* (2020) Genomic evidence for recurrent genetic admixture during the domestication of Mediterranean olive trees (*Olea europaea* L.). *BMC Biology* 18(1), 1–25.

Kaya, H.B., Akdemir, D., Lozano, R., Cetin, O., Kaya, H.S. *et al.* (2019) Genome wide association study of 5 agronomic traits in olive (*Olea europaea* L.). *Scientific Reports* 9, 1–14.

Klepo, T., Toumi, A., de la Rosa, R., León, L and Belaj, A. (2014) Agronomic evaluation of seedlings from crosses between the main Spanish olive cultivar 'Picual' and two wild olive trees. *Journal of Horticultural Science and Biotechnology* 89, 508–512.

Korte, A. and Farlow, A. (2013) The advantages and limitations of trait analysis with GWAS: a review. *Plant Methods* 9, 29.

Lavee, S. (2013) Evaluation of the need and present potential of olive breeding indicating the nature of the available genetic resources involved. *Scientia Horticulturae* 161, 333–339.

Lavee, S. and Avidan, B. (2011) Heredity diversity in populations of free-, self-, and specific cross-pollinated progenies of some olive (*Olea europaea* L.) cultivars. *Israel Journal of Plant Sciences* 59, 29–37.
Lavee, S. and Zohary, D. (2011) The potential of genetic diversity and the effect of geographically isolated resources in olive breeding. *Israel Journal of Plant Sciences* 59, 3–13.
Lavee, S., Haskal, A. and Wodner, M. (1986) 'Barnea' a new olive cultivar from first breeding generation. *Olea* 17, 95–99.
Lavee, S., Avidan, N., Haskal, A. and Ogrodovich, A. (1996) Acortamiento del periodo juvenil en los plantones de olivo obtenidos de semillas. Un instrumento de revalorizacion de la mejora genetica. *Olivae* 60, 33–41.
Lavee, S., Harshemesh, H., Haskal, A., Avidan, B., Ogrodovich, A., Avidan, N. and Trapero, A. (1999) 'Maalot' a new cultivar for oil extraction resistant to *Spilocaea oleagina* (Cast). *Acta Horticulturae* 474, 125–128.
Lavee, S., Avidan, B. and Meni, Y. (2003) 'Askal', a new high performing oil variety for intensive and super-intensive olive orchards. *Olivae* 97, 53–59.
Lavee, S., Avidan, B., Meni, Y., Haskal, A. and Wodner, M. (2004) Three new semi-dwarf varieties of olive tree for table use. *Olivae* 102, 33–41.
León, L., Martin, L.M. and Rallo, L. (2004a) Phenotypic correlations among agronomic traits in olive progenies. *Journal of the American Society for Horticultural Science* 129, 271–276.
León, L., Garrido-Varo, A. and Downey, G. (2004b) Parent and harvest year effects on near-infrared reflectance spectroscopic analysis of olive (*Olea europaea* L.) fruit traits. *Journal of Agricultural and Food Chemistry* 52, 4957–4962.
León, L., de la Rosa, R. and Barranco, D. (2007) Breeding for early bearing in olive. *HortScience*, 42, 499–502.
León, L., Beltran, G., Aguilera, M.P., Rallo, L., Barranco, D. and de la Rosa, R. (2011) Oil composition of advanced selections from an olive breeding program. *European Journal of Lipid Science and Technology* 113, 870–875.
León, L., Velasco, L. and de la Rosa, R.R. (2015) Initial selection steps in olive breeding programs. *Euphytica* 201, 453–462.
León, L., de la Rosa, R., Velasco, L. and Belaj, A. (2018) Using wild olives in breeding programs: implications on oil quality composition. *Frontiers in Plant Science* 9, 232.
León, L., de la Rosa, R. and Arriaza, M. (2021) Prioritization of olive breeding objectives in Spain: analysis of a producers and researchers survey. *Spanish Journal of Agricultural Research* 19, e0701–e0701.
Leyva-Pérez, M., Valverde-Corredor, A., Valderrama, R., Jiménez-Ruiz, J., Muñoz-Merida, A. *et al.* (2015) Early and delayed long-term transcriptional changes and short-term transient responses during cold acclimation in olive leaves. *DNA Research* 22, 1–11.
Li, J., Ji, X., Wang, Z., Zeng, Y. and Zhang, J. (2021) Morphological, molecular and genomic characterization of two inter-subspecific hybrids between olive cultivars and olive subspecies. *Horticulturae* 7, 138.
Lopez-Escudero, F.J., Blanco-Lopez, M.A., del Rio Rincón, C. and Caballero, J. (2007) Response of olive cultivars to stem punture inoculation with a defoliating pathotype of *Verticillium dahliae*. *HortScience* 42, 294–298.
Manai, H., Mahjoub-Haddada, F., Oueslati, I., Daoud, D. and Zarrouk, M. (2008) Characterization of monovarietal virgin olive oils from six crossing varieties. *Scientia Horticulturae* 115, 252–260.
Mariotti, R., Pandolfi, S., de Cauwer, I., Saumitou-Laprade, P., Vernet, P. *et al.* (2021) Diallelic self-incompatibility is the main determinant of fertilization patterns in olive orchards. *Evolutionary Applications* 14, 983–995.
Mikhail, E.G. (2015) Behaviour of some olive accessions resulting from an olive improvement program. *Annals of Agricultural Science, Moshtohor* 53(1), 99–114.
Montes-Osuna, N. and Mercado-Blanco, J. (2020) Verticillium wilt of olive and its control: What did we learn during the last decade? *Plants* 9, 1–31.
Moral, J., Alsalimiya, M., Roca, L.F., Díez, C.M., León. L, de la Rosa, R. *et al.* (2015) Relative susceptibility of new olive cultivars to *Spilocaea oleagina*, *Colletotrichum acutatum* and *Pseudocercospora cladosporioides*. *Plant Disease* 99, 58–64.
Moral, J., de la Rosa, R., León, L., Barranco, D., Michailides, T.J. and Trapero, A. (2008) High susceptibility of olive cultivar FS-17 to *Alternaria alternata* in Southern Spain. *Plant Disease* 92, 1252.
Navas-López, J.F., Cano, J., de la Rosa, R., Velasco, L and León, L. (2020) Genotype by environment interaction for oil quality components in olive tree. *European Journal of Agronomy* 119, 126115.
Navas-Lopez, J.F., León, L., Rapoport, H.F., Moreno-Alías, I., Lorite, I.J. and de la Rosa, R. (2019a) Genotype, environment and their interaction effects on olive tree flowering phenology and flower quality. *Euphytica* 215, 1–13.

Navas-Lopez, J.F.F., León, L., Trentacoste, E.R.R. and de la Rosa, R. (2019b) Multi-environment evaluation of oil accumulation pattern parameters in olive. *Plant Physiology and Biochemistry* 139, 485–494.

Osman, I. (2015) Agronomic and qualitative evaluation of some olive selections derived from a breeding program. *Annals of Agricultural Science, Moshtohor* 53(1), 123–138.

Padula, G., Giordani, E., Bellini, E., Rosati, A., Pandolfi, S. (2008) Field evaluation of new olive (*Olea europaea* L.) selections and effects of genotype and environment on productivity and fruit characteristics. *Advances in Horticultural Science* 22, 87–94.

Pérez, A.G., León, L., Pascual, M., Romero-Segura, C., Sánchez-Ortiz, A. *et al.* (2014) Variability of virgin olive oil phenolic compounds in a segregating progeny from a single cross in *Olea europaea* L. and sensory and nutritional quality implications. *PLOS ONE*, 9, e92898–e92898.

Pérez, A.G., León, L., Sanz, C. and de la Rosa, R. (2018) Fruit phenolic profiling: a new selection criterion in olive breeding programs. *Frontiers in Plant Science* 9, 1–14.

Pérez, A.G., León, L., Pascual, M., de la Rosa, R., Belaj, A. and Sanz, C. (2019) Analysis of olive (*Olea europaea* L.) genetic resources in relation to the content of Vitamin E in virgin olive oil. *Antioxidants* 8, 242.

Porras Soriano, A., Soriano Martín, M.L. and Porras Piedra, A. (2003) Grafting olive cv. Cornicabra on rootstocks tolerant to *Verticillium dahliae* reduces their susceptibility. *Crop Protection* 22, 369–374.

Pritsa, T.S., Voyiatzis, D.G., Voyiatzi, C. and Sotiriou, M.S. (2003) Evaluation of vegetative growth traits and their relation to time to first flowering of olive seedlings. *Australian Journal of Agricultural Research* 54, 371–376.

Rallo, L., Barranco, D., de la Rosa, R. and León, L. (2008) 'Chiquitita' olive. *HortScience* 43, 529–533.

Rallo, L., Barranco, D., Castro-García, S., (2014) High-density olive plantations. In: Janick, J. (ed.) *Horticultural Reviews* vol. 41. Wiley Blackwell., Hoboken, New Jersey, pp. 303–384.

Ripa, V., de Rose, F., Caravita, M.A., Parise, M.R., Perri, E. *et al.* (2008) Qualitative evaluation of olive oils from new olive selections and effects of genotype and environment on oil quality. *Advances in Horticultural Science* 22, 95–103.

Rosati, A., Paoletti, A., Caporali, S. and Perri, E. (2013) The role of tree architecture in super high density olive orchards. *Scientia Horticulturae* 161, 24–29.

Rosati, A., Marchionni, D., Mantovani, D., Ponti, L. and Famiani, F. (2021) Intercepted photosynthetically active radiation (PAR) and spatial and temporal distribution of transmitted par under high-density and super high-density olive orchards. *Agriculture* 11(4), 351.

Sales, H., Nunes, J. and Vaz Patto, M.C. (2020) Achievements and challenges towards a sustainable conservation and use of 'Galega vulgar' *Olea europaea* variety. *Agronomy* 10, 1467.

Santos-Antunes, F., León, L., de la Rosa, R., Alvarado, J., Mohedo, A. *et al.* (2005) The length of the juvenile period in olive as influenced by vigor of the seedlings and the precocity of the parents. *HortScience* 40, 1213–1215.

Serrano, A., de la Rosa, R., Sánchez-Ortiz, A. and León, L. (2020) Genetic and environmental effect on volatile composition of extra virgin olive oil. *European Journal of Lipid Science and Technology* 122, 2000162.

Serrano, A., de la Rosa, R., Sánchez-Ortiz, A., Cano, J., Pérez, A.G. *et al.* (2021a) Chemical components influencing oxidative stability and sensorial properties of extra virgin olive oil and effect of genotype and location on their expression. *LWT Food Science and Technology* 136, 110257.

Serrano, A., Rodríguez-Jurado, D., Román, B., Bejarano-Alcázar, R. de la Rosa, R. and León, L. (2021b) Verticillium wilt evaluation of olive breeding selections under semi-controlled conditions. *Plant Disease* 105, 1781–1790.

Torres-Sánchez, J., de la Rosa, R., León, L., Jiménez-Brenes, F.M., Kharrat, A. and López-Granados, F. (2022) Quantification of dwarfing effect of different rootstocks in 'Picual' olive cultivar using UAV-photogrammetry. *Precision Agriculture* 23(1), 178–193.

Trapero, C., Díez, C.M., Rallo, L., Barranco, D. and López-Escudero, F.J. (2013) Effective inoculation methods to screen for resistance to Verticillium wilt in olive. *Scientia Horticulturae* 162, 252–259.

Trapero, C., Rallo, L., López-Escudero, F.J., Barranco, D. and Díez, C.M. (2015) Variability and selection of verticillium wilt resistant genotypes in cultivated olive and in the *Olea* genus. *Plant Pathology* 64, 890–900.

Unver, T., Wu, Z., Sterck, L. and Van der Peerr, Y. (2017) Genome of wild olive and the evolution of oil biosynthesis. *Proceedings of the National Academy of Sciences* 114, E9413–E9422.

Valverde, P., Trapero, C., Arquero, O., Serrano, N, Barranco, D. *et al.* (2021) Highly infested soils undermine the use of resistant olive rootstocks as a control method of verticillium wilt. *Plant Pathology* 70, 144–153.

Velasco, L., Fernández-Cuesta, A., de la Rosa, R., Victoria Ruiz-Méndez, M. and León, L. (2014) Selection for some olive oil quality components through the analysis of fruit flesh. *Journal of the American Oil Chemists' Society* 911731–1736.

Velasco, L., de la Rosa, R. and León, L. (2015) Advanced olive selections with enhanced quality for minor constituents. *Grasas y Aceites* 66, e100–e100.

Yilmaz-Düzyaman, H., de la Rosa, R. and León, L. (2022) Seedling selection in olive breeding progenies. *Plants* 11(9), 1195.

3.7 Breeding Strategies and Achievements for Table Olive Varieties

G. Ben-Ari[1], D. Namdar[1] and P. Rallo[2]
[1]*Institute of Plant Sciences, ARO, The Volcani Center, Rishon LeZion 7528809, Israel;* [2]*Departamento de Agronomía, Universidad de Sevilla, 41013 Sevilla, Spain*

3.7.1 Introduction

The selection of olive varieties has been based on the local selection of high-performance trees from the natural olive forests in the east Mediterranean, representing the primary center of domestication (Namdar *et al.*, 2015; Galili *et al.*, 2021), then from the hybrids derived from crosses among varieties brought to center and west Mediterranean and by introgressions from local wild populations (Besnard *et al.*, 2018; Julca *et al.*, 2020). It has been argued that fruit trees, and olive among them, were consciously rather than randomly selected (Barazani *et al.*, 2014). This empirical selection continued along the time until now, when the demand for new olive cultivars is dramatically increasing during the last decades.

A marked worldwide shift from traditional forms of cultivation to intensively mechanized methods of olive growing created a need for new cultivars suitable for these technologies (De Castro *et al.*, 2019; Ibtissem *et al.*, 2014). Other initiatives were directed to a targeted selection from traditional cultivars, well adapted to their agro-environment (Lavee *et al.*, 2014). However, the production level of local cultivars is often rather low and does not meet modern needs and market requirements, as many of the native cultivars are not suited to modern methods of intensification. New mechanical harvest technologies require further adaptations. Thus, empirical selection cannot be relied upon to develop cultivars suitable to the needs of the ongoing technological progress in olive growing (Ben-Ari *et al.*, 2014, 2021). Therefore, in the last decades, large-scale olive-breeding programs have been undertaken, primarily in Mediterranean olive-producing countries, as well as in other areas where olive growing has been recently introduced, by the use of different approaches. In Australia, there is a program working on the selection of feral progenies escaped from cultivation (Sedgley and Wirthensohn, 2000; Guerin *et al.*, 2003), in China, there are inter-cultivar and inter-subspecies *cuspidata* crosses under selection (Li *et al.*, 2021), and in Argentina, local genetic materials with unknown origin are under an agronomic evaluation (Torres *et al.*, 2014).

Within the Mediterranean area, most of the current breeding programs are based on inter-cross between commercial cultivars that bear specific desired traits, in order to produce and select within multiple progenies. The extremely wide genetic diversity, typical of the progenies produced by most olive cultivars (de la Rosa *et al.*, 2016; Dridi

DOI: 10.1079/9781789247350.0011

et al., 2018; Lavee and Avidan, 2011), may be used as a source for selection of new cultivars. F1 or succeeding progenies are exposed to several selection cycles, in order to identify superior types for one or multiple traits. At present, newly released commercial olive cultivars are still the result of classical breeding methodologies based on direct agronomical evaluation (Ben-Ari and Lavi, 2012; Goldental-Cohen *et al.*, 2017; Lavee *et al.*, 2014), because genetic information available on the identity of markers linked to traits or specific genes involved in phenotypic traits of agricultural importance is still rather limited.

As a consequence of the delay in the release of knowledge and genomic tools, classic olive-breeding programs are still extended over several decades and require large plots of land, involving high maintenance costs. Genomic information is required for enabling a rational and driven choice of cross parents suited to the aims of the breeding. Developing a large genomic database is crucial, particularly as most characters under selection are under a multi-locus control.

3.7.2 Breeding for Table Olive, Principles and Perspectives

For table olives production, olive fruit characters are of particular importance, under a morphological, mechanical and metabolic point of view. The epicarp, while representing a very low percentage of the whole fruit, is of great technological importance for table olives and consists mainly of cellulose and cutin; mesocarp represents 70–90% of the entire fruit and stone accounts for 10–30%. Main metabolites of olive fruit pulp are water (60–75%), lipids (10–25%), reducing sugars (2–5%) and phenols (1–3%) (Karagiannis *et al.*, 2021). Olives also have significant amounts of sterols and flavonoids, such as tocopherols, carotenoids and minerals (Conte *et al.*, 2020; Evangelou *et al.*, 2018; Guo *et al.*, 2018; Johnson and Mitchell, 2018; Montaño *et al.*, 2010). Table olives hold a high nutritional value also, due to their richness in monounsaturated fatty acids, vitamin E and phenolic compounds (Rocha *et al.*, 2020).

The fruit weight for table olives is usually higher than that for olive oil production and ranges between 2 g and 8 g. The characteristic strong bitter taste decreases with fruit ripening, as the peel color changes from green to yellow, to purple and, finally, black, but it usually remains very high at harvesting, mainly for green olives; thus, an industrial debittering process is usually needed before consumption.

According to the International Olive Council (IOC), table olives are defined as:

> ...the product prepared from the fruits of varieties of the cultivated olive tree chosen for their production of olives whose weight, shape, flesh-to-stone ratio, taste, firmness and ease of detachment from the stone, make them suitable for processing and preservation by natural fermentation, or by heat treatment with or without the addition of preservatives.

IOC recently reported on the importance of table olives in the European diet, being the leading fermented food consumed in Europe and accounting for a worldwide production close to 3 million tons (IOC, 2021).

Table olives are classified by the degree of ripeness – determined by their skin color, which can range from green to black; the debittering treatments – i.e. natural, dry and/or wrinkled, chemically or physically treated (brine-cured, dry-cured, lye-cured, pickled, baked, etc.), or darkened by oxidation; and by preparation styles – which can keep fruits intact or depitted, stuffed, dressed etc.

Table olives are generally classified according to three main categories: (i) treated green olives (also termed Spanish-style green olives); (ii) olives darkened by oxidation (ripe olives); and (iii) natural black olives (Cillidag, 2013).

Fruit size, shape, chemical composition and flavor of table olives are mainly influenced by the cultivar, even if agricultural practices and environmental conditions may slightly modify the final fruit characteristics (Boskou, 2015; Ghanbari *et al.*, 2012).

World table-olive production has tripled in the last 30 years; in fact, in 1990, it was nearly 950,000 tons, and by 2020 it reached 3134 million tons (IOC, 2021). Table olive consumption has increased proportionately in the last 25 years, as consumption levels rose in correlation to production level over this period (Fig. 3.7.1a).

The main table olive producer in the 2020/2021 season was Egypt, with a production reaching 800,000 tons. Spain followed in second place with 590,000 tons, and third, fourth and fifth were Turkey, Algeria and Greece, respectively. The main table olive consumer in 2020/2021 was Egypt with a consumption of 700,000 tons and export surplus availability of 100,000 tons. The second country was Turkey with a consumption of 345,000 tons; third, fourth and fifth were Algeria, USA and Spain, respectively.

The main table varieties grown in the five major producing countries are 'Aggezi Shami', 'Hamed' and 'Toffahi' in Egypt; 'Gordal Sevillana', 'Manzanilla de Sevilla' and 'Hojiblanca' in Spain; 'Gemlik', 'Memecik' and 'Memely' in Turkey; 'Azeraj' and 'Sigoise' in Algeria, and 'Konservolia', 'Chalkidiki' and 'Kalamon' in Greece (Conte *et al.*, 2020).

3.7.3 Methods for Breeding of New Olive Cultivars

The most relevant desired traits for table olives are disease resistance, high fruit size, high flesh:stone ratio, low tree height and low fruit damage from mechanical harvesting (Bevilacqua *et al.*, 2015; Rejano *et al.*, 2010).

The two main approaches that have been practiced for the development of new table olive cultivars are clonal selection and cross-breeding. Clonal selection focusses on the search for intra-cultivar variability caused by spontaneous somatic mutations that may lead to new and improved clones. Clonal selection has been practiced mainly during the 1970s and 1980s within some of the main olive cultivars, most of them intended for oil production. Field surveys and selection of outstanding trees have been conducted for some table olive and double-purpose cultivars as well, namely: 'Manzanilla de Sevilla' in Spain (Suarez *et al.*, 1990); 'Nocellara del Belice' in Italy (Caruso *et al.*, 2014; D'Hallewin *et al.*, 1990); 'Negrinha', 'Galega', 'Carrasquenha', 'Cordovil de Serpa', 'Santulhana' and 'Redondil' in Portugal (Leitao *et al.*, 1999; Serrano, 1990); 'Picholine Marocaine' in Morocco (Zaher *et al.*, 2011); and 'Souri' in Israel (Lavee *et al.*, 2008), among others. Nevertheless, few of these clones have been exploited commercially. No significant differences have been observed in long-term experiments between the clones and the original cultivars in terms of agronomical performance (Rallo *et al.*, 2018a) with only a few exceptions (Ben-Ari *et al.*, 2014). Actually, there are still doubts about the presence of somatic mutations in the investigated clones, and to clarify this issue, the only effective strategy is represented by the resequencing of the entire genome of the clones of interest, as observed, for example, for the clones of some grape varieties (Calderón *et al.*, 2021).

Classical cross-breeding has been more successful in developing new olive cultivars. The pioneer program for olive breeding was initiated in the 1960s in Israel (Lavee, 1990), leading to the distribution of new commercial cultivars, including table olive varieties, such as 'Kadesh' and 'Kadeshon' (Rymon and Lavee, 1990; Biton *et al.*, 2012). Other breeding programs focussed on table olives are being conducted in Turkey (Ersoy *et al.*, 2008; Ozdemir *et al.*, 2020), Tunisia (Dridi *et al.*, 2018), Italy (Sorrentino *et al.*, 2016), South Africa (Costa, 2014) and Spain (Rallo *et al.*, 2011; Rallo *et al.*, 2018b). Olive breeding involves the crossing of traditional cultivars and/or selected bred genotypes, followed by seedling germination and their further field evaluation, usually following two or three selection stages: at first, a rough seedling evaluation, followed by a preliminary selection and, as a final step, an advanced selection of the outstanding genotypes clonally propagated from the previous selection stage (Rallo *et al.*, 2018a).

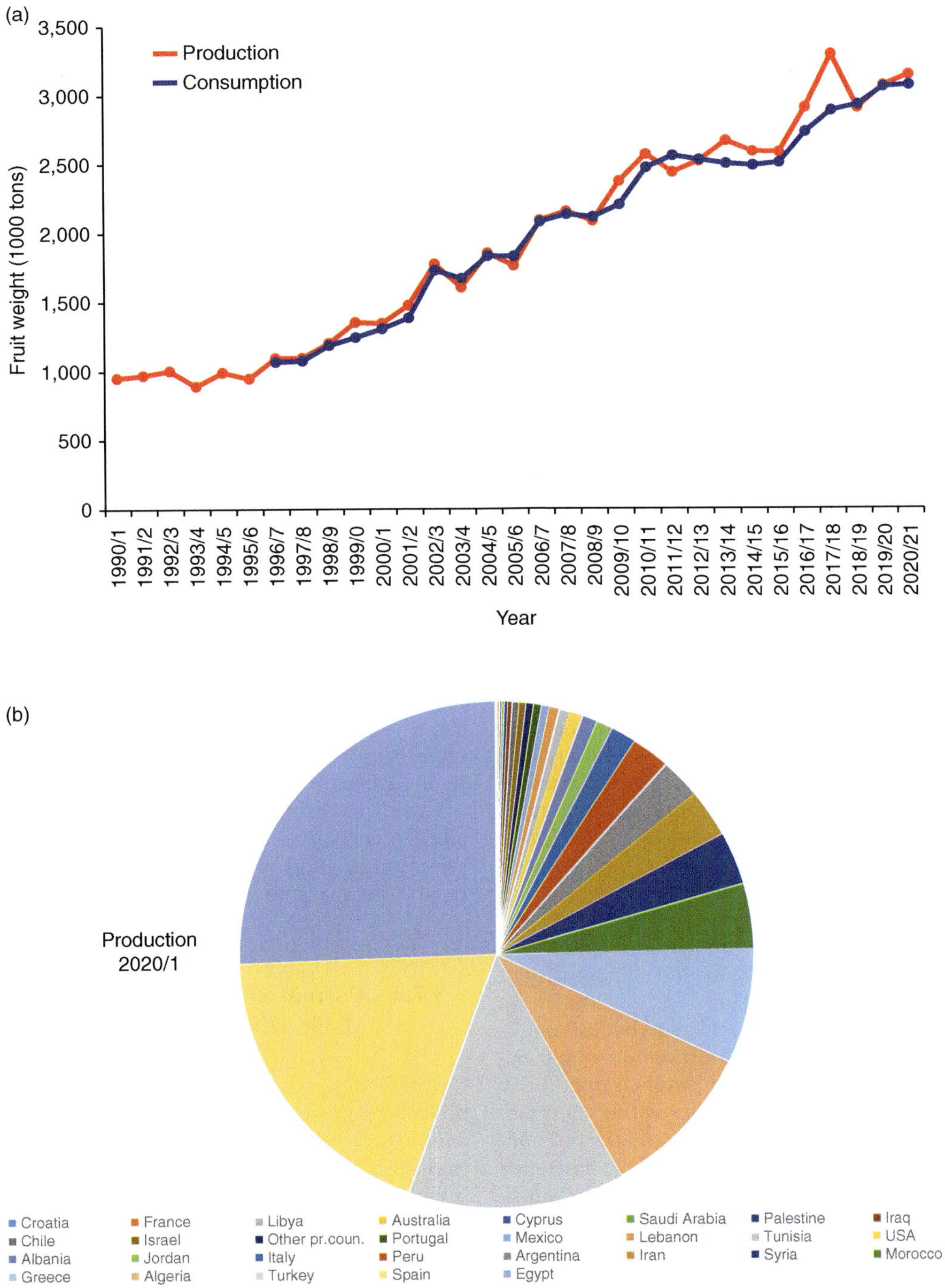

Figure 3.7.1. (a) table olive world production (red line) and consumption (blue line) over the last three decades, as reported by the International Olive Council; (b) table olive world production by countries during the 2020/2021 season; (c) table olive world consumption by countries during the 2020/2021 season.

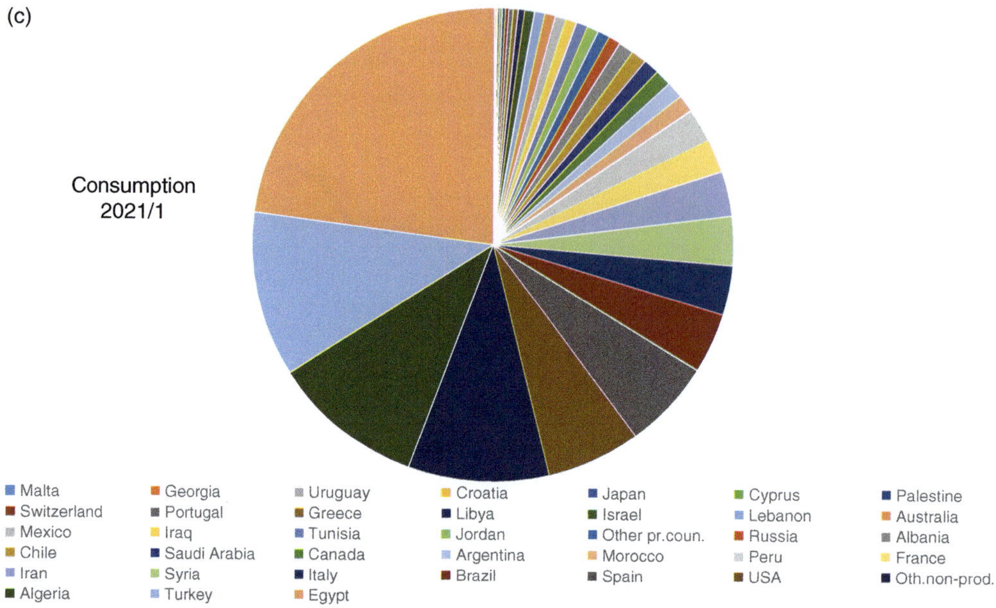

Figure 3.7.1. Continued.

Crosses among cultivars have been carried out in different breeding programs without a preliminary survey regarding their inter-compatibility or self-compatibility. Although emasculation of flowers was attempted in some first crossings, the most common procedure consisted in bagging perfect flowers before anthesis, in order to avoid contamination with alien pollen, followed by the application of pollen of the other parental cultivar, without emasculation, taking into account the self-sterility of most cultivars (Mariotti *et al.*, 2020a). In fact, paternity tests performed by SSR markers revealed that self-pollination rarely occurs in olive and that inter-compatibility between different cultivars is the main factor assuring successful crosses (Díaz *et al.*, 2007). The recent discovery of the existence of a diallelic self-incompatibility system in olive, with only two incompatibility groups (Mariotti *et al.*, 2020 a, b), will help to establish inter-compatibility relationships for future selection of genitors and crosses design in breeding programs.

Cross-breeding is strongly hampered by the juvenility period of seedlings, i.e. the time elapsed from germination to the appearance of the first flowers in a new plantlet, which may last up to 15 years. Juvenility represents one of the major drawbacks in cross-breeding programs. Recent advances in the identification of markers and genes involved in the transition from juvenile to adult stage open new opportunities for the selection of short juvenile genotypes based on specific markers (Haberman *et al.*, 2017).

3.7.4 Current Requirements of New Table Olive Cultivars

Cultivars designated for production of table olives are required to produce high-quality fruits. Among olive pests, the most harmful is certainly *Bactrocera oleae*, which causes severe damage to the pulp, making fruits unsuitable for processing. Some cultivars are less susceptible than others to the olive fruit fly and suffer less damage (Malheiro *et al.*, 2015; Grasso *et al.*, 2017). The mechanisms of resistance still remain elusive (Corrado *et al.*, 2012), but disruption of insect symbiosis with bacteria (Sinno *et al.*, 2020), modifying the fruit volatile route to interact with insect chemotaxis (Giunti *et al.*, 2020), or using low-susceptible cultivars as parents for inter-cross breeding (Quesada-Moraga *et al.*, 2018) could represent new routes towards the fight against *Bactrocera*.

Fruit size is another important aspect taken into account to produce new high-quality table olive varieties. Optimal table olives are characterized by relatively large-sized fruits and a high flesh-to-stone ratio. The two traits are linearly related, as fruit size increases, as flesh-to-stone ratio increases, when mesocarp is significantly larger than endocarp (Hammami *et al.*, 2011).

Also, the low injury of fruits during harvesting and storage before processing represent another challenging aspect to take into consideration in evaluating the characteristics of new table olive cultivars, in addition to the ability, for green table olives, to abscise immature fruit from the tree by mechanical shaking, without loss of foliage and without damaging the fruit skin (Corell *et al.*, 2020; Zipori et al., 2021). Manual harvesting still remains the most suitable method to fulfill these requirements; however, it represents the most expensive component of table olive production costs worldwide (Zipori *et al.*, 2014). For this reason, new table olive cultivars suitable for mechanical harvesting are necessary. At present, a common way to apply mechanical harvesting of green table olives is the pre-treatment of trees with ethylene (Biton *et al.*, 2022), which enables fruit abscission with minimal defoliation, even when harvested by a trunk shaker. Besides, since the possibility of using straddle harvesters in super-high-density table olive orchards was first studied (Morales-Sillero *et al.*, 2014), a few commercial initiatives are taking place in Spain involving mainly cv. 'Manzanilla Cacereña'.

The second factor affecting fruits at harvesting is the external fruit color change or browning as a result of fruit injury. The development of cultivars with firm skin and higher resistance to bruising during fruit detachment by mechanical harvest is required (Jiménez *et al.*, 2011, 2016, 2017; Goldental-Cohen *et al.*, 2017, 2019). Also, bruise resistance, as many other agronomical traits, should be a quantitative trait controlled by multiple genes (Biton *et al.*, 2012; Atienza *et al.*, 2014; Ipek *et al.*, 2017; Sebastiani and Busconi, 2017).

Fruit appearance is another important aspect to consider for table olive breeding. In fact, aesthetic attributes such as fruit size, shape, flesh-to-stone ratio and skin color are currently the main selection criteria for table olive breeders (Rallo *et al.*, 2011).

Other olive fruit qualitative traits that are gaining attention as breeding goals are those associated with the nutraceutical properties of the final product. Consumers are aware of the relationship between food and health and the olive fruit is considered an important source of biologically active compounds. Olive products contain multiple bioactive molecules such as phenols, α-tocopherol, triterpenes and pigments, among others, some of which are exclusive to olives (Mousavi *et al.*, 2022; Ozdemir *et al.*, 2020). Focussing on these biologically potent constituents in table olive breeding will allow the development of new cultivars with enhanced and distinct nutritional properties.

Another trait which should be considered while developing new table olive cultivars is the texture of green olives achieved during processing, for increased acceptance by a broader range of consumers (Fadda *et al.*, 2014).

3.7.5 Main Achievements

At the University of Seville, a breeding program exclusively intended for the improvement of table olives has been undertaken since the early 2000s, by the cross between the most distinguished table olive varieties. After a long evaluation process, 33 new genotypes were selected based on their outstanding tree and fruit traits and their adaptability to standard processes, namely Spanish-style and black ripe olives (Medina *et al.*, 2012; Rallo *et al.*, 2018b). At present, 20 advanced selections are under evaluation in the field for suitability to mechanical harvesting by means of trunk shakers and straddle harvesters (Rallo *et al.*, 2020), and three of them are about to be registered. Recently, new table olive programs involving the collaboration of the table olive industry and public universities are taking place in Spain (BALAM and the University of Cordoba; Aceitunas Guadalquivir and the University of Seville).

At Izmir Olive Research Institute (Turkey), from the cross of 'Gemlik' and 'Memecik' varieties, ten new genotypes are currently under evaluation for fruit processing as natural green split olive, Spanish style green olive and traditional Turkish-style natural turning-black olive (Irmak *et al.*, 2022).

In Egypt, from a breeding program of the Horticulture Research Institute in Giza, six new genotypes obtained from the cross- or open pollination of Egyptian and Spanish cvs 'Aggizi', 'Hamed', 'Toffahi', 'Picual', 'Manzanillo' and 'Arbequina' are currently under evaluation for morphological and flowering traits (El-Badawy *et al.*, 2019).

An olive breeding program has been conducted in Tunisia since 1994 using controlled crosses of cv. 'Meski', mainly used for table olives in northern Tunisia, with other named table olive cultivars such as 'Agezzi', 'Chetoui', 'Chemlali', 'Besbessi', 'Manzanillo', 'Picholine' and 'Ascolana Tenera'. After the pre-selection step, 13 progenies are currently under evaluation for the morphological traits (Dridi *et al.*, 2018).

Below, we list some of the main new table olive cultivars recently released:

Lavee. Dual-purpose, medium-sized tree with controlled vigorous growth habit; high fruit yield; resistant to *Spilocaea oleagina*; good eating quality with an organoleptic character that is of well-balanced fruit-grassy taste. (ARO, Israel)

Masepo. Particularly characterized by a small to medium-sized tree facilitating hand-picking; suitable for intensive orchard growing with harvesting from ground level, with only minimal pruning required; good resistance to *Spilocaea oleagina*; round fruit with attractive black color when ripe and desirable taste when processed at full black maturation. (ARO, Israel)

Sepoka. Particularly characterized as a small to medium-sized tree facilitating hand picking, suitable for intensive orchard growing with harvesting from the ground level with only minimal pruning required, good resistance to *Spilocaea oleagina* and elongated shaped fruit, with attractive, shiny appearance, light-green-yellow colored skin covered with large greenish mottles which is suitable for both green and black pickling (particularly desirable for green processing). (ARO, Israel)

Kadeshon. Particularly characterized by a dwarf-sized tree facilitating hand-picking; suitable for intensive orchard growing with harvesting from ground level with only minimal pruning required; good resistance to *Spilocaea oleagina*; large, slightly elongated shaped fruit, suitable for both green and black pickling. (ARO, Israel)

Dolce di Sicilia (Sweet of Sicily). Obtained from a cross between 'Nocellara del Belice' and 'Tonda Dolce di Partanna'. Showed remarkable vigor, rather rapid growth and, at the level of phenological phases, a very evident advance in the vegetative restart. Fruits of excellent characteristics, low fibrousness, good pulp:stone ratio and a medium-high oil percentage (22–24%). (CREA, Italy)

Rosso di Sicilia (Sicilian red). Originates from the same parents as the previous one, but rather different because it has a lower vigor, less dense foliage, less rapid growth and, at the level of phenological phases, an average vegetative restart with early ripening. The drupes are excellent with an average weight of 4–6 g and excellent organoleptic characteristics. Has a medium oil percentage. (CREA, Italy)

Don Carlo. Dual-purpose cultivar; low vigor; self-fertility; early entry to production; high productivity; high oil yield; late ripening; good resistance to cold; tolerant of most common plant diseases. (CREA, Italy)

3.7.6 Future Perspectives

Several cost-effective and rapid solutions to speed up olive breeding, both for table and for oil production, are beginning to emerge. The first and most promising tools refer to the possibility to apply genomics-assisted

Fig. 3.7.2. Examples of some new cultivars selected for table olives at ARO – Volcani Center (Israel).

breeding. Fast screening methods based on different advanced technologies may contribute to accelerate breeding schemes by reducing the time and cost required for the evaluation stage. Identification of major genes or quantitative trait loci (QTLs) associated with relevant traits to improve olive varieties is still at its initial stage (Baldoni *et al.*, 2016). Nevertheless, the publication of genome sequences of numerous olive varieties (Cruz *et al.*, 2016; Jiménez-Ruiz *et al.*, 2020; Rao *et al.*, 2021; Vatansever *et al.*, 2022), along with the evolution of massive and cost-effective sequencing techniques, will allow development of new functional DNA markers to be applied at the seedling selection stage.

Another avenue that should represent a breakthrough for the fast and non-destructive analysis of fruits directly on the plant is the near-infrared spectrometry (NIRS), which showed high correlation between given spectra and important traits for table olives, such as fruit morphology (Morales-Sillero *et al.*, 2011), firmness, pigments and phenolic content (Cirilli *et al.*, 2016) and external fruit damage (Guzmán *et al.*, 2013). Furthermore, the use of portable NIRS devices may allow an on-site, cheap, rapid and *ad hoc* monitoring of genotypes.

Finally, digital phenotyping technologies, such as airborne imagery based on unmanned aerial vehicles (UAV), are becoming powerful tools for fast and affordable screening of cross-progenies or individuals at last selection phases and have proved very efficient in the selection of olive genotypes based on architectural traits (De Castro *et al.*, 2019; Rallo *et al.*, 2020).

Some important traits, such as fungal resistance, flowering and lipid composition have been successfully manipulated in olive through genetic transformation of somatic embryos (Palomo-Ríos *et al.*, 2021); thus, genetic transformation and gene editing should represent other alternative biotechnologies that could be useful in generating novel genotypes with individually chosen improved traits. The development of new genetic modification (nGM) techniques will create new opportunities.

3.7.7 Conclusions

In the coming years, due to the length of each stage of traditional breeding programs and the high costs associated with growing and evaluating many genotypes, genomics-assisted breeding is expected to become a widespread tool to develop new olive cultivars. For this purpose, research should focus on delivering markers linked to genes involved in economically and agriculturally relevant traits, to be applied to ensure fast genotype screening of seedlings. Only seedlings carrying the expected genomic markers should be kept for further selection, reducing the size of seedling populations and time for phenotyping.

The future optimal table olive cultivar will ensure high and stable yield, improved fruit texture and fully mechanical harvest. To ensure suitability to climate change, the future optimal table olive cultivar should be resistant to high temperatures during spring and summer and with low cold requirements. In addition, it should be resistant to several diseases, especially to the olive fruit fly.

References

Atienza, S.G., de la Rosa, R., León, L., Martín, A. and Belaj, A. (2014) Identification of QTL for agronomic traits of importance for olive breeding. *Molecular Breeding* 34, 725–737.

Baldoni, L., Khadari, B. and de la Rosa, R. (2016) Genetic mapping and detection of quantitative trait loci. In: Rugini, E., Baldoni, L., Muleo, R. and Sebastiani, L. (eds) *The Olive Tree Genome*. Springer, Cham, Switzerland, pp. 65–74.

Barazani, O., Westberg, E., Hanin, N., Dag, A., Kerem, Z. *et al.* (2014) A comparative analysis of genetic variation in rootstocks and scions of old olive trees: a window into the history of olive cultivation practices and past genetic variation. *BMC Plant Biology* 14(1), 1–8.

Ben-Ari, G. and Lavi, U. (2012) Marker-assisted selection in plant breeding. In: Altman, A. and Hasegawa, P.M. (eds) *Plant Biotechnology and Agriculture*. Academic Press, San Diego, California, pp. 163–184.

Ben-Ari, G., Biton, I., Mani, Y., Avidan, B. and Lavee, S. (2014) The diversity in performance of commercial olive clones selected from the autochthonous cv. Souri population for intensive irrigated cultivation. *HortScience* 49, 425–429.

Ben-Ari, G., Biton, I., Many, Y., Namdar, D. and Samach, A. (2021) Elevated temperatures negatively affect olive productive cycle and oil quality. *Agronomy* 11(8), 1492.

Besnard, G., Terral, J.F. and Cornille, A. (2018) On the origins and domestication of the olive: a review and perspectives. *Annals of Botany* 121(3), 385–403.

Bevilacqua, A., de Stefano, F., Augello, S., Pignatiello, S., Sinigaglia, M. *et al.* (2015) Biotechnological innovations for table olives. *International Journal of Food Sciences and Nutrition* 66, 127–131.

Biton, I., Shevtsov, S., Ostersetzer, O., Mani, Y., Lavee, S. *et al.* (2012) Genetic relationships and hybrid vigour in olive (*Olea europaea* L.) by microsatellites. *Plant Breeding* 131, 767–774.

Biton, I., Namdar, D., Mani, Y. and Ari, G.B. (2022) Table olives: toward mechanical harvesting. In: Yonar, T. (ed.) *Olive Cultivation*. InTech. 978-1-80355-442-6.

Boskou, D. (2015) Olive fruit, table olives, and olive oil bioactive constituents. In: Boskou, D. (ed.) *Olive and Olive Oil Bioactive Constituents*. AOCS Press, pp. 1–30.

Calderón, L., Mauri, N., Muñoz, C., Carbonell-Bejerano, P., Bree, L. *et al.* (2021) Whole genome resequencing and custom genotyping unveil clonal lineages in 'Malbec' grapevines (*Vitis vinifera* L.). *Scientific Reports* 11(1), 1–11.

Caruso, T., Marra, F.P., Costa, F., Campisi, G., Macaluso, L. *et al.* (2014) Genetic diversity and clonal variation within the main Sicilian olive cultivars based on morphological traits and microsatellite markers. *Scientia Horticulturae* 180, 130–138.

Cillidag, S.I. (2013) Table olive processing technologies. In: Caballero, J., D'Andria, R., Fernández, M., Garrido, A., Rallo, L. *et al.* (eds) *Present and Future of the Mediterranean Olive Sector*. CIHEAM/IOC, Zaragoza, Spain, 64–67.

Cirilli, M., Bellincontro, A., Urbani, S., Servili, M., Esposto, S. *et al.* (2016) On-field monitoring of fruit ripening evolution and quality parameters in olive mutants using a portable NIR-AOTF device. *Food Chemistry* 199, 96–104.

Conte, P., Fadda, C., Del Caro, A., Urgeghe, P.P. and Piga, A. (2020) Table olives: an overview on effects of processing on nutritional and sensory quality. *Foods* 9, 514.

Corell, M., Martín-Palomo, M.J., Girón, I., Andreu, L., Galindo, A. *et al.* (2020) Stem water potential-based regulated deficit irrigation scheduling for olive table trees. *Agricultural Water Management* 242, 106418.

Corrado, G., Alagna, F., Rocco, M., Renzone, G., Varricchio, P. *et al.* (2012) Molecular interactions between the olive and the fruit fly *Bactrocera oleae*. *BMC Plant Biology* 12, 86.

Costa, C. (2014) Evaluation and development potential of olive cultivars in South Africa. *VII International Symposium on Olive Growing* 1057, 525–532.

Cruz, F., Julca, I., Gomez-Garrido, J., Loska, D., Marcet-Houben, M. *et al.* (2016) Genome sequence of the olive tree, *Olea europaea*. *Biology of Sex Differences* 5.

De Castro, A.I., Rallo, P., Suárez, M.P., Torres-Sánchez, J., Casanova, L. *et al.* (2019) High-throughput system for the early quantification of major architectural traits in olive breeding trials using UAV images and OBIA techniques. *Frontiers in Plant Science* 10, 1472.

De la Rosa, R., Arias-Calderón, R., Velasco, L. and León, L. (2016) Early selection for oil quality components in olive breeding progenies. *European Journal of Lipid Science and Technology* 118, 1160–1167.

D'Hallewin, G., Mulas, M. and Schirra, M. (1990) Characteristics of eleven table-olive clones selected from Nera cultivar. *International Symposium on Olive Growing* 286, 49–52.

Díaz, A., Martín, A., Rallo, P., and de la Rosa, R. (2007) Cross-compatibility of the parents as the main factor for successful olive breeding crosses. *Journal of the American Society for Horticultural Science* 132, 830–835.

Dridi, J., Fendri, M., Breton, C.M. and Msallem, M. (2018) Characterization of olive progenies derived from a Tunisian breeding program by morphological traits and SSR markers. *Scientia Horticulturae* 236, 127–136.

El-Badawy, H., EL-Gioushy, S., Ikram-Saadeldin and Reem-Abo El-Ata (2019) Evaluation of some morphological and flowering traits in new six olive genotypes grown under Egypt conditions. *Asian Journal of Agricultural and Horticultural Research* 3(1), 1–16.

Ersoy, N., Arsel, A.H., Sefer, F., Guloglu, U. and Kaya, H. (2008) The first findings on the promising individuals from hybridization, In: Ozkaya, M.T., Lavee, S. and Ferguson, L. (eds) *Proceedings of the Fifth International Symposium on Olive Growing* 1, 49–53.

Evangelou, E., Kiritsakis, K., Sakellaropoulos, N. and Kiritsakis, A. (2018) Table olives production, postharvest processing, and nutritional qualities. In: Siddiq, M. and Uebersax, M.A. (eds) *Handbook of Vegetables and Vegetable Processing*. John Wiley & Sons Ltd., pp. 727–744.

Fadda, C., Caro, A.D., Sanguinetti, A.M.M and Piga, A. (2014) Texture and antioxidant evolution of naturally green table olives as affected by different sodium chloride brine concentrations. *Grasas Y Aceites* 65, 002.

Galili, E., Langgut, D., Terral, J.F., Barazani, O., Dag, A. *et al.* (2021) Early production of table olives at a mid-7th millennium BP submerged site off the Carmel coast (Israel). *Scientific Reports* 11, 2218.

Ghanbari, R., Anwar, F., Alkharfy, K.M., Gilani, A.-H. and Saari, N. (2012) Valuable nutrients and functional bioactives in different parts of olive (*Olea europaea* L.) – a review. *International Journal of Molecular Science* 13, 3291–3340.

Giunti, G., Campolo, O., Laudani, F., Algeri, G.M. and Palmeri, V. (2020) Olive fruit volatiles route intraspecific interactions and chemotaxis in *Bactrocera oleae* (Rossi) (Diptera: Tephritidae) females. *Scientific Reports* 10(1), 1–10.

Goldental-Cohen, S., Burstein, C., Biton, I., Ben Sasson, S., Sadeh, A. *et al.* (2017) Ethephon induced oxidative stress in the olive leaf abscission zone enables development of a selective abscission compound. *BMC Plant Biology* 17, 87–87.

Goldental-Cohen, S., Biton, I., Many, Y., Ben-Sason, S., Zemach, H., Avidan, B. and Ben-Ari, G. (2019) Green olive browning differ between cultivars. *Frontiers in Plant Science* 10, 1260.

Grasso, F., Coppola, M., Carbone, F., Baldoni, L., Alagna, F. *et al.* (2017) The transcriptional response to the olive fruit fly (*Bactrocera oleae*) reveals extended differences between tolerant and susceptible olive (*Olea europaea* L.) varieties. *PLOS ONE* 12(8), e0183050.

Guerin, J., Mekuria, G., Burr, M., Collins, G. and Sedgley, M. (2003) Selection of improved olive cultivars. *Acta Horticulturae* 622, 231–234

Guo, Z., Jia, X., Zheng, Z., Lu, X., Zheng, Y. *et al.* (2018) Chemical composition and nutritional function of olive (*Olea europaea* L.): a review. *Phytochemistry Reviews* 17, 1091–1110.

Guzmán, E., Baeten, V., Pierna, J.A.F. and Garcia-Mesa, J.A. (2013) Infrared machine vision system for the automatic detection of olive fruit quality. *Talanta* 116, 894–898.

Haberman, A., Bakhshian, O., Cerezo-Medina, S., Paltiel, J., Adler, C. *et al.* (2017) A possible role for flowering locus T-encoding genes in interpreting environmental and internal cues affecting olive (*Olea europaea* L.) flower induction. *Plant Cell and Environment* 40, 1263–1280.
Hammami, S.B.M., Manrique, T. and Rapoport, H.F. (2011) Cultivar-based fruit size in olive depends on different tissue and cellular processes throughout growth. *Scientia Horticulturae* 130, 445–451.
Ibtissem, L., Mezghani, A.M. and Messaoud, M. (2014) Phenotypic diversity of some olive tree progenies issued from a Tunisian breeding program. *European Scientific Journal* 10(6) e – ISSN 1857-7431.
IOC (2021) Table olive figures. Available at: https://www.internationaloliveoil.org/what-we-do/economic-affairs-promotion-unit/#figures (accessed 6 May 2023).
Ipek, A., Ipek, M., Ercjişli, S. and Tangu, N.A. (2017) Transcriptome-based SNP discovery by GBS and the construction of a genetic map for olive. *Functional & Integrative Genomics* 17, 493–501.
Irmak, Ş., Sefer, F. Öztürk Güngör, F., Susamcı, E., Güloğlu, U., Yıldırım, A. and Tusu, G. (2022) Determination of table olive characteristics of new olive varieties obtained by crossbreeding of Gemlik and Memecik variety. *Journal of Agriculture Faculty of Ege University* 59(2), 195–208.
Jiménez, R., Rallo, P., Suarez, M.-P., Morales-Sillero, A., Casanova, L. *et al.* (2011) Cultivar susceptibility and anatomical evaluation of table olive fruit bruising. *Acta Horticulturae* 924, 419–424.
Jiménez, M.R., Rallo, P., Rapoport, H.F. and Suárez, M.P. (2016) Distribution and timing of cell damage associated with olive fruit bruising and its use in analyzing susceptibility. *Postharvest Biology and Technology* 111, 117–125.
Jiménez, M.R., Casanova, L., Suárez, M.P., Rallo, P. and Morales-Sillero, A. (2017) Internal fruit damage in table olive cultivars under superhigh-density hedgerows. *Postharvest Biology and Technology* 132, 130–137.
Jiménez-Ruiz, J., Ramírez-Tejero, J.A., Fernández-Pozo, N., Leyva-Pérez, M.D.L.O., Yan, H. *et al.* (2020) Transposon activation is a major driver in the genome evolution of cultivated olive trees (*Olea europaea* L.). *The Plant Genome* 13(1), e20010.
Johnson, R.L. and Mitchell, A.E. (2018) Reducing phenolics related to bitterness in table olives. *Journal of Food Quality* 2018, 3193185.
Julca, I., Marcet-Houben, M., Cruz, F., Gómez-Garrido, J., Gaut, B.S. *et al.* (2020) Genomic evidence for recurrent genetic admixture during the domestication of Mediterranean olive trees (*Olea europaea* L.). *BMC Biology* 18(1), 1–25.
Karagiannis, E., Michailidis, M., Skodra, C., Stamatakis, G., Dasenaki, M. *et al.* (2021) Proteo-metabolomic journey across olive drupe development and maturation. *Food Chemistry* 363, 130339.
Lavee, S. (1990) Aims, methods, and advances in breeding of new olive (*Olea-europaea* L.) cultivars. *Proceedings of the I International Symposium on Olive Growing* 286, 23–36.
Lavee, S. and Avidan, B. (2011) Heredity diversity in populations of free-, self-, and specific cross-pollinated progenies of some olive (*Olea europaea* L.) cultivars. *Israel Journal of Plant Sciences* 59, 29–37.
Lavee, S., Singer, A., Haskal, A., Avidan, B., Avidan, N. *et al.* (2008) Diversity in performance between trees within the traditional Souri olive cultivar (*Olea europaea* L.) in Israel under rain-fed conditions. *Olivae* 109, 33–45.
Lavee, S., Avidan, B. and Ben-Ari, G. (2014) Trends in breeding new olive varieties in Israel for quality and economic management. *Agricultural Sciences* 5, 9.
Leitao, F., Serrano, J.F., Potes, M.F., Clara, M.I., Felix, M.R. *et al.* (1999) Preliminary results on clonal and sanitary selection of *Olea europaea* L. cv. 'Santulhana' in north-east of Portugal. In: Metzidakis, I. (ed.) *Third International Symposium on Olive Growing* Vol. 1, pp. 163–166.
Li, J., Ji, X., Wang, Z., Zeng, Y. and Zhang, J. (2021) Morphological, molecular and genomic characterization of two inter-subspecific hybrids between olive cultivars and olive subspecies. *Horticulturae* 7(6), 138.
Malheiro, R., Casal, S., Baptista, P. and Pereira, J.A. (2015) A review of *Bactrocera oleae* (Rossi) impact in olive products: from the tree to the table. *Trends in Food Science & Technology* 44, 226–242.
Mariotti, R., Pandolfi, S., De Cauwer, I., Saumitou-Laprade, P., Vernet, P. *et al.* (2020a) Diallelic self-incompatibility is the main determinant of fertilization patterns in olive orchards. *Evolutionary Applications* 1–14.
Mariotti, R., Fornasiero, A., Mousavi, S., Cultrera, N.G.M., Brizioli, F. *et al.* (2020b) Genetic mapping of the incompatibility locus in olive and development of a linked sequence-tagged site marker. *Frontiers in Plant Science* 10, 1760.
Medina, E., Morales-Sillero, A., Ramírez, E.M., Rallo, P., Brenes, M. and Romero, C. (2012) New genotypes of table olives: profile of bioactive compounds. *International Journal of Food Science & Technology* 47(11), 2334–2341.
Montaño, A., Sánchez, A.H., López-López, A., de Castro, A. and Rejano, L. (2010) Chemical composition of fermented green olives: acidity, salt, moisture, fat, protein, ash, fiber, sugar, and polyphenol. In: Preedy,

V.R. and Watson, R.R. (eds) *Olives and Olive Oil in Health and Disease Prevention*. Academic Press, San Diego, California, pp. 291–297.

Morales-Sillero, A., Fernandez-Cabanas, V.M., Casanova, L., Jimenez, M.R., Suarez, M.P. *et al.* (2011) Feasibility of NIR spectroscopy for non-destructive characterization of table olive traits. *Journal of Food Engineering* 107, 99–106.

Morales-Sillero, A., Rallo, P., Jimenez, M.R., Casanova, L. and Suarez, M.P. (2014) Suitability of two table olive cultivars ('Manzanilla de Sevilla' and 'Manzanilla Cacereña') for mechanical harvesting in super-high density hedgerows. *HortScience* 49(8), 1028–1033.

Mousavi, S., Stanzione, V., Mariotti, R., Mastio, V., Azariadis, A. *et al.* (2022). Bioactive compound profiling of olive fruit: the contribution of genotype. *Antioxidants* 11(4), 672.

Namdar, D., Amrani, A., Getzov, N. and Milevski, I. (2015) Olive oil storage during the fifth and sixth millennia BC at Ein Zippori, Northern Israel. *Israel Journal of Plant Sciences* 62, 65–74.

Ozdemir, Y., Ozturk, A., Tangu, N.A., Akcay, M.E., Ozyurt, U. and Ercisli, S. (2020) Determination of raw and processed black olive characteristics of six cultivars. *Journal of Animal and Plant Sciences* 30, 1567–1573.

Palomo-Ríos, E., Narváez, I., Pliego-Alfaro, F. and Mercado, J.A. (2021) Olive (*Olea europaea* L.) genetic transformation: current status and future prospects. *Genes* 12, 386.

Quesada-Moraga, E., Santiago-Álvarez, C., Cubero-González, S., Casado-Mármol, G., Ariza-Fernández, A. and Yousef, M. (2018) Field evaluation of the susceptibility of mill and table olive varieties to egg-laying of olive fly. *Journal of Applied Entomology* 142(8), 765–774.

Rallo, L., El Riachy, M. and Rallo, P. (2011) The time and place for fruit quality in olive breeding. In: Jenks, M.A. and Bebeli, J.B. (eds) *Breeding for Fruit Quality*. Wiley-Blackwell, Chichester, UK, 323–347.

Rallo, L., Barranco, D., Díez, C.M., Rallo, P., Suárez, M.P. *et al.* (2018a) Strategies for olive (*Olea europaea* L.) breeding: cultivated genetic resources and crossbreeding. In: Al-Khayri J., Jain S. and Johnson D. (eds) *Advances in Plant Breeding Strategies: Fruits*. Springer, Cham, Switzerland, pp. 535–600.

Rallo, P., Morales-Sillero, A., Brenes, M., Jimenez, M.D., Sanchez, A.H. *et al.* (2018b) Elaboration of table olives: assessment of new olive genotypes. *European Journal of Lipid Science and Technology* 120, 1800008–1800019.

Rallo, P., de Castro, A.I., López-Granados, F., Morales-Sillero, A., Torres-Sánchez, J. *et al.* (2020) Exploring UAV-imagery to support genotype selection in olive breeding programs. *Scientia Horticulturae* 273, 109615.

Rao, G., Zhang, J., Liu, X., Lin, C., Xin, H., Xue, L. and Wang, C. (2021) De novo assembly of a new *Olea europaea* genome accession using nanopore sequencing. *Horticulture Research* 8.

Rejano, L., Montaño, A., Casado, F.J., Sánchez, A.H. and de Castro, A. (2010) Table olives: varieties and variations. In: Preedy, V.R. and Watson, R.R. (eds) *Olives and Olive Oil in Health and Disease Prevention*. Academic Press, San Diego, California, pp. 5–15.

Rocha, J., Borges, N. and Pinho, O. (2020) Table olives and health: a review. *Journal of Nutritional Science* 9, e57.

Rymon, D. and Lavee, S. (1990) Extension of the marketed product-line of table olives by marketing a new low-fat variety – 'Kadesh', 286 ed. *International Society for Horticultural Science* (ISHS), Leuven, Belgium, pp. 481–484.

Sebastiani, L. and Busconi, M. (2017) Recent developments in olive (*Olea europaea* L.) genetics and genomics: applications in taxonomy, varietal identification, traceability and breeding. *Plant Cell Reports* 36, 1345–1360.

Sedgley, M. and Wirthensohn, M. (2000) The Australian olive improvement programme. *Olivæ* (83), 27–30.

Serrano, J.M.F. (1990) Clonal selection in Portuguese olive varieties. *Acta Horticulturae* 286, 53–56.

Sinno, M., Bézier, A., Vinale, F., Giron, D., Laudonia, S., Garonna, A.P. and Pennacchio, F. (2020) Symbiosis disruption in the olive fruit fly, *Bactrocera oleae* (Rossi), as a potential tool for sustainable control. *Pest Management Science* 76(9), 3199–3207.

Sorrentino, G., Muzzalupo, I., Muccilli, S., Timpanaro, N., Russo, M.P. *et al.* (2016) New accessions of Italian table olives (*Olea europaea*): characterization of genotypes and quality of brined products. *Scientia Horticulturae* 213, 34–41.

Suarez, M.P., Lopezrivares, E.P., Cantero, M.L. and Ordovas, J. (1990) Clonal selection on Manzanilla-de-Seville. *Proceedings of the I International Symposium on Olive Growing* 286, 117–119.

Torres, L.E., Taborda, R., Costero, B., Degutis, L., Teich, I., Prenol, L.V. and Conci, L. (2014) Genetic relationships between olive (*Olea europaea* L.) cultivars and promissory genotypes in Catamarca, Argentina. *Acta Horticulturae* 1057, 495–499.

Vatansever, R., Hernandez, P., Escalante, F.J., Dorado, G. and Unver, T. (2022) Genome-wide exploration of oil biosynthesis genes in cultivated olive tree varieties (*Olea europaea*): insights into regulation of oil biosynthesis. *Functional & Integrative Genomics* 22(2), 171–178.

Zaher, H., Boulouha, B., Baaziz, M., Sikaoui, L., Gaboun, F. *et al.* (2011) Morphological and genetic diversity in olive (*Olea europaea* subsp. *europaea* L.) clones and varieties. *Plant Omics* 4, 370–376.

Zipori, I., Dag, A., Tugendhaft, Y. and Birger, R. (2014) Mechanical harvesting of table olives: harvest efficiency and fruit quality. *HortScience* 49, 55–58.

Zipori, I., Fishman, A., Zelas, Z.B.B., Subbotin, Y. and Dag, A. (2021) Effect of postharvest treatments of mechanically harvested 'Manzanilla' table olives on product quality. *Postharvest Biology and Technology* 174, 111462.

3.8 Biotechnological Approaches in Olive Breeding

I. Narváez, L. Ribalta, J.A. Mercado, F. Pliego-Alfaro and E. Palomo-Ríos
Instituto de Hortofruticultura Subtropical y Mediterránea 'La Mayora' (IHSM-UMA-CSIC), Departamento de Botánica y Fisiología Vegetal, Universidad de Málaga, Spain

3.8.1 Introduction

Olives cultivars used nowadays are the result of seedling selection from wild or cultivated trees; however, the urgent need to increase production, resistance to biotic and abiotic stress, as well as adaptability to new planting systems, has encouraged researchers to start olive-breeding programs for obtaining cultivars with these traits (Rallo *et al.*, 2018). As in most woody species, rapid evaluation of the sexual progeny is not possible due to the long time required for olive seedlings to reach the adult phase. In this chapter, the use of other approaches, such as mutagenesis, somaclonal variation and genetic transformation, will be explored to evaluate their feasibility for incorporation into olive-breeding programs.

3.8.2 Mutagenesis

Induced mutations have been used for over 70 years to improve sexually and asexually propagated crops. According to Jain (2002), chemical (ethyl-methane-sulphonate, EMS; sodium azide, SA) as well as physical (X-rays, gamma rays) mutagens are among the most widely used. However, physical mutagens are preferable since tissue penetration is more uniform and results are easier to reproduce. These traits are particularly relevant in the case of gamma rays (Jain, 2005).

In olive, RAPD and ISSR analysis of callus exposed to chemical mutagen SA in the range 1–5 mM resulted in genetic dissimilarity values in the range 0.14–0.21 (Sheikh and Moradnejad, 2014). Interestingly, no damage could be observed in the callus, even at the highest dosage; hence, this approach could be useful to generate variability in olive embryogenic cells.

Oražem *et al.* (2013) exposed *in vitro* grown shoots to X-radiation observing that rooting capacity was inhibited at 30 Gy; a lower dosage (10 Gy) did not affect shoot growth, although following AFLP analysis, a high number of absent/novel fragments could be detected. Moreover, a decrease in nuclear DNA content was also noticeable, indicating that mutations would occur even in the absence of morphological alterations. Since most selected varieties would just require minor modifications, this strategy could be useful for incorporation into olive-breeding programs.

Other types of radiation have also been used in olive, e.g. a new cultivar, 'Briscola', was identified in 1970 from a group of 441

DOI: 10.1079/9781789247350.0012

self-rooted plants of cv. 'Ascolana', previously exposed to gamma rays. The new cultivar derived from one of those plants, after successive pruning and multiplication through cuttings of the new growth, showed bushy habit and dense vegetation being only used for ornamental purposes since fruit size and oil content were lower than the original mother plant 'Ascolana' (Roselli and Donini, 1982). Also in the 1970s, using gamma irradiation in self-rooted plants, three somatic mutants were obtained, two with compact growth habit, LC and FC from 'Leccino' and 'Frantoio' cultivars, respectively, and a third with dwarf growth habit (LD), from 'Leccino' (Petruccioli *et al.*, 1974, cited by Panneli *et al.*, 1990). These mutants showed lower number of fruits per fruiting shoot than the control. Interestingly, leaf area and thickness were higher than the control in both compact mutants while the opposite was observed in the case of the dwarf LD mutant (Panneli *et al.*, 1990, 1992). Karyological analysis of these materials showed that while FC and LC mutants were mixoploids, LD mutant showed the same ploidy level (2n) as the original 'Leccino' mother plant (Rugini *et al.*, 1996). FC and LC mutants produced abnormal drupes and, when cultured *in vitro*, shoot-tip fragmentation occurs and different growth patterns appeared; e.g. in the case of the FC line, shoots with ovate, large and thick leaves and others with lanceolate, narrow and thin leaves were identified. Following rooting, plants derived from the shoots with ovate/thick leaves had the same phenotype as the original mixoploid FC mother plant and root-tip analysis showed tetraploid cells, while root tips derived from shoots with lanceolate leaves mainly contained diploid cells. Obtainment of triploid plants was possible following germination of seeds from large drupes of FC and LC mutants. Triploid level was confirmed by chromosome counting of root meristems (Rugini *et al.*, 1996). Further studies on floral biology of LD confirmed that despite the larger cell size of tetraploid flowers, no drastic effects on fruit size could be detected (Caporali *et al.*, 2014).

As result of the shoot-tip fragmentation approach previously described, several tetraploid and diploid mutant lines FRM5-4n and FRM5-2n from 'Frantoio' and LM3-4n and LME-2n from 'Leccino' could be obtained for further evaluation under field conditions (Rugini *et al.*, 2016a). Tetraploid plants showed a clear size decrease, and when used as rootstocks a noticeable reduction in canopy volume and trunk diameter was obtained, particularly in the case of the LM3-4n line. Interestingly, the diploid mutant LME-2n also showed a good performance in reduction of scion vigor when used as rootstock. These results led authors to conclude that this material could be of interest for using in high-density plantings (Rugini *et al.*, 2016a). More recently, Bashir (2021) exposed *in vitro* shoots of a 'Frantoio' tetraploid and a diploid 'Moraiolo' lines to a 3 Gy X-ray dosage to select salt-tolerant mutants. The 'Frantoio' tetraploid line showed more plasticity to radiation allowing the obtainment of more stable mutants in comparison to diploid 'Moraiolo'. Moreover, when exposed to a high (4%) PEG (8000) concentration, both types of mutants showed a lower malondialdehyde (MDA) content than controls, indicating a faster recovery of membrane injury.

Chemical agents with antimitotic activity have also been used in olive; e.g. Ozair *et al.* (2014) evaluated the effect of oryzalin (0–400 mg/l) in *in vitro*-grown shoots, observing a general increase in shoot length and leaf area although no further attempt to identify possible mutants was subsequently reported. In cv. 'Canino', fruits derived from shoots exposed to oryzalin in the *in vitro* proliferation phase showed variation in size in relation to those of the untreated control (Rugini *et al.*, 2016b). More recently, Ribalta *et al.* (2021) obtained promising results when using colchicine to induce tetraploidy in olive embryogenic cells derived from a seed of cv. 'Picual'. After two assessments through flow cytometry, it was observed that some calli sections were tetraploids. Shoots regenerated from these sections were also tetraploids. Currently, rooted plants are growing in the greenhouse. Chromosome counting and morphological analysis will be carried out to confirm the tetraploid nature of this material.

3.8.3 Somaclonal Variation

Somaclonal variation (SV) is associated to *in vitro* culture and several factors; e.g., genotype, type of explant, type and concentration of growth regulators, number of subcultures, regeneration pathway and the stress generated during the *in vitro* process – contribute to appearance of variability in regenerated plants (Bairu *et al.*, 2011). Generally, most new traits are undesirable since alterations are spontaneous and occur at random; however, sometimes plants with improved agronomical characteristics can be obtained.

Olive micropropagation takes place through sprouting and elongation of axillary buds (Lambardi *et al.*, 2013), a regeneration pathway in which resulting plants are more genetically stable than those obtained via indirect organogenesis or somatic embryogenesis (George and Debergh, 2008). However, several investigations have reported the occurrence variation in olive micropropagated material; e.g. Farahani *et al.* (2011), using RAPD analysis, observed an increase in polymorphism with the number of subcultures (until seven times) in micropropagated cvs. 'Dezful Baghe-Malek' and 'Dezful Safiabad', while Hassan *et al.* (2016), using the same type of markers in cvs. 'Manzanillo', 'Koroneiki' and 'Picual', multiplied *in vitro* only during three subcultures, obtained a high similarity range with the mother plant (0.82–0.94).

Kangarloo *et al.* (2015) also found genetic alterations, by chloroplast *rpl16* intron sequences and RAPD markers, in rooted shoots of cv. 'Koroneiki', which had been maintained during one year *in vitro*. Allatif and Hmmam (2022) observed a noticeable genotype effect using the start codon targeted (SCoT) analysis in shoots of different cvs maintained *in vitro* during six subcultures, being 'Picual' and 'Toffahi' less genetically stable than others. However, there are also reports where no genetic alterations had been found, e.g. micropropagated shoots of cv. 'Dezful' maintained *in vitro* during four subcultures, were 100% identical to the mother plant following RAPD analysis (Peyvandi *et al.*, 2013). Similarly, Brito *et al.* (2010), working with *in vitro* shoots of two wild olive species, found no variations in relation to the donor field trees using flow cytometry and SSR markers.

Somaclonal variation has been shown to affect field behavior of micropropagated material; e.g., in cv. 'Nocellara Etnea', yield was similar to grafted plants but fruit weight showed a significant reduction; in cv. 'Carolea', level of flower induction was similar to that of conventionally propagated plants but fruit set was much lower, resulting in poor yield (Briccoli-Bati *et al.*, 2006). These results confirm that genetic make-up plays a key role in occurrence of somaclonal variation (Bairu *et al.*, 2011).

Type of explant also affects variation. Cañas and Benbadis (1988) observed dichotomy and double leaves in shoots regenerated from cotyledons, via adventitious organogenesis; however, Mencuccini and Rugini (1993) did not find any morphological difference in plants regenerated from petiole explants. More recently, Mirzaei *et al.* (2021) compared, through AFLP markers, the genetic stability of shoots regenerated from callus produced at basal part of bud explants cultured in liquid medium with zeatin (CIS: callus induced shoots) with shoots derived from meristems (MIS: meristem induced shoots). In two of the cvs used, 'Amin' and 'Meshkat', similar variations were observed in the two types of material with the highest degree of similarity with the mother plant (97%) found in MIS of cv. 'Amin', while in the case of 'Arbequina', 62% and 80% genetic similarity were found for CIS and MIS materials, respectively.

In plants regenerated via somatic embryogenesis, genetic and phenotypic variations have also been found, especially in long-term cultures. Rugini and de Pace (2016) observed phenotypical differences in plants derived from embryogenic cultures maintained in proliferation over three years. Leva *et al.* (2012) identified two somaclones, BOS (bush olive somaclone) and COS (columnar olive somaclone), in plants regenerated from three-year-old embryogenic callus derived from cotyledons of cv. 'Fragivento'. Both somaclones showed differences related to growth rate, leaf, fruit and pit traits. Molecular analysis through RAPDs revealed polymorphism between these somaclones and the donor plant, and between both of them. Authors suggested that these two variants could be used as rootstocks. Bradaï *et al.* (2016a) characterized embryogenic lines obtained from radicles of zygotic embryos of cv. 'Picual', maintained for two and eight years *in vitro*, observing a lower maturation and germination capacity in embryos derived from aged lines. Four-and-a-half years after acclimatization, these plants were analyzed to detect differences in vegetative and reproductive traits; plants obtained from the eight-year-old lines generally showed higher variation, although it was genotype-dependent (embryogenic line) (Bradaï *et al.*, 2016b). These results were later confirmed by SSR and RAPD markers (Bradaï *et al.*, 2019). In contrast, Narváez *et al.* (2019) did not find any variation, by RAPD and SSR markers, in shoots regenerated from somatic embryos of a wild olive genotype in comparison to micropropagated shoots and the mother plant; however, a polymorphism could be detected in a callus sample through RAPD markers. Working with *Olea* spp., Lopes *et al.* (2009) did not detect any alteration using SSR markers in the material analyzed through the somatic embryogenesis process (from non-embryogenic tissue to somatic embryos) and the donor plant.

Induction of somaclonal variation is also being used in olive as a tool to increase resistance to the fungal pathogens *Rosellinia necatrix* and *Verticillium dahliae* following exposure of embryogenic cultures to several cycles of increasing concentrations of crude filtrate of the pathogen culture. Some lines survived exposures to 40–60% *R. necatrix* or *V. dahliae* culture filtrates. Plants have been regenerated and acclimatized; after one year in the greenhouse their response to these pathogens will be evaluated.

3.8.4 Genetic Transformation

Genetic transformation has raised as a valuable tool for improving woody perennials, such as olive, with long juvenile periods and a high level of heterozygosis. Although olive is recalcitrant to *in vitro* manipulation, several studies have reported the obtainment of transgenic plants with modified traits, e.g. stress resistance, plant architecture, flowering behavior and volatile profiles. Most olive transformation protocols use embryogenic structures as explants (Rugini *et al.*, 2020; Palomo-Ríos *et al.*, 2021). Somatic embryos are bipolar structures similar to zygotic embryos, but derived from somatic cells; in olive, somatic embryogenesis has been developed for juvenile, e.g. immature zygotic embryos (Rugini, 1988), radicles from mature embryos (Orinos and Mitrakos, 1991; Cerezo *et al.*, 2011), cotyledon fragments (Leva *et al.*, 1995), seedling roots (Shibli *et al.*, 2001), adult types of explants such as petioles (Rugini and Caricato, 1995) and leaf sections (Mazri *et al.*, 2013).

3.8.4.1 *Agrobacterium tumefaciens*-mediated transformation

The main system used for olive genetic transformation is via *Agrobacterium tumefaciens*. Rugini *et al.* (1999, 2000) were able to transform somatic embryos derived from adult petiole explants of cv. 'Canino', using *A. tumefaciens* LBA4404 strain carrying the pBIN19 binary vector with *nptII* gene for kanamycin selection. Embryogenic masses were inoculated with the bacterial suspension for 48 h in flasks under darkness, at 80 rpm agitation. Afterwards, embryogenic masses were washed, blotted dry over sterile paper and cultured onto solid medium supplemented with 250 mg/l cefotaxime to control bacterial growth. After 30 days, the selection agent kanamycin, at 100 mg/l, was added to the medium. Later on, isolated embryos were cultured in liquid medium supplemented with zeatin, and those turning green cultured on solid medium with higher kanamycin concentration, 150 mg/l, in darkness. The new embryos obtained in this selection medium were cultured without antibiotics for maturation and germination; obtained plants were acclimated and transferred to pots in the greenhouse.

Torreblanca *et al.* (2010) developed an alternative *A. tumefaciens*-mediated transformation methodology using 1–2 mm-diameter somatic embryos from an embryogenic callus line derived from a mature zygotic embryo of cv. 'Picual'. These embryos were inoculated with the bacterial suspension for 20 minutes under mild agitation; afterwards, embryos were blotted dry over sterile paper and cultured onto a solid medium (embryogenic medium) in darkness for a co-culture period of 48 h. Embryos were washed for two hours with a liquid embryogenic medium supplemented with cefotaxime and timentin at 250 mg/l each, to eliminate the bacteria, and cultured onto a solid medium supplemented with those antibiotics and paromomycin at 200 mg/l for selection (selection medium). The embryos were re-cultured weekly during the first month and bi-weekly thereafter. The aminoglycoside antibiotic paromomycin was chosen instead of the most frequently used kanamycin, since it showed a better selection effect on olive somatic embryos (Pérez-Barranco *et al.*, 2009). Nevertheless, to avoid the appearance of chimeric embryos, after three months of culture onto a selection medium, the embryogenic callus derived from inoculated embryos was disaggregated and incubated into a liquid medium supplemented with 50 mg/l paromomycin at 120 rpm for three weeks in darkness. Then, the suspensions were filtered through a 3 x 3 mm mesh pore and the isolated embryos were cultured again onto a selection medium, supplemented with cefotaxime and timentin at 250 mg/l and paromomycin at 200 mg/l for two additional months. Transgenic embryo maturation and germination were performed as described by Cerezo *et al.* (2011), maintaining the selection pressure in these phases. Later on, Palomo-Ríos *et al.* (2017), after checking the transgenic nature of embryogenic masses by PCR, proceeded to maturation of putative transgenic embryos without using antibiotics for selection. According to Torreblanca *et al.* (2010), *A. tumefaciens* strain AGL1 provided higher transformation rates than EHA105 and LBA4404, yielding 7–14% of paromomycin-resistant calli, depending on the binary vector employed.

Palomo-Ríos *et al.* (2017) studied the effect of genotype on transformation rate using four embryogenic lines derived from zygotic embryos of cv. 'Picual'. For some of these genotypes, the use of an increasing selection pressure from 50 to 200 mg/l paromomycin was recommended; afterwards, embryogenic callus was cultured in liquid medium supplemented with 25–50 mg/l paromomycin. Narváez *et al.* (2020) also found this approach useful in the case of binary vectors with large transgenes within the T-DNA, which may hinder regeneration of transformed embryos.

The use of fluorescent reporter genes has been proposed as an alternative for early detection of transformed explants using lower selection pressure to speed up the regeneration process. Green (GFP) and red (DsRed) reporter genes were tested in olive by Cerezo *et al.* (2019). Both fluorescence signals could be detected, using an epifluorescence microscope, six weeks after embryo inoculation with *A. tumefaciens*; however, DsRed signal was higher and easily detected in somatic embryos. DsRed could also be observed in roots and leaves of regenerated plants using a confocal microscope.

3.8.4.1.1 Modification of Agronomical Traits

3.8.4.1.1.1 PLANT ARCHITECTURE AND FLOWERING PATTERN. In olive, somatic embryos from adult cv. 'Canino' were transformed with *A. tumefaciens* LBA4404 strain carrying the pBIN19 binary vector with rolABC genes from *A. rhizogenes* (Rugini *et al.*, 1999, 2000). Transgenic plants showed a hairy root phenotype and dwarf characteristics, such as short internode length and axillary shoots growth. A long juvenile phase was also observed (Rugini *et al.*, 2008).

The floral integrator FLOWERING LOCUS T (FT) is known to be responsible for flower induction in *Arabidopsis* (Notaguchi *et al.*, 2008; Turck *et al.*, 2008); in woody species such as citrus and apple, overexpression of FT-homolog genes has been shown to shorten the juvenile period (Endo *et al.*, 2005; Tränkner *et al.*, 2010). To gain understanding of factors affecting olive flower induction, Haberman *et al.* (2017) obtained transgenic plants overexpressing the FT gene from *Medicago truncatula* under the control of the CaMV35S promoter. The juvenile phase was extremely short in plants of three lines that showed early flowering *in vitro*. Following acclimatization, these lines presented an altered growth habit, with a profuse axillary branching and reduced size; moreover, plants of one line showed flower formation throughout the year (Fig. 3.8.1), although the amount of flowering was higher after winter (Haberman *et al.*, 2017). An in-depth study of these transgenic lines with altered flowering pattern and growth habit is being carried out (Elena Palomo-Ríos, University of Málaga, November 2022, personal communication).

3.8.4.1.1.2 BIOTIC AND ABIOTIC STRESS TOLERANCE. Osmotin is known as PR (pathogenesis-related) protein playing a key role in biotic and abiotic stress tolerance (Hakim *et al.*, 2018). Olive plants were transformed with *A. tumefaciens* strain LBA4404, using the pKYLX71 plasmid harboring the tobacco osmotin gene under the control of 35S promoter (Rugini *et al.* 1999, 2000). Transgenic plants showed increased resistance to infection by *Spilocaea oleagina*, causing the olive leaf spot disease, but also higher susceptibility to *Otiorhynchus cribricollis* G. and *Lychtensia viburnii* S. pests (Rugini and de Pace, 2016). Osmotin transgenic plants displayed a higher resistance to drought stress in the presence of polyethylene glycol (1–4%) (Silvestri *et al.*, 2017). These lines were also more tolerant to salinity and this behavior was associated with sulphur metabolism, with increasing activities of O-acetyl-serine-(thiol)lyase (OASTL) at higher NaCl levels (Bashir *et al.*, 2021). Tolerance to *Verticillium dahliae*, causal agent of verticillium wilt, and *Rosellinia necatrix*, the necrotrophic fungus that causes white root rot in fruit trees, has also been addressed. To achieve this goal, the *afp* gene, from *Aspergillus giganteus*, coding for antifungal protein AFP, and the *npr1* from *Arabidopsis thaliana* were overexpressed in olive (Narváez *et al.*, 2018, 2020). AFP disturbs the integrity of the plasma membrane and inhibits chitin biosynthesis in sensitive fungi (Leiter *et al.*, 2017). Theis *et al.* (2003) hypothesised that AFP acts in a species-specific manner, being effective to generate tolerance against some ascomycetes but not others. On the other hand, *npr1* is a key regulator of the systemic acquired resistance (SAR) (Cao *et al.*, 1997) and it has been used in other species to enhance resistance to a broad spectrum of pathogens (Lin *et al.*, 2004, Parkhi *et al.*, 2010a). In both cases, the transformation protocol described by Torreblanca *et al.* (2010) was used and transgenic plants were tested against *V. dahliae* and *R. necatrix* infection in the greenhouse. Plants of the NPR1 transgenic line with the highest transgene expression showed less severe symptoms than control when inoculated with a non-defoliating (ND) *Verticillium* pathotype; however, it did not show a positive response following inoculation with the defoliating (D) pathotype. Similar results had previously been reported by Parkhi *et al.* (2010b) in cotton. In a study of the Verticillium wilt resistance in tomato, Fradin *et al.* (2009) observed that the mechanism of *V. dahliae* infection depended on the fungus race,

Fig. 3.8.1. Olive plant transformed with *FT* gene from *Medicago truncatula* growing in the greenhouse.

which may explain the different results obtained with D and ND pathotypes in transformed olive plants. As regards *R. necatrix* infection, two transgenic lines overexpressing *afp* and one of the *npr1* lines showed partial tolerance to this pathogen.

3.8.4.1.1.3 VOLATILE PROFILES. The role of the 13-hydroperoxide lyase gene (13-HPL) in the lipoxygenase pathway has been studied by transgenic approaches (Cerezo *et al.*, 2021); this gene plays a key role in the synthesis of volatile compounds involved in olive oil aroma (Padilla *et al.*, 2010). Two constructs were used, one of them to induce overexpression of the gene and the other one for silencing through RNAi. As expected, a higher 13-HPL enzymatic activity and C6 volatiles content were detected in leaves of overexpresser plants, when compared with control non-transformed leaves, while RNAi plants displayed lower activity and higher amounts of C5 compounds. The olive oil profile was not described, but similar results would be expected since it has been shown that leaves, fruit mesocarp and calli derived from cotyledons displayed a similar lipid composition (Williams *et al.*, 1998). Interestingly, transformed plants with low 13-HPL activity showed reduced growth and vigor (Fig. 3.8.2). Similar results have previously been reported in *hpl* rice mutant plants (Liu *et al.*, 2012; Tong *et al.*, 2012).

Fig. 3.8.2. Olive plants transformed with *13-HPL* gene. *Left to right*: untransformed control, plant overexpressing the gene, plant with gene silenced showing reduced growth.

3.8.4.2 *Agrobacterium rhizogenes*-mediated transformation

The inoculation of shoots of recalcitrant woody species with *A. rhizogenes* for rooting is a common strategy. The chimeric plants obtained, composed of a wild-type stem with transgenic roots, could be very useful for studying soilborne pathogens and their interactions with plants (Taylor *et al.*, 2006). *In vitro* shoots of cvs 'Dolce Agogia' and 'Moraiolo' were infected with the *A. rhizogenes* strain 1855, wounding the stems with an inoculated blade knife; then they were cultured in ½ MS medium. Fifty percent of shoots developed roots after ten days (Rugini, 1986); however, the agropine gene of the TR-DNA could only be detected in 20% of stems. Later on, 1 mM putrescine was added to the bacterial suspension and the rooting rate was increased (Rugini, 1992). Strobel *et al.* (1988) suggested the use of the super-virulent strain 232 for inoculation of greenhouse-rooted shoots to improve their growth. The proof of concept was carried out with cv. 'Manzanillo' shoots.

3.8.4.3 Biolistic-mediated Transformation

The approximations carried out using biolistic have yielded only transient transformation; however, these protocols are easier and faster than other transformation methodologies, and they are suitable for studies on promoter activity or protein subcellular localization. Lambardi *et al.* (1999) bombarded large somatic embryos of cv. 'Canino', > 5 mm length; two constructs were employed, pZ085 and pCGUδ0, both containing the β-*glucuronidase* (GUS) reporter gene. The efficiency of different particles, tungsten vs. gold, and two different devices, Particle Inflow Gun and PDS-1000/He, were studied, although no differences were found. GUS signal was detected using a bombardment pressure of 580-KPa. Later on, Pérez-Barranco *et al.* (2009) analyzed the optimal transient transformation conditions of an embryogenic callus line obtained from a zygotic embryo of cv. 'Picual' using a PDS-1000/He device. They tested the particle nature and size, DNA concentration per shot, target distance and bombardment pressure; optimal parameters were chosen as 900 psi at a target distance of 6 cm, using gold particles of 1 μm coated with 1μg/μl of DNA. GUS signal could be detected up to 12 weeks after the bombardment event. The use of three plasmids, pGUSInt, pJGUS5, and pCGUΔ1, harboring the reporter gene β-*glucuronidase* under CaMV35S, 35S with enhancer and the sunflower ubiquitin promoters, respectively, was tested. Efficiency of transient transformation was estimated as the number of GUS-positive spots in embryo surface, with the ubiquitin promoter yielding the highest values.

References

Allatif, A.A. and Hmmam, I. (2022) Shoot proliferation, rooting, and genetic stability of six micropropagated olive (*Olea europaea* L.) genotypes. *Horticultural Science and Technology* 40, 432–445.

Bairu, M.W., Aremu, A.O. and Van Staden, J. (2011) Somaclonal variation in plants: causes and detection methods. *Plant Growth Regulation* 63, 147–173.

Bashir, M.A. (2021) Mutation induction for development of olive breeding material and its evaluation for drought tolerance. PhD thesis, Arid Agriculture University Rawalpindi, Pakistan.

Bashir, M.A., Silvestri, C., Coppa, E., Brunori, E., Cristofori, V. *et al.* (2021) Response of olive shoots to salinity stress suggests the involvement of sulfur metabolism. *Plants* 10, 350.

Bradaï, F., Pliego-Alfaro, F. and Sánchez-Romero, C. (2016a) Long-term somatic embryogenesis in olive (*Olea europaea* L.): influence on regeneration capability and quality of regenerated plants. *Scientia Horticulturae* 199, 23–31.

Bradaï, F., Pliego-Alfaro, F. and Sánchez-Romero, C. (2016b) Somaclonal variation in olive (*Olea europaea* L.) plants regenerated via somatic embryogenesis: influence of genotype and culture age on phenotypic stability. *Scientia Horticulturae* 213, 208–215.

Bradaï, F., Sánchez-Romero, C. and Martín, C. (2019) Somaclonal variation in olive (*Olea europaea* L.) plants regenerated via somatic embryogenesis: influence of genotype and culture age on genetic stability. *Scientia Horticulturae* 251, 260–266.

Briccoli-Bati, C., Godino, G., Monardo, D. and Nuzzo, V. (2006) Influence of propagation techniques on growth and yield of olive trees cultivars 'Carolea' and 'Nocellara Etnea'. *Scientia Horticulturae* 109, 173–182.

Brito, G., Lopes, T., Loureiro, J., Rodriguez, E. and Santos, C. (2010) Assessment of genetic stability of two micropropagated wild olive species using flow cytometry and microsatellite markers. *Trees* 24, 723–732.

Cañas, L.A. and Benbadis, A. (1988) In vitro plant regeneration from cotyledon fragments of the olive tree (*Olea europaea* L.). *Plant Science* 54, 65–74.

Cao, H., Glazebrook, J., Clarke, J.D., Volko, S. and Dong, X.N. (1997) The Arabidopsis NPR1 gene that controls systemic acquired resistance encodes a novel protein containing ankyrin repeats. *Cell* 88, 57–63.

Caporali, S., Hammami, S.B.M., Moreno-Alías, I., Rapoport, H.F., Chiancone, B. *et al.* (2014) Effects of tetraploidy on olive floral and fruit biology. *Scientia Horticulturae* 179, 198–203.

Cerezo, S., Mercado, J.A. and Pliego-Alfaro, F. (2011) An efficient regeneration system via somatic embryogenesis in olive. *Plant Cell, Tissue and Organ Culture* 106, 337–344.

Cerezo, S., Palomo-Ríos, E., Ben Mariem, S., Mercado, J.A. and Pliego-Alfaro, F. (2019) Use of fluorescent reporter genes in olive (*Olea europaea* L.) transformation. *Acta Physiologiae Plantarum* 41, 49.

Cerezo, S., Hernández, M.L., Palomo-Ríos, E., Gouffi, N., García-Vico, L. *et al.* (2021) Modification of 13-hydroperoxide lyase expression in olive affects plant growth and results in altered volatile profile. *Plant Science* 313, 111083.

Endo, T., Shimada, T., Fujii, H., Kobayashi, Y., Araki, T. and Omura, M. (2005) Ectopic expression of an FT homolog from Citrus confers an early flowering phenotype on trifoliate orange (*Poncirus trifoliata* L. Raf.). *Transgenic Research* 14(5), 703–712.

Farahani, F., Yari, R. and Masoud, S. (2011) Somaclonal variation in Dezful cultivar of olive (*Olea europaea* subsp. *europaea*). *Geneconserve* 40, 216–233.

Fradin, E.F., Zhang, Z., Juarez Ayala, J.C., Castroverde, C.D.M. *et al.* (2009) Genetic dissection of Verticillium wilt resistance mediated by tomato Ve1. *Plant Physiology* 150, 320–332.

George, E.F. and Debergh, P.C. (2008) Micropropagation: uses and methods. In: George, E.F., Hall, M.A. and De Klerk, G.J. (eds) *Plant Propagation by Tissue Culture*, 3rd edn. Vol. 1: *The Background*. Springer, Dordrecht, The Netherlands, pp. 29–65.

Gutièrrez-Pesce, P., Taylor, K., Muleo, R. and Rugini, E. (1998) Somatic embryogenesis and shoot regeneration from transgenic roots of the cherry rootstock 'Colt' (*Prunus avium* x *P. pseudocerasus*) mediated by pRi 1855 T-DNA of *Agrobacterium rhizogenes*. *Plant Cell Reports* 17, 574–580.

Haberman, A., Bakhshian, O., Cerezo-Medina, S., Paltiel, J., Adler, C. *et al.* (2017) A possible role for flowering locus T-encoding genes in interpreting environmental and internal cues affecting olive (*Olea europaea* L.) flower induction. *Plant, Cell and Environment* 40(8), 1263–1280.

Hakim, U.A., Hussain, A., Shaban, M., Khan, A.H., Alariqi, M. *et al.* (2018) Osmotin: a plant defense tool against biotic and abiotic stresses. *Plant Physiology and Biochemistry* 123, 149–159.

Hassan, S.A.M., Allatif, A.M.A. and Mahfouze, H.A. (2016) Assessment of genetic stability of micropropagated olive (*Olea europaea* L.) cultivars using RAPD marker. *International Journal of PharmTech Research* 9, 816–825.

Jain, S.M. (2002) A review of induction of mutations in fruits of tropical and subtropical regions. *Acta Horticulturae* 575, 295–302.

Jain, S.M. (2005) Major mutation-assisted plant breeding programs supported by FAO/IAEA. *Plant Cell, Tissue and Organ Culture* 82, 113–123.

Kangarloo, B., Noormohammadi, Z., Farahani, F., Ghasemzadeh-Baraki, S. and Sheidai, M. (2015) An assessment of tissue culture-induced genetic variability in olive *(Olea europaea* L.) using chloroplast rpl16 intron sequences and single primer amplification reaction (SPAR) markers. *The Journal of Horticultural Science and Biotechnology* 90, 719–723.

Lambardi, M., Amorosi, S., Caricato, G., Benelli, C., Branca, C. and Rugini, E. (1999) Microprojectile-DNA delivery in somatic embryos of olive (*Olea europaea* L.). *Acta Horticulturae* 474, 505–509.

Lambardi, M., Ozudogru, E.A. and Roncasaglia, R. (2013) In vitro propagation of olive (*Olea europaea* L.) by nodal segmentation of elongated shoots. In: Lambardi, M., Ozudogru, E.A. and Jain, S.M. (eds) *Protocols for Micropropagation of Selected Economically-Important Horticultural Plants*. Springer, New-York, pp. 33–44.

Leiter, E., Gall, T., Csernoch, L. and Pocsi, I. (2017) Biofungicide utilizations of antifungal proteins of filamentous ascomycetes: current and foreseeable future developments. *Biocontrol* 62, 125–138.

Leva, A., Muleo, R. and Petruccelli, R. (1995) Long-term somatic embryogenesis from immature olive cotyledons. *Journal of Horticultural Science* 70, 417–421.

Leva, A.R., Petruccelli, R. and Rinaldi, L.M.R. (2012) Somaclonal variation in tissue culture: a case study with olive. In: Leva, A. and Rinaldi, L. (eds) *Recent Advances in Plant In Vitro Culture*. IntechOpen. pp. 123–129.

Lin, W.C., Lu, C.F., Wu, J.W., Cheng, M.L., Lin, Y.M. *et al.* (2004) Transgenic tomato plants expressing the Arabidopsis NPR1 gene display enhanced resistance to a spectrum of fungal and bacterial diseases. *Transgenic Research* 13, 567–581.

Liu, X., Li, F., Tang, J., Wang, W., Zhang, F. *et al.* (2012) Activation of the jasmonic acid pathway by depletion of the Hydroperoxide Lyase OsHPL3 reveals crosstalk between the HPL and AOS branches of the oxylipin pathway in rice. *PLOS ONE* 7(11), e50089.

Lopes, T., Capelo, A., Brito, G., Loureiro, J. and Santos, C. (2009) Genetic variability analyses of the somatic embryogenesis induction process in *Olea* spp. using nuclear microsatellites. *Trees* 23, 29–36.

Mazri, M.A., Belkoura, I., Pliego-Alfaro, F. and Belkoura, M. (2013) Somatic embryogenesis from leaf and petiole explants of the Moroccan olive cultivar Dahbia. *Scientia Horticulturae* 159, 88–95.

Mencuccini, M. and Rugini, E. (1993) In vitro shoot regeneration from olive cultivar tissue. *Plant Cell, Tissue and Organ Culture* 32, 283–288.

Mirzaei, L., Yadollahi, A., Kermani, M.J., Naderpour, M., Zeinanloo, A.A., Farsi, M. and Davoodi, D. (2021) Evaluation of genetic stability in olive callus induced and meristem induced shoots using flow cytometry and amplified fragment length polymorphism techniques. *Plant Methods* 17, 31.

Narvaez, I., Khayreddine, T., Pliego, C., Cerezo, S., Jiménez-Díaz, R.M. *et al.* (2018) Usage of the heterologous expression of the antimicrobial gene *afp* from *Aspergillus giganteus* for increasing fungal resistance in olive. *Frontiers in Plant Science* 9, 680.

Narváez, I., Martín, C., Jiménez-Díaz, R.M., Mercado, J.A. and Pliego-Alfaro, F. (2019) Plant regeneration via somatic embryogenesis in mature wild olive genotypes resistant to the defoliating pathotype of *Verticillium dahliae*. *Frontiers in Plant Science* 10, 1471.

Narváez, I., Pliego Prieto, C., Palomo-Ríos, E., Fresta, L., Jiménez-Díaz, R.M. *et al.* (2020) Heterologous expression of the AtNPR1 gene in olive and its effects on fungal tolerance. *Frontiers in Plant Science* 11, 308.

Notaguchi, M., Abe, M., Kimura, T., Daimon, Y., Kobayashi, T. *et al.* (2008) Long-distance, graft-transmissible action of Arabidopsis FLOWERING LOCUS T protein to promote flowering. *Plant Cell Physiology* 49(11), 1645–1658. Epub 2008 Oct. 11. Erratum in: *Plant Cell Physiology* 2008 December, 49(12), 1922. PMID: 18849573.

Oražem, P., Štajner, N. and Bohanec, B. (2013) Effect of X-ray irradiation on olive shoot culture evaluated by morphological measurements, nuclear DNA content and SSR and AFLP markers. *Trees* 27(6), 1587–1595.

Orinos, T. and Mitrakos, K. (1991) Rhizogenesis and somatic embryogenesis in calli from wild olive (*Olea europaea* var. *sylvestris* (Miller) Lehr) mature zygotic embryos. *Plant Cell, Tissue and Organ Culture* 27, 183–187.

Ozair, M., Hafiz, I.A., Abbasi, N.A., Mahmood, M.A. and Bibi, N. (2014) Effect of different concentrations of Oryzalin on in vitro growth of explants of olive cv. Moraiolo. *Scholarly Journal of Agricultural Science* 4(1), 51–59.

Padilla, M.N., Hernández, M.L., Pérez, A.G., Sanz, C. and Martínez-Rivas, J.M. (2010) Isolation, expression, and characterization of a 13-Hydroperoxide lyase gene from olive fruit related to the biosynthesis of the main virgin olive oil aroma compounds. *Journal of Agricultural and Food Chemistry* 58(9), 5649–5657.

Palomo-Ríos, E., Cerezo, S., Mercado, J.A. and Pliego-Alfaro, F. (2017) Generation and selection of transgenic olive plants. *Bio-protocol* 7(22).

Palomo-Ríos, E., Narváez, I., Pliego-Alfaro, F. and Mercado, J.A. (2021) Olive (*Olea europaea* L.) genetic transformation: current status and future prospects. *Genes* 12, 386.

Panneli, G., Famiani, F., Rugini, E., Bignami, D. and Natali, S. (1990) Preliminary characteristics of olive somatic mutants from gamma irradiated 'Frantoio' and 'Leccino' plantlets. *Acta Horticulturae* 286, 77–80.

Panneli, G., Famiani, F. and Rugini. E. (1992) Effects of the change on ploidy levels on anatomical, cytological, reproductive, growth performance and polyamine contents, in mutants of gamma irradiated olive plants. *Acta Horticulturae* 317, 209–218.

Parkhi, V., Kumar, V., Campbell, L.M., Bell, A.A., Shah, J. and Rathore, K.S. (2010a) Resistance against various fungal pathogens and reniform nematode in transgenic cotton plants expressing Arabidopsis NPR1. *Transgenic Research* 19, 959–975.

Parkhi, V., Kumar, V., Campbell, L.M., Bell, A.A., Shah, J. and Rathore, K.S. (2010b) Expression of Arabidopsis NPR1 in transgenic cotton confers resistance to non-defoliating isolates of *Verticillium dahliae* but not the defoliating isolates. *Journal of Phytopathology* 158, 822–825.

Pérez-Barranco, G., Torreblanca, R., Padilla, I.M.G., Sánchez-Romero, C., Pliego-Alfaro, F. and Mercado, J.A. (2009) Studies on genetic transformation of olive (*Olea europaea* L.) somatic embryos. I. Evaluation of different aminoglycoside antibiotics for nptII selection. II. Transient transformation via particle bombardment. *Plant, Cell Tissue and Organ Culture* 97(3), 243–251.

Peyvandi, M., Monsef, M. and Hosseini-Mazinani, M. (2013) Assessment of genetic stability of olive in vitro propagated by RAPD marker. *Journal of Cell and Molecular Research* 5, 81–86.

Rallo, L., Barranco, D., Díez, C.M., Rallo, P., Suárez, M.P., Trapero, C. and Pliego-Alfaro, F. (2018) Strategies for olive (*Olea europaea* L.) breeding: cultivated genetic resources and crossbreeding. In: Al-Khayri, J.M., Jain, S. and Johnson, D. (eds) *Advances in Plant Breeding Strategies: Fruits*. Springer, pp. 535–600.

Ribalta, L., Narváez, I., Palencia, F., Mercado, J.A., Palomo-Ríos, E. *et al.* (2021) Obtaining tetraploid olive plants by colchicine treatments. In: XIV Reunión de la Sociedad Española de Cultivo in vitro de Tejidos Vegetales. El Ejido (Almería), Libro de Resúmenes, p. 82.

Roselli, G. and Donini, B. (1982) 'Briscola'. Nuova cultivar di olivo a sviluppo compatto. *Rivista di Ortoflorofrutticoltura Italiana* 66(2), 103–114.

Rugini, E. (1986) Olive. In: Bajaj, Y.P.S. (ed.) *Biotechnology in Agriculture and Forestry*, Vol. 10. Springer, Berlin, pp. 253–267.

Rugini, E. (1988) Somatic embryogenesis and plant-regeneration in olive (*Olea europaea* L). *Plant, Cell Tissue and Organ Culture* 14, 207–214.

Rugini, E. (1992) Involvement of polyamines in auxin and *Agrobacterium rhizogenes*-induced rooting of fruit trees in vitro. *Journal of the American Society for Horticultural Science* 117, 532–536.

Rugini, E. and Caricato, G. (1995) Somatic embryogenesis and plant recovery from mature tissues of olive cultivars (*Olea europaea* L.) 'Canino' and 'Moraiolo'. *Plant Cell Reports* 14, 257–260.

Rugini, E. and de Pace, C. (2016) Olive breeding with classical and modern approaches. In: Rugini, E., Baldoni, L., Muleo, R. and Sebastiani, L. (eds) *The Olive Tree Genome*. Springer, New York, pp. 163–191.

Rugini, E., Pannelli, G., Ceccarelli, M. and Muganu, M. (1996) Isolation of triploid and tetraploid olive (*Olea europaea* L.) plants from mixoploid cv. 'Frantoio' and 'Leccino' mutants by in vivo and in vitro selection. *Plant Breeding* 115(1), 23–27.

Rugini, E., Gutierrez-Pesce, P. and Spampinato, P.L. (1999) New perspective for biotechnologies in olive breeding: morphogenesis, in vitro selection and gene transformation. *Acta Horticulturae* 474, 107–110.

Rugini, E., Biasi, R. and Muleo, R. (2000) Olive (*Olea europaea* var. *sativa*) transformation. In: Jain, S.M. and Minocha, S.C. (eds) *Molecular Biology of Woody Plants*, Vol. 2. Kluwer Academic Publishers, Dordrecht, The Netherlands, pp. 245–279.

Rugini, E., Gutierrez-Pesce, P. and Muleo, R. (2008) Olive. In: Kole, C. and Hall, T.C. (eds) *Compendium of Transgenic Crop Plants: Transgenic Temperate Fruits and Nuts*. Blackwell Publishing, Oxford, UK, pp. 233–258.

Rugini, E., Silvestri, C., Ceccarelli, M., Muleo, R. and Cristofori, V. (2016a) Mutagenesis and biotechnology techniques as tools for selecting new stable diploid and tetraploid olive genotypes and their dwarfing agronomical characterization. *Hortscience* 51, 799–804.

Rugini, E., Cristofori, V. and Silvestri, C. (2016b) Genetic improvement of olive (*Olea europaea* L.) by conventional and in vitro biotechnology methods. *Biotechnology Advances* 34(5), 687–696.

Rugini, E., Baldoni, L., Silvestri, C., Mariotti, R., Narváez, I. *et al.* (2020) *Olea europaea* Olive. In: Litz, R.E., Pliego-Alfaro, F. and Hormaza, J.I. (eds) *Biotechnology of Fruit and Nut Crops*, 2nd edn. CAB International, Wallingford, UK and Boston, Massachusetts, pp. 343–376.

Sheikh, A.A.D. and Moradnejad, M. (2014) Mutagenesis in olive (*Olea europaea* L.) calli caused by sodium azide and detection of mutants using ISSR and RAPD markers. *The Journal of Horticultural Science and Biotechnology* 89, 153–158.

Shibli, R.A., Shatnawi, M., Abu, E. and Al-Juboory, K.H. (2001) Somatic embryogenesis and plant recovery from callus of 'Nabali' olive (*Olea europaea* L.). *Scientia Horticulturae* 88, 243–256.

Silvestri, C., Celletti, S., Cristofori, V., Astolfi, S., Ruggiero, B. and Rugini, E. (2017) Olive (*Olea europaea* L.) plants transgenic for tobacco osmotin gene are less sensitive to in vitro-induced drought stress. *Acta Physiologiae Plantarum* 39(10), 229.

Strobel, G.A., Nachmias, A. and Hess, W.M. (1988) Improvements in the growth and yield of olive trees by transformation with the Ri plasmid of *Agrobacterium rhizogenes*. *Canadian Journal of Botany* 66, 2581–2585.

Taylor, C.G., Fuchs, B., Collier, R. and Lutke, W.K. (2006) Generation of composite plants using *Agrobacterium rhizogenes*. *Methods in Molecular Biology* 343, 155–167.

Theis, T., Wedde, M., Meyer, V. and Stahl, U. (2003) The antifungal protein from *Aspergillius giganteus* causes membrane permeabilization. *Antimicrobial Agents and Chemotherapy* 47, 588–593.

Tong, X., Qi, J., Zhu, X., Mao, B., Zeng, L. *et al.* (2012) The rice hydroperoxide lyase OsHPL3 functions in defense responses by modulating the oxylipin pathway. *Plant Journal* 71(5), 763–775.

Torreblanca, R., Cerezo, S., Palomo-Ríos, E., Mercado, J.A. and Pliego-Alfaro, F. (2010) Development of a high throughput system for genetic transformation of olive (*Olea europaea* L.) plants. *Plant, Cell Tissue and Organ Culture* 103(1), 61–69.

Tränkner, C., Lehmann, S., Hoenicka, H., Hanke, M.V., Fladung, M. *et al.* (2010) Over-expression of an FT-homologous gene of apple induces early flowering in annual and perennial plants. *Planta* 232(6), 1309–1324.

Turck, F., Fornara, F. and Coupland, G. (2008) Regulation and identity of florigen: FLOWERING LOCUS T moves center stage. *Annual Review of Plant Biology* 59, 573–594.

Williams, M., Morales, M.T., Aparicio, R. and Harwood, J.L. (1998) Analysis of volatiles from callus cultures of olive *Olea europaea*. *Phytochemistry* 47(7), 1253–1259.

4 Olive Propagation and Nursery

M. Lambardi[1], A. Fabbri[2], M. Micheli[3] and A. Vitale[4]
[1]*Institute of Bioeconomy, CNR, Sesto Fiorentino, Italy;* [2]*Department of Food and Drug, University of Parma, Italy;* [3]*Department of Agricultural, Food and Environmental Sciences, University of Perugia, Italy;* [4]*Olive Technical Consultant, Pistoia, Italy*

4.1 Olive Propagation Today

Plant propagation consists of the application of specific biological principles and particular techniques for the multiplication of plants. The plants obtained in the process should also be identical, or as similar as possible, to the plants from which they are derived. Olive trees are today mainly reproduced by cutting propagation and micropropagation (i.e. *in vitro* propagation), but grafting propagation is still largely utilized, especially in some Mediterranean countries, such as Italy and Greece. In Italy, for instance, grafting propagation still represents about 65% of the 5 million trees produced by the olive nurseries (Catalano and Sonnoli, 2007).

Olive-cutting propagation is based on the 'misting technique', which spread widely between 1950 and 1960 and is the most common propagation method where there are no financial or technical constraints. Under such circumstances, and, at any rate, wherever the nurseries are medium to large in size, they can afford fairly high investment in terms of buildings (greenhouses), instruments (for the management and control of humidity levels, temperature, light etc.) and skilled personnel. As regards the material used for propagation, this comes from the nursery itself (stock plants), neighboring nurseries and plants in production. The result of all this is that, as a rule, the material is indigenous in origin; only where stock plants of exotic cultivars are collected and grown by nurseries or institutions, genotypes of other regions or countries can be made available to local producers.

Starting from the 1990s, the propagation of the olive tree has begun to make use of *in vitro* technology that guarantees the production of plants of high genetic-sanitary quality, through a process that consists of, essentially, two phases: the phase of multiplication of shootlets in the tissue culture laboratory; and the phase of rooting-acclimatization in the greenhouse. Today, an increasing yearly production of olive micropropagated plants is observed in the Mediterranean area, mainly in Italy, Tunisia, Turkey and Algeria.

Finally, there is still a substantial demand for plants propagated by grafting, particularly in Italy. This fact is due to an ancient belief, still alive, of many olive growers that grafted olive trees have a better root system that goes deeper in the soil, thus favoring a superior anchoring of the plants and a better supply of water. On these aspects, the debate between proponents of propagation by

 The Olive: Botany and Production (eds A. Fabbri *et al.*)

DOI: 10.1079/9781789247350.0013

cuttings/micropropagation and those of grafting persists (see 4.3.2). On the other hand, the grafting technique could soon regain great importance in olive trees if, as many hope, new selected clonal rootstocks should begin to spread.

The success of propagation, whatever the technique adopted, depends on the knowledge of the many aspects controlling the formation of a new individual:

- genetical characteristics of the cultivars which regulate their ability to be propagated;
- anatomical structure and physiology of the whole plant and the organs to be propagated;
- methods most suited to propagate the various cultivars and rootstocks; and
- techniques and structures, and their optimization to obtain the best technical and economical results.

4.2 Propagation by Cuttings

With varying results, every part of the olive tree has been used for propagation (Fabbri *et al.*, 2004). Shoots and branches are, as a rule, excellent propagation material, when the proper techniques are applied. The olive tree is also very efficient at differentiating new meristems at the collar, where the transition from trunk to root system takes place, as well as on old wood (limbs) where latent buds are present. This peculiarity explains the relative ease in propagating the olive from large plant parts, which enabled ancient growers to asexually propagate this tree and to domesticate it far earlier than most other fruit trees. Root cuttings are the only organs of the olive not able to produce new plants; indeed, roots do not possess adventitious buds, and are incapable of producing such structures when separated from the parent plant.

4.2.1 Ancient systems of propagation

Branches

The use of branches, four years old or more, is a very ancient method of olive propagation (Phoenicians, Romans, Arabs). The method makes use of the possibility of producing shoots (from latent or adventitious buds) and roots from branches. Cuttings (20–50 cm long, with a diameter of 5–10 mm or more) are obtained from branches during the cold season (autumn/winter). The cuttings are kept until spring in a cool and not excessively humid place, in layers of sand. They are then positioned in an outdoors plot horizontally, vertically or inclined. The horizontal branch cuttings were the most used. In spring, among the new shoots, the most vigorous is kept for growth and the others are eliminated, unless the cutting is of such a size to support more than one plantlet. After six to eight months the plantlets have reached a height varying between 60 cm and 80 cm, and at their bottom end, near the insertion on old wood, a small root system should be present. The rooted shoots are then separated from the branch, after which they receive normal nursery care.

Branch material is easily obtained from pruning operations, which in this case should be made in winter. As a rule, the cuttings should not be left in the open after collection; they should either be put to root or stratified in sand. Treatments with root-promoting chemicals are useful, and rooting performance is directly proportional to stem diameter. Rooting is also obtained from very old trunks, when uprooted, cut into sections 20–30 cm long; the new shoots and roots are produced from the living tissues that are in the cortex, when the piece of wood is buried a few centimeters down.

Examples of traditional propagation by branch cutting are the ‘Cormoni’ (Apulia and southern Italy in general), 40–60 cm-long branches with their foliage removed; the ‘Estacas’ (Spain), big cuttings, as long as 2–3 m, more than 6 cm thick, obtained during pruning; the ‘Garrotes’ (Spain), cuttings 50–100 cm in length, and 3–5 cm in diameter (Fabbri *et al.*, 2004).

Good results are obtained with these methods but they are not sufficient to justify nursery use. On a commercial scale, it is almost impossible to find the large quantities of material necessary for this kind of propagation. However, this technique can be useful for small olive growers, when they need to replant a small percentage of the orchard each year and are not willing to invest in propagation equipment.

Ovules

Ovules are characteristic swellings, hyperplasias, which appear as protuberances and are found at the level of the collar (Baldini, 1986). At the ovule site, a build-up of sap due to a slowing down of its circulation occurs. This is most often found at the base of the trunk where the root structure joins the trunk, producing torsions in the vessels, thus slowing sap circulation. This determines hypernutrition of cambium cells, which actively proliferate and produce the extroflection of tissues, particularly parenchyma tissues, that constitute most of the ovule. In some cultivars these structures can also be found in the lower trunk parts of five- to six-year-old trees. Accumulated starchy substances may cause the formation and emission of adventitious shoots and roots, the type of structure depending on whether the ovule is exposed, or not, to light.

The ovules can be removed with special cutting tools, taking great care of exposed tissues, which must be disinfected and protected. Ovules are taken from healthy trees at the beginning of spring or in autumn; they are then kept in slightly damp sand until the time comes to put them under the earth in the ovulary, or ovule bed. The weight of an ovule varies, on average, from 100 g to 3 kg. If they are rooted in nurseries their weight is usually 500–800 g.

Suckers (or pollards)

Suckers develop at the base of the trunk and have their origin in ovules. Generally, they are taken from the tree when they have acquired their ovule masses, and have produced sufficiently developed roots (they cannot, therefore, be considered cuttings in a strict sense). To help these processes, the base of the tree is covered with a thin layer of earth which encourages the initiation of roots. Girdling can further enhance adventitious root formation. Subsequently, the rooted suckers are removed in the spring from the stock plant with some old, collar wood, to be grown in the nursery before being planted out.

This method of multiplication can be used for the replacement of small numbers of trees, it cannot be used at the nursery level because it is slow and costly. The use of suckers is part of common practice to re-establish an olive orchard after a strong frost, as happened, for instance, in the hill of Florence after the terrible frost of 1966. Once the death of the upper part of the tree is ascertained, it is cut down and removed to avoid the presence of wood that can easily be attacked by parasites (fungi, bacteria, insects), while the stump is left so that suckers can develop freely (the stump usually survives the cold because it is covered by a few centimeters of soil). After about one year, the less favorably positioned suckers are removed so that in the second or third year there is one sucker, or at most two or three suckers, from which to grow a new canopy.

4.2.2 Propagation by shoot cuttings

Shoot collection and cutting preparation

For the preparation of cuttings, scions can be taken all year round, but, in practice, this is done during two annual peak points of concentration of rooting, corresponding to the times of maximum intensity of vegetative growth (i.e. March–April and September–October in the Mediterranean basin). In temperate climates, at the borderline of the growing area for olive (such as central Italy), it is possible to get from stock plants shoots that are long enough to make more than one cutting (Fig. 4.1a), but not more than once a year. From these shoots, cuttings are obtained by cutting them in 10–15-cm-long pieces and 4–6 mm in diameter, with four to six nodes. The four to six leaves at the distal end are retained, while the two to three basal nodes are left without leaves (Fig. 4.1b). The basal cut must be made just under a node, and care must be taken to avoid leaving a part of the internode, which can be a point of necrosis. Generally, new roots do

Fig. 4.1. (a) medium-sized, well-developed shoots from which cuttings and microcuttings can be prepared. (b) semi-hardwood cutting ready for planting in the rooting bed. (c) olive microcutting, cv. 'Arbequina'. (d) olive cuttings tied in bunches, ready for hydro-alcoholic IBA treatment by quick dipping.

not grow from internode tissues, they usually develop within node tissues. Once the cuttings are prepared, basal treatment with a rooting hormone (see below) must be carried out within 20–40 minutes.

Microcuttings

In recent years, an innovative propagation technique by cutting has become widespread, based on the use of short cuttings (7–8 cm, i.e. about half of a traditional cutting), called 'microcuttings'. The microcutting is prepared with one node, in the middle of its length, which maintains the two full leaves (Fig. 4.1c). The microcuttings are propagated without any treatment with auxin or with treatments similar to the traditional cuttings, after which they are rooted in multi pots. The main advantage of this technique, which is already well affirmed in Spanish nurseries and is widely spreading also in other olive-growing countries, such as Italy, is that of doubling the material available for propagation compared to the traditional use of longer cuttings.

Self-prepared auxin treatments to promote rooting of cuttings

Among the available auxins, the indole-3-butyric acid (IBA) is by far the most used plant growth regulator (PGR) for olive propagation, because it is not toxic to hardwood and semi-hardwood cuttings over a wide concentration range, and is very effective in the root promotion of a large number of plant species. Alternatively, the 1-naphthaleneacetic acid (NAA), which is stronger than IBA in terms of stimulation of olive

Fig. 4.2. (a) A rooting bench with copper pipes for bottom heating of cuttings; (b) cutting propagation into a perlite substrate.

adventitious rooting, is sometimes preferred. However, because uniform results are difficult to obtain with NAA, its use is mainly restricted to cultivars which respond unsatisfactorily to IBA. Some commercial preparations are composed of a mixture of IBA and NAA.

Hydro-alcoholic solutions are the most commonly used IBA preparation, as the nursery workers can easily prepare their rooting treatment at a low cost by buying the IBA as a pure product. In this method the cuttings are treated by quick-dipping (five seconds) their basal parts (0.5–1 cm) in an IBA hydro-alcoholic solution. In rooting treatments, the concentration of auxin is usually expressed in parts per million (ppm). In hydro-alcoholic treatments, it refers to the weight of the auxin compound per volume of the solution (w/v), e.g. 1000 ppm = 0.1 g of auxin in 100 ml of solution. However, to test the effectiveness of different growth regulators (e.g. IBA vs. NAA) in rooting promotion, equimolar concentrations (M) of the compounds should be compared. Effective concentrations of IBA in hydro-alcoholic solutions are in the range of 1000–4000 ppm: 1000 ppm when cuttings are collected from cultivars with a high rooting potential or from vigorous vegetative shoots of the stock plant, 4000 ppm for cuttings collected from difficult-to-root cultivars, or during the rest period or bloom. Table 4.1 reports the rooting potential of some of the most commonly cultivated olive cultivars. With respect to NAA, treatments with this PGR are generally in the range of 1000–2000 ppm. As a rule, ethanol concentrations in the rooting solution should be not higher than 30%, to limit the risks of strong dehydration and injury of basal cutting tissue.

As the quantity of IBA that can be dissolved is related to ethanol concentration, 30% ethanol may not dissolve the large amounts of auxin required to produce concentrated treatment solutions (e.g. over 4000 ppm). Salts of some auxins (e.g. K^+-IBA) are available, and they can be simply dissolved in distilled or deionized water. However, the cost of K^+-IBA formulation is much higher than IBA. Cuttings are most efficiently prepared in bundles of 10–15 (Fig. 4.1d) and treated this way, instead of one-by-one. Immediately after the treatment with IBA, the cuttings are inserted in the rooting medium under misting. Cuttings are rooted in benches with a bottom heating system, generally made of water radiators of plastic or copper pipes (Fig. 4.2a), and filled with a bed of perlite (Fig. 4.2b), while microcuttings are generally rooted in multi pots containing mixtures of different components (such as peat, perlite, soil, coconut fibers), under misted tunnels.

In root-promoting talc formulations, auxin is dispersed in the inert talcum powder. Here, the concentration of auxin refers to the weight per weight (w/w), that is grams of auxin per grams of talc (e.g. 1000 ppm = 0.1 g of auxin in 100 g of talc). Auxin-talcum powder mixtures can be purchased commercially (see below) or prepared by the nursery workers, using reagent grade auxin and talcum powder. To get a homogeneous dispersion, the auxin is previously dissolved in a solvent (generally diethyl ether; alternatively, ethanol can be used, but

Table 4.1. Rooting potential of some of the most commonly cultivated olive cultivars. The reported percentages are the average of results obtained by different authors, after IBA treatments. (From Fabbri *et al.*, 2004)

High (100–66%)		Medium (66–33%)		Low (33–0%)	
Cultivar	**Country***	**Cultivar**	**Country***	**Cultivar**	**Country***
AGLANDAU	F	AGGEZI SHAMI	ET	AZÉRADJ	DZ
ARBEQUINA	E	AZAPA	RCH	BELLA DI SPAGNA	I
ASCOLANA TENERA	I	BARDHE I TIRANE	AL	BIANCHERA	I
BARNEA	IL	BELLA DI CERIGNOLA	I	BIANCOLILLA	I
BOUTEILLAN	F	BICAL CASTELO		BÜYÜK TOPAK	
		BRANCO	P	ULAK	TR
CORATINA	I	BIDH EL HAMMAM	TN	CARRASQUENHA	P
CORDOVIL		CAILLETIER	F	CHEMLAL	DZ
CASTELO BRANCO	P				
FRANTOIO	I	ÇAKIR TR		CHEMLALI DE SFAX	TN
GORDAL		CARRASQUEÑO	E	DOMAT	TR
DE GRANADA	E				
LECCINO	I	CHALKIDIKI	GR	EMPELTRE	E
LECHIN DE SEVILLA	E	CHEMCHALI	TN	FARGA	E
LUCQUES	F	ERKENCE	TR	GORDAL	
				SEVILLANA	E
MANZANILLA		GALEGA VULGAR	P	LECCIO DEL	
DE SEVILLA	E			CORNO	I
MISSION	USA	KALAMATA	GR	LIANOLIA	
				KERKYRAS	GR
MIXAN	AL	PICHOLINE	F	NABALI BALADI	IL
MORAIOLO	I	PICHOLINE		NOCELLARA	
		MAROCAINE	MA	ETNEA	I
NOCELLARA		PICUAL	E	OGLIAROLA	
MESSINESE	I			MESSINESE	I
OBLICA	HR	SIGOISE	DZ	OUESLATI	TN
PENDOLINO	I	TAGGIASCA	I	SALONENQUE	F
VERDAL	E	VERDALE DE		VERDEAL	
		L'HÉRAULT	E	ALENTEJANA	P

*AL, Albania; DZ, Algeria; E, Spain; EG, Egypt; HR, Croatia; F, France; GR, Greece; I, Italy; IL, Israel; P, Portugal; RCH, Chile; TN, Tunisia; MA, Morocco; TR, Turkey; USA, United States of America

it will require more time to evaporate). which is then used to wet totally the talc. When the solvent is completely evaporated, the powder is mixed and used for the treatment of cuttings. Cuttings are dipped in the powder with their 1–2 mm basal parts. It may be beneficial to pre-wet cutting bases with water so that powder adheres better. Cuttings are then lightly tapped to remove excess powder. To avoid brushing off the powder during the insertion of cuttings in the rooting medium, a stick can be used to make a 5 mm hole in the medium before each cutting is inserted.

Talc formulations are advantageous in allowing treatments with high IBA concentrations (up to 10.000), which are very useful when propagation of very difficult-to-root cultivars (Table 4.1) is pursued. Another advantage is that talc formulations are easy to use. On the other hand, rooting results are less uniform with talc formulations than with hydro-alcoholic treatments, due to the variability in the amount of powder adhering to the cuttings.

Commercial auxin preparations

A number of commercial rooting preparations are available, differing in formulations and concentrations. Essentially, they contain one or more auxins in an inert matrix (talc or, less frequently, gel or paste, often mixed with lanolin to increase adherence), plus some minor additives (such as fungicides and microelements). These products are formulated to be effective for a large number of species. Usually, complete directions and a list of plants tested come with each product. However, there is very poor information on the activity of commercial preparations in inducing adventitious root formation in olive stem cuttings, as olive is rarely among the tested species. The only guideline that can be suggested is to check carefully the composition of preparations, in order to choose the one in which the active ingredient concentration (auxin) is in the range indicated above.

Soaking and wounding the cuttings

Although they have never become routine in commercial olive propagation, soaking and wounding the cuttings, in combination with auxin treatments, can improve their rooting performance.

As regards soaking, with the immersion of their bottom (proximal) part in water, cuttings lose by diffusion a number of natural substances, some of which have an inhibiting effect on the rooting process. At the same time, mechanisms that enable the cutting to root occur more readily. Soaking does not alter the annual pattern of the rooting response of cuttings; however, it enhances and stimulates an already existing ability. This effect is shown especially in hard-to-root olive cultivars. For instance, a positive effect on both rooting percentage, and quality of roots from olive cuttings, has been reported when cuttings were soaked for 24 hours at a pH of 8.5 before treatment with 4000 ppm IBA (Bartolini *et al.*, 1977).

Basal wounding is often reported as beneficial to stimulate rooting of difficult-to-root species or cultivars. It consists in making thin longitudinal incisions at the base of cuttings with a sharp knife, before auxin treatment. Wounded cuttings display greater absorption of growth regulators at the base of cuttings, especially when the quick-dip method is used. Following wounding and auxin treatment, callus and adventitious root production are greater, particularly along the edges of the incisions. With regard to the olive, there is no specific information on the subject. Considering the increase of propagation costs due to this practice, its application is justified in commercial olive nurseries only when rooting of very difficult-to-root cultivars is attempted.

Polyamines

It has been reported that polyamines can interfere in different ways with adventitious rooting of several species (Hausman *et al.*, 1997). As regards olive, putrescine was shown to promote adventitious rooting of cuttings, in synergism with IBA treatments (Rugini *et al.*, 1997). Therefore, putrescine could be considered for use with cuttings of olive cultivars which respond poorly to auxin treatments alone.

Fungicides

With many woody species, treatment of cuttings with fungicides, whether incorporated into the auxin-talcum powder or used alone, has been shown to protect newly formed roots from fungal attack, as well as increase survival and overall quality of the rooted cuttings. It should be noted that the use of cuttings affected by fungi and insects should always be avoided to limit the spread of diseases, and also because cuttings suffering biotic stresses have lower rooting capacity.

4.3 Grafting Propagation

Although propagation by cutting, microcutting and micropropagation is by far the most used approach worldwide in olive propagation, some countries, such as Italy, have still a quite large production of olive trees by grafting. The main purpose of grafting is the reproduction of genotypes (cultivars, selections, elite trees) that cannot be propagated by cutting or other methods, or that can be propagated with such means at exceedingly high costs. As reported above, rooting ability differs markedly among olive cultivars (e.g. several table cultivars are very hard to root or do not root at all) and also micropropagation is often poorly effective, so that propagation by grafting turns out to be the only viable approach. Furthermore, studies are underway to identify and select clonal rootstocks that can give desired characteristics to olive cultivars, such as the resistance to important diseases (e.g. *Xylella fastidiosa*) or the reduction of the vegetative development of the plant in semi-intensive and intensive orchards. It is presumed that the advent of clonal rootstocks will lead to a return to a wider use of graft propagation.

On the other hand, grafting has some disadvantages, i.e. (i) the 'seed-to-orchard' procedure is far more lengthy and laborious than the 'cutting-to-orchard' one; moreover, grafting propagation is by far more expensive than cutting propagation; (ii) seedlings used as rootstocks are a genetically varied population of individuals, and their effects on the grafted cultivar may likewise be varied; (iii) the germination percentages of several genotypes are very low; moreover, in many olive cultivars the germination of healthy and perfect seeds may be heavily delayed, as they must overcome a strong double dormancy (chemical and mechanical); (iv) grafting is a complex operation which requires long practice; as a consequence, the production of grafted trees is usually restricted to specialized nurseries, where skilled labor is present.

4.3.1 Production of olive seedlings

Seed propagation in the olive is undertaken mainly for the purpose of producing seedlings to be used as rootstocks for olive trees. The advantage of seed propagation lies mostly in the possibility to produce a very large number of high-quality virus-free rootstocks. Moreover, the seeds are cheap and the production of seedlings can be carried out with little skill and equipment. One important drawback of rootstock production by seed is that the seedlings are heterogeneous in terms of vigor and root development, for which growth characteristics of grafted plants can differ, even markedly. Proper handling of olive seeds – from fruit collection up to seed germination and seedling development – is fundamental to produce fast-growing homogeneous rootstocks to be used in grafting propagation.

The method of grafting scions on seedlings is still largely used in Italy where, in the highly specialized nurseries of Pescia (Pistoia province, Italy), the large numbers of plants required for establishing the new industrial olive plantations have been produced over the last half-century; this would not have been possible if only branch cuttings and ovules had been available. The method is also used widely in Argentina, where it allowed a rapid diffusion of the species. On the other hand, nurseries of other important olive-growing countries, such as Spain, use, almost exclusively, stem cuttings for olive propagation (Fabbri *et al.*, 2004).

Sources of stones, quality of seeds and harvesting time

What is commonly called 'olive seed' is, in fact, a stone, i.e. the 'true' olive seed is enclosed in the fruit stony endocarp (the pit). The best sources of stones for seed propagation are trees of specific cultivars, selected from olive orchards with plants in good conditions of productivity. When possible, it is advisable to select plants of the same cultivar in the inner part of the orchard, in order to reduce as much as possible cross-pollination and, therefore, heterogeneity of seedlings. Once trees producing good-quality seeds have been located, it is recommended to collect the stones every year from the same plants. Cultivars used as sources of stones for seedling rootstock production are generally a result of personal experience of nurserymen, who have acquired information on seed germinability and superior characteristics of derived seedlings in terms of vigor, quality of root apparatus, diameter reached at grafting time, and tolerance to biotic and abiotic stresses. Examples of cultivars used as stone sources are 'Canino', 'Mignolo', 'Maurino', 'Moraiolo', 'Frangivento' and 'Americano' in Italy, 'Arbequina' and 'Verdal' in Spain, and 'Allegra' and 'Oblonga' in the USA. A limited use of stones from wild olive trees is still sporadically made in Italy and Spain.

In general, the best fruits for seed propagation are those harvested at their full ripening, when the peel and the pulp have the same color. This condition is reached between middle autumn and early winter (November–December in the Northern Hemisphere), depending on the cultivar and the climatic characteristics of the growing area. Progress in olive fruit ripening coincides with total carbohydrate accumulation in the seed, and this parameter seems positively correlated with germination ability (Rinaldi *et al.*, 1994). Moreover, an anticipated collection (e.g. at the moment of the color change of fruits) increases the percentage of empty seeds (seeds not containing embryos), as the fruits with empty seeds drop during ripening. The percentage of empty seeds at ripening is generally in a range of 10–30%.

Sowing

Before sowing, it is worthwhile to carry out some operations aimed at increasing the germination percentage and at synchronizing, as much as possible, seed swelling and embryo development. Olive seed germination is hastened by softening the stony endocarp before sowing with one of the following practices: (i) by soaking the stones, performed by placing them in frequent changes of water (e.g. every one to three days for a period of 10–20 days); during this process, floating seeds, which are empty, must be discarded; (ii) by chemical scarification, which consists in placing the stones in containers and covering them with 5% sodium hydroxide (NaOH) in water for 24 h. The mixture should be stirred cautiously at intervals during the treatment to produce uniform results, and the progress of the treatment should be monitored by periodical drawing out of samples to check the thickness of the seedcoat. In the end, the NaOH solution is poured off and the stones are washed for about a week, renewing the water daily.

Sowing is undertaken in cold frames, germination benches or simple plastic trays like the ones used for fruits and vegetables, under a glasshouse or plastic tunnels. Alternatively, sowing can be done in the open on well-drained nursery soil, in rows which are mulched during the winter months (this system is quite common in Californian olive nurseries). A good olive seedbed should have a loose but fine physical texture, delivering adequate moisture to the seeds and good, but not excessive, aeration. The quantity of stones required is 1.5–3 kg per m^2 of seedbed, depending on the size of the stones. A precautionary fungicide treatment of the substrate (e.g. with TMTD 50 or Captan, 250 g/hl) is convenient and should be repeated at least monthly during seed germination, in order to avoid or reduce collar rots due to *Pythium* and/or *Rhizoctonia* contamination. Moreover, a wire net around the seedbed should protect the stones from rodent attacks.

Transplant and growth of seedlings

Olive seed germination in nurseries is considered to be in the order of 40–60% of the harvested seeds. For instance, the 'Mignolo' is a cultivar with a high proportion of empty seeds; therefore, its final seedling production is generally no more than 50% of the harvested seeds, as about 30% are empty seeds and about 20% are non-germinating seeds or seeds lost during the first stages of germination.

The spring after sowing, when the seedlings have four to six months growth and are 10–15 cm tall, they have their roots slightly trimmed and are transplanted into well-worked and manured soil, or in pots (2.5 lt) filled with a common olive substrate. To obtain uniform plants, seedlings should be selected at this time, rejecting the thin, the knelt and the contaminated ones. The 'grafting area' is a plot of the nursery specifically dedicated to graftage, where the transplanted seedlings, in pots or in the soil, are grown up to grafting time. Plant density is about 150 units/m^2. By the end of the year the young plants are 30–50 cm tall, and they have the thickness of a pencil at collar level. In the following spring (i.e. 18 months after sowing, one year after transplanting in the 'grafting area') the plants are ready for grafting.

4.3.2 Grafting technique

In countries with an ancient tradition of olive grafting, the most commonly used grafting technique is by far a type of bark grafting (called in Italy *innesto a penna*). It was developed in Pescia (Italy), which has been for a century, and still is today, the world's most important center for olive propagation. Pescia nurserymen have tested and perfected this grafting procedure over the years, and the result, therefore, derives from a widespread experimentation. Hence, bark grafting is both biologically suitable, and economically viable.

Bark grafting in olive

Olive scion wood is obtained from one-year-old vigorous shoots collected from mature and healthy stock plants. Stock plants can be located in productive orchards or in-field collections. However, in order to achieve a maximum uniformity of plants, scion wood should be collected yearly from the same plants, better if grown in a specific area of the nursery (the 'stock plants plot') where they should be periodically checked for their maintenance in good sanitary and genetic conditions. To avoid dehydration, the shoots can be placed in vertical position inside a pot, containing 1–2 cm of water, remembering that these basal ends of the shoots must not be used for grafting.

The graft, since it depends on the bark separating readily from the wood, can be done in spring (April–May in central Italy, one month earlier in southern Italy), i.e. after active growth of the seedling rootstocks has started. In general, this optimal condition for grafting olive plants lasts about one month. Grafting olive trees should be avoided during hot and windy days. If rootstocks are grown in the greenhouse, it is possible to anticipate the grafting time by about one month.

In olive, bark grafting is generally performed on 18-month-old seedlings in pots or in the field, provided they have attained a thickness of 6–8 mm (all plants below such size are discarded). Only one scion per rootstock is grafted. To prepare the rootstock to be grafted, the stem is cut 10–12 cm above the soil, or at 5–10 cm if in a container (Fig. 4.3a). After eliminating the leaves, a vertical knife cut (2 cm long) is then made at the top end of the stub through the bark to the wood (Fig. 4.3b).

The best scion wood is taken from the middle part of shoots, rather than from the apical parts (where the wood is still immature) or from the basal parts (too thick). In general, only the portion of scion wood having the right thickness (3–4 mm) should be used to obtain scions. The scion is 6–8 cm long and is prepared in such a way as to contain a node in its upper third. The two leaves of the node are cut to 1/4–1/3 of their size to limit scion dehydration. One oblique cut is made with the knife along one side at the base of the scion, extending about one-third of the way into the scion. The bark is lifted slightly along both sides of this cut. The scion is then inserted between the bark and the wood of the rootstock, positioned under the vertical cut of the bark, and gently pushed downward to penetrate a bit deeper below the bark (Fig. 4.3c). The graft zone is then fastened with raffia or rubber bands in order to produce a firm contact between the scion and the rootstock. After the scion has been fastened by tying, all cut surfaces, including the top of the scion, are covered thoroughly with grafting wax (Fig. 4.3d).

In general, the best productivity in terms of grafts per hour is achieved with a team of three persons, i.e. one grafter and two helpers (one devoted to the preparation of plants and materials, one to tie the grafted plant and apply the wax). Three experienced workers can make 200–300 grafts per hour (1500–2500 grafts per day).

Both in pots or in the field, it is always better to avoid irrigating the grafted plants for the first ten days after grafting or, if irrigation is necessary, attention should be paid in maintaining dry the grafting point. Callusing takes place within 15–20 days when petioles naturally abscise from successfully grafted scions. If a correct grafting technique has been applied, the rate of healing should be over 75%, but rates above 90% are always the goal. In well-healed grafts, two new shoots develop on the scion within

Fig. 4.3. (a) Seedling cut back to 5–10 cm length; leaves are still present on the stub, but will be excised. (b) A longitudinal slit is made at the top end of the stub. (c) Scion inserted under the lifted bark of the seedling (*left*); detail of the oblique cut at the base of the scion (*right*). (d) Bark grafting tied with a rubber band and waxed top and at the grafting zone.

30 days. When these have reached 5–10 cm in length, the best shoot is chosen and the other one is removed. In the summer, the young shoots are supported with stakes to ensure vertical growth.

Grafting on suckers

This practice can be pursued when grave damage (due, e.g., to exceptional cold spells, mechanical damages, and severe pathogen attacks) has involved olive trees, and it is deemed convenient to reconstruct the orchard by utilizing the trees' existing root system, rather than replanting new ones. In this case, the tree should be cut back to soil level, and a variable number of suckers (as many as possible in the first year) are left to grow freely until the canopy is reconstituted and the plant starts producing again. This can take a few years. If the tree had been propagated by grafting, some suckers (or all of them) may originate from the rootstock, which, as a rule, was a seedling. Such a canopy is not suitable for production, both because the characteristics of the new tree are going to be different from the desired cultivar (most likely worse) and because the tree will maintain for a long period juvenile features. Hence, it is suggested to graft the selected suckers with a suitable cultivar, which of course can be different from the pre-existing one. The most vigorous and better-positioned suckers will be selected for grafting, no more than three or four. Grafts should be made as low as possible in order to reduce the likelihood of the development of latent buds from the rootstock.

Grafted cuttings

This technique is also named 'bench grafting' because it is performed at benches by skilled grafters as a large-scale operation. In the olive it has been proposed in the past (Jacoboni and Fontanazza, 1976), having as its principal object the multiplication of cultivars that do not root readily under mist. Cuttings and scions are taken from one-year-old shoots, the cuttings from selected plants (generally grown from seed and chosen for their high rooting capacity), the scions from the cultivar to be propagated. The cuttings are 15–18 cm long, prepared so as to leave two pairs of leaves. The scions (of the same thickness as the cuttings) are prepared with two nodes and one pair of leaves, and grafted at the bench with the 'wedge' or 'whip-and-tongue' technique. The base of the grafted cuttings is then treated with IBA at 2000–4000 ppm, and set in a substrate of peat and perlite, suitably moistened and maintained under plastic cover to preserve humidity. The bottom should be maintained at approximately 28°C. In these conditions of high temperature and humidity, rooting of the cutting and healing of the graft occur almost simultaneously within 60–90 days. Although the technique has never found any commercial application, it can be a useful method for propagating small numbers of plants when little space is available.

Production of clonal rootstocks in olive

The use of clonal rootstocks is of fundamental importance for improving the productivity of many fruit species, making it possible to control important characteristics, such as vigor, precocity of production, resistance to parasites, diseases, and adverse environmental and soil conditions. Unlike most fruit species, information on olive rootstocks is scattered and fragmentary, not only because in many olive-producing countries the use of grafting is limited, but also because, even where it is generally practiced, the rootstocks used are produced from seed, either of cultivars or of wild types. Although there has been no systematic research on best combinations of scion/rootstock that would allow any control of the characteristics of the olive tree, in recent years, some clonal rootstocks have been proposed as a result of studies carried out mainly in California, Italy and Spain (Bartolini *et al.*, 1998; Petruccelli *et al.*, 2012). Worthy of mention are:

- 'Oblonga' (California), employed because of its resistance to *Verticillium dahliae*;
- 'FS-17®' (Italy), which induces a reduction of vigor in the grafted cultivar; recently, a variant of the 'FS-17', named 'Favolosa', is proving to be of great interest as it is able to induce tolerance to *Xylella fastidiosa* in the cultivar that is grafted onto it (see, e.g., *Favolosa, // it.wikipedia.org/w/index.php?title=Favolosa&oldid=120438544; accessed 28/1221*);
- 'Lechin de Sevilla' (Spain), tolerant to dry and calcareous soils and inducing cold resistance in the grafted cultivars ('Gordal sevillana', 'Manzanilla de Sevilla', 'Morona');
- 'Leccino Compatto', 'Leccino Dwarf' (Italy), inducing reduction of vigor of the grafted cultivar, and tolerance to cold;
- 'Verdial de Badajoz' (Spain), tolerant to dry and calcareous soils (grafted with 'Manzanilla de Sevilla');
- 'Hojiblanca' (Spain), tolerant to dry and calcareous soils and inducing cold resistance in the grafted cultivar ('Aloreña');
- 'Urano®', 'Tosca07®' (Italy), rootstocks selected at the Vivai Sonnoli of Pescia, able to consistently reduce the growth of the grafted cultivar.

It has to be noted that the above reported information is either generally agreed on in the various countries or merely declared by breeders; however, definitive experimental data are not always available.

Grafted plants or self-rooted plants?

One of the first questions the olive grower has to face is which type of planting material should he utilize – grafted or self-rooted plants? It must be remembered that not all cultivars can be obtained self-rooted; in this case, as for many table cultivars, the choice of the grafted tree has no alternative. When both types of material are available at comparable prices, the problem arises as to which is best. The grafted versus self-rooted trees issue is an old one. A discrete amount of research was undertaken on this topic in the 1960s and 1970s, both in Italy and California. Generally speaking, no significant or generalizable differences were noticed in specific trials with respect to productivity (Guerriero *et al.*, 1972, 1974). Thus, in central Italy (Tuscany), grafted trees proved to be more productive; in the south (Sicily), the advantage was for the self-rooted trees. But a closer examination of the results shows different behaviors in different experimental fields and, above all, different performances depending on the cultivar being evaluated. Most growers are interested in a better vigor in the first years after plantation; even in this case differences exist among cultivars, which cannot be generalized: 'Moraiolo' rootlings appear weaker than grafted plants, at least in the first years of culture; 'Frangivento' seems to have an opposite behavior; while 'Leccino' shows no differences. Self-rooted 'Frantoio' resulted weaker if compared to plants grafted on 'Frantoio' seedlings, but more vigorous if compared to 'Frantoio' grafted on 'Lea' seedlings. Other cultivars ('Manzanillo', 'Mission', 'Gordal') have quite different behaviors in the grafted/self-rooted comparison (Hartmann, 1958).

Substantial differences occur with respect to the development of the root system (Scaramuzzi, 1963). Rooted cuttings develop their roots initially at a unique level in the soil, usually depending on the depth of planting. Usually, the root apparatus is quite superficial, with a wide geotropic angle. Seedlings, on the other hand, irrespective of the fact, of course, that they are or not grafted, develop their roots from an axis of a discrete length, usually around 20 cm, and, having a narrower geotropic angle, tend to grow deeper into the soil. This means that in arid, sandy, windy areas, self-rooted plants need, in the beginning, to be staked and irrigated regularly in the dry periods, and require greater care than usual. Practical experience not always confirms this point of view: in certain arid zones (southern Tunisia), the olive has long been propagated by ovules without any adverse effect on resistance to drought or longevity. Grafted trees, on the other hand, have a few drawbacks; the main one is that whenever it is necessary to reconstitute a damaged canopy (e.g. by frost, diseases or senescence), all suckers coming from the stock that are not suitable for production need to be eliminated. Grafted trees should be planted with the grafting point below the soil level; however, it must be remembered that, with time, soil levels subside due to erosion, and the rootstock may show up and produce its own suckers.

In addition to the above, and related to the container technology that is dominating in almost all nurseries, it should be remembered that all types of plants undergo transplants in containers of different sizes, at least twice. This operation means breakage (often voluntary cuts, called transplant pruning) of longer roots, and regeneration of a root system, more or less compact, of a cylindric shape, and uniformly distributed. This means that, morphologically speaking, two-year-old root systems coming from the two types of propagation are not easily distinguished, and consequently an analogous behavior is to be expected in the soil. Finally, a known truth among plant physiologists is that roots, in their development, are far more influenced by the edaphic environment than by their genetic characteristics. This means that the alleged innate characteristic of better survival in the soil of the taproot, were it true, can hold true only in the very first period of root development. Then, in sandy soils, roots go deep in search of water; in clay soils, they stay closer to the surface; that is, they respond to environmental stimuli in the same way. The opposite has never been demonstrated.

This is not the final word on the issue; more research is needed on a wide scale to ascertain with precision the effect of grafting and of the different genotypes utilized as stocks, although self-rooted plants are, at least in certain countries and planting systems, becoming far more widespread than grafted trees.

The search for clonal rootstocks able to control plant behavior is still ongoing, although the results so far cannot be compared with those of deciduous fruit trees.

4.4 Micropropagation

Micropropagation represents an efficient technique for the aseptic culture of the plant explants (cells, tissues and organs), under environmentally controlled conditions, using artificial nutritive media. It exploits the ability of the initial explants of regenerating whole plantlets. The potentiality of differentiated and specialized cells to form complete plants is referred to as *cellular totipotency* (Bhojwani and Dantu, 2013). Micropropagation is based on this ability and has now become a common practice in horticulture and forestry to preserve the desirable characters of selected genotypes or varieties. Therefore, it is applied for the mass propagation of true-to-type plant material.

At the end of the aseptic path, rooted *vitro*-derived shoots are ready to be acclimated to recover the original autotrophic condition. In general, the scheme of the technique is articulated in five phases as follows:

- Phase 0: selection, collection and preparation of the mother plants.
- Phase 1: taking from the donor plant of initial explants, their decontamination and *in vitro* establishment.
- Phase 2: *in vitro* multiplication (proliferation) of new shoots.
- Phase 3: *in vitro* rooting of proliferated shoots.
- Phase 4: acclimatization of *vitro*-derived plantlets.

The differences with the main traditional systems of vegetative propagation (cutting and grafting) consist in the following considerations: (i) many plantlets can be produced starting with a few initial explants within a shorter period of time and in more limited spaces; (ii) recalcitrant-to-root genotypes can be cloned; (iii) propagation is ideally carried out in aseptic conditions (avoiding contaminations) and pathogen-free high-quality plantlets are produced in controlled conditions; (iv) a suitable adjustment of factors influencing the regeneration process is possible, such as nutrient- and growth-regulator levels, light and temperature; (v) propagation can be continued all year round, regardless of seasonal changes; (vi) plant material can be stored over long periods when needed; (vii) less time and space are required for propagation; (viii) the diffusion and the commercial exchange of germplasm are simplified as regards the quarantine systems to control the plants' import; and (ix) the elimination of micro-organisms (particularly viruses) affecting plant material is possible by specific *in vitro* techniques such as micrografting, meristem culture and thermotherapy.

Propagation of plants from small parts has changed the concept of production on a large scale and over a billion commercially important plants are today clonally produced by tissue culture every year in the world (Bhojwani and Dantu, 2013). It is mainly about important horticultural, medicinal and forestry species for which the *in vitro* protocols have been developed and optimized.

Not all the species can be micropropagated, but one of them is the olive (*Olea europaea* L.), although the success of the technique is strongly dependent on the aptitude of each genotype to be cultured in aseptic conditions (Fabbri *et al.*, 2004). Micropropagation of olive by shoot tip and nodal culture is available for many olive cultivars, as reported by Rugini *et al.* (2020). The specifications of olive micropropagation are described below, through the different phases of the technique.

4.4.1. Phase 0: selection, collection and preparation of the mother plants

Regarding the initial step, the use of adult trees as mother plants poses several problems such as oxidation of tissues from field-grown, microbial contamination, and unresponsive cultivars (Rugini *et al.*, 2020). On the contrary, rooted cuttings or potted young stock plants grown in the greenhouse (Sghir *et al.*, 2005) or hardwood cuttings sprouted in a controlled environment (Vidoy-Mercado *et al.*, 2012) are the ideal material to initiate *in vitro* propagation of olive, particularly when segments are excised from vigorous shoots after sprouting (Rugini and Fedeli, 1990).

Additional brief information

a) Age of the mother plant can affect the response of explants.
b) Time of collection or vegetative conditions of initial explants can influence the decontamination protocol.
c) Microbe contamination can be reduced when slight water stress is induced on potted stock plants by diminishing the amount of watering a few days before explant collection.

4.4.2 Phase 1: *in vitro* establishment of explants

After excision and isolation of initial explants, their sterilization is fundamental before the introduction of *in vitro* conditions. Pre-soaking initial explants in PPM™ (Plant Preservative Mixture™) or ethanol (Fig. 4.4a) could be necessary for surface cleaning, even if the use of ethyl alcohol can cause tissue dehydration. Treatments with calcium or sodium hypochlorite, at different concentrations, are useful for obtaining sterile explants, free from exogenous spores of fungi or bacteria. The use of very low concentrations of mercuric chloride is also contemplated (Lambardi and Rugini, 2003), but this salt is noxious and toxic and should be used only if really necessary.

Sometimes initial explants of olive can release phenols after excision, which get oxidized and cause browning of the medium. The oxidation products of the phenols could be phytotoxic. A simple method to protect the explants consists of the addition of antioxidants (ascorbic or citric acid) to the nutritive medium.

Fig. 4.4. (a) Pre-soaking of uninodal segments of olive excised from young shoots of mother plants. (b) Shoots of olive proliferated at the end of Phase 2. (c) Elongated *vitro*-derived shoots of olive. (d) Initial bi-nodal explants placed on the proliferation agarized medium. (e) Shoot of olive rooted on agarized medium. (f) Shoot of olive rooted on soil mixture.

In vitro establishment is achieved when explants are clean from contaminations and can produce new shoots after a period of incubation on the culture medium (see Phase 2).

Additional brief information

a) The mother plant should be cultivated under optimal conditions (in the field or in the greenhouse) to minimize contaminations in the *in vitro* culture (Cassells and Doyle, 2005).
b) Some surfactants (Tween 20/80) play an important role to improve the effect of disinfectants.
c) The addition of 1ml l^{-1} of PPM™ within the nutritive medium has proved suitable for counteracting the fungi contaminations in olive cultures. PPM™ can be considered the ultimate solution against endogenous contamination.

4.4.3 Phase 2: *in vitro* multiplication (proliferation) of new shoots

After the *in vitro* establishment, the multiplication of olive explants is usually achieved through the stimulation of nodal axillary buds to develop new elongated shoots. In this step, the medium composition plays an essential role and can determine a satisfactory multiplication rate, as well as the development of well-structured and vigorous shoots. The MS medium (Murashige and Skoog, 1962) has been the most extensively used for the vegetative propagation of many plant species *in vitro*, but the development of Olive Medium (OM) for axillary bud stimulation and subsequent shoot multiplication (Fig. 4.4b) marked an important step forward in the improvement of olive micropropagation. It was formulated by Rugini (1984) on the basis of the analysis of the main mineral elements found in shoot apices of vigorously growing plants and in mature zygotic embryos. The OM is richer in calcium, magnesium, sulfur, phosphorus, boron, copper and zinc than MS, and has a low calcium:nitrogen ratio (1:11). It also contains glutamine as a nitrogen source (Lambardi and Rugini, 2003).

As regards the carbon source, mannitol (one of the major carbohydrates of olive metabolism) proved to be better than sucrose, as it increases shoot proliferation, improves the general quality and uniformity of shoot cultures, and reduces the basal callus formation (Leva *et al.*, 2013).

In vitro proliferation of olive is rarely obtained with the production of adventitious shoots due to the effect of a strong apical dominance, and the removal of the apical meristem may not be an efficient solution. However, the exogenous application of PGRs, particularly cytokinins, can force the axillary buds to grow even in the presence of the terminal bud (Bhojwani and Dantu, 2013). Zeatin proved to be the most effective at inducing maximum stimulation of axillary buds in many olive varieties, although it is very expensive. As a consequence, many studies were carried out with the aim of testing the effectiveness of alternative cytokinins or cytokinin-like substances with different results in relation to the genotypes evaluated: 6-benzyl adenine (BA), 6-γ,γ-dimethylallylaminopurine (2iP), kinetin, thidiazuron (TDZ), metatopolin or dikegulac (Rugini *et al.*, 2020).

Shoot proliferation of olive explants is accomplished by segmentation of elongated shoots (Fig. 4.4c) in uni-/bi-nodal propagules which represent the initial explants for a new subculture (Lambardi *et al.*, 2013). Micropropagation is usually carried out in heterotrophic or mixotrophic conditions. In contrast, if the shoots of olive can be cultured under autotrophic-like conditions, in which the culture vessel is ventilated with CO_2 through a microporous filter and the cultures are maintained under high light intensity (150 μmol m^{-2} s^{-1}), growth and quality of the shoots is considerably improved (Bhojwani and Dantu, 2013). Moreover, in the absence of carbohydrates in the medium, contamination of the cultures can be minimized.

Additional brief information

a) In consideration of the strong dominance exercised by the apex, the multiplication rate is given by the total number of nodes formed for each olive explant at the end of the proliferation phase.
b) Micropropagated olive shoots have a thin cuticle, low stomata density, and a single palisade layer (Cozza *et al.*, 1997).
c) The performance of olive during multiplication can be influenced by the size of initial explants (uninodal or binodal segments) and by their orientation on the agarized medium (vertical or horizontal placement; Fig. 4.4d).
d) The addition to the medium of natural plant substances (coconut milk, natural extracts or seed oils) can improve the proliferation and the vegetative activity of olive explants as biostimulants (Micheli *et al.*, 2018).
e) Proliferating olive cultures are usually maintained at 22±2°C and 40 μmol $m^{-2}s^{-1}$ of light intensity with a 16 h photoperiod.
f) After proliferation, a passage in a medium containing filter-sterilized GA_3 can induce a rapid elongation of shoots, which are more suitable for subsequent rooting (Rugini *et al.*, 2001).
g) Dikegulac was tested in adult olives for its capacity to increase fruit set due to its effects on plant growth reduction and an increase in the number of short shoots (Moradnezhad *et al.*, 2017).

4.4.4 Phase 3: *in vitro* rooting of proliferated shoots

In olive, the rooting aptitude is strongly affected by the genotype. In addition, other variables that can affect the process of adventitious root formation are: excessive osmolarity of the culture medium, low availability of reserve substances in the tissues, lack of adequate nutrients, contamination by pathogens, juvenility of the materials used, and the thickness and lignification of the cell wall (Cançado *et al.*, 2013). Bi- or tri-nodal explants excised from the proliferated shoots are usually utilized. Medium formulation and auxin treatment are the most studied factors to improve rooting performance. The proliferation medium can be used generally by reducing to half the strength of the original macroelement formulation (Lambardi and Rugini, 2003). As regards the use of PGR, both IBA and NAA have been indicated as most effective for the stimulation of rooting; however, other factors can be usefully modulated to increase the rooting activity: the use of putrescine seems to promote rooting ability, improved production of difficult-to-root cultivars can be achieved by etiolation of shoots, basal treatments of explants with H_2O_2 also stimulate rooting, and the carbon source may affect root development (Rugini *et al.*, 2020).

To reduce the costs of micropropagation, many laboratories now remove unrooted shoots from the *in vitro* environment and root them outside the culture vessel (George *et al.*, 2008). In fact, alternatively to the traditional use of rooting agarized media (Fig. 4.4e), basal semi-solid substrates or soil mixtures have also given positive results (*in vitro* or *in vivo* conditions) after the exposure of the olive shoots to a basal dipping in inductive auxin solutions. The formed roots show a structure more similar to the definitive ones and seem to be more functional than those developed in the agarized media (Fig. 4.4f).

Additional brief information

a) Basal etiolation of shoots can be performed by the black painting of the outside of the jars and/or by covering the agarized rooting medium with black sterile polycarbonate granules (Mencuccini, 2003).
b) Lucchesini and Vitagliano (2002) emphasized the importance of rooting olives in aerated systems to obtain vigorous plants for better acclimatization.

c) The *in vitro* rooting ability can depend on the ontogenetic phase of the mother plant.
d) In specific genotypes, NAA can produce a stronger rooting stimulation than IBA, as a lower hormone concentration is required to obtain rhizogenesis.
e) The seasonal effect on the *in vivo* rooting ability has been shown also *in vitro* when some olive cultivars were rooted without darkening the medium.

4.4.5 Phase 4: acclimatization of *vitro*-derived plantlets

Acclimatizing is a critical point in olive micropropagation because of the change of climatic conditions (humidity, light, asepsis) that characterize the passage to the *ex vitro* environment. As a consequence, this step should be carried out under high humidity with continuous light to obtain the development of plants with good lateral branches (Leva, 2011). The rooting of vigorous proliferated shoots in mixed *vitro-vivo* conditions allows the development of functional roots and the reduction of acclimatization stresses, accelerating the growth of the plantlets.

Interesting possibilities exist for combining the use of Arbuscular Mycorrhizal Fungi (AMF) with micropropagation of olive to improve plant acclimatization: mycorrhizae are known to positively affect plant growth by improving nutrient uptake, especially that of phosphorus, increasing the rhizosphere volume, and protecting plants against biotic and abiotic stresses (Binet *et al.*, 2007).

Following acclimation, a one-year hardening period is required before the final plantation of young plants in the orchard.

Additional brief information

a) Rugini *et al.* (2016) suggested that a liquid medium containing a high amount of mannitol or sucrose (4–5%), putrescine and GA_3 should be added to the semi-solid medium (double-phase) to accelerate hardening after the emergence of adventitious roots.
b) During the acclimatization phase, the plantlets must be potted in appropriate compost substrates (e.g. peat moss, perlite and polystyrene granules, 2:2:1) and in fog conditions to reduce as much as possible the stress resulting from their transfer to an *in vivo* environment (Lambardi and Rugini, 2003).

4.5 Olive Nursery

The modern olive nursery, where the various propagation and acclimatization operations are carried out, does not differ substantially from the ordinary tree nurseries. Therefore, only general considerations on the equipment of the nursery, plus some more specific notions on the olive nursery, are reported below.

4.5.1 Greenhouse

There are several factors behind the planning of greenhouse systems: climate, crop and bottom line (the market value of what the greenhouse produces). Concerning the latitude where the structures have to be built, it is necessary to plan the orientation of the greenhouse to guarantee good lighting, in view of the microclimate of the area. In function of the greenhouse proprieties, only about 60–80% of the PAR (photosynthetically active radiation) outside actually reaches the plants, depending on the cover material (plastic or glass) and construction parts that intercept the light. Once inside the greenhouse, the radiation has two effects: it is used in photosynthesis (the PAR component), and the energy it carries (both the PAR and near-infrared component) heats everything, i.e. the plants and all parts of the greenhouse and equipment included. The control zone is a centralized system that contains the greenhouse environmental controls of humidity, temperature and light, using a different system in the function of the technology level: windows opening and closure, water regulation for mist or fog system, temperature control (both in the beds and in the ambient), humidity control (by use of a fan and/or heat pump), and water treatment, in function of the electrical conductivity (EC).

Rooting zone

As for cutting and microcutting propagation, the structuring of this area depends on the type of propagation utilized: in fact, nowadays it is possible to consider useful the classic 'rooting beds' with a raised edge (10–12 cm), fixed or movable, filled with an inert substrate like perlite and covered with plastic tunnels (height 120–180 cm), according to latitude and greenhouse structure (Fig. 4.5). Mobile pallets are usually preferred by growers because, despite the utilization of more expensive materials (aluminum), they allow an increase of the number of plants per square meter and provide better material handling and automation ('robotic greenhouse'). In this system, usually the pallet is filled with an inert material, like perlite, that covers 5–8 cm of draining material (gravel or crushed stone), utilized to improve the uniformity of temperature at the bottom of the cuttings; the heating system is placed at the bottom of the bed. The use of an inert substrate like perlite certainly reduces the risk of the accumulation of salinity, mainly sodium, that can harm rooting more than what is possible to have in peat.

It is important to remember that the EC for the propagation of plants has to be below 500 μS cm^{-1} to avoid this problem, and the use of a substrate that improves drainage and preserves the bottom of cuttings from high or low temperatures is positive. A similar concept of propagation is the 'closed frame technique' (Fig. 4.6), largely used in olive-cutting propagation. It consists of a heated bench, which is placed inside a metal-and-plastic box and covered with an airtight lid. The bench, 30–40 cm high, is raised to about 1 m and contains a rooting substrate or small plastic containers in which olive cuttings are placed. No extra watering is required in the rooting period (50–60 days), during which time the lid should never be opened. In the last decade, the use of rooting areas without tunnels has increasingly spread, where the control of temperature and humidity is entrusted to a fog system (high and low pressure), facilitating the possibility of managing large quantities of material with less labor.

The use of units equipped with the solar supplement allows the management to intervene to control sunlight (directly correlated with the warming of the environment). However, to improve the rooting of olive

Fig. 4.5. Tunnel with mist system which allows maintenance of a high level of humidity, reducing to the minimum the risk of cutting stress.

Fig. 4.6. Closed frames, a relatively old system for rooting olive cuttings; it consists of a raised box with basal heating and covered with glass or plastic.

cuttings, it is necessary to keep the lower ends of cuttings at a constant temperature range (18–25°C) by setting up the correct heating system, starting from distribution pumps and heating diffusers to the sizing of the burner or other heat generators. The most common diffusers are the water radiators (plastic or copper pipes), the low-voltage electric resistances, and the heating sheets made of electrical-resistance material. Among them, the most used are the water radiators (usually plastic pipes), with a distance between the pipes that has to be not more than 15 cm and covered with an inert material.

Climatic chamber

The rooting of olive cuttings in a climatic chamber is a possibility that must be evaluated for difficult-to-root cultivars that can benefit from a more precise environmental control. Morini *et al.* (1990) suggested that the propagation with a low level of light is possible using the most appropriate light spectra for rooting: a test on the cv. 'Leccino' under fluorescent light at 39–41 μmol m^{-2} s^{-1} compared different qualities of light (red, yellow, white, green and blue) and demonstrated that yellow light gave the greater effects in rooting of cuttings.

This type of propagation requires a chamber with fog system at high pressure, small tunnels or containers, closed by a plastic film, where one or few plateaux with preforma systems are allocated (e.g. a plateau with 240–360 holes). The use of peat as substrate is not recommended for phytopathological problems, as well as paper pots, although the latter create fewer problems for the best oxygenation and drainage of the substrate. High rooting percentages can be obtained by this technique (up to 80%), maintaining for 40–60 days the cuttings under LED light (Gregoriu *et al.*, 2007) (intensity at 54 μmol m^{-2} s^{-1}) for 16 hours/day, without any operation other than the control of the water status of preforms (one or two water interventions per day is usually necessary in 60 days).

Greenhouse and environmental conditions (light, humidity, temperature) for rooting

The environmental conditions that should be considered for olive rooting are light, humidity and temperature. Light is directly connected with plant photosynthesis, but at the same time, under covered area, it can increase the temperature too much and may induce thermal stress on cuttings, hence negatively affecting rooting. Hartmann studied the olive tree response to shading and showed that the plant maintains a relatively high photosynthetic capacity under low light intensity (Hartmann *et al.*, 2010). This characteristic gives the possibility to make an environment with low irradiance during spring/summer without affecting the photosynthetic activity required at the root stage, reducing the high temperature and giving good results of rooting also without artificial lights during autumn/winter. These considerations follow the basic principle of propagation by cutting for most ornamental and fruit plants (Hartmann *et al.*, 2010), where it has been shown that a high light is not needed before the rooting and may even determine a stress until the cuttings start to produce roots.

Regarding humidity, in the first two weeks it is suggested a high level of RH (90–95%) be maintained to preserve the material in rooting. Since the various technologies for water regulation are not perfect – different control devices, such as timers, electronic leaves, scales, photoelectric cells and integrated computer systems – usually the farmer tends to an excessive use of water to avoid leaf fall, thus causing a condition in which the substrate gets exceedingly moist. Because of the difficulty in regulating the ambient humidity, it is possible to understand why the use of mist/fog systems under the tunnel is very common, as well as the use of high tunnels to reduce the rise of temperature around the cuttings. It is possible to root the cuttings even in the absence of a tunnel by means of the use of mist or fog systems, but only with a high level of experience in the management of the protected environment. Olive cuttings are sensible to the air temperature above 30–35°C, which increases the transpiration and causes leaves to fall, as well as to basal temperature below 15°C, which increases the rooting times. Temperature stresses, such as excessive heat or cold, increase cellular ROS (reactive oxygen species) levels and render membranes and proteins dysfunctional.

4.5.2 Material and substrates for rooting cuttings and microcuttings

The choice of the substrate for the rooting phase is linked to obtaining the right level of moisture, aeration and correct temperature at the bottom of the cutting. The newly formed roots are not able to absorb nutrients, and therefore the substrates are not required to contain available mineral nutrients.

To achieve the right rooting conditions, several materials have been used, alone or in mixtures, the most common being perlite, peat, vermiculite and sand (Table 4.2). The optimum rooting quality for several cultivars was obtained on the perlite substrate (Awan *et al.*, 2003), because it is recognized to have a unique capillary action, and increase aeration and drainage within the container for its uniformity and lightness (Paradiso and De Pascale, 2008). It is possible to compare different aeration capacities of several substrates, with an ideal substrate in the range 20–25% (De Boodt and Verdonck, 1972). These values were suitable for perlite, but inadequate for peat moss. In addition, Isfendiyaroglu *et al.* (2009) showed that overwatering can cause anaerobic conditions, leading to less rooting rate as it may be signaled on peat moss. It is well known that more granular inorganic substrates, such as perlite, have low water-retention capacity and are used for their porosity, while peat moss, used for nutrient content and high water-holding capacity, makes favorable conditions for fungal growth (Awan *et al.*, 2003) and needs incorporation of perlite material to improve the aeration status of the substrate.

However, considering the labor cost required for the transplant of bare-root cuttings, a new propagation system is based on the use of paper pots and preforma (Fig. 4.7): the first consists of peat mixed with perlite, surrounded by paper to form a cylinder; the second is a new substrate composed of peat mixed with a special adhesive to form a spongy compound. These systems may be used in all the structures described above by substituting the inert material, and reducing transplant problems and the cost of management. Moreover, this type of material permits increased waiting time than do traditional systems for transplanting, as the roots wrapping around the paper pot receive water and mineral salts for several months, allowing the nursery to transplant without the stresses that characterize bare-root transplants. A typical preforma is made of a polymer prepared in the form of cubes and having physical characteristics that guarantee a high level of porosity and good drainage.

Fig. 4.7. Type of substrate used for olive rooting. *Left to right:* perlite, preforma, paperpot and peat moss mixture. Preforma and paperpot are largely used for rooting microcuttings.

Table 4.2. Physical and chemical characteristics of perlite, vermiculite, sand, peat and preforma. (From Zaccheo and Cattivello, 2009)

Physical characteristics								
					Peat decomposition classes			
Parameters	Unit of measure	Perlite	Vermiculite	Sand	1–3	4–6	7–10	Preforma
Actual density	Kg m^{-3}	520–2.540	2650	2610				1582
Relative density	Kg m^{-3}	42–147	90–100	1700	40–79	80–159	160–250	111.4
Total porosity	% (v v^{-1})	94–98	95	35	93–97	90–95	85–93	93
Air at pF1	% (v v^{-1})	46–76	41	24	10–55	10–40	5–25	
Water at pF1	% (v v^{-1})	19–51	39	11	42–83	50–85	68–80	47.2
Water retention	g H_2O g^{-1}s.s.	3–3.8	6	2.2	6–15	4–8	3–5	5.08
Water easily available	% (v v^{-1})	4–18	1.3	10				8.7
Reserve water	% (v v^{-1})	2 – 4	1,1	2				
Chemical characteristics								
pH		5.2 – 7.7	7	6	3–5	3–5	3–5	4.5 – 6.2
EC	dS m^{-1}	0 – 0.4	0.02	0.12	0.1–0.3	0.2–0.5	0.3–0.7	0.4 – 0.6
CSC	cmol L^{-1}	0.2	12	0	8–18	15–25	24–34	?

EC, electrical conductivity; CSC, cation exchange capacity

In olive cuttings, both systems are used mainly for microcuttings.The choice of the paper pot or preforma systems permits reduction of the costs of propagation, improving at the same time the handling of material, starting from the transplant as the extraction of plants from the trays without any stress is facilitated. Another consideration is that, with paper pot or preforma systems, the nurserymen can sell and move the material after rooting easily, at lower costs and in a shorter time. However, the management must be more careful, as both systems require tunnels, with coverage in polythene, placed on pallets on the ground or raised, commonly humidified with a fog system (high-pressure) to reduce substrate wetting. The sand is not a good substrate when used alone, as it has a very poor water-retention capacity, is very heavy and its particles may cause injuries to the cuttings when they are inserted into the substrate. Moreover, it may contain weed seeds and various harmful pathogens.

4.5.3 Pathogen control in greenhouse

Recent studies were made on the effect in greenhouse of nitrogen (N) on the incidence of olive leaf spot (*Fusicladium oleagineum*) and showed that disease incidence was significantly higher in plants subjected to a high N treatment than in those subjected to a low N treatment under different conditions (Roca *et al.*, 2018). These results suggest that N excess increases susceptibility to olive leaf spot. Because of the increase in climate change and the increasing restrictions on pesticides, it is also interesting to study the mycorrhization of young olive plants with *Glomus intraradices*, a choice that may be useful to increase plant tolerance to biotic and abiotic stresses (Porras-Soriano *et al.*, 2009). The management of fertilization should consider that the interaction between plants and mycorrhizae, with high levels of phosphorus (P) in a substrate, reduces plant growth (Jiménez-Moreno *et al.*, 2018), while when the substrate contains low P concentrations, root growth increases compared to a control without mycorrhization (P should not be supplied to the substrate if plants are to be mycorrhized).

Another element that is considered to have beneficial effects on plant tolerance to biotic and abiotic stresses is silicon (Si), the second most abundant element in the earth's crust after oxygen. All plants growing in soil contain Si in their tissues, and the concentration depends on the soil and plant species (Debona *et al.*, 2017). The Si taken up by plants is transported from the roots to shoots, either actively or passively, via the transpiration stream. In the shoot, silicic acid is concentrated through loss of water due to transpiration and is polymerized into insoluble silica, forming a silica gel layer between the cuticle and epidermal cells and preventing Si translocation to new, growing leaves. This silica layer constitutes a physical barrier that is believed to reduce the incidence of pests and diseases, as well as to improve photosynthetic rates and tolerance to water stress, drought conditions and other abiotic stresses. Silicon also forms a chemical barrier, inducing the production of phenolic compounds, phytoalexins and other products that activate plant defense mechanisms (Debona *et al.*, 2017). In olive propagation, Si can be applied by foliar sprays, directly to the substrate, or through irrigation water, although Savvas and Ntatsi (2015) indicated that application through the soil is more effective than foliar application in increasing Si levels in plant tissues. Cultivar differences were evident in responses to Si application for olive leaf spot (Nascimento-Silva *et al.*, 2019), where the effect of pathology reduction was more pronounced in 'Arbequina', a moderately susceptible cultivar to olive leaf spot, than in the more susceptible cv. 'Picual'.

4.5.4 Plant hardening and growing in the nursery

The process of hardening prepares the plants for the final transplant in the field. In this phase, it is necessary to reduce the basal temperature maintained under the rooting zone, preferably in a gradual manner; hence, it is suggested to use a well-aerated greenhouse, with a mechanized opening in the roof and on the sides. Usually, when the cuttings are rooted in autumn or late winter/early spring, the plants are moved under a closed greenhouse to finish the development of the roots before transplant (April–May in the Mediterranean area). Otherwise, if cuttings are placed to root in the late spring/early summer, it is important to control temperature and humidity during the first weeks by the use of nets to avoid stress and the ensuing negative impact on rooting and health of the cuttings. The operations to be carried out for plant growth in the nursery will depend on the choice of rooting substrate: if perlite is utilized, it is necessary to transplant bare root plants in pots with an organic substrate (not less than 500 ml of volume), taking care to keep the roots moist during transplant. In the first few days after transplant, the leaves should also be kept moistened, just like the substrate, without being waterlogged. In the case of use of paper pot or preforma systems, it is not recommended to preserve the plants as in the bare root system, since the roots always remain preserved in the substrate. A final pot, not less than 2000 ml (usually 2500 ml), can be used from the start but this procedure is not optimal for the growth rate of the canopy, which is lower (Fig. 4.8). The plant tends to develop its root system first, before the growth of the crown, causing in this way the reduction of foliage, a lower photosynthesis and a slower plant development. Anyway, it has been demonstrated in olive (Cimato and Franchini, 2002) that all roots (thinner and thicker) contribute, albeit to different degrees, to water and mineral absorption: water, P, K and Ca are absorbed by even the most lignified zones, while in adjacent areas, not only the absorption but also the transport may be greater.

By observing root development on different substrates, it emerges that in the fertile substrate the absorbing mass amounts to half of the aerial part, while in the non-fertile substrate the root tends to ramify intensively (root mass greater than aerial mass). These considerations suggest use in the pots of a substrate with a high level of drainage, avoiding excess salinity due to excessive fertilization by the grower. In the history of olive propagation, it is common to use sphagnum peat blonde (0–40 mm) in addition to pumice (different grain size, as a function of the vessel), usually in a ratio in favor of peat. In recent years, several nurseries in Pescia (Italy) prefer a more draining substrate, increasing the percentage of pumice (up to 70%), obtaining excellent rooting results. As concerns plant fertilization, it is commonly a good practice to supply few elements during cultivation (mainly N, P and Ca) to increase root development. In this way, an approach could be to mix in substrate 1–3 kg m^{-3} of slow-release fertilizer (12 months) plus powdered manure (5–7 kg m^{-3}) to avoid nutritional deficits during cultivation. In particular, the important role of Ca in the Mediterranean basin should be considered where water resources are scarce and irrigation with saline water is commonplace, although olive plants are considered moderately tolerant to peat salinity (Marin *et al.*, 1985). As in other glycophytic species, salt tolerance in the olive is associated with ion-exclusion mechanisms located in the roots (Benlloch *et al.*, 1991; Melgar *et al.*, 2006), which reduce translocation and accumulation of Na^+ and Cl^- to the aerial part. It has been shown that olive plants are less sensitive to leaf Cl^- than Na^+ (Bongi and Loreto, 1989), and that the Cl^- ion does not cause toxicity (Melgar *et al.*, 2009). In contrast, Na^+ toxicity is usually a concern when Ca^{2+} concentrations are relatively low. Indeed, when Ca^{2+} was applied

Fig. 4.8. Comparison of the radical development between olive plants, rooted *ex vitro* after *in vitro* rooting induction (*left*), and rooted cuttings (*right*) in 2.5-liter pots. During the cold season, the vegetative growth and the development of root system is better in olive trees from micropropagation.

to saline irrigation water, leaf Ca^{2+} concentrations increased with the amount of Ca^{2+} applied, but with a concomitant decrease in leaf Na^{+} concentrations (Melgar *et al.*, 2006). This indicates that Ca^{2+} plays an important role in the Na^{+} exclusion mechanism, and that maintaining adequate Ca^{2+} nutrition can protect olive plants against Na^{+} toxicity (Tattini and Traversi, 2009).

The radical temperature to avoid physiological issues should be maintained between 2°C and 27°C, considering the possibility of using greenhouses in cold periods and shadow areas in the late spring/summer.

4.6 Stock Plants

Nurseries producing olive trees need to use 'controlled' material for propagation which corresponds to the genetic and morphological characteristics set out in the certification procedure (see 4.7). The methods used in the growing of stock plants vary according to the 'product' to be supplied for propagation: seed, shoots for grafting and cuttings, buds and microcuttings for micropropagation. Since these plants are the source of material for propagation (cuttings, buds, microcuttings, scions, seeds), controls must be carried out to ensure that stock plants maintain good health and genetic true-to-typeness.

Material for cutting propagation, micropropagation and grafting should never be taken from seedlings, as the resulting trees would be of unknown characteristics, and all positive features fixed by cloning would be lost. On the other hand, when it is necessary to collect material for propagation by cutting, microcutting and *in vitro* culture from adult and old trees, the selection of organs more reactive to 'regeneration' is fundamental, such as suckers, pollards and vertically growing shoots.

Specific training systems can induce a sort of 'rejuvenation' in organs used for collecting propagation material, involving an improvement of the rooting ability of cuttings and microcuttings, as well as an easier development of explants for micropropagation and establishment of shoot cultures. All these systems involve vigorous pruning to stimulate strong and rapid growth of new shoots. Experience acquired up to the present in managing olive trees as a source of propagation material is neither abundant nor definitive. It would seem that it is best to place the plants at minimum distances of about 2.0 m x 2.5 m, to make mechanized operations possible. The lowest branch height is established at 80–90 cm since the first year. From the second year, when propagation material is being taken, shoots are excised while leaving the two basal buds (nodes) as spurs, and leaving untouched all the shoots that have not grown sufficiently. Thus, future development is aided by avoiding, as far as possible, an imbalance between development below and above ground. If there is excessive vegetative growth, it is advisable to carry out normal cutbacks. Olive trees pruned this way acquire their definitive shape after four to five years, at which point they will each produce 200–400 traditional cuttings annually, and more than double microcuttings.

Specialized nurseries are structured to use, as a source of high-quality material, mainly for microcutting and *in vitro* propagation, young clonal potted plants maintained in greenhouses under controlled conditions of temperature, light and humidity. These stock plants are treated with fungicides before being used for the collections of microcuttings or explants for micropropagation.

The role of light on olive rhizogenesis has not been sufficiently studied, but the results obtained in some woody plants under relatively low irradiance, which showed most consistent effect of shade on leaf area and photosynthesis activity, motivated similar studies in olive, as reported, e.g., by Cartechini and Tombesi (1986) and Proietti *et al.* (1988). Ninety percent shade caused an 81% increase in olive leaf area (Kappel and Flore, 1983), and the greater leaf size in shaded leaves can be caused by an increase in the contents of auxins and gibberellins within the leaves under low irradiance (Salisbury and Ross, 1978). Thus, shaded olive leaves had a 7.0%, 16.2% and 27.0% reduction in stomatal density in 30%, 60% and 90% shade, respectively) in relation to the control. The smaller reduction in both stomatal density and thickness of palisade tissues, as well as the greater increase in leaf area under 90% shade of olive, suggests a good adaptation of olive to low irradiance. This may give the olive tree the advantage of maintaining a relatively high photosynthetic capacity under the same low irradiance (Gregoriou *et al.*, 2007). Considering these studies, therefore, it is possible to affirm that during spring/summer, stock plants need to be covered to increase the rooting ability of cuttings; in addition, the high intensity of light during these seasons would suggest the possibility to study the effects of photoselective nets for the shadow of stock plants, considering that in some woody plants they have led to carbohydrate accumulation.

As concerns water management, it is important that there is the correct substrate in the function of the irrigation system and in the function of the pot size, guaranteeing a very good drainage to avoid water stagnation. In addition to that, the plants usually are in a covered area where a localized irrigation is present, frequently used for the fertilization of plants, so that it is possible to properly manage water as well as fertilization. Stock plants need a low level of N in order to increase the rhizogenesis of cuttings, while the effect of P_2O_5 and K_2O is negligible (Dag *et al.*, 2012).

4.7 Olive Certification: The Example of Italy

The guarantees of genetic origin and good sanitary conditions represent the key points of the certification of olive plants produced in the nurseries and intended for marketing. Compliance with these conditions is a prerequisite to having good results in terms of production and longevity of the olive grove. A large number of olive varieties obtained by natural and artificial selection have generated, over time, a great deal of confusion in relation to the huge genetic heritage of this species. This problem is increased by the expansion of olive cultivation outside the traditional Mediterranean areas (Fabbri *et al.*, 2004). In addition, concerning the sanitary characteristics, the olive plants may be mainly affected by fungal, bacterial, phytoplasmic or viral diseases of greater or lesser economic importance (Petruccelli *et al.*, 2012). The latency state of many of these parasites could cause their diffusion in the propagation material produced by contaminated mother plants. So, in an attempt to limit the potential impact of the spread of pests and pathogens through the movement of infected plants, specific laws have been issued at regional, national and European levels (Catalano *et al.*, 2016). In recent years, many efforts have been conducted to bring the highest number of Italian olive cultivars to the genetic and sanitary status suitable for their inclusion in the Italian and European certification programs (Faggioli *et al.*, 2017).

Currently, in Italy, the nursery production and the marketing of olive plants are subjected to such mandatory standards as Plant Passport (EU Reg. 2031/2016) and CAC (Conformitas Agraria Communitatis) (EU Council Directive 2020/177). Moreover, the plant propagation material is qualified by voluntary standards, as an EU Certification (EU Council Directive 2020/177) and National Voluntary Certification (MiPAAFT Decree, 19 March 2019). The different categories of materials (National Voluntary, Community Voluntary, Community Mandatory) are also identified through labels of different colors, as required by specific EU regulations.

The Italian model of certification begins with (i) the transfer of plant propagating material (plants) from the breeder (or whoever intends to introduce a new accession for a specific cultivar in the certification system) to an institution, with the aim of carrying out an evaluation and eligibility of the registration request by a technical certification committee authorized by the ministry (Fabbri *et al.*, 2004). Once the nuclear stock has been registered, (ii) the subjects responsible for its production and selection (public institutions or authorized research centers) must maintain the material healthily in a screenhouse as pre-basic plants for conservation and pre-multiplication. The following step (iii) is the pre-multiplication of basic material carried out in the open field (repository for pre-multiplication) by the same subjects. Cuttings and seeds collected from the basic material (iv) are used to produce the certified mother plants for the establishment of the multiplication mother blocks of nursery associations. Certified mother plants obtained represent the source for nursery-certified olive plant production (Albanese *et al.*, 2012).

The Regional Phytosanitary Service (LD 536/1992) is in charge of the issue of authorizations required in this field, monitoring phytosanitary and genetic conditions of plant species by visual inspections and instrumental examinations, and the issue of related certificates.

References

Albanese, G., Saponari, M. and Faggioli, F. (2012) Phytosanitary certification. In: Muzzalupo I. (ed.) *Olive Germplasm: The Olive Cultivation, Table Olive and Olive Oil Industry in Italy*. InTech OpenEditors, London.

Awan, A.A., Iqbal, A., Rehman, M.J. and Idris, G. (2003) Response of olive hard wood cuttings to different growth media and basal injuries for propagation. *Asian Journal of Plant Science* 2, 883–886.

Baldini, E. (1986) *Arboricoltura Generale*. CLUEB, Bologna, Italy.

Bartolini, G., Fiorino, P. and Bouzar, M. (1977) Ricerche sull'influenza dell'immersione in acqua delle talee. 3. Effetto della bagnatura a pH diversi sulla capacità rizogena in talee di olivo, cv. 'Frantoio'. *Rivista Ortoflorofrutticoltura Italiana* 6, 409–417.

Bartolini, G., Prevost, G., Messeri, C. and Carignani, G. (1998) *Olive Germplasm: Cultivars and World-wide Collections*. FAO, Rome.

Benlloch, M., Marin, L. and Fernández-Escobar R. (1991) Salt tolerance of various olive varieties. *Acta Horticulturae* 356, 215–217.

Bhojwani, S.S. and Dantu, P.K (2013) *Plant Tissue Culture: An Introductory Text*. Springer, New Delhi.

Binet, M.N., Lemoine, M.C., Martin, C., Chambon, C. and Gianinazzi, S. (2007) Micropropagation of olive (*Olea europaea* L.) and application of mycorrhiza to improve plantlet establishment. *In Vitro Cellular and Developmental Biology – Plant* 43, 473–478.

Bongi, G. and Loreto, F. (1989) Gas-exchange properties of salt-stressed olive (*Olea europea* L.) leaves. *Plant Physiology* 90(4), 1408–1416.

Cançado, G.M.A., Setotaw, T.A. and Ferreira, J.L. (2013) Applications of biotechnology in olive. *African Journal of Biotechnology* 12(8), 767–779.

Cartechini, A. and Tombesi, A. (1986) Influenza del contenuto in carboidrati e in grassi sulla differenziazione delle gemme a fiore dell'olivo. *Rivista di Ortoflorofrutticoltura Italiana* 70(6), 287–295.

Cassells, A.C. and Doyle, B.M. (2005) Pathogen and biological contamination management: the road ahead. In: Loyola-Vargas, V.M. and Vázquez-Flota, F. (eds) *Plant Cell Culture Protocols*. Humana Press, New York, pp. 35–50.

Catalano, L. and Sonnoli, A. (2007) Le prospettive del vivaismo olivicolo: la parola ai vivaisti. *Proceedings of the National Congress 'Ricerca e trasferimento delle innovazioni tecnologiche del vivaismo olivicolo'*. Bari, Italy, 27 February 2007.

Catalano, L., Barba, M. and Benedettini, G. (2016) The qualification of nursery production: from regional programs to the National Voluntary Certification Service. *Georgofili* 13(2), 63–75.

Cimato, A. and Franchini E. (2002) In: Cimato, A. and Franchini E. (eds) *La fertilizzazione dell'oliveto*. ARSIA, Florence, Italy, pp. 28–36.

Cozza, R., Turco, D., Briccoli Bati, C. and Bitonti, M.B. (1997) Influence of growth medium on mineral composition and leaf histology in micropropagated microplants of *Olea europaea*. *Plant Cell Tissue and Organ Culture* 51, 215–233.

Dag, A., Erel, R., Ben-Gal, A., Zipori, I. and Yermiyahu, U. (2012) The effect of olive tree stock plant nutritional status on propagation rates. *HortScience* 47, 307–310.

De Boodt, M. and Verdonck, O. (1972) The physical properties of the substrates in horticulture. *Acta Horticulturae* 26, 37–44.

Debona, D., Rodrigues, F.A. and Datnoff, L.E. (2017) Silicon's role in abiotic and biotic plant stresses. *Annual Review of Phytopathology* 55, 85–107.

Fabbri, A., Bartolini, G., Lambardi, M. and Kailis, S.G. (2004) *Olive Propagation Manual*. CSIRO Publications, Canberra.
Faggioli, F., Roschetti, A., Luigi, M., Leo, G. and Albanese, G. (2017) Sanitary selection of olive (*Olea europaea* L.) cultivars for worldwide distribution. *Asian Journal of Plant Pathology* 11(1), 28–34.
George, E.F., Hall, M.A. and De Klerk, G. (2008) Plant growth regulators I: Introduction; auxins, their analogues and inhibitors. In: George, E.F., Hall, M.A. and De Klerk, G. (eds) *Plant Propagation by Tissue Culture*. Springer, Dordrecht, The Netheralands, pp. 175–204.
Gregoriu, K., Pontikis, K. and Vemmos, S. (2007) Effects of reduced irradiance on leaf morphology, photosynthetic capacity, and fruit yield in olive (*Olea europaea* L.). *Photosynthetica* 45(2), 172–181.
Guerriero, R., Scaramuzzi, F., Crescimanno, F.G. and Sottile I. (1972) Ricerche comparative tra olivi innestati ed autoradicati. Osservazioni nei primi anni di impianto. *Tecnica Agraria* 4.
Guerriero, R., Scaramuzzi, F., Crescimanno, F.G. and Sottile, I. (1974) Confronto tra diverse combinazioni di innesto dell'olivo su piante autoradicate. *Tecnica Agraria* 4.
Hartmann, H.T. (1958) Rootstock effects in the olive. *Proceedings of the American Society of Horticultural Science* 72, 242–251.
Hartmann, H.T., Kester, D.E., Davies Jr, F.T. and Geneve, R.L. (2010) *Plant Propagation: Principles and Practices*. 8th edn. Prentice-Hall, Englewood Cliffs, New Jersey.
Hausman, J.F., Kevers, C., Evers, D. and Gaspar, T. (1997) Conversion of putrescine to 1-aminobutyrric acid, an essential pathway for root formation by poplar shoots *in vitro*. In: Altman, A. and Waisel, Y. (eds) *Biology of Root Formation*. Plenum Press, New York, pp. 133–139.
Isfendiyaroglu, M., Özeker, E. and Başer, S. (2009) Rooting of 'Ayvalik' olive cuttings in different media. *Spanish Journal of Agricultural Research* 7(1), 165–172.
Jacoboni, N. and Fontanazza, G. (1976) Un nuovo tipo di propagazione dell'olivo: l'innesto-talea in cassone riscaldato alla base. *Italia Agricola* 133, 104–112.
Jiménez-Moreno, M.J., del Carmen Moreno-Márquez, M., Moreno-Alías, I., Rapoport, H. and Fernández-Escobar, R. (2018) Interaction between mycorrhization with *Glomus traradices* and phosphorus in nursery olive plants. *Scientia Horticulturae* 233, 249–255.
Kappel, F. and Flore, J.A. (1983) Effect of shade on photosynthesis, specific leaf weight, leaf chlorophyll content and morphology of young peach trees. *Journal of American Society for Horticulture Science* 108, 541–544.
Lambardi, M. and Rugini, E. (2003) Micropropagation of olive (*Olea europaea* L.). In: Jain, S.M. and Ishii, K. (eds) *Micropropagation of Woody Trees and Fruits*. Kluwer Academic Publishers, Springer, Dordrecht, The Netherlands, pp. 621–646.
Lambardi, M., Ozudogru, E.A. and Roncasaglia, R. (2013) *In vitro* propagation of olive (*Olea europaea* L.) by nodal segmentation of elongated shoots. In: Lambardi, M., Ozudogru, E.A. and Jain, S.M. (eds) *Protocols for Micropropagation of Selected Economically-Important Horticultural Plants*. Springer, New York, pp. 33–44.
Leva, A.R. (2011) Innovative protocol for *ex-vitro* rooting on olive micropropagation. *Central European Journal of Biology* 6, 352–358.
Leva, A.R., Sadeghi, H. and Petruccelli, R. (2013) Carbohydrates modulate the *in vitro* growth of olive microshoots. I. The analysis of shoot growth and branching patterns. *Journal of Plant Growth Regulation* 32, 53–60.
Lucchesini, M. and Vitagliano, C. (2002) Micro-environment effects in *in vitro* rooting of Frantoio and Leccino cultivars. Atti delle VI Giornate Scientifiche SOI, Italy.
Marin, L., Benlloch, M. and Fernández-Escobar, R. (1985) Screening of olive cultivars for salt tolerance. *Scientia Horticulturae* 64, 113–116.
Melgar, J.C., Benlloch, M. and Fernàndez-Escobar R. (2006) Calcium increases sodium exclusion in olive plants. *Scientia Horticulturae* 109(3), 303–305.
Melgar, J.C., Guidi, L., Remorini, D., Agati, G., Degl'Innocenti, E. *et al.* (2009) Antioxidant defences and oxidative damage in salt-treated olive plants under contrasting sunlight irradiance. *Tree Physiology* 29(9), 1187–1198.
Mencuccini, M. (2003) Effect of medium darkening on *in vitro* rooting capability and rooting seasonality of olive (*Olea europaea* L.) cultivars. *Scientia Horticulturae* 97, 129–139.
Micheli, M., Fernandes da Silva, D., Farinelli, D., Agate, G., Pio, R. and Famiani, F. (2018) Neem oil used as a 'complex mixture' to improve *in vitro* shoot proliferation in olive. *Hortscience* 53(4), 531–534.
Moradnezhad, M., Hosseini, R., Zarrabi, M.M. and Golmohammadi, F.G. (2017) A new approach for olive (Arbequina cv.) micropropagation: effect of dikegulac, light and carbon source. *International Journal of Horticultural Science and Technology* 4(1), 79–87.

Morini, S., Loreti, F. and Sciutti, R. (1990) Effect of light quality on rooting of 'Leccino' olive cutting. *Acta Horticulturae* (286), 73–76.

Murashige, T. and Skoog, F. (1962) A revised medium for rapid growth and bioassays with tobacco tissue cultures. *Physiology of Plants* 15, 473–497.

Nascimento-Silva, K., Roca-Castillo, L., Benlloch-González, M. and Fernández-Escobar, R. (2019) Silicon reduces the incidence of *Venturia oleaginea* (Castagne). Rossman & Crous in potted olive plants. *HortScience* 54, 1962–1966.

Paradiso, R. and De Pascale, S. (2008) Effects of coco fiber addition to perlite on growth and yield of cut gerbera. *Acta Horticulturae* 779, 529–534.

Petruccelli, R., Micheli, M., Proietti, P. and Ganino, T. (2012) Moltiplicazione dell'olivo e vivaismo olivicolo. *Italus Hortus* 19(1), 3–22.

Porras-Soriano, A., Soriano-Martin, M.L., Porras-Piedra, A. and Azcon, R. (2009) Arbuscular mycorrhizal fungi increased growth, nutrient uptake and tolerance to salinity in olive trees under nursery conditions. *Journal of Plant Physiology* 166, 1350–1359.

Proietti, P., Preziosi, P. and Tombesi, A. (1988) Influence of shading on olive leaf photosynthesis. In: *Proceedings of 2nd International Meeting on Mediterranean Tree Crops*, Crete, 2–4 November, pp. 334–342.

Rinaldi, L.M.R., Menabeni, D., Lambardi M. and Cimato, A. (1994) Changes in carbohydrates in olive seeds (*Olea europaea* L.) during fruit maturation and their correlation with germination. *Acta Horticulturae* 356, 58–61.

Roca, L.F., Romero, J., Bohórquez, J.M., Alcántara, E., Fernández-Escobar, R. and Trapero, A. (2018) Nitrogen status affects growth, chlorophyll content and infection by Fusicladium oleagineum in olive. *Crop Protection* 109, 80–85.

Rugini, E. (1984) In vitro propagation of some olive (*Olea europaea sativa* L.) cultivars with different rootability, and medium development using analytical data from developing shoots and embryos. *Scientia Horticulturae* 24, 123–134.

Rugini, E. and Fedeli, E. (1990) Olive (*Olea europaea* L.) as an oilseed crop. In: Bajaj. Y.P.S. (ed.) *Biotechnology in Agriculture and Forestry*. Springer, Berlin, pp. 593–641.

Rugini, E., Di Francesco, G., Muganu, M., Astolfi, S. and Caricato, G. (1997) The effect of polyamines and hydrogen peroxide of root formation in olive and the role of polyamines as an early marker for rooting ability. In: Altman, A. and Waisel, A. (eds) *Biology of Root Formation*. Plenum Press, New York, pp. 65–73.

Rugini, E., Biasi, R., Muganu, M., Pannelli, G., Aversa, G. *et al.* (2001) L'organizzazione di un moderno vivaismo olivicolo alla base della produzione di piante certificate. *Frutticoltura* 5, 11–24.

Rugini, E., Cristofori, V. and Silvestri, C. (2016) Genetic improvement of olive (*Olea europaea* L.) by conventional and *in vitro* biotechnology methods. *Biotechnology Advances* 34, 687–696.

Rugini, E., Baldoni, L., Silvestri, C., Mariotti, R., Narvaez, I. *et al.* (2020) *Olea europaea Olive*. In: Litz, R.E., Pliego-Alfaro, F. and Hormaza, J.I. (eds) *Biotechnology of Fruit and Nut Crops*. CAB International, Boston, Massachusetts, pp. 343–376.

Salisbury, F.B. and Ross C.W. (1978) Photomorphogenesis. *Plant Physiology*, 2nd edn. Wadsworth Publishing Co., Belmont, California, pp. 438–463.

Savvas, D. and Ntatsi, G. (2015) Biostimulant activity of silicon in horticulture. *Scientia Horticulturae* 196, 66–81.

Scaramuzzi, F. (1963) Osservazioni comparative sull'apparato radicale degli olivi in vivaio. *L'Italia Agricola* n. 12.

Sghir, S., Chatelet, P., Ouazzani, N., Dosba, F. and Belkoura, I. (2005) Micropropagation of eight Moroccan and French olive cultivars. *HortScience* 40, 193–196.

Tattini, M. and Traversi, M.L. (2009) On the mechanism of salt tolerance in olive (*Olea europea* L.) under low or high Ca+ supply. *Environmental and Experimental Botany* 65(1), 72–81.

Vidoy-Mercado, I., Imbroda-Solano, I., Barceló-Muñoz, A. and Pliego-Alfaro, F. (2012) Differential *in vitro* behavior of the Spanish olive (*Olea europaea* L.) cultivars 'Arbequina' and 'Picual'. *Acta Horticulturae* 949, 27–30.

Zaccheo, P. and Cattivello, C. (2009) *I substrati di coltivazione. Aspetti teorici ed applicativi di un fattore chiave delle produzioni in contenitore.* Edagricole-New Business Media, Bologna, Italy.

5.1 Planting Models for New Olive Growing

T. Caruso[1], R. Lo Bianco[1], L. Ferguson[2], G. Marino[2] and F. Famiani[3]
[1]*Dipartimento di Scienze Agrarie, Alimentari e Forestali, Università degli Studi di Palermo, Palermo, Italy;* [2]*Department of Plant Sciences, University of California, Davis, California, USA;* [3]*Dipartimento di Scienze Agrarie, Alimentari e Ambientali, Università degli Studi di Perugia, Perugia, Italy*

5.1.1 Current Olive Growing and New Objectives

In many areas of the Mediterranean basin, olive growing has a very long tradition, is located in marginal sites, and primarily consists of medium- to large-size trees, managed with a high use of labor. Traditional olive orchards are usually trained to globe system or to vase (which has regional variations). The different situations can be: low planting density, from 17–100 trees/ha in environments with low water availability to 200–300 trees/ha in areas with relatively high rainfall and a short growing season; irregularly spaced trees; highly sloping land; the presence of terraces and lunettes; trees with phytosanitary problems (Figs 5.1.1–5.1.6) (Famiani and Gucci, 2011; Rallo *et al.*, 2013; Famiani, 2019; Lo Bianco *et al.*, 2021). These conditions lead to low and alternating production, difficulties in using machines (low level of mechanization) and high management costs (especially for harvesting), resulting in olive groves that generate marginal profits.

For a more complete description of traditional olive groves, it is important to consider that, in some cases, they are multi-product systems, i.e. olive is associated with other crops or livestock productions (Fig. 5.1.7). In addition to the productive function, they may also perform other roles, such as landscaping, carbon sequestration, biodiversity, recreational, historical, monumental and/or environmental (Proietti *et al.*, 2014, 2017; Regni *et al.*, 2017; Famiani, 2021). This means that traditional olive groves are often multi-functional agricultural systems (Figs 5.1.8, 5.1.9). For further details, this topic is covered in Chapter 9.

The structural conditions of traditional groves that limit productivity and mechanization, along with the increasing worldwide consumption of olive oil (Famiani, 2021), are pushing for a renewal of olive orchards and an increase in cultivated areas. For this to happen, new plantings that meet the requirements of modern olive growing need to be established, i.e. rapid and abundant fruiting and large-scale use of mechanization. Therefore, it is very important to shorten, as much as possible, the time required for trees to reach full production. In this regard, planting systems with a high number of small trees per unit area are more efficient than those based on a few large trees. Moreover, new plantings should ensure high/full mechanization of cultural practices, particularly fruit harvesting

DOI: 10.1079/9781789247350.0014

Fig. 5.1.1. Traditional rainfed olive orchards in southern Tunisia, with trees spaced 24 m x 24 m (17 trees/ha). Trees are widely spaced because of water shortage (rainfall 100–250 mm/year). (Photo: Franco Famiani)

Fig. 5.1.2. Traditional rainfed olive orchard in northern Tunisia with 100–150 trees/ha (rainfall 250–400 mm/year). (Photo: Franco Famiani)

Fig. 5.1.3. Traditional rainfed olive orchard in central Italy (rainfall 700–1000 mm/year), comprised of irregularly spaced, medium-sized trees (200–300 trees/ha). (Photo: Franco Famiani)

Fig. 5.1.4. Traditional rainfed olive orchard in southern Italy with interventions to improve the productive potential: increase in planting density by adding trees and applying irrigation. (Photo: Franco Famiani)

Fig. 5.1.5. Traditional rainfed olive orchard in Lebanon on highly sloping land with terraces. (Photo: Franco Famiani)

Fig. 5.1.6. Traditional rainfed olive orchard in central Italy on highly sloping land with terraces and lunettes. (Photo: Franco Famiani)

Fig. 5.1.7. Multi-product, traditional rainfed olive orchard in central Italy: olive cultivation associated with sheep grazing. (Photo: Franco Famiani)

Fig. 5.1.8. Multi-functional olive orchards. Besides the productive function, olives have a landscape importance: view of part of the belt of olive orchards from Assisi to Spoleto in central Italy, candidate for a UNESCO heritage site. (Photo: Franco Famiani)

Fig. 5.1.9. Multi-functional olive orchards. Besides the productive function, olives may have historical/monumental importance: large, old trees in southern Italy. (Photo: Franco Famiani)

(Famiani and Gucci, 2022). Among other cultural practices, the simplification of pruning (execution from the ground with hand-held devices) and/or its mechanization is essential (Famiani *et al.*, 2009; Connor *et al.*, 2014). In some situations, in establishing new olive orchards, it is very important to use cultivars, which, in the environment considered, enable unique and distinctive quality attributes to be obtained for both table and oil olives. This will greatly contribute to the enhancement of the product value. The high environmental heterogeneity has been a major factor in the spread of numerous and different cultivars in the various olive-growing countries. The enhancement of the olives/oils value through product differentiation, is, in turn, favored by the different and numerous environment/variety combinations, even within individual countries (Famiani and Gucci, 2011). In some countries, it is possible to certificate the link between the territory/environment and the quality of olives and oils (i.e. use of trademark certification, such as Protected Designation of Origin (PDO) and Protected Geographic Indication (PGI), which implies the use of specific varieties. Finally, new modern plantings (selected environments, cultivars and planting systems) must ensure low environmental impact and take into consideration the climatic change. A good choice of cultivation sites and varieties is at the basis of organic growing (environments less favorable to the main adversities and use of tolerant varieties) and can allow for better adaptation to conditions determined by climate change.

At present, the reference planting models are intensive olive orchards with planting densities ranging from 200 to 1000 trees/ha and super-high-density (SHD) olive orchards, with planting densities higher than 1000 trees/ha (Lo Bianco *et al.*, 2021).

Herewith, information is presented on the requirements and technical aspects related to the establishment and management of intensive and SHD olive orchards. The main advantages and disadvantages are also presented for the purpose of aiding growers in selecting the best option based on the environmental conditions where the orchard is to be established, farm size, availability of labor, the technical level that can be ensured, and commercial targets.

5.1.2 Environmental Factors and Planting Sustainability

Numerous studies have shown that the amount of radiant energy intercepted by the canopy is greater in small than in large trees, as small ones have a higher ratio of exposed leaf surface to canopy volume than large ones (Russell *et al.*, 1989; Mariscal *et al.*, 2000; Villalobos *et al.*, 2006; Connor *et al.*, 2014). Hence, the amount of light intercepted by many small trees is greater than that of a few large ones (Poorter *et al.*, 2006). In addition, light penetrates and is distributed more evenly within the canopy of small trees; conversely, in large trees, there is a marked difference in light intensity between the periphery and the internal parts of the canopy. Small trees have a greater proportion of photosynthetically active organs (especially leaves) and fruiting wood (one- and two-year-old shoots and branches). Indeed, in large trees, the frame mainly consists of woody parts, such as stems and branches of different ages (Poorter *et al.*, 2006; Niinemets, 2010). Although they serve as permanent supporting, transporting and reserve structures, they are not metabolically productive because they primarily consume carbon by respiration and by assimilate immobilization (Connor *et al.*, 2011). Respiration and carbon immobilization by old wood inevitably withdraws photoassimilates from the growth of new roots and, above all, from the growth of the reproductive organs: flower buds, inflorescences and fruits (Turktas *et al.*, 2013).

Furthermore, small-sized trees can be set out in the field at closer distances and, therefore, it is possible to cover a larger portion of the allotted space in a shorter time (Musacchi *et al.*, 2021). Thus, the overall light interception per unit of soil surface by the olive orchard in the first years of planting is greater with small than with large trees, producing positive implications on the photosynthetic efficiency of the planting system (Connor *et al.*, 2014). Under low light intensities (less than 30% of full radiation), flower bud induction and morphogenesis are reduced or do not take place at all, resulting in little or no flower formation (Sadok *et al.*, 2013). In low light conditions, even oil accumulation and synthesis of polyphenols, tocopherols, sterols and volatile compounds do not proceed regularly, negatively affecting the quantity and quality of the final product (Conde *et al.*, 2008).

Based on these considerations of radiation interception, small trees are the most efficient in high-density plantings, while medium-size ones in intensive low-density plantings (densities of 200–400 trees/ha) if canopies are structured to favor light interception. This is the case of the vase training system, where the empty center and the partitioning of vegetation among the primary branches enables good light interception (Famiani *et al.*, 2009).

As far as other environmental factors are concerned, the slope of the land is an important factor for mechanization. For intensive orchards harvested with trunk shakers, there are no problems up to slopes of 20–30% (especially with tracked tractors) (Fig. 5.1.10). New straddle machines for SHD orchards are able to operate up to slopes of 20% (Fig. 5.1.11) (Famiani and Gucci, 2022).

Water availability is another important factor. Small trees with a proportionally smaller root system adapt less to conditions of limited water availability compared to large trees (Marino *et al.*, 2016).

The economic lifespans of intensive and SHD orchards are > 40 years and > 20 years, respectively (Rallo *et al.*, 2013). This creates differences in the economical sustainability

Fig. 5.1.10. In intensive olive orchards on sloping lands, trunk shakers can be used up to slopes of 30%, especially when the shaker is mounted on tracked tractors. (Photo: Franco Famiani)

in case of severe adversities, such as freezing temperatures that could damage the whole aerial part of the trees. The economical sustainability is usually greater when the economic lifespan is longer, because the negative effects of adversities on yields and costs can be spread over more years.

5.1.3. Planting Models and Cultivar Characteristics

Management practices have been tailored to the agronomic features of cultivars that have played a fundamental role in the development of traditional olive growing in the various olive-growing areas. Without irrigation, fertilizers and pesticides, and considering that, in some cases, other crops were associated with olive, often the most vigorous and productive cultivars were preferred for their greater ability to grow and produce in these conditions. These cultivars, favored by the extraordinary longevity of the olive, are still used today and there is a tendency to enhance their typicality, along with their growing models. We could say that traditional olive growing has been 'cultivar-centric', i.e. it has primarily focused on cultivars able to give good productive results in a specific environment despite the limited cultural practices that were used in the past (Lo Bianco *et al.*, 2021). Current olive growing, as well as the growing of nearly all other agricultural crops in industrialized countries, demands an integral or a high level of mechanization of management practices, pushing to a 'mechano-centric' approach (Salomone and Ioppolo, 2012; Connor *et al.*, 2014; Díez *et al.*, 2016; Lo Bianco *et al.*, 2021).

Unfortunately, the old cultivars, which are considered family assets and were selected as described above, rarely adapt

Fig. 5.1.11. In SHD orchards, straddle harvesters can be used up to slopes of 20%. (Photo: Franco Famiani)

well to SHD planting systems that include full mechanization. Of the 600 olive cultivars certified and listed in the World Bank of Olive Germplasm established in Spain (Trujillo *et al.*, 2014; Belaj *et al.*, 2020), only three or four are suitable for SHD plantings.

While waiting for genetic improvement programs aimed at obtaining cultivars suitable for new growing models, and considering the recent world climate changes, it is essential to evaluate the germplasm pool (including local cultivars) in the different olive-growing areas. Cultivars should be selected on the basis of the management needs of the modern planting systems. In particular, they should be selected according to the management criteria of 'pedestrian systems' (managed from the ground), i.e. based on low-vigor trees, which are more eco-physiologically efficient, easier to prune and harvest mechanically, and easier to protect against diseases and pests (Lo Bianco *et al.*, 2021).

A great improvement in olive growing will take place if the evaluation of the existing olive germplasm and the implementation of genetic improvement programs increase the number of varieties suitable for pedestrian olive orchards. This should also enable full mechanization (beneficial for low-price competitiveness) and the pursuit of other production objectives, such as typical oil and olive characteristics, tolerance to adversities and self-compatibility, ultimately leading to economic, environmental and social sustainability. Furthermore, the great opportunities offered by pedestrian olive systems would also benefit the table-olive industry. Indeed, table olives, in several cases, must be harvested manually and delicately placed in baskets to avoid skin and/or pulp damage that could reduce quality during processing and preservation (Pérez-Ruiz *et al.*, 2018). Therefore, the possibility of harvesting from the ground without ladders would greatly improve harvesting efficiency, especially for green olives with high commercial value.

In the selection of existing varieties and the production of new ones, the numerous possibilities to enhance the products and byproducts of the olive/oil supply chain in the cosmetics and energy sectors must also be considered (Rodrigues *et al.*, 2015).

The cultivars that are most prevalent in the different olive-growing areas, besides being well adapted to the specific environment, often have a high tolerance to specific abiotic/biotic adversities. Furthermore, frequently, they give high-quality olives and oils, which can also be certified for their typicality (i.e. PDO and PGI), and they have features that allow for efficient mechanized harvesting with trunk shakers. However, as mentioned above, they often have a medium/high vigor, which is not suitable for high planting densities. In these situations, the particular features of these cultivars can be exploited with the intensive model, with densities of 200–400 trees/ha. Also in this case, the evolution in the management of this kind of orchard has brought to the development of a 'pedestrian' model, where all cultural practices which directly regard the trees (harvesting and pruning) can be done from the ground (see chapters 6.4 and 6.5).

5.1.4. Planting Systems

5.1.4.1 Intensive systems

The planting density in intensive olive orchards varies from 200 to 1000 trees/ha, with trees arranged in a square or rectangle, depending on the planting density and the training system. The 'volume' or three-dimensional (3D) training forms are the most used (Scaramuzzi, 2006). Several studies have also demonstrated the validity of two-dimensional (2D) training systems, also called hedgerow or 'planar' olive plantings (Russo *et al.*, 2015; Díez *et al.*, 2016; Mairech *et al.*, 2020; Gurrero-Casado *et al.*, 2021), which create a continuous canopy 'wall' along the row. There are three types of intensive systems, based on planting density: low, medium and high.

Low-density intensive systems

In these systems, trees are arranged in 5–7 m × 5–7 m square/rectangular spacings (200–400 trees/ha) and are trained to 3D form, particularly according to the vase form (Godini and Bellomo, 2002; Sola-Guirado *et al.*, 2014; Castillo-Ruiz *et al.*, 2017; Famiani, 2021). The choice of the distances depends on the vigor of the cultivar (larger for vigorous cultivars), amounts of rainfall, soil fertility and cultural practices, with particular regard to irrigation (larger when rainfall is low and in rainfed conditions) (Figs 5.1.12–5.1.14).

For low-density plantings, two-year-old trees are the best for establishing new plantings. They should have a single trunk 1–1.2 m high, with several lateral shoots in the apical portion (Fig. 5.1.15). However, one-year-old trees can also be used, but a longer time is needed to reach the adult stage/full production and they require more careful training soon after planting (Fig. 5.1.16).

The economic lifetime of intensive plantings is greater than 40 years; that is, as long as the trees respond well to cultural practices and a large amount of wood has not accumulated. Up to this age, the trees also respond well to the use of trunk shakers for harvesting. The orchard economic life differs greatly from the biological life of olive trees, which can be much longer.

In low-density intensive olive orchards, the trees begin to produce significantly after about four years, reaching full production after six to ten years, with 4–12 tons of olives/ha, depending on the environment (especially rainfall amounts) and the cultivation practices (especially irrigation). The main problem with this planting system is that the trees take six to ten years to occupy

Fig. 5.1.12. Young intensive low-density olive orchard in central Italy (rainfall 700–1000 mm/year), with drip irrigation and trees spaced 5 m × 5 m. (Photo: Franco Famiani)

Fig. 5.1.13. Young intensive low-density olive orchard in southern Tunisia (rainfall 100–250 mm/year), with drip irrigation and relatively wide tree spacing. (Photo: Franco Famiani)

Fig. 5.1.14. Adult intensive low-density olive orchard in central Italy. (Photo: Franco Famiani)

Fig. 5.1.15. Newly established intensive low-density olive orchard in central Italy, with two-year-old trees, drip irrigation and 6 m × 5 m spacing. (Photo: Franco Famiani)

Fig. 5.1.16. Newly established intensive low-density olive orchard in central Italy, with one-year-old trees and 6 m × 6 m spacing. (Photo: Franco Famiani)

all the space available and reach full production, and for the trunks to reach the size suitable for trunk shakers for olive harvesting. Therefore, it is important to reduce this time, by accelerating the growth of trees in the first years after planting. Recently, good results have been obtained by deflowering young trees to avoid fruiting and hence to concentrate the use of assimilates toward tree growth (Famiani *et al.*, 2022). Moreover, an increase in growth has also been obtained by adding biostimulants to fertigation (Almadi *et al.*, 2020) or using an organic fertilizer also having a biostimulant action (Mazeh *et al.*, 2021). Further investigations would be very useful to set up protocols which produce synergic effects through the combination of deflowering, fertigation and biostimulants. An important role on tree growth is also exerted by pruning, which must be minimized, by limiting it to the elimination of vegetative formations which, if not removed, would damage the tree structure. The use of these techniques could also be useful to speed up initial tree growth in SHD orchards (first and second year).

All cultivars can be used for the establishment of intensive olive orchards. This allows for the pursuit of all possible production objectives chosen by the farm, including those aimed at the enhancement of the product value. This can be done by obtaining high quality levels and differentiating production by exploiting the numerous environment/cultivar combinations. In fact, intensive olive orchards, through an appropriate choice of cultivars, allow the following oils to be obtained: (i) certified with PDO or PGI; (ii) those with particular compositional and sensory attributes (typical oils) and/or health benefits (e.g. high content in antioxidant substances), which can be monovarietal or declared mixtures of different varieties ('blend'); (iii) organic, in suitable environments and choosing the most resistant varieties to abiotic and biotic adversities (in particular to the olive fly and peacock eye); (iv) those of standard quality aimed at maximizing oil production, through an appropriate choice of the harvesting period. It is therefore a system that favors flexibility (Famiani and Gucci, 2011).

Considering that most of the cultivars are self-sterile, when choosing the variety, it is important to plan for pollinizers, unless two to three interfertile varieties are used as main cultivars. In order to have good pollination, if trees are uniformly distributed as single trees, 10–15% of pollinizers is a good reference. If they are planted out in rows the incidence should be 20–25% (one row every four or five rows).

In low-density intensive plantings, vase is the best training form. Usually, the free-vase version is used (for details on this training form, see the chapter 6.4) because it allows for rapid growth and fruiting of the trees, good lighting and ventilation of the foliage (also useful for discouraging pathogen and pest attacks), and maximizes the efficiency of mechanical harvesting with trunk shakers (Famiani and Gucci, 2011, 2022). Indeed, structuring the trees to a single trunk 1 m high and with four to five primary branches with an angle of insertion on the trunk not too open, an overall height of 4–5 m, relatively short (rigid) and numerous secondary branches, and without long pendants, allows for high harvesting efficiency: percentage of detached fruits with respect to the total > 85–90% (Famiani and Gucci, 2022). Furthermore, this vase structure (especially if less than 4.5 m in height) allows the use of facilitating equipment for pruning, i.e. pruning from the ground with tools mounted on rods.

Intensive low-density orchards can be established on both flat and sloping areas (up to 30% slope). They require good water availability, but can also give good results in rainfed conditions. It is important to adjust the canopy volume and density in relation to water availability. In rainfed conditions, with rainfall of 500–600 mm/year, especially on light soil, the canopy volume should not be greater than 8000–10,000 m^3/ha, while in irrigated orchards, it can reach 12,000–13,000 m^3/ha, which, in the best conditions, can be increased up to 15,000 m^3/ha. However, in intensive olive orchards, the application of irrigation is always advisable, especially in areas with limited rainfall. The methods to be used are the drip system or the microsprinkler system. When fertigation is used in young orchards where the trees still have a small root system, the use of lines with the emitters placed near the trees (new ones will be added as the trees grow), instead of continuous drip lines, enables efficient use of water and nutrients and other compounds. This is particularly important considering the climate-change scenario, which demands efficient use of these resources, particularly water. The drip lines can be suspended at a height of 0.6–0.8 m when the trees are young and not able to be harvested mechanically. Later, they can be laid on the ground (managing the soil along the row with weeding), buried (sub-irrigation) or lifted to a height that does not cause problems to the trunk shaker, even when an inverted umbrella is used. Irrigation is especially important in young orchards, as it is one of the most effective factors in increasing tree growth.

If rainfall is greater than 600–700 mm or if irrigation is applied, soil management by grassing is recommended. Along with the advantages to soil fertility (increases organic matter, improves availability of nutrients, etc.) and environmental aspects (reduces erosion to the point of eliminating it, favors a better balance between useful and harmful organisms etc.), the presence of a grass cover also improves the bearing capacity of the soil and therefore facilitates mechanical harvesting, especially if rains occur and if treatments against pests and diseases are needed (Famiani and Gucci, 2011). Furthermore, it also makes pruning easier, when carried out with facilitating equipment or, possibly, with mechanical pruners. Under low-rainfall conditions, temporary grassing can be used by covering the soil with grass from autumn to the beginning of spring and tilling it in the spring/summer period. In intensive olive groves, with a rectangular arrangement of trees in rows, it is possible to practice integrated management, differentiating the management of the row from that of the inter-row. The use of green covers in intensive plantings is always very useful to improve both environmental and economic sustainability (simplification/mechanization of harvesting and pruning). In arid environments, tillage is usually applied, but in these conditions, rainfall and related difficulties for mechanization are not frequent during the harvesting and pruning periods.

Fertilization, especially with nitrogen, is one of the most important factors for accelerating initial growth in young trees. In adult trees, along with adequate pruning and irrigation, fertilization is critical for obtaining balanced trees that are more easily and quickly manageable with pruning and less prone to alternate fruit bearing and disease/pest attacks. For intensive cultivations, indications in different countries are available for proper fertilization protocols which often include Integrated Production Regulations. This helps to specify schemes for fertilizer application that allows the impact on the environment to be minimized, while ensuring optimal nutrition to the trees.

Harvesting of low-density intensive olive orchards is carried out mechanically with self-propelled or towed trunk shakers, which can also be equipped with an inverted umbrella to collect the detached fruit, enabling complete harvest mechanization (Fig. 5.1.17). For a good efficiency of the machines, the trunk must have a diameter between 10–15 cm and 40 cm at the gripping point of the shaker head and the canopy volume of each tree should be less than 50 m^3 (Famiani and Gucci, 2022).

Medium-density intensive systems

Two training systems are used: vase and monocone (Fontanazza, 1984; Palliotti et al., 1997; Famiani *et al.*, 2009; Massenti *et al.*, 2022). The vase form can be used in orchards of up to 500 trees/ha with cultivars of modest vigor in poor, dry soils. Trees are trained to three main branches

Fig. 5.1.17. Self-propelled trunk shaker with an inverted umbrella for collecting the detached fruit. (Photo: Franco Famiani)

with a narrow angle with respect to the trunk to avoid excessive opening of the canopy (Pannelli and Alfei, 2019). The monocone training system has a central axis, generally about 4–4.5 m high, with primary branches starting at about 1 m from the ground and with decreasing length from the bottom to the top of the canopy (Palliotti *et al.*, 1997; Famiani *et al.*, 2009; Pannelli and Alfei, 2019). The medium-density system is used with planting distances of 5–6 m between rows and 2.5–3.0 m within the row (planting density of 550–800 trees/ha), mostly up to about 650 trees/ha. Spacing varies in relation to water availability, cultivar vigor and length of the growing season. The monocone was designed for establishing continuous hedgerows by arranging the trees in the field according to a rectangle (Fontanazza, 1984; Fig. 5.1.18). Harvesting can be done with 'side-by-side' trunk shakers (Fig. 5.1.19; Fridley *et al.*, 1971; Ferguson, 2006).

Pruning can be partially mechanized with disc saws mounted on mobile bars, operated by a tractor, and manual finishing to balance the structure of the canopy, particularly on the top of the tree (Lodolini *et al.*, 2018).

The monocone training system has been mainly used in Italy and also tried in other countries, but it has not spread. This is likely because, as the canopy grows, there is progressive shading of the internal parts and the vegetation tends to grow laterally and in height. This makes resizing necessary, which imbalances the trees, leading to excessive vegetative activity.

Wide-hedgerow orchards

These are plantings with 250–450 trees/ha with a high level of irrigation and a long growing season in subtropical areas and plantings with 450–800 trees/ha and 3D

Fig. 5.1.18. Intensive medium-density olive orchard in central Italy, with trees trained to monocone and spaced 6 m × 3 m. (Photo: Franco Famiani)

Fig. 5.1.19. Side-by-side trunk shaker. (Photo: Franco Famiani)

canopies that form a wide hedgerow with their canopies (Fig. 5.1.20). These are especially found in new areas of olive cultivation in the southern hemisphere and are not easy to manage in the long term because of reciprocal shading (Rallo *et al.*, 2013). Therefore, to limit these problems, it is beneficial to choose low-vigor and early-bearing cultivars and control the growth by applying deficit irrigation and proper pruning and fertilization (Rallo *et al.*, 2013). These orchards are usually harvested by using a large straddle harvester (Colossus) or side-by-side trunk shakers (Rallo *et al.*, 2013). Pruning, especially on large farms, is executed mechanically or with a combination of mechanical and manual operations (Fig. 5.1.20).

High-density intensive systems

The potential of high planting densities (650–1000 trees/ha) has only been slightly explored, but the advent of SHD planting systems at the beginning of the current millennium prompted some research groups to evaluate its potential (Tous *et al.*, 1999). Therefore, cultivars native to the various olive-growing areas and having low to medium vigor might be re-evaluated under these planting densities. The reference planting distance is 5 m × 2–2.5 m (in few cases 5 m × 3 m) (Massenti *et al.*, 2022). The proposed training system is the 'free palmette' to form continuous hedgerows, 3–4 m high and about 1.5–2.5 m thick depending on the vigor of the cultivar and type of harvesting machine: trunk shakers with side-by-side frames or, better, straddle harvesters (Figs. 5.1.21 and 5.1.22; Moutier *et al.*, 2010). New straddle machines have been produced that are able to harvest trees up to 4.0–4.5 m in height with canopies up to 2.0–2.5 m thick, and this increases the use of this kind of harvester greatly (Fig. 5.1.23;

Fig. 5.1.20. Wide-hedgerow orchard where mechanical pruning is applied. (Photo: Pierluigi Pierantozzi)

Fernández-Escobar *et al.*, 2013; Famiani and Gucci, 2022). Moreover, continuous lateral canopy shakers have been produced, which can harvest one side or, if operated in pairs, both sides of a tree row, and can work up to a height of 5 m (Famiani and Gucci, 2022).

The free palmette training form requires a supporting structure (trellis) with poles and wires (2–3, depending on the final tree height). To obtain the free palmette, the main axis of the young plant, which should be at least 0.8 m (better, 1.0–1.2 m) high at planting, is tied to a stake. Then, it must be supported by adding ties as it grows, generally every 0.3–0.4 m. If damaged, it must be replaced with an alternative shoot and any competing branches must be removed. At the same time, the lower shoots are progressively removed up to a height of about 0.8 m from the ground. The lateral branches growing on the main axis are initially left to grow freely and only the most vigorous ones growing in the inter-row direction are removed. Those growing along the row are trained to form a vegetation wall. Usually six to eight branches (3–4 pairs irregularly spaced) form the main structure of the free palmette. Production pruning must maintain the canopy structure, the correct vegetation density and ensure the renovation of aged branches. Pruning can be partially mechanized, by doing mechanical topping in the summer and manual hedging in the winter.

Some recent studies have shown positive results with this system, both for the high yields that can be achieved and for the number of varieties that can be used at densities of 650–1000 trees/ha (Lo Bianco *et al.*, 2021). In fact, the larger planting distances with respect to those used for SHD plantings (> 1000 trees) significantly increase the number of cultivars that can adapt to this system.

In Italy, an autochthonous cultivar planted at a density of 1000 trees/ha, irrigated with about 1300 m^3 water/ha, gave cumulated fruit yields of approximately 100 t/ha in the first seven years from planting (Marino *et al.*, 2019). After more than 15 years of trials using a large pool of genotypes with different vigor, vegetative and fruiting habits, the results obtained with this new high-density pedestrian system are definitely interesting. Therefore, this growing system can be considered ready for testing in the various olive-growing areas of the

Fig. 5.1.21. Tree trained to free palmette. (Photo: Franco Famiani)

world, where local cultivars of high agronomic value might be re-evaluated (Marino *et al.*, 2017, 2019).

Overall, this high-density intensive system may represent a further possibility for the establishment of new olive orchards. Its strength lies in the fact that, with respect to SHD, a large number of cultivars can adapt to free-palmette hedgerow training. Therefore, this system might be a good compromise between mechanized management and the possibility of producing high amounts of oil per hectare with unique flavors that can be linked to a specific territory. The latter is possible thanks to the greater possibility of using selected combinations of cultivars and environments (Massenti *et al.*, 2022). This would also enable the social and economic benefits that derive from the uniqueness and typicality of an olive product to be obtained. As for cultivars having bio-reproductive traits in line with intensive HD pedestrian systems, it is important to draw attention to FS-17, recently renamed 'Favolosa'®, which, together with 'Leccino', is the only cultivar tolerant to *Xylella fastidiosa* subsp. *Pauca* (Hanani *et al.*, 2021).

Super-high-density systems

The SHD orchard system (1200–2225 trees/ha) was developed in Spain in the 1990s.

Fig. 5.1.22. Row of adult trees trained to free palmette. (Photo: Franco Famiani)

Fig. 5.1.23. Straddle harvester with a relatively large harvesting tunnel. (Photo: Franco Famiani)

The objective was to facilitate mechanical harvesting with over-the-row straddle harvesters (Tous *et al.*, 2011). The trees are planted at 3.5–4 m between rows and 1–2 m apart in the row to quickly fill the space between trees and create an uninterrupted fruit-bearing wall for continuous harvesting (Connor *et al.*, 2014). Hence, the management unit in the SHD hedgerow system is not the single tree but the hedgerow (Figs 5.1.24 and 5.1.25).

The primary advantage of the hedgerow system is the economic viability, due to the high productivity per surface area and the possibility of mechanizing harvesting, the greatest expense in olive production. A mature SHD olive orchard can produce 10–14 t of olives per ha, and the straddle harvester can harvest one hectare in 2–2.5 hours (Tous *et al.*, 2012). As a comparison, a trunk shaker with reverse umbrella takes 6–14 hours to harvest a single hectare (Tous *et al.*, 2012; Famiani and Gucci, 2022).

An additional advantage of the SHD system is the early entry into production. The first yield is two years after planting (YAP). Full production can be achieved as early as three to five YAP.

The high profitability of the SHD system is very appealing, particularly in countries new to olive production. In fact, the USA, Argentina, Chile and Australia have all developed olive industries based on HD or SHD systems. In contrast, countries with existing olive oil industries have been slow to adopt the SHD. The primary reason for this low adoption rate is that their cultivars are too vigorous to be grown in a SHD system and harvested with a straddle harvester. Low vigor and early high productivity are the key cultivar traits for SHD systems. The best cultivars for the SHD system have

Fig. 5.1.24. SDH hedgerow olive orchard in California: harvesting with an over-the-row straddle harvester. (Photo: Giulia Marino)

Fig. 5.1.25. SDH hedgerow olive orchard in central Italy. (Photo: Franco Famiani)

highly dissected and thin branches (Rosati *et al.*, 2013). In other crops commonly grown in the SHD hedgerow system, for example apple and cherry, dwarfing rootstocks are used to control scion vigor. Growers can then increase planting density with their favored scion cultivars. However, despite intensive research and some promising results, there are no commercially available olive rootstocks that control scion cultivar vigor (Romero *et al.*, 2014; Rugini and Pace 2016; Díaz-Rueda *et al.*, 2020; Torres-Sánchez *et al.*, 2022). Currently, three own-rooted olive cultivars, 'Arbequina', 'Arbosana' and 'Koroneiki' dominate global SHD olive production. A survey by California's UC Davis Olive Center reported 4600 hectares of SHD olives planted in three years, 2005–2008. Over 78% was 'Arbequina', 16% 'Arbosana' and 6% 'Koroneiki'.

A consequence of the widespread adoption of the SHD cultivars is the loss of local cultivars that could potentially produce a more diverse olive oil industry with differing sensory and phenolic profiles. For example, a trial in Italy demonstrated that the 'international' cultivars 'Arbequina', 'Arbosana' and 'Koroneiki' produced oils with lower phenolic contents than the local cultivars (Fig. 5.1.26; Marino *et al.*, 2017). Researchers and breeders have worked on selecting and developing other cultivars for SHD, and more options are now available. Some are traditional cultivars, like 'Maurino' (Tombesi *et al.*, 2011) and 'Calatina' (Marino *et al.*, 2017), selected for their performance in SHD systems. Other cultivars have been developed as crosses between local cultivars and international cultivars like 'Lecciana', a cross between 'Arbosana' and 'Leccino' (Camposeo *et al.*, 2021) and 'Sikitita', a cross between 'Picual' and 'Arbequina' (Rallo *et al.*, 2008). Other new cultivars are hybrids between 'Arbequina' and 'Arbosana', such as 'Oliana' ('Agromillora').

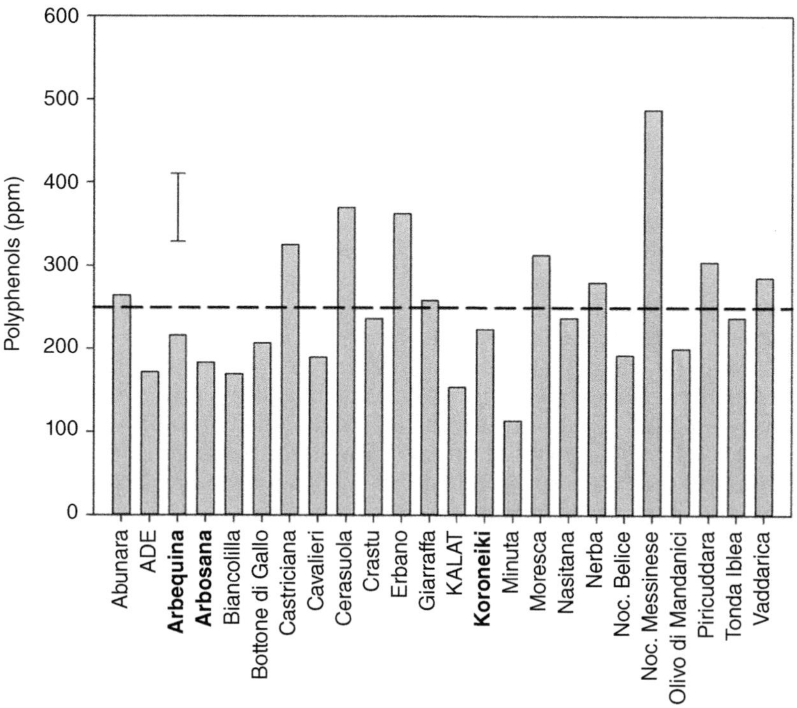

Fig. 5.1.26. Total phenol content of the three international cultivars (in bold) in comparison to 21 Sicilian cultivars. Horizontal line shows the average values of all the cultivars. (From Marino *et al.*, 2017)

SHD orchards are expensive to establish and require higher irrigation, fertilizer and pest-management inputs than more spaced orchards. For this reason, SHD orchards are generally planted in the best soils. The high tree density and the small soil volume explored by the root system make the trees more susceptible to water stress. For example, trees in a rainfed low-density olive orchard maintained the same water status as trees in an adjoining HD olive orchard that was irrigated weekly (Marino *et al.*, 2016). In an agricultural framework threatened by climate change, particularly drought, the SHD system could potentially be less sustainable than lower-density systems.

The SHD system has a shorter productive life than lower-density orchards. The lower canopy light penetration due to shading has been suggested as a factor limiting long-term productivity (Pastor *et al.*, 2007). Decreased light penetration could result from excessive vigor in response to over-irrigation and -fertilization. An equilibrium between orchard density, irrigation, fertilizer inputs and productivity, growth and longevity is the site-specific goal which can allow productivity to be maintained over a longer time. For example, Díez *et al.* (2016) measured yields for 14 years after planting in an experimental SHD orchard and found no yield reduction for the cultivars 'Arbequina', 'Arbosana' and 'Koroneiki', averaging 12–14 t per ha. For the cultivar 'Arbequina', the yield actually increased from 16.1 t ha^{-1} to 29.9 t ha^{-1} as tree density increased from 780 trees per ha to 2254 trees per ha. Interestingly, the increased productivity was correlated to shorter distances between rows more than to distance between trees in the row.

The SHD system was developed for over-the-row mechanical harvesting. This requirement determines the planting spacing and training. Connor *et al.* (2014) specified the structural parameters of a SHD system as follows: row height, row distance, canopy width and depth, alley width, canopy slope and the height above ground level, maintained foliage-free, also called the skirt, as seen in Fig. 5.1.27.

At planting, the first decision is the distance between rows. This is generally 3.5–4 m for the size of the mechanical harvester. The tree height, canopy depth plus skirt height is generally 2.5–3.5 m to accommodate the harvester. The 0.6–0.9 m trunk depth (height) below the canopy skirt must be pruned to remove shoots, allowing the straddle harvester plates to close in about the tree trunks to collect the harvested fruit.

Within these space limitations and to facilitate harvesting, the next planting decision is to select the tree density and training system that will generate maximum hedgerow productivity. The first decision, row spacing, discussed above, affects light interception and therefore shading and potential flowering. Shading has been reported to change the light quality within the canopy in favor of the far red, with consequences on flowering patterns (Tombesi and Cartechini, 1986), fruit size, oil content (Connor *et al.*, 2012) and the resulting chemical profile of the oil (Grilo *et al.*, 2021). Specifically, Grilo *et al.* (2019) demonstrated that oil extracted from fruit obtained from the upper layer of the hedgerow receiving more light represented 60% of the overall production and had higher fat content and more total phenols. Connor and Gómez-del-Campo (2013) modeled light interception and penetration within the canopy and concluded that the canopy depth should equal the alley width for maximum cropping potential. They also showed that different pruning methods that improve light penetration between adjacent rows, such as increasing the hedgerow slope from top to bottom, would allow closer spacing between rows and increase yield potential.

The most commonly used row orientation to date has been north to south, as it maximizes summer light interception. However, Gomez del Campo *et al.* (2022) recently reported a seven-year study demonstrating that both the N–S and E–W hedgerows produced similar fruit and oil yields per ha. They also reported that the NE–SW

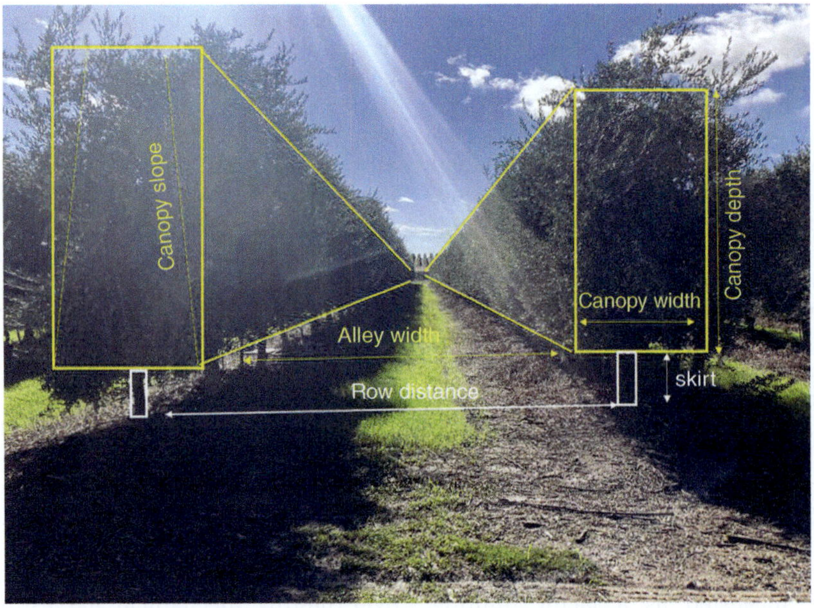

Fig. 5.1.27. Definition of structural parameters of hedgerow orchards as described by Connor *et al.* (2014). (Photo: Giulia Marino)

hedgerow produced a significantly higher oil yield (12% average) than the E–W hedgerow. This yield was higher than all the yields of the other hedgerow orientations in two of the seven years. The NE–SW orientation increased fruit number per ha, but not fruit size or oil concentration.

Pollinizers should be placed within 30 m of the producing cultivar, to ensure high yield. They can be distributed as single trees within rows of the primary cultivar, or as entire rows or groups within other rows. The primary parameter to consider when selecting the pollinizer is flowering time, i.e. it must bloom simultaneously with the primary cultivar. Pollinizers should not be highly alternate bearing. It is recommended to use more than one variety as pollinizer. Research on compatible pollinizers is limited.

A trellis system at planting is required to ensure precise tree alignment for mechanical harvesting. The most common trellis systems have posts at both ends of the row. The 13–15-cm-diameter treated wood or metal posts are usually installed at a 65–75° angle (Fig. 5.1.28). Within the row, 1.8 m-long metal T-posts are inserted every 10–20 m and a bamboo stake or an iron rod is used for each tree (Fig. 5.1.29). A single galvanized wire is placed at 0.6–0.8 m above ground, or two galvanized wires, at 0.6–0.8 m and 1.6–1.8 m. The irrigation system and trellis end posts for the trellis are installed prior to planting. The T-posts, bamboo tree stakes and wire can be installed immediately after planting. The tree's central leader should be tied loosely every 30 cm to the stake to avoid girdling as the tree grows. In the first two years the primary training is limited to: (i) maintaining the central leader, only replacing it with an alternative shoot if damaged and removing competing limbs as the leader approaches the top wire; (ii) removing lower shoots; and (iii) tying the central leader to the stake as it grows. When the central leader grows beyond the top wire by 30–40 cm, generally two YAP, the trees need to be trained for the harvester. Over-the-row straddle canopy contact harvesters remove olives located on the outer 0.5 m of the

Fig. 5.1.28. Detail of an end post of a trellis system in a mature SDH orchard in Stockton, California. (Photo: Emily Santos)

Fig. 5.1.29. One-year-old olive trees in an SHD system tied to bamboo stakes every 30 cm. (Photo: Emily Santos)

canopy. Some straddle harvesters have a series of horizontal bars under tension, bow rods that detach olives by agitating in multiple directions while pressing against the canopy's fruiting wall. Other straddle harvesters have heads with vertically stacked rings of radiating rods that extend into the canopy and detach fruit with a horizontal whipping motion. The canopies most efficiently harvested are narrow with the primary scaffolds parallel to the tree row and bearing annually renewed shoots in a fruiting wall accessible with bow rods or radiating tines (Connor *et al.*, 2014). For efficient mechanical harvesting, all the large lateral branches that extend into the middle of the row should be removed. This will limit canopy damage that makes the tree susceptible to olive knot, the bacterial pathogen *Pseudomonas savastanoi* pv. *savastanoi* spread by rain, which enters the tree through cracks in the bark. The lowest meter of trunk should be kept free of suckers.

The final objective of spacing and training of trees in SHD orchards is a sloped or vertical hedgerow wall that receives sufficient light to flower and mature fruit that can be successfully mechanically harvested and caught in a catch frame as shown in Fig. 5.1.30.

In recent years, the rainfed version has also been proposed. In this case, the tree density usually ranges from 800 to 1000 trees/ha, with distances of 5–7 m between rows and 1.5–2 m between trees within the row. Densities of 4 m × 2 m have also been proposed. Usually, the cultivars that are used are the same as used in irrigated orchards. Because of the larger distances between the rows, the canopy volume per ha is reduced correspondingly; orchards require four to six years to reach full production (4–6 t/ha) (Rallo *et al.*, 2013).

Recently, a new system has been proposed. It is based on the use of trees whose apex is cut in the nursery to promote ramification (Smarttree), and no support structure (or a very limited one) is needed (Fig. 5.1.31). After planting, trees are trained to obtain canopies with a globe/fan shape. Afterwards, the canopies are widened along the row and then the growth is promoted upwards, until reaching the final height of 2.5–3.0 m. This system is proposed with most of the pruning executed mechanically, but, up to now, very little knowledge is available. Therefore, further evaluations would be useful to acquire more information on the pros and cons of this system.

Fig. 5.1.30. Receiver attached to the mechanical olive harvester in the adjacent row (unseen in this picture). (Photo: Louise Ferguson)

Fig. 5.1.31. New SHD system based on trees whose apex is cut in the nursery to promote ramification, and no support structure is needed (or a very limited one). (Photo: Franco Famiani)

5.1.5. Comparison and Choice of Different Planting Systems

Establishing new olive plantings, in addition to enabling a high degree of mechanization, promotes easier use of the most modern management techniques. Furthermore, the whole productive potential of the trees can be expressed.

At the present time, there are mainly two planting models: intensive and SHD

(Table 5.1.1) (Famiani and Gucci, 2011; Famiani *et al.*, 2021; Lo Bianco *et al.*, 2021; Massenti *et al.*, 2022).

Concerning the intensive system, there are basically two different options in relation to the availability of water and the varieties chosen: low density, up to 400 trees/ha, with trees trained to vase; high density, with 650–1000 trees/ha trained to free palmette.

The low-density intensive system is well known in terms of management needs, productivity and duration, and its establishment is highly flexible. In fact, it can be established both in the plains and hills, even with slopes of up to 30%. Establishing the orchard does not require large investment and all cultivars can be used. The knowledge and technical experience for managing it are available. The occurrence of adversities that damage trees and/or reduce production (e.g. frost, drought, rain or high temperatures during flowering) is more acceptable due to the relatively long economic life of the orchard (more than 40 years). All the production objectives of the farm can be pursued (PDO, PGI, organic olives or oils, etc.). Furthermore, it has a low environmental impact. On the other hand, it reaches full production only six to ten years after planting and requires more labor for pruning and harvesting compared to the SHD system.

The high-density intensive system is less known, but the results obtained from recent studies are promising. Being intermediate, it can combine the advantages of low-density intensive orchards with those of SHD ones. Indeed, there are more varieties that can potentially be cultivated with this system. With respect to low-density intensive orchards, it improves the possibility for mechanization of both harvesting and pruning (Massenti *et al.*, 2022). Further studies to evaluate its potential in the different olive-growing areas would be very useful.

The SHD system requires flat terrain or not very steep slopes and minimal risk of abiotic adversities (e.g. frost). Greater investments are necessary for establishment (however, a second-generation SHD system has been proposed without a support frame, making it much cheaper), as well as good water availability (irrigation), even though it is also proposed in rainfed conditions. Management of cultivation techniques is more complex and high technical skill is required; it can also be managed for organic production. At the present time, it is based on the use of a few very fertile and productive cultivars with limited vigor and compact habit. This system has important advantages, such as rapid entry into full production (three to five years from planting) and the total mechanization of harvesting with a straddle harvester.

In the future, new orchards (intensive low-density and high-density and SHD) will complement traditional ones, which need to be maintained for their landscape, monumental, historical and/or environmental importance. They also produce valuable oils and at least partial mechanization of harvesting is possible.

The choice of planting systems must be based on: size and orography of the farm; possibility of joining associations or co-operatives or of using third-party services for mechanized harvesting (if the farm is small); availability of capital and the need to have or not a rapid turnover of the same; availability and degree of skilled labor and technical management; production/commercial objective of the farm. The use of several planting systems instead of just one will allow the predominant requirements of each farm to be met. Furthermore, the increasing need to respond to the diversified and changing demands of consumers in terms of price and quality can be met (Famiani and Gucci, 2011; Massenti *et al.*, 2022). Finally, the best combinations between environmental conditions and planting systems will also minimize the environmental impact of new orchards.

Regarding research, efforts need to be directed at optimizing SHD and high-density growing systems in relation to the different environmental conditions, also considering the effect of ongoing climate change. For both planting systems, research aimed at producing/selecting dwarfing rootstocks and evaluating the adaptation of the numerous varieties, also minor ones present in the various olive-growing countries, must be encouraged. Moreover, efforts need to be

Table 5.1.1. Environmental requirements, yield, management and costs of intensive low-density and high-density orchards and super-high-density (SHD) ones.

Parameter	INTENSIVE – Low-density	INTENSIVE – High-density	SHD
Soil orography	Plantations in the plains and hills, with slopes up to 30%	Plantations in the plains and hills, with slopes up to 20% (15% if large straddle machines are used)	Plantations in the plains and hills, with slopes up to 10–20%
Environmental requirements	Low economic impact due to damage from abiotic stress	Medium economic impact due to damage from abiotic stress	High economic impact due to damage from abiotic stress
Cultivars	All	Several	Few
Planting densities	200–400 trees/ha	650–1000 trees/ha	1200–2500 trees/ha
Nursery material	Two-year-old trees or one-year-old trees	One-year-old, self-rooted trees 0.8 m tall	Six–eight-month-old self-rooted/ micropropagated trees, 0.6 m tall
Tree training form	Vase	Free palmette	Central axis (SHD 1.0), Multi-axis (Smarttree) (SHD 2.0)
Cost of establishment*	4000–8000 €/ha	6000–10,000 €/ha	SHD 1.0 7000–12,000 €/ha; SHD 2.0: 6000-8000 €/ha
Full production	6–10 years from planting	4–6 years from planting	3–5 years from planting
Economic life of the orchard	> 40 years	30–40 years	> 20 years
Average yield (lower values in rainfed conditions)	4000–12,000 kg of olives/ha	4000–14,000 kg of olives/ha	4000–14000 kg of olives/ha
Pruning	Facilitated by using poled devices. Sometimes integrated mechanically	Facilitated by using poled devices, integrated mechanically	Mechanically integrated by using facilitating devices
Irrigation	1000–1200 m^3/ha/year	1500–2000 m^3/ha/year	2000–2500 m^3/ha/year
Soil management	Soil tillage or total/partial green cover. Sometimes with weeding.	Soil tillage or total/partial green cover. Often, tillage or weeding within the rows.	Soil tillage or total/partial green cover. Usually, weeding or tillage along the rows.
Fertilization			
Control of biotic stress	Easy to manage	Low difficulty	Medium difficulty
Harvesting method	Mechanical with trunk shakers equipped with reversed umbrella interceptor (0.5–1 ha/day)	Mechanical with straddle machine, lateral canopy shaker or side-by-side trunk shaker interceptor (1–2 ha/day)	Mechanical with straddle machine (2.5–4 ha/day)
Harvesting cost- €/Kg olives	0.07–0.21	0.05–0.18	0.04–0.16
Degree of management skills	Medium	Medium–high	High
Level of acquired experience	High	Low	Medium

*European countries

made to produce new varieties suitable for these planting systems through plant-breeding programs. In addition, especially for low-density intensive orchards, techniques should be set up that promote faster tree growth in the first years after planting, to reduce the time required to enter into full production.

In all olive-growing systems, i.e. traditional, intensive and SHD, the application of environmentally friendly cultivation techniques to improve environmental sustainability is a key objective. Olive growing has important positive characteristics. Indeed, all types of olive groves have a good carbon sequestration capacity (Proietti *et al.*, 2014; Mairech *et al.*, 2020). Furthermore, the olive tree is a species, which, overall, needs fewer inputs for cultivation than other fruit tree species. In addition, CO_2 emissions can be reduced by applying cultivation schemes capable of minimizing them (Proietti and Regni, 2020). Interesting indications have also emerged from recent studies that evaluated the water footprint of different types of olive orchards. Results gave good water footprint values for both intensive and high-density olive orchards (Pellegrini *et al.*, 2016; Rossi *et al.*, 2020). Comparisons of CO_2 emissions per unit of product among different olive-growing systems have given contrasting results (De Gennaro *et al.*, 2012; Romero Gámez *et al.*, 2017; Camposeo *et al.*, 2022). However, CO_2 emissions (carbon footprint) and water consumption (water footprint) per unit of product depend on the management protocols used, as well as on the production levels achieved. Since the latter are maximized in new plantings, all new orchards might tend to go in the direction of environmental sustainability.

Finally, improvements in the production efficiency and environmental sustainability of olive orchards must also derive from research based on the application of smart technologies for monitoring the conditions of the trees and implementing precise management practices.

References

Almadi, L., Paoletti, A., Cinosi, N., Daher, E., Rosati, A., Di Vaio, C. and Famiani F. (2020) A biostimulant based on protein hydrolysates promotes the growth of young olive trees. *Agriculture* 10(12), 618.

Belaj, A., de la Rosa, R., León, L., Gabaldon-Leal, C., Santos, C. *et al.* (2020) Phenological diversity in a World Olive Germplasm Bank: potential use for breeding programs and climate change studies. *Spanish Journal of Agricultural Research* 18(1), 701.

Camposeo, S., Vivaldi, G.A., Montemurro, C., Fanelli, V. and Cunill Canal, M. (2021) Lecciana, a new low-vigour olive cultivar suitable for super high density orchards and for nutraceutical EVOO production. *Agronomy* 11(11), 2154.

Camposeo, S., Vivaldi, G.A., Russo, G. and Melucci, F.M. (2022) Intensification in olive growing reduces global warming potential under both integrated and organic farming. *Sustainability* 14, 6389.

Castillo-Ruiz, F.J., Sola-Guirado, R.R., Castro-Garcia, S., Gonzalez-Sanchez, E.J., Colmenero-Martinez, J.T. and Blanco-Roldán, G.L. (2017) Pruning systems to adapt traditional olive orchards to new integral harvesters. *Scientia Horticulturae* 220 122–129.

Conde, C., Delrot, S. and Gerós, H. (2008) Physiological, biochemical and molecular changes occurring during olive development and ripening. *Journal of Plant Physiology* 165(15), 1545–1562.

Connor, D.J., Gómez-del-Campo, M. and Comas, J. (2012) Yield characteristics of N–S oriented olive hedgerows, cv. Arbequina. *Scientia Horticulturae* 133, 31–36.

Connor, D.J. and Gómez-del-Campo, M. (2013) Simulation of oil yield and quality of N–S oriented olive hedgerow orchards in response to structure. *Scientia Horticulturae* 150, 92–99.

Connor, D.J., Gómez-del-Campo, M., Rousseaux, M.C. and Searles, P.S. (2014) Structure, management and productivity of hedgerow olive orchards: a review. *Scientia Horticulturae* 169, 71–93.

Connor, D.J., Loomis, R.S. and Cassman, K.G. (2011) Crop ecology: productivity and management in agricultural systems. Cambridge University Press, Cambridge, UK.

De Gennaro, B., Notarnicola, B., Roselli, L. and Tassielli, G. (2012) Innovative olive-growing models: an environmental and economic assessment. *Journal of Cleaner Production* 28, 70–80.

Díaz-Rueda, P., Franco-Navarro, J.D., Messora, R., Espartero, J., Rivero-Núñez, C.M. *et al.* (2020) SILVOLIVE, a Germplasm collection of wild subspecies with high genetic variability as a source of rootstocks and resistance genes for olive breeding. *Frontiers in Plant Science* 11, 629.

Díez, C.M., Moral, J., Cabello, D., Morello, P., Rallo, L. and Barranco, D. (2016) Cultivar and tree density as key factors in the long-term performance of super high-density olive orchards. *Frontiers in Plant Science* 7, 1226.

Famiani, F. (2019) Mantenere le funzioni produttive in oliveti di interesse ambientale e paesaggistico. *Proceedings 'La multifunzionalità dell'olivicoltura umbra'*. Spoleto (PG), Italy, 13 December 2020, pp. 15–29.

Famiani, F. (2021) Stato attuale e linee di sviluppo per il rilancio dell'olivicoltura italiana. In: Alpi, A., Nanni, P. and Vicenzini, M. (eds) *Olivo, Olivicoltura, olio di Oliva – Guardando al Futuro*. Accademia dei Georgofili, Firenze, Italy, pp. 75–102.

Famiani, F. and Gucci, R. (2011) Moderni modelli olivicoli. Accademia Nazionale dell'Olivo e dell'Olio, Spoleto (PG), Italy. Available at: www.accademiaolivoeolio.com.

Famiani, F. and Gucci, R. (2022) La raccolta delle olive. Edagricole – New Business Media, Bologna, Italy.

Famiani, F., Camposeo, S. and Gucci, R. (2021) *Tipologie di impianto per una moderna olivicoltura*. Accademia Nazionale dell'Olivo e dell'Olio, Spoleto (PG), Italy. Available at: www.accademiaolivoeolio.com

Famiani, F., Proietti, P., Lodolini, E.M. and Neri, D. (2009) Gestione della chioma. In: Pisante, M., Inglese, P. and Lercker, G. (eds) *L'ulivo e l'Olio*. Bayer CropScience – Collana Coltura & Cultura, Milan, Italy, pp. 390–411.

Famiani, F., Cinosi, N., Paoletti, A., Farinelli, D., Rosati, A. and Lodolini, E.M. (2022) Deflowering as a tool to accelerate growth of young trees in both intensive and super-high-density olive orchards. *Agronomy* 12(10), 2319.

Ferguson, L. (2006) Trends in olive fruit handling previous to its industrial transformation. *Grasas y Aceites* 57(1), 9–15.

Fernández-Escobar, R., de la Rosa, R., León, L., Gomez, J.A., Testi, L. *et al.* (2013) Evolution and sustainability of the olive production systems. *Options Mediterraneennes* 106, 11–42.

Fontanazza, G. (1984) *Allevamento e potatura dell'olivo*. Universale Edagricole, Bologna, Italy.

Fridley, R.B., Hartmann, H.T., Melschau, J.J., Chen, P. and Whisler, J. (1971) *Olive Harvest Mechanization in California*. University of California, Berkeley, California.

Godini, A. and Bellomo, F. (2002) Olivicoltura superintensiva in Puglia per la raccolta meccanica con vendemmiatrice. *Proceedings of the International Congress of Oliveculture*, Spoleto (PG), Italy, pp. 230–234.

Gómez-del-Campo, M., Trentacoste, E.R. and Connor, D.J. (2022) Long-term effects of row orientation on oil yield and oil yield components of hedgerow olive orchards cv. Arbequina. *Scientia Horticulturae* 294, 110770.

Grilo, F., Caruso, T. and Wang, S.C. (2019) Influence of fruit canopy position and maturity on yield determinants and chemical composition of virgin olive oil. *Journal of the Science of Food and Agriculture* 99(9), 4319–4330.

Grilo, F., Sedaghat, S., Di Stefano, V., Sacchi, R., Caruso, T. and Lo Bianco, R. (2021) Tree planting density and canopy position affect 'Cerasuola' and 'Koroneiki' olive oil quality. *Horticulturae* 7(2), 11.

Guerrero-Casado, J., Carpio, A.J., Tortosa, F.S. and Villanueva, A.J. (2021) Environmental challenges of intensive woody crops: the case of super high-density olive groves. *Science of the Total Environment* 798, 149212.

Hanani, A., Valentini, F., Sanzani, S.M., Santoro, F., Minutillo, S.A. *et al.* (2021) Community analysis of culturable sapwood endophytes from Apulian olive varieties with different susceptibility to *Xylella fastidiosa*. *Agronomy* 12(1), 9.

Lo Bianco, R., Proietti, P., Regni, L. and Caruso, T. (2021) Planting systems for modern olive growing: strengths and weaknesses. *Agriculture* 11(6), 494.

Lodolini, E.M., Polverigiani, S., Grossetti, D. and Neri, D. (2018) Pruning management in a high-density olive orchard. *Acta Horticolturae* 1199, 385–390.

Mairech, H., Lopez-Bernal, A., Moriondo, M., Dibari, C., Regni, L., Proietti, P., Villalobos, F.J. and Testi, L. (2020) Is new olive farming sustainable? A spatial comparison of productive and environmental performances between traditional and new olive orchards with the model OliveCan. *Agricultural Systems* 181, 102816.

Marino, G., Pernice, F., Marra, F.P. and Caruso, T. (2016) Validation of an online system for the continuous monitoring of tree water status for sustainable irrigation managements in olive (*Olea europaea* L.). *Agricultural Water Management* 177, 298–307.

Marino, G., Macaluso, L., Marra, F.P., Ferguson, L., Marchese, A. *et al.* (2017) Horticultural performance of 23 Sicilian olive genotypes in hedgerow systems: vegetative growth, productive potential and oil quality. *Scientia Horticulturae* 217, 217–225.

Marino, G., Macaluso, L., Grilo, F., Marra, F.P. and Caruso, T. (2019) Toward the valorization of olive (*Olea europaea* var. *europaea* L.) biodiversity: horticultural performance of seven Sicilian cultivars in a hedgerow planting system. *Scientia Horticulturae* 256, 108583.

Mariscal, M.J., Orgaz, F. and Villalobos, F.J. (2000) Modelling and measurement of radiation interception by olive canopies. *Agricultural and Forest Meteorology* 100(2–3), 183–197.

Massenti, R., Ioppolo, A., Veneziani, G., Selvaggini, R., Servili, M., Lo Bianco, R. and Caruso, T. (2022) Low tree vigor, free palmette training form, and high planting density increase olive and oil yield efficiency in dry, sloping areas of Mediterranean regions. *Horticulturae* 8(9), 817.

Mazeh, M., Almadi, L., Paoletti, A., Cinosi, N., Daher, E. *et al.* (2021) Use of an organic fertilizer also having a biostimulant action to promote the growth of young olive trees. *Agriculture* 11(7), 593.

Moutier, N., Ricard, J.M. and Le Verge, S. (2010) Vigour control of the olive tree in a high density planting system: two experimental approaches. *Proceedings, XXVIII International Horticultural Congress on Science and Horticulture for People (IHC2010): Olive Trends Symposium: From the Olive Tree to Olive Oil: New Trends and Future Challenges. Acta Horticulturae* 924, 185–193.

Musacchi, S., Iglesias, I. and Neri, D. (2021) Training systems and sustainable orchard management for European pear (*Pyrus communis* L.) in the Mediterranean area: a review. *Agronomy* 11(9), 1765.

Niinemets, Ü. (2010) A review of light interception in plant stands from leaf to canopy in different plant functional types and in species with varying shade tolerance. *Ecological Research* 25(4), 693–714.

Palliotti, A., Famiani, F., Proietti, P., Boco, M., Antognozzi, E. *et al.* (1997) Effects of training system on tree growth, yield and oil characteristics in different olive cultivars. *Proceedings III International Symposium on Olive Growing,* Chania, Greece 474, 189–192.

Pannelli, G. and Alfei, B. (2019) 'L'olivo a Vaso Policonico'. Terroir e Sostenibilità. Edagricole: Milan, Italy.

Pastor, M., García-Vila, M., Soriano, M.A., Vega, V. and Fereres, V. (2007) Productivity of olive orchards in response to tree density. *The Journal of Horticultural Science and Biotechnology* 82(4), 555–562.

Pellegrini, G., Ingrao, C., Camposeo, S., Tricase, C., Contò, F. and Huisingh D. (2016) Application of water footprint to olive growing systems in the Apulia region: a comparative assessment. *Journal of Cleaner Production* 112, 2407–2418. DOI: 10.1016/j.jclepro.2015.10.088

Pérez-Ruiz, M., Rallo, P., Jiménez, M.R., Garrido-Izard, M., Suárez, M.P. *et al.* (2018) Evaluation of over-the-row harvester damage in a super-high-density olive orchard using on-board sensing techniques. *Sensors* 18(4), 1242.

Poorter, L., Bongers, L. and Bongers, F. (2006) Architecture of 54 moist-forest tree species: traits, trade-offs, and functional groups. *Ecology* 87(5), 1289–1301.

Proietti, P. and Regni L. (2020) Mitigazione del cambiamento climatico attraverso una filiera olivicolo-olearia sostenibile. *Proceedings 'La multifunzionalità dell'olivicoltura umbra'*. Spoleto (PG), Italy, 13 December, pp. 31–41.

Proietti, S., Sdringola, P., Desideri, U., Zepparelli, F., Brunori, A. *et al.* (2014) Carbon footprint of an olive tree grove. *Applied Energy* 127, 115–124.

Proietti, S., Sdringola, P., Regni, L., Evangelisti, N., Brunori, A. *et al.* (2017) Extra virgin olive oil as carbon negative product: experimental analysis and validation of results. *Journal of Cleaner Production* 166, 550–562.

Rallo, L., Barranco, D., de la Rosa, R. and León, L. (2008) 'Chiquitita' olive. *HortScience* 43, 529–531.

Rallo, L., Barranco, D., Castro-Garcia, S., Connor, D.J., Gómez del Campo, M. and Rallo, P. (2013) High-density olive plantations. *Horticultural reviews* 41, 303–383.

Regni, L., Nasini, L., Ilarioni, L., Brunori, A., Massaccesi, L., Agnelli, A. and Proietti, P. (2017) Long term amendment with fresh and composted solid olive mill waste on olive grove affects carbon sequestration by prunings, fruits, and soil. *Frontiers in Plant Science* 7, 2042.

Rodrigues, F., Pimentel, F.B. and Oliveira, M.B.P. (2015) Olive by-products: challenge application in cosmetic industry. *Industrial Crops and Products* 70, 116–124.

Romero, A., Hermoso, J.F. and Tous, J. (2014) Olive rootstocks to control 'Arbequina IRTA-I18' clone vigour: results from a second one comparative trial. *Acta Horticulturae* 1057, 577–584.

Romero-Gámez, M., Castro-Rodríguez, J. and Suárez-Rey, E.M. (2017) Optimization of olive growing practices in Spain from a life cycle assessment perspective. *Journal of Cleaner Production* 149, 25–37.

Rosati, A., Paoletti, A., Caporali, S. and Perri, E. (2013) The role of tree architecture in super high density olive orchards. *Scientia Horticulturae* 161, 24–29.

Rossi, L., Regni, L., Rinaldi, S., Sdringola, P., Calisti, R. *et al.* (2020) Long-term water footprint assessment in a rainfed olive tree grove in the Umbria region, Italy. *Agriculture* 10, 8.
Rugini, E. and Pace, C.D. (2016) Olive breeding with classical and modern approaches. In: Rugini, E., Baldoni, L., Muleo, R. and Sebastiani, L. (ed.) *The Olive Tree Genome*. Springer, pp. 163–193.
Russell, G., Jarvis, P.G. and Monteith, J. L. (1989) Absorption of radiation by canopies and stand growth. In: Russell, G., Marshall, B. and Jarvis, P.G. (eds) *Plant Canopies: Their Growth, Form and Function* 31, 21–39.
Russo, G., Vivaldi, G.A., De Gennaro, B. and Camposeo, S. (2015) Environmental sustainability of different soil management techniques in a high-density olive orchard. *Journal of Cleaner Production* 107, 498–508.
Sadok, I.B., Celton, J.M., Essalouh, L., El Aabidine, A.Z., Garcia, G. *et al.* (2013) QTL mapping of flowering and fruiting traits in olive. *PLOS ONE* 8(5), e62831.
Salomone, R. and Ioppolo, G. (2012) Environmental impacts of olive oil production: a life cycle assessment case study in the province of Messina (Sicily). *Journal of Cleaner Production* 28, 88–100.
Scaramuzzi, F. (2006) The landscape planning policy in Italy constrains olive growing competitiveness: lecture of the session of the Accademia Dei Georgofili held in Padova, 17 May, 2006.
Sola-Guirado, R.R., Castro-García, S., Blanco-Roldán, G.L., Jiménez-Jiménez, F., Castillo-Ruiz, F.J. and Gil-Ribes, J.A. (2014) Traditional olive tree response to oil olive harvesting technologies. *Biosystems Engineering* 118, 186–193.
Tombesi, A. and Cartechini, A. (1986) L'effetto dell'ombreggiamento della chioma sulla differenziazione delle gemme a fiore dell'olivo. *Rivista di Ortoflorofrutticoltura Italiana* 70, 277–285.
Tombesi, A., Proietti, P., Iacovelli, G., Tombesi, S. and Farinelli, D. (2011) Vegetative and productive behaviour of four olive italian cultivars and 'Arbequina' according to super intensive olive training system in central Italy. *Acta Horticulturae* 924, 211–218.
Torres-Sánchez, J., de la Rosa, R., León, L., Jiménez-Brenes, F. M., Kharrat, A. and López-Granados, F. (2022) Quantification of dwarfing effect of different rootstocks in 'Picual' olive cultivar using UAV-photogrammetry. *Precision Agriculture* 23(1), 178–193.
Tous, J., Romer, A., Plana, J. and Baiges, F. (1999) Planting density trial with 'Arbequina' olive cultivar in Catalonia (Spain). *Acta Horticulturae* 474, 177–180.
Tous, J., Romero, A., Hermoso, J.F. and Ninot, A. (2011) Mediterranean clonal selections evaluated for modern hedgerow olive oil production in Spain. *California Agriculture* 65(1), 34–40.
Tous, J., Romero, A., Hermoso, J.F., Msallem, M. and Larbi, A. (2012) Olive orchard design and mechanization: present and future. *Proceedings of VII International Symposium on Olive Growing.* San Juan, Argentina, 1057, 231–246.
Trujillo, I., Ojeda, M.A., Urdiroz, N.M., Potter, D., Barranco, D., Rallo, L. and Díez, C.M. (2014) Identification of the Worldwide Olive Germplasm Bank of Córdoba (Spain) using SSR and morphological markers. *Tree Genetics & Genomes* 10(1), 141–155.
Turktas, M., Inal, B., Okay, S., Erkilic, E.G., Dundar, E. *et al.* (2013) Nutrition metabolism plays an important role in the alternate bearing of the olive tree (*Olea europaea* L.). *PLOS ONE* 8(3), e59876.
Villalobos, F.J., Testi, L., Hidalgo, J., Pastor, M. and Orgaz, F. (2006) Modelling potential growth and yield of olive (*Olea europaea* L.) canopies. *European Journal of Agronomy* 24(4), 296–303.

5.2 Site Selection and Olive Orchard Design and Establishment

F. Famiani[1], R. de la Rosa[2], N. Cinosi[1], L. Regni[1] and P. Proietti[1]
[1]*Dipartimento di Scienze Agrarie, Alimentari e Ambientali, Università degli Studi di Perugia, Perugia, Italy;* [2]*Centro IFAPA "Alameda del Obispo", Córdoba, Spain*

5.2.1 Introduction

An olive orchard must be designed to ensure abundant and high-quality production, economical management and environmental sustainability. This means that light interception, temperature and availability of nutrients and water, which influence the physiological and biological processes at the basis of vegetative growth and production (quantity and quality), need to be optimized. Moreover, production costs must be minimized. Therefore, the mechanization of cultivation operations, particularly harvesting, must be ensured.

Replanting, that is replacement of a pre-existing olive orchard, should be considered when: the trees are old and deteriorated, due to decay, and do not respond satisfactorily to crop care; the tree density is low, due to the death of some trees following frosts and/or parasitic attacks; the arrangement of the trees and/or their structure and/or the cultivars present do not allow easy use of machines for harvesting; it is necessary to change the cultivated variety to meet the requirements of production certifications (PDO, PGI) and/or new production objectives of the farm. Early bearing in the new olive-growing systems makes replanting more economically viable.

In this chapter, the main references that were consulted are reported in the reference list. Only very few specific ones are also reported directly in the text.

5.2.2 Site Selection

To be profitable, olive tree cultivation requires suitable climatic, soil, sanitary and socio-economic conditions. Furthermore, anthropogenic inputs and environmental impact must be minimal. High economic sustainability must be ensured along with an abundant, genuine, high-quality product.

The presence of healthy and productive olive orchards near the area where the new orchard is to be established is an important positive indication of the suitability of the site for olive growing. However, in many countries, olives are being planted outside the traditional growing areas. Indications are given below on the main features for selecting a suitable site for olive growing.

5.2.2.1 Climatic needs

The climatic suitability of an area should be obtained from historical series of at least

DOI: 10.1079/9781789247350.0015

30 years of the main climatic parameters (minimum and maximum temperature, precipitation, wind speed, frequency of hailstorms etc.). This information can be obtained from meteorological stations located near the area under consideration. Sometimes, useful information can be obtained from websites that have databases on the climatic parameters of different areas. However, models predicting future climate changes should also be taken into consideration. For many Mediterranean areas, those models predict higher temperature and water scarcity. These changes could severely hamper the normal reproductive development of olive.

Temperature

Winter temperatures should not drop below –6°C to –7°C, especially for a relatively long time (days). It is not recommended to plant olive trees where winter temperatures drop below –11°C, even for a short time, for more than once every 20–30 years (Table 5.2.1). Snow, fog and wind exacerbate the effects of frost. However, during the winter, a period of relatively low temperatures (between 0°C and 7°C, with values between 5°C and 7°C being the best) seems to be necessary for optimal flower bud differentiation. This cold requirement varies in relation to cultivar, but 50–60 hours is enough for most cultivars. However, there are recent reports of olive trees flowering and fruiting in the subtropical climate of Tenerife, Canary Islands, where the minimum winter temperature ranges from 10°C to 15°C; the effect of the lack of winter-chilling temperatures was a desynchronization on flowering bud bloom, but the final olive productivity was only partially affected. More research on the real chilling needs of olive is needed to determine the climatic conditions where good productivity can be expected.

Table 5.2.1. Minimum temperature for the different phenological stages.

Phenological stage	Minimum temperature
Beginning of inflorescence development/bloom	10°C
Fruit set	15°C
Fruit growth – veraison	20°C
Ripening	5°C
Vegetative (winter) rest	–4°C to –5°C

A period of relatively low temperatures (around 0°C) is also beneficial for reducing some parasites, such as black scale, olive fly and olive moth. The risk of late frosts (April–May in the northern hemisphere), which damage the flower buds and therefore the production and/or early frosts (autumn, which damage the fruit not yet harvested and therefore the extracted oil) should be very low (no more than one frost every 10–15 years).

During the growing season, the optimal temperature is around 25°C. The olive tree can withstand very high temperatures (> 40–45°C), but if these temperatures persist for a long time, the vegetative-productive responses of the trees are compromised, especially if associated with water scarcity. High daily temperature variations could also damage olive flowers or fruits even in the case where temperatures do not go below 0°C.

In areas located in the northern limits of cultivation, hilly areas between 200 m and 500 m asl are less susceptible to risks of low-temperature damage, compared to those of the plains.

Regarding oil quality, temperature can affect the acidic composition and phenol content. High temperatures during oil synthesis tend to reduce the amount of oleic acid and increase that of palmitic and/or linoleic acid; when temperatures are very high there is also an increase in linolenic acid. Generally, in hot environments, the oil tends to have a lower amount of phenolic substances.

Relatively high temperatures during harvesting accelerate degradation of the olives during storage, thus reducing the quality of the extracted oil; hence, the work must be organized in such a way as to eliminate or reduce the storage time of the olives to a minimum (very few hours).

Rainfall

In areas where rainfall exceeds 800–900 mm/year, the olive tree can be grown with good responses in terms of production and vegetative growth, without irrigation.

However, irrigation can improve the vegetative-productive activity in the event of particularly hot and dry periods during the growing season. With 500–700 mm/year of rainfall, there can be good levels of growth and production, but significant improvements can be achieved with irrigation. With rainfall around 400 mm/year or lower, irrigation promotes very great increases in vegetative growth and production. Significant and prolonged water shortages during the growing phase of the fruit during which the oil accumulates (August–September in the northern hemisphere) can cause the onset of the dry/wood sensorial defect in the oil.

Air humidity

High air humidity, which can occur more frequently in valley bottoms, near lakes and in coastal areas, favoring the attack of pathogens (in particular peacock eye) and phytophagous pests (in particular olive fly and black scale), leads to serious conditions for the olive tree. However, in areas where winter temperatures can reach critical levels, large bodies of water play a positive role in mitigating the thermal conditions. New harvest systems such as straddle harvester machines (see subchapter 6.5) could favor the spread of airborne pathogens in the case of high air humidity or fog during harvesting (Fig. 5.2.1). In such situations it is recommended to choose varieties most resistant to the aforementioned pathogens/pests.

Wind, hail and snow

Problems for the olive tree occur in areas subject to: strong and/or salty, very hot or cold winds; frequent hailstorms. If hail occurs close to harvesting, the quality of the oil can be affected. In all the situations, it promotes olive knot attacks, especially in cultivars most susceptible to this pathogen. In the case of hedgerow orchards, support structures (see 5.2.4) are mandatory in the case of strong winds in perpendicular direction to the tree row. To counteract the effect of strong winds, sturdy stakes (poles) must be used for young plants and, in areas most at risk, windbreaks should be set up, made with eucalyptus or cypress trees or with nets supported by special poles.

Regarding snow, there can be problems in environments where heavy snowfalls occur, causing the branches to break, especially if the foliage is dense and the snowfalls are late (heavier snow), and/or if the main branches are trained with an excessively open angle to the vertical.

Fig. 5.2.1. The damage caused to trees by the straddle machine can facilitate the spread of diseases, especially in cases of high air humidity or fog during harvesting. (Photo: Franco Famiani)

5.2.2.2 Pedological needs

The profile, composition and drainage of the soil should be analyzed.

Profile and depth of cultivation

The soil profile is evaluated by digging pits (more than one if the land is not uniform), at least 1 m deep, to visually determine the depth available for root growth and variations in texture, presence of stones, rocks, salt and carbonates and the level of the water table. This should be done after two to three days of dry weather from digging the pit, when it is easier to analyze the different characteristics of the soil. It is important that the plants have a volume of soil that: does not restrict root growth, ensures anchoring of the tree; and provides sufficient water and nutrients. Generally, it is possible to distinguish the topsoil, the more active layer suitable for root development, which is soft, porous and has a good amount of nutrients, organic matter and aerobic microorganisms. Below that, a more inert layer can be found (the subsoil), not as suitable for root development, which may be more compact and less porous and permeable. When evaluating the soil profile, it is important to evaluate the thickness of the layer where roots can grow (root-growing depth), which is the distance between the upper limit of a soil layer that creates an obstacle to root growth and the soil surface. This depth can also be limited by a high water table. The root-growing depth should not be less than 0.8–1 m, as most of the roots of the olive tree develop in the first 0.8 m (Table 5.2.2). Lower depths could hamper the plant growth; however, in the case of a hedgerow system, restriction in plant growth may not be very important due to the high plant density and the low plant vigor needed to form the hedgerows.

Table 5.2.2. Best soil characteristics for olive.

Soil characteristics	Optimal conditions
Thickness of the growing depth where roots can grow	> 1 m
Drainage	Good
Texture	Medium
pH	6.8–7.5
Salinity	< 4 dS/m

Physical and chemical soil characteristics

Soil analysis provides information on texture, pH, organic matter content, cation exchange capacity, amount of total and active limestone, nutrients, salts and sodium.

The olive grows best on medium-textured soils (35–50% sand, 25–45% silt, 20–25% clay), but also grows well in clay loam, silt loam and silty clay loam soils. The soil should be deep, fertile, wet and well drained and preferably have a pH of 6.8–7.5. However, olive is widely adaptable, being able to grow and produce also in soils rich in very fine gravel or limestone (up to 50–60% of total limestone), rocky, poor and dry, with a pH between 5.5 and 8.5, relatively saline and/or sodic. In loose, sandy soils, if water is available or if there is a water table that the roots can reach, and nutrients are supplied, the olive can grow and produce well. Problems can occur in heavy clay soils (clay > 40%) in the plains, as there may be water stagnation, to which the olive is very sensitive. Prior to planting, drainage systems are sometimes advisable to lower the water table (see 5.2.4).

The olive is a species that is one of the most resistant to salinity, which is measured as specific electrical conductivity of the saturated soil extract (CEes) and is expressed in dS/m. A 10% reduction in production may occur if the CEes reaches values around 4 dS/m, about 25% with values around 5 dS/m and 50% or more with values around 8 dS/m. However, olive cultivars have a very different resistance to salinity. The olive is a species that is one of the most tolerant to sodium (expressed as exchangeable sodium percentage - ESP): a decrease in production of about 25% occurs when the ESP reaches values of 20–40.

Regarding pH, when the soil pH drops to values around 5.5, toxicity problems begin to arise due to a high release of ions such as aluminum and manganese and to the immobilization of important nutrients, such as calcium, magnesium, potassium, phosphorus, molybdenum, boron and copper. When the pH rises to values above 8.5, the soils are generally rich in sodium and if there is a high content of active limestone, iron chlorosis may occur. The olive has a lower sensitivity to excess boron and chlorides in the soil than most fruit tree species. It tolerates soils with boron and chlorides up to 2 ppm and 10–15 meq/l, respectively, showing at these values a limited reduction in production (about 10%).

5.2.2.3 Modification/correction of soil characteristics

Texture modification/correction

In general, modification/correction of soil texture is not practical. The only case in which it can be done is in those soils that have layers with very different textures (e.g. sandy and clayey). In this case, in the pre-planting phase, using a burglary plow, the layers can be mixed, leading to a better final texture for olive tree cultivation.

Modification/correction of the root-growing depth

Economically sustainable adjustments of the root-growing depth of the soil are feasible only when it is possible to break up compact soil layers with the burglary plow and when it is possible to drain soils that have an excessively high water table.

Soil reaction modification/correction (pH)

Although the soil reaction (pH) can be adjusted, it is expensive and often does not resolve the problem and therefore treatments must be evaluated very carefully. The pH of acid soils can be increased by adding quicklime or slaked lime or carbonates, such as dolomite and calcium and magnesium carbonate. The amount to be applied depends on the starting pH and on the amount of clay and organic matter. In general, applications must be repeated every two to three years on sandy soils and every five to ten years on clay ones. The pH of alkaline soils can be lowered by adding calcium sulphate. Iron chlorosis may occur if the pH is too high due to excessive limestone, so varieties resistant to iron chlorosis should be selected. The application of iron-based chelates is a valuable aid in overcoming the drawbacks caused by excessive alkalinity; however, this must be evaluated economically, especially in dry farming conditions.

Correction of excess salinity, sodium, boron or chlorides

Excess salinity, sodium, boron and chlorides can be due to using irrigation water excessively high in these compounds. In these cases, alternative water sources should be sought. Within certain limits, these soils can be corrected by washing (abundant irrigations that move the elements/compounds to a greater depth) with good-quality water; i.e. water with low amounts of these compounds. This practice requires a large amount of water and well-drained soil. It can also lead to the leaching of beneficial elements.

Correction of sodic soils is particularly difficult.

5.2.2.4 Sanitary aspects of the soil

The presence of pathogenic fungi, such as *Verticillium dahliae* and *Armillaria mellea*, are very dangerous to the olive tree and must be carefully monitored. As for Verticillium, if the pre-existing olive orchard had been attacked or if susceptible species (e.g. Solanaceae, such as tomato, pepper, aubergine, Cucurbitaceae, cotton, strawberry etc.) had been planted there previously, it is recommended to analyze the soil for the presence of inoculum. If the amount of inoculum is high, then the plot should not be used to establish an olive orchard or measures should be taken to lower the inoculum, such as solarization. The same precautions should be taken with plants with root rot caused by *Armillaria mellea*. If the risk of such pathogens is limited and the area is situated in a position that does not cause particular problems in managing the rest of the plot, it is possible to proceed with the establishment, excluding the part of the land at risk.

In any case, when preparing the ground for the new plantation, to reduce health risks it is very important to remove as much plant residue as possible, particularly pieces of stumps and roots. Furthermore, in case of replanting, it is good to set out the new rows at the center of the inter-row of the old olive orchard.

5.2.2.5 Orography and exposure of the terrain

In the case of hilly terrain, for easy use of machines for cultivation and harvesting, the slope must not exceed 15–30% (with steeper slopes, track-type tractors must be used).

In the northern hemispere, southern exposure, ensuring high lighting, is the best. Eastern and western exposures and south-east and south-west are also good. Northern exposures cause a significant reduction in lighting, negatively affecting the vegetative-productive activity of plants. In areas at risk of frost, an eastern exposure can be risky due to a rapid change in temperature in the early morning.

5.2.2.6 Socio-economic aspects

For the choice of the planting site, socio-economic aspects must also be taken into account, such as the presence of mills nearby, the availability of manpower to carry out the work, particularly pruning, the presence of technical assistance, and access to markets that allow for an adequate enhancement of the product.

5.2.3. Orchard Design

5.2.3.1 Choice of cultivar

Over the centuries, as a result of the selective pressure exerted by olive growers as well as the significant environmental differences found in the areas where the olive tree has developed, more than 1600 varieties of olive have originated and spread throughout the world. Over 600 have been described in Italy. The existing olive heritage is rather stable, as it is difficult to obtain new varieties through genetic improvement using classical methods (crossing, use of mutagenic agents etc.). Therefore, only very few new varieties have been obtained by systematic breeding.

The choice of cultivars for establishing the new olive orchard is crucial and therefore must be done very carefully. In fact, in the event of poor choices, there will be negative repercussions on the production and management of the olive orchard for its entire duration.

5.2.3.2 Parameters/requirements in choosing cultivars

In choosing cultivars well, it is necessary to identify where the new olive orchard is to be established and observe the varieties that are present or are in the surrounding area and have been shown to respond well to the environmental conditions. The following characteristics must be considered to choose the most suitable cultivar:

- Abundance and regularity of production (quantity of fruit, oil yield and alternation of production).
- Adaptation to adverse pedo-climatic conditions (soil salinity, drought, frosts etc.) and sensitivity to pathogens (peacock eye, olive knot etc.) and pests (olive fly, black scale etc.) detected in the area.
- Suitability for mechanized harvesting: for optimal use of harvesting machines, medium or high weight fruits (> 2.0 g) and not high resistance to detachment are necessary. If several cultivars are to be planted, it might be beneficial to choose them with progressive ripening times, so that they can be harvested sequentially at the optimal time.
- Vigor, which expresses the growth capacity of the variety, and the growth habit, which can be upright/compact,

expanded or weeping. These factors must be considered in determining the planting distances and the training system. Generally, less vigorous varieties tend to have a higher fructification and come into production earlier than vigorous ones.

- Pulp consistency and fruit pigmentation: olives with high pulp firmness and limited or late pigmentation, generally, have greater resistance to bruising and wounding during harvesting and/or transport and/or storage, which can negatively affect oil quality (breakdown of the liposomes in which the oil is contained in the pulp cells which causes the oil to come into contact with enzymes that favor acidification and oxidation).
- Self-incompatibility (self-sterility), which characterizes most olive cultivars. When self-incompatible cultivars are used, it is advisable to plant pollinator varieties (pollinizers) and/or to use several inter-compatible cultivars in the same orchard. This is especially important in areas where olive tree is not very common and there is little airborne pollen during the flowering period.
- Oil characteristics, particularly the fatty acid composition (oils with oleic acid content > 73%, linoleic acid < 10%, oleic/linoleic ratio higher than 7 are considered excellent), the amount of antioxidant compounds (especially phenolic substances and tocopherols), the oil stability and the sensory profile.

Information on the main olive cultivars worldwide is available in the *World Catalog of Olive Varieties*, published by the International Olive Council (Barranco *et al.*, 2000).

5.2.3.3 Cultivar selection based on the production/commercial objectives, the cultivation method and the presence of particular adversities

Once the characteristics of potential cultivars and their compatibility with the environment and cultural needs have been evaluated, they should be selected according to the production/commercial objectives to be pursued and the cultivation method to be adopted. Since it is difficult to have cultivars that meet all the requirements, priorities based on specific objectives should be established.

Production/commercial objective

If the goal is to obtain a standard quality oil (extra-virgin), the cultivars must have high productivity and adaptability to mechanized harvesting (for high production and low costs, given that the price of the standard product is usually not high). If the goal is to obtain a product with a Protected Designation of Origin (PDO) or Protected Geographical Indication (PGI) mark, the varietal choice must correspond to the rules set out in the production regulations, which specify both the cultivars to be used and their respective percentages. If the objective is to expand an existing olive orchard, the varieties already present must be considered, so that the final proportions are in line with the PDO or PGI specifications. If the goal is to obtain an oil with particular compositional, sensory and/or health characteristics (e.g. high content of antioxidants), cultivars having these characteristics should be chosen. In this regard, monovarietal oils are important. In this context, it is very important to refer to the cultivars that have been established over time in the various areas and which give rise to variety/environment combinations that produce particular oils, with easily recognizable compositional and organoleptic characteristics (high degree of typicity). Sometimes, minor varieties are important. Often, they have been maintained because they give high-quality production (oils with particular sensory or nutritional features) or have a high degree of resistance to specific adversities. Oils with peculiar characteristics can also be obtained by mixing oils of different varieties ('blends'). For establishing hedgerow or super-high-density (SHD) olive orchards, limited vigor is the fundamental requisite. Another very important feature is the high branching and small diameters of branches/shoots. For this type of orchard, the varieties that have shown the best adaptation are the Spanish 'Arbequina' and 'Arbosana', while the Greek 'Koroneiki' is presently less used. In recent years, some new varieties have been produced by genetic improvement and are being used in the new SHD orchards, such as 'Sikitita' ('Picual' × 'Arbequina'), 'Sikitita2' ('Picual' × 'Arbequina'), 'Lecciana' ('Arbosana' × 'Leccino') or I-15 ('Arbosana' × 'Koroneiki'). Another new variety is 'Oliana' ('Arbequina' × 'Arbosana'), but its use has been reducing because of its low oil content.

In recent years, intensive high-density olive orchards have been developed in Italy with the main purpose of using Italian varieties at densities higher than those of intensive low-density orchards, but less than hedgerow ones. The preliminary available information indicates that the varieties 'FS17', 'Leccio del Corno', 'Maurino', 'Nociara', 'Piantone di Mogliano' and, to a lesser extent, 'Piantone di Falerone' are the most promising.

Cultivation method

If the olive orchard will be managed organically, resistance to abiotic and biotic adversities is of primary importance. Low susceptibility to olive fly and peacock eye and olive knot diseases is very important. There are significant varietal differences regarding susceptibility to these adversities. As for the olive fly, there is a negative correlation between susceptibility to attack and fruit size and the content of oleuropein in the fruit. Also, early ripening/veraison varieties are less susceptible to late attacks (the fly favors green olives for egg laying). Generally, the varieties with fruits with relatively firm pulp are less susceptible to attack. In organic cultivations, the presence of cultivars that ripen at different times, or of oil and table cultivars together, is not recommended, as this would favor the succession of several generations of olive fly.

Resistance to abiotic adversities is also important. These adversities, in addition to causing direct damage, make plants more susceptible to biotic adversities due to the debilitation and/or injuries they cause (e.g. frost damage favors the spreading of olive knot, especially in cultivars that are highly sensitive to this pathogen).

Choice of cultivars in the presence of particular adversities

When a new orchard is to be established in an area having a high risk of a severe adversity, the resistance to such adversity should be the main factor to keep in mind when choosing cultivars.

5.2.3.4 Choice of cultivars and pollination

In all situations, when the main cultivar is self-incompatible (like most olive cultivars), particular attention must be given to the choice of pollinizers. If the pollinizers are already known, the choice must focus on these. Otherwise, the cultivars in the surrounding area that are usually cultivated with the one being considered should be selected. The pollinating cultivars to be favored are those that have abundant blooms and high pollen production, and that coincide as much as possible with the blooming of the cultivar to be pollinated. Even the few existing self-compatible cultivars take advantage of cross-pollination. Therefore, it is recommended that pollinizers be included even when choosing self-compatible varieties.

If there are several cultivars with good characteristics and that are inter-compatible to be chosen for planting, it is beneficial to use several of them to set up multi-varietal orchards. Ideally, two or, even better, three or four could be used, if in line with the production objective, to avoid that any low-flowering cultivar could significantly affect the overall production of the olive orchard. This will avoid any problems due to 'mismatches' in flowering of the different cultivars as a result of the seasonal trend. Furthermore, the risk of serious production losses is reduced in case of adversities, due to the different sensitivities of the different cultivars. In large olive orchards, combining cultivars that ripen at different times makes it easier to harvest each variety at the optimal time.

In any case, it is important to confirm the cross-compatibility of the chosen cultivars. Recent studies have shown that olive cultivars are divided into two large groups of incompatibility (G1 and G2), on the basis of which all the cultivars of one group are cross-incompatible with each other and fertile with all those of the other group and *vice-versa* (Saumitou-Laprade *et al.*, 2017). Then there would seem to be another level of interaction between the pollen tube and the ovary which would make it possible to explain the self-compatibility of some cultivars (Breton *et al.*, 2021). Therefore, to have efficient pollination it is necessary to ensure the simultaneous presence of cultivars of group G1 and cultivars of group G2.

5.2.3.5 Choice of a cultivar widespread in an environment different from that in which the new olive orchard is to be established

When choosing a cultivar taken from another environment, this often involves modifications in the vegetative and productive behavior of the trees, in the quality of the product (e.g. variations in the acidic composition and in the amount of phenolic substances) and in the resistance to adversities. Therefore, it is recommended to introduce the cultivar only in situations where experiments have been carried out that have confirmed its adaptability to the environment in which it is to be introduced. In other cases, the risk involved must be considered. Accordingly, the agronomic information available for a cultivar should always be considered cautiously. It is possible that the available data could correspond to environmental conditions or orchard management that differ from the one to be established.

5.2.3.6 Choice of training system

For new olive orchards it is important to choose training systems that favor rapid growth and early time for full production, good lighting of the entire canopy, high and constant production, a microclimate within the canopy unfavorable to the development of pathogens and harmful insects, and make easy the application of cultivation practices, in particular pruning and harvesting.

Usually, the vase training system is recommended for intensive low-density olive orchards, while the central axe system or the multi-axis system (Smarttree) are recommended for hedgerow olive orchards and, finally, the free palmette system is recommended for intensive high-density orchards.

For more information on the choice of training systems, see the chapters 5.1 and 6.4.

5.2.3.7 Choice of planting density, spacing and distances

Intensive low-denisty olive orchards

Planting density (number of trees/ha) should be based on the size that the trees will reach as adults (which depends on the vigor of the cultivars, the fertility of the soil, the climatic conditions, the training system and the cultivation technique with particular reference to irrigation) and the need to mechanize cultivation practices, particularly harvesting. At full development, the crowns of the trees must be well lit and ventilated and, therefore, the arrangement and planting distances must be designed so as to avoid mutual shading among neighboring trees.

For the olive tree, unlike other fruit species, rootstocks able to reduce the vigor of all grafted varieties are not available. In addition, very few varieties have low vigor. Hence, in most cases relatively high planting distances are needed. The growth rate of the olive is rather slow and therefore the time required to reach full development is relatively long (6–10 years). Consequently, in the first years after establishment, there is not optimal use of the available light energy. If tree density is increased (and therefore the amount of ground covered by foliage), a greater amount of solar radiation is intercepted and thus production in the first years after planting is increased, but, after a few years, this may cause an excessive competition for light among the trees. This would create the need to periodically carry out drastic pruning to bring the crowns back into the space allocated for them, with consequent vegetative-productive unbalance.

In most situations, with new intensive orchards, the best results are obtained with 200–400 trees per hectare. The most common planting designs are the square, at 5 × 5 m to 7 × 7 m spacing and the rectangle, at 6 × 5 m or 7 × 4 m to 7–8 × 6–7 m spacing (Fig. 5.2.2). The wider spacings are indicated for the most vigorous cultivars and/or for particular production needs (e.g. inclusion of other plants/animals) and/or in the most favorable environmental and cultural conditions or in case of low availability of water (low rainfall and no irrigation). Only in particular situations, i.e. non-vigorous cultivars and trees trained to have a compact crown, they can be reduced to 5 × 4 m (500 trees/ha).

The square layout favors more uniform illumination of the foliage and ease in carrying out cultivation operations according to the two directions of planting. For the same number of plants/ha, the rectangle gives greater distances between the rows and this can benefit cultivation practices. Moreover, it facilitates different soil management of the row compared to the inter-row and can increase the efficiency of water distribution for irrigation.

Fig. 5.2.2. An intensive low-denisty olive orchard, after two years from planting, with trees spaced 5 m x 5 m. The irrigation lines are supported by the tree poles. (Photo: Franco Famiani)

The choice of planting distances that are too close is a very serious mistake, which is difficult to correct (periodically, intense pruning must be done to bring the canopy back into the available space, with consequent alteration in the vegetative–productive balance of the trees). In general, planting at less than 6 m between trees should only be done when it is certain that with the cultivar(s) used, in the considered environment and cultivation conditions, there will be no problems of excessive development.

To overcome the problem of not fully using the land in the first years after the planting, a layout has been proposed, which is the use of a double number of trees compared to the final number in order to have increased production in the first years. Overall, this is not economical, because the increase in production in the first years does not compensate for the higher planting costs and the expense necessary for uprooting the excessive trees, which must be done when shading problems begin to occur between neighboring trees (usually 10–12 years after planting). Transplanting the uprooted trees into another plot to create a new olive orchard is technically feasible if done at the right time (autumn) but, considering the cost and the time required for the complete productive recovery of the trees, the operation is not economically sustainable. There could be economic benefits if the uprooted trees were sold for ornamental purposes, but this is seldom the case, as the demand for these trees is limited and occasional.

For further information on intensive olive orchards also see chapters 5.1 and 6.4.

Tree density in environments with limited water

In environments with limited water availability (300–450 mm rainfall/year or less) and where it is impossible to irrigate, which is a recurring situation in the Mediterranean basin, the use of dry-farming techniques is required. The number of trees/ha may go down to 100–150 or even less, with trees arranged in a square at 8 m × 8 m or 10 m × 10 m; in cases where rainfall is extremely low, such as in southern Tunisia (about 200 mm/year or less), very low densities of 15–20 trees/ha (24 m × 24 m) are used. Reducing the number of trees/ha increases the volume of soil available for each tree and therefore the amount of water and nutrients available.

Planting density and mechanization

Regarding the relationship between planting density and mechanization, particularly harvesting, a distance of at least 5 m between rows (6 m is even better) is needed for optimal handling of shakers. If using an inverted umbrella interceptor frame, the distance between the trees along the rows should not be less than 5 m.

Planting density and cultivation method

In planting olive orchards that will be managed organically, relatively wider planting distances (+ 5–10%) than those of the conventional method are recommended, to ensure optimal lighting and ventilation of the canopy, a prerequisite for lower susceptibility to parasites.

5.2.3.8 Hedgerow olive orchards

Recently, in Spain, a hedgerow olive-growing model has been developed, which is spreading to most olive-growing countries. It consists of the use of a high number of trees/ha (1200–2500) of low-vigor varieties, in order to reach full production in three to five years after planting. The trees are trained to form a continuous hedgerow for straddle harvesting (modified grape harvesters), which significantly reduces harvesting time (2–3 h/ha) and the labor needed, and, hence, the cost of harvesting.

The planting distances are 3.5–4.5 m between rows and 1.2–2.0 m between plants within the rows; the 4.0 m × 1.5 m spacing is the most used (Fig. 5.2.3). Shorter distances are used in environments where soil fertility is lower and/or the growing season is shorter and/or less vigorous varieties are used.

The hedgerow system described is defined as first generation. The trees are trained with a central axis and, usually, a support structure with poles and wires is used. Another type, defined as second generation, has a free-shaped canopy (with or without early pruning of the central axis). Usually, for this system, trees whose apex is cut in the nursery to promote ramification are used (Smarttree) (Fig. 5.2.4). This second type does not always require the use of support structures (with savings on establishment costs), and favors the use of a higher level of mechanization for pruning.

A particular case is hedgerow orchards in dry-farming conditions. In recent years, this system has become popular in several countries such as Spain. There is little information on its viability, the cultivars that are suitable and the best tree density and canopy management. In the first orchards planted with this system, densities from 7 m × 2 m to 4 m × 2 m were used and 'Arbosana' and 'Sikitita' were among the most suitable cultivars (Figs 5.2.5 and 5.2.6). However, more research is needed given the future scarcity of water forecasted for the Mediterranean area under the climate-warming scenario.

For further information on SHD olive orchards, see the chapters 5.1 and 6.4.

Fig. 5.2.3. A hedgerow olive orchard with trees spaced 4 m x 1.5 m. (Photo: Franco Famiani)

Fig. 5.2.4. A second-generation hedgerow olive orchard with trees spaced 4 m x 1.5 m, without the use of supporting structures. (Photo: Franco Famiani)

5.2.3.9 Intensive high-density olive orchards

In recent years, intensive high-density olive orchards have been proposed in Italy. They mainly consist of 650–1000 trees/ha, which is a density between that of intensive and SHD olive orchards. Generally, planting distances are 5 m × 2–2.5 m, sometimes 5 m × 3 m (Fig. 5.2.7). This system has been proposed to increase the number of varieties that could be cultivated according to a hedgerow system (larger distances allow a larger number of cultivars to adapt to this system). The concept is to create a system that allows for full production in four to six years, the partial mechanization of pruning and the use of machines for continuous harvesting. The harvesters are large straddle machines or machines that can operate on one side of the row at a time or on both simultaneously, up to 4–5 m in height; prototypes and versions of these harvesters are currently being improved. Machines for continuous harvesting also include 'side-by-side' shakers.

For further information on intensive high-density olive orchards, see also the chapters 5.1 and 6.4.

In Spain there have been several attempts to set up orchards at 6 m × 2 m with irrigation, where the trees are trained to form large hedgerows that are harvested with a large straddle machine pulled by a tractor (Fig. 5.2.8). The wide spacing and bigger hedgerows of this system could increase the number of suitable cultivars, but more research is needed to determine its long-term viability.

5.2.3.10 Arrangement of rows and trees of different cultivars

Row orientation

In all new orchards, rows must be oriented in such a way as to optimize light interception and facilitate cultivation practices. As for the light, the north-south (N-S) orientation is the best (it is important to get as close to this orientation as possible). Long rows are best for cultivation practices. However, if this involves an unsuitable orientation for good lighting of the canopy, the latter should be favored, unless in particular situations the area consists of very narrow tracks in the N-S

Fig. 5.2.5. A low-density hedgerow on a dry farm: planting distances 7 m x 2 m. (Photo: Raúl de la Rosa)

Fig. 5.2.6. A high-density hedgerow on a dry farm: planting distances 4 m x 2 m. (Photo: Raúl de la Rosa)

Fig. 5.2.7. Intensive high-density olive orchard with trees spaced 5 m x 2.5 m. (Photo: Franco Famiani)

direction and large ones in the opposite direction. In sloping terrain with a steep slope (> 10%), the rows must be arranged vertically, i.e. according to the maximum slope, because this makes the use of machines safer.

In intensive olive orchards, in all situations which deviate significantly from the possibility of using a N-S orientation, it could be convenient to use a square layout.

In the hedgerow and intensive high-density olive orchards, the N-S orientation is more binding because of the continuous walls of vegetation that are formed. In those cases where a straddle harvester is used, the total length of the tree rows should be adjusted also considering the maximum crop load of this machine.

Arrangement of trees of different cultivars

If more than one main cultivar is planted, the different cultivars should be set out according to blocks (e.g. blocks of four rows per cultivar, which are repeated if the plot is large). In this way, if necessary, it is possible to manage the trees (e.g. pesticide treatments only for the cultivars that need them) and the olives (e.g. harvested at different times if ripening occurs at different times; separation of the olives for the production of monovarietal oils) of the different cultivars

Fig. 5.2.8. A large tractor-towed straddle machine for harvesting large hedgerows. (Photo: Raúl de la Rosa)

separately. In areas at risk of cold damage, if the new orchard is established on a hill with vertical rows and if the cultivars used have a different susceptibility to low temperatures, they should be divided by altitude, placing the most resistant ones in areas at greatest risk of cold damage.

In designing the arrangement of the rows and trees (squaring), in order to have a space for easy maneuvering of the machines, headlands should be planned with a width of 6 m plus half the distance of the plants along the rows (however, a total width of at least 8 m must always be ensured), and a distance of the rows from the lateral borders equal to that between the rows.

Density and distribution of pollinizers

The pollinizers, which are mandatory in orchards where the main cultivar is self-incompatible and there might be a lack of airborne pollen of compatible cultivars, can be uniformly distributed within the plantation. In this case, they should consist of at least 10% of the total number of trees. Alternatively, they may be arranged in rows that alternate with those of the main cultivars. In the latter case, they should not be less than 20% (one row every five). In any case, the maximum distance between the pollinators and the trees to be pollinated should not exceed 30 m. The arrangement of the pollinizers in separate rows makes it possible to achieve the advantages described in relation to arranging in blocks of rows in the case of using different main cultivars to create the olive orchard.

If the plot is located in an area where there is a predominant wind during flowering, pollinizers should be placed in such a way as to be well disposed to the wind, therefore ensuring the best distribution of pollen. If there is a self-compatible cultivar among those used, it should be placed in the first part of the olive orchard hit by the predominant wind.

When several main inter-compatible cultivars are used for establishing the olive orchard, it is not necessary to have trees with the sole function of pollinizers; however, the inclusion of pollinizers is always a useful choice to create the best conditions for obtaining the maximum productive potential of the olive orchard.

When using a cultivar where it is not known to which group of incompatibility it belongs to, it is important to use two cultivars as pollinizers, one of the group G1 and one of the group G2. In this case, after a group of rows of the main cultivar, for example four rows, it is advisable to use one row of the cultivar G1 and one row of the cultivar G2, which pollinate each other, and one of them ensures the pollination of the main cultivar.

If the new plantation is established in an area where there are other olive orchards, these, too, can contribute to pollinating the new one.

Arrangement of rows in case of replanting

In case of replanting, to minimize sanitary risks, it is very important to carefully remove the residues (stumps and roots) of the previous trees and to plant the new rows in the center of the inter-row of the old olive orchard.

5.2.3.11 Choice of trees

Trees obtained by grafting or cutting

Trees derived from grafting or from cuttings (self-rooted) are available in the nursery. In both cases they are grown and marketed in pots and therefore with potting soil. There are no substantial differences between the two types of trees, as rootstocks capable of conferring particular advantages to grafted plants are not available. However, those derived from cuttings seem to have a greater susceptibility to water stress and in anchoring in the early stages of development, but they are more uniform and enter into production sooner. Furthermore, they do not have problems in case it is necessary to cut at the base following a frost that has damaged the aerial part of the tree, whereas in grafted ones, cutting under the grafting point, the suckers that develop derive from the rootstock and therefore must be grafted. The differences in sensitivity to water stress and in anchoring are due to the different structures of the root systems in the early stages of development of the two types of trees. The trees from cuttings have a very voluminous root system with slender roots, whereas the grafted plants have fewer roots, but are more robust and able to immediately penetrate deeply. However, after two to four years of growth in the field, these differences are no longer noticeable. Grafted plants are often preferred in environments with limited rainfall and no irrigation. Instead, those from cuttings are preferred in areas at risk of frost. In cold areas, if no cutting plants are available, grafted ones can be used with the precaution of burying the grafting point at the time of planting. In any case, the choice depends on the nursery production, where propagation by grafting is preferred over that by cuttings and *vice versa* for varieties with reduced rhizogenic capacity.

Age and structure of plants

Regardless of the method of propagation, the plants are marketed at an age ranging from six months to two years (in the case of grafted ones, the age means from the grafting).

In the case of intensive low-density olive orchards with pot-grown trees, it is recommended to use two-year-old plants with the main axis free from vigorous lateral branches up to a height of 1.0–1.2 m, and with branches that form an already well-developed crown in the apical portion (plants already structured in the nursery) (Fig. 5.2.9). The use of two-year-old trees implies that they have grown well in the nursery. In this regard, it is important that they had been transplanted into larger pots at the end of the first year. This is necessary to avoid aging and twisting of the roots inside the pots. Well-grown two-year-old plants, compared with one-year-old ones, allow the time to reach full production to be reduced. Furthermore, with these trees, management difficulties resulting from insect attacks, such as *Margaronia* or olive moth, are avoided/strongly reduced. They are much more harmful in plants that have not yet reached the height at which the main branches develop, because by damaging the apexes, the tree's growth in height and the

Fig. 5.2.9. Two-year-old tree suitable for intensive low-denisty olive orchards. (Photo: Franco Famiani)

times of training pruning can be significantly prolonged. In addition, with younger plants (one year old, 50–100 cm in height), it is necessary to ensure greater initial care and pay particular attention to the control of weeds along the row and to the training pruning. In the case of older plants (three years old), there is the risk of a greater transplant crisis, especially if the plants have been raised in pots of inadequate size and shape (too small, round and without internal ribs). In this case the roots may have developed around the walls of the pot (helical development), creating a slowdown in initial growth and reduction in stability when planted out. It is always good to make sure that in the nursery the pot has not been in contact with the ground. This could promote the development of roots outside the pot, which, when the plants are sold, must be cut (if this has been done you can see the white of the roots cut through the drainage holes of the pot), with consequent reduction of the root system and increased risk of promoting transplant crisis.

For SHD and intensive high-density plantings, in general, six- to eight-month-old plants are used, 0.4–0.6 m tall (Fig. 5.2.10). Also, in this case, the use of the tall plants (0.8 m in height) facilitates the initial management, in terms of pruning, and reduces the negative effects resulting from any attacks of *Margaronia*. For the second-generation superintensive orchards ramified trees (Smarttree) are used. Obviously, taller trees are more expensive. In the case of organic orchards, due to the lack of inorganic nitrogen supply, the initial growth of the plant is much slower. Therefore 1.0–1.3 m-tall plants are recommended even for SHD and intensive high-density orchards (Fig. 5.2.11).

Fig. 5.2.10. Trees 0.4–0.6 m tall are often used in hedgerow orchards. (Photo: Franco Famiani)

Certified plants

In the European Union, plants must meet the requirements imposed by Community standards, and therefore be accompanied by the plant passport and the marketing document with the indication 'CE Quality', which require varietal compliance and the absence of the main parasites. In the case of an organic olive orchard, the plants must come from nurseries that use the organic production method (Fig. 5.2.12).

Certification systems have also been developed in extra-EU countries. It is always recommended to use certified trees for establishing new orchards.

5.2.4 Establishment of the Olive Orchard

5.2.4.1 Soil preparation

Before starting to clean up and reorganize the land, it is important to ensure that there are no hydrogeological and landscape restrictions to be respected and to ask for any authorizations from the authorities (e.g. for the movement of high volumes of soil).

Soil clearing

The soil must be cleared of plant residues from any previous crops or, in case of abandoned, uncultivated land, shrubs, trees etc., must be removed. It is possible to use a special brush cutter with a horizontal axis that cuts and shreds them so that they can be used to enrich the soil in organic matter; however, remains of stumps and larger roots must be removed. Bulldozers, skid steer loaders, grubbers or excavators can be used. The same operation must be performed in case of replanting, which can be done immediately after grubbing out the old olive orchard, because there is no problem of soil exhaustion (soil sickness) with the olive tree.

If there are large stones in the plot, they must be removed. If the stones are small, stone-crushing machines can be used

Fig. 5.2.11. Trees about 1 m tall. In hedgerow orchards, the use of tall plants aids the initial management of the trees and reduces the negative effects from *Margaronia* attacks. (Photo: Franco Famiani)

to crush them, but this should be evaluated economically.

Surface soil leveling and drainage

If the ground surface is uneven, it should be leveled (with bulldozers and skid loaders), eliminating depressions and bumps. This is important to avoid possible water stagnation (which, in addition to creating problems of root asphyxia, can favor collar rot and Verticillium wilt) and to aid maneuvering of equipment and the application of irrigation.

If major excavations are needed for surface leveling that go well beyond moving the topsoil, it is recommended that topsoil be stockpiled then spread on the surface once the leveling operation has been completed. This allows the maintenance of the most fertile layers of the soil where most of the roots develop, resulting in a better and more regular growth of the plants. If this is not cost-effective, at least abundant organic amendments should be carried out in the area(s) where the subsoil has been brought to the surface. In hilly areas, excavation and filling in must be carried out with prior authorization from the competent authorities, if required, and in such a way as not to destabilize the slope. It is important to ensure good cohesion between the soil brought in and that on the spot. This can be achieved by scarifying the soil on the spot and then making progressive fillings, 30–40 cm thick, which must be well compacted by repeated passages with the machines, up to the desired height.

Drainage in flat areas

In flat areas, if there is a risk of water stagnation (e.g. poor permeability of clay soils,

Fig. 5.2.12. An organic hedgerow orchard. (Photo: Raúl de la Rosa)

presence of high groundwater due to intense and long-lasting rains or in situations in which water flows with difficulty due to the low grade), it is necessary to install ditches (drain ways – drawbacks: production reduction, frequent maintenance, obstacle to maneuvering equipment) or tile drains (drains – drawbacks: occlusion of the holes, problem that can be mitigated by using drains coated with coconut fiber). Ditching is effective in eliminating water in all situations that cause stagnation, while tile drains are not good in the case of soils with particularly low permeability (e.g. very clayey), as the flow of water towards the drains would be so slow as to make them ineffective. Tile drains, on the other hand, are particularly recommended in the case of a high water table. Ditching or tile draining must ensure a cultivation clearance of at least 50–60 cm. In situations most at risk of stagnation, the distance between ditches should not be greater than 20 m, while in rather permeable soils they can reach 35–50 m. Tile drains with corrugated and perforated PVC tiles can be installed with tiling machines at a depth of 1–1.5 m and 5–10 m apart in clayey soils and 20–25 m in relatively loose soils. In the plains, whenever possible, the use of tile drains rather than ditches is always better, because ditches cause significant management problems, in particular for the movement of machines.

The grade of the drains must not be less than three per 1000. The drains must outlet the water into natural areas, such as ditches, rivers, lakes etc. or into artificial collection ditches (ditches or collectors). The drains must flow into collectors capable of disposing of the collected water.

Surface soil shaping and drainage in hilly areas

In hilly areas, up to a 5% grade, the rows can be arranged perpendicularly with respect to the slope, so that soil-management operations are carried out sideways. This allows surface run-off (and therefore erosion) to be reduced and favors the accumulation of water in the soil. Working the soil sideways can give rise to problems only in the case of clayey soils at risk of landslides. Alternatively, with grades greater than 5%, the rows and cultivation operations are carried out in the direction of the maximum slope. In this case, to limit erosion, the length of the rows should not exceed 100–150 m if the slope is between 10% and 20% and 50–100 m if greater than 20%.

When the grade exceeds 10%, especially if the plot is long, it is useful to install temporary ditches along the slope, according to the contour lines, at a distance between them from 30 m to 70–80 m depending on the grade and permeability of the soil. The ditches must be 5–10 cm deeper than that of the main working depth, a grade ranging from 1% to 2.5%, depending on the nature of the soil, and a length not exceeding 200 m. These ditches must be restored every time work is carried out. Obviously, it is possible to install temporary ditches only if trees are spaced enough and do not form hdgerows.

The water from the ditches is collected and outlet into natural drains or artificial waterways, preferably with grassing, laying stones or concrete slabs on the bottom, which run downstream.

If grassing (permanent green cover) is used, it is not necessary to create the aforementioned temporary ditches. Therefore, in hilly plots it is always advisable to use grassing especially if the olive orchard is irrigated. If grassing is used, it is advisable to arrange the rows in the direction of the maximum slope also when the grade of the soil is lower than 5%.

In the case of clayey soils, if deep stagnation occurs (with the risk of landslides), due to insufficient natural drainage, tile drains should be installed; this is also important to do in case of impluvium where water tends to accumulate.

If water consistently arrives from the land above, it is advisable to create a guard ditch upstream of the plot at least 50–70 cm deep to collect the water from the upstream slope and drain it into the existing hydraulic network.

When the grade of the plot exceeds 30%, terraces must be built. Considering the cost and difficulties terraces create for mechanization, it is justifiable only in particular cases such as production of typical oils that get very high prices and/or where there is the need to consolidate the slopes, making the olive tree an important environmental and/or landscape feature.

Service infrastructures

At the time of surface shaping, it is also important to plan service infrastructures, such as farm roads, excavations for burying irrigation pipes, ponds or tanks for collecting water, wells and, if it is in an extremely windy area, windbreak barriers.

5.2.4.2 Basic fertilization

Basic fertilization has the purpose of bringing the fertility to suitable levels for a good development of the trees and eventually creating reserves of non-mobile nutrients. Soil testing should be done in the top 60–80 cm. The values should be compared with reference values, in order to establish the amount of fertilizer to be applied. Nitrogen is not included as it is soluble and subject to leaching. Basic fertilization is not needed if the organic matter and nutrient content of the soil is at or above the reference levels. In the latter case, fertilization should be resumed when values drop below the reference ones.

In medium-fertile soils, generally, 40–60 t/ha of aged manure is needed (up to 100 t/ha can be spread), 150–250 kg/ha of phosphorus and 200–300 kg/ha of potassium. If the amount of organic matter to be added is very high because the organic matter in the soil is low, it should be applied gradually over several years. In loose soils, the organic matter is mineralized rapidly and the nutrients are very mobile (including phosphorus and potassium), therefore it is preferable to divide the amount of organic matter and fertilizers over the years, in order to avoid depletion. Aged bovine manure is the most widely used organic fertilizer because it is excellent and relatively easy to find, but manure from other animals (horses, sheep, rabbits, chickens etc.) and fertilizers obtained from composting of various substances (pomace, straw, stalks, manure of various origins etc.) can also be used. Composts found on the market, also derived from the recycling of municipal solid waste (MSW), provided that they are of high quality and without significant amounts of heavy metals, can be used.

Generally, composts have a relatively high dry-matter content (from 35% to 55%), compared to manure; therefore, they are used in doses reduced by 30–50%. The use of amendments available on the market should be evaluated economically since large amounts must be used in order to affect the balance of organic matter.

In the European Union, in the case of organic production, fertilizers allowed by Regulation (EU) 2018/848 must be used.

If the fertilizers mentioned above are not available, a valid alternative to provide organic matter is the use of green manuring carried out with mixtures of grasses and legumes with autumn sowing (barley (75 kg/ha) + vetch (*Vicia villosa* 25 kg/ha or *Vicia sativa* 50 kg/ha), or, alternatively, barley (50 kg/ha) + field bean (*Vicia faba* 100 kg/ha)) or spring sowing (with the same species and amounts of autumn sowing or replacing barley with oats (70–80 kg/ha) or with Italian ryegrass (15–20 kg/ha)). The green manure is then worked into the soil. Green manuring can be done before deep plowing or after planting. In the latter case, the incorporation is more superficial. Green manure generally brings 4–6 t/ha of dry organic matter equivalent to 0.4–1.8 t of humus (the isohumic coefficient, which expresses the amount of humus that can be obtained from the organic substance considered, can vary from 0.1 to 0.3). It is good to remember that, with green manuring practiced as a basic fertilization, the main objective is to have a good humus yield. Therefore, mowing should be done at a relatively advanced stage, after the grasses have eared out and the legumes have flowered, when there is a relatively high content of lignin and cellulose in the biomass.

Regarding phosphorus (P) and potassium (K), simple perphosphate (19–21% P_2O_5), triple superphosphate (46% P_2O_5), Thomas slag (16–18% of P_2O_5 – byproducts of steel processing), potassium sulphate (48–52% of K_2O), potassium saline (40% of K_2O) can be used. For trees grown according to the organic method, fertilization is based on the use of organic matter, but there are also simple natural fertilizers that can be used. For P it is possible to use soft natural phosphate (obtained from the grinding of phosphorites), aluminum calcium phosphate (Phospal) and Thomas slag. For K, raw potassium salt and potassium sulphate containing magnesium salt can be used. In acid soils, natural phosphate and Thomas slag are used, while in alkaline soils (calcareous), aluminum calcium phosphate (Phospal) is used.

Basic fertilization is done before plowing so that fertilizers are incorporated in the layer of the root zone. However, if very deep plowing is done, the fertilizers should be incorporated with a shallower process to prevent them from being buried at an excessive depth.

5.2.4.3 Deep-till plowing

Deep-till plowing consists of working the soil deeply. The objectives are: to favor root development and water percolation and eliminate any mechanical barriers; to improve soil aeration; to bury soil amendments; to correct the chemical composition and the pH; to improve the availability of nutrients; to mix any layers of soil with different textures if this leads to an improvement in the final texture; to complete the removal of root residues (this operation should be done immediately after the uprooting when it is easier to remove the roots because they are still fresh and not crumbly).

The use of deep tilling is particularly important in compact soils where it is necessary to reach a depth of 80–100 cm. If not done, the trees could have a stunted development. In loose soils, a depth of 50–70 cm is sufficient; in these soils, deep tilling is less important because they have a good degree of aeration, do not give rise to water stagnation and do not hinder root development.

Method of deep tilling

Deep-till plowing is advisable when useful for mixing the subsoil with the topsoil (e.g. clay and sand) to obtain a texture more suitable for olive cultivation, or when the soil is particularly compact. In other cases, it is better to use a two-layer operation, which consists of using a ripper which cuts and raises the soil up to a depth of 80–100 cm, followed by a plowing at a depth of 40 cm, incorporating the organic residues and fertilizers (Fig. 5.2.13). The best results are obtained with cross-ripping using a distance between the teeth of the ripper of 40–50 cm.

The two-layer operation is particularly good when the subsoil is rocky or stony, as it allows the soil to be tilled without bringing too many stones to the surface. If stones are brought to the surface, they must be removed or crushed. Other advantages of two-layer working are: the incorporation of fertilizers in the topsoil where most of the roots develop; the absence of forming a tillage sole, which can cause difficulties in infiltration of excess water with possible consequences also for the stability of the slopes; reducing erosion that often occurs when hilly soils are worked in the direction of the slope, especially if the ripping is performed first in the direction of the slope and then in the transverse direction and also if plowing is done according to this latter sense. Generally, the two-layer operation is also less expensive than classic deep-till plowing.

Fig. 5.2.13. The use of a ripper to cut and raise the soil up to a depth of 0.8–1.0 m. (Photo: Franco Famiani)

Time to do deep-till plowing

The best time to carry out deep-till plowing is summer, but it can also be done at other times as long as the ground is in good tilth (not too wet and not too dry). When the two-layer operation is used, it is also possible to do ripping in spring and plowing from the end of summer to the beginning of autumn.

5.2.4.4 Refinement of the soil surface

After deep-till plowing, a finishing operation must be carried out to break down clods and even out the ground. A harrow (with discs or teeth) or grubber is used. This also takes care of uprooting any remaining crop residues.

5.2.4.5 Squaring

Squaring, the process of establishing the position of the rows and therefore the trees, is done by 'staking', that is, by driving stakes where the trees will be planted. Squaring is done using various tools, such as tacheometers, surveying squares, tape measures, poles, wires etc. A basic reference alignment is traced with poles, generally along the longer side of the plot, after which, using a tacheometer or, more simply, a surveyor square or a metric cord, main alignments are set perpendicularly to this at regular distances (multiples of the distance between the rows). When using the metric cord (for small orchards/farms), the system of the right-angle triangle is used with the sides in a ratio of 3:4:5. In practice, the first 12 m of rope are used and a right-angle triangle is built on the ground by placing and holding the 0 m and the 12 m notch of the tape at the point where you want to trace the alignment perpendicular to the baseline (basic reference alignment). Then, the rope is run along the baseline and, at a distance of 4 m, a pole is placed around which the tape is rotated and then pulled with another pole so as to form a triangle with sides of 4 m (along the basic reference alignment), 5 m and 3 m, with the 3 m side forming an angle of 90° with that of the 4 m side, and therefore the alignment perpendicular to the reference baseline is identified.

Once the main alignments have been set, squaring is completed using a tape measure, poles and stakes, identifying the position of all the rows and along these the points where the plants will be placed.

When possible, rectangular plots are created with the longer side in a N-S direction to optimize lighting.

Squaring and use of transplanters

Machines have been used to plant the trees by performing only a partial squaring or by completely eliminating the need for squaring. These are transplanter machines that use two different guiding systems to automatically or semi-automatically identify the point where to plant the tree.

The oldest system, developed about 40 years ago, is laser-guided. In this system, partial squaring is needed to identify the beginning and the direction of each row. On the edges, at the point where the row begins, a laser beam emitter is placed which is directed towards a receiver placed at the point where the row ends. In this way, the alignment of the row is precisely identified. A receiver installed on the tractor guides the transplanter, enabling the correct direction of each row to be maintained. The transplanter can mechanically measure the planting distance between the trees with precision, placing them at regular intervals. After each row has been transplanted, the emitter and receiver are moved to the next row. This system is more suitable for flat and non-rolling plots.

More recently, satellite-guided transplanters have been produced, which completely eliminate the need for squaring. These machines use a receiver of the global positioning signals emitted by the GPS (USA system) and GLONASS (Russian system) satellites, which allows automatic, precise identification of the point where to plant each tree. The system uses a receiving station (base station) which is fixed at a point on the edge all the time. This station communicates with the satellites and with a regional position correction network, and 'centimeter' precision is obtained. A computer is installed on the tractor or transplanter, connected to another receiver which takes both the satellite signal and that of the base station, in order to obtain the same centimeter accuracy. The computer automatically calculates all the points at which to plant the trees. A control system installed on the transplanter, communicating with the computer, allows all alignment and inclination errors to be corrected in real time, in order to carry out planting with extreme precision. This system is suitable for any plot, even with an accentuated grade and undulation. It is used both on transplanters and on pile-driven machines. Usually, all the SHD and intensive high-density orchards are planted with this system, which, however, can also be used for the plantation of intensive low-density olive orchards, especially if planted over a large area.

5.2.4.6 Planting

Time of planting

Planting in mild winter climates, where the risk of cold damage is negligible, especially if spring rainfall is limited, is preferable to autumn, otherwise in late winter/early spring. In the latter case, in mild environments, planting can be done starting about a month before the end of winter, while in those at risk of freezing it is better to postpone until after the beginning of spring (up to 20–30 days later). With potted trees, it is also possible to plant them later, as long as a good availability of water is ensured. In hot, arid environments, where aridity is more problematic than low temperatures, planting can also be done during winter.

5.2.4.7 Planting of intensive low-density olive orchards

To plant the trees, it is necessary to dig holes by hand or with a small excavator or an auger operated by a tractor or with a motor drill, 40 cm wide and deep (Fig. 5.2.14). The soil must be dry, especially if the soil is clayey, to avoid compaction of the walls, which would then hinder water flow, and poor placement and/or excessive compaction of the soil close to the roots if planting is carried out immediately. The holes could also be made a few days before planting to improve the structure of the walls and the soil which will then be used to fill them. A pole should be driven into the bottom of the hole, which should be placed on the north side of the plant (northern hemisphere), especially if it has a large diameter, to avoid shading of the leaves along the stem (Fig. 5.2.15). In areas with heavy winds, the pole should be placed on the side where the main wind blows, to avoid damage of the plant. If basic fertilization has not been carried out, fertilizer or organic matter can be put in the hole (e.g. manure, which should be well matured to prevent its volume from being excessively reduced, which would result in a serious sinking of the plant) and then covered with a layer of soil. Recently, it has also been proposed to use leonhardite, which contains humic and fulvic acids (that have biostimulant effects) and microbial biostimulant, which contains microorganisms able to develop mycorrhizae, distributed around the potting soil with the aim of increasing the initial growth of the trees (Figs 5.2.16 and 5.2.17). Research is in progress to evaluate their potential.

The trees should be removed from the pot, taking care not to break the root ball, after which they must be positioned so that the collar is at ground level or slightly lower (no more than 5 cm) and the stem near the pole. Only in particular cases should trees be planted deeper. With grafted plants in areas at risk of freezing, it is advisable to bury the grafting point (5–10 cm) to facilitate rooting and, therefore, in case of damage to the aerial part, allow regrowth of the plant, cut at the base, by means of a sucker without having to re-graft it.

To avoid breaking the root ball when the trees are unpotted, it is important to have the right degree of humidity, so watering the potted trees the day before planting is advisable.

Fig. 5.2.14. The use of a small excavator to dig the holes in a new olive orchard. (Photo: Franco Famiani)

To put the tree in the exact position where the pole is, a 1–1.2 m-long plank can be used on which three notches have been made, two near the ends of the plank and one in its center. Before digging the hole, make the central notch coincide with the pole and place two sticks in correspondence to the two notches at the ends, then remove the central pole and the plank and dig the hole. When setting the plant in the hole, put the plank back by making the two extreme notches coincide with the sticks and position the plant at the central notch. After that, the hole is filled by placing the soil under and around the root ball, compressing it progressively so as to make it adhere well to the ball and thus create good continuity to favor root development.

The tree is then tied to the pole and about 10 liters of water are given to facilitate contact between the soil and the roots.

5.2.4.8 Support poles in intensive low-density olive orchards

The poles, driven into the ground to a depth of about 0.5 m, must be about 1.5 m above ground for trees trained according to the vase system and 2–2.5 m for those trained according to the monocone system. They can be made up of different types of materials. Wooden ones are usually chestnut, but they can also be flowering ash, hornbeam or pine. Pine poles are generally treated to avoid rot, prolonging their life. The others are generally

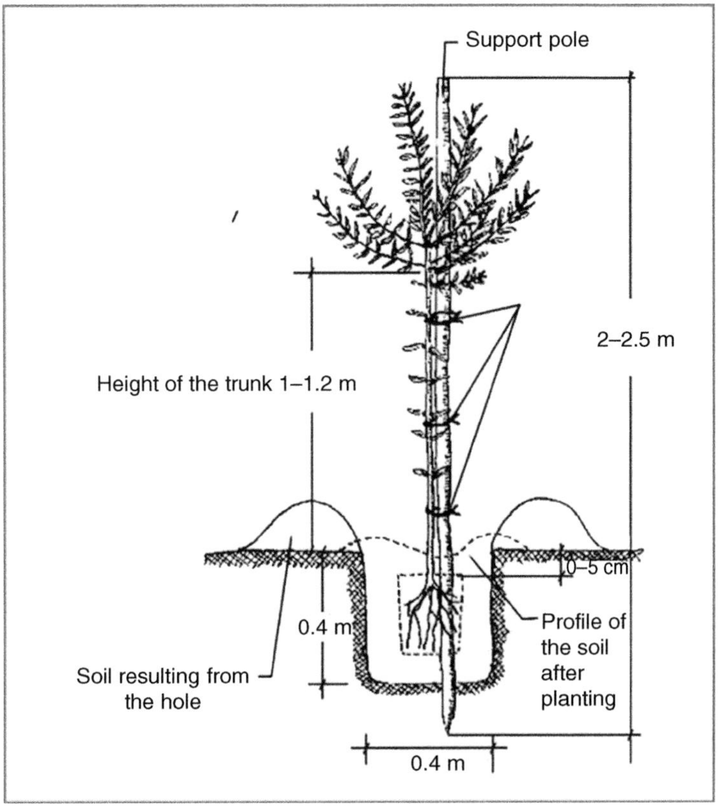

Fig. 5.2.15. Scheme of planting of a two-year-old tree in an intensive low-density olive orchard. (Drawing: Franco Famiani)

marketed naturally and therefore it could be useful to treat the part that will be buried by soaking them in a 1% copper sulphate solution for three to four days, in order to avoid rotting in a short time. The wooden poles should be no less than 6 cm in diameter to provide support for the trees until they are able to stand on their own (after four to six years of age) (Fig. 5.2.18). For installing these poles in the ground, a post driver can be used that is attached laterally to normal tractors (Fig. 5.2.19). In this case, the poles can first be installed and then the holes for planting adjacent to them can be made, by hand if the ground is soft or with the specific equipment previously indicated. If plastic poles are used, they must be of the reinforced type (with a hexagonal or octagonal starry section, that is, with internal reinforcement) and be no less than 6 cm in width. If the new olive orchard is equipped with a drip-irrigation system supported by a structure along the row consisting of headed and intermediate poles and wire, on which the dripping line is attached, the poles can be tied to the wire itself. This increases their supporting capacity and therefore less robust poles than those indicated can also be used. However, with respect to this solution, an interesting possibility, which is simple and cheap, is to fix the irrigation lines directly on the plant posts, especially if they are sturdy (Fig. 5.2.2).

It is important to consider that, in all cases, the installation of suspended irrigation lines can cause problems for harvesting with a shaker equipped with a reverse umbrella interceptor. Therefore, they can be suspended at a height of about 0.6 m from the soil up to the starting of mechanical harvesting. Then it is necessary to arrange the drip lines at a height that is not reached by the umbrella, or on the ground along the row (managing them by weeding) or underground (sub-irrigation).

Fig. 5.2.16. Leonhardite put in the hole at planting time. (Photo: Franco Famiani)

Fig. 5.2.17. A microbial biostimulant, potentially able to develop mycorrhizae, put in the hole at planting time. (Photo: Franco Famiani)

Fig. 5.2.18. A sturdy support pole for two-year-old trees in an intensive low-density olive orchard. (Photo: Franco Famiani)

Tying the trees to the poles

Tying of the plants to the poles, two to three ties per plant, should be neither too tight nor too loose, as in the first case it could cause choking of the growing stem and in the latter case the wind could cause the plants to rub on the pole, resulting in damage to the bark. The ties should be made in a figure of eight to reduce the risk of choking and to avoid direct contact between the pole and the tree (Fig. 5.2.20). The ties should be loosened/retied periodically as the tree grows. Plastic tubular ties or other non-rigid materials should be used to make the ties.

5.2.4.9 Structures for supporting the tree rows and/or the irrigation system in hedgerow orchards

In hedgerow systems, header and intermediate poles are used to support the rows, to which 1–3 wires are attached. In correspondence to each tree, iron rods (diameter

Fig. 5.2.19. Installation of poles in the ground by using a post driver that is attached laterally to a tractor. (Photo: Franco Famiani)

Fig. 5.2.20. Example of a figure-of-eight tie to support the tree. (Photo: Franco Famiani)

6–8 mm), tied to wires, are used to support the trees. Bamboo (around 2 cm in diameter) or plastic (3 cm in diameter) poles can also be used. In all cases, the height from the soil to the top of rods/poles supporting the trees is usually of 2 m.

For constructing the support structure of the irrigation system and in the case of SHD systems, it is possible to use wooden, prestressed, reinforced concrete or galvanized iron poles and iron wire (2.5 mm in diameter) with triple galvanization or steel (the best for durability, ease of installation and reduced stretching, but more expensive), while those made of polyamide are not recommended as they could be inadvertently cut during pruning. A valid alternative, in terms of cost, duration and resistance, to galvanized wire is the use of steel wire protected with a zinc and aluminum alloy (crapal wire), which considerably prolongs its life.

The head poles should be tilted towards the headlands by 15–20° and should be attached to special anchors, which are made up of concrete plates with a 10–12 mm iron rod, which are buried at a depth from 0.8–1.0 m depending on the nature of the ground (greater in loose soils), letting the rod with a loop at its end protrude from the ground, or by galvanized iron anchors that screw into the ground, even those with a loop that remain above ground. These anchors are easy to install, but in sandy soils they do not always guarantee adequate anchoring.

Attaching the wire to the poles is best done using special metal collars that are attached to the poles. To keep the wire at the right tension, wire tighteners must be attached near the head poles and, if the rows are longer than 100 m, also along the row. Collars used on head poles can also be equipped with a spool tightener.

The end poles are secured to the anchors by a wire that connects the loop of the latter to a collar.

The distance between the poles should be established according to the density of the plants, windiness and size of the poles to which the plants are tied: generally, this is every 10–20 m in SHD and high-density systems.

When harvesting the firsts crops, the shaking movement of the straddle machine can be transmitted by the wire of the supporting structure to the trees that are located a few meters in advance of the machine itself. Therefore, the fruits of those shaken trees could fall some meters ahead of the machine thus producing some yield loss. One solution for this is to use only a wire at 60–80 cm from the ground or to eliminate the supporting structure three to four years after planting when the trees do not need any additional support.

5.2.4.10 Use of transplanters

The transplanters, both laser- and satellite-guided, can be used for all types of olive orchards (intensive low-density and high-density and SHD). They are attached to the rear of the tractor and operate by making a groove inside to which the plants and poles are inserted at a predetermined distance (Fig. 5.2.21). When support poles are to be used, they can also be installed with the transplanter, which can also enable the irrigation line to be laid. The transplanter allows a very high working capacity, enabling planting from 3–4 to 6–8 ha/day, depending on the conditions (hill/plain, length of rows etc.). Therefore, it is particularly useful when large areas have to be planted. There are stations on the transplanter for the operators who place the trees and poles in the devices for planting.

5.2.4.11 Application of mulch

At the time of planting, to avoid competition from weeds, mulching can be applied along the rows, at a width of 1–1.5 m, with plastic film (polyethylene, polypropylene, polyester) or with non-woven fabric that allows air and rainwater to pass (Fig. 5.2.22). Under the mulching film, irrigation lines and possibly fertilization (fertigation) can be placed. Mulching favors the initial development of the trees (greater growth of the root system and the aerial part) by inhibiting weed development along the row and improving soil conditions. However, the costs related to the purchase and installation of the mulching material as well as its disposal limit the use of this technique. Another drawback is the spread of rodents and moles, which in some areas can cause damage to the trees.

5.2.4.12 Use of tree shelters

In areas susceptible to rodent attacks (voles, hares etc.), tree shelters should be used. This is more likely to occur in olive orchards established near wooded areas. Shelters are tubes, generally made of polypropylene, placed around the trees. They reduce the risk of damaging plants by mechanical means or when weeding with herbicides. If weeding is done, neither nets nor those with holes in the lower portion (to facilitate

Fig. 5.2.21. Use of a tractor-towed transplanter. (Photo: Franco Famiani)

internal ventilation) should be used. The growth of suckers on the stem inside the shelter can cause problems, and so it is important that the shelter can be moved up to eliminate the weeds and suckers. Therefore, the shelter should be shorter than the height of the stem between the soil and the first ramifications (Figs 5.2.11 and 5.2.12).

In SHD plantings, shelters are mainly used to protect the tree stems during soil-management operations (Figs 5.2.6 and 5.2.12). The cost of shelters must be considered. In some cases, as a low-cost solution, an insulating sheath, cut at the desired length, can be used as a shelter (Fig. 5.2.11).

5.2.5 Post-planting Operations (First Year)

After planting, starting from the vegetative restart, or in the case of planting in the spring, 10–15 days after planting of the trees, the following operations are recommended:

- Localized nitrogen fertilization (2–4 applications during the spring, for a total of about 50–100 g/tree (depending on tree age/size), avoiding direct contact of the fertilizer with the stem).
- If no ordinary irrigation is carried out, emergency irrigation in case of drought,

Fig. 5.2.22. Use of mulching. (Photo: Franco Famiani)

especially if self-rooted plants have been used; application of water also makes it possible to improve the absorption of the nitrogen applied.

- If mulching has not been done, weeds should be eliminated (weeding or chemical weeding), as they exert strong competition towards water and nutrients, negatively affecting the growth of young olive trees.
- Elimination of vigorous shoots that develop along the stem of the two-year-old trees used in intensive low-density orchards and, in hedgerow orchards, starting of the removal of the lowest shoots.
- At the beginning of autumn, in environments where there could be damage from low temperatures, carrying out a treatment with 1% Bordeaux mixture to stop the growth of the shoots and favor their hardening.
- Monitoring of pathogens and phytophages that can attack and cause serious damage to the trees, with particular regard to moth, *Margaronia* and weevil, and application of pesticide treatments when needed; by damaging the apexes

these pests cause the interruption of growth and the development of lateral shoots, with consequent slowdown in growth and greater difficulties in forming the crown; in case of hailstorms, treatments with copper-based products should be made to avoid the spread of olive knot.

- Replacement of dead trees.

In the first one to two years, it is very important to promote the vegetative growth in order to shorten the time required by the trees to occupy their available space. As known, significant effects on growth are obtained by minimizing pruning and applying fertigation. Recently, notable effects have been obtained, in both intensive and superintensive olive orchards by deflowering young trees by spraying a solution of naphthaleneacetic acid at 160 (NAA) at full bloom (open flowers > 80–90%) (Famiani *et al.*, 2022). This eliminates the competition for assimilates due to fruits, significantly favoring vegetative growth. In superintensive orchards, where very fertile cultivars are normally used, NAA might be best applied during the first and, possibly, the second year in the case of abundant flowering. In intensive olive orchards, the repeated use of NAA to prevent fruit set in the first few years may be a strategy to promote a significant reduction in the time required by trees to reach the trunk size that enables trunk shakers to be used (minimum diameter of 10–15 cm) and the time required to reach final tree size and full production (Famiani *et al.*, 2022). Positive effects on the growth of young trees have also been obtained by using biostimulants, in the first part of the vegetative season, or organic fertilizers also having a biostimulant action (Almadi *et al.*, 2020; Mazeh *et al.*, 2021). Experiments are in progress to evaluate the effects of humic and fulvic acids (contained, for example, in leonhardite) and microbial biostimulant applied at planting and used as starters.

References

Alfei, B., Pannelli, G. and Ricci A. (2003) Progettazione e realizzazione del nuovo oliveto. In: *Olivicoltura di qualità*. Edagricole, Bologna, Italia, pp. 61–89.

Almadi, L., Paoletti, A., Cinosi, N., Daher, E., Rosati, A., Di Vaio, C. and Famiani F. (2020) A biostimulant based on protein hydrolysates promotes the growth of young olive trees. *Agriculture* 10(12), 618.

Bandino, G., Dettori, S. and Sedda P. (2003) L'Impianto dell'oliveto. In: *Manuale di olivicoltura*. Regione Autonoma della Sardegna - Consorzio Interprovinciale per la Frutticoltura, Cagliari-Oristano-Nuoro, Italia, pp. 47–92.

Barranco, D., Cimato, A., Fiorino, P., Rallo, L., Touzani, A. *et al.* (2000) *Catalogo Mondiale delle varietà di olivo*. Consiglio Oleicolo Internazionale, Madrid.

Boulouha, B. and Gregoriou, C. (2010) Design of new groves, germplasm (varieties) and propagation – nursery. In: Good Agricultural Practices for the near east & north Africa countries. FAO, Regional Office for the Near East, Cairo, Egypt, pp. 94–108.

Breton, C.M., Farinelli, D., Koubouris, G., Famiani, F., Raymond, M. and Bervillé, A. (2021) A dual-successive-screen model at pollen/stigma and pollen tube/ovary explaining paradoxical self-incompatibility diagnosis in the olive tree: an interpretative update of the literature. *Plants* 10, 1938.

Connor, D.J., Gómez-del-Campo, M., Rousseaux, M.C., Searles, P.S. (2014) Structure, management and productivity of hedgerow olive orchards: A review. *Scientia Horticulturae* 169, 71–93.

Fabbri, A. and Ganino T. (2003) Impianto di un oliveto. In: *Note tecniche sulla coltivazione dell'olivo in Emilia*. Progetto per lo Sviluppo dell'olivicoltura da olio nelle province emiliane, pp. 4–11.

Famiani F. and Gucci R. (2022) *La raccolta delle olive*. Edagricole – New Business Media, Bologna, Italy.

Famiani, F. and Proietti P. (2007) Speciale Nuovi Impianti: tutte le tecniche che portano al nuovo oliveto. *Olivo e Olio* 6, 44–50.

Famiani, F., Camposeo, S. and Gucci, R. (2021) *Tipologie di impianto per una moderna olivicoltura*. Accademia Nazionale dell'Olivo e dell'Olio, Spoleto (PG), Italy, pp. 77.

Famiani, F., Cinosi, N., Paoletti, A., Farinelli, D., Rosati, A. and Lodolini, E.M. (2022) Deflowering as a tool to accelerate growth of young trees in both intensive and super-high-density olive orchards. *Agronomy* 12(10), 2319.

Famiani, F., Proietti, P., Inglese, P. (2021) *Progettazione e impianto dell'oliveto*. Accademia Nazionale dell'Olivo e dell'Olio, Spoleto (PG), Italy, pp. 75.

Famiani, F., Proietti, P., Micheli, M., Boco, M., Standardi, A., Ferranti F. and Reale, L. (2007) Effects of tree shelters on young olive tree growth and physiology. *New Zealand Journal of Crop and Horticultural Science* 35, 303–312.

Fontanazza, G. (1993) Criteri per l'impianto dell'oliveto in coltura intensiva, Le cultivar ed il miglioramento genetico. In: *Olivicoltura intensiva meccanizzata*. Edagricole, Bologna, Italia.

Gucci, R., Mancuso, S. and Sebastiani L. (2003) Resistenza agli stress ambientali. In: Fiorino, P. (ed.) *Olea - Trattato di Olivicoltura*. Edagricole, Bologna, Italia, pp. 91–111.

Hidalgo, J., Hidalgo, J.C. and Vega V. (2022) Evolucón de diferentes modelos productivos de olivar en los primeros años tras la plantación. *Vida Rural*.

Inglese, P., Famiani, F., Galvano, F., Servili, M., Esposto, S. and Urbani, S. (2011) Factors affecting extra-virgin olive oil composition. *Horticultural Reviews* 38, 83–147.

Lopez-Escudero FJ, Mercado-Blanco J (2011) Verticillium wilt of olive: A case study to implement an integrated strategy to control a soil-borne pathogen. *Plant and Soil* 344, 1–50.

Mazeh, M., Almadi, L., Paoletti, A., Cinosi, N., Daher, E. *et al.* (2021) Use of an organic fertilizer also having a biostimulant action to promote the growth of young olive trees. *Agriculture* 11(7), 593.

Medina-Alonso MG, Navas JF, Cabezas JM, *et al* (2020) Differences on flowering phenology under Mediterranean and Subtropical environments for two representative olive cultivars. *Environmental and Experimental Botany* 180, 104239.

Navarro, C. and Parra M.A. (2017) Plantacion. In: Barranco, D., Fernández-Escobar, R. and Rallo, L. (eds.) *El cultivo del olivo*. Coedición Junta de Andalusia – Consejería de Agricoltura y pesca; Ediciones Mundi-Prensa, Madrid, Spain, pp. 155–187.

Pannelli, G. and Alfei, B. (2008). I parametri importanti per scegliere le cultivar. *Olivo e Olio* 6, 46–56.

Rallo, L., Barranco, D., Castro-Garcia, S., Connor, D.J., Gómez del Campo, M. and Rallo, P. (2013) High-density olive plantations. *Horticultural reviews* 41, 303–383.

Rosati, A., Paoletti, A., Caporali, S., & Perri, E. (2013) The role of tree architecture in super high density olive orchards. *Scientia Horticulturae* 161, 24–29.

Saumitou-Laprade, P., Vernet, P., Vekemans, X., Billiard, S., Gallina, S. *et al.* (2017) Elucidation of the genetic architecture of self-incompatibility in olive: evolutionary consequences and perspectives for orchard management. *Evolutionary Applications* 10, 867–880.

Sibbett, G.S. and Osgood J. (2005) Site selection and preparation, tree spacing and design, planting, and initial training. In: Sibbett, G.S., Ferguson, L., Coviello, J.L. and Lindstrand, M. (eds.) *Olive production manual. University of California – Agriculture and Natural Resource*, California, USA, pp. 27–33.

Tombesi, A. and Tombesi, S. (2007) Orchard planning and planting. In: Production techniques in olive growing. International Olive Council, Madrid, Spain, pp. 15–40.

Vitagliano, C. and Sebastiani L. (2003) Impianto. In: Fiorino, P. (ed.) *Olea – Trattato di olivicoltura*. Edagricole, Bologna, Italy, pp. 195–209.

6.1 Soil Management

S. Camposeo[1] and J.A. Gómez[2]
[1]*Dipartimento di Scienze del Suolo, della Pianta e degli Alimenti, Università di Bari Aldo Moro, Bari, Italy;* [2]*Instituto de Agricultura Sostenible, Consejo Superior de Investigaciones Científicas, Córdoba, Spain*

6.1.1 Introduction

6.1.1.1 Why soil management?

Soil is the 'home' of plant roots, so growers must make it suitable for the life of hypogeal organs. This is the fundamental of soil management, especially for olive orchards, the roots of which live in the same soil, whether for a few decades or many centuries. Nevertheless, other reasons should be considered in a sustainable agriculture scenario. First of all, sustainable use of soil starts from soil conservation. Soil is a non-renewable natural resource, and as such it must be protected from erosion, contamination and other forms of degradation. Moreover, as soil management influences strongly all physical, chemical and biological processes, it is linked with water storage and dry-farming measures, frost-damage prevention, enhancement of on-farm biodiversity, carbon storage/climate-change mitigation and, not least, access to improved olive yield and, in general, to food security (Fig. 6.1.1). Good soil management practices are needed to improve olive growth and productivity by providing adequate nutrients and water to the trees. The soil has to maintain a good structure, ensuring adequate root growth and activity. The soil should be well aerated, with proper ratios of water and oxygen, regulated by infiltration/evaporation processes; not too much water to induce erosion and waterlogging, and not too little to safeguard the olive tree functionality, especially during the critical phenological phases of olive vegetation and fructification.

6.1.1.2 Interactions with other agronomical techniques

It is clear that most orchard agronomic and environmental problems are due to poor soil management: tree health and root health depend on good soil conditions. Indeed, irrigation, fertilization, pest and disease control and fruit harvesting are agronomical techniques strongly related to it. There are strong interactions among different aspects of soil, water and nutrient processes related to soil management. For instance, cover crops in olive orchards reduce nutrient losses and increase soil water retention capacity. Analogous effects, that is modifying infiltration and soil–water balance, might be achieved by minimum tillage. Minimum tillage is traditionally suggested in dry-farming olive orchards as a useful

DOI: 10.1079/9781789247350.0016

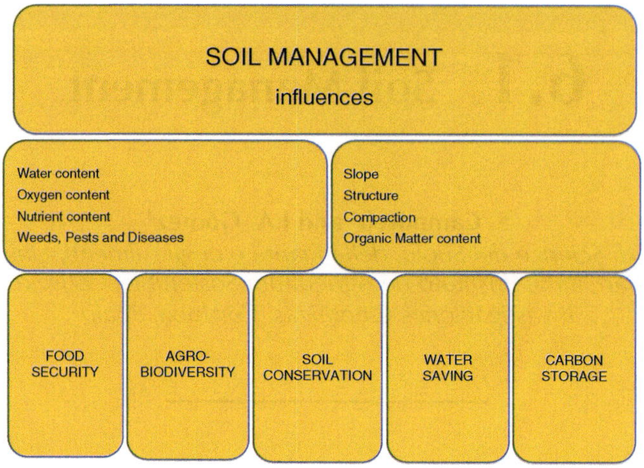

Fig. 6.1.1. Factors affected by soil management decisions.

practice: (i) to break the continuity of the soil capillaries and decrease water losses by evaporation; and (ii) for weed control, both in integrated as well as organic agriculture. Cover crops, by spontaneous vegetation or species sowed on purpose, provide ground cover, reducing erosion risk and decreasing soil evaporation, but at the same time they consume water through transpiration introducing a potential risk for competition with the olive tree if not managed properly (Castellini *et al.*, 2019). Although water is scarce in the majority of Mediterranean olive orchards, rainfall occurs in autumn/winter, when this water can be used to maintain a plant cover to improve soil characteristics and reduce erosion (Kavvadias *et al.*, 2018). However, at a specific date, which will vary according to the climatic conditions in each locality each year, the cover will have to be eliminated by tillage, or chemical or mechanical mowing to minimize this risk of competition for water (Xiloyannis *et al.*, 2018). Simultaneously, soil management needs to fulfill additional regulations. In the European Union, for instance, current regional legislation requires that soil management practices ensure control of the *Xylella fastidiosa pauca* vectors.

The macronutrient balance and the macronutrients to be applied will depend on the soil management techniques used. It should be highlighted that tillage causes mineralization of organic matter and provides readily assimilated nutrients, most importantly nitrogen, while live plant covers immobilize them. Mulching with olive pomace enhances potassium soil availability; considering the great importance of soil potassium content for olive growing, olive pomace mulching could reduce the amount of K fertilizers (Camposeo and Vivaldi, 2011). As for micronutrients such as iron, most soils have adequate contents, but they could be found in forms which cannot be assimilated by the olive tree. However, the soil organic molecules form compounds, such as chelates, which can be taken up by the tree. Consequently, it is always wise to increase organic matter content, which is poor in most olive orchards, because it makes fertilization management easier and more effective (Zipori *et al.*, 2020). Moreover, tillage allows for fertilizers landfilling, just before the sprouting phase.

Finally, appropriate soil management is needed to improve olive orchard mechanization level and efficiency, with particular emphasis on olive mechanical harvesting. Natural green covers make nets shifting easier when trunk-shaker machines are used and increase the soil-bearing capacity when heavy machines, such as straddle harvesters, need to be used. On the contrary, bare soil technique is still being used when preparing the ground areas under the large-size olive trees for mechanical picking of the olives naturally dropped, a practice still in use, although it will reduce the quality of the harvested fruit and, as a result, of the extracted oil.

6.1.1.3 Sustainable agriculture scenario

In recent years, the olive sector is shifting towards intensification by new cropping system implementation, causing a strong positive economic impact. Indeed, the olive-growing systems are moving from traditional low-density (< 250 trees ha^{-1}) to intensive medium-density (300–600 trees ha^{-1}) and, mostly, to super-high-density or superintensive (> 1200 trees ha^{-1}) ones. Making olive growing environmentally sustainable is the main challenge for the future; but sustainable intensification needs efficient orchard management (Camposeo *et al.*, 2022). As a first step, efficient use of agronomical inputs and natural resources is required, involving soil, water and nutrient management, precious resources that, in the case of soil, are non-renewable (Bechara *et al.*, 2018). When it comes to soil management techniques, it is essential to consider the existence of different cropping systems, which can be summarized as traditional, intensive and superintensive, each one having different technical solutions, also determined by the specific farm conditions (e.g. climate, soil, topography etc.). In fact, there is no such thing as a single technique, a good idea being to integrate different management techniques, each having specific advantages and disadvantages.

Traditional olive-growing soil management has long been considered a minor technical issue and, until the last decades of the 20th century, soil tillage and chemical weeding were chiefly used, especially under dry-farming conditions (Agriculture 1.0 and 2.0). After the introduction of significant changes in planting systems, together with deficit irrigation, soil management started to arouse increased interest among olive growers, as it represents one of the aspects to be considered in order to reach specific goals not only from a quantitative and qualitative perspective but also for environmental protection (Agriculture 3.0).

6.1.1.4 Integrated or organic farming

Intensification and professionalization of olive cultivation worldwide is leading to a sophistication of the decision-making process of farm operations (Fig. 6.1.2), illustrated by the expansion of certified systems. For instance, in Europe, olive growers have to choose mainly between two possible certified farming methods: integrated or organic. Indeed, EU directives 2005/1698 and 2009/128 officially launched integrated agriculture, aimed at reducing the dominance of the traditional and conventional systems, towards production systems in which any farm operation is registered and based on technical grounds. This trend goes beyond certified systems, and the European Union, through its Common Agricultural Policy, makes it compulsory for each farm to keep an official registry of farm operations. Adopting an integrated method means applying all available technical tools for crop production and pest management, including them in the same orchard in different phenological phases; the integrated method reduces significantly the use of synthetic chemicals in compliance with agronomic, ecological, toxicological and economic principles. For these reasons, integrated agriculture is also referred to as sustainable agriculture (Agriculture 3.0) and its application for soil management results in soil conservation and in a better agro-ecosystem quality as well. At the same time, with Directive 834/2007, the EU promoted organic agriculture that is based on the complete elimination of synthetic chemicals.

6.1.2 Factors Conditioning Soil Management

To be successful, any decision regarding soil management needs to take into account the conditions of the olive orchard and the farm. These conditions are described in Table 6.1.1, and they can be classified into

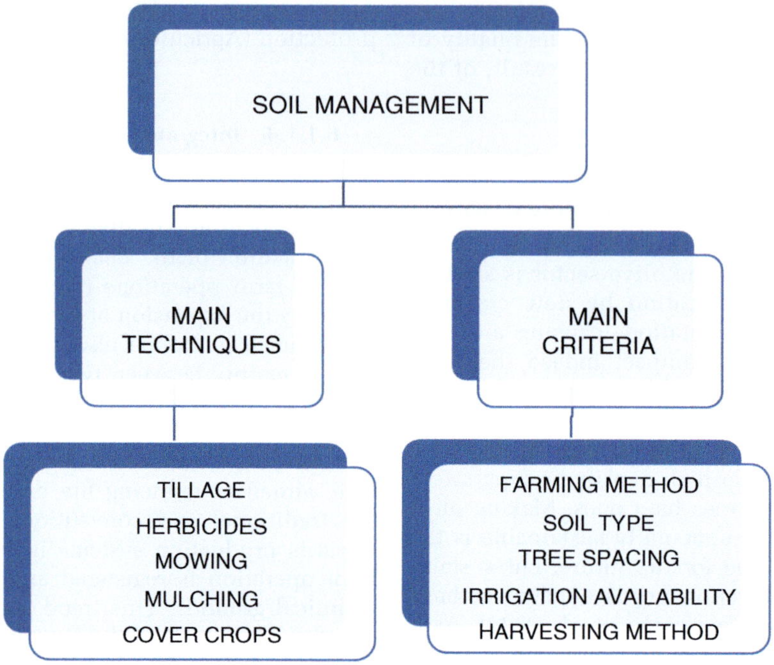

Fig. 6.1.2. Main types of soil management techniques and criteria involved in their selection.

geographical features (slope, soil, climate), agronomic (planting system, water availability) and farming situations (access to machinery and equipment, production system, investment capacity). Any skilled olive grower will attempt to implement a soil management process best suited to their edaphoclimatic conditions within the constraints imposed by agronomic conditions (e.g. tree density or access to irrigation, or lack thereof) and farming possibilities (e.g. available equipment).

This can be easily illustrated if a comparison was made among the different kinds of orchards shown in Fig. 6.1.3. The extensive olive orchard on rainfed conditions on a very steep slope, combined with olive grazing, calls for a soil management process where the soil has temporary cover crops from natural vegetation, which is managed with periodic grazing by the sheep at low density. This orchard is tilled every few years and seeded with species that can enrich the seedbank to provide better grazing material. Given the rough terrain, the pruning residues are burned on the spot for sanitary reasons given the high cost of taking them out. The second orchard is a rainfed orchard with higher tree density and canopy size on a very steep slope. Here, although the management process shown in the picture concerns bare soil, the best strategy will be to seed a dense cover crop perpendicular and parallel to the maximum slope. Given the steep slope, if the certified production system allows it, using herbicide may ensure better control. Chopping the pruning residues to create a mulch can complement soil protection. Regarding the third situation, i.e. a superintensive orchard in a relatively flat land, the main difficulty will be to ensure trafficability and proper management of the pruning residues. A no-tillage system creating a mulch of chopped pruned residues will be the best solution, unless sanitary conditions require elimination of the pruning. Given the shadowing and high compaction in the lanes, the installation of a cover crop may prove rather complicated, but probably some herbaceous vegetation from the seed bank will emerge in the bare spots across the mulch and tree

Table 6.1.1. Factors affecting soil management strategies.

#	Factor	Reason
1.a	Soil type	It conditions rainfall infiltration, soil stability against compaction and erosion, and water-holding capacity.
1.b	Climate	It regulates water availability and erosion risk. It conditions how to use cover crops.
1.c	Topography: slope length and steepness	They affect trafficability and water and tillage erosion risk.
2.a	Tree density	It conditions traffic direction, equipment to use and orientation of the lanes, particularly in super-intensive production.
2.b	Access to irrigation	It diminishes risk of water competition, and increases yields and farm profitability.
2.c	Pest, diseases and weeds present at the farm and in the area	It needs to be evaluated locally to see how to implement specific uses, like cover crop or mulch of pruning residues, to avoid increasing infestation.
2.c	Harvesting method	It conditions timing of maximum traffic in the orchard and it might affect implantation of cover crops.
2.d	Pruning	It is necessary to know timing and amount of pruning residues to evaluate feasibility of use as mulch.
2.e	Fertilization	To evaluate whether there is the need of some operation to incorporate fertilizer (e.g. compost) and also to define and manage cover crops, if this is used.
3.a	Certified production systems	To determine what operations are allowed in the production system used on the farm.
3.b	Agricultural and environmental regulations applying to the farm	Some operations are forbidden or limited depending on the farm conditions. For instance, in orchards receiving Common Agricultural Policy subsidies in Europe, or orchards in the catchment of reservoirs used for human consumption.
3.c	Access to staff, equipment and investment	To determine if an intended management can be feasibly applied on the farm.

lines. They can be controlled through mowing or herbicide depending on the production systems and equipment available on the farm.

6.1.3 Approaches to Soil Management

6.1.3.1 Types of soil management

Given the large variety of olive cultivation systems and environmental conditions on which they are grown, there are multiple strategies to carry out soil management in an olive orchard, as summarized in Table 6.1.2 and Fig. 6.1.4. Broadly speaking, we can distinguish between strategies designed to keep a bare soil, using tillage, herbicide or a combination of both, and strategies designed to provide a ground cover to the soil consisting of living vegetation (cover crops) or inert material (usually chopped pruning material). Nevertheless, multiple combinations of these basic options can be seen in a given olive orchard; for instance, cover crop in lanes and herbicide use in the tree line. Historically, bare soil management was recommended in olive orchards because it provided a straightforward way to minimize the risk of competition of the olive tree with vegetation naturally present at the orchard (weeds), to incorporate fertilizer and/or manure, and to increase temporarily infiltration rates. However, it should be noted that for most of the history of olive cultivation, until the extension of mechanization and herbicides, the ability of farmers to till was limited by the cost of animal power. As such, prior to the advent of mechanization,

Fig. 6.1.3. Different types of olive orchards. (a) low-intensity grazed orchard (courtesy of Mª Auxiliadora Soriano); (b) rainfed orchard on very steep slope at 120 trees ha^{-1}; (c) super-intensive deficit-irrigated orchard in flatland.

tillage intensity in olive orchards was low and we can envision a situation in which tillage was used to keep vegetation under control but the orchard presented some kind of ground cover during much of the year. All this changed with the adoption of mechanization and herbicides, which allowed bare soil to be maintained all year round (see Fig. 6.1.4a). As a result, severe problems related to soil degradation (erosion, loss of soil organic carbon and nutrient, off-site contamination) arose (Figs 6.1.5, 6.1.6), and, as late as the 1960s, the use of ground-covered soil management was started to be promoted in olives (Gómez *et al.*, 2014). Nowadays, the promotion of these soil management options, particularly cover crops, is encouraged not only for soil conservation purposes but also for the provision of other ecosystem services such as maintenance of biodiversity or improvement of landscape values.

Table 6.1.2 is intended only as an outline of the basic possibilities for implementation of soil management strategies. They can be combined to implement the one that is needed in a specific orchard depending on conditions (see section 6.1.3.2) and objectives.

Table 6.1.2. Main soil management options with brief definitions.

Objective	Definition
Bare soil tillage	Soil kept clean of vegetation by regular tillage
Bare soil herbicide	Soil kept clean of vegetation by herbicide use
Temporary cover crop	Cover crop in the orchard, usually in the lanes, during the rainy season terminated chemical or mechanically at the onset of the dry season
Permanent cover crop	Cover crop in the orchard kept permanently, usually in the lanes, regularly controlled by mowing or grazing
Green manure	Cover crop, usually with legumes species, incorporated into the soil at the time of maximum biomass to enhance nutrient content
Mulching	Mulch on the lanes surface by chopped pruning residues or, occasionally, with straw from outside the orchard
Combined systems	Combination of the basic types (above); for instance, bare soil in the lanes by regular tillage and cover crop under the tree canopies, or alternated cover with one lane with cover crop and another with mulch of pruning residues

Fig. 6.1.4. Examples of different kinds of soil managements. (a) bare soil using tillage; (b) minimum tillage (tilled only in lanes); (c) bare soil using herbicide; (d) temporary cover with mix for biodiversity; (e) mulch of chopped pruning residues; (f) temporary cover crop with mix for biodiversity.

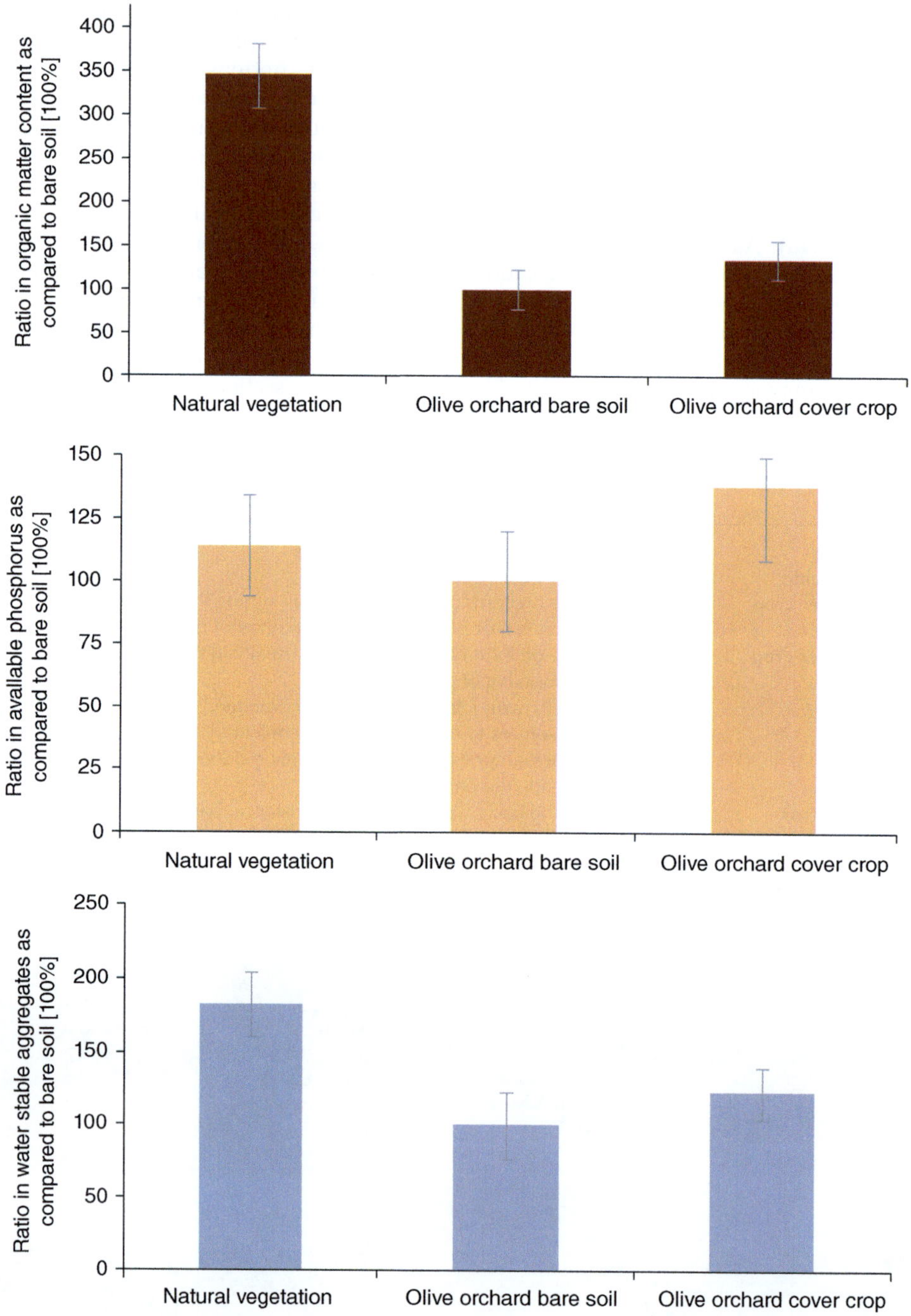

Fig. 6.1.5. Example of the effect of different soil management on top-soil quality in the lanes of olive orchards. (Adapted from Gomez *et al.*, 2014)

Given their environmental benefits, ground cover-based soil management options are usually preferable, but great attention needs to be paid to potential risks (see below) in order to make sure they are avoided. To prevent these risks, particularly when transitioning

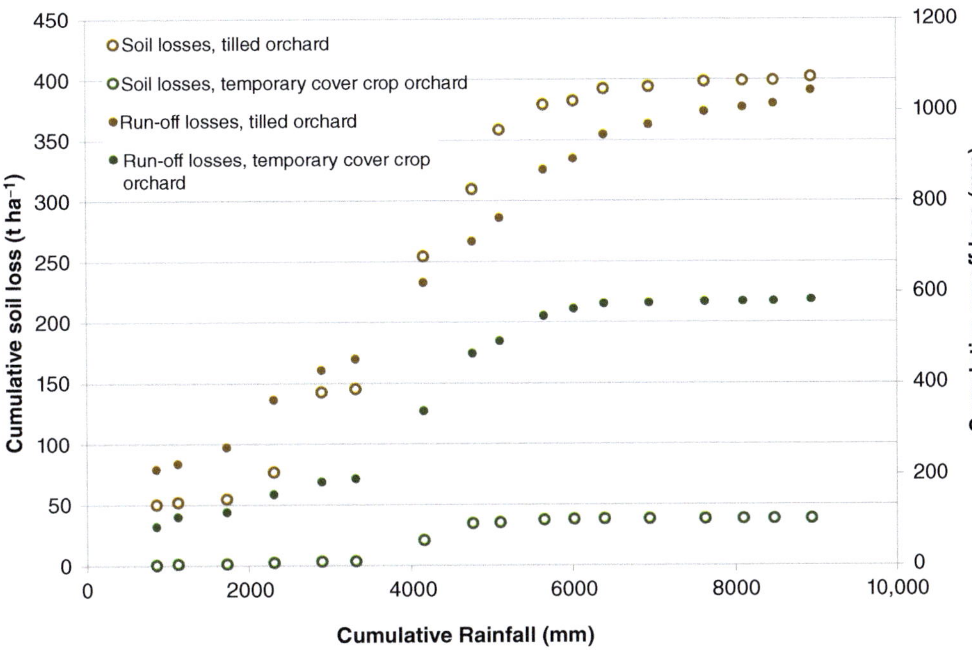

Fig. 6.1.6. Comparison of bare soil by tillage vs. temporary cover crop cumulative long-term run-off and soil losses in a long-term experiment. (Adapted from Gomez *et al.*, 2022)

from a bare-soil management process, planning for a gradual learning of the technique using a conservative approach might be a good strategy. For instance, if you plan to use a cover crop, you will start in only one part of your orchards with a relatively early date of control. This will allow you to learn how to tune the best control date based on the rainfall of the year.

6.1.3.2 Impact of different soil management strategies

Table 6.1.3 shows a comparison of the beneficial effects of the different soil management options, with Table 6.1.4 showing a similar comparison of the risks. It is important to highlight a few major points in order that these tables can be interpreted correctly. Firstly, be critical of what you actually achieved in the orchard and do not be deceived by the soil management strategy that you are attempting to deploy. For instance, an orchard in which you attempt a cover crop and you have very little vegetation on the lanes is closer to bare soil with herbicides (Fig. 6.1.4c) than to a cover crop (Fig. 6.1.4d). Secondly, remember that the benefits that you might achieve with a cover crop or mulch are positively correlated with the amount of biomass and ground cover that you actually achieve. For instance, if you have a low tree density and produce a small amount of pruning, you will achieve a moderate benefit when using a mulch of chopped pruning residues, since they will cover a small fraction of the orchard ground. Thirdly, be aware of the pest and disease risk in your orchard and region when planning soil management strategy. For instance, never bring inert mulch from outside of your orchard if there is a risk of such material coming from an area infested with *Verticillium dahliae*. Moreover, in the presence of *Xylella fastidiosa*, you should manage your cover crop with species that cannot host vectors for this disease. Fourthly, there is always a risk for competition with water between the cover crop and the olive, and management of cover crop is aimed to control this. One of the reasons for this is that the moderate increase in infiltrating rainfall water as compared to bare soil (Fig. 6.1.6) cannot compensate for the potential losses of water transpired by the cover crop if it is not controlled once the rainy season ends. There are several technical publications which can expand on the basic ideas outlined in this section, like Tombesi *et al.* (2007) or Alcántara *et al.* (2017).

6.1.3.3 Examples of soil management strategies

Fig. 6.1.7 illustrates some soil management strategies adapted to different kinds of olive orchards. For instance, option A of a bare soil through tillage, although not being the one that we would recommend to achieve the best provision of ecosystem services, might be an alternative used by farmers in flatland areas where there is little risk of soil erosion,

Table 6.1.3. Potential benefit effects of different soil management on olive orchards: ++, high benefit; +, moderate benefit; ±, moderate but temporary benefit; 0, no benefit; (leg), benefit only if legumes are used. Cover crops or mulch will provide an effect proportional to the actual ground cover achieved and biomass returned to the soil. Layout adapted from Grant *et al.* (2006) with information from Grant *et al.* (2006), Gómez *et al.* (2014), Alcantara *et al.* (2017) and Bombino *et al.* (2021).

Potential benefit	Bare soil with tillage	Bare soil using herbicide	Permanent cover crop	Temporary cover crop	Green manure	Mulch of pruning residues or straw
Increase of soil organic carbon	0	0	+	+	++	+
Increase of biological activity	0	0	+	+	++	+
Nitrogen incorporation into soil	0	0	+ (leg)	+ (leg)	++ (leg)	0
Infiltration improvement	±	0	++	+	+	+
Run-off reduction	±	0	++	+	+	+
Erosion reduction	±	0	++	++	+	++
Reduction of dust	0	+	++	+	+	++
Reduction of soil compaction	±	0	+	+	+	0
Weed control	±	±	++	++	++	++
Reduce excess of soil water	0	0	++	+	++	0
Improvement of trafficability	0	++	++	+	+	++

Table 6.1.4. Incidence of potential risks associated with different types of soil management: --, high risk; -, moderate risk; 0, no or negligible risk. Layout adapted from Grant *et al.* (2006) with information from Grant *et al.* (2006), Gómez *et al.* (2014), Alcantara *et al.* (2017), Sastre *et al.* (2016), Song *et al.* (2020) and Bombino *et al.* (2021).

Potential risk	Bare soil with tillage	Bare soil using herbicide	Permanent cover crop	Temporary cover crop	Green manure	Mulch of pruning residues or straw
Competition for soil water	0	0	- -	-	-	0
Increase of soil erosion (water)	- -	- -	0	0	-	0
Run-off losses	-	- -	0	0	0	0
Increase of frost damage in spring	0	0	-	-	- -	-
Interfere in eliminating some plagues	0	0	-	-	-	-
Increase of rodents	0	-	-	-	-	-
Interfere in chopping/ withdrawing pruning residues	0	0	0	-	- -	0
Increase of nematodes	0	0	- -	-	-	-

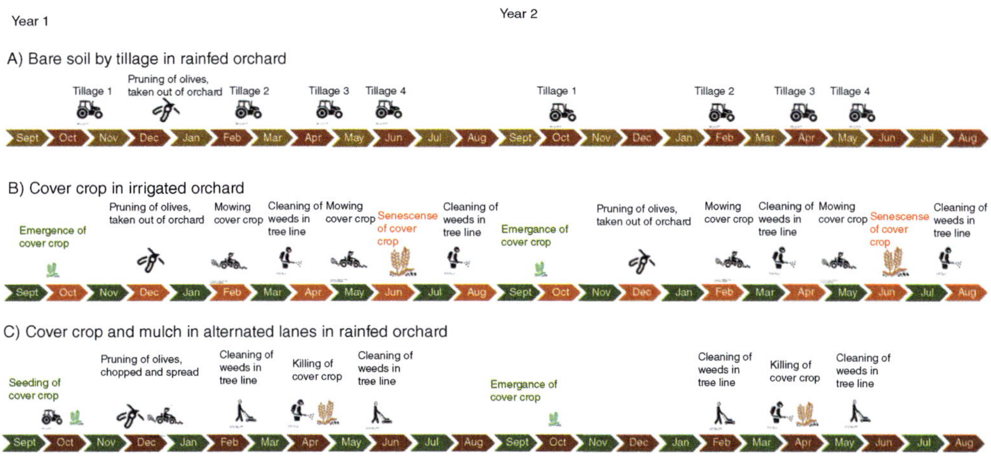

Fig. 6.1.7. Example of operations for three different kinds of soil management strategies in olive orchards: (A) bare soil through tillage; (B) permanent cover crop controlled by mowing in the lanes in an irrigated olive orchard from orchard seed bank; (C) seeded cover crops and mulch of chopped pruning residues in alternate lanes in a rainfed olive orchard, for the conditions of southern Spain with onset of the rainy season in October and of the dry season in late April.

and the option of using a cover crop or a mulch is prevented by the lack of equipment and seed availability. Option B is very much used in olive orchards on sloping areas where we want to achieve a higher degree of protection against erosion but the use of irrigation calls for control of weeds in the tree lane where the irrigation lanes are located. Option 3 shows a viable alternative in case we have difficulties implanting a cover crop due to damage by traffic in our orchard. The use of alternate lanes of a mulching makes it possible to concentrate heavier traffic, particularly on moist soil, in those lanes, allowing a well-established cover crop to be implanted in every other lane. Weed cleaning in the lane area will be dependent on the techniques allowed by the certification system used (e.g. organic vs. certified), and the technical constraints of the orchard: weed species present, available equipment, environmental regulations, farmer´s choice. There are some publications which give further insight, some of them listed in the reference section.

6.1.3.4 Comments on use of cover crops

The use of cover crops in olive orchards has evolved from being mostly oriented to soil conservation to a broader provision of ecosystem services. Table 6.1.5 summarizes the main reasons for using a cover crop, raises some questions regarding its design, and explains choice of species and major management decisions that the olive farmer will face. This is a dynamic area in which innovation – particularly regarding increasing seed availability of species best suited to fully develop its phenological cycle before the onset of the dry period and use of mixes with different species – is happening. It is not unusual to face difficulties when trying to establish a cover crop for the first time in an orchard. To minimize these difficulties, Table 6.1.6 summarizes the major problems that we might face and suggests some ideas to address them, with Fig. 6.1.8 showing some field examples.

6.1.4 Complementary Techniques

There are several elements that need to be considered in order to achieve a sustainable use of the olive farm complementarily to soil management, particularly in relation to trafficability, soil conservation and improvement of biodiversity and landscape values. For all of them, we need a landscape

Table 6.1.5. Cover crop composition and some key questions and management decisions based on the main soil management objective. (From Gómez *et al.*, 2021)

Objective	Best type of cover	Key questions	Major management decisions
Enhancement of biodiversity, landscape values	Mix of plants with flowers	What species to use to achieve a cover crop with staggered flowering periods and moderate costs	How to control cover crop to maintain composition. When and how to use mowing
Improvement of soil fertility	Legumes or Legumes/ grasses	Species composition	When to incorporate the cover crop. Soil preparation to maximize biomass production
Erosion control	Grasses	Species with best-adapted phenology to climate conditions When to reseed	When to kill the cover crop if it is temporary cover crop
Grazing	Legumes or Legumes/ grasses	Species composition When to reseed	Grazing density; time of the year when grazing is allowed
Trafficability	Grasses	Species composition. Time of seeding in relation to farm operations	When to control the cover crops, mechanical (mowing) or chemically?

vision of our orchard visualizing it as a combination of hillslopes, traversed by paths where traffic is concentrated and connected by areas of run-off concentration, where ephemeral streams and gullies might appear (Fig. 6.1.9).

Soil conservation can be greatly facilitated if — when planted — we try to adapt the tree lines to the contours (Fig. 6.1.10a). In flat areas, or areas with moderate slope, this might be relatively easy to implement, even building broad-based terraces (Fig. 6.1.10b). However, as the slope increases the cost of terracing might be prohibitive or an obstacle to full mechanization (Fig. 6.1.10c). In this case, it is common to see orchards planted with the lanes in the direction of the maximum slope (Fig. 6.1.9). If this situation happens, in addition to using a cover crop across the lanes, it is good practice to fractionate the slope with vegetated barriers perpendicular to the maximum slope, such barriers consisting of herbaceous plants or shrubs, as they can prevent the formation of rills (Fig. 6.1.11).

The areas of flow convergences among hillslopes, thalwegs, act as ephemeral streams during large rainfall events. It should be noted that in orchards, cover crops will reduce run-off, even though this will still be significant in large rainfall events, particularly in rainy years when the soil profile is close to saturation. The practice of not planting trees in these thalweg areas is increasingly relied upon since it allows excess run-off to be delivered safely, preventing the formation of gullies if these areas are left permanently protected (Fig. 6.1.12a, b). If, despite our best efforts, gullies appear in our orchard, we need to act promptly to restore them, since they can grow rapidly, producing loss of trees and fractioning the orchard and expanding further. It is beyond the scope of this chapter to provide a detailed description of gully restoration in olive orchards, in respect of which reference should be made to several technical publications (e.g. Geyik, 1986; Gómez *et al.*, 2019, 2021). Nevertheless, the basic principles and guidelines are summarized in Table 6.1.7. It is important to note that the final purpose of gully restoration is, through a combination of check dams and revegetation, to transform our gully into a well-protected ephemeral stream where the energy of the concentrated flow during large storms can be dissipated without additional erosion (Fig. 6.1.12c, d). The vegetated areas of

Table 6.1.6. Summary of problems that might arise when implanting a cover crop and suggestions to address them.

Problem	Reasons	Suggestions
No establishment of natural herbaceous vegetation when we stop tilling or applying herbicide	1 Deplenishment of soil seed bank 2 Over compaction 3 Low nutrient content in the soil 4 Too-arid conditions 5 Overgrazing by rabbits	1 Seed a cover crop. 2 Till before seeding. 3 Fertilize to help implanting the cover crop. 4 If conditions are too arid you might find it impossible to establish a proper cover crop. 5 See below.
Seeded cover crops are overgrazed by rabbits	If a high density of rabbits are present they can eat the cover crop preventing it from establishing.	Sometimes use of non-palatable species (e.g. *Anthemis Arvensis*) can overcome this problem. Success is not assured, so first try a small area. If it does not work, consider use of an inert mulch.
High cost of seed	Seeds for cover crops adapted to olive conditions are usually expensive to harvest and process.	Consider first the option of natural vegetation present in the orchard. If not, look for species that propagate easily to seed a moderate amount across your orchard; let it propagate over multiple years.
I can´t achieve self-seeding.	1 Phenological cycle of your cover crop is too long, can´t produce enough seed before wilting in dry season. 2 Need to control cover crop before maturing for seeds due to risk for water competition. 3 Overgrazing or premature date for controlling cover crop	1 Change the species seeded for one(s) with a better-adapted cycle. 2 Leave a narrow strip to allow self-seeding. 3 Reduce grazing density, adjust date for controlling the cover crop after maturity.
Infestation by weeds	Weeds present at the farm are more competitive that the one that seeds as a cover crop.	1 If weeds need to be eradicated, reduce their level of infestation as much as possible before seeding a cover crop. Consider seeding species that can be competitive with the ones you try to eradicate. 2 If weeds are not dangerous and can act as a cover crop, manage them as a cover crop instead of fighting against nature.
Machine traffic prevents establishment of cover crop	Damage and compaction by machine tires kept the soil bare.	Try to reduce traffic, concentrating it in a few lanes and avoiding operations on wet soil. Combine in alternate lanes inert mulch (where to concentrate traffic) with others with a cover crop.

Fig. 6.1.8. Examples of problems, and solutions to address these problems, in implanting a cover crop; leaving a strip for self-seeding, if needed, to control the cover crop (a); use of less-palatable species on areas with high rabbit density (b).

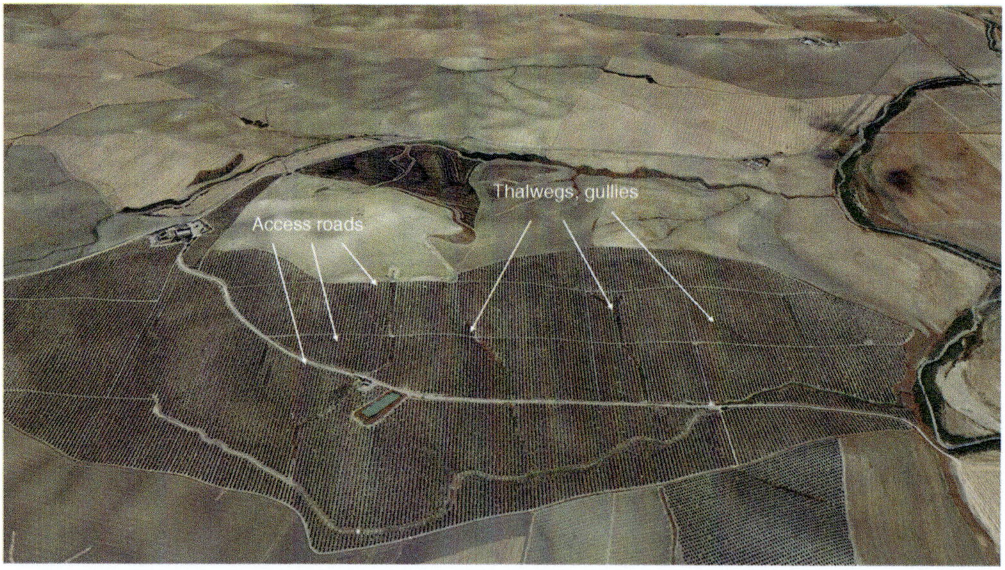

Fig. 6.1.9. Elements to consider for a landscape approach complementary to soil management in olive orchards.

these protected streams or restored gullies can become islands of biodiversity, greatly improving the provision of ecosystem services by our orchard.

The soil in the lanes where the traffic within the orchard tends to concentrate, like areas of access or where trailers are loaded and unloaded, can become rapidly deteriorated. Avoiding trafficking on wet soil and making a wise distribution of traffic across the orchard can minimize this problem, which can become acute in years where harvest coincides with a rainy winter. If erosion becomes extremely intense in these areas, we may want to create perpendicular obstacles to water, e.g. with embedded stones or pruning residues that can divert run-off to the side of the trafficked lanes, acting like water bars (NRCS, 2010).

Fig. 6.1.10. Examples of contouring (a); and terracing (b, c) in olive orchards.

6.1.5 Environmental Sustainability of Different Soil Management Techniques

6.1.5.1 Life cycle assessment analysis

The environmental burdens of different soil management techniques in an olive orchard could be evaluated and compared using the LCA (Life Cycle Assessment) analysis. LCA is the most appropriate instrument for assessing the sustainability of an agricultural process because it makes it possible to highlight both the environmental loads due to field activities and those related to production,

Fig. 6.1.11. Example of crossed vegetated barriers (along and perpendicular to the main slope) to reduce run-off and sediment connectivity.

Table 6.1.7. Basic principles and guidelines for gully erosion control. (Adapted from Gómez *et al.*, 2019)

#	Principle
1	Enhance infiltration in contributing catchment to reduce run-off flow.
2	Divert, if necessary and always in an area protected against concentrated flow, all or part of the run-off flowing to the gully head.
3	Stabilize the gully section against incision and widening using structural methods and vegetation.

#	Guideline	Reason
1	Check dams not too high, < 1.5 m if possible.	They are safer to build and more stable. The system is more resilient if a single check dam fails.
2	Check dams properly spaced.	Prevents scoring between consecutive dams and dissipates the stream energy.
3	Good structural design	As well as being structurally stable, a spillway and toe protection should be included to prevent scouring the check dam base.
4	Complement with revegetation.	Revegetate with shrubs and herbaceous plants the gully boundaries to stabilize against erosion and mass movements.

installation and disposal of the components used for soil management. The LCA results of five different techniques of soil management for weed control are shown in Fig. 6.1.13.

The results obtained suggest the convenience of promoting the use of mulching for weed control in olive orchards as a sustainable practice, not only in organic orchards (Russo *et al.*, 2015). The comparison between techniques of mulching with plastic materials revealed, in general, a greater environmental burden of non-woven tissue compared to the polyethylene film. Mechanical weeding had greater environmental burdens for both abiotic depletion and ozone layer depletion potentials indexes and lower values for both acidification and global warming potentials. Chemical weed control had the major environmental burdens

Fig. 6.1.12. Example of protected thalwegs to prevent gully formation (a and b; courtesy of Jose Mora); and restored gullies (c and d).

for all indexes. Moreover, in the plots mulched with organic material, two bio-indicators, such as the orchid *Serapias lingua* and the fungus of the genus *Coprinus* spp., were observed. Therefore, organic mulching could be considered a suitable technique for greater environmental sustainability for soil management in intensive plantings as well (Fig. 6.1.14).

There are also other risks and less beneficial effects of plastic mulching with regard to waste treatment, life cycle balance and impact on soil quality. These include the persistence of unrecovered plastic mulch in

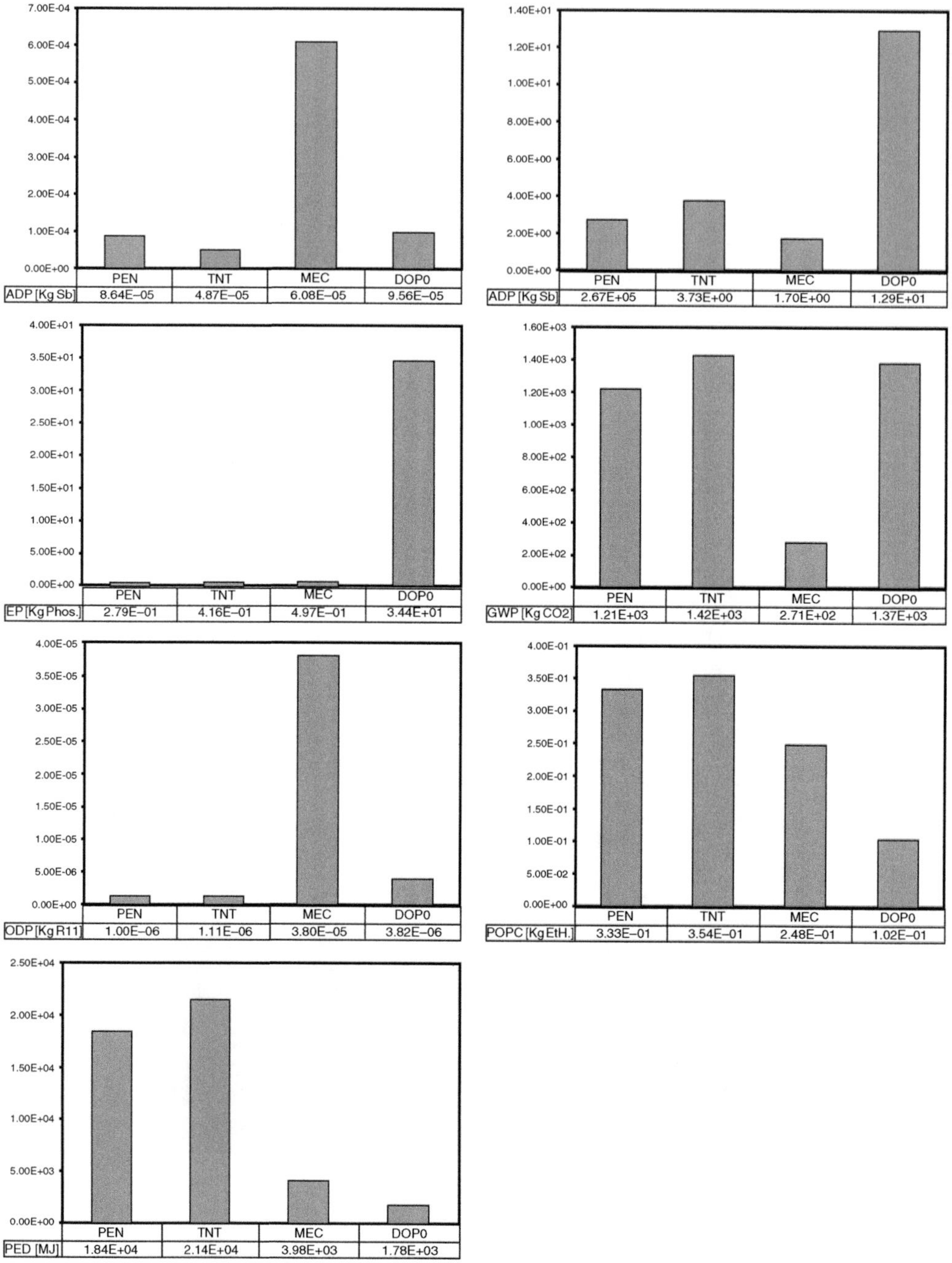

Fig. 6.1.13. LCA analysis of different soil management in a super-high-density olive orchard: de-oiled olive pomace mulching (DOP), non-woven tissue mulching (TNT), polyethylene film mulching (PEN), chemical weeding (CHI), mechanical weeding (MEC). Indexes quantified: abiotic depletion potential (ADP), acidification potential (AP), eutrophication potential (EP), global warming potential (GWP), ozone layer depletion potential (ODP), photochemical ozone creation potential (POPC), primary energy demand (PED). (From Russo *et al.*, 2015)

Fig. 6.1.14. Row of an olive orchard mulched with organic material. (From Russo *et al.*, 2015)

soil, their potential to alter soil quality by shifting the edaphic biocoenosis, accelerating carbon and nitrogen metabolism, as well as potentially degrading soil organic matter. This further includes inducing soil water repellency, increasing the risk of mycotoxin formation in soil and an enhanced release of climate-relevant gases. Although several attempts have been made to extend the life cycle of plastic mulches and to reduce their input to soil by recovering and recycling used mulching films, in some cases the cost and effort of these options seem to outweigh their benefits from yield increases, water savings and facilitated pest control. If, in addition, external effects of mulching on (public) goods or resources, like soil quality, biodiversity or recreation, are taken into account, a holistic ecosystem services assessment could make plastic mulching uneconomical and unsustainable.

6.1.5.2 Other environmental analysis

In comparison with tillage, soil management practices involving recycling of resources, such as sowing a mixture of winter cover crops, soil mulching with chopped pruning residue and composted byproducts without soil mechanical disturbance, have great potential for enhancing soil biodiversity and fertility. Cover crops stimulate and select naturally soil microorganisms such as those involved in the carbon and nitrogen cycles, and those that promote plant growth and/or that act as a deterrent against pathogenic microorganisms. Sustainable farming practices increased significantly soil microbial communities: ten times the fungi CFU and three times the bacteria CFU (Sofo *et al.*, 2010, 2014). The results on biological indicators of soil quality, such as mycorrhizal colonization of spontaneous weed flora, actinobacteria, Azotobacter and proteolytic bacteria, suggest favorable effects on soil biology and ecosystem complexity in the sustainable treatments.

Moreover, cover crops could hardly increase soil vertical infiltration of rainwater to about 160 mm day^{-1}), so reducing losses through run-off (Palese *et al.*, 2014). Greater accumulation of rainwater in the ground brings several advantages, such as help to increase production of olive-growing systems in dry farming, less sensitivity to attack by pathogens, reduction of soil erosion and consequently less transport of sediments and pollutants (fertilizers, herbicides, pesticides) in water basins, and reduced flow of sediment-laden run-off in rivers, ensuring control of floods. With sustainable techniques, the carbon content in the surface layer of soil (0–15 cm) may increase significantly after seven years.

The arthropod community population was also relatively high in cover-crop treatments, but the seeded cover crops provided the additional benefit of increasing the number of predators for olive pests. Seeding of a mix of selected species was shown to be particularly effective in accelerating the increase in plant biodiversity, which was achieved naturally in the system of natural cover crop only after two decades.

At a traditional olive grove in southern Italy, the change of management practices from tillage and burning of pruning residues to no-tillage, the adoption of spontaneous cover crops and the mulching of pruning residues had a beneficial effect on soil carbon concentration (from 1.1% up to 1.4%) and, in turn, on the stock of carbon (0–30 cm depth) at a mean rate of 2.4 t C ha^{-1} yr^{-1} (equal to 8.8 t CO_2 ha^{-1} yr^{-1}).

6.1.6 Does Soil Management Affect Olive/Oil Quality?

As previously reported, soil management is an agricultural technique that can have a large effect on olives and oil yield. But what about qualitative aspects? Although it is a matter of discussion in the olive sector, there are few published studies on this question. It seems, from available literature, that there is an indirect effect of soil management on quality due to the soil water and nutrient availabilities (nitrogen especially) that influence leaf water and nutrient status.

Although tillage may lead to higher yield by controlling competition for water and nutrients, the introduction of cover crops in olive orchards can lead to higher polyphenol contents, enhancing VOO quality; moreover tillage treatment decreased maturity index (García-González *et al.*, 2019). Available studies in rainfed conditions found small, and in many cases, statistically negligible differences in fruit ripening, pigments and several fatty acids as well as in sensory parameters in tilled soil vs. grass-covered soil (Sastre *et al.*, 2016). Sastre *et al.* (2016) noted how differences tend to appear only in dry years, in which most of the phenolic compounds, especially oleuropein, mainly accumulated in olive fruits of the cover-cropped rainfed olive groves, suggesting a potential improvement in the nutritional value of these fruits which requires further research. This is in line with other studies that reported that cover-cropped orchards presented an increase in the amounts of some aromatic compounds (hexanal and hexenal, if irrigated; hexenyl and hexenol, if rainfed) responsible for the distinctive flavor notes of olive oil (Tekaya *et al.*, 2016).

A particular case is the use of wastewater irrigation, a relevant water source under water scarcity conditions. In this situation, Tekaya *et al.* (2017) evaluated the use of wastewater for two different soil managements, tillage and bare soil, with herbicide cover with a mulch of chopped pruning residues. Their results suggest that, in this situation, the best combination should be to combine use of wastewater with soil tillage, with wastewater recommended for use only in the 'off' year to support vegetative growth and improve physiological performance of olive trees. In the 'on' year, tilling the soil under rainfed conditions proved to be the most favorable cropping strategy for enhancing both physiological performance of olive trees and olive oil quality. It is clear that soil management, irrigation and fertilization are strictly linked. These findings must be taken into consideration when developing strategies for oil quality enhancement

6.1.7 New Frontiers

6.1.7.1 Soil management in a climate change scenario

Olive cultivation has experienced an astonishing transformation in the last decades, but it is currently facing additional challenges. Among them are climate change, with drier and warmer scenarios predicted for the olive-growing areas, and increasing fresh-water scarcity, along with a trend towards the intensification of olive-planting systems, all leading to an increase in tree crop water requirements. Innovative and environmentally friendly soil management strategies need to be developed for olive orchards, combining profitability, optimum use of natural resources and resilience under a changing climate. Different kinds of innovative mulching techniques, both alive as green cover crops and biodegradable films, start to be available, but further developments are expected in the coming years (Montanaro *et al.*, 2018). Mulching made with new biodegradable films or 'live mulching', made with innovative green ground covers, could be more environmentally friendly techniques (Fracchiolla *et al.*, 2021). The idea is that this green ground cover enriching the soil with dry matter should contribute to minimize the negative effects of increasingly erratic and scarce rainfall. In addition, biodegradable films could be capable of reducing evaporation losses and promoting vegetative and productive tree activities (Gucci *et al.*, 2012; Fracchiolla *et al.*, 2013). Unfortunately, till now, this kind of innovative mulching has been applied successfully in vineyards, but little is known in olive cultivation (Camposeo and Vivaldi, 2011; Russo *et al.*, 2015).

Knowledge on mulching impact on soil water, nutrients and temperature dynamics and on olive tree performance would be very useful to improve environmental sustainability of olive orchards and to better cope with the climate change scenario. Finally, certain agricultural practices (e.g. no tillage, cover crops, supply of organic raw material, recycling of pruning residuals) might increase soil carbon stock and, in turn, enhance the climate change mitigation potential of agricultural soils (Fig. 6.1.15).

Fig. 6.1.15. Inter-row of an organic olive orchard sowed with a mixture of leguminous species for green manure. The rows are naturally grassed.

Given the trade-off between the use of water and vegetation for the provision of ecosystem services, further research on vegetative barriers along the orchard boundaries might provide alternatives for sustainable olive cultivation in areas too arid for a successful use of cover crops in the lanes. The beneficial increase of soil organic carbon (SOC) due to certain soil management practices strengthens the relevance of orchards as managed ecosystems for climate change mitigation (Liu *et al.*, 2014).

6.1.7.2 Soil management in the precision agriculture scenario

Precision agriculture (PA), known also as Agriculture 4.0 or digital agriculture, is based on the field management of spatial and temporal variability, using ICT (information and communications technologies). PA application started over the last 25 years, when GPS (global positioning systems) and sensors were made available for crops. In particular, fruit tree crops pose an emerging and challenging sector for PA technology and management. Today, digital agriculture is not only geared towards easier access to PA, but also offers new opportunities for the decision-making process at farm level with access to technical knowledge and decision rules for olive farmers.

Examples of these possible PA applications for soil management in olive orchards include:

- monitoring of evolution of our orchard from satellite images, some of them freely available (see, for instance, Fast tool (European Commission, 2021));
- support tools for deciding type, dose and dates for sowing of cover crops and green manure; and
- site-specific weed control with herbicides, mowing or tillage.

Fig. 6.1.16. New generation of super-high-density olive orchard, where the application of precise agriculture is the main challenge for its environmental sustainability.

Some farm experiences are available, but in all cases georeferencing, prescription maps, real-time systems and specialized technical assistance are needed.

However, with respect to traditional current methods, spreading of PA could reduce significantly herbicide quantity, waste, fuels and traffic load, with strong positive economic and environmental impacts (Pellegrini *et al.*, 2016; Camposeo *et al.*, 2022). PA and digital agriculture represent the main opportunity as well as challenge for environmental sustainability of olive cultivation, including the new generation of super-high-density olive-planting systems (Fig. 6.1.16).

Finally, very recent studies (Popolizio *et al.*, 2022) open a window to the possibility of using retentates extracted from olive mill wastewater as 'bioherbicides' in order to integrate weed-control techniques. These compounds of natural origin acting as pre-emergence herbicides represent an opportunity to reduce the environmental impact of the olive oil sector.

References

Alcántara, C., Soriano, M.A., Saavedra, M. and Gómez, J.A. (2017) Sistemas de manejo del suelo. In: Barranco Navero, D., Fernández-Escobar, R. and Rallo Romero, L. (eds) *El cultivo del olivo*, 7th edn. Mundiprensa: Madrid. Available at: http://hdl.handle.net/10261/246622 (accessed 12 May 2023).

Bechara, E., Papafilippaki, A., Doupis, G., Sofo, A. and Koubouris, G. (2018) Nutrient dynamics, soil properties and microbiological aspects in an irrigated olive orchard managed with five different management systems involving soil tillage, cover crops and compost. *Journal of Water and Climate Change* 9(4), 736–747. DOI: 10.2166/wcc.2018.082

Bombino, G., Denisi, P., Gómez, J.A. and Zema, D.A. (2021) Mulching as best management practice to reduce surface runoff and erosion in steep clayey olive groves. *International Soil and Water Conservation Research* 9, 26–36. DOI: 10.1016/j.iswcr.2020.10.002

Camposeo, S. and Vivaldi, G.A. (2011) Short-term effects of de-oiled olive pomace mulching application on a young super high-density olive orchard. *Scientia Horticulturae* 129, 613–621. DOI: 10.1016/j.scienta.2011.04.034

Camposeo, S., Vivaldi, G.A., Russo, G. and Melucci, F.M., (2022) Intensification in olive growing reduces global warming potential under both integrated and organic farming. *Sustainability* 14, 6389. DOI: 10.3390/su14116389

Castellini, M., Stellacci, A.M., Mastrangelo, M., Caputo, F. and Manici, L.M. (2019) Estimating the soil hydraulic functions of some olive orchards: soil management implications for water saving in soils of Salento Peninsula (southern Italy). *Agronomy* 10, 177. DOI: 10.3390/agronomy10020177

European Commission (2021) Farm sustainability tool for nutrients. Available at: https://fastplatform.eu/whyfast (accessed 12 May 2023).

Fracchiolla, M., Caramia, D., Lasorella, C. and Montemurro, P. (2013) Ground cover management strategies in an Apulian oil-producing olive grove: agronomic and ecological assessment proposals. *Advances in Horticultural Science* 27, 44–54. Available at: https://www.jstor.org/stable/42882922 (accessed 12 May 2023).

Fracchiolla, M., Cazzato, E., Lasorella, C., Camposeo, S. and Popolizio, S. (2021) Mulching with almond hull and olive leaves for weed control in fennel (*Foeniculum vulgare* Mill.) cultivation and flower beds. *Italus Hortus* 28, 59–68. DOI: 10.26353/j.itahort/2021.3.5968

García-González I., García-Díaz A., Sastre B., Teutscherova N., Pérez M.Á., Bienes R., Espejo R. and Hontoria C. (2019) Mycorrhizal, nutritional and virgin olive oil parameters affected by groundcovers. *Journal of Plant Nutrition and Soil Science* 182(5), pp. 815–823. DOI: 10.1002/jpln.201800439

Geyik, M.P. (1986) FAO watershed management field manual: gully control. FAO Conservation Guide 13/2. Food and Agriculture Organization of the United Nations, Rome. Available at: http://www.fao.org/3/AD082E/AD082e00.htm (accessed 12 May 2023).

Gómez, J.A., Infante-Amate, J., González de Molina, M., Vanwalleghem, T., Taguas, E.V. and Lorite, I. (2014) Review: Olive cultivation, its impact on soil erosion and its progression into yield impacts in southern Spain in the past as a key to a future of increasing climate uncertainty. *Agriculture* 4, 170–198. DOI: 10.3390/agriculture4020170

Gómez, J.A., Castillo, C., Mora, J., Lora, A., Taguas, E.V. *et al.* (2019) Criterios técnicos para el diseño y evaluación de cárcavas, revegetación para diversificación del paisaje, muros de contención, mejora ambiental de fuentes y abrevaderos y construcción de charcas artificiales. Junta de Andalucía, Seville, Spain.

Gómez, J.A., Dostal, T., Klik, A., Fereres, E., Intrigliolo, D. *et al.* (2021) *Best Management Practices for Optimized Use of Soil and Water in Agriculture*. DOI: 10.20350/digitalCSIC/13964

Gómez, J.A., Montoliu, J. and Guzmán, G. (2022) Long-term hydrologic effect of temporary cover crops in an olive orchard on a sandy-loamy soil. EGU General Assembly 2022, Vienna, 23–27 May. EGU22-2252 Available at: https://doi.org/10.5194/egusphere-egu22-2252 (accessed 12 May 2023).

Grant, J., Anderson, K.E., Prichard, T., Hasey, J., Bugg, R.L., Thomas, F. and Johnson, T. (2006) Cover crops for walnut orchards. University of California, Publication 21627. Available at: https://anrcatalog.ucanr.edu/pdf/21627e.pdf (accessed 12 May 2023).

Gucci, R., Caruso, G., Bertolla, C., Urbani, S., Taticchi, A. *et al.* (2012) Changes of soil properties and tree performance induced by soil management in a high-density olive orchard. *European Journal Agronomy* 41, 18–27. DOI: 10.1016/j.eja.2012.03.002.

Kavvadias, V., Papadopoulou, M., Vavoulidou, E., Theocharopoulos, S., Koubouris, G. *et al.* (2018) Effect of sustainable management of olive tree residues on soil fertility in irrigated and rain-fed olive orchards. *Journal of Water and Climate Change* 9(4), 764–774. DOI: 10.2166/wcc.2018.143

Liu, H., Li, Y., Josef, T., Zhang, R. and Huang, G. (2014) Quantitative estimation of climate change effects on potential evapotranspiration in Beijing during 1951–2010. *Journal of Geographical Science* 24, 93–112. DOI: 10.1007/s11442-014-1075-5

Montanaro, G., Nuzzo, V., Xiloyannis, C. and Dichio, B. (2018) Climate change mitigation and adaptation in agriculture: the case of the olive. *Journal of Water and Climate Change* 9(4), 633–642. DOI: 10.2166/wcc.2018.023

NRCS (Natural Resources Conservation Service) (2010) Practice Standard. ACCESS ROAD. Code 560.

Palese, A.M., Vignozzi, N., Celano, G., Agnelli, A.E., Pagliai, M. and Xiloyannis, C. (2014) Influence of soil management on soil physical characteristics and water storage in a mature rainfed olive orchard. *Soil Tillage Research* 144, 96–109. DOI: 10.1016/j.still.2014.07.010

Pellegrini, G., Ingrao, C., Camposeo, S., Tricase, C., Contò, F. and Huisingh, D. (2016) Application of water footprint to olive growing systems in the Apulia region: a comparative assessment. *Journal of Cleaner Production* 112, 2407–2418. DOI: 10.1016/j.jclepro.2015.10.088

Popolizio, S., Fracchiolla, M., Leoni, B., Cazzato, E. and Camposeo, S. (2022) Phytotoxic effects of retentates extracted from olive mill wastewater suggest a path of bioherbicide development. *Agronomy* 12, 1378. DOI: 10.3390/agronomy12061378

Russo, G., Vivaldi, G.A., De Gennaro, B. and Camposeo, S. (2015) Environmental sustainability of different soil management techniques in a high-density olive orchard. *Journal Cleaner Production* 107, 498–508.

Sastre, B., Pérez-Jiménez, M.Á., Bienes, R., García-Díaz, A. and De Lorenzo, C. (2016) The effect of soil management on olive yield and VOO quality in a rainfed olive grove of central Spain. *Journal of Chemistry* 2016, art. no. 4974609. DOI: 10.1155/2016/4974609

Sofo, A., Palese, A.M., Casacchia, T., Celano, G., Ricciuti, P., Curci, M., Crecchio, C. and Xiloyannis, C. (2010) Genetic, functional, and metabolic responses of soil microbiota in a sustainable olive orchard. *Soil Science* 175(2), 81–88. DOI: 10.1097/SS.0b013e3181ce8a27

Sofo, A., Ciarfaglia, A., Scopa, A., Camele, I., Curci, M. *et al.* (2014) Soil microbial diversity and activity in a Mediterranean olive orchard using sustainable agricultural practices. *Soil Use Management* 30(1), 160–167. DOI: 10.1111/sum.12097

Song, D., Tariq, A., Pan, K., Chen, W., Zhang, A., Sun, X., Ran, Y. and Zeng, F. (2020). Effects of straw mulching practices on soil nematode communities under walnut plantation. *Scientific Reports* 10: 15351. https://doi.org/10.1038/s41598-020-72530-5

Tekaya, M., Mechri, B., Dabbaghi, O., Mahjoub, Z., Laamari, S. *et al.* (2016) Changes in key photosynthetic parameters of olive trees following soil tillage and wastewater irrigation, modified olive oil quality. *Agricultural Water Management* 178, 180–188. DOI: 10.1016/j.agwat.2016.09.023

Tekaya, M., Chehab, H., Flamini, G., Gharbi, I., Mahjoub, Z. *et al.* (2017) Modification of pomological characteristics and flavour components of fruits and virgin olive oil following wastewater irrigation and soil tillage. *Journal of the Science of Food and Agriculture* 98(8), 2942–2952. DOI: 10.1002/jsfa.8791

Tombesi, A., Tombesi, S., Saavedra, M.M., Fernández, R., d'Andria, R. *et al.* (2007) Production techniques in olive growing. International Olive Council. Available at: https://www.internationaloliveoil.org/wp-content/uploads/2019/12/Olivicultura_eng.pdf (accessed 12 May 2023).

Xiloyannis, C., Palese, A.M., Sofo, A., Mininni, A.N. and Lardo, E. (2018) The agro-ecosystemic benefits of sustainable management in an Italian olive grove. *Acta Horticulturae* 1199, 303–307. DOI: 10.17660/ActaHortic.2018.1199.47

Zipori, I., Erel, R., Yermiyahu, U., Ben-Gal, A. and Dag, A. (2020) Sustainable management of olive orchard nutrition: a review. *Agriculture* 10, 11. DOI: 10.3390/agriculture10010011

6.2 Water Requirements and Irrigation

A. Ben-Gal[1], G. Caruso[2], R. Gucci[2], R. Lo Bianco[3] and F.P. Marra[4]
[1]*Department of Soil, Water and Environmental Sciences, Agricultural Research Organization – Volcani Institute, Gilat Research Center, Israel;* [2]*Department of Agriculture, Food and Environment, University of Pisa, Italy;* [3]*Department of Agricultural, Food and Forest Sciences, University of Palermo, Italy;* [4]*Department of Architecture, University of Palermo, Italy*

6.2.1 Introduction

Olive (*Olea europaea* L.) is one of the most widely cultivated fruit species in the arid and semi-arid regions of the Mediterranean basin characterized by dry summers, cold and wet winters, and a high evaporative demand throughout the growing season. In these areas, the olive tree has been traditionally grown at low planting densities under rainfed conditions mainly for its ability to withstand drought. However, despite the high resistance of the species to water deficit, it has been shown that the use of irrigation increases production even in traditional olive groves (less than 300 trees ha^{-1}), becomes highly recommended in new high-density plantings (300–600 trees ha^{-1}), and is virtually indispensable in high-density orchards (over 600 trees ha^{-1}). At high densities, considering the olive tree's confined root system and the high leaf area index (LAI), a correct and precise application of water is crucial to match water requirements, reduce water consumption, promote vegetative growth, and maximize yields and oil quality (Beede and Goldhamer, 1994; Moriana *et al.*, 2003; Rapoport *et al.*, 2004). The effect on product quality is nowadays very important, since significant economic benefits are derived from the improvement and diversification of olive oil (Caruso *et al.*, 2014; Gucci *et al.*, 2019; Servili *et al.*, 2007).

Olive trees have evolved different strategies to cope with drought stress (Bacelar *et al.*, 2004; Boussadia *et al.*, 2008; Faraloni *et al.*, 2011). Some of them are anatomical, like the tomentosity of the abaxial side of leaves; the cuticle thickness; the position of stomata, localized in small depressions thus limiting transpiration; and the small diameter of xylem vessels that reduce the likelihood of embolism within the xylem system. Other physiological mechanisms include osmotic adjustment, turgor maintenance, and stomatal closure (for a detailed description of the water relations see subchapter 7.1).

Under conditions of limited water resources and irregular supply, deficit-irrigation strategies have been developed to save water and control excessive vegetative growth in fruit tree species. Deficit-irrigation scheduling requires reliable monitoring of plant-water status to avoid excessive water stress, which may have severe impact on tree growth and/or yield.

Determining the correct water needs of olive orchards is essential for productivity and sustainability in any irrigation strategy. Moreover, knowledge of crop-water requirements

DOI: 10.1079/9781789247350.0017

is crucial for water resources management and planning in order to improve water-use efficiency.

It is well known that water consumption occurs mainly through crop evapotranspiration, which may be estimated from reference evapotranspiration and crop coefficients (Fereres *et al.*, 2012) such that the measurement of actual crop evapotranspiration (ETc) for the whole vegetative cycle is equal to the crop water requirement. Although the calculation of ETc is a fair estimator of olive water requirement, it is difficult to determine the level of water stress of the plant.

Recently, the use of plant-based water status indicators has become popular to study plant water relations and for planning irrigation programs. The measurement of plant water status may be useful for irrigation scheduling because of its dynamic nature, which is directly related to climatic and soil conditions, as well as with crop productivity (Fereres *et al.*, 2012).

This chapter provides basic information on methods for estimating the irrigation needs of the olive tree (irrigation frequency and volume) based on both agrometeorological approaches and plant-based methods.

6.2.2 Water Status Assessment

6.2.2.1 Monitoring soil moisture

Soil water content can be monitored through direct or indirect methodologies. Among direct methods, the most used and reliable is the thermo-gravimetric method, which consists of weighing the sample at a certain moisture content, then drying it in an oven at 105°C until it reaches a stable weight. The difference between initial weight and dry weight is the amount of the water present in the original sample. It is a simple method, requires no expensive equipment, but it is destructive and time-consuming.

Indirect methods are based on the measurement of physical parameters, which are closely correlated with the volumetric soil-water content. These methods allow continuous, non-destructive, automatic monitoring and are often used in the field. Electrodes embedded in porous material like gypsum or fiberglass measure the electrical conductivity of the substrate, which is closely correlated with the volumetric soil moisture. These sensors have been used for decades and are still common as they are quite reliable and inexpensive. However, they need to be calibrated frequently and readings are affected by soil type and salinity.

Methods based on the measurement of soil relative dielectric permittivity are widely used. Permittivity is highly correlated with soil water content. Even a slight variation in the soil water content of the soil changes its dielectric permittivity. The wetter the soil, the higher the electrical permittivity. At room temperature, permittivity values are 80 for water, 4–5 for the solid phase and about 1 for air. Time Domain Reflectometry (TDR) and Frequency Domain Reflectometry (FDR) are technologies based on the emission of electromagnetic waves into the ground (Fig. 6.2.1). TDR and FDR sensors measure the transmission time of an electromagnetic wave between the electrodes or changes in its frequency, respectively. TDR sensors typically consist of two or three metal rods or nails that are driven into the ground. One of the rods transmits the electromagnetic wave, while the others

Fig. 6.2.1. FDR sensor installed at 0.30 m depth in a commercial olive orchard for the continuous monitoring of the volumetric water content of the soil.

receive the wave that propagated through the substrate. The time taken by the wave to reach the receivers is used to calculate the transmission speed in the soil, and consequently its dielectric permittivity and the volumetric water content.

An FDR sensor is very similar to TDR, also consisting of two to three electrodes that emit and receive high-frequency electromagnetic waves. Instead of measuring time, FDR sensors measure the dielectric constant of the substrate generated between two electrodes. The dielectric constant of water for dry (non-organic) soils ranges between 2 and 5. The soil water content is easy to calculate, and it is generally expressed as a percentage of volume. In recent decades, TDR sensors have been widely used and many companies produce them. This type of sensor is inexpensive, and the remote acquisition and transmission of data can be automated. The weak point of both TDR and FDR sensors is that they are not very reliable or precise in soils rich in organic matter, clay or rock fragments, and in compacted soils. Salinity can also be a problem as it may generate the overestimation of soil moisture. For this reason, it is strongly recommended that electrical conductivity measurements be carried out both in the laboratory and in the field to obtain good correspondence.

Sensors measuring the matrix potential of the water present in the soil (ψ_m) are also common, with tensiometers among the best known and most widespread. They consist of a microporous ceramic capsule connected to a pressure sensor through a duct full of water. When the tensiometer is inserted in non-saturated soil ($\psi < 0$), water exchanges occur through the porous barrier between the capsule and the soil. The exchange stops when the pressure inside the capsule, measured by a pressure gauge, equals water pressure inside the soil. Tensiometers can generally measure up to about –85 kPa (–0.85 bar), therefore water content can be measured only within a limited range and values become unreliable in very dry soils, which can be a problem for rainfed or severely deficit irrigated olive orchards.

6.2.2.2 Plant-based measurements

Plant water status can be assessed directly as the relative water content of different organs or their water potential, or indirectly via measures of plant responses to changes in water status.

6.2.2.3 Leaf and stem water potential

Water potential is the most widely used parameter to describe plant water status. Leaf and stem water potential vary diurnally and seasonally with transpiration, the difference between them being that while leaf water potential can vary between leaves of a single tree depending on their specific location and age, stem water potential is measured in a non-transpiring plant or leaf in equilibrium with and representing the average potential of the entire tree. Leaf (or stem) water potential is often used to monitor water status for irrigation scheduling.

There are several ways to measure water potential. The most widely used is the Scholander pressure chamber (Fig. 6.2.2) (Scholander *et al.*, 1965). It is a relatively simple instrument composed of a cylinder where leaves or shoot tips are placed with

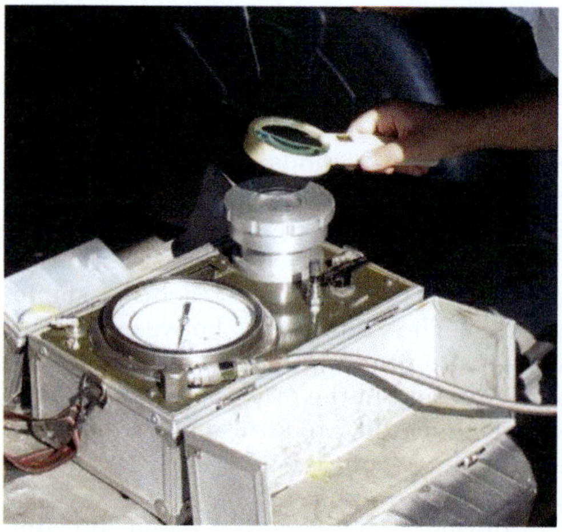

Fig. 6.2.2. Custom-built pressure chamber to measure leaf water potential of olive trees.

the petiole end protruding out of the chamber, which is then pressurized at a slow flow rate until sap appears on most of the cut end surface. To get a good description of plant water status the leaf water potential should be determined at different times of the day: in the dark at the end of the night period (pre-dawn) and at mid-day, when it reaches its minimum value because of transpiration (Caruso *et al.*, 2017; Corell *et al.*, 2020; Gómez-del-Campo, 2013).

Pre-dawn leaf water potential (Ψ_{pd}) determinations require leaf samples taken before sunrise because at the end of the night period the water potential reaches an equilibrium with the soil water potential. Mid-day leaf water potential (Ψ_{md}) should be measured within an hour of solar noon, under clear sky conditions under full sunlight, and leaves should be sampled from the outer part of the canopy. Stem water potential (Ψ_{stem}) is determined by enclosing single leaves or shoot tips in a dark, plastic bag to stop transpiration for a sufficiently long period so that the leaf water potential equilibrates with that of the xylem (Begg and Turner, 1970). For stem water potential measurements, leaves should be selected from the inner part of the canopy, close to the trunk or to a main branch to prevent the effect of pressure drops due to hydraulic constrictions at the branch-shoot level of lateral ramifications.

Regardless of the type of measurements (Ψ_{pd}, Ψ_{md}, Ψ_{stem}), once the sample is placed in the chamber, the flow of gas inside should be adjusted at a low, constant rate of 0.02 $MPa \cdot s^{-1}$ for maximum accuracy (Naor and Peres, 2001). The wetting of the tissue can be observed using a magnifying lens.

To measure the water potential of olive trees the pressure chamber should be capable to read very low values because stress can be severe. A good magnifying lens is also needed to detect the appearance of sap on the small petiole cut surface. The pressure chamber has been proposed, and is, in some cases, used as an instrument for monitoring water status in commercial olive orchards, but it is time-consuming and therefore its commercial use is still unpopular.

Nowadays, pressure transducers and microtensiometers that measure continuously are available commercially for monitoring tree water status in the orchard. These sensors are reliable and relatively quick to install, but the initial investment may be high. They also need to be checked periodically and adjustments to reposition them made during the seasonal operation. Data interpretation also requires qualified staff. The values generally correlate well with water potentials measured by the pressure chamber, but there is more variability than the pressure chamber due to anatomical and environmental factors affecting the measurements, so that calibration and periodic checks need to be done.

6.2.2.4 Relative water content (RWC)

The RWC (%) is the water content of a leaf with respect to the maximum water content at full turgor and can be calculated according to the following expression:

$$RWC = \left[(FW - DW) / (TW - DW) \right] * 100 \quad (1)$$

where FW, TW and DW are the sample fresh, turgid and dry weight, respectively. This method is relatively easy and only requires a precision scale, but measurements are destructive, discontinuous and time-consuming.

6.2.2.5 Leaf gas exchange

Tree water status is also related to stomatal conductance (the degree of opening of the stomata), which decreases as water deficit increases (Fernandes-Silva *et al.*, 2016; Moriana *et al.*, 2002; Marino *et al.*, 2018; Shackel *et al.*, 2021). Gas exchange parameters can be measured non-destructively in the field using either portable infrared gas analyzers (IRGAs) or porometers. Both types of devices include a chamber enclosing part of the leaf during measurements. An IRGA measures the concentration of CO_2 and H_2O in an air sample by detecting the absorption of an emitted infrared light source through that sample. Porometers measure the difference in relative humidity between incoming and outgoing air that is correlated with leaf stomatal conductance. However, IRGAs are expensive and require highly qualified operators for calibrations, measurements and data interpretation for irrigation scheduling. Consequently, the use of this equipment is limited to experimental purposes. Porometers are less expensive than IRGAs but are similarly unsuitable for irrigation applications in commercial orchards.

6.2.2.6 Dendrometers

Plant water status can also be determined by measuring the variations in diameter of different organs, such as stems, branches, roots and fruit induced by diurnal variations of transpiration using linear variable differential transformer (LVDT) dendrometers (Fig. 6.2.3). When transpiration begins early in the morning, a tension occurs in the xylem forcing part of the water stored in the plant tissues during the night to move to the xylem to be lost through transpiration. This allows the plant to respond quickly to changes in transpirative demand when root water uptake is not sufficient to cope with it. Changes in the water content of extensible tissues of the stem induce diurnal diameter variations, which are readily reversible. A number of indices have been developed to measure stem diameter variations including stem growth rate, maximum daily stem shrinkage, daily minimum stem diameter and daily maximum stem diameter and relate them to tree water status (Fig. 6.2.4) (Girón *et al.*, 2015; Corell *et al.*, 2019; Moriana and Fereres, 2002; Moriana *et al.*, 2010).

Fig. 6.2.3. Linear variable differential transformer (LVDT) dendrometer installed on an olive tree trunk for the automatic recording of trunk diameter variations.

6.2.2.7 Sap flow sensors

Sap flow sensors have been used for detecting water stress and assessing crop water needs of olive trees (Giorio and Giorio, 2003; Nadezhdina *et al.*, 2007; López-Bernal *et al.*, 2010; Marino *et al.*, 2021). The sap flow technique is used to assess transpiration rates of plants by measuring the rate at which the sap ascends stems. In this approach, the thermal dissipation rate from a heated needle (Granier type) or the velocity of a short pulse of heat moving along the stem carried by convection (heat pulse method) are measured to estimate the transpiration stream. The velocity of the heat convection is used as a proxy for plant transpiration rate determined by changes in stomatal opening. The sap flow velocity is diminished when transpiration is reduced (for example, at night and during water-deficit conditions). Therefore, pairing sap flow with environmental measures could represent a suitable way to optimize irrigation scheduling. Benefits of sap flow sensors include continuous measurements and low-cost options. The great advantage of sap flow sensors, compared with gas exchange and water potential measurements, is that data can be automatically acquired. The main limits of sap flow systems are that they require technical expertise and need normalization to functional cross-sectional area of the stem or trunk to quantify transpiration water losses and estimate irrigation volumes.

6.2.2.8 Canopy temperature

Leaf temperature is inversely proportional to leaf transpiration because the phase

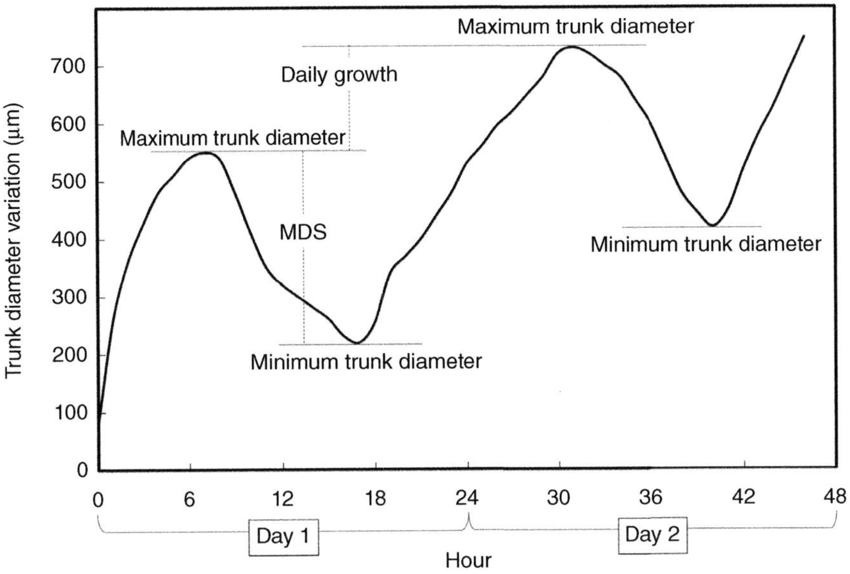

Fig. 6.2.4. Trunk diameter variation and the derived indices measured in July on a fully productive eight-year-old olive tree in Tuscany. The data were acquired using a linear variable differential transformer (LVDT) dendrometer. (From Caruso *et al.*, unpublished)

transition of liquid water into water vapor results in a reduction in temperature. Measurements of canopy temperature to assess water stress began about 50 years ago and can be used for irrigation scheduling in commercial orchards as a result of the development of both affordable image acquisition systems and user-friendly data-image processing software (Fig. 6.2.5). Canopy temperature is related to leaf water status, but it is also strongly influenced by ambient weather. Therefore, the use of canopy temperature as a water stress indicator requires normalization against a reference. The most common leaf temperature index for irrigation scheduling is the crop water stress index (CWSI) (Idso *et al.*, 1981). The calculation of the CWSI relies on two thresholds: the non-water-stressed baseline (T_{wet}), which represents a 'healthy' unstressed plant that transpires at the potential rate for a given evaporative demand, and the maximum stressed baseline (T_{dry}), which corresponds to a plant that is not transpiring. Due to the difficulty of having fully transpiring and non-transpiring crop references, different artificial and natural wet and dry surfaces have been used for empirical determination of T_{dry} and T_{wet} and analytical methods for CWSI calculation have been also proposed (Ben-Gal *et al.*, 2009; Maes and Steppe, 2012; Egea *et al.*, 2017; Bian *et al.*, 2019). The main advantage of temperature-based methods is that they can be scaled up to field-scale and used with remote-sensing platforms.

An alternative to manually acquired measurements, canopy temperature can be measured remotely. Aerial images from thermal cameras can be acquired from unmanned aerial vehicles (UAVs), aircraft or satellites. The high flexibility in flight scheduling and the high spatial ground resolution of UAV platforms allow images to be taken at the most appropriate time of the day (crucial for thermal remote sensing) and to discriminate between olive tree canopies and the bare soil or grass cover in orchard alleys. Conversely, their limited payload and short flight time restrict their application to small or medium-sized orchards (Matese *et al.*, 2015). Another disadvantage is related to the regulations applied by aviation authorities for UAVs, which prevent their use in certain areas. Satellite and airborne surveys allow mapping of large areas with a single image or flight,

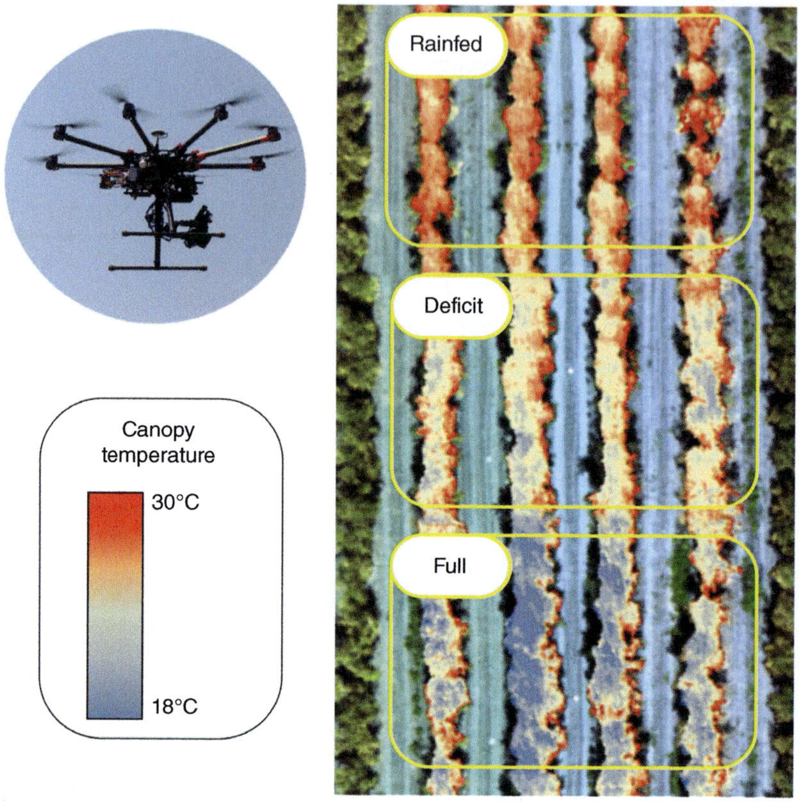

Fig. 6.2.5. Canopy temperature of olive trees subjected to full irrigation (100% ET0), deficit irrigation (40–50% of full) and rainfed condition (only rainfall) measured at 50 m above ground level by a thermal camera mounted on an unmanned aerial vehicle. (From Caruso *et al.*, unpublished)

respectively, but present lower image resolution and suffer from cloud cover. UAV imagery has a high potential for application to precision irrigation, since it allows scanning of entire fields to better understand which sections are over- or underirrigated and take into account the spatial variability in the orchards. The main constraint with these methods is that they may not be dynamic enough for real-time management and currently are not fully automatable.

6.2.3 Determining Water Requirements

6.2.3.1 Evapotranspiration processes

Water moving from terrestrial systems to the atmosphere is termed evapotranspiration (ET). The individual components of ET include evaporation from the soil, in some instances evaporation of water intercepted by the plant canopy and litter layer (E), and transpiration through the stomata of plants (T). The function of E and T within ecosystems is distinctly different: T is usually associated with plant productivity, whereas E does not directly contribute to production (Kool *et al.*, 2014). Evapotranspiration represents the major water requirement for crops including olive trees. Irrigation is often designed to return ET losses in an orchard.

The evaporation of water is fundamentally a phase transition from liquid to gas driven by available liquid water, available energy to drive the transition (evaporate the water) and a gradient in potential energy (vapor pressure) between the evaporating surface and the atmosphere. Measurement of ET is therefore possible either by closing a water balance or energy balance.

Both energy and water balances can be used to solve for ET. Both are conservation equations where the evaporation components are computed as the residual of the budget of a defined system of interest. The components included in the budget equations will depend on the size and timescale of the system (Kool *et al.*, 2014). For example, the water balance in an olive orchard can be simply described as the sum of inputs and outputs:

$$I + P = ET + D + RO + \Delta S \qquad (2)$$

where I is irrigation, P precipitation, ET evapotranspiration, D deep drainage or lateral flow out of the orchard, RO surface run-off/on, and ΔS change in water stored in the root zone of the soil. Minor water balance components that are routinely neglected include changes in plant storage, lateral flow and capillary rise. If all the other variables in the equation are known, ET can be simply calculated. If ET is known, the water balance can be used to calculate I (irrigation). The water balance is only practical as a tool to estimate ET when this is relatively large, otherwise small errors in measurements of other components can result in large errors. Usually, the limiting aspect in solving for ET using a water balance is the difficulty in calculation or measurement of the variables D and ΔS.

The most direct method to close the water balance is via lysimeters. A lysimeter is a soil-plant system isolated from its surroundings in such a way to allow control or measurement of all water fluxes entering or exiting. Lysimeters in field conditions have been used to quantify ET for many crops including young olives (Puppo *et al.*, 2014, 2019). Lysimeters are expensive to build and challenging to manage and are therefore limited to experimental determination of crop factors to be used in estimations and models of ET. Systems of experimental lysimeters have also been used to quantify the relative effects of particular variables, such as drought or salinity, on ET (Ben-Gal *et al.*, 2010, 2017).

The surface energy balance ties into the water balance through the evaporation component and is defined as:

$$R_n - \lambda ET - H - G = 0 \qquad (3)$$

where Rn is net radiation, H is sensible heat and G is soil heat flux (all in W m^{-2}). Extended versions of the surface energy balance include a thermal conversion factor for CO_2 fixation; energy advection into the canopy air layer, and the rate of energy storage per unit area in the layer (Kool *et al.*, 2014).

A simple general equation can be derived for transport within a gas by either molecules, particles, or eddies, sometimes defined as 'carriers', capable of transporting water vapor. Even when the carriers are moving randomly, net transport in a specific direction may occur provided that the scalar decreases with distance in that direction (Monteith and Unsworth, 2013). Based on the diffusion analogy, the turbulent exchange can be described as:

$$ET = \rho\overline{w'q'} \qquad (4)$$

with ET in kg m^{-2} s^{-1}, where ρ is air density (kg m^{-3}) and $\overline{w'q'}$ is the mean product of turbulent fluctuation velocity in the vertical direction (m s^{-1}) and turbulent fluctuating specific humidity (kg kg^{-1}). This equation is based on the assumption that the surface boundary layer is a fully turbulent region where the vertical distribution of vapor fluxes is relatively constant. This is usually valid except when there is condensation or, under stable atmospheric conditions, when stratification in the vertical distribution of vapor fluxes predominates (Brutsaert, 1982).

Acquisition and quantification of either the necessary energy balance or gas transport components to enable calculation of ET is possible but requires meteorological and detection equipment which is costly and currently impractical to include in agricultural fields for day-to-day decision making.

In light of the impracticality of any of the water or energy balance techniques for determination of ET, methods for its estimation have been developed. The most popular methods for estimating ET are calculations of potential maximum ET, which are then multiplied by a crop factor (Kc) or number of factors to consequently estimate the actual ET of a given crop (ETc).

$$ETc = Kc * ET_0 \quad (5)$$

Reference evapotranspiration (ET_0) is the ET of a reference crop surface, typically full cover grass at 20 cm above the soil surface. There are several models used to calculate ET_0, the most popular for irrigation scheduling is the Penman-Monteith (Monteith, 1965) equation as developed and presented by the FAO (Allen *et al.*, 1998). Penman-Monteith calculates ET_0 using standard weather station data including radiation, air temperature, soil temperature, relative humidity and wind speed. It can be determined daily or hourly. Converting ET_0 to ETc (crop ET), as mentioned, requires multiplying by a single crop coefficient or different coefficients (Allen *et al.*, 1998). One coefficient will reflect canopy cover or leaf area as well as crop phenological stage. Another coefficient is often used to estimate the E from the soil. Coefficients can also be included to adjust ET_0 for stress-causing conditions (i.e. salinity) (Minhas *et al.*, 2020) or to allow for deficit irrigation (Pereira *et al.*, 2020).

According to the FAO 56 paper, the basic procedure for calculating ETc consists of (i) determining crop growth stages and selecting a corresponding coefficient (Kc) for each stage; (ii) adjusting Kc for frequency of wetting or climatic conditions; (iii) building a Kc curve; and (iv) calculating ETc. For oil olive orchards, where optimization of oil yield and quality has been found for deficit irrigation strategies (Ben-Gal *et al.*, 2021; Fernández *et al.*, 2013; Gucci *et al.*, 2019) the first step will include identification of crop stages for deficit supply and corresponding Kcs. Example Kc curves for olive from FAO 56 (Allen *et al.*, 1998) and Ben-Gal *et al.*, (2021), who suggested different strategies for 'on' and 'off' years, are shown in Fig. 6.2.6.

The values of crop coefficients or factors (Kc) for olives from the literature need to be carefully examined before using them in places other than where they were developed. This is due to their tendency to represent different things. For example, some have separate factors for canopy cover and for phenological growth stages, others represent a single factor for ET and others separate factors for E and T. The Kc will likely change with tree spacing which is associated to the intensity of the cropping system; this is mostly due to differences in relative canopy cover by the trees in the orchard. In growing areas where biennial bearing (differences in yields between subsequent 'on' and 'off' years) is significant, water consumption and therefore Kc values for olives should reflect the bearing status. Trees with high fruit load have been found to require up to 30% more water compared to trees with low to no fruit (Bustan *et al.*, 2016; Naor *et al.*, 2013).

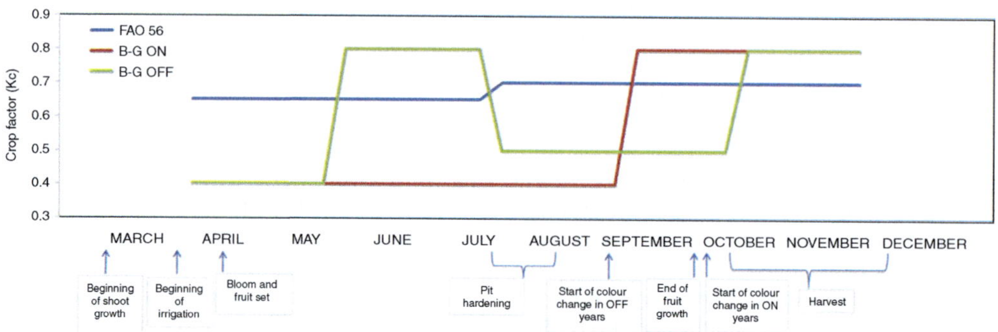

Fig. 6.2.6. Example of crop factor (Kc) course for oil olive trees from the literature. FAO 56 (Allen *et al.*, 1998) and Ben-Gal *et al.* (2021) for 'on' years (B-G ON) and 'off' years (B-G OFF).

6.2.3.2 Water use and tree productivity

The transpiration (T) component of ET reflects a tree's productivity since opening of stomata to allow water loss to the atmosphere as vapor is directly tied to carbon acquisition, needed for photosynthesis, growth and production. In general, transpiration and growth are linearly related. A plant that transpires more also grows more and a plant that becomes larger will concurrently transpire more water. In any given place and time, maximizing production is accompanied by maximized water use by a plant. For example, olives (cv. 'Barnea') in a lysimeter experiment in Israel were found to reduce their growth and water consumption as a function of increasing root zone salinity levels caused by both increased irrigation water salinity and reduced leaching level (Ben-Gal *et al.*, 2017). Fig. 6.2.7 demonstrates the strong relationship between tree-scale vegetative growth and water use found in the experiment.

Furthermore, water use and requirements of olive orchards are difficult to determine since they are influenced by many factors including pruning strategy, variety and irrigation system, in addition to planting density and meteorological conditions.

Water-use efficiency (WUE) is an ambiguous term, and one should be warned to confirm the intended specific definition before trying to interpret any use of it. WUE is used by biologists to evaluate carbon sequestration (photosynthesis) as a function of transpiration. This is usually measured at a scale of a leaf or part of a leaf. Another of the many definitions for WUE is biomass production as a function of water consumption (Heydari, 2014). For the objectives of this chapter in which our major interest is irrigation, the term **water productivity** (WP) is less ambiguous, defined as yield (usually economic yield) as a function of water applied, either as precipitation or irrigation (Van Halsema and Vincent, 2012; Fereres *et al.*, 2012). Water productivity can be important to olive growers since it can be increased either by increasing yield at a rate greater than the increase in irrigation application or by reducing irrigation at a rate greater than the reductions of yield. One should be cautious here as well since WP (and most sorts of WUE) can increase to potentially infinite levels in the case where both irrigation and yield approach zero. The most obvious practical use for interest in WP is to ultimately calculate the relative

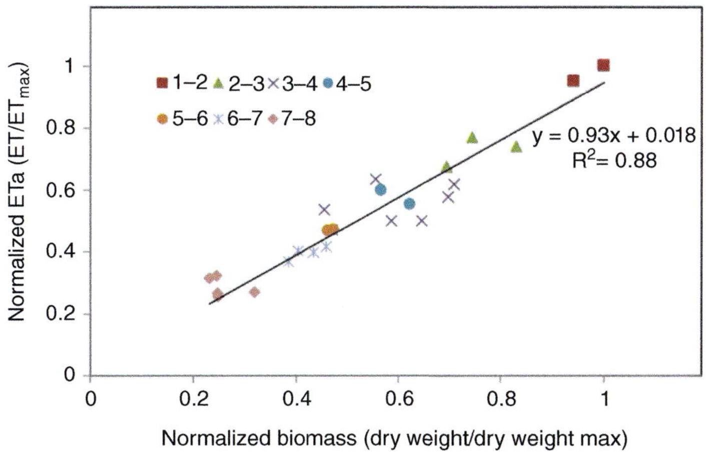

Fig. 6.2.7. Linear relationship between olive biomass production and tree-scale evapotranspiration. ETa is actual tree-scale evapotranspiration (ET), biomass is tree-scale total biomass production. Both variables were measured during a three-year lysimeter experiment in Israel evaluating irrigation water salinity and irrigation (leaching) amounts (Ben-Gal *et al.*, 2017). ECe is time-profile averaged electrical salinity of saturated paste solution from soil in the lysimeters. Each point shows data from a single olive tree (cv. 'Barnea').

yield value and irrigation costs and determine if increased WP is, or is not, accompanied by increased profit and therefore economically convenient.

Water productivity of olives is sensitive to location (especially weather and atmospheric demand for water) as well as variety, fruit load in locations where biennial bearing is prominent, and irrigation scheduling. Regulated deficit irrigation, discussed later in the chapter, attempts to increase WP by limiting water during specific stages of crop phenology when the resulting stress is expected to cause no or relatively low negative effect to fruit and oil yield.

6.2.3.3 Irrigation volumes and scheduling

Once ET_0 and Kc have been determined, actual water volume for irrigation is a function of converting from units of length (mm water) to cubic meters and multiplying by orchard area.

$$I = ET_0 * K_c * A * 10, \text{ or, } I = ET_c * A * 10 \quad (6)$$

where ET_0 and ETc are in mm, Kc is unitless and A is in Ha. The 10 is to correct for units, bringing I to cubic meters (m^3).

6.2.3.4 Irrigation frequency

Irrigation scheduling includes determination of timing of water application in addition to amount of water applied. Strategies for determining frequency of application events is highly dependent on irrigation method. In flood and other surface delivery methods, as well as in sprinkler systems wetting the entire soil surface, each irrigation event can be costly. In these cases, following an irrigation event, due to lack of oxygen, moisture in the soil tends to temporarily be higher than optimal for tree health and growth. Irrigation therefore should be applied only when necessary, and in the minimum number of events. In these cases, the concept of PAW or plant available water can be utilized to determine which irrigation volume fills the soil's storage capacity without wasting water to deep drainage. Actual PAW is the difference between the crop lower limit for water uptake and the volumetric soil water content (volume water/volume soil). PAW is a function of soil texture and crop rooting depth and pattern. Estimation of PAW can be made based on knowledge of depth of wet soil and rooting pattern (root density as a function of depth in soil profile). Water content can be measured and monitored with soil water sensors. Irrigation scheduling uses PAW by calculating maximum allowable depletion (MAD). MAD specifies the maximum amount of soil water a crop is allowed to extract from the active rooting zone between irrigation events. Typically, irrigation events are planned to prevent soil water from being reduced to half of the PAW. Using this concept, irrigation events will be more frequent for light soils, shallow root zones and high ET_0 and less frequent for heavier soils with high water holding capacity, deep root zones and mild weather with low ET_0.

Microirrigation techniques, including drip irrigation, enable growers to be much less sensitive to PAW and allow frequent water application. Since microirrigation wets only part of the soil, there is less sensitivity to too much water. Also, microirrigation systems are typically permanently installed in an orchard and automated with little or no added costs involved with their operation concerning individual water application events. In these cases, irrigation can be scheduled to provide actual water requirements without consideration of MAD or PAW since microirrigation typically involves water volumes much lower than the root zone soil's water-holding capacity. Increased frequency, up to daily application in fruit tree crops and even higher in vegetable crops, based on return of ETc for the given period, has been found to optimize uptake of both water and nutrients by roots and be beneficial for growth and production (Ben-Gal and Dudley, 2003; Segal *et al.*, 2006; Silber *et al.*, 2003). High-frequency irrigation of olives additionally benefits accuracy in regulated deficit strategies by making it easier to quickly change irrigation amounts in order to reach desired water-stress levels.

6.2.4 The Concept of Deficit Irrigation in Olive Orchards

In several fruit crops, irrigation has been used as a tool for improving production. When pursuing this task, it is essential to establish a series of objectives, such as controlling tree vigor, preventing occasional periods of water stress, managing fruit development or modifying its quality. When tree water deficit, the difference between transpiration and root water uptake, is less than 10%, the damage is minimal and easily reversible as the balance is restored; while if water deficit reaches about 30–40%, the damage may become irreversible even after trees recover the optimal water status (Gucci *et al.*, 2012).

Irrigation techniques using controlled water deficit (Regulated Deficit Irrigation, RDI) arise from the need to make the most of the increasingly scarce water resources as well as to improve product (oil) quality (Ben-Gal *et al.*, 2021). During transpiration through the stomates, the uptake of CO_2 from the atmosphere is associated with water vapor loss. The limitation of transpiration has often been associated with production losses by farmers, since it has been repeatedly pointed out that plants give up water to obtain carbon and, if carbon intake must be maximized to achieve high yields, potential transpiration must be satisfied.

However, in trees, water deficit reduces vegetative growth, but not necessarily the production of fruit, giving opportunity to increase irrigation water productivity (WP). It is indeed possible to reduce the transpiration of woody plants without reducing fruit yields, by varying water supply according to the phenological phase. For example, the Spanish olive cultivar 'Arbequina' did not exhibit significant yield (olives and oil) losses when irrigation volumes were reduced to 50% and 75% of ETc during the fruit growing period (from three weeks after fruit set to before the beginning of veraison) (Alegre *et al.*, 2000). Other experiments conducted in Italy, Spain, Turkey and Morocco have shown that the reduction of irrigation volumes to 66% of ETc does not determine any decrease in production compared to fully irrigated trees; even supplying only 33% of the irrigation needs has shown significant production advantages compared to rainfed management, due to a greater number and size of fruits (Patumi *et al.*, 1999; d'Andria and Morelli, 2002; Marra *et al.*, 2016). It is important that there is no water deficit during active vegetative growth and flowering, since it would negatively affect both flower fertility (and consequently the number of fruits) and shoot growth, necessary to increase leaf area of the current year and the production of the following year.

Another strategy to improve tree WP is the application of deficit irrigation throughout the irrigation season. In a semi-arid climate, olive trees periodically suffer from water shortages, which may or may not reduce their yields depending on the cultivar and length of the dry period (Lambers and Oliveira, 2019; Boyer, 1985). To ensure suitable and regular growth rates, complementary irrigation is generally applied with quantities of water that, however, do not fully satisfy the crop water needs, but rather are able to ensure a minimum sustainable production.

Some studies carried out in Tunisia with reduced irrigation volumes of up to 20% of ETc on local varieties, such as 'Chemlali' or 'Chetoui', as well as on 'Picholine', 'Manzanilla' and 'Coratina', have shown increases of WP even for the lowest water supplies (Mezghani *et al.*, 2012). In general, in semi-arid conditions, with very low average annual rainfall, even water supplies reduced to 20–50% of ETc can guarantee a minimum of vegetation and production at least in medium- and low-density plantings. Under these conditions, adjusting irrigation volumes according to the expected crop load is highly recommended, i.e. a more modest volume in the off years and *vice versa*.

The variety of experiments and results suggest that the specific levels and timing of water deficit applied with RDI are a function of geographic location and weather conditions, cultivar and growing system (planting density, type of pruning, training form etc.).

6.2.4.1 Intensive olive orchards and irrigation management

High-density (HD) and super-high-density (SHD) olive-growing systems have become real business opportunities as they are increasingly utilized to increase yields and reduce production costs. These planting systems need high-irrigation inputs. Indeed, unlike traditional olive-growing systems, irrigation is essential to obtain an optimal and constant level of production in HD and SHD olive growing (Camposeo *et al.*, 2008; Strippoli *et al.*, 2013). There are basically three cultivars on which these systems (mainly SHD) have been calibrated: two Spanish, ‘Arbequina’ and ‘Arbosana’, and one Greek, ‘Coroneiki’. The introduction of new cultivars into new environments raises the need to evaluate their eco-physiological behavior in response to irrigation. Several studies have been carried out in the attempt to fine-tune the irrigation management of HD and SHD systems and improve their sustainability. It has been shown that, with irrigation, the olive tree can recover both the optimal water status and leaf gas exchanges within two to three days, if the water stress imposed in the pre-irrigation phase determines values of leaf water potential greater than –1.0 and –2.5 MPa before sunrise and noon, respectively; at values lower than these thresholds, recovery is much slower (20 days), and below –2.0 and –4.0 MPa, respectively, it may not occur at all (Vivaldi *et al.*, 2013). There are, however, some varietal differences. For example, ‘Arbequina’ showed a better response to irrigation than ‘Coratina’, which, instead, showed a higher WUE together with a lower leaf transpiration (Vivaldi *et al.*, 2013).

There is no consensus on the best approach to irrigation for SHD systems. RDI may represent one of the most valid strategies, as it manages to offer the right combination of water saving, vegetative growth and oil production. For example, it has been shown that the optimal amount of irrigation water for an SHD ‘Arbequina’ orchard is around 30% of ETc with savings of 70% or more and a reduction in yield of about 26% (Fernández *et al.*, 2013). But, again, this may depend on irrigation water availability, irrigation management expenses, local environment (soil type, rainfall amount and distribution etc.), market and social issues.

6.2.4.2 Productive responses to RDI

As mentioned above, irrigation can increase yield, fruit size, pulp-pit ratio, mesocarp cell size and oil content (Caruso *et al.*, 2014; Gucci *et al.*, 2009). The sensitivity to water deficit of each cultivar in a given environment is critical for the development of an adequate RDI strategy; one of the basic principles of RDI is, in fact, to reduce the irrigation volumes during specific phenological phases, so as not to influence fruit yield. RDI strategies have been used for some years in tree crops to control vegetative growth, modify dry-matter partitioning, save water and, in many cases, improve fruit quality (Li *et al.*, 1989).

In olive, a period of water deficit during the first stage of fruit growth decreases the final fruit size, acting mainly on the cell-division process; while water deficit after pit hardening acts mainly on the cell-expansion process (Beede and Goldhamer, 1994). More recent observations have shown that cell division and expansion occur simultaneously and tend to overlap during fruit development (Gucci *et al.*, 2009). In non-limiting environmental conditions (irrigated trees), growth does not follow a double sigmoid but an almost linear trend. Cell division is characterized by high metabolic activity, strong nitrogen absorption, and intense respiratory activity; hence, an adequate water supply is important, to avoid a slowdown in cell formation, ultimately leading to smaller fruit size and increased fruit drop.

The production of oil per tree tends to increase in irrigated plantings, even if the oil measured as a percentage of the fruit fresh weight is lower than that found in olives obtained without irrigation (Ramos and Santos, 2010). However, the increase in yield often occurs at the expense of the oil quality (Alegre *et al.*, 2000). The responses in terms of oil yield or number of olives per tree may vary both in relation to the amount of water applied and to variations in evapotranspiration.

Early approaches using Ψ_{stem} thresholds were successful in managing RDI to increase WP and olive oil quality. Responses to irrigation were more evident in 'on' than in 'off' years, indicating that crop load, rather than irrigation, was the primary determinant of oil yields (Naor *et al.*, 2013). Attempts to irrigate according to pre-set target values of Ψ_{stem} also significantly increased WP as yield decreased by less than 15% under ~40% reduction of water supply (Ben-Gal *et al.*, 2021). From both the above studies, it was evident that the choice of optimal Ψ_{stem} threshold needs to consider crop load, environment and cultivar. While Ψ_{stem}-based irrigation scheduling has proved to be effective in driving RDI, its high labor demand and cost make its use in commercial olive orchards questionable.

Recent approaches to plan an RDI strategy were tested providing 45% of the olive water needs and using patch clamp pressure (ZIM) probes to estimate and study the daily leaf turgor pressure curves. The association between the shape of those turgor pressure curves and the water stress level of the tree was found to be valid for many olive genotypes and cultivation conditions (Padilla-Díaz *et al.*, 2016). The approach with ZIM probes allowed for a considerable saving of water, without compromising in any way either the tree water status or fruit yields, even in comparison with water consumption estimates using the crop coefficient.

Despite a clear increase of fruit weight in fully irrigated trees, as mentioned above, percent of oil yield decreases; this is explained by an increase in terms of fresh weight but a decrease of about 10% in dry weight (Girona *et al.*, 2002). The decrease in percent of oil yield in well-irrigated olive plantings seems to depend not on the fruit oil content *per se* but on the reduced extractability of the oil, as water contained in the fruit would determine the formation of emulsions during the extraction process (Berenguer *et al.*, 2006).

Opinions regarding the amount of oil accumulated in the olive in relation to irrigation are contrasting; some studies have shown that RDI decreases the overall tree crop load to the advantage of higher oil quality and quantity inside the olive, while other studies have found only minimal differences (Lavee and Wodner, 2004). Such discrepancies may be attributed to the balance between sink and source in olive trees during fruit development as well as to environmental factors.

Ripening of the drupes plays a key role in determining olive oil quality, and some changes in quality parameters induced by changes in ripening patterns can be attributed to irrigation. In general, irrigating beyond veraison is not beneficial when the goal is the production of high-quality oil. It is therefore clear that RDI represents the best compromise for the production of quality oil, as it allows for a considerable increase of the oil yield compared to rainfed trees and a minimal decrease in yield along with substantial water savings compared to fully irrigated trees. In this regard, monitoring tree water status is also becoming increasingly important for a precise irrigation strategy that targets quality, a challenging opportunity in a rapidly evolving market for high-quality olive oils (Gucci *et al.*, 2009).

6.2.4.3 Effects of irrigation management on the composition of olive oil

Various investigations have recently targeted the effects of irrigation regimes on the chemical and sensory characteristics of olive oil. As mentioned above, this is because today's needs include both saving water resources and producing quality oil that can be competitive on the market.

Both in high-density and low-density planting systems, obviously with the necessary differences, the producer must always find the optimum balance between quality and costs. Despite this, the results obtained were not always concordant.

Regarding the analytical parameters that allow olive oil classification in the different commercial categories, such as degree of acidity, absorbance in the ultraviolet (K270) and number of peroxides, only for this last parameter several authors agree that there is no influence of irrigation (Patumi *et al.*, 1999; Faci *et al.*, 2000).

As irrigation increases, a number of parameters decrease, including ripening index, total polyphenols and oxidative stability; also monounsaturated fatty acids and the oleic:linoleic acid ratio decrease, while polyunsaturated fatty acids generally increase, although sometimes the differences are minimal and therefore negligible. In SHD systems, a decrease in irrigation may cause a reduction of free fatty acids, negatively affecting oil quality (Dag *et al.*, 2015).

Sensory attributes like fruity, bitter and spicy are absent or very low in the oils obtained from fully irrigated olive trees, while they may be excessively accentuated by severe water deficits. Irrigation levels between 40% and 60% of the ETc enable balanced olive oils to be obtained. These irrigation regimes can favor the development of artichoke, grass, green apple, ripe fruit and flower flavor notes that enrich the sensory profile of the oils (Berenguer *et al.*, 2006).

An important aspect that has been often pointed out by several studies is the convenience of using some specific cultivars for the production of oil under RDI regimes. This must be carefully considered because high-yielding cultivars are very often preferred to those producing high-quality oils, i.e. quantity over quality. Furthermore, clear and generalized indications for precise regulation of irrigation regimes are still lacking, so growers prefer to sacrifice part of the quality of the oil in favor of its quantity. This is the case, for example, with the 'Cordovil' cultivar in Portugal, appreciated for the production of an oil with a high amount of free fatty acids and unique sensory properties. However, as there are no regulations governing the irrigation management of PDO 'Cordovil', farmers tend to sacrifice part of quality in favor of quantity (Ramos and Santos, 2009). Similar issues can be found in other olive-growing areas with other olive cultivars.

6.2.4.4 Partial root-zone drying (PRD)

Often the RDI strategy is complex to develop and apply, particularly regarding the identification of the most suitable period to impose the deficit as well as its degree or intensity. Therefore, an alternative strategy to RDI was developed, the PRD (Partial Root-zone Drying) strategy, i.e. the irrigation of a portion of the root system. PRD was first developed from experiments on water stress in grapevine (Dry and Loveys, 1998) conducted in Australia. This strategy consists in irrigating one half of the root system of the plant, leaving the other half dry; after a preset time based on the characteristics of the soil and sensitivity of the plant, most commonly every two weeks for trees in loamy soils, the irrigated and the dry sides are exchanged. The main motivation in irrigating plants with the PRD technique consists in the potential production of chemical signals (abscisic acid, cytokinins, pH changes) in the dry roots which, thanks to the xylem flow from wet roots, are transported to the leaves; here they may cause a reduction in stomatal conductance (Gowing *et al.*, 1990). In particular, PRD is claimed to induce a partial stomatal closure, reducing water losses due to transpiration without significantly affecting CO_2 intake, and consequently yields (Lo Bianco, 2013). Through this mechanism, the root-derived chemical signal modifies the physiology of the canopy, improving WP, i.e. determining a lower water consumption per unit of dry matter produced. Knowing soil water potential in the drying root-zone (or a specifically studied time for switching dry and wet sides) is critical to avoid root death and loss of the chemical signal, which would compromise the success of PRD and plant growth itself.

Several PRD experiments have been carried out using different irrigation regimes (from 60% to 40% of ETc) on various species and cultivars, in most cases using localized irrigation. Based on the results of these experiments, different models have been developed, such as a dry-irrigated side alternation every four weeks instead of two, or the beginning of the alternation after full bloom, or only during fruit ripening.

6.2.5 Irrigation Water Quality and Salinity Effects in Olive

Due to freshwater scarcity in much of olive's popular growing areas, use of unconventional sources of water for irrigation is becoming prevalent. These alternative water sources include recycled wastewater from industrial and municipal sources and brackish groundwater.

6.2.5.1 Irrigation with recycled wastewater

Issues of irrigation with recycled water include possible long-term effects to the soil and concerns regarding fertilizer requirements (Pedrero *et al.*, 2020). Wastewaters and especially municipal wastewater effluents tend to be high in sodium and therefore contribute to elevated sodium adsorption ratio (SAR) in soils. The SAR is a measure of the effect of sodium relative to the effects of calcium and magnesium on soil physical conditions. High SAR indicates risk for permeability hazard on soil structure. When SAR in soils is high, there is risk for reduced infiltration and water movement, particularly following rain or irrigation with low-salinity water (Erel *et al.*, 2019; Pedrero *et al.*, 2020).

Heavy soils have been found to become less productive over time when irrigated with recycled wastewater (Assouline *et al.*, 2016). The reasons for the decline in fruit tree productivity are not completely clear but seem to involve combined sodicity and organic components causing issues of soil hydrophobicity and hypoxia and related plant stress.

Another issue of irrigation with recycled water lies in nutrient management. Nutrients delivered with the water should be monitored and considered in fertilizer decision-making (Erel *et al.*, 2019). This is particularly true for drip-irrigated orchards where fertilizers are provided as solutes with the water (fertigation). Level of nutrients, including the macronutrients (N, P and K) and micronutrients, depend on the wastewater source as well as the methods of its treatment and storage.

Other contaminants potentially contained in recycled wastewaters include human pathogens that can promote disease in people coming in direct contact or digesting the water (Pedrero *et al.*, 2020). In olives, where the commercial products are processed and free of pathogens, the major concern for human health lies with workers in contact with the irrigation water itself.

6.2.5.2 Olive response to salts

Common to recycled wastewater and brackish groundwater is a relatively high concentration of salts. The presence of salts in water changes calculation of irrigation requirements due to the need for leaching. If salts in irrigation water are allowed to accumulate in the root zone, they quickly become detrimental to water uptake and plant growth and functioning.

Salts in an olive tree's root zone cause both osmotic stress and toxicity and consequential responses (Ben-Gal, 2011; Ben-Gal *et al.*, 2017). Osmotic stress, caused by lowering of the potential energy of the soil solution due to dissolved salts, is felt by the tree much as drought is sensed, essentially lowering the energy gradient between soil, roots, leaves and atmosphere and reducing water uptake. Toxic effects occur as ions of the salts, particularly sodium and chloride, are taken up into the roots. These effects begin with the roots themselves, which are the tree's first defense in protection from salinity (Ben-Gal *et al.*, 2017; Soda *et al.*, 2017; Tan *et al.*, 2020). Olives exposed to salt have been found to change root-growing patterns including altered architecture and morphology (Tan *et al.*, 2020). One response mechanism to salt is a high mortality rate of young roots (Soda *et al.*, 2017). Under high-enough exposure, ions of salts enter the plant, and eventually accumulate in sensitive above-ground tissues where their effects can be related to lowered photosynthesis, growth and production.

Olives are considered relatively tolerant to salinity but this should be regarded cautiously due to (i) salinity responses being very cultivar-specific, with different varieties behaving differently to the same exposure to salinity (Gucci and Tattini, 2010; Ben-Gal, 2011); and (ii) controlled studies consistently showing that tolerance reported in the literature is too often shown as survival of young seedlings. Response to salt can have tremendous negative effects on growth and production of mature olive trees long before mortality is an issue.

Olive tree response to irrigation water salinity depends both on the concentration of salts in the water and on irrigation rate, the combination of which determines soil solution and root-zone salinity. Deficit irrigation with water containing salts will lead to high concentrations of salts while excess irrigation will allow lower concentrations as some of the salts are leached beyond the root zone. Root-zone salinity is typically expressed as electrical conductivity of soil sample extract solution (ECe). Response curves of olive production as a function of normalized root-zone salinity found in the literature vary greatly (Fig. 6.2.8). Work by Aragüés *et al.* (2004, 2005) suggested dynamic responses changing over the three years of their experiment with an average threshold value of 4.2 dS/m below which the 'Arbequina' olives were not affected. While their threshold was the same as that given by Maas and Hoffman (1977) for a generic 'moderately sensitive' olive crop, the response to rising salinity in Aragüés *et al.* (2004, 2005) was greater, with 16% compared to 12% decrease in yield for every increase of 1 dS/m in ECe. A study on bearing 'Barnea' trees in large (2.5 m^3) lysimeters (Ben-Gal *et al.*, 2017) suggested a sigmoidal response curve (Van Genuchten and Gupta, 1993) with a much lower threshold, if any, indicating even greater sensitivity to salinity (Fig. 6.2.8).

Regarding salinity and olive oil quality (Tietel *et al.*, 2019), there is no clear trend: increased salinity has been found to cause increase in some desirable quality parameters including polyphenol and tocopherol levels but to cause contrasting non-desirable decrease in a number of important compounds, including 16:1 and 18:3 fatty acids.

6.2.5.3 Irrigation management with water containing salt

To minimize the negative effect of root-zone salinity when irrigating with water containing salts, irrigation must be supplied at high rates, such that some of it carries the excess salts out of the root zone to deeper soil. The leaching requirement (LR) is a function

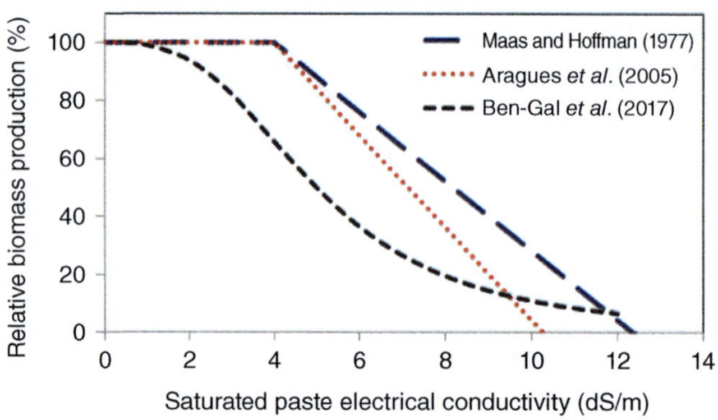

Fig. 6.2.8. Salinity response curves for olive biomass as a function of averaged saturated paste soil extract electrical conductivity. Blue dashed line indicates responses from Maas and Hoffman (1997) for generic olives categorized as 'moderately tolerant'. Red dotted line indicates average response over three years for 'Arbequina' olives (Aragues *et al.*, 2004, 2005). Black dashed line is from two-year controlled lysimeter experiment on 'Barnea' olives (Ben-Gal *et al.*, 2017).

of the salinity of the irrigation water, the sensitivity of the crop and the level of reduced yield due to salts that the grower is willing to accept. The United Nations Food and Agriculture Organization (FAO) (Ayers and Westcot, 1985) recommends computing LR as follows:

$$LR = \frac{EC_{iw}}{5 \cdot EC_e^{ac} - EC_{iw}}$$

where, EC_{iw} is irrigation water EC, and EC_e^{ac} is soil-saturated paste extract EC of average soil salinity tolerated by the crop at a predetermined acceptable level. EC_e^{ac} values used to determine LR are usually either the EC_e of the threshold value (EC_e –0%), reflecting a 0% yield-loss decrease due to salinity, or EC_e –10%, reflecting a 10% yield-loss level. The LR calculations assume that all salinity is potentially negative; ignoring 'positive' EC from nutrients, which, in fertigated agriculture, may contribute up to 1 dS/m to EC_{iw} and which may be taken up by plants at rates much greater than for other dissolved salts. Chemical reactions in the soil that could alter the assumed conservative relationship between EC_e and EC_{iw} are also ignored. Olives are regarded as being 'moderately tolerant' to salinity, and as such would have approximate threshold of 4 dS/m EC_e and slope of further yield decrease of 12–16% for every increase of 1 dS/m EC_e (Fig. 6.2.8; Aragüés *et al.*, 2004; Maas and Hoffman, 1977). Examples of LRs for olives for EC_{iw} ranging from 1 dS/m to 8 dS/m are shown in Table 6.2.1 (Ayers and Westcot, 1985). The LRs are given for two different scenarios, EC_e –0% and EC_e –10%. The relative water application rates (irrigation relative to potential ET: I/ETp) corresponding to the calculated LRs are also shown in Table 6.2.1.

It has been suggested (Dudley *et al.*, 2008; Meiri and Plaut, 1985) that calculations of LR, like those presented by the FAO, are problematic because they do not consider soil type, climate, or salinity-induced reduction in transpiration by the plants. Such omissions could possibly result in underestimation of actual leaching and overestimation of leaching requirements.

Table 6.2.1. Leaching requirement (LR) according to FAO (Ayers and Westcot, 1985) for olives at increasing irrigation water salinity (EC_{iw}). Based on EC_e –0% threshold of 4.0 dS/m (zero yield decrease due to salinity: 100% potential yield) and EC_e –0% of 4.8 dS/m (up to 10% yield decrease due to salinity: 90% potential yield). I/Ep is the equivalent irrigation rate (irrigation relative to potential evaporation) at each LR.

	90% potential yield		100% potential yield	
EC_{iw}	LR	I/Ep	LR	I/Ep
1	0.05	1.07	0.05	1.07
2	0.09	1.09	0.11	1.10
4	0.19	1.24	0.25	1.35
6	0.33	1.52	0.43	1.75
8	0.50	1.96	0.67	3.01

Approaches for evaluating crop-production functions that consider salinity and water application rate, crop sensitivity, soil type, climate and plant feedback responses have recently been developed (Shani *et al.*, 2007). Yield as a function of irrigation water salinity and water application rate from a lysimeter study in Israel (Ben-Gal *et al.*, 2017) and as predicted by Shani *et al.* (2007) are found in Fig. 6.2.9. The nature of the irrigation-water-salinity response is dependent upon the amount of water applied, on crop sensitivity, on climate, and on the soil in which the water is applied and the trees grow. As irrigation salinity increases, growers can minimize decreased yields by increasing the leaching fraction (LF), but they must carefully consider the costs versus the expected returns. Theoretically, if one were irrigating with salt-free water, crops would respond linearly, at least in terms of vegetative growth and ET, to water quantity up to the point where climatic demand was satisfied. Additional water in this case would not have any beneficial effect on growth or yield and would therefore be superfluous (in the context of this discussion, potential negative effects of excess water are ignored). When irrigation water contains salts, the plant takes up and transpires less water, such that even at low application rates (targets close to or less than ETp), some leaching of salts out of the root zone occurs. Lower transpiration is accompanied by reduced plant growth and yield (Fig. 6.2.9). For olive cultivation in well-drained soils in Israel, irrigation water of EC = 0.5 dS/m does not limit production. At EC = 2 dS/m and higher there is clear added benefit to increased irrigation that leaches salts and increases production. Irrigation with EC = 2 dS/m limits production to ~80% of the potential yield with non-salty water. Increasing irrigation water salinity to 5 dS/m, which is close to the lowest salinity of available water in the region, limits production to 50% of the potential yield. EC_{iw} has a substantial influence on actual LF. Due to lowered ET caused by higher salinity conditions, the drainage component of the water balance increases as EC_{iw} increases and I/ETp remains constant. Significant volumes of the water applied for the irrigation water containing salts, but necessary to maximize yields, are lost from the root zone.

Again, both LRs and yield loss due to irrigation water salinity are sensitive to weather (demand for water), variety, and soil environment. These, as well as the environmental costs of irrigation with non-conventional water sources, as leachate water pollutes deep soils and groundwater with salts and other agro-sourced or natural contaminants, will be lower in olive-growing areas with some summer rainfall events and relatively moderate climate and in regions with natural hydrological leaching of salts to the sea, while they will be greater in more arid areas.

6.2.6 Irrigation Methods

Irrigation methods are defined as the way water is distributed in the field. Irrigation practices are usually grouped into pressurized and gravity methods. In the first, water is transported in closed pipeline networks whereas in gravitational methods the transport and distribution take place through networks of free-surface channels. In olive, the distribution of irrigation inputs is generally rationalized adopting microirrigation methods, such as microsprinkler, drip and subsurface irrigation.

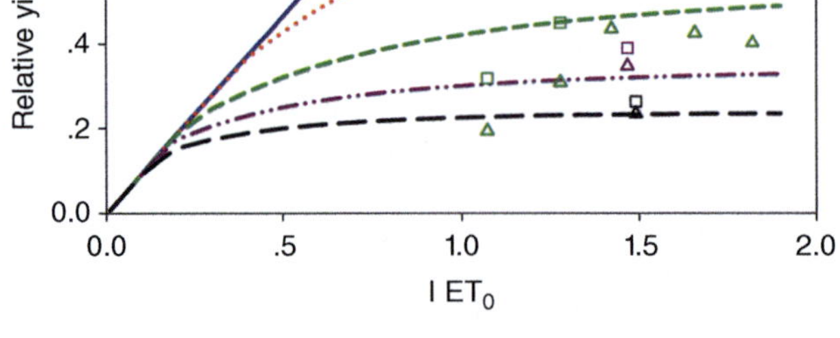

Fig. 6.2.9. Water production functions calculated by the Shani *et al.* (2007) ANSWER model for irrigation water salinity levels evaluated experimentally in Ben-Gal *et al.* (2017). Symbols are values of tree size measured as trunk cross-sectional area (TCSA, squares) and total accumulated above-ground dry weight of biomass (AGDW, triangles). Lines are simulated results using input values found in Ben-Gal *et al.* (2017), simulating semi-arid Mediterranean conditions with no summer rains. Line style and color of lines and symbols correspond to given irrigation water salinity levels.

The use of microsprinkler systems has the benefit of distributing water over a larger area of soil than drip systems and is a method applied mostly in traditional olive groves. This method has the disadvantage of having greater losses due to evaporation and favor the development of weeds that can compete with the olive trees, both aspects that result in a lower irrigation efficiency.

Drip irrigation is a cost-effective and efficient method of directly watering the soil near plant rows or individual plants. The many important agronomic and economic advantages of drip make it highly efficient, potentially reducing irrigation requirements and operating costs. It reduces weed growth because most of the soil surface is not wetted.

The main disadvantage of localized irrigation is the risk of clogging of the emitters mainly due to the very small diameter of the nozzles used and the low operating pressures; hence, filtering the water in advance is critical. Clogging is indeed a serious threat for the crop if the water is not properly filtered and the equipment not properly maintained. In addition, localized irrigation systems, characterized by low flow rates, are not very suitable for complementary irrigation when intervals between periods of water availability are long.

More recently, the possibility of irrigating olive groves is spreading using subsurface irrigation, which involves distributing water directly to the roots. The main features of this system are: high irrigation efficiency; low percolation losses; absence of losses due to run-off and evaporation during irrigation; possibility to irrigate with modest volumes of water and low operation pressures and it does not create any interference to crop operations on the soil surface. This last aspect is important, as there are no external constraints for the mechanical harvesting of olives, or soil cultural care such as tillage. In addition, it is particularly easy to manage the application of fertilizer in irrigation water with advantages both for the olive grove and for the environment. The benefits are achievable only with an accurate design of the system and a regular maintenance. In fact, for an adequate distribution of water, particular attention must be paid to the choice of the flow rate and the arrangement (distance) of the drippers.

References

Alegre, S., Marsal, J., Mata, M., Arbones, A., Girona, J. and Tovar, M.J. (2000) Regulated deficit irrigation in olive trees (*Olea europaea* L. cv. Arbequina) for oil production. *IV International Symposium on Olive Growing* 586, 259–262.

Allen, R.G., Pereira, L.S., Raes, D. and Smith, M. (1998) Crop evapotranspiration – guidelines for computing crop water requirements. FAO irrigation and drainage paper 56. FAO, Rome 300(9), D05109.

Aragüés, R., Puy, J. and Isidoro, D. (2004) Vegetative growth response of young olive trees (*Olea europaea* L., cv. Arbequina) to soil salinity and waterlogging. *Plant and Soil* 258(1), 69–80.

Aragüés, R., Puy, J., Royo, A. and Espada, J.L. (2005) Three-year field response of young olive trees (*Olea europaea* L., cv. Arbequina) to soil salinity: trunk growth and leaf ion accumulation. *Plant and Soil* 271(1), 265–273.

Assouline, S., Narkis, K., Gherabli, R. and Sposito, G. (2016) Combined effect of sodicity and organic matter on soil properties under long-term irrigation with treated wastewater. *Vadose Zone Journal* 15(4).

Ayers, R.S. and Westcot, D.W. (1985) *Water Quality for Agriculture* (Vol. 29). Food and Agriculture Organization of the United Nations, Rome.

Bacelar, E.A., Correia, C.M., Moutinho-Pereira, J.M., Gonçalves, B.C., Lopes, J.I. and Torres-Pereira, J.M. (2004) Sclerophylly and leaf anatomical traits of five field-grown olive cultivars growing under drought conditions. *Tree Physiology* 24(2), 233–239.

Beede, R. and Goldhamer, D. (1994) Olive irrigation management. *Olive Production Manual* 3353, 61–68.

Begg, J.E. and Turner, N.C. (1970) Water potential gradients in field tobacco. *Plant Physiology* 46(2), 343–346.

Ben-Gal, A. (2011) Salinity and olive: from physiological responses to orchard management. *Israel Journal of Plant Sciences* 59(1), 15–28.

Ben-Gal, A. and Dudley, L.M. (2003) Phosphorus availability under continuous point source irrigation. *Soil Science Society of America Journal* 67(5), 1449–1456.

Ben-Gal, A., Agam, N., Alchanatis, V., Cohen, Y., Yermiyahu, U. *et al.* (2009) Evaluating water stress in irrigated olives: correlation of soil water status, tree water status, and thermal imagery. *Irrigation Science* 27(5), 367–376.

Ben-Gal, A., Kool, D., Agam, N., van Halsema, G.E., Yermiyahu, U. *et al.* (2010) Whole-tree water balance and indicators for short-term drought stress in non-bearing 'Barnea' olives. *Agricultural Water Management* 98(1), 124–133.

Ben-Gal, A., Beiersdorf, I., Yermiyahu, U., Soda, N., Presnov, E. *et al.* (2017) Response of young bearing olive trees to irrigation-induced salinity. *Irrigation Science* 35(2), 99–109.

Ben-Gal, A., Ron, Y., Yermiyahu, U., Zipori, I., Naoum, S. and Dag, A. (2021) Evaluation of regulated deficit irrigation strategies for oil olives: a case study for two modern Israeli cultivars. *Agricultural Water Management* 245, 106577.

Berenguer, M.J., Vossen, P.M., Grattan, S.R., Connell, J.H. and Polito, V.S. (2006) Tree irrigation levels for optimum chemical and sensory properties of olive oil. *HortScience*, 41(2), 427–432.

Bian, J., Zhang, Z., Chen, J., Chen, H., Cui, C., Li, X., Chen, S. and Fu, Q. (2019) Simplified evaluation of cotton water stress using high resolution unmanned aerial vehicle thermal imagery. *Remote Sensing* 11(3), 267.

Boussadia, O., Mariem, F.B., Mechri, B., Boussetta, W., Braham, M. and El Hadj, S.B. (2008) Response to drought of two olive tree cultivars (cv Koroneki and Meski). *Scientia Horticulturae* 116(4), 388–393.

Boyer, J.S. (1985) Water transport. *Annual Review of Plant Physiology* 36(1), 473–516.

Brutsaert, W. (1982) *Evaporation into the Atmosphere*. Springer, Dordrecht, The Netherlands. https://doi.org/10.1007/978-94-017-1497-6

Bustan, A., Dag, A., Yermiyahu, U., Erel, R., Presnov, E. *et al.* (2016) Fruit load governs transpiration of olive trees. *Tree Physiology* 3636(3), 380–391.

Camposeo, S., Ferrara, G., Palasciano, M. and Godini, A. (2008) Varietal behaviour according to the superintensive oliveculture training system. *Acta Horticulturae* 791, 271–274. https://doi.org/10.17660/ActaHortic.2008.791.38

Caruso, G., Gucci, R., Urbani, S., Esposto, S., Taticchi, A. *et al.* (2014) Effect of different irrigation volumes during fruit development on quality of virgin olive oil of cv. Frantoio. *Agricultural Water Management* 134, 94–103. https://doi.org/10.1016/j.agwat.2013.12.003

Caruso, G., Gucci, R., Sifola, M.I., Selvaggini, R., Urbani, S. *et al.* (2017) Irrigation and fruit canopy position modify oil quality of olive trees (cv. Frantoio). *Journal of the Science of Food and Agriculture* 97(11), 3530–3539.

Corell, M., Martín-Palomo, M.J., Girón, I., Andreu, L., Trigo, E. *et al.* (2019) Approach using trunk growth rate data to identify water stress conditions in olive trees. *Agricultural Water Management* 222, 12–20. https://doi.org/10.1016/j.agwat.2019.05.029

Corell, M., Martín-Palomo, M., Girón, I., Andreu, L., Galindo, A., Centeno, A., Pérez-López, D. and Moriana, A. (2020) Stem water potential-based regulated deficit irrigation scheduling for olive table trees. *Agricultural Water Management* 242, 106418.

Dag, A., Naor, A., Ben-Gal, A., Harlev, G., Zipori, I. *et al.* (2015) The effect of water stress on super-high-density 'Koroneiki' olive oil quality. *Journal of the Science of Food and Agriculture* 95(10), 2016–2020. https://doi.org/10.1002/jsfa.6913

d'Andria, R. and Morelli, G. (2002) Irrigation regime affects yield and oil quality of olive trees. *Acta Horticulturae* 586, 273–276. https://doi.org/10.17660/ActaHortic.2002.586.52

Dry, P.R. and Loveys, B. (1998) Factors influencing grapevine vigour and the potential for control with partial rootzone drying. *Australian Journal of Grape and Wine Research* 4(3), 140–148.

Dudley, L.M., Ben-Gal, A. and Shani, U. (2008) Influence of plant, soil, and water on the leaching fraction. *Vadose Zone Journal* 7(2), 420–425.

Egea, G., Padilla-Díaz, C.M., Martinez-Guanter, J., Fernández, J.E. and Pérez-Ruiz, M. (2017) Assessing a crop water stress index derived from aerial thermal imaging and infrared thermometry in super-high density olive orchards. *Agricultural Water Management* 187, 210–221.

Erel, R., Eppel, A., Yermiyahu, U., Ben-Gal, A., Levy, G. *et al.* (2019) Long-term irrigation with reclaimed wastewater: implications on nutrient management, soil chemistry and olive (*Olea europaea* L.) performance. *Agricultural Water Management* 213, 324–335.

Faci, J., Berenguer, M., Espada, J. and Gracia, S. (2000) Effect of variable water irrigation supply in olive (*Olea europaea* L.) cv. Arbequina in Aragon (Spain). II. Extra virgin oil quality parameters. *IV International Symposium on Olive Growing* 586, 649–652.

Faraloni, C., Cutino, I., Petruccelli, R., Leva, A., Lazzeri, S. and Torzillo, G. (2011) Chlorophyll fluorescence technique as a rapid tool for in vitro screening of olive cultivars (*Olea europaea* L.) tolerant to drought stress. *Environmental and Experimental Botany* 73, 49–56.

Fereres, E., Goldhamer, D.A. and Sadras, V. (2012) Yield response to water of fruit trees and vines: guidelines. FAO Irrigation and Drainage Paper 66. FAO, Rome.

Fernandes-Silva, A.A., López-Bernal, Á., Ferreira, T.C. and Villalobos, F.J. (2016) Leaf water relations and gas exchange response to water deficit of olive (cv. Cobrançosa) in field grown conditions in Portugal. *Plant and Soil* 402(1), 191–209.

Fernández, J.E., Perez-Martin, A., Torres-Ruiz, J.M., Cuevas, M.V., Rodriguez-Dominguez, C.M. *et al.* (2013) A regulated deficit irrigation strategy for hedgerow olive orchards with high plant density. *Plant and Soil* 372(1), 279–295.

Giorio, P. and Giorio, G. (2003) Sap flow of several olive trees estimated with the heat-pulse technique by continuous monitoring of a single gauge. *Environmental and Experimental Botany* 49(1), 9–20.

Girón, I., Corell, M., Martín-Palomo, M., Galindo, A., Torrecillas, A., Moreno, F. and Moriana, A. (2015) Feasibility of trunk diameter fluctuations in the scheduling of regulated deficit irrigation for table olive trees without reference trees. *Agricultural Water Management* 161, 114–126.

Girona, J., Luna, M., Arbonés, A., Mata, M., Rufat, J. and Marsal, J. (2002) Young olive trees responses (*Olea europea*, cv 'Arbequina') to different water supplies: water function determination. *Acta Horticulturae* 586, 277–280. https://doi.org/10.17660/ActaHortic.2002.586.53

Gómez-del-Campo, M. (2013) Summer deficit-irrigation strategies in a hedgerow olive orchard cv. 'Arbequina': effect on fruit characteristics and yield. *Irrigation Science* 31(3), 259–269.

Gowing, D., Davies, W. and Jones, H. (1990) A positive root-sourced signal as an indicator of soil drying in apple, Malus x domestica Borkh. *Journal of Experimental Botany* 41(12), 1535–1540.

Gucci, R. and Tattini, M. (2010) Salinity tolerance in olive. In: Janick, J. (ed.) *Horticultural Reviews* 21, 177–214.

Gucci, R., Lodolini, E.M. and Rapoport, H.F. (2009) Water deficit-induced changes in mesocarp cellular processes and the relationship between mesocarp and endocarp during olive fruit development. *Tree Physiology*, 29(12), 1575–1585.

Gucci, R., Fereres, E. and Goldhamer, D.A. (2012) Olive. In: Steduto, P., Hsiao, T.C., Fereres, E. and Raes, D. (eds) *Crop Yield Response to Water*. Irrigation and Drainage Paper 66. FAO, Rome, pp. 300–313.

Gucci, R., Caruso, G., Gennai, C., Esposto, S., Urbani, S. and Servili, M. (2019) Fruit growth, yield and oil quality changes induced by deficit irrigation at different stages of olive fruit development. *Agricultural Water Management*, 212, 88–98.

Heydari, N. (2014) Water productivity in agriculture: challenges in concepts, terms and values. *Irrigation and Drainage* 63(1), 22–28.

Idso, S., Jackson, R., Pinter Jr, P., Reginato, R. and Hatfield, J. (1981) Normalizing the stress-degree-day parameter for environmental variability. *Agricultural Meteorology* 24, 45–55.

Kool, D., Agam, N., Lazarovitch, N., Heitman, J., Sauer, T. and Ben-Gal, A. (2014) A review of approaches for evapotranspiration partitioning. *Agricultural and Forest Meteorology*, 184, 56–70.

Lambers, H. and Oliveira, R.S. (2019) Plant water relations. In: Lambers, H., Chaplin III, F.S. and Pons, T.L. (eds) *Plant Physiological Ecology*. Springer, Cham, Switzerland, pp. 187–263.

Lavee, S. and Wodner, M. (2004) The effect of yield, harvest time and fruit size on the oil content in fruits of irrigated olive trees (*Olea europaea*), cvs. Barnea and Manzanillo. *Scientia Horticulturae* 99(3–4), 267–277.

Li, S.-H., Huguet, J.-G., Schoch, P. and Orlando, P. (1989) Response of peach tree growth and cropping to soil water deficit at various phenological stages of fruit development. *Journal of Horticultural Science* 64(5), 541–552.

Lo Bianco, R. (2013) Responses of apple to partial root-zone drying: a review. In: Ali, M.H. (ed.) *Irrigation management, Technologies, and Environmental Impacts*. Nova Science Publishers, pp. 71–86.

López-Bernal, Á., Alcántara, E., Testi, L. and Villalobos, F.J. (2010) Spatial sap flow and xylem anatomical characteristics in olive trees under different irrigation regimes. *Tree Physiology*, 30(12), 1536–1544.

Maas, E.V. and Hoffman, G.J. (1977) Crop salt tolerance current assessment. *Journal of the Irrigation and Drainage Division* 103(2), 115–134.

Maes, W. and Steppe, K. (2012) Estimating evapotranspiration and drought stress with ground-based thermal remote sensing in agriculture: a review. *Journal of Experimental Botany* 63(13), 4671–4712.

Marino, G., Caruso, T., Ferguson, L. and Marra, F.P. (2018) Gas exchanges and stem water potential define stress thresholds for efficient irrigation management in olive (*Olea europaea* L.). *Water* 10(3), 342.

Marino, G., Scalisi, A., Guzmán-Delgado, P., Caruso, T., Marra, F.P. and Lo Bianco, R. (2021) Detecting mild water stress in olive with multiple plant-based continuous sensors. *Plants* 10(1), 131.

Marra, F., Marino, G., Marchese, A. and Caruso, T. (2016) Effects of different irrigation regimes on a super-high-density olive grove cv. 'Arbequina': vegetative growth, productivity and polyphenol content of the oil. *Irrigation Science* 34(4), 313–325.

Matese, A., Toscano, P., Di Gennaro, S.F., Genesio, L., Vaccari, F.P. *et al.* (2015) Intercomparison of UAV, aircraft and satellite remote sensing platforms for precision viticulture. *Remote Sensing* 7(3), 2971–2990.

Meiri, A. and Plaut, Z. (1985) Crop production and management under saline conditions. In: Pasternak, D. and San Pietro, A. (eds) *Biosalinity in Action: Bioproduction with Saline Water*. Springer, pp. 253–271.

Mezghani, M.A., Charfi, C.M., Gouiaa, M. and Labidi, F. (2012) Vegetative and reproductive behaviour of some olive tree varieties (*Olea europaea* L.) under deficit irrigation regimes in semi-arid conditions of Central Tunisia. *Scientia Horticulturae* 146, 143–152.

Minhas, P., Ramos, T.B., Ben-Gal, A. and Pereira, L.S. (2020) Coping with salinity in irrigated agriculture: crop evapotranspiration and water management issues. *Agricultural Water Management* 227, 105832.

Monteith, J.L. (1965) Evaporation and environment. *Symposia of the Society for Experimental Biology* 19, 205–234.

Monteith, J.L. and Unsworth, M. (2013) *Principles of Environmental Physics: Plants, Animals, and the Atmosphere*. Academic Press.

Moriana, A. and Fereres, E. (2002) Plant indicators for scheduling irrigation of young olive trees. *Irrigation Science* 21(2), 83–90.

Moriana, A., Villalobos, F. and Fereres, E. (2002) Stomatal and photosynthetic responses of olive (*Olea europaea* L.) leaves to water deficits. *Plant, Cell and Environment* 25(3), 395–405.

Moriana, A., Orgaz, F., Pastor, M. and Fereres, E. (2003) Yield responses of a mature olive orchard to water deficits. *Journal of the American Society for Horticultural Science* 128(3), 425–431.

Moriana, A., Girón, I., Martín-Palomo, M., Conejero, W., Ortuño, M., Torrecillas, A. and Moreno, F. (2010) New approach for olive trees irrigation scheduling using trunk diameter sensors. *Agricultural Water Management* 97(11), 1822–1828.

Nadezhdina, N., Nadezhdin, V., Ferreira, M.I. and Pitacco, A. (2007) Variability with xylem depth in sap flow in trunks and branches of mature olive trees. *Tree Physiology* 27(1), 105–113.

Naor, A. and Peres, M. (2001) Pressure-increase rate affects the accuracy of stem water potential measurements in deciduous fruit trees using the pressure-chamber technique. *The Journal of Horticultural Science and Biotechnology* 76(6), 661–663.

Naor, A., Schneider, D., Ben-Gal, A., Zipori, I., Dag, A. *et al.* (2013) The effects of crop load and irrigation rate in the oil accumulation stage on oil yield and water relations of 'Koroneiki' olives. *Irrigation Science* 31(4), 781–791.

Padilla-Diaz, C., Rodriguez-Dominguez, C., Perez-Martin, A., Montero, A., Garcia, J. and Fernandez, J. (2016) Scheduling a deficit irrigation strategy from leaf turgor measurements: impact on water status, gas exchange and oil yield. *VIII International Olive Symposium* 1199, 267–272.

Patumi, M., d'Andria, R., Fontanazza, G., Morelli, G., Giorio, P. and Sorrentino, G. (1999) Yield and oil quality of intensively trained trees of three cultivars of olive (*Olea europaea* L.) under different irrigation regimes. *The Journal of Horticultural Science and Biotechnology* 74(6), 729–737.

Pedrero, F., Grattan, S., Ben-Gal, A. and Vivaldi, G.A. (2020) Opportunities for expanding the use of wastewaters for irrigation of olives. *Agricultural Water Management* 241, 106333.

Pereira, L., Paredes, P. and Jovanovic, N. (2020) Soil water balance models for determining crop water and irrigation requirements and irrigation scheduling focusing on the FAO56 method and the dual Kc approach. *Agricultural Water Management* 241, 106357.

Puppo, L., García, C., Girona, J. and García-Petillo, M. (2014) Determination of young olive-tree water consumption with drainage lysimeters. *Journal of Water Resource and Protection* 6, 841–851.

Puppo, L., García, C., Bautista, E., Hunsaker, D.J., Beretta, A. and Girona, J. (2019) Seasonal basal crop coefficient pattern of young non-bearing olive trees grown in drainage lysimeters in a temperate sub-humid climate. *Agricultural Water Management* 226, 105732.

Ramos, A.F. and Santos, F.L. (2009) Water use, transpiration, and crop coefficients for olives (cv. Cordovil), grown in orchards in Southern Portugal. *Biosystems Engineering* 102(3), 321–333.

Ramos, A. F. and Santos, F.L. (2010) Yield and olive oil characteristics of a low-density orchard (cv. Cordovil) subjected to different irrigation regimes. *Agricultural Water Management* 97(2), 363–373.

Rapoport, H.F., Costagli, G. and Gucci, R. (2004) The effect of water deficit during early fruit development on olive fruit morphogenesis. *Journal of the American Society for Horticultural Science* 129(1), 121–127.

Scholander, P.F., Bradstreet, E.D., Hemmingsen, E. and Hammel, H. (1965) Sap pressure in vascular plants: negative hydrostatic pressure can be measured in plants. *Science* 148 (3668), 339–346.

Segal, E., Ben-Gal, A. and Shani, U. (2006) Root water uptake efficiency under ultra-high irrigation frequency. *Plant and Soil* 282(1), 333–341.

Servili, M., Esposto, S., Lodolini, E., Selvaggini, R., Taticchi, A. *et al.* (2007) Irrigation effects on quality, phenolic composition, and selected volatiles of virgin olive oils cv. Leccino. *Journal of Agricultural and Food Chemistry* 55(16), 6609–6618.

Shackel, K., Moriana, A., Marino, G., Corell, M., Pérez-López, D. *et al.* (2021) Establishing a reference baseline for midday stem water potential in olive and its use for plant-based irrigation management. *Frontiers in Plant Science* 2715.

Shani, U., Ben-Gal, A., Tripler, E. and Dudley, L.M. (2007) Plant response to the soil environment: an analytical model integrating yield, water, soil type, and salinity. *Water Resources Research* 43(8).

Silber, A., Xu, G., Levkovitch, I., Soriano, S., Bilu, A. and Wallach, R. (2003) High fertigation frequency: the effects on uptake of nutrients, water and plant growth. *Plant and Soil* 253(2), 467–477.

Soda, N., Ephrath, J.E., Dag, A., Beiersdorf, I., Presnov, E. *et al.* (2017) Root growth dynamics of olive (*Olea europaea* L.) affected by irrigation induced salinity. *Plant and Soil* 411(1), 305–318.

Strippoli, G., Vivaldi, G., Camposeo, S. and Conto, F. (2013) Sprouts seasonal elongation of two olive cultivars in a high-density orchard. *Agricultural Sciences* 4(8), 376.

Tan, J., Ben-Gal, A., Shtein, I., Bustan, A., Dag, A. and Erel, R. (2020) Root structural plasticity enhances salt tolerance in mature olives. *Environmental and Experimental Botany* 179, 104224.

Tietel, Z., Dag, A., Yermiyahu, U., Zipori, I., Beiersdorf, I., Krispin, S. and Ben-Gal, A. (2019) Irrigation-induced salinity affects olive oil quality and health-promoting properties. *Journal of the Science of Food and Agriculture* 99(3), 1180–1189.

Van Genuchten, M.T. and Gupta, S. (1993) A reassessment of the crop tolerance response function. *Journal of the Indian Society of Soil Science* 41, 730.

Van Halsema, G.E. and Vincent, L. (2012) Efficiency and productivity terms for water management: a matter of contextual relativism versus general absolutism. *Agricultural Water Management* 108, 9–15.

Vivaldi, G., Strippoli, G. and Camposeo, S. (2013) Ecophysiological response to irrigation of two olive cultivars grown in a high-density orchard. *Agricultural Sciences* 4(8a).

6.3 Nutrition and Fertilization

A. Dag, R. Erel, I. Zipori and U. Yermiyahu
Gilat Research Center, Volcani Institute, Ministry of Agriculture, Israel

6.3.1 The Importance of Balanced Nutrient Availability for Olive Cultivation

Plant nutrition is an essential part of crop cultivation and production. Plants absorb most nutrients from the soil and if not replenished properly, these nutrients will eventually be depleted. The plant might then be exposed to different deficiencies, impairing its vegetative and reproductive performance. The absorbed nutrients can be replenished by supplying them in the form of mineral fertilizers, organic manures or green manures, or by recycling certain plant parts.

Like most plants, the olive tree is composed of 20 elements: carbon (C), hydrogen (H), oxygen (O), nitrogen (N), phosphorus (P), potassium (K), calcium (Ca), magnesium (Mg), sulphur (S), iron (Fe), zinc (Zn), manganese (Mn), molybdenum (Mo), copper (Cu), boron (B), nickel (Ni), silicon (Si), cobalt (Co), sodium (Na) and chlorine (Cl) (López-Villalta and Muñoz-Cobo, 1996; Kirkby, 2012). All of these elements are considered essential nutrients, and therefore a deficiency in or overexposure to any one of them might impair growth, production or product quality.

The non-mineral nutrients C, H and O are absorbed from the air, in the form of carbon dioxide (CO_2) and from the soil, in the form of water (H_2O) during photosynthesis and transpiration, to form carbohydrates which compose approximately 95% of the plant tissue's dry weight. All of the other nutrients are absorbed by the root system from the soil solution as ions or ionic oxides (López-Villalta and Muñoz-Cobo, 1996; Kirkby, 2012). The mineral nutrients are divided into two major groups, according to their concentrations in the plant tissue and the amounts absorbed by the plants. N, P, K, Ca, Mg and S belong to the group of macronutrients. They are absorbed in relatively large amounts, and their concentrations in the plant tissues are relatively high and usually expressed in percentage of dry weight. All of the other nutrients belong to the group of micronutrients. They are absorbed in relatively small amounts, their concentrations in the plant tissues are low, and they are usually expressed in milligrams per kilogram dry weight (or in parts per million [ppm]). Overexposure to some nutrients is more critical than to others; for example, overexposure to Ca or Mg will rarely impair plant performance, whereas overexposure to Na, Cl or B will markedly affect it.

Mineral nutrient availability strongly depends on soil pH, as illustrated in Fig. 6.3.1: within the common range of pH values in

DOI: 10.1079/9781789247350.0018

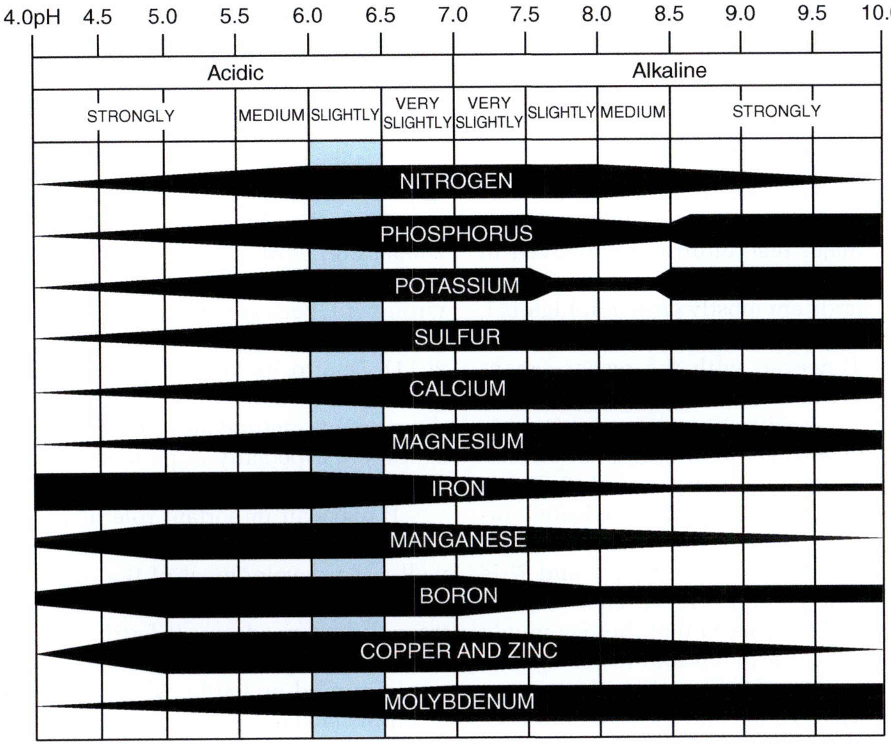

Fig. 6.3.1. Relations between soil pH and nutrient availability. Blue bar represents optimal soil pH. (From Wikimedia Commons)

agricultural soils, macroelement availability is barely affected by soil pH, whereas microelement availability depends strongly on soil pH, sometimes leading to toxicity in acidic soils (Marschner and Rengel, 2012). Olives are generally grown in alkaline soils and not in acidic ones.

Olives are one of the most resistant species to harsh growing conditions and are therefore often cultivated on non-fertile, poor and even marginal soils, calling for special attention to orchard nutrition. The growers have at their disposal two major tools to achieve a proper and balanced nutrient supply to the orchard: soil analysis and leaf analysis. In addition, olive cultivation occurs in two major and distinct systems: (i) traditional, low-density, extensive, rainfed orchards; and (ii) modern, high-density, intensive, irrigated orchards. Fertilization considerations are different in these two systems, and will be discussed in detail in this chapter. When a fertilization plan for a given orchard is elaborated, the environmental impact of fertilizer application must be considered as well, such as the generally negative effects of over-fertilization.

Certain amounts of nutrients are removed from the orchard and must be replenished by fertilizers, foliar applications, green manure or organic substances. However, there is a certain level of natural recycling of nutrients in an olive orchard. The lifespan of an olive leaf is approximately two years. During senescence, the leaves translocate some of their nutrients for incorporation into the tissues of newly formed leaves. Moreover, the falling leaves remain on the ground and decompose to form organic matter which, eventually, undergoes mineralization and can account for part of the trees' mineral nutrition (Fig. 6.3.2). However, it is very difficult to assess the quantitative contribution of this process to orchard nutrition.

Pruned material is also a source of nutrient removal from the orchard. If the pruned material is chopped *in situ*, at least part of the nutrient content in the material will be recycled. In this case as well, it is difficult to estimate the contribution of this material to plant nutrition. Chopped pruned material has additional advantages, serving as mulch and thus reducing water loss due to direct evaporation from the soil surface, and soil erosion by wind and water.

The major macronutrients involved in olive orchard nutrition are N, P and K. As olive orchards are mostly grown in high-pH, calcareous soils, Ca, Mg and S deficiencies are rare. However, with the recent expansion of olive cultivation to new regions, including those with low-pH soils, Ca, Mg and S deficiencies might occur. An estimation of the annual amounts of N, P and K removed from an intensive, irrigated orchard are presented in Table 6.3.1.

The annual removal of N, P and K is approx. 100, 11 and 139 kg ha^{-1}, respectively. It can be assumed that these amounts will be about five- to ten-fold lower in rainfed, extensive orchards due to lower tree density, lower yields and lower amounts of plant material removed by pruning. This emphasizes the link between water availability and nutrient uptake. As most nutrient movement into the roots is associated with water uptake, for which the driving force is transpiration, high water consumption will usually be associated with higher nutrient uptake (Zipori *et al.*, 2015).

N is the most significant macronutrient for most crops, including olives. Vegetative growth is primarily governed by N availability and uptake, and vegetative growth is important in olives as it produces the fruit-bearing branches of the following year. However, over-fertilization with N reduces fruit yield due to competition between vegetative and reproductive organs (Erel *et al.*, 2008; Haberman *et al.*, 2019a), and impairs oil quality, especially free fatty acid content and polyphenol levels (Dag *et al.*, 2009a; Erel *et al.*, 2013a). Over-fertilization with N also has a negative impact on the environment, because of the risk of nitrate migration into the ground water (Segal *et al.*, 2011). Due to its high mobility in the soil profile and its effect on oil yield and quality, the amount, method and timing of N application are especially important and will be further detailed in this chapter.

P plays an important role in photosynthetic and reproductive plant processes, as well as in development of the root system. Until recently, it was believed that olive trees do not need any P fertilization. This notion was based on experiments carried out in rainfed orchards, where no response to supplemental P was found, and on the fact that olive roots have symbiotic mycorrhizal

Table 6.3.1. Annual amounts of N, P and K removed from an intensive, irrigated olive orchard. (Adapted from Zipori *et al.*, 2020)

Nutrient	N	P	K
Removed with fruit (kg ha^{-1})	40.4	4.4	78.1
Removed with pruned material (kg ha^{-1})	59.3	6.9	60.5
Total amount removed (kg ha^{-1})	99.7	11.3	138.6

Fig. 6.3.2. Falling olive leaves remain on the ground. (Photo: Tali Nachshon Dag)

Table 6.3.2. Comparison of fertilization in rainfed vs. irrigated olive orchards.

	Rainfed	Irrigated
Timing of fertilizer application	Towards the end of winter	During spring/ summer
Formulation of fertilizers	Solids	Liquid forms or dissolved solids
Fertilization efficiency	Low	High
Tree nutrient requirement	Low	High

relationships, enhancing P uptake (Dag *et al.*, 2009c). However, in recent studies, it was found that in intensive, irrigated orchards, fruit and oil production respond positively to P fertilization (Erel *et al.*, 2016; Haberman *et al.*, 2021). There is insufficient information regarding over-fertilization with P. In container experiments, where trees were grown in an inert medium, over-fertilization with P had a negative effect on oil quality, but this response was not found in field experiments, probably due to the high rate of P fixation in the soil, leading to reduced availability to the plants. P mobility in the soil is very limited and therefore over-fertilization with this nutrient is not likely to have a significant negative impact on the environment.

The mineral element that is removed by olive trees in the largest amounts is K (Table 6.3.1). K plays an important role in controlling the plants' water balance. Hence, extreme K deficiency resulted in yield reduction (Haberman *et al.*, 2019a) but had no negative effect on oil quality, whereas over-fertilization with K had no negative effect on yield or on oil quality (Dag *et al.*, 2009a; Erel *et al.*, 2013a). K is less mobile in the soil than N, due to its interaction with the soil's clay minerals, but it is more mobile than P. Extreme over-fertilization with K might result in its migration below the root zone into the ground water, posing an environmental hazard.

There is very limited information regarding the response of olive trees to microelement application. Recent practice in irrigated, intensive orchards has adopted foliar application of Zn to improve fruit set, although its contribution has not been fully proven (Zipori *et al.*, 2016). Attempts to improve olive yields by applying Fe chelates in high-pH, calcareous soils resulted in higher chlorophyll content in the leaves but had no effect on yield (Zipori *et al.*, 2011).

Balanced nutrient availability is important not only with respect to yield, oil quality and environmental considerations, but also from an economic point of view. The cost of mineral fertilizers has been increasing steadily in the last decades and their application must be carried out intelligently. Soil and leaf analyses, as well as expected fruit yield, are useful tools in determining the fertilization regime in olive orchards and will be further discussed in this chapter.

6.3.2 Fertilization of Traditional and Modern Olives

Olive is one of the world's oldest fruit trees, and has been grown for thousands of years around the Mediterranean basin as a rainfed crop (Zohary and Spiegel-Roy, 1975). The traditional orchards' fruit and yield production is mainly a function of the amount of rain in the winter prior to new crop development (Lavee, 2009). In the Mediterranean – the largest and most important olive-growing region – summers are often dry and trees are therefore typically exposed to temporary water stress, which causes reduced growth rates and varying availability of nutrients (Lahiri, 1980). In traditional olive cultivation (Fig. 6.3.3), drought is probably the dominant limiting factor. Most of the fertilization studies available in the literature were conducted in agro-systems where drought was inherent and therefore, the olive's response to fertilizer might have been minor and inconsistent, depending on water availability. Once water is removed as a key limiting factor by proper irrigation, the next limiting factor (in our case mineral nutrients) will take its place.

In some regions, there is a relatively long tradition of using irrigation for table olives to accelerate fruit growth. However, until about 40 years ago, the use of irrigation was not common practice in oil olive cultivation. Nevertheless, the general trends of orchard intensification, including increasing tree density and the need to increase yields to achieve profitable orchards, have gradually led to the introduction of irrigation in oil olive orchards as well (Lavee, 2009). One of the first reports on irrigation in oil olives originated from Israel in the early 1960s (Samish and Spiegel, 1961).

Fig. 6.3.3. Traditional rainfed olive orchard, Alma, Galilee, Israel. (Photo: Ayala Noy Meir)

Adequate water supply is a prerequisite for plant growth and the driving force for plant nutrient uptake (da Silva *et al.*, 2011). Therefore, the shift from rainfed to irrigated cultivation systems has had a tremendous effect on fertilization management. Despite the importance of intensive growing conditions, most of the studies available in the literature have been conducted under rainfed conditions. According to Liebig's Law of the Minimum, it is only when the dominant limiting factor (in this case, water) has been removed (by applying suitable irrigation) that the next limiting factor can be manifested. Therefore, when fertilization experiments were carried out under rainfed cultivation, the effect of water stress may well have overshadowed any possible role of nutrients. While in rainfed orchards, solid fertilizers are generally applied towards the end of winter to allow the rain to dissolve the fertilizers and enable their movement into the soil, in irrigated orchards, fertilizers are generally applied during the summer with the irrigation water (fertigation), either as dissolved solid fertilizer or in liquid form (Table 6.3.2.). This application system enables much better control of nutrient availability throughout the growing season (Bar-Yosef, 1999) to meet the olive tree's needs. Furthermore, wet soil during the summer in the irrigated orchards enhances the availability of nutrients, especially P and K, compared to the relatively dry soil in rainfed orchards (Zipori *et al.*, 2015). It might therefore be speculated that in intensive, irrigated olive orchards, due to the improved nutrient application and, consequently, improved uptake efficiency, the fertilization requirement might be reduced. However, in reality, this is not often the case; in intensive, irrigated orchards, tree density is much higher than in rainfed ones. Tree density in rainfed orchards is 100–120 trees ha^{-1}; in irrigated orchards, it is about three-fold higher (300–400 trees ha^{-1})

(Fig. 6.3.4) and in super-intensive orchards, it is about ten-fold higher (1200–1400 trees ha^{-1}) (Fig. 6.3.5). The much higher vegetative growth and fruit yield per hectare raise the nutritional demand. It is therefore obvious that intensive or super-intensive irrigated orchards require different fertilization protocols than rainfed ones.

Fig. 6.3.4. Intensive, irrigated olive orchard (Israel). (Photo: Tali Nachshon Dag)

Fig. 6.3.5. Super-high-density, intensive olive orchard (Israel). (Photo: Tali Nachshon Dag)

6.3.3 Management of Specific Nutrients

6.3.3.1 Fertilization considerations

Nutrient cycles in olive agro-ecosystems

The overall purpose of tree nutrition is to restore the nutrients removed by harvesting and pruning, in addition to those required for adequate vegetative development of root, stem, branches, leaves, flowers and fruit. Nonetheless, external inputs and outputs should be considered. For example, significant losses can be related to leaching/run-off, immobilization, irreversible sorption, or vaporization, while significant contributions of nutrients can be derived from atmospheric deposits (dust), rain, irrigation water, mineralization, desorption, and weathering of soil minerals. Nutrients can be recycled within the agro-system from soil organic pools found mostly from defoliated leaves/mulch back to the plants and from the plant to organic and mineral pools in the soil. Nutrient recycling in olive agro-systems has been demonstrated for natural mulch (Palese *et al.*, 2009; Michalopoulos *et al.*, 2020) or by restoring olive mill water (Zipori *et al.*, 2018). Soil microorganisms and weeds (when applicable) can also preserve a significant pool of nutrients that can release (mineralization) or absorb (immobilization) minerals. These processes should be taken into account when developing a sustainable fertilization strategy and to avoid over-fertilization. For example, Fernández-Escobar *et al.* (2009a, 2012) reported that in traditional rainfed orchards, the natural N cycle (i.e. mineralization and deposition from rain) can sustain optimal N levels for many years without any reduction in yield. These examples highlight the need to account for the natural services of the agro-ecosystem to prevent over-fertilization, which has a disruptive environmental impact (Tilman *et al.*, 2002; Segal *et al.*, 2011) as well as inherent economic waste. Specifically, over-fertilization leads to severe water contamination, eutrophication, and excessive greenhouse gas emission (Tellez-Rio *et al.*, 2017). The aim of this section is to promote the use of diagnostic tools to develop advanced fertilization strategies that will elevate nutrient utilization efficiency and minimize environmental impact while optimizing oil yield and oil quality.

Factors affecting nutrient demand

Massive work has been done regarding olive mineral nutrition, mostly with respect to N nutrition. Several reviews and book chapters related to olive fertilization are available (López-Villalta and Muñoz-Cobo, M., 1996; Freeman and Carlson, 2005; Therios, 2009; Fernández-Escobar, 2010). The olive tree is native and well-adapted to the Mediterranean basin where it is most often grown in nutrient-poor, shallow, rocky hillside soils and where extended periods of drought during the summer are common (Lavee, 1996; Wiesman, 2009). It is not surprising, then, that conventional wisdom considers the olive tree as having low nutrient demand (Therios, 2009; Wiesman, 2009). This consensus is not supported by solid experimental evidence, but is based on observations of fruiting trees under rainfed conditions. In poor soils, especially during the long dry summers, we may speculate scenarios of deficient macroelement availability to olives. This is particularly important under modern and intensive agro-systems, where productivity is substantially increased. Modern varieties and advanced agronomic practices have increased olive oil yields to 2.0–2.5 t ha^{-1}. Nutrient demand of these productive agro-systems differs from that of traditional agro-systems where oil yield is typically much lower (in the range of 0.3–0.6 t ha^{-1}). An in-depth discussion of the nutritional demands of modern, irrigated and fertigated olive can be found in Erel *et al.* (2018). In addition, fertilization efficiency also differs in traditional vs. modern agro-systems (see section 6.3.2). It is therefore recommended that, in addition to the efficiency of nutrient application methods, the expected yield and vigor of the orchard be considered.

Nutrient budget in olive orchards

As already noted, nutrient demand is derived from the offtake by the yield and other plant material removed from the orchard after subtracting the nutrients supplied from natural sources. It is therefore essential to evaluate the magnitude of the nutrient removal. In a controlled experiment, Dag *et al.* (2009a) reported fruit mineral concentrations in response to irrigation solution mineral concentrations. Fruit N concentration (dry weight) increased from 0.3% to 1.4% when N in the irrigation water increased from 6 to 200 mg/L; P content increased from 0.03% to 0.15% when P in the irrigation water increased from 0.2 to 20 mg/L; and K increased from 1% to 2.2% when K in the irrigation water increased from 10 to 210 mg/L. A comparable range of fruit mineral concentrations was reported by Jordao and Lietao (1989) in a survey of 50 cultivars growing in Portugal.

The vast range of fruit mineral concentrations implies that enhanced fertilization may create 'luxury consumption' and a consequential significant increase in mineral accumulation in fruit. Thus, the high concentrations in fruit may give a false impression of the fertilization requirements due to the massive amounts removed at harvest. The same trend was reported by Fernández-Escobar *et al.* (2006) who found that excessive N application results in higher N removal without any increase in either total yield or vegetative growth. Moreover, excessive N fertilization might impair fruit set (Erel *et al.*, 2008, 2013b) and oil quality (Fernández-Escobar *et al.*, 2006; Morales-Sillero *et al.*, 2007; Dag *et al.*, 2009a; Erel *et al.*, 2013a).

To estimate mineral removal from olive orchards by fruits, a calculation was performed using average fruit yield results from Dag *et al.* (2009a) and Connor and Fereres (2005) (Table 6.3.3).

Xiloyannis *et al.* (2000) estimated annual fruit demand in a six-year-old irrigated orchard (cv. 'Coratina') as: 56 kg N, 4.8 kg P and 44 kg K per ha. The authors estimated that fruit mineral removal represents 33–67% of total N uptake. Therios (2006) estimated annual mineral uptake (per ha) as: 15–35 kg N, 8 kg P and 10–50 kg K. For intensive agrosystems with a typical yield of 10 t ha^{-1}, nutrient offtake by fruit is approximately: 40, 5 and 60 kg N, P and K, respectively (11 kg P_2O_5 and 70 kg K_2O). These theoretical levels were later validated in a field trial (average yield of 10 t ha^{-1}) that resulted in offtake of 40, 4.4 and 78 kg ha^{-1} N, P and K, respectively (Zipori *et al.*, 2020). Combining the data provided by Xiloyannis *et al.* (2000) we can estimate that total N and P demand for annual vegetative growth is roughly double that removed with the fruit, while K demand is lower. These estimates were confirmed by Zipori *et al.* (2020) who reported pruning offtake of 59, 6.9 and 61 kg ha^{-1} N, P and K, respectively.

Table 6.3.3. Average nutrient removal per ton of fresh fruit.

Nutrient	kg t^{-1}	Source
N (kg)	3.6	Connor and Fereres (2005)
	4.2	Dag *et al.* (2009b)
	2.9	Fernández-Escobar *et al.* (2015)
	5.6	Zipori *et al.* (2020)
P (kg)*	0.55	Connor and Fereres (2005)
	0.42	Dag *et al.* (2009b)
	0.49	Fernandez-Escobar *et al.* (2015)
	0.65	Zipori *et al.* (2020)
K (kg)**	5.5	Connor and Fereres (2005)
	6.3	Dag *et al.* (2009b)
	4.4	Fernandez-Escobar *et al.* (2015)
	5.7	Zipori *et al.* (2020)

*To convert to P_2O_5, multiply by 2.29.
**To convert to K_2O, multiply by 1.21.

6.3.3.2 Nitrogen

Introduction

N is absorbed by the root mainly in the form of ammonium (NH_4^+) or nitrate (NO_3^-). The crucial role of N is a result of its high concentration in the plant's organs; it is usually the most abundant element among the nutrients (Marschner, 2012). The N form affects the pH in the rhizosphere and, consequently, affects the availability of other nutrients. N is the most commonly – and sometimes only – nutrient applied regularly in olive orchards as fertilizer. Similar to many other plants, olives favor the ammonium form of N (Therios and Sakellariadis, 1988; Therios, 2009). Furthermore, ammonium nutrition enhances P and Fe uptake, but may impair that of cations such as Ca and Mg, although K is not affected by NH_4 application (Therios and Sakellariadis, 1988; Tattini *et al.*, 1989).

In most traditional orchards, N is applied once a year, during late winter to early spring, in the hopes that rain will be sufficient to drive the fertilizer into the active root zone (Freeman *et al.*, 2005; Therios, 2006). Because of the complex nature of N chemistry in the soil, the applied N may be lost through various different mechanisms. Being a negatively charged ion, nitrate is easily leached beneath the active root zone (Segal *et al.*, 2011). Furthermore, after a heavy rain, anaerobic conditions promote denitrification and N loss to the atmosphere. When N application is carried out through the irrigation system (fertigation), concentrations and amounts can be closely monitored and the leaching of excess N and contamination of ground water minimized (Neilsen *et al.*, 1999).

Nitrogen fertilization recommendations

Recommendations for N fertilization are largely speculative and frequently excessive (Fernández-Escobar *et al.*, 2009a; Fernández-Escobar, 2010). Recommendations for traditional orchards cover a wide range, from 0.45 kg N to 2 kg N per tree per year (for example: 0.45–0.9 kg in Freeman *et al.*, 2005; 0.5–1 kg in López-Villalta and Muñoz-Cobo, (1996); 0.75 in Jasrotia *et al.*, 1997). Therios (2006) recommended taking into account the amount of precipitation, giving a range of 0.4 kg tree^{-1} when precipitation is less than 400 mm, to 1.5 kg tree^{-1} in regions with more rain. Recommendations for fertigation are scarce. A group led by Morales-Sillero studied the effect of fertigation on fruit, oil and soil in intensive, irrigated olive orchards (Morales-Sillero *et al.*, 2007, 2008 2009). They recommended an annual amount of 0.4 kg N per tree for oil production and 0.6 kg N per tree for table olives, although the authors did indicate a risk of ground water contamination. In Israel's intensive orchards, the common practice is 200 kg N ha^{-1}. However, in an eight-year study in a super-intensive orchard (900 trees ha^{-1}) in Israel, when trees received the recommended N amount, massive annual N build-up in the root zone was reported. The excessive N was almost completely leached during the winter (Segal *et al.*, 2011; Erel *et al.*, 2019) . In a field trial in an intensive olive orchard, optimal yield was found in response to 150 kg N ha^{-1}, whereas higher and lower N rates resulted in reduced yield (Haberman *et al.*, 2019b). Furthermore, whereas low N levels reduced vegetative growth and accelerated alternate bearing, a high N rate (300 kg ha^{-1}) reduced flowering intensity and quality. In a controlled pot experiment, optimal productivity was recorded at an N concentration of 45 mg l^{-1} (Erel *et al.*, 2008, 2013b). For propagation purposes, N fertilization needs to be adjusted for the mother plants because increases in N level have been found to negatively affect rooting rate and cutting survival. Propagation success increased threefold when N in the irrigation water was reduced from the highest to lowest treatments. Enhanced root development under low N concentrations resulted in higher root weight compared to that under high N concentrations (Dag *et al.*, 2012).

Reproduction

Reports on reproductive responses of olive trees to N fertilization are inconsistent. One explanation for this is differences in the initial nutritional status of the trees and soil fertility. In the 1950s, Hartmann (1958) reported a positive effect of N fertilization in olive only in poor shallow soil, whereas no effect was observed in deep fertile soil, even when no N was applied for three years. In the shallow soil, when no N was added, leaf N was low (1.3%). This was the only case in which fruit set, and therefore yield, were well correlated with leaf N concentration. Cimato *et al.* (1989) reported an increase in fruit set and yield when N was sprayed onto the foliage. The control trees in this experiment were below the deficiency threshold (approximately 1.3%), which might explain the positive effect of N application.

Severe N deficiency was reported to increase the proportion of staminate (male) flowers compared to hermaphroditic (perfect) flowers (Uriu, 1953; Erel *et al.*, 2013b; Haberman *et al.*, 2019b). Similarly, lower ratios of perfect flowers were noted in one out of three seasons in trees grown in containers under low environmental N (Erel *et al.*, 2013b). Furthermore, low N resulted in a low number of flowers per inflorescence and lower flowering intensity. When N was limiting, fruit set increased gradually as leaf N increased from 0.8% to 1.6%. In contrast, high N levels resulted in a decrease in fruit set (Erel *et al.*, 2008, 2013b). High leaf N concentration was correlated with low ovule viability (Fernández-Escobar *et al.*, 2008), which might explain the lower fruit set under those conditions. These studies indicate some of the hazards of N over-fertilization additional to economic waste and environmental damage. In an extended study in a rainfed orchard, Fernández-Escobar *et al.* (2009a) found no decrease in yield, even when N was withheld for nine years, compared to well-fertilized trees. During this period, leaf N did not drop below the N threshold for deficiency (1.4%).

Vegetative growth

Vegetative growth responded positively to N fertilization when N nutritional status was low, whereas no further response was observed when higher amounts of N were given (Sánchez-Zamora and Fernández-Escobar, 2002; Therios, 2006; Erel *et al.*, 2013b; Haberman *et al.*, 2019b). Therios and Sakellariadis (1988) reported an increase in plant fresh weight when N in the irrigation water was increased from 14 ppm to 112 ppm, whereas no further increase was measured when N was raised to 224 ppm. Erel *et al.* (2008) found no increase in trunk circumference or branch extension when N in the irrigation solution was increased from 25 ppm to 197 ppm. When N in the irrigation solution was increased from 6 ppm to 25 ppm, a significant increase in both growth parameters was noted.

The effect of N level on oil composition and quality is discussed in section 6.3.8.

6.3.3.3 Phosphorus

Introduction

P is an essential element for plant development that is frequently applied in many agro-systems to maintain productivity. Plants acquire P as orthophosphate (typically H_2PO_4). Due to complex chemical reactions, applied P rapidly reacts with soil components and is subsequently immobilized in near-surface soil layers (topsoil). Portions of the 'fixed P' may be adsorbed to Fe/Al oxides or precipitate as insoluble Ca–P complexes, which have low availability to plants (Simpson *et al.*, 2011). To acquire the sparingly available P, plants have developed complex mechanisms that manipulate chemical and physical conditions in the rhizosphere to increase P solubility (Hinsinger *et al.*, 2009). Processes such as acidification, exudation, root exploration and root–fungus associations have considerable metabolic costs (Lynch and Ho, 2005). P over-fertilization results in its accumulation in the topsoil where it is susceptible to transport to aquatic ecosystems by means of soil erosion or run-off and thus represents a considerable environmental threat. Development of algae and other aquatic plants in natural water bodies is frequently limited by P. Hence, transport of P from arable soils accelerates algal bloom (eutrophication), with large detrimental ecological effects.

Plants acquire P through two main pathways: direct P uptake by roots and indirect uptake via arbuscular mycorrhizal fungi (AMF) (Fig. 6.3.6). The presence of AMF in arable soils improves P acquisition, primarily because of their extensive extra-radical mycelial networks, which extend the roots' absorbing area beyond the P-depletion zone (Smith and Read, 2010). In its natural habitats, olive roots are spontaneously colonized by AMF (Estaún *et al.*, 2003). In nurseries, where natural AMF are frequently absent, inoculation with AMF significantly enhances plant growth and, specifically, P uptake (Estaún *et al.*, 2003; Dag *et al.*, 2009c).

The plant's nutritional level affects the intensity of root colonization with AMF (Deng *et al.*, 2017). Specifically, high soil P and N levels have a negative impact on the symbiotic root–fungus association. For example, in a controlled pot experiment described by Erel *et al.* (2013b), root AMF colonization was negatively correlated to N and P concentration in the irrigation solution (Fig. 6.3.7).

Fig. 6.3.6. The effect of AMF inoculation on olive plant growth (cv. 'Koroneiki'). (For more details, see Dag *et al.*, 2009c.)

For a long time, knowledge about the role of P in olive cultivation was very limited (Freeman *et al.*, 2005), likely due to the common belief that olive trees take up P very efficiently (López-Villalta and Muñoz-Cobo, 1996; Freeman *et al.*, 2005; Therios, 2006) as a result of extensive root systems and symbioses with mycorrhizal fungi (Estaún *et al.*, 2003; Dag *et al.*, 2009c). P deficiency is indeed rare, and P is not commonly applied as a fertilizer (López-Villalta and Muñoz-Cobo, 1996; Freeman *et al.*, 2005; Therios, 2006; Fernández-Escobar, 2010). Nevertheless, in the last decade, several studies relating directly to olive P nutrition have been published (Erel *et al.*, 2013b; Erel *et al.*, 2016; Jiménez-Moreno and Fernández-Escobar, 2017; Ferreira *et al.*, 2018; Haberman *et al.*, 2021) indicating the significance of regular P fertilization in intensive, irrigated olive agro-systems compared to little effect in traditional agro-systems or when soil P level is high (Jiménez-Moreno and Fernández-Escobar, 2017; Ferreira *et al.*, 2018). This indicates the need for routine leaf and/or soil analyses to predict the trees' response to P. In an intensive, irrigated orchard, when Olsen-P dropped below 10 mg kg^{-1}, yield was significantly reduced, even when leaf P was near (Haberman *et al.*, 2021) or above the deficiency threshold. This calls for a revision of the P threshold values for irrigated orchards (Erel *et al.*, 2019). This was not the case in traditional rainfed orchards (Ferreira *et al.*, 2018).

Reproduction

Only a few studies have tested the reproductive responses of olives to P fertilizer. However, recently, a series of studies was published

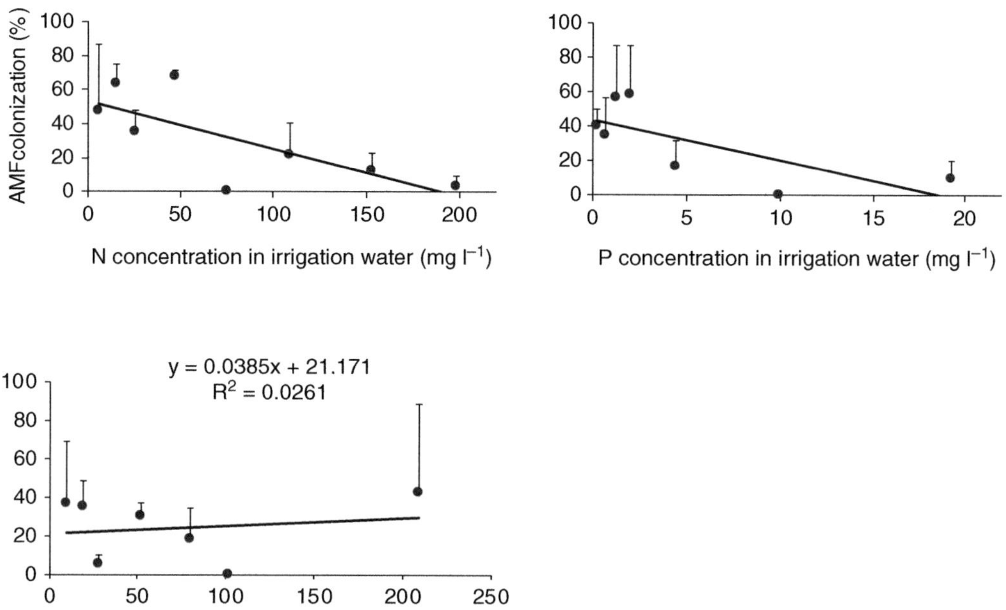

Fig. 6.3.7. Mycorrhizal colonization rate in response to N (*left panel*) and P (*right panel*) concentration in the irrigation water in a large container experiment (Erel *et al.*, 2013b). Samples were taken two years after trees had been exposed to the different nutrient concentrations. Each point on the graph is the average + SD of three trees.

reporting on investigations of olive's response to P and P fertilization (Erel *et al.*, 2013b, 2016; Jiménez-Moreno and Fernández-Escobar, 2016, 2017). López-Villalta and Muñoz-Cobo (1996) quoted several studies conducted in Spain, which found a positive yield response to P fertilization. However, this response was only found when P was applied for a few years. Freeman *et al.* (2005) reported no positive response to P fertilization in California olive orchards. Similarly, Jiménez-Moreno and Fernández-Escobar (2017) found no response to foliar application of P to young olive trees. P fertilization has traditionally been recommended only when P leaf concentration decreases below a deficiency threshold of 0.1%. We recently found a pronounced positive effect of P on fruit set, and consequently on yield, for three successive seasons (Erel *et al.*, 2008, 2013b). Fruit set and yield were both linearly correlated to leaf P concentration. Leaf P concentration ranged from 0.06% to 0.19%, indicating a positive effect of P even when leaf P concentration is well above the deficiency threshold of 0.1%. A possible explanation for the improved fruit set is an increase in the ratio of perfect flowers to male flowers (Erel *et al.*, 2013b, 2016). Increased P was also found to increase the rate of reproductive buds in relation to dormant and vegetative buds, inflorescence weight, pistil weight and fruitlet persistence. In a long-term field study (6 years), withholding P resulted in a significant decline in yield due to reduced flowering and fruit set. In addition, low P accelerated fluctuations in alternate bearing, whereas vegetative growth was hardly affected by P-fertilizer level (Haberman *et al.*, 2021). Combining the results obtained in irrigated olive orchards (Haberman *et al.*, 2021; Erel *et al.*, 2019) indicated that the common sufficiency threshold of 0.1% is not sufficient for intensive agro-systems, and we suggest 0.13% as new threshold (see section 6.3.4).

6.3.3.4 Potassium

Introduction

Potassium (K) is an essential macronutrient found at relatively high concentrations in most terrestrial plants. It plays a central role in the plant's water economy, especially via osmoregulation and maintenance of cell turgor. Appropriate K nutrition is known to raise the plant's tolerance to water stress (Cakmak, 2005; Römheld and Kirkby, 2010), and it is the main osmoticum responsible for stomatal movement (Marschner, 1995). Despite widespread traditional cultivation, knowledge on the role of K nutrition in olive physiology is scarce. In fact, K is considered to be particularly important in olive cultivation (Fernández-Escobar, 2010). However, the assumption of a high K requirement for tree health and productive capacity lacks solid scientific evidence and probably stems from the high K concentration in the fruit flesh (Hartmann *et al.*, 1966; Freeman *et al.*, 2005; Bustan *et al.*, 2013) and the relatively common occurrence of K deficiencies in olive orchards (Fernández-Escobar, 2010). In the 1960s, some reports relating K to yield were published (not all of them in English) and were subsequently used by a few authors (for example, López-Villalta and Muñoz-Cobo, 1996; Klein and Lavee, 1977) as a basis for understanding and for recommendations. In general, K has a crucial role in the plant's water balance and K deficiency has been claimed to impair carbohydrate assimilation rate (Cakmak, 2005). However, in olive, we found no correlation between K level and drought-tolerance mechanisms (Erel *et al.*, 2014). Moreover, stomatal control was not significantly impaired by K deficiency, regardless of water availability.

Interactions between potassium and sodium

Contrary to K, which is acknowledged as an essential macronutrient, Na is generally considered in plant nutrition for its adverse toxic effects on plants and its antagonistic relationship with K. However, in some plant species, Na seems to be able to at least partially replace K. Beneficial effects of Na on plant growth and yield have long been acknowledged, specifically under conditions of low K availability (Subbarao *et al.*, 2003; Wakeel *et al.*, 2011; Kronzucker *et al.*, 2013). Olive is considered to be salt-tolerant, with the extent of this tolerance depending on the variety (Gucci and Tattini, 1997). Salt tolerance in general (Marschner, 1995) and the ability to compartmentalize Na in cell vacuoles in particular, is crucial for Na to function as a nutrient (Wakeel *et al.*, 2011). The tolerance mechanisms in olive make it a promising candidate for the utilization of Na as a nutrient. Interestingly, Na was able to replace K and restore carbohydrate assimilation rate to nearly optimum level (Fig. 6.3.8; Erel *et al.* 2014). Potassium deficiency was found to diminish photoprotection mechanisms due to reduced photosynthetic and photorespiratory capacity. The lower rates of CO_2 and O_2 assimilation in K-deficient trees resulted in elevated levels of excitation energy (Erel *et al.*, 2015).

When soil moisture is high, K becomes more available to the olive tree (Restrepo-Diaz *et al.*, 2008a; Zipori *et al.*, 2015) (Fig. 6.3.9). Beutel *et al.* (1983) stated that K deficiencies may be corrected solely by irrigation.

The mobility, and therefore availability, of K for uptake by the roots under fertigation is significantly greater than under broadcast application (Neilsen *et al.*, 1999). Reports on K deficiencies are common in the Mediterranean basin, even in relatively deep fertile soils. Therefore, regular K application is recommended.

Availability, growth, and production

As already noted, K deficiency is common in rainfed orchards. In Andalusia in southern Spain, low K is considered to be the main nutritional disorder (Fernández-Escobar *et al.*, 1994; Fernández-Escobar, 2010), despite high soil K and K application levels. In a field trial under rainfed conditions, Restrepo-Diaz *et al.* (2008b) studied the effect of K application using two different salts – KCl and K_2SO_4 – compared to an untreated control. They found a positive correlation between K uptake and water availability (annual precipitation), irrespective of the form of K applied. The unfertilized control bore the same fruit load for the three examined seasons even though leaf K in those trees was below the level that is considered

Fig. 6.3.8. Typical appearance of olive trees (cv. 'Barnea') with no K and no Na (*left*), no K and 2 mM Na (*center*), and 2 mM K and Na (*right*) in the irrigation solution. Photo was taken one year after treatment initiation. (Adapted from Erel *et al.*, 2014)

Fig. 6.3.9. Drip-irrigation system in intensive olive orchard in Israel. (Photo: Tali Nachshon Dag)

sufficient (0.8%), but higher than the accepted deficiency level of 0.4%. While K application significantly increased leaf K, no effect on vegetative growth was found in two out of the three seasons of the study. In a different report, the same group found that even foliar absorption of K (using rubidium as a K analog) is limited when water stress is induced. K fertilization enhanced branch extension when K concentration in the solution increased from 2 mg l^{-1} to 4 mg l^{-1}, and further increased it when K concentration was elevated to 98 mg l^{-1} (Restrepo-Diaz *et al.*, 2008b). Freeman *et al.* (2005) reported that K uptake increases when the applied amount is increased, but no increase in yield was noted in trees showing adequate levels of K. Foliar spray of K on trees with leaf K below the sufficiency threshold also did not result in increased vegetative growth, even though leaf K level increased significantly (Mimoun *et al.*, 2004). Furthermore, under soilless culture, we found no effect of K on growth or yield, even when K was withheld for three years and leaf K values decreased to below the sufficiency level of 0.8% (Erel *et al.*, 2013b). In a modern irrigated orchard, Haberman *et al.* (2019a) reported that withholding K fertilization for six years leads to a significant reduction in yield attributed to reduced flowering and fruit set, but not growth. Apparently, the low K levels accelerated alternate bearing. Withholding K fertilization led to lower soil-exchangeable and soluble K concentrations and, therefore, the authors suggested using soil tests as a diagnostic tool to complement leaf analyses.

K-application rates are generally similar or higher than those for N. The annual amounts of K range between 0.5 and 1.5 kg $tree^{-1}$ or approx. 300 kg ha^{-1} (Morales-Sillero *et al.*, 2007; Hussein, 2008), depending on soil type, tree size, irrigation practice, and results of the leaf analysis (Therios, 2009; Fernández-Escobar, 2010). While in irrigated orchards, fertigation is the recommended application method, under rainfed conditions, a combination of soil and foliar application is the best approach. In modern irrigated orchards, Haberman *et al.* (2019a) calculated the net K balance and concluded that K fertilization of 180 kg ha^{-1} is sufficient for a balanced fertilization scheme.

Diagnostic leaf dynamics

Leaf K concentration tends to be high at the beginning of the growing season and then decreases along the season to minimum values at the end of the winter (Fernandez-Escobar *et al.*, 1999; Freeman *et al.*, 2005; Bustan *et al.*, 2013). These annual dynamics point out the importance of time of sampling and the standards used. In the past, in Israel, for various reasons, leaf sampling was conducted in the winter, when leaf K is at minimum values, although the international standards used for comparison were based on July sampling. As a result, deficiency was inaccurately diagnosed and remarkable amounts of K were used to correct it with no effect, which is not surprising because the trees were actually well within the sufficient range for K. Leaf K concentration is highly dependent on fruit load, with lower values following high yields (Klein and Lavee, 1977; Fernández-Escobar *et al.*, 1999; Restrepo-Diaz *et al.*, 2008b; Bustan *et al.*, 2013). The high K concentration found in fruit explains the strong relationship between fruit load and leaf K. We found K concentrations in fruit flesh ranging between 1.7% and 3.5% in response to various K concentrations in the irrigation solution (Dag *et al.*, 2009a). However, the role of K in fruit remains unknown and we found no major effect of K on oil-quality parameters or fatty acid composition (Dag *et al.*, 2009a; Erel *et al.*, 2013a).

Salinity

Under saline conditions, antagonism between K and Na is common, and therefore K uptake decreases when environmental Na concentration increases and *vice versa* (Klein *et al.*, 1993; Loupassaki *et al.*, 2002; Erel *et al.*, 2014). As a result, under high salinity, K deficiency may occur even when soil K is high. Moreover, high K (and Ca) levels can moderate the negative effects of salinity (Larbi *et al.*, 2020). The ratio of K to Na is considered to be a good indicator of a tree's salinity status (Benlloch *et al.*, 1991).

6.3.3.5 Calcium, magnesium and sulphur

The macronutrient Ca is essential for normal plant development and is a significant component of plant cell walls (Gilliham *et al.*, 2011). Ca is often considered immobile in the phloem, and is transported via the xylem to highly transpiring plant organs (e.g. leaves). As a result of its low mobility, Ca accumulates in plant tissues with time. We therefore expect to find high Ca concentrations in mature leaves, followed by young leaves and then fruit (Fernández-Escobar *et al.*, 1999). The macroelement Mg is required for several metabolic processes, such as chlorophyll synthesis. Since both Ca and Mg are present as divalent cations, they share several chemical properties, reflected in highly antagonistic uptake patterns (Tang and Luan, 2017). Thus, high Ca availability will impair Mg uptake and *vice versa*. Mg also tends to accumulate in mature leaves (Fernández-Escobar *et al.*, 1999), although plants remobilize Mg to a certain extent.

S is the sixth most abundant macroelement (after N, K, P, Ca and Mg), required for the synthesis of many proteins. It has a structural role in several proteins and amino acids and participates in catalytic or electrochemical reactions in the plant. Because olive cultivation is found mostly in calcareous soils rich in Ca, Mg and S, deficiencies are rare, and generally fertilization is not required or practiced. That said, Ca and Mg can have an indirect effect on the olive's nutritional status via antagonistic uptake with K or NH_4, or a synergetic effect in the case of salinity (Larbi *et al.*, 2020).

In many plants, water transport to the fruit is predominantly via the phloem. Hence, Ca concentration in the fruit is often low compared to that in the leaves. This poses a challenge for adequate Ca transport to the fruit, even under high Ca availability in the soil solution (Ho and White, 2005). Indeed, olive fruit Ca concentration is extremely low (0.1–0.2%), ten times lower than that found in the leaves. Despite being of high interest, the effect of fruit Ca concentration on fruit properties has not been thoroughly studied.

6.3.3.6 Microelements

Iron

Fe is the most abundant, and probably most important micronutrient in olive nutrition. It is involved in many redox reactions in plants and plays a specific role in chlorophyll metabolism. It is therefore not surprising that Fe-deficiency symptoms are characterized by leaf chlorosis (Fig. 6.3.10).

Cultivation areas with alkaline soil have huge amounts of Fe; however, in oxidized soil, the Fe is in an insoluble form (Fe-III) and therefore presents very low bioavailability (Lindsay and Schwab, 1982). The availability of Fe is most limited in highly buffered calcareous soil, resulting in what is often called 'lime-induced chlorosis'. Fe-induced chlorosis is common in olives growing in calcareous soils and is often related to a lack of extractable Fe (Benítez *et al.*, 2002) rather than total Fe.

Chemical plant analysis, in particular leaf analysis, is still the most common method used for diagnostic purposes in trees and is based on the relationship between plant growth rate and nutrient content. However, the use of leaf analysis has some limitations when applied to lime-induced chlorosis, because in many field-grown plants, there is no correlation between leaf Fe concentration and the degree of chlorosis (Römheld, 2000). This lack of correlation results from inactivation of Fe in the leaves or from inhibition of leaf growth due to Fe chlorosis, especially when bicarbonate concentration is high (Marschner, 2011). Therefore, several alternative diagnostic approaches were tested to estimate the 'active' fraction of Fe in the leaf (Bavaresco *et al.*, 1993). These methods seem promising, but results in the field are inconsistent (Abadía, 1992; Pestana *et al.*, 2003). In essence, there is no practical diagnostic tool to identify Fe-induced chlorosis in olives. Alternatives include soil test using chelate (Benítez *et al.*, 2002) or leaf analysis for active Fe (Pestana *et al.*, 2003). For both approaches, standards for olives are needed. Fe deficiencies generally can be handled by soil application of iron chelate (Zipori *et al.*, 2011).

Boron

B is an essential microelement for plant development and is required for several structural and metabolic functions, and, in particular, for successful fruit set. In olives,

Fig. 6.3.10. Typical iron deficiency in leaves of olive is 'Souri' cv. Israel. (Photo: Younis Mujira)

both deficiencies and toxicities are common in many cultivation areas (Benlloch *et al.*, 1991; Tsadilas and Chartzoulakis, 1997). The olive is considered moderately tolerant to B toxicity (Chatzissavvidis *et al.*, 2005), and it is therefore rare in natural environments but may occur when B concentration in the irrigation water or the soil is excessive. Irrigation water with > 2 mg l^{-1} B may be hazardous in the long term, depending on the variety and soil properties.

Soils with low B availability are abundant in olive orchards. The olive has a unique soluble sugar (mannitol) that allows B remobilization from mature to developing tissues when its level is low (Liakopoulos *et al.*, 2005). When the olive tree B level is low in the spring, fruit set might be impaired; foliar application of B has been shown to increase flower viability and fruit set (Perica *et al.*, 2001).

Additional micronutrients (zinc, manganese, copper, molybdenum)

All of these microelements are essential nutrients and required in low quantities by all plants. Deficiencies in Zn, Mn, Cu and Mo are rare, even in poor sandy soils (Hartmann and Brown, 1953; Freeman *et al.*, 2005). Cu is occasionally sprayed in orchards attacked by the pathogenic fungus *Spilocaea oleaginea* (peacock spot), and therefore leaf concentration is high.

6.3.4 Diagnosing Nutritional Status

Good fertilization strategy should rely on accurate diagnoses that can account for the complexity of nutrient cycling in the olive agro-ecosystem (see 6.3.3). Fertilization-management strategy that overlooks this complexity is likely to yield inaccurate fertilization recommendations. Specifically, generalization of nutrient-application rates ('one size fits all') will over- or underestimate the actual orchard needs. That said, in the name of simplicity, this is one of the most common fertilization approaches. For example, until ~2010, olive-fertilization recommendations by the Israeli extension services were fixed at: 200–0–300 units of NPK, respectively, regardless of soil fertility traits, tree productivity or vigor, and leaf nutrient concentrations. This strategy was later shown to lead to excessive N and K fertilization (Segal *et al.*, 2011; Erel *et al.*, 2019; Haberman *et al.*, 2019a, b, 2021). In contrast, a customized fertilization strategy can increase fertilization-use efficiency without affecting yield (Fernández-Escobar *et al.*, 2012). For example, in deep, fertile soils, olive can thrive on little to no N, K or P fertilization (López-Villalta and Muñoz-Cobo, 1996) because natural cycling processes, such as precipitation, soil pools, mineralization and weathering of soil minerals, can meet the olives' demand for nutrients. This is especially true in traditional agro-systems where growth and yields are typically low, limited by water availability. Fernández-Escobar *et al.* (2012) reported no response to N fertilization for seven years. A similar finding related to K fertilization was reported by Restrepo-Diaz *et al.* (2008b). We therefore conclude that **fertilization using a fixed recipe (as commonly practiced in many areas) will inevitably lead to unbalanced nutrition in the plant**. As a rule of thumb, more fertilizers are required in sandy, shallow, highly productive and vigorous orchards, and less in deep fertile soils with low olive productivity. Specific attention should be paid to nutrient concentrations in the irrigation with reclaimed wastewater which may supply most or all of the olives' requirement (Segal *et al.*, 2011; Erel *et al.*, 2019). To add to the complexity, olives are often cultivated in rugged terrain with large spatial variability in soil type and depth and, consequently, soil fertility. It is therefore essential to develop a precise, site-specific fertilization strategy.

6.3.4.1 Visual symptoms

Visual deficiency or toxicity symptoms are an important diagnostic tool with a major drawback: symptoms typically occur after damage to potential yield has been incurred (Table 6.3.4).

Table 6.3.4. Thresholds from the literature for first observation of visible deficiency symptoms in olive leaves.

Nutrient			Source
Macronutrients			
N	%	1.28	Fernández-Escobar *et al.* (2016)
		1.20	Erel *et al.* (2008)
		0.90	Hartmann and Brown (1953)
P	%	0.03	Fernández-Escobar *et al.* (2016)
		0.03	Erel *et al.* (2008)
		0.03	Hartmann and Brown (1953)
K	%	0.19	Fernández-Escobar *et al.* (2016)
		0.30	Erel *et al.* (2014)
		0.10	Hartmann and Brown (1953)
Ca	%	0.68	Fernández-Escobar *et al.* (2016)
Mg	%	0.10	Fernández-Escobar *et al.* (2016)
		0.06	Hartmann and Brown (1953)
Micronutrients			
Fe	mg kg^{-1}	NA	
Mn	mg kg^{-1}	17	Fernández-Escobar *et al.* (2016)
Zn	mg kg^{-1}	8.8	Fernández-Escobar *et al.* (2016)
B	mg kg^{-1}	33	Fernández-Escobar *et al.* (2016)
		15	Tsadilas and Chartzoulakis (1997)
		13	Hartmann and Brown (1953)

NA, not applicable

Therefore, we always prefer to diagnose nutritional status before the visible effects are manifested. One major exception is the diagnosis of Fe deficiency, which cannot be identified easily by today's methods (Pestana *et al.*, 2003). Furthermore, analysis of the 14 essential elements is costly, and many farmers will focus on the main macronutrients – N, P, and K – to save on laboratory expenses, and rely on visual diagnosis for marginal deficiencies, such as of B. It is therefore important to be familiar with the typical visual symptoms of deficiency and toxicity. Leaf-deficiency symptoms for most nutrients are presented in Hartmann and Brown (1953), Fernández-Escobar *et al.* (2016) and de Souza *et al.* (2019).

Fig. 6.3.11. Visual symptoms for olive leaves (var. 'Barnea') with deficient, adequate and excess N concentrations in the irrigation solution. Photo was taken approximately two years after exposure to the specified N concentration. N concentrations (*left to right*): 200, 70, 25, 15 and 6 ppm N in the irrigation water. (Photo: Ran Erel)

Nitrogen-deficiency symptoms

N-deficiency symptoms are presented in Fig. 6.3.11. Under severe N deficiency, growth is drastically inhibited; the leaves are smaller and light-green in color. However, leaf drop is less significant when N is low compared to other mineral deficiencies. Holding N to below 1.0% in leaves may result in shoot dieback, but not necrosis (Hartmann and Brown, 1953). In young cuttings, symptoms occurred when leaf N dropped to 1.28%

(Fernández-Escobar *et al.*, 2016), in agreement with our previous finding (Erel *et al.*, 2008).

Phosphorus-deficiency symptoms

Visual P-deficiency symptoms are very rare and difficult to distinguish *(Fernández-Escobar et al.*, 2016). P deficiency is characterized by smaller leaves with green color (in contrast to N deficiency). Leaf texture is thicker and stiffer. Under severe deficiency, leaf P drops to 0.03% and irregular necrotic patches appear (Fig. 6.3.12) (Jiménez-Moreno and Fernández-Escobar, 2016). On the whole-tree scale, P deficiency is characterized by reduction to cessation of growth followed by abscission of young leaves and eventual shoot dieback. Visual symptoms start to appear when leaf P drops below 0.05%.

Potassium-deficiency symptoms

K deficiency symptoms are presented in Fig. 6.3.13. Withholding K for two years under sand culture did not cause any visual symptoms in the leaves or on a whole-tree basis, although K concentration in young leaves dropped to 0.3% and in mature leaves further decreased to 0.1% (Hartmann and Brown, 1953). Trees deprived of K were somewhat reduced in size, mostly due to decreased root weight, but to a much lesser extent than responses to N, Mg and P deficiencies. In soilless culture for three years, we found no reduction in growth or yield when no K was added to an irrigation solution with an average K concentration of only 5–10 ppm (Erel *et al.*, 2013a). Leaf K in trees deprived of K decreased continuously, reaching as low as 0.25% after 2.5 years. When trees were grown with no K and no Na, visual symptoms of K deficiency were observed at leaf K concentration lower than 0.3% (Fig. 6.3.8). Symptoms of K deficiency started with chlorosis due to low chlorophyll concentration (Erel *et al.*, 2015). As deficiency progressed, necrosis starting at the leaf tips and later spreading to the leaf edges was observed. The visual symptoms were more pronounced in mature leaves. Severely necrotic leaves dropped rapidly.

Salinity symptoms

Salinity stress in agro-systems commonly results from irrigation with saline, recycled or other marginal water. This is particularly important in olive cultivation because use of marginal water is on the rise and olive has only moderate resistance to salinity stress (Ben-Gal, 2011). Poor irrigation management

Fig. 6.3.12. Visual symptoms for olive leaves (var. 'Barnea') with adequate to deficient P concentrations in the irrigation water. Photo was taken approximately two years after exposure to the specified P concentration. P concentrations (*left to right*): 20, 10, 0.7 and 0.2 ppm P in the irrigation water. (Photo: Ran Erel)

Fig. 6.3.13. Development of K-deficiency symptoms in leaves from healthy (left) to severely K-deficient (far right) plants. Leaf K concentration at the visible deficiency level was lower than 0.25% when no Na was present in the irrigation solution. (Photo: Ran Erel)

or natural salinization processes can result in significant salinity stress. If leaf analysis does not indicate salt stress in time, a visual assessment can identify it. Stressed olive trees are characterized by retarded growth, small, thick leaves and overall poor development. Under extreme stress conditions, plants suffer from major leaf defoliation and tip burn (Fig. 6.3.14). That said, in most cases, moderate salinity stress will not have pronounced visual symptoms, but evapotranspiration and yield will drop considerably (Ben-Gal *et al.*, 2017).

6.3.4.2 Plant and tissue analyses

To account for the many variables governing nutrient requirements, leaf analysis is considered the best diagnostic tool for assessing nutritional status and guiding olive fertilization (López-Villalta and Muñoz-Cobo, 1996; Fernández-Escobar *et al.*, 2009b; Fernández-Escobar, 2010). For a complete diagnosis, both leaf and soil analyses should be performed, as the latter can indicate excessive concentrations or toxic elements in the soil solution (Reuter and Robinson, 1997). However, foliar

Fig. 6.3.14. Young olive trees growing under desert conditions and irrigated with saline water (EC 4.2 dS m^{-1}) showing extreme salinity symptoms following rain that leached the accumulated salts to the root zone. (Photo: Nitzan Shatzkin (Shatz))

analysis must be used with care to achieve proper diagnosis of olive tree nutritional status. Two major limitations of leaf mineral analysis should be considered. The first is the asymptotic shape of the response curve to mineral availability in the soil, which makes the interpretation and diagnosis of excess nutrients somewhat problematic, specifically for N (Fig. 6.3.15). The second is that leaf mineral concentration may not reflect the concentration in the soil as a result of a different limiting factors, such as water shortage, disease etc. (Mengel and Kirkby, 2001). Fig. 6.3.15 illustrates the challenges involved in diagnosing N over-fertilization in olive leaves. While N accumulates in a logarithmic fashion, saturation is reached at the relatively low environmental concentration of ~25 mg l^{-1}. When N concentration increases further, up to ~150 mg l^{-1}, no significant change in leaf N is noted. However, high levels of N are associated with impaired productivity (Erel *et al.*, 2008, 2013b; Haberman *et al.*, 2019b) and reduced oil quality (Fernández-Escobar *et al.*, 2006; Dag *et al.*, 2009a; Erel *et al.*, 2013a). In contrast, N deficiency is detected in leaf samples and, indeed, the low N

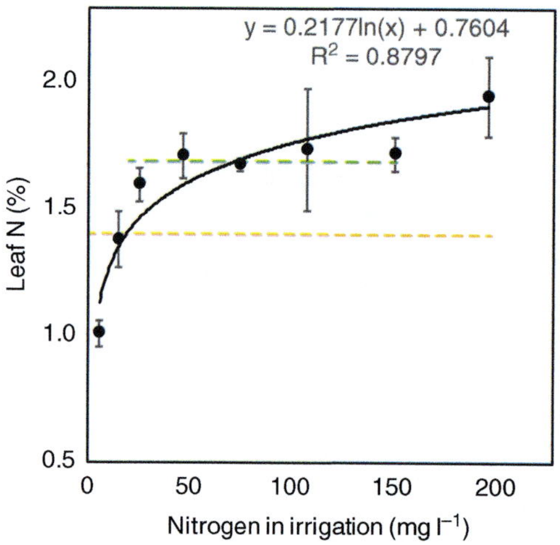

Fig. 6.3.15. Leaf N content in response to N concentration in the irrigation solution. Data were obtained from the experimental set-up described in Erel *et al.* (2013b) two years after treatment initiation. The green line indicates the range in which increased N in irrigation solution does not reflect in leaf N. The yellow line indicates the deficiency threshold of 1.4% N.

treatments with low productivity were reflected in leaf N levels below the commonly reported (Erel *et al.*, 2013b) threshold of 1.4%. Thus, while foliar analysis is a good diagnostic tool for N deficiency, excess N is not manifested in leaf N concentrations (Erel *et al.*, 2013b), as also suggested by Molina-Soria and Fernández-Escobar (2008, 2012).

In a long-term study, Fernández-Escobar *et al.* (2009a,b) observed that the use of leaf analysis enabled a drastic reduction in the amount and expense of olive fertilization without reducing yield. This approach should be examined and validated in intensive agro-systems.

Leaf P threshold was set at 0.1% in a traditional agro-system (Fernandez-Escobar, 2010). Results of a series of controlled and field experiments suggested that the threshold for leaf P be elevated to 0.13% for intensive agro-systems (Zipori *et al.*, 2020; Haberman *et al.*, 2021), as yield significantly increased with P application when leaf P was near or above 0.1%.

Procedure

The defined diagnostic leaf should be sampled in July (northern hemisphere) from a non-bearing branch. The diagnostic leaves are the youngest fully developed leaves (Fig. 6.3.16) sampled from around the tree canopy. Leaves should be collected from 10–50 representative trees in a homogeneous plot (in terms of variety, soil type, irrigation etc.), typically covering 1–10 ha depending on the spatial variability. Abnormal trees and those growing in border rows should be avoided.

International standards

International standards of mineral content for foliar diagnosis were determined and published by Chapman (1966), Childers (1966) and Beutel *et al.* (1983) for July sampling, and by Bouat (1987) for winter sampling. Those references have since been used by many researchers and published in well-acknowledged books such as *El cultivo del olivo* (later translated to *Olive Cultivation* (Fernández-Escobar, 2010)), *World Olive Encyclopaedia* (López-Villalta and

Fig. 6.3.16. Illustration of the diagnostic leaves (youngest fully developed leaves). Leaves should be sampled in July (northern hemisphere) from a non-bearing branch. (Photo: Tali Nachshon Dag)

Muñoz-Cobo, 1996), *Olive Production Manual* (Freeman *et al.*, 2005) and others. These standards are widely accepted and currently used by many growers worldwide. However, the original sources for the standards were based mainly on observations, trials and surveys in traditional rainfed orchards carried out more than half a century ago (for examples, Hartmann and Brown, 1953; Fahmy, 1958; Hartmann, 1958; Fahmy and Nasrallah, 1959; Samish *et al.*, 1961). Therefore, the relevance of this knowledge for intensive olive orchards should be validated. Based on recent studies, in Table 6.3.5 we suggest some revised thresholds suitable for modern, irrigated agro-systems.

SEASONAL DYNAMICS. The seasonal changes in minerals in the diagnostic leaves of mature 'Picual' trees grown under dryland cultivation were discussed in detail by Fernández-Escobar *et al.* (1999), who emphasized the importance of sample timing and fruit load on mineral concentrations. Perica (2001) found a significant difference in leaf N concentration as a result of leaf location in the canopy, with an apparent response to light exposure. The effects of time and fruit load on leaf mineral concentration have been reported in several studies (Bouat, 1987; Jordao *et al.*, 1993; Sibbett and Ferguson, 2000).

N demand in the spring and early summer during vegetative and productive development is high. Leaves may be an important N pool for new and growing fruit and branches. When N is limited, remobilization from leaves to growing organs is enhanced (Klein and Weinbaum, 1984; Fernández-Escobar *et al.*, 2004). This phenomenon explains the depletion in leaf N during the summer in non-irrigated orchards. Proper fertigation enables sufficient N supply and there is therefore no N depletion in fertigated orchards. Similar phenomena have been found for P and K, but with less supporting data.

Time of sampling

One of the most important parameters essential for proper interpretation of leaf mineral data is the timing of the sampling (Beutel *et al.*, 1983). The optimal time for leaf sampling is controversial. Many researchers recommend sampling in the summer (López-Villalta and Muñoz-Cobo, 1996; Freeman *et al.*, 2005; Fernández-Escobar, 2010) because of the relatively minor changes in leaf mineral concentration expected between mid-June and late August, and because most of the standards were established for this period. However, some researchers claim that winter, when tree metabolism is at its lowest, is the best time to sample leaves (Bouat, 1987; Lavee, 1996; Therios, 2006). This disagreement may result in misinterpretation of analysed data due to confusion over the different sampling dates with the others' standards.

Fruit load

Fruit load may play a significant role in the dynamics of leaf mineral concentration

Table 6.3.5. Critical element concentrations in diagnostic leaves sampled in July (northern hemisphere). New suggested standards for modern agro-systems are indicated in ***bold italics***.

Nutrient		Deficiency	Sufficiency	Over-fertilization/ toxicity	Source
N	%	< 1.4	1.5-2.0		Fernández-Escobar (2010)
			> 1.2-1.3		Fernández-Escobar *et al.* (2009b)
			1.4-1.8	***> 1.9***	Erel *et al.* (2013a); Haberman *et al.* (2019b)
P	%	0.05	0.1-0.3		Fernández-Escobar (2010)
			< 0.13		Zipori *et al.* (2020); Haberman *et al.* (2021)
K	%	0.4	> 0.8		Fernández-Escobar (2010)
			> 0.9		Zipori *et al.* (2020); Haberman *et al.* (2019a)
Ca	%	0.30	> 1.0		Fernández-Escobar (2010)
Mg	%	0.08	> 0.1		Fernández-Escobar (2010)
Na	%			> 0.2	Fernández-Escobar (2010)
Fe	**mg kg^{-1}**		none		
Mn	**mg kg^{-1}**		> 20		Fernández-Escobar (2010)
Zn	**mg kg^{-1}**		> 10		Fernández-Escobar (2010)
			> 8		Zipori *et al.* (2016)
B	**mg kg^{-1}**	14	19–150	185	Fernández-Escobar (2010)
Cu	**mg kg^{-1}**		> 4		Fernández-Escobar (2010)
Cl	%			> 0.5	Fernández-Escobar (2010)

(Jordao *et al.*, 1993; Fernández-Escobar, 2010; Bustan *et al.*, 2013), because growing fruit are a major sink for minerals (Klein and Weinbaum, 1984; Cimato *et al.*, 1989; Fernández-Escobar *et al.*, 2004). Considerable quantities of minerals are then removed at harvest and, consequently, may affect tree nutritional status. Fernández-Escobar *et al.* (2004) found that during an on year, 60–70% of the N in the branches is allocated to fruit. The influence of biannual bearing on leaf mineral concentration was studied under dryland cultivation (Fernández-Escobar *et al.*, 1999), under a supplemental irrigation regime (Ulger *et al.*, 2004), and in a modern intensive orchard (Bustan *et al.*, 2013), results were contradictory. In the rainfed traditional orchard, a pronounced effect of fruit load on N, P, K and Mg was reported, with lower levels of these nutrients following an on season (Fernández-Escobar *et al.*, 1999). In the semi-intensive agro-system, Ulger *et al.* (2004) found no effect of fruit load on P, Ca or Mg, while N was higher in the on year and K was higher in the off year. In the intensive agro-system, no effect of fruit load on N, P or Mg was found, whereas K decreased significantly in trees bearing a heavy crop (Bustan *et al.*, 2013). These different results may indicate an agro-system effect on mineral dynamics.

6.3.4.3 Soil and water analyses

Soil and water analyses are good complementary approaches to leaf analysis. These are especially important for diagnosing excessive mineral/salt accumulation (Fig. 6.3.14). Unlike leaf analysis, there is no accepted protocol or procedure to determine soil nutrient availability, and, most importantly, there are no thresholds. Furthermore, soil tests vary greatly with location and soil type. This makes interpretation challenging and traditional recommendations by experienced agronomists valuable. The total elemental concentration in the soil is in large part irrelevant because most nutrients in the soil are found in forms that have very little bioavailability. Hence, various procedures have been developed to assess nutrient availability rather than content.

Due to its complex and rapid chemistry, the most challenging nutrient to assess in soil samples is N. Under high-N fertilization, its main form is NO_3 (Segal *et al.*, 2011), whereas under low environmental N, it is NH_4 (Fernández-Escobar *et al.*, 2009a). Whereas soil NO_3 is simply extracted with water, NH_4 adsorbs to the soil and must be chemically extracted, typically with 2 M KCl solution. The timing of the soil N test is very important as N rapidly leaches through the soil profile following rain events. In mid-summer, excessive accumulation in the soil (e.g. over 20–30 mg kg^{-1}) indicates N over-fertilization (Fernández-Escobar *et al.*, 2009a). However, to the best of our knowledge, no standards have been developed for optimal olive cultivation.

Soil P tests vary from place to place, but the Olsen P extraction (0.5 M $NaHCO_3$ at pH 8.5) is prevalent for calcareous soils. Other soil P tests (e.g. Bray, Mehlich, resin) are also in use. Good olive productivity was obtained when Olsen P was 12–20 mg kg^{-1}, whereas values below 10 were associated with reduced yields (Haberman *et al.*, 2021).

For K, both water extraction (soluble K) and extraction with ammonium acetate (extractable K) are in wide use. Clearly, the extractable pool is much larger and is found in equilibrium with the soil solution. Good olive productivity results were obtained when extractable K was 200–300 mg kg^{-1} (Haberman *et al.*, 2019a). In a wide range of Algerian soils, Mouas Bourbia *et al.* (2013) reported a linear correlation between exchangeable K and leaf K, indicating that ~150 mg extractable K kg^{-1} soil is needed to obtain a leaf K content of > 0.8%. That said, soil texture has a large effect on the soil's ability to accumulate K in the exchange phase and therefore should be considered when running soil tests.

Soil tests are very important in the case of salinity/sodicity hazard (Aragüés *et al.*, 2005), especially when irrigation with saline water or high B concentration is practiced. In the case of salinity hazard, saturated paste extraction is a good diagnostic tool to assess the expected damage from salinity (Maas and Hoffman, 1977). To identify B availability, extraction with hot water has been reported to be well correlated with olive uptake (Tsadilas and Chartzoulakis, 1997).

To conclude from the scarce available information, Table 6.3.6 presents rough estimations of soil nutrient thresholds required for olive cultivation. These standards should be validated for a wide range of soils and conditions, as has been done for leaf analyses. Soil tests, especially for N, K and P, can give valuable information when collected routinely, because they can indicate trends of depletion or accumulation of these important nutrients.

6.3.5 Sustainable Management of Orchard Nutrition

6.3.5.1 Introduction

The plant material (fruit, pruned material) that is removed from olive orchards carries

Table 6.3.6. Suggested soil criteria for loamy to clay soils.

Nutrient	Extraction method	Low	Sufficient	Excessive	Source
P (mg kg^{-1})	**Olsen**	**< 10**	**15–25**	**> 30**	Haberman *et al.* (2021)
K (mg kg^{-1})	**Ammonium acetate**	**< 100**	**150–300**	**> 350**	Haberman *et al.* (2019a).
Fe (mg kg^{-1})	**Oxalate (Fe_{ox})**		**> 0 .35**		Benítez *et al.* (2002)
B (mg kg^{-1})	**Hot water**		**0.33**		Tsadilas and Chartzoulakis (1997)
Salinity (EC dS m^{-1})	**Saturated paste**		**< 1.2**	**> 4.2***	Ben-Gal *et al.* (2017)
			< 4.0		Aragüés *et al.* (2005)
N		**Not applicable**			

*EC_{50} (reduction of 50% in biomass compared to fresh water)

with it nutrients that have to be constantly replenished to avoid the development of nutritional deficiencies which will, in turn, impair plant vegetative growth, fruit yield and oil quality. The replenishment of plant nutrients can be carried out in different ways, using different nutrient sources. Sustainable management of olive orchard nutrition aims to provide trees with the required nutrients, with as little negative impact as possible on the environment. The major potential environmental hazard resulting from soil fertilization is the pollution of underground water and above-ground water bodies with fertilizer residues. In practice, when fertilizers are applied, there is never 100% uptake efficiency by the crop; there is always a certain residual fraction. Rodrigues *et al.* (2012) estimated fertilization recommendations for extensive, rainfed orchards to be more than 20-fold higher than the amounts removed by the trees. Erel *et al.* (2018) estimated this amount in intensive, irrigated orchards to be at least fivefold higher than that actually removed. Zipori *et al.* (2020) found that in an intensive, drip-fertigated orchard, the residual amounts of N, P and K were 30%, 60% and 40%, respectively. In some cases, e.g. high-density fertigated orchards, sustainability considerations are different from those of extensive, rainfed, widely spaced orchards. There is a difference in the potential impact of macro- and micronutrient fertilization on the environment. Olive trees are not regularly fertilized with micronutrients; when they are, it is very often done by foliar application, which has a lower potential impact on the environment compared to soil application. Therefore, the discussion in this section will concentrate on the sustainable nutrition of olive orchards with the macroelements N, P and K.

The potential impact of fertilization on tree performance and on the environment is not the same for all macronutrients. As already noted, the most important nutrient for agricultural crops is N. However, N also has the largest potential for impacting the environment. Regardless of the form applied to the orchard, most N will end up as nitrate through mineralization and nitrification processes. Nitrate mobility in the soil is very high and most NO_3-N residues will eventually reach the ground water. A small fraction of NO_3-N will be converted to N_2 or N_2O by denitrification and will be emitted to the atmosphere. Therefore, when it comes to N fertilization, the grower has to provide the trees with the required amounts but also adopt a fertilization regime that will minimize N migration below the root zone.

P is the least mobile nutrient in the soil in most olive-growing regions in the world. Even when applied in soluble form, P undergoes rapid fixation due to its interactions with Ca, Mg and Fe, which reduce its mobility to practically zero. Plants take up most P by root interception and not by movement of soluble P with the water. In very acidic soils, with no or low Ca and Mg, P might remain in a soluble form and migrate below the root zone. However, olive trees are usually grown in neutral or high-pH soils. Therefore, migration of P below the root zone into the ground water is quite rare. On the other hand, due to its fixation to the upper soil layers, it might contaminate above-ground water bodies when soil-erosion events occur.

The potential impact of K fertilization on the environment lies between those of N and P. K is less mobile than N in the soil due to its adsorption to clay minerals and the organic fraction. On the other hand, unlike P, K does not undergo chemical fixation. Thus, K poses a moderate hazard to the environment. In very sandy soils, where clay content is very low, the risk of K migration below the root zone into the ground water is higher than in heavy-textured soils. Therefore, K fertilization must also be related to soil texture when addressing environmental considerations.

6.3.5.2 Chemical fertilizers

Chemical fertilizers are widely used for olive fertilization in both rainfed, extensive orchards and intensive, irrigated ones. There are distinct differences in fertilization considerations for rainfed vs. irrigated olive orchards. In the latter, the application of chemical fertilizers can be done quite precisely and can be more properly adapted to the trees' phenological stages, because the irrigation water carries the applied chemicals to the root zone (fertigation). On the other hand, in rainfed orchards, the application of chemical fertilizers must take place before or during rainfall, usually by broadcast application, so that the rain will transport the nutrients to the root zone. However, this does not always coincide with the corresponding phenological stage and tree uptake rates, and the dissolved fertilizers may therefore migrate below the root zone. The common water-application practice in irrigated olive orchards is regulated or unregulated deficit irrigation, with potentially low migration of nutrients, especially N, below the root zone during the irrigation season. Over-fertilization will be accompanied by nutrient migration below the root zone in the case of significant rain events occurring during or after the fertilization season. The fertilization regime has to be adapted to the prevailing conditions in the growing area; for example, in Mediterranean climates, characterized by wet, cool winters and dry, hot summers, the irrigation and fertilization seasons are spring and summer, coinciding with the trees' active period. If fertilization is completed one to two months before the end of the irrigation season, the trees will exhaust the residual soil N during the remaining irrigation period, dramatically reducing the potential for N migration below the root zone with the rains. On the other hand, in regions with abundant summer rains, fertilization has to be adapted to the climatic conditions to avoid heavy nutrient leaching below the root zone. This can be done through frequent applications of small amounts of fertilizer when solid fertilizers are broadcasted, or by frequent, technical fertigation events in which the water is the only fertilizer carrier.

6.3.5.3 Organic material (manure, compost)

Organic manure, either composted or crude, is often used in olive orchards as a source of nutrients (Fig. 6.3.17). The use of manure is more common in traditional, rainfed orchards than in intensive, irrigated ones, although sometimes the latter are fertilized with manure as well. The manure is usually broadcasted around the trees in widely spaced orchards or over the entire soil surface in more dense ones. In some rare specific cases, organic manure is applied in trenches below the drip-irrigation lines in irrigated orchards (Fig. 6.3.17). When organic material is broadcasted, tillage is required in most cases to avoid decomposition of the organic material by solar radiation. The release of nutrients through mineralization of the organic matter can only take place when water is available. In rainfed orchards, this means during the winter. However, during the winter, the olive trees' nutrient uptake is minimal so that a significant part of the nutrients, especially N, might leach below the root zone (Dahan *et al.*, 2014). On the other hand, mineralization processes are temperature-dependent and relatively slow during the winter. There is a specific period, in the spring and early summer, when the temperatures are high enough and rainfall events are still occurring, during which mineralization of organic compounds can take place. This coincides with increased tree activity and nutrient uptake. It is practically impossible to control the mineralization rates of the applied organic matter, and therefore the risk of polluting underground water with mineralized nutrients, especially N, is high. On the other hand, organic manure increases soil organic content and improves water-retention properties, cation-exchange capacity and hydraulic conductivity. An important environmental aspect of compost application is the recycling of animal manure. Animal husbandry produces large amounts of manure that might become an environmental pollutant if not treated properly. The incorporation of this material into compost is beneficial for both parties: animal husbandry and horticulture (Beltran *et al.*, 2004).

6.3.5.4 Cover crops

Cover crops (Fig. 6.3.18) affect both physical and nutritional soil properties in orchards. Their use is usually related to their ability to reduce the risk of soil erosion and

Fig. 6.3.17. Compost trench installation in mature olive orchards. (Photo: Isaac Zipori)

run-off, especially on sloped terrains (Hernández *et al.*, 2001). Their contribution to plant nutrition is debated, despite their contribution to soil fertility, as they compete with the trees for nutrients and water (Ferreira *et al.*, 2013). Cover crops can be either perennial or annual and can consist of natural vegetation or be purposefully sown. In non-tilled orchards, annual cover crops are usually maintained on the soil surface during the rainy season. They can then be mowed and/or treated with herbicides, or left to dry out during the summer, thus protecting the soil surface from wind erosion and improving rainfall penetration in the subsequent rainy season. In tilled orchards, they can be incorporated into the soil by tilling. The common concept is that leguminous cover crops contribute N. However, this contribution depends on the C:N ratio in the vegetative material (Aulakh *et al.*, 1991). When the ratio is high, N immobilization occurs, and *vice versa*. A C:N ratio of 24 has been suggested as a threshold indicator for the potential contribution of plant residues to mineral soil N (Trinsoutrot *et al.*, 2000).

Most studies regarding cover crops and olive orchards show a positive effect on the environment through reduction of nutrient losses, as well as improved soil fertility through improvement of soil physical properties, increased water retention and increased organic matter content (Nieto *et al.*, 2012). The use of cover crops is often part of a non-tillage policy, which has additional advantages, such as soil conservation and

Fig. 6.3.18. Natural cover between rows in a high-density, intensive olive orchard (Israel). (Photo: Tali Nachshon Dag)

reduction of CO_2 emission into the atmosphere. The practice of using cover crops combined with no-tillage has more advantages than disadvantages and can be considered a key component in sustainable olive orchard management (Gómez, 2017).

6.3.5.5 Plant material recycling and nutrition

Some nutrient recycling occurs naturally. The lifespan of an olive leaf is about two years (Lavee, 1996). In senescing leaves (> 1 year old), nutrients are translocated to the newly formed, young tissues (Weinbaum, 1988). The senescent leaves eventually drop and form a soil layer with higher organic matter content below the trees (Fig. 6.3.2). These leaves, together with dropped fruit, are a potential source of nutrients as they eventually decompose and undergo mineralization. However, it is difficult to assess the quantitative contribution of this material to the orchard's nutrition. When the dropped leaves and fruit are left on the soil surface, i.e. in non-tilled orchards, their decomposition rates are lower than when they are incorporated into the soil by tillage.

Large amounts of nutrients are removed from olive orchards with the pruned material. The pruned material can either be removed from the orchard or chopped *in situ* (Fig. 6.3.19). Removal from the orchard creates a challenge for waste disposal, and sometimes phytosanitary problems. Chopping this material *in situ* is therefore the preferred solution. The chopped material forms soil mulching, reducing water loss and soil erosion. It increases soil organic matter content, improves soil structure and reduces weed growth. The chopped pruned material eventually decomposes, undergoes mineralization and can contribute to the orchard's nutrition. Again, it is quite difficult to quantitatively assess the contribution of this material to the orchard's nutrition because decomposition rates are inconsistent and depend on factors such as temperature, humidity and microorganism activity.

Fig. 6.3.19. A mature olive orchard after chopping of pruned material (a) and a close-up view of the chopped material (b). Gilat Reseach Center, Israel. (Photo: Arnon Dag)

Moreover, some of the mineralized nutrients, especially N, are subjected to denitrification processes and find their way into the atmosphere as N_2.

6.3.5.6 Application of olive pomace and olive mill wastewater

The olive oil extraction industry produces large amounts of byproducts that have to be dealt with. Three-phase olive mills produce low-water-content olive pomace and olive mill wastewater (OMW), whereas two-phase olive mills produce high-water-content, semi-liquid olive pomace. These byproducts are potential environmental pollutants. However, when properly treated, they can become a valuable source of nutrients. Olive pomace from two-phase systems is easiest to handle (Castellani *et al.*, 2017). It can be incorporated into composts and contribute significant amounts of nutrients. It can also serve as livestock feed after proper treatment, or can be dried and serve as high-energetic-value burning material for heating. OMW from three-phase systems is more difficult to handle. This byproduct cannot be introduced into domestic-sewage-treatment plants because of its high biochemical oxygen demand, and fat and polyphenol concentrations, which might cause the treatment plant's collapse. In most olive-oil-producing countries, the most common solution is to spread OMW on the soil surface of olive orchards in limited permissible amounts of 50–80 m^3 ha^{-1} (Zipori *et al.*, 2018). There is a potential risk in this practice, as it can have negative effects on soil properties or soil fauna due to toxicity. However, recent studies (Levy *et al.*, 2018; Zipori *et al.*, 2018; Kurtz *et al.*, 2021) have shown that controlled application of OMW to olive orchards over several years has no long-term adverse effects on soil properties or tree performance. Moreover, controlled application of OMW can provide most of the trees' K requirements and significant amounts of their P needs.

Whereas the controlled application of OMW to olive orchards has become common practice, permitted by the official authorities of most olive-oil-producing countries, the application of two-phase olive pomace to agricultural soils in general, and olive orchards in particular, has been quite limited. Nevertheless, the small amount of literature available on the subject indicates a positive effect of two-phase olive pomace amendment on soil properties and tree performance when application rates are limited to 30 t ha^{-1} pomace (dry weight). Application of double this amount resulted in decreased yield (López-Piñeiro *et al.*, 2008). In another study, the application of OMW at a rate of 50 t ha^{-1} significantly improved tree performance and some soil properties (Regni *et al.*, 2017). It seems that when properly applied, i.e. in limited amounts that are evenly distributed over the soil surface, two-phase olive pomace can be beneficial to olive orchards.

6.3.5.7 Irrigation with recycled wastewater as a source of nutrients

The globally increasing demand for water on the one hand, and the increasing quantities of wastewater produced by the world's growing population on the other, stress the importance of using alternative water sources for the irrigation of agricultural crops, including olives. One such source is recycled wastewater (RWW), which has become an alternative in regions with a limited supply of fresh water for agriculture (Muyen *et al.*, 2011; Grant *et al.*, 2012). Irrigation with RWW carries agronomic and environmental risks that require special consideration, but it also reduces the need to invest energy and funds for complete purification of this water, and allows recycling of water and nutrients that would otherwise be disposed of into the environment and potentially contaminate water bodies. In a case study carried out in an olive orchard over eight years, irrigation with RWW provided most of the orchard requirements for N, P and K and had no adverse effects on tree performance (Erel *et al.*, 2019) and oil quality (Basheer *et al.*, 2019). The amounts of nutrients supplied by RWW depended on water composition and the amount of water applied. A negative impact of the use of RWW in that case was a steady increase in soil sodium adsorption ratio (SAR) values. RWW contains relatively large amounts of Na, which can alter the ratio of Na to Ca+Mg and increase the soil SAR, leading to negative effects on the soil's physical properties (Ayoub *et al.*, 2016; Erel *et al.*, 2019). However, this can be dealt with by periodic soil liming or application of other compound containing Ca and Mg. When irrigating olive orchards with RWW, the contribution of the irrigation water to the orchard's NPK supply has to be subtracted from that of the fertilizers added to the crop. Where very-low-quality RWW is used for olive irrigation, over-fertilization might occur due to elevated N, P or K levels, negatively impacting oil yield and quality. In such cases, specific measures have to be considered, such as improving the water quality or diluting the RWW with fresh water.

6.3.5.8 Additional environmental considerations

Direct environmental pollution of the soil, air and ground and surface water bodies due to over-fertilization of agricultural crops is quite obvious and straightforward. However, over-fertilization also has an indirect negative impact. The fertilizer industry and its supporting systems (transport and application of fertilizers) emit CO_2 and other polluting products into the environment. If the applied fertilizers are not taken up by the plants, the damage is even greater. Raw materials for the production of some fertilizers, such as P and K, are exhaustible resources. Adoption of precise fertilization approaches, as well as recycling of waste and plant material, will reduce the rate of exhaustion of these raw materials.

6.3.5.9 Sustainable olive nutrition in organic orchards

The source of nutrients in organic orchards is generally organic materials. The most common organic nutrient source in olive orchards is composted animal manure (Larney and Hao, 2007). Macronutrient concentrations in compost vary according to the source materials, but they are usually in the range of 1.4–2.5%, 0.3–1.0% and 2.1–2.9%, for N, P and K, respectively. The relatively low N concentrations in composts imply the application of large amounts in olive orchards to satisfy the plants' N requirements, but this may lead to over-fertilization of P (Cayuela *et al.*, 2004). Nevertheless, apart from being a nutrient source, compost has a positive effect on microelement availability, soil organic matter content, soil hydraulic conductivity, soil water-retention capacity and soil structure (Whalen *et al.*, 2003; Cayuela *et al.*, 2004).

In chemically fertigated olive orchards, nutrient control can be carried out quite precisely, adjusting application to optimal rates and timing and minimizing losses. In organic orchards, the mineralization rates of the applied organic material cannot be controlled by the grower and depend mainly on temperature, soil water content and the soil's microbiome. If compost application is followed by heavy rain or irrigation, N migration might occur, either below the root zone or with the run-off (Dahan *et al.*, 2014). On the other hand, most of the compost N is in organic form, which is less liable to migrate with water.

In general, it can be stated that under organic orchard management it is more difficult to satisfy the plants' needs and control negative environmental repercussions. On the other hand, the use of animal manure as a source material for composts helps treat and eliminate a waste that otherwise is a severe pollutant (Beltran *et al.*, 2004).

6.3.6 Selection and Types of Fertilizers

The strategy for achieving adequate and optimal plant nutrition should take into consideration the plant's needs, as well as the soil's and irrigation water's ability to provide nutrients. These two parameters can be assessed quantitatively quite precisely and the difference between them is then applied as fertilizer. The basic principles of optimal fertilization should be based on the following guidelines:

1. Application of each nutrient in the required amount at the right time. Under-fertilization might result in low yield and/or quality, whereas over-fertilization might cause yield reduction (Haberman *et al.*, 2019b), lower product quality (Erel *et al.*, 2013a) and environmental pollution (Segal *et al.*, 2011), especially of water bodies such as underground water aquifers, rivers and water reservoirs. Long-term under-fertilization of olive trees might also result in enhanced tree susceptibility to pests and diseases (Rodrigues *et al.*, 2019; Fernandez-Escobar, 2019). Fertilization timing refers to both plant and climatic considerations; e.g. under Mediterranean conditions, olive tree activity during the cold, wet winter is low, resulting in very low nutrient-uptake rates (Fernandez-Escobar *et al.*, 2004). Therefore, orchard fertilization with soluble fertilizers during this period is neither useful nor environmentally friendly. On the other hand, in regions with heavy summer rains, types of fertilizers and timing of applications must be carefully considered to minimize nutrient leaching below the root zone.

2. To select suitable fertilizers, the grower should know which nutrients need to be applied and the most suitable application method for the growing conditions. Soil and leaf analyses are useful tools that can help the grower reach the right decision, thereby avoiding deficiencies as well as unnecessary costs.

3. Factors such as fertilizer availability, cost and effectiveness should also be considered.

Fertilizers can be divided into two major groups: solid fertilizers and soluble/liquid fertilizers. In each group, there are organic fertilizers and chemical ones. Slow-release fertilizers form a separate subgroup within the chemical fertilizers. The common fertilizer-application methods are broadcast application, application with the irrigation water (fertigation), and foliar application.

6.3.6.1 Chemical fertilizers

Chemical fertilizers can be mononutrients, such as urea or ammonium nitrate (N), potassium chloride (K) and superphosphate (P). They can contain more than one nutrient, such as potassium nitrate (K and N), potassium sulphate (K and S), monopotassium phosphate (K and P), monoammonium phosphate (N and P) and polyhalite (K, Mg, Ca and S). There is a wide range of solid, soluble compound fertilizers containing all three macronutrients (N, P and K) in various ratios, with or without microelements, such as 20:20:20, 17:10:27 ($N:P_2O_5:K_2O$, respectively) and many more. In general, the more compounds there are in the fertilizer, the higher its cost.

For rainfed olives, the most common fertilization practice is broadcast application of solid chemical fertilizers. This fertilization practice is strongly climate-dependent as the agent driving the fertilizer into the soil and down to the root zone is rainfall. As already noted in this chapter, the most susceptible nutrient in this case is N, which is the most mobile of the three macroelements. When N is applied at the beginning of winter, winter rains will likely leach most of it below the root zone and its uptake will be minimized, due to both leaching and low tree activity. In this case, it is more efficient to apply N fertilizers towards the end of the rainy season, when sufficient rain to carry the N to the root zone is still expected, and tree activity is on the rise. P and K fertilizers are less susceptible to leaching and can be applied at any time during the winter. Fertilizer broadcasting is almost the only way to apply chemical fertilizers to rainfed olive orchards. The number of applications per year is limited and can be manual or mechanical. The amount of nutrients required to satisfy the needs of rainfed olives is relatively small due to the low number of trees per unit area, low water availability over long periods, and limited vegetative and reproductive production.

Irrigated olives benefit from the fact that they are almost always drip-irrigated and thus fertilizers can be injected into the irrigation system and transported directly to the root zone (fertigation). This method enables the grower to apply fertilizers with any required combination of nutrients at the most suitable times (Erel *et al.*, 2018). The only usable fertilizers in this case are soluble ones. The preparation of a fertilizer solution from soluble, solid fertilizers for fertigation requires some basic knowledge in order not to mix substances that will lead to precipitation of insoluble salts. For example, mixing monoammonium phosphate with calcium nitrate will result in the precipitation of calcium phosphate, which will reduce the availability of both nutrients to the trees and also clog the drippers. In some countries, the grower can obtain compound liquid fertilizers at the orchard level. This option is more expensive than preparing a solution from solid, soluble fertilizers, but it saves on labor. In intensive, irrigated olive orchards, fertigation is the most recommended fertilization method. There are different fertigation techniques. The simplest one consists of a fertilization tank, installed as a bypass within the irrigation system, which ensures a constant flow of nutrients during irrigation. However, fertilization control with this method is limited. In computer-controlled irrigation systems, fertilizer injectors are installed, enabling precise proportional fertilization during irrigation.

Foliar nutrient application is relatively common in olive orchards, especially rainfed ones (Zipori *et al.*, 2020). The advantage of this method is the ability to provide nutrients directly to the canopy, where they are rapidly taken up. It also gives the grower flexibility with respect to application timing. The disadvantage of this method is that high nutrient concentrations in the sprayed solution can cause leaf burn, restricting the application to small amounts per tree or unit area. Therefore, this method is more suitable for the application of microelements, although it is sometimes used for K application as well. For the other macroelements, a few annual applications might be required to satisfy the plants' needs. This is an expensive method. It requires good-quality spraying equipment, it is time-consuming and labor-intensive, and the substances for foliar application are usually expensive.

6.3.6.2 Organic fertilizers

Organic plant nutrition aims to enhance plant ion uptake from the soil, by either direct ion transport from the soil solution or indirect transport. The latter pathway is based on symbiosis with microorganisms that extract soil ions and translocate them into the plant roots. Improving soil fertility, that is enhancing the soil biome and organic matter, is a fundamental goal for organic growers (Y. Tugendhaft, personal communication). N is the major macronutrient supplied. N sources are legumes, compost and amendments such as fermented chicken manure, guano and hydrolysed or fermentation products of the food industry. Growing legumes alongside olive trees (Pardini *et al.*, 2002) can provide N needs in low-productivity olive orchards (Rodrigues *et al.*, 2013), if there is sufficient rain or irrigation water for their vegetative development. Legumes must be chopped at first flowering to retain the fixed N in the soil.

Under water scarcity or when higher yields are desired, applying compost or other organic amendments will satisfy the crop's demand for N (Baldi *et al.*, 2010). These must be well-irrigated to dissolve nutrients and allow roots and microorganisms to develop in the added organic matter, and thus increase soil organic N uptake (Sánchez-García *et al.*, 2016). Compost application can increase mycorrhizal activity, which provides P for the olive tree's needs (Dag *et al.*, 2009c; Thirkell *et al.*, 2016).

Application of solid amendments may cause nitrate leaching below the root zone (Dahan *et al.*, 2014), and reduced risk of environmental pollution is fundamental to organic farming. Composts and solid amendments should therefore be applied in spring, when olive roots are active, the tree is in the vegetative phase and nutrient uptake is at its peak. Nutrient composition of the organic amendment should be analysed prior to application and the amounts applied should address crop needs. Leaf analysis must be conducted annually, to detect nutrient deficiencies. Adequate application of mature compost will provide macro- and microelements. KCl or B can be applied to address K or B deficiencies. In rare cases of Mg or Ca deficiencies, incorporating polyhalite in the root zone is a good practice. Mature compost application increases the soil's humic fraction, which, in turn, increases metal ion uptake by plant roots.

6.3.7 Plant Biostimulants

A promising and environmentally friendly approach is the use of natural plant biostimulants (PBs) that enhance flowering, plant growth, fruit set, productivity and nutrient-use efficiency, and can also improve tolerance to a wide range of abiotic stressors (Rouphael and Colla, 2020). PBs are substances and materials, with the exception of nutrients and pesticides, which, when applied to plants, seeds or growing substrates in a specific formulation, have the capacity to modify plants' physiological processes in a way that provides potential benefits for growth, development and/or stress responses. Based on this definition, PBs are categorized on the basis of their touted agricultural functions, and include diverse natural bioactive substances: (i) humic and fulvic acids; (ii) animal and vegetal protein hydrolysates; (iii) macroalgal seaweed extracts; and (iv) beneficial microorganisms: (i) AMF; and (ii) N-fixing bacteria of strains belonging to the genera *Rhizobium*, *Azotobacter*, and *Azospirillum*.

There is a clear need to improve our understanding of PB functioning, so that their efficacy can be improved and industrial production processes optimized (Brown and Saa, 2015). This is hard to achieve due to the diversity of sources of these materials and the complexity of the products, which often contain poorly characterized molecules (Calvo *et al.*, 2014). Hence, despite several publications on the positive impact of PBs on olive growth (Almadi *et al.*, 2020) and on resistance to biotic (López-Moral *et al.*, 2021) and abiotic (Graziani *et al.*, 2022) stresses, it is hard to draw clear recommendations for the use of PBs in olive plantations. Furthermore, some organic fertilizers also act as PBs (Mazeh *et al.*, 2021), complicating the picture even more. Still, olive growers often use them, depending on the qualifications of the marketing people and their ability to convince the growers of PBs' beneficial effects, even without any good experimental data to back up their claims. Growers who use these compounds are therefore advised to ask for results of controlled experiments in which the compounds were used, or to leave some untreated control trees so that they will be able to evaluate the PBs' contribution to their crop.

6.3.8 Olive Nutrition and Oil Quality

Olive oil has a central role in the Mediterranean diet; its protective effects against major human diseases, such as coronary heart disease, cancer and health-related cognitive decline, have been related mainly to two fundamental components: monounsaturated fatty acids (MUFAs) and polyphenols (Gorzynik-Debicka *et al.*, 2018). Increasing awareness of the health and gastronomic properties of olive oil has led to a steady increase in the global demand for this product in the last few decades. Fertilization, irrigation, higher tree density, advanced pruning and mechanical harvesting serve to satisfy this increased demand. However, there is also increasing awareness of olive oil quality. Therefore, while considering the appropriate fertilization regime for the olive orchard, in addition to productivity, environment and costs, oil quality should also be taken into account.

Free fatty acid content was found to increase linearly with increasing N in olive fruit in a controlled experiment (Erel *et al.*, 2013a) and in a wide survey in Israel (Bustan *et al.*, 2014). In addition, polyphenol content in the oil decreased linearly with increasing N availability and tree N status (as reflected by N level in diagnostic leaves) (Erel *et al.*, 2013a; Centeno *et al.*, 2017). Aside from their health and gastronomic properties, polyphenols protect the oil against oxidation and elongate its shelf life (Gutfinger, 1981). The decrease in polyphenols can be explained by the accumulation of N in the fruit as a result of N over-fertilization (Fernández-Escobar *et al.*, 2006). In this situation, phenylalanine preferentially flows to the synthesis of proteins rather than phenylpropanoids, resulting in reduced polyphenols due to the protein competition model (Jones and Hartley, 1999). With respect to fatty acid profiles, MUFAs decreased with increasing N availability (Dag *et al.*, 2009a). Hence, there seems to be a clear negative correlation between N availability/tree N status and the major oil-quality parameters. To produce high-quality olive oil, the grower therefore needs to avoid N over-fertilization (Fernández-Escobar *et al.*, 2006) and, when possible, supply the N to the plant throughout the spring till mid-summer to avoid N fertilization during the oil-accumulation stage, which occurs towards the end of the summer until the beginning of winter. Furthermore, while using RWW for irrigation, it is crucial to assess its nutrient content and subtract it from the fertilization program to avoid N over-fertilization (Basheer *et al.*, 2019). Whereas N availability has a significant effect on olive oil quality, no such clear effect has been found for the other nutrients (Dag *et al.*, 2009a), and their application should strive to balance productivity, tree health and fertilization costs.

Acknowledgment

We would like to thank Yizhar Tugendhaft for providing the information on organic fertilizers.

References

Abadía, J. (1992) Leaf responses to Fe deficiency: a review. *Journal of Plant Nutrition* 15, 1699–1713.

Almadi, L., Paoletti, A., Cinosi, N., Daher, E., Rosati, A. *et al.* (2020) A biostimulant based on protein hydrolysates promotes the growth of young olive trees. *Agriculture* 10(12), 618. https://doi.org/10.3390/agriculture10120618

Aragüés, R., Puy, J., Royo, A. and Espada, J. (2005) Three-year field response of young olive trees (*Olea europaea* L., cv. Arbequina) to soil salinity: trunk growth and leaf ion accumulation. *Plant and Soil* 271, 265–273.

Aulakh, M.S., Walters, D.T., Doran, J.W., Francis, D.D. and Mosier, A.R. (1991) Crop residue type and placement effects on denitrification and mineralization. *Soil Science Society of America Journal* 55(4), 1020–1025.

Ayoub, S., Al-Shdiefat, S., Rawashdeh, H. and Bashabsheh, I. (2016) Utilization of reclaimed wastewater for olive irrigation: effect on soil properties, tree growth, yield and oil content. *Agricultural Water Management* 176, 163–169.

Baldi, E., Toselli, M., Marcolini, G., Quartieri, M., Cirillo, E. *et al.* (2010) Compost can successfully replace mineral fertilizers in the nutrient management of commercial peach orchard. *Soil Use and Management* 26(3), 346–353.

Bar-Yosef, B. (1999) Advances in fertigation. *Advances in Agronomy* 65, 1–77.

Basheer, L., Dag, A., Yermiyahu, U., Ben-Gal, A., Zipori, I. and Kerem, Z. (2019) Effects of reclaimed wastewater irrigation and fertigation level on olive oil composition and quality. *Journal of the Science of Food and Agriculture* 99(14), 6342–6349.

Bavaresco, L., Fraschini, P. and Perino, A. (1993) Effect of the rootstock on the occurrence of lime-induced chlorosis of potted *Vitis vinifera* L. cv. 'Pinot blanc'. *Plant and Soil* 157, 305–311.

Beltran, E.D., Loresco, M.M., Dizon, R.C. and Zuñiga, A.J.M. (2004) Waste management system at DTRI university farm: utilization of manure, urine and farm residues as organic fertilizer at the dairy farm, its effects in the environment and sustainable livestock production system. DTRI 2004 Annual Report (Philippines). University of the Philippines at Los Baños.

Ben-Gal, A. (2011) Salinity and olive: from physiological responses to orchard management. *Israel Journal of Plant Sciences* 59, 15–28. https://doi.org/10.1560/IJPS.59.1.15

Ben-Gal, A., Beiersdorf, I., Yermiyahu, U., Soda, N., Presnov, E. *et al.* (2017) Response of young bearing olive trees to irrigation-induced salinity. *Irrigation Science* 35, 99–109.

Benítez, M., Pedrajas, V., Del Campillo, M. and Torrent, J. (2002) Iron chlorosis in olive in relation to soil properties. *Nutrient Cycling in Agroecosystems* 62, 47–52.

Benlloch, M., Arboleda, F., Barranco, D. and Fernández-Escobar, R. (1991) Response of young olive trees to sodium and boron excess in irrigation water. *HortScience* 26, 867–870.

Beutel, J., Uriu, K. and Lilleland, O. (1983) Leaf analysis for California deciduous fruits. In: *Soil and Plant Tissue Testing in California*. University of California Bulletin 1879. University of California, Berkeley, California, pp. 15–17.

Bouat, A. (1987) Olives: introduction and general. In: Martin-Prével, P., Gagnard, J. and Gautier, P. (eds) *Plant Analysis as a Guide to the Nutrient Requirements of Temperate and Tropical Crops*. Lavoisier Publishing Inc., Paris, pp. 299–309.

Brown, P. and Saa, S. (2015) Biostimulants in agriculture. *Frontiers in Plant Science* 6, 671. https://doi.org/10.3389/fpls.2015.00671

Bustan, A., Avni, A., Yermiyahu, U., Ben-Gal, A., Riov, J. *et al.* (2013) Interactions between fruit load and macroelement concentrations in fertigated olive (*Olea europaea* L.) trees under arid saline conditions. *Scientia Horticulturae* 152, 44–55.

Bustan, A., Kerem, Z., Yermiyahu, U., Ben-Gal, A., Lichter, A. *et al.* (2014) Preharvest circumstances leading to elevated oil acidity in 'Barnea' olives. *Scientia Horticulturae* 176, 11–21.

Cakmak, I. (2005) The role of potassium in alleviating detrimental effects of abiotic stresses in plants. *Journal of Plant Nutrition and Soil Science* 168, 521–530.

Calvo, P., Nelson, L. and Kloepper, J.W. (2014) Agricultural uses of plant biostimulants. *Plant and Soil* 383, 3–41

Castellani, F., Vitali, A., Bernardi, N., Marone, E., Palazzo, F., Grotta, L. *et al.* (2017) Dietary supplementation with dried olive pomace in dairy cows modifies the composition of fatty acids and the aromatic profile in milk and related cheese. *Journal of Dairy Science* 100(11), 8658–8669.

Cayuela, M.L., Bernal, M.P. and Roig, A. (2004) Composting olive mill waste and sheep manure for orchard use. *Compost Science & Utilization* 12(2), 130–136.

Centeno, A., García, J.M. and Gómez-del-Campo, M. (2017) Effects of nitrogen fertilization and nitrification inhibitor product on vegetative growth, production and oil quality in 'Arbequina' hedgerow and 'Picual' vase-trained orchards. *Grasas y Aceites* 68(4), e215. https://doi.org/10.3989/gya.0441171

Chapman, H.D. (1966) *Diagnostic Criteria for Plants and Soils*. University of California, Division of Agricultural Sciences, Berkeley, California.

Chatzissavvidis, C.A., Therios, I.N. and Molassiotis, A.N. (2005) Seasonal variation of nutritional status of olive plants as affected by boron concentration in nutrient solution. *Journal of Plant Nutrition* 28, 309–321.

Childers, N.F. (1966) *Mineral Nutrition of Fruit Crops*. Horticultural Publications, Rutgers, New Jersey.

Cimato, A., Marranci, M. and Tattini, M. (1989) The use of foliar fertilization to modify sinks competition and to increase yield in olive (*Olea europaea* cv Frantoio). *Acta Horticulturae* 286, 175–178.

Connor, D. and Fereres, E. (2005) The physiology of adaptation and yield expression in olive. *Horticultural Reviews* 31, 155–229.

da Silva, E.C., Nogueira, R.J.M.C., da Silva, M.A. and de Albuquerque, M.B. (2011) Drought stress and plant nutrition. *Plant Stress* 5(1), 32–41.

Dag, A., Ben-David, E., Kerem, Z., Ben-Gal, A., Erel, R. *et al.* (2009a) Olive oil composition as a function of nitrogen, phosphorus and potassium plant nutrition. *Journal of the Science of Food and Agriculture* 89, 1871–1878.

Dag, A., Bustan, A., Avni, A., Lavee, S. and Riov, J. (2009b) Fruit thinning using NAA shows potential for reducing biennial bearing of 'Barnea'and 'Picual' oil olive trees. *Crop and Pasture Science* 60(12), 1124–1130.

Dag, A., Yermiyahu, U., Ben-Gal, A., Zipori, I. and Kapulnik, Y. (2009c) Nursery and post-transplant field response of olive trees to arbuscular mycorrhizal fungi in an arid region. *Crop and Pasture Science* 60, 427–433.

Dag, A., Erel, R., Ben-Gal, A., Zipori, I. and Yermiyahu, U. (2012) The effect of olive tree stock plant nutritional status on propagation rates. *HortScience* 47, 307–310.

Dahan, O., Babad, A., Lazarovitch, N., Russak, E.E. and Kurtzman, D. (2014) Nitrate leaching from intensive organic farms to groundwater. *Hydrology and Earth System Sciences* 18, 333–341.

de Souza, F.B.M., Coelho, V.A.T., Pio, R., Rodas, C.L., da Silva, I.P. *et al.* (2019) Visual symptoms and nutritional deficiencies in olive plants subjected to nutrient deprivation. *Acta Scientiarum: Agronomy* 41. https://doi.org/10.4025/actasciagron.v41i1.39582

Deng, Y., Feng, G., Chen, X. and Zou, C. (2017) Arbuscular mycorrhizal fungal colonization is considerable at optimal Olsen-P levels for maximized yields in an intensive wheat-maize cropping system. *Field Crops Research* 209, 1–9. https://doi.org/10.1016/j.fcr.2017.04.004

Erel, R., Dag, A., Ben-Gal, A., Schwartz, A. and Yermiyahu, U. (2008) Flowering and fruit set of olive trees in response to nitrogen, phosphorus, and potassium. *Journal of the American Society for Horticultural Science* 133, 639–647.

Erel, R., Kerem, Z., Ben-Gal, A., Dag, A., Schwartz, A. *et al.* (2013a) Olive (*Olea europaea* L.) tree nitrogen status is a key factor for olive oil quality. *Journal of Agricultural and Food Chemistry* 61, 11261–11272.

Erel, R., Yermiyahu, U., Van Opstal, J., Ben-Gal, A., Schwartz, A. *et al.* (2013b) The importance of olive (*Olea europaea* L.) tree nutritional status on its productivity. *Scientia Horticulturae* 159, 8–18.

Erel, R., Ben-Gal, A., Dag, A., Schwartz, A. and Yermiyahu, U. (2014) Sodium replacement of potassium in physiological processes of olive trees (var. Barnea) as affected by drought. *Tree Physiology* 34, 1102–1117. https://doi.org/10.1093/treephys/tpu081

Erel, R., Yermiyahu, U., Ben-Gal, A., Dag, A., Shapira, O. *et al.* (2015) Modification of non-stomatal limitation and photoprotection due to K and Na nutrition of olive trees. *Journal of Plant Physiology* 177, 1–10. http://dx.doi.org/10.1016/j.jplph.2015.01.005

Erel, R., Yermiyahu, U., Yasuor, H., Chamus, D.C., Schwartz, A. *et al.* (2016) Phosphorus nutritional level, carbohydrate reserves and flower quality in olives. *PLOS ONE* 11, e0167591. https://doi.org/10.1371/journal.pone.0167591

Erel, R., Yermiyahu, Y., Ben-Gal, A. and Dag, A. (2018) Olive fertilization under intensive cultivation management. *Acta Horticulturae* 1217, 207–224. https://doi.org/10.17660/ActaHortic.2018.1217.27

Erel, R., Eppel, A., Yermiyahu, U., Ben-Gal, A., Levy, G. *et al.* (2019) Long-term irrigation with reclaimed wastewater: implications on nutrient management, soil chemistry and olive (*Olea europaea* L.) performance. *Agricultural Water Management* 213, 324–335. https://doi.org/10.1016/j.agwat.2018.10.033

Estaún, V., Camprubí, A., Calvet, C. and Pinochet, J. (2003) Nursery and field response of olive trees inoculated with two arbuscular mycorrhizal fungi, *Glomus intraradices* and *Glomus mosseae*. *Journal of the American Society for Horticultural Science* 128, 767–775.

Fahmy, I. (1958) Changes in carbohydrate and nitrogen content of Souri olive leaves in relation to alternate bearing. *Proceedings of the American Society for Horticultural Science* 72, 252–256.

Fahmy, I. and Nasrallah, S. (1959) Changes in macro-nutrient elements of Souri olive leaves in alternate bearing years. *Proceedings of the American Society for Horticultural Science* 74, 372–377.

Fernández-Escobar, R. (2010) Fertilization. In: Barranco, D., Rallo. L. and Fernández-Escobar, R. (eds) *Olive Growing*, 5th edn. RIRDC, Pendle Hill, Australia.

Fernández-Escobar, R. (2019) Olive nutritional status and tolerance to biotic and abiotic stresses. *Frontiers in Plant Science* 1151.

Fernández-Escobar, R., García-Barragán, T. and Benlloch, M. (1994) Estado nutritivo de las plantaciones de olivar en la provincia de Granada. *Informacion Tecnica Economica Agraria* 90, 39–49.

Fernández-Escobar, R., Moreno, R. and Garcıa-Creus, M. (1999) Seasonal changes of mineral nutrients in olive leaves during the alternate-bearing cycle. *Scientia Horticulturae* 82, 25–45.

Fernández-Escobar, R., Moreno, R. and Sánchez-Zamora, M. (2004) Nitrogen dynamics in the olive bearing shoot. *HortScience* 39, 1406–1411.

Fernández-Escobar, R., Beltrán, G., Sánchez-Zamora, M., García-Novelo, J., Aguilera, M. *et al.* (2006) Olive oil quality decreases with nitrogen over-fertilization. *HortScience* 41, 215–219.

Fernández-Escobar, R., Ortiz-Urquiza, A., Prado, M. and Rapoport, H. (2008) Nitrogen status influence on olive tree flower quality and ovule longevity. *Environmental and Experimental Botany* 64, 113–119.

Fernández-Escobar, R., Marin, L., Sánchez-Zamora, M., García-Novelo, J., Molina-Soria, C. *et al.* (2009a) Long-term effects of N fertilization on cropping and growth of olive trees and on N accumulation in soil profile. *European Journal of Agronomy* 31, 223–232.

Fernández-Escobar, R., Parra, M., Navarro, C. and Arquero, O. (2009b) Foliar diagnosis as a guide to olive fertilization. *Spanish Journal of Agricultural Research* 7, 212–223.

Fernández-Escobar, R., García-Novelo, J.M., Molina-Soria, C. and Parra, M.A. (2012) An approach to nitrogen balance in olive orchards. *Scientia Horticulturae* 135, 219–226. https://doi.org/10.1016/j.scienta.2011.11.036

Fernández-Escobar, R., Sánchez-Zamora, M.A., Garcia-Novelo, J.M. and Molina-Soria, C. (2015) Nutrient removal from olive trees by fruit yield and pruning. *HortScience* 50, 474–478.

Fernández-Escobar, R., Guerreiro, M., Benlloch, M. and Benlloch-González, M. (2016) Symptoms of nutrient deficiencies in young olive trees and leaf nutrient concentration at which such symptoms appear. *Scientia Horticulturae* 209, 279–285. https://doi.org/10.1016/j.scienta.2016.07.002

Ferreira, I.Q., Arrobas, M., Claro, A.M. and Rodrigues, M.A. (2013) Soil management in rainfed olive orchards may result in conflicting effects on olive production and soil fertility. *Spanish Journal of Agricultural Research* 11(2), 472–480.

Ferreira, I.Q., Rodrigues, M.Â., Moutinho-Pereira, J.M., Correia, C.M. and Arrobas, M. (2018) Olive tree response to applied phosphorus in field and pot experiments. *Scientia Horticulturae* 234, 236–244.

Freeman, M. and Carlson, R. (2005) Mineral nutrient availability. In: Sibbett, G. and Ferguson, L. (eds) *Olive Production Manual*. University of California Division Agriculture and Natural Resources (UCANR Publications), Berkeley, California, pp. 75–82.

Freeman, M., Uriu, K. and Hartmann, H. (2005) Diagnosing and correcting nutrient problems. In: Sibbett, G. and Ferguson, L. (eds) *Olive Production Manual*. University of California Division Agriculture and Natural Resources (UCANR Publications), Berkeley, California, pp. 77–86.

Gilliham, M., Dayod, M., Hocking, B.J., Xu, B., Conn, S.J. *et al.* (2011) Calcium delivery and storage in plant leaves: exploring the link with water flow. *Journal of Experimental Botany* 62, 2233–2250. https://doi.org/10.1093/jxb/err111

Gómez, J.A. (2017) Sustainability using cover crops in Mediterranean tree crops, olives and vines – challenges and current knowledge. *Hungarian Geographical Bulletin* 66(1), 13–28.

Gorzynik-Debicka, M., Przychodzen, P., Cappello, F., Kuban-Jankowska, A., Marino Gammazza, A. *et al.* (2018) Potential health benefits of olive oil and plant polyphenols. *International Journal of Molecular Sciences* 19(3), 686. https://doi.org/10.3390/ijms19030686

Grant, S.B., Saphores, J.D., Feldman, D.L., Hamilton, A.J., Fletcher, T.D. *et al.* (2012) Taking the 'waste' out of 'wastewater' for human water security and ecosystem sustainability. *Science* 337(6095), 681–686.

Graziani, G., Cirillo, A., Giannini, P., Conti, S., El-Nakhel, C. *et al.* (2022) Biostimulants improve plant growth and bioactive compounds of young olive trees under abiotic stress conditions. *Agriculture* 12(2), 227. https://doi.org/10.3390/agriculture12020227

Gucci, R. and Tattini, M. (1997) Salinity tolerance in olive. *Horticultural Reviews* 21, 177–214.

Gutfinger, T. (1981) Polyphenols in olive oils. *Journal of the American Oil Chemists' Society* 58(11), 966–968.

Haberman, A., Dag, A., Shtern, N., Zipori, I., Erel, R. *et al.* (2019a) Long-term impact of potassium fertilization on soil and productivity in intensive olive cultivation. *Agronomy* 9, 525. https://doi.org/10.3390/agronomy9090525

Haberman, A., Dag, A., Shtern, N., Zipori, I., Erel, R. *et al.* (2019b) Significance of proper nitrogen fertilization for olive productivity in intensive cultivation. *Scientia Horticulturae* 246, 710–717. https://doi.org/10.1016/j.scienta.2018.11.055

Haberman, A., Dag, A., Erel, R., Zipori, I., Shtern, N. *et al.* (2021) Long-term impact of phosphorous fertilization on yield and alternate bearing in intensive irrigated olive cultivation. *Plants* 10, 1821. https://doi.org/10.3390/plants10091821

Hartmann, H. (1958) Some responses of the olive to nitrogen fertilizers. *Proceedings of the American Society for Horticultural Science* 72, 257–266.

Hartmann, H. and Brown, J. (1953) The effect of certain mineral deficiencies on the growth, leaf appearance, and mineral content of young olive trees. *Hilgardia* 22, 119–130.

Hartmann, H.T., Uriu, K. and Lilleland, O. (1966) Olive nutrition. In: Childers, N.F. (ed.) *Fruit Nutrition*. Horticultural Publications, Rutgers, New Jersey, pp. 252–261.
Hernández, A.J., Prieto, N. and Pastor Piñeiro, J. (2001) Management of an olive crop in a semiarid environment using sown or resident leguminous covers. In: *First World Congress on Conservation Agriculture*, Madrid, 1–5 October 2001, pp. 419–423.
Hinsinger, P., Bengough, A.G., Vetterlein, D. and Young, I.M. (2009) Rhizosphere: biophysics, biogeochemistry and ecological relevance. *Plant and Soil* 321, 117–152.
Ho, L.C. and White, P.J. (2005) A cellular hypothesis for the induction of blossom-end rot in tomato fruit. *Annals of Botany* 95, 571–581. https://doi.org/10.1093/aob/mci065
Hussein, A. (2008) Response of Manzanillo olive (*Olea europaea*, L.) cultivar to irrigation regime and potassium fertigation under Tabouk conditions, Saudi Arabia. *Journal of Agronomy* 7, 285–296.
Jasrotia, A., Singh, R., Singh, J. and Bhutani, V. (1997) Response of olive trees to varying levels of N and K fertilizers. *Acta Horticulturae* 474, 337–340.
Jiménez-Moreno, M.J. and Fernández-Escobar, R. (2016) Response of young olive plants (*Olea europaea*) to phosphorus application. *HortScience* 51, 1167–1170.
Jiménez-Moreno, M.J. and Fernández-Escobar, R. (2017) Influence of nutritional status of phosphorus on flowering in the olive (*Olea europaea* L.). *Scientia Horticulturae* 223, 1–4. https://doi.org/10.1016/j.scienta.2017.05.028
Jones, C.G. and Hartley, S.E. (1999) A protein competition model of phenolic allocation. *Oikos* 86, 27–44.
Jordao, P. and Lietao, F. (1989) The olive's mineral composition and some parameters of quality in fifty olive cultivars grown in Portugal. *Acta Horticulturae* 286, 461–464.
Jordao, P., Dias, J., Calouro, F. and Duarte, M. (1993) Effect of fertilization on the leaf macronutrient concentrations of olive tree. *Acta Horticulturae* 356, 197–201.
Kirkby, E. (2012) Introduction, definition and classification of nutrients. In: Marschner, P. (ed.) *Marschner's Mineral Nutrition of Higher Plants*. Academic Press, London, pp. 3–5.
Klein, I. and Lavee, S. (1977) Effect of nitrogen and potassium fertilizers on olive production. In: *Proceedings of the 13th Colloquium of the International Potash Institute: Fertilizer Use and Production of Carbohydrates and Lipids*. International Potash Institute, Zug, Switzerland, pp. 295–304.
Klein, I. and Weinbaum, S. (1984) Foliar application of urea to olive: translocation of urea nitrogen as influenced by sink demand and nitrogen deficiency. *Journal of the American Society for Horticultural Science* 109, 356–360.
Klein, I., Ben-Tal, Y., Lavee, S., De Malach, Y. and David, I. (1993) Saline irrigation of cv. Manzanillo and Uovo di Piccione trees. *Acta Horticulturae* 356, 176–180.
Kronzucker, H.J., Coskun, D., Schulze, L.M., Wong, J.R. and Britto, D.T. (2013) Sodium as nutrient and toxicant. *Plant and Soil* 369, 1–23.
Kurtz, M.P., Dag, A., Zipori, I., Laor, Y., Buchmann, C., Saadi, I. *et al.* (2021) Toward balancing the pros and cons of spreading olive mill wastewater in irrigated olive orchards. *Processes* 9, 780.
Lahiri, A.N. (1980) Interaction of water stress and mineral nutrition on growth and yield. In: Turner, N.C. and Cramer, P.J. (eds) *Adaptation of Plants to Water and High Temperature Stress*. John Wiley, New York, pp. 341–352.
Larbi, A., Kchaou, H., Gaaliche, B., Gargouri, K., Boulal, H. *et al.* (2020) Supplementary potassium and calcium improves salt tolerance in olive plants. *Scientia Horticulturae* 260, 108912. https://doi.org/10.1016/j.scienta.2019.108912
Larney, F.J. and Hao, X. (2007) A review of composting as a management alternative for beef cattle feedlot manure in southern Alberta, Canada. *Bioresource Technology* 98(17), 3221–3227.
Lavee, S. (1996) Biology and physiology of the olive. In: *World Olive Encyclopedia*. International Olive Oil Council, Madrid.
Lavee, S. (2009) The revolutionary impact of introducing irrigation-intensification to the olive oil industry. *Acta Horticulturae* 888, 21–30.
Levy, G.J., Dag, A., Raviv, M., Zipori, I., Medina, S. *et al.* (2018) Annual spreading of olive mill wastewater over consecutive years: effects on cultivated soils' physical properties. *Land Degradation & Development* 29(1), 176–187.
Liakopoulos, G., Stavrianakou, S., Filippou, M., Fasseas, C., Tsadilas, C. *et al.* (2005) Boron remobilization at low boron supply in olive (*Olea europaea*) in relation to leaf and phloem mannitol concentrations. *Tree Physiology* 25, 157–165. https://doi.org/10.1093/treephys/25.2.157
Lindsay, W.L. and Schwab, A.P. (1982) The chemistry of iron in soils and its availability to plants. *Journal of Plant Nutrition* 5, 821–840. https://doi.org/10.1080/01904168209363012

López-Moral, A., Agustí-Brisach, C. and Trapero, A. (2021) Plant biostimulants: new insights into the biological control of Verticillium wilt of olive. *Frontiers in Plant Science* 12, 782. https://doi.org/10.3389/fpls.2021.662178

López-Piñeiro, A., Albarrán, A., Nunes, J.R. and Barreto, C. (2008) Short and medium-term effects of two-phase olive mill waste application on olive grove production and soil properties under semiarid Mediterranean conditions. *Bioresource Technology* 99(17), 7982–7987.

López-Villalta, L. and Muñoz-Cobo, M. (1996) Production techniques. In: *World Olive Encyclopedia*. International Olive Oil Council, Madrid, pp. 145–194.

Loupassaki, M., Chartzoulakis, K., Digalaki, N. and Androulakis, I. (2002) Effects of salt stress on concentration of nitrogen, phosphorus, potassium, calcium, magnesium, and sodium in leaves, shoots, and roots of six olive cultivars. *Journal of Plant Nutrition* 25, 2457–2482.

Lynch, J.P. and Ho, M.D. (2005) Rhizoeconomics: carbon costs of phosphorus acquisition. *Plant and Soil* 269, 45–56.

Maas, E.V. and Hoffman, G.J. (1977) Crop salt tolerance – current assessment. *Journal of the Irrigation and Drainage Division* 103, 115–134.

Marschner, H. (1995) *Mineral Nutrition of Higher Plants*, 2nd edn. Academic Press, London, UK.

Marschner, H. (2011) *Marschner's Mineral Nutrition of Higher Plants*, 3rd edn. Academic Press, London.

Marschner, P. (2012) *Marschner's Mineral Nutrition of Higher Plants*. Academic Press, London.

Marschner, P. and Rengel, Z. (2012) Nutrient availability in soils. In: Marschner, P. (ed.) *Marschner's Mineral Nutrition of Higher Plants*. Academic Press, London, pp. 315–330.

Mazeh, M., Almadi, L., Paoletti, A., Cinosi, N., Daher, E. *et al.* (2021) Use of an organic fertilizer also having a biostimulant action to promote the growth of young olive trees. *Agriculture* 11(7), 593. https://doi.org/10.3390/agriculture11070593

Mengel, K. and Kirkby, E.A. (2001) *Principles and Plant Nutrition*. Panima Publishing, Dordrecht, The Netherlands.

Michalopoulos, G., Kasapi, K.A., Koubouris, G., Psarras, G., Arampatzis, G. *et al.* (2020) Adaptation of Mediterranean olive groves to climate change through sustainable cultivation practices. *Climate* 8, 54. https://doi.org/10.3390/cli8040054

Mimoun, M.B., Loumi, O., Ghrab, M., Latiri, K. and Hellali, R. (2004) Foliar potassium application on olive tree. In: Badraoui, M., Bouabid, R., Ait-Houssa, A. and Hassan II, I.A.V. (eds) *IPI Regional Workshop on Potassium and Fertigation Development in West Asia and North Africa Region*. International Potash Institute, Zug, Switzerland.

Molina-Soria, C. and Fernández-Escobar, R. (2008) The reliability of the established critical leaf nitrogen concentration in olive orchards. *Acta Horticulturae* 868, 209–212.

Molina-Soria, C. and Fernández-Escobar, R. (2012) A proposal of new critical leaf nitrogen concentrations in olive. *Acta Horticulturae* 949, 283–286.

Morales-Sillero, A., Jiménez, R., Fernández, J., Troncoso, A. and Beltrán, G. (2007) Influence of fertigation in 'Manzanilla de Sevilla' olive oil quality. *HortScience* 42, 1157–1162.

Morales-Sillero, A., Jimenez, R., Fernandez, J., Troncoso, A. and Rejano, L. (2008) Effect of fertigation on the 'Manzanilla de sevilla' table olive quality before and after 'Spanish-style' green processing. *HortScience* 43, 153–158.

Morales-Sillero, A., Fernández, J., Ordovás, J., Suárez, M., Pérez, J. *et al.* (2009) Plant-soil interactions in a fertigated 'Manzanilla de Sevilla' olive orchard. *Plant and Soil* 319, 147–162.

Mouas Bourbia, S., Barré, P., Boudiaf Naït Kaci, M., Derridj, A. and Velde, B. (2013) Potassium status in bulk and rhizospheric soils of olive groves in North Algeria. *Geoderma* 197–198, 161–168. https://doi.org/10.1016/j.geoderma.2013.01.007

Muyen, Z., Moore, G.A. and Wrigley, R.J. (2011) Soil salinity and sodicity effects of wastewater irrigation in South East Australia. *Agricultural Water Management* 99(1), 33–41.

Neilsen, G., Neilsen, D. and Peryea, F. (1999) Response of soil and irrigated fruit trees to fertigation or broadcast application of nitrogen, phosphorus, and potassium. *HortTechnology* 9, 393–401.

Nieto, O.M., Fernández-Ondoño, E. and Castro, J. (2012) Sustainable agricultural practices for Mediterranean olive grove: effect of soil management on soil properties. *Spanish Journal of Soil Science* 2(1). https://doi.org/10.3232/SJSS.2012.V2.N1.05

Palese, A., Pasquale, V., Celano, G., Figliuolo, G., Masi, S. *et al.* (2009) Irrigation of olive groves in Southern Italy with treated municipal wastewater: effects on microbiological quality of soil and fruits. *Agriculture, Ecosystems & Environment* 129, 43–51.

Pardini, A., Faiello, C., Snowball, R., Mancuso, S. and Longhi, F. (2002) Cover crop species and their management in vineyards and olive groves. *Advances in Horticultural Science* 16(3–4), 225–234.

Perica, S. (2001) Seasonal fluctuation and intracanopy variation in leaf nitrogen level in olive. *Journal of Plant Nutrition* 24, 779–787.

Perica, S., Brown, P.H., Connell, J.H., Nyomora, A.M., Dordas, C. *et al.* (2001) Foliar boron application improves flower fertility and fruit set of olive. *HortScience* 36, 714–716.

Pestana, M., de Varennes, A. and Faria, E.A. (2003) Diagnosis and correction of iron chlorosis in fruit trees: a review. *Journal of Food Agriculture and Environment* 1, 46–51.

Regni, L., Nasini, L., Ilarioni, L., Brunori, A., Massaccesi, L. *et al.* (2017) Long term amendment with fresh and composted solid olive mill waste on olive grove affects carbon sequestration by prunings, fruits, and soil. *Frontiers in Plant Science* 7, 2042. https://doi.org/10.3389/fpls.2016.02042

Restrepo-Diaz, H., Benlloch, M. and Fernández-Escobar, R. (2008a) Plant water stress and K+ starvation reduce absorption of foliar applied K+ by olive leaves. *Scientia Horticulturae* 116, 409–413.

Restrepo-Diaz, H., Benlloch, M., Navarro, C. and Fernández-Escobar, R. (2008b) Potassium fertilization of rainfed olive orchards. *Scientia Horticulturae* 116, 399–403.

Reuter, D. and Robinson, J.B. (1997) *Plant Analysis: An Interpretation Manual.* CSIRO Publishing, Clayton, Australia.

Rodrigues, M.Â., Ferreira, I.Q., Claro, A.M. and Arrobas, M. (2012) Fertilizer recommendations for olive based upon nutrients removed in crop and pruning. *Scientia Horticulturae* 142, 205–211.

Rodrigues, M.Â., Correia, C.M., Claro, A.M., Ferreira, I.Q., Barbosa, J.C. *et al.* (2013) Soil nitrogen availability in olive orchards after mulching legume cover crop residues. *Scientia Horticulturae* 158, 45–51.

Rodrigues, M.Â., Coelho, V., Gouveia, M.E., Arrobas, M., Raimundo, S., Correia, C.M. and Bento, A. (2019) Nitrogen fertilization can significantly reduce the incidence of the olive fruit fly. *VI Jornada Nacional del Grupo de Olivicultura de la Sech.*

Römheld, V. (2000) The chlorosis paradox: Fe inactivation as a secondary event in chlorotic leaves of grapevine. *Journal of Plant Nutrition* 23, 1629–1643. http://dx.doi.org/10.1080/01904160009382129

Römheld, V. and Kirkby, E. (2010) Research on potassium in agriculture: needs and prospects. *Plant and Soil* 335, 155–180. http://dx.doi.org/10.1007/s11104-010-0520-1

Rouphael, Y. and Colla, G. (2020) Biostimulants in agriculture. *Frontiers in Plant Science* 11, 40. https://doi.org/10.3389/fpls.2020.00040

Samish, R.M. and Spiegel, P. (1961) The use of irrigation in growing olives for oil production. *Israel Journal of Agricultural Research* 11, 87–95.

Samish, R., Moscicki, W., Kessler, B. and Hoffman, M. (1961) A nutritional survey of Israel vineyards and olive groves by foliar analyses. Institute of Horticulture: Scientific activities 1981–1984, Department of Oleiculture and Viticulture, special publication no. 232. Agricultural Research Organization, Israel Ministry of Agriculture, Rishon Lezion, Israel.

Sánchez-García, M., Sánchez-Monedero, M.A., Roig, A., López-Cano, I., Moreno, B. *et al.* (2016) Compost vs biochar amendment: a two-year field study evaluating soil C build-up and N dynamics in an organically managed olive crop. *Plant and Soil* 408(1), 1–14.

Sánchez-Zamora, M. and Fernández-Escobar, R. (2002) The effect of foliar vs. soil application of urea to olive trees. *Acta Horticulturae* 594, 675–678.

Segal, E., Dag, A., Ben-Gal, A., Zipori, I., Erel, R. *et al.* (2011) Olive orchard irrigation with reclaimed wastewater: agronomic and environmental considerations. *Agriculture, Ecosystems & Environment* 140, 454–461. https://doi.org/10.1016/j.agee.2011.01.009

Sibbett, G. and Ferguson, L. (2000) Nitrogen, boron, and potassium dynamic in 'on' vs 'off' cropped Manzanillo olive trees in California, USA. *Acta Horticulturae* 586, 369–373.

Simpson, R.J., Oberson, A., Culvenor, R.A., Ryan, M.H., Veneklaas, E.J. *et al.* (2011) Strategies and agronomic interventions to improve the phosphorus-use efficiency of farming systems. *Plant and Soil* 349, 89–120.

Smith, S.E. and Read, D.J. (2010) *Mycorrhizal Symbiosis*. Academic press.

Subbarao, G.V., Ito, O., Berry, W.L. and Wheeler, R.M. (2003) Sodium: a functional plant nutrient. *Critical Reviews in Plant Sciences* 22, 391–416. https://doi.org/10.1080/07352680390243495

Tang, R.-J. and Luan, S. (2017) Regulation of calcium and magnesium homeostasis in plants: from transporters to signaling network. *Current Opinion in Plant Biology* 39, 97–105.

Tattini, M., Cimato, A. and Bertoni, P. (1989) Nitrogen nutrition of self-rooted olive in sand culture: effect of NH_4-N/NO_3-N on growth and nutritional status. *Acta Horticulturae* 286, 311–318.

Tellez-Rio, A., Vallejo, A., García-Marco, S., Martin-Lammerding, D., Tenorio, J.L. *et al.* (2017) Conservation agriculture practices reduce the global warming potential of rainfed low N input semi-arid agriculture. *European Journal of Agronomy* 84, 95–104.

Therios, I. (2006) Mineral nutrition of olive trees. In: *Proceedings of the 2nd International Seminar on Recent Advances in the Olive Industry*, pp. 403–410.

Therios, I.N. (2009) *Olives*. CAB International, Wallingford, UK.

Therios, I.N. and Sakellariadis, S.D. (1988) Effects of nitrogen form on growth and mineral composition of olive plants (*Olea europaea* L.). *Scientia Horticulturae* 35, 167–177. http://dx.doi.org/10.1016/0304-4238(88)90110-0

Thirkell, T.J., Cameron, D.D. and Hodge, A. (2016) Resolving the 'nitrogen paradox' of arbuscular mycorrhizas: fertilization with organic matter brings considerable benefits for plant nutrition and growth. *Plant, Cell & Environment* 39(8), 1683–1690.

Tilman, D., Cassman, K.G., Matson, P.A., Naylor, R. and Polasky, S. (2002) Agricultural sustainability and intensive production practices. *Nature* 418, 671–677.

Trinsoutrot, I., Recous, S., Bentz, B., Lineres, M., Cheneby, D. *et al.* (2000) Biochemical quality of crop residues and carbon and nitrogen mineralization kinetics under nonlimiting nitrogen conditions. *Soil Science Society of America Journal* 64(3), 918–926.

Tsadilas, C. and Chartzoulakis, K. (1997) Boron deficiency in olive trees in Greece in relation to soil boron concentration. *Acta Horticulturae* 474, 341–344.

Ulger, S., Sonmez, S., Karkacier, M., Ertoy, N., Akdesir, O. *et al.* (2004) Determination of endogenous hormones, sugars and mineral nutrition levels during the induction, initiation and differentiation stage and their effects on flower formation in olive. *Plant Growth Regulation* 42, 89–95. https://doi.org/10.1023/B:GROW.0000014897.22172.7d

Uriu, K. (1953) *Pistil Abortion in the Olive (Olea europaea L.) as Influenced by Certain Physiological Conditions*. University of California, Davis, California.

Wakeel, A., Farooq, M., Qadir, M. and Schubert, S. (2011) Potassium substitution by sodium in plants. *Critical Reviews in Plant Sciences* 30, 401–413. https://doi.org/10.1080/07352689.2011.587728

Weinbaum, S.A. (1988) Foliar nutrition of fruit trees. In: Neumann, P.M. (ed.) *Plant Growth and Leaf-Applied Chemicals*. CRC Press, Boca Raton, Florida, pp. 81–100.

Whalen, J.K., Hu, Q. and Liu, A. (2003) Compost applications increase water-stable aggregates in conventional and no-tillage systems. *Soil Science Society of America Journal* 67(6), 1842–1847.

Wiesman, Z. (2009) *Desert Olive Oil Cultivation: Advanced Biotechnologies*. Elsevier, New York.

Xiloyannis, C., Celano, G., Palese, A., Dichio, B. and Nuzzo, V. (2000) Mineral nutrient uptake from the soil in irrigated olive trees, cultivar Coratina, over six years after planting. *Acta Horticulturae* 586, 453–456.

Zipori, I., Yermiyahu, U., Ben-Gal, A. and Dag, A. (2011) Response of oil-olive trees to iron application. *Acta Horticulturae* 888, 295–300.

Zipori, I., Yermiyahu, U., Erel, R., Presnov, E., Faingold, I. *et al.* (2015) The influence of irrigation level on olive tree nutritional status. *Irrigation Science* 33, 1–11. https://doi.org/10.1007/s00271-015-0465-5

Zipori, I., Yermiyahu, U., Tugendhaft, Y., Ben-Gal, A. and Dag, A. (2016) The response of olive (*Olea europaea*) trees to zinc nutrition. *Acta Horticulturae* 1199, 351–356.

Zipori, I., Dag, A., Laor, Y., Levy, G.J., Eizenberg, H. *et al.* (2018) Potential nutritional value of olive-mill wastewater applied to irrigated olive (*Olea europaea* L.) orchard in a semi-arid environment over 5 years. *Scientia Horticulturae* 241, 218–224. https://doi.org/10.1016/j.scienta.2018.06.090

Zipori, I., Erel, R., Yermiyahu, U., Ben-Gal, A. and Dag, A. (2020) Sustainable management of olive orchard nutrition: a review. *Agriculture* 10, 11. https://doi.org/10.3390/agriculture10010011

Zohary, D. and Spiegel-Roy, P. (1975) Beginnings of fruit growing in the Old World: olive, grape, date, and fig emerge as important Bronze Age additions to grain agriculture in the Near East. *Science* 187(4174), 319–327.

6.4 Olive Training Systems and Pruning

F. Famiani[1], N. Cinosi[1], P. Proietti[1], T. Caruso[2], D. Neri[3], P. Pierantozzi[4], M. Msallem[5] and E.M. Lodolini[3]
[1]*Dipartimento di Scienze Agrarie Alimentari e Ambientali, Università degli Studi di Perugia, Italy;* [2]*Dipartimento di Scienze Agrarie, Alimentari e Forestali, Università degli Studi di Palermo, Italy;* [3]*Dipartimento di Scienze Agrarie, Alimentari e Ambientali, Università Politecnica delle Marche, Ancona, Italy;* [4]*Instituto Nacional de Tecnología Agropecuaria, San Juan, Argentina;* [5]*International Consultant – Olive Development, Tunis, Tunisia*

6.4.1 Importance of Pruning

Pruning is one of the most important cultural practices in olive growing. Correct pruning results in rapid growth of young plants, and abundant and steady production of high-quality fruit and oil in adult trees. After harvesting, it is the second most expensive cultural practice, leading to the need to reduce its cost through simplified/mechanized operations.

Modern pruning must adapt to the changes taking place in olive growing in terms of new plantations, with the number of trees per hectare ranging from 200–400 (intensive olive orchards) to 1200–2500 (super-high-density, SHD, olive orchards), and increasing application of irrigation, mechanization (particularly harvesting) and treatments against diseases and pests (operation and effectiveness of spray must be guaranteed). Moreover, the maximum safety of pruners must be ensured.

Pruning is also used to reshape and rejuvenate (old) trees and, in cold areas, for recovery from frost injuries.

In this chapter, the main references are reported in the reference list. Only very few specific ones are reported directly in the text.

6.4.2 General Principles for Olive Pruning

Pruning is the cultural technique which has the greatest impact on production of assimilates, through photosynthesis, and the vegetative-reproductive balance of trees. Hence, it has a great effect on their potential growth and fruiting.

In young trees with a small-sized canopy, assimilate production is maximized by minimizing pruning intensity (allowing maximum leaf surface area), by only eliminating vegetative formations, which, if not eliminated, would damage their structure.

In adult trees, assimilate production through photosynthesis per unit of surface area (i.e. hectare) is maximized by achieving the best combination between total leaf surface area and light interception, temperature and water and nutrient availability. Pruning has a direct effect on leaf surface area and light interception, whereas temperature and water availability depend on the environment (amounts and distribution of rainfall) and the possibility of irrigation.

The best agronomic results (abundant and regular yield) are obtained when the orchard is designed with a planting density and

DOI: 10.1079/9781789247350.0019

tree size (obtained with pruning) that enables the maximum canopy volume per hectare that can be sustained by the environmental conditions and cultural practices applied.

The choice of training system and pruning must aid mechanization, particularly mechanical harvesting. This is done by choosing the training system and by structuring/shaping the trees in such a way that they will be suitable for the chosen harvesting system.

Finally, safety of pruners must be maximized by eliminating the use of ladders and not climbing on the trees. Pruning must be done directly from the ground by using hand-held tools and evaluating possibilities for mechanization.

6.4.3 Main Training Systems

The efficiency of the canopy increases with a high light-exposed surface over the occupied volume and this depends on the shape and size of the canopies.

It is important to note that every training system has advantages and disadvantages. Several factors need to be considered in selecting the most suitable one before planting the orchard: the harvesting system (manual, facilitated with hand-held tools or mechanical), the fruit destination (oil or table), the size of the farm, the variety (vigor and growth habit), the plant density, and the labor availability. In choosing the training system for new orchards, the planting model is very important: intensive, which can be at low, medium or high density, or super-high-density. For more information on planting models see chapters 5.1 and 5.2.

The main training systems for olive are summarized in Table 6.4.1 with the main features, advantages and disadvantages.

6.4.3.1 Considerations on the main training systems

Vase, together with globe, is the most widespread training system in traditional orchards. In the vase, the central part of the canopy is relatively empty, and the vegetation is then spread around the different main (primary) branches, ensuring high canopy lighting. For this reason and the fact that, if the canopy is well structured, trunk shakers can be used for very efficient mechanical harvesting, it is the most common training system in intensive olive orchards.

There are different ways to shape the canopy within the vase training system: the inverted cone vase, the cylindrical cone vase, the multi-cone vase and the free vase. The inverted cone vase encourages growth of vegetation in the upper part of the trees. This is sometimes used in environments very susceptible to peacock eye and with varieties very sensitive to this disease. It is known that peacock eye especially infects the lower portions of the canopy, which are reduced in this system. It is sometimes used in young orchards in the Middle East with varieties very sensitive to peacock eye and environments favorable to such disease. The canopies tend to escape upwards. Therefore, pruning must contrast this tendency and maintain the tree in balance.

In the cylindrical vase, secondary branches are trained to obtain a continuous cylindrical canopy. This shape, with respect to other kinds of vases, enables a larger volume of the canopy to be obtained at parity of tree height. However, for good lighting of the canopy it entails that trees be widely spaced (Fig. 6.4.12). If the trees are close, the canopy forms a continuous vertical external wall, creating reciprocal shading and difficulty for the light to reach the basal portions of the canopy (Fig. 6.4.13). For this reason, the cylindrical vase is usually used in traditional orchards with widely spaced trees. In intensive olive orchards it can be used where the space between canopies avoids reciprocal shading. Considering its limited height, it is sometimes used with table olive varieties which are manually harvested (i.e. cultivar Ascolana Tenera in central Italy).

The multi-cone (or polyconic) vase is structured in order to divide the vegetation into cones (one for each main branch). This is obtained by training secondary branches with decreasing length from the bottom to the top of the main branches (Fig. 6.4.14). The different length of the secondary branches (shorter at the top and longer at the bottom) creates conditions for good lighting of the whole external parts of the canopy,

Table 6.4.1. Olive training systems with the description of the main features and advantages and disadvantages.

Training system	Main features	Advantages	Disadvantages
Vase (traditional and intensive orchards) (Figs 6.4.1 and 6.4.2)	Single trunk and 3–5 primary branches Less vegetation in the central portion of the canopy Increasing length of the secondary branches from the apical to the basal portion of the canopy	Uniform light distribution Suitable for different growth habits Suitable for olive and table varieties Simplified and selective pruning Suitable for harvest with trunk shakers (important: single trunk 1.0–1.2 m high, primary branches inserted with 30–45° angle with respect to the vertical, secondary branches relatively short and rigid)	Less suitable for mechanical pruning
Vasebush (traditional orchards) (Figs 6.4.3 and 6.4.4)	Primary branches originating directly from the stump/ ground (absence of a single trunk) Used for restructuring pruning through coppicing (especially in frost-damaged trees) and starting from new emitted suckers Secondary branches as in the vase	Limited height Suitable for different growth habits Suitable for olive and table varieties	Unsuitable for mechanical harvesting with trunk shakers
Globe (traditional orchards) (Figs 6.4.5 and 6.4.6)	As in the vase but without empty space in the central part of the canopy It can also be the consequence of bad or no pruning.	Very common in climates with high sunlight level to prevent sunburning of the branches	Vegetation concentrated in the peripheral portions of the canopy, with heavy shading of the internal parts, leading to defoliation Requires skilled labor for pruning to prevent self-shading
Monocone (intensive orchards) (Fig. 6.4.7)	Single-axis 4–5 m high First 1–1.2 m of the axis free of ramifications Primary branches horizontally inserted, helicoidally arranged and with increasing length from the top to the base of the central leader (conical shape)	Suitable for integrated partial mechanical pruning Suitable for mechanical harvesting with trunk shakers	Less suitable for varieties with expanded or pendulous growth Requires skilled labor for pruning to maintain the size of the canopy and prevent self-shading in adult trees Difficulty in maintaining a good balance between vegetative and reproductive activities, especially if trees are closely spaced along the row

Continued

Table 6.4.1. Continued.

Training system	Main features	Advantages	Disadvantages
Central leader (super-high-density orchards) (Figs 6.4.8 and 6.4.9)	Central leader free of vegetation for the first 50–60 cm from the ground Tree height and canopy width limited to the size of the harvesting machine Lateral branches selected among those developing naturally on the central leader Branches are mainly selected along the row to create the final skeletal structure of the canopy, while long and rigid lateral branches developing into the inter-row are removed creating a flattened canopy (hedgerow) along the row.	Suitable for harvesting with over-the-row machines Simplified and selective pruning can be easily applied. Suitable for mechanical pruning (especially topping) integrated with simplified and selective interventions	Less/not suitable for varieties with medium/ high vigor, expansive growth, low branching ability and rigid branches Requires skilled labor to manage pruning intensity, frequency and integration with/ application of mechanical pruning
Multi-axis - Smarttree (super-high-density orchards) (Fig 6.4.10)	Absence of a central leader Several branches are trained in order to obtain a small globe initially and then a flattened fan conformation along the row. Periodical hedging and topping to keep the canopy size within the harvester limits	Suitable for harvest with over-the-row machines Suitable for mechanical pruning	Excessive clustering of the vegetation due to repeated mechanical pruning
Free palmette (intensive high-density orchards) (Fig. 6.4.11)	Central leader free of vegetation for the first 50–60 cm from the ground The growth of stronger branches must be directed along the row to form a continuous wall (hedgerow) Up to 3–4 m in height and 1.5–2.5 m in width of the hedgerow depending on the vigor and growth habit of the cultivar and the type of harvesting machinery used	Suitable for harvesting with large over-the-row or side-by-side machines Simplified and selective pruning can be easily applied Suitable for mechanical pruning (especially topping) integrated with simplified and selective interventions	Less/not suitable for varieties with high vigor Requires skilled labor to manage pruning intensity, frequency and integration with mechanical pruning

Fig. 6.4.1. A tree trained to vase in an intensive olive orchard in central Italy. (Photo: Franco Famiani)

Fig. 6.4.2. Trees trained to vase in a traditional olive orchard in central Italy. (Photo: Franco Famiani)

Fig. 6.4.3. Trees trained to bush vase in a traditional olive orchard in central Italy. (Photo: Franco Famiani)

Fig. 6.4.4. Trees trained to bush vase in a traditional olive orchard in Andalusia, Spain. (Photo: Franco Famiani)

Fig. 6.4.5. Trees trained to globe in a traditional olive orchard in the southern Tunisia. (Photo: Franco Famiani)

Fig. 6.4.6. A tree trained to globe in a traditional olive orchard in Syria. (Photo: Franco Famiani)

Fig. 6.4.7. Trees trained to monocone in an intensive olive orchard in central Italy. (Photo: Franco Famiani)

Fig. 6.4.8. A tree trained to central leader in a super-high-density olive orchard in central Italy. (Photo: Enrico Maria Lodolini)

Fig. 6.4.9. A row of trees trained to central leader in a super-high-density olive orchard in Spain. (Photo: Enrico Maria Lodolini)

Fig. 6.4.10. Trees trained according to a multi-axis system in a super-high-density olive orchard in central Italy. (Photo: Franco Famiani)

Fig. 6.4.11. A tree trained to free palmette in an intensive high-density olive orchard in Sicily, Italy. (Photo: Franco Famiani)

Fig. 6.4.12. A tree trained to cylindrical vase in a traditional olive orchard in Abruzzo Region (Italy), where trees are widely spaced. (Photo: Franco Famiani)

especially the lower portions, even when the trees are close. Moreover, the partitioning of the vegetation among the different main branches, creating discontinuity, greatly contributes to increasing the external canopy surface-to-volume ratio, favoring good lighting of the whole canopy, especially when the trees are closely spaced (Fig. 6.4.15).

Although the free vase is the most recommended in intensive olive orchards worldwide, there is no clear definition and description of this kind of vase. Herein, we will attempt to fill this gap. A free vase has the full functionality of the vase shape but has less geometrical regularity. In fact, it is obtained and maintained by only pruning parts, which, if not pruned, would reduce the functionality of the canopy. Aesthetic aspects are not taken into account. Functionality includes the structure of the scaffold and secondary branches that maximize the efficiency of mechanical harvesting with trunk shakers and the maintenance of good vitality of all the components of the tree, included the internal and dorsal parts of the main branches. As the leading criterion is functionality, in training pruning, supernumerary main branches are kept until the time when, if not removed, they would damage the structure of trees. In production pruning, functionality is maximized and maintained by ensuring a good structure of the main branches, the right volume and lighting of the whole canopy, and good rigidity of the scaffold to enable efficient harvesting with the trunk shaker. Good functionality and lighting are mainly guaranteed by the correct control of the apical portion of the canopy, which must be narrower than the lower part. Likewise, it is important there is good inclination of the main branches which must favor the growth of productive

Fig. 6.4.13. Trees trained to cylindrical vase in an intensive olive orchard in central Italy, where trees are closely spaced. (Photo: Franco Famiani)

Fig. 6.4.14. A tree trained to multi-cone or polyconic vase in an intensive olive orchard in central Italy. (Photo: Franco Famiani)

Fig. 6.4.15. Comparison of the cylindrical vase (*top*) and polyconic vase (*bottom*). The shape of the canopy of the polyconic vase creates conditions for good lighting of the canopy, especially the lower portions, even with closely spaced trees. It also determines a high external canopy surface-to-volume ratio, which also favors good lighting of the canopy. (Drawing: Franco Famiani)

vegetation and prevent excessive vegetative growth (vigorous suckers) in the central part of the vase. However, some vegetation (weak suckers and small branches) must be kept in the central portion of the canopy in order to shade the upper part of main brunches to avoid sun-burning. The rigidity and suitability of the canopy for using trunk shakers is ensured by a not-too-wide inclination of the main branches (which, as seen above, is also useful for good functionality of the canopy) and by avoiding the formation of pendant branches. This description of the free vase resembles the multi-cone vase in terms of structure, but has a less regular (aesthetical) shape (Fig. 6.4.16). It has high functionality, especially with relatively close planting distances, as is the case with intensive olive orchards. The free vase must also enable rapid and simplified pruning, which is an indispensable aspect on large farms, also to reduce manpower.

The vase training system is also often used in traditional orchards, where large size trees can be found. As the space between trees increases, the size of the trees also increases, through progressive enlargement of the main branches. This causes a progressive increase in the angle of inclination of the main branches, creating conditions for increasing risk of burning of the bark of the upper parts of the main branches and the emission of vigorous suckers, with consequent waste of sap and excessive vegetative activity. To some extent, these problems can be mitigated by allowing some small branches to grow in the internal part of these vases to shade the bark and reduce the emission of suckers. However, if the size becomes very large, the vegetation inside the vase becomes very abundant leading toward the globe system, which is characterized by a spherical canopy, which protects the main branches from burning. For this reason, the globe system is used in traditional olive orchards, especially in very sunny areas and with very large trees.

The multi-trunk and bush vases have multi-stems, which, respectively, derive from the system of planting in Spain (Andalusia) and, in most cases, in Italy and other countries, from coppicing, after the aerial

Fig. 6.4.16. Trees trained to free vase in an intensive olive orchard in central Italy. (Photo: Franco Famiani)

part of the tree has been completely damaged, usually by freezing in cold areas of cultivation. These training systems have spread because a large canopy can develop in a short time. This is especially important when the distance among trees is large. Nowadays, these systems are still present in traditional olive orchards. In some cases, restructuring pruning is used to bring the trees to two or one trunk in order to enable mechanical harvesting with trunk shakers (with a single trunk, the trunk shaker can also be combined with an inverted umbrella to also mechanize the collection of detached fruits).

The monocone training system has been used mainly in Italy and also tried in other countries, but it has not spread. This is likely due to the fact that as the canopy grows there is progressive shading of its internal parts and the vegetation tends to grow laterally as well as in height. This makes resizing (intense pruning) necessary, which imbalances the trees, leading to excessive vegetative activity and irregular production.

The central axis, the multi-axis (Smarttree) and the free palmette are training systems used for hedgerow systems. They are used for intensive high-density or SHD orchards. The size and the shape of the canopy ensure good lighting. Hedgerows enable the use of straddle machines, which are very efficient for harvesting. For the free palmette, the use of lateral continuous harvesting can also be considered. The success of these training systems largely depends on the variety, which should have low vigor, especially for SHD orchards. For medium-vigorous cultivars, slightly wider spacing of the trees along and between rows is used (650–1000 trees/ha). For low-vigor cultivars, spacing is reduced and, as a result, the number of trees per hectare is increased (1200–2500 trees/ha). Once harvesting is mechanized with a straddle machine, pruning is the most expensive operation, accounting for up to half of the total production costs. Pruning must ensure a good balance along the entire height of the hedgerow and a size of the vegetation wall that enable the efficient use of straddle machines. With hedgerow training systems pruning can be, at least partially, mechanized.

6.4.4 Physiological Basis of Pruning

6.4.4.1 Growth habit

The olive has a bush-like growth habit and single-trunk tree training derives from modification of the natural habit through a series of pruning operations. Growth is basitonic with the basal portion of the canopy growing more vigorously than the apical part, so that in combination with the presence of multiple stems originating from the stump, the canopy appears wider than high.

The root system of the olive has a high transmigration ability to access favorable portions of soil to adsorb water and nutrients. This ancestral trait enables its survival in adverse climatic conditions (arid or semiarid climates) but represents a high energetic cost for the tree.

6.4.4.2 Canopy-to-root ratio

There is a close correlation between canopy size and root system (canopy-to-root ratio), which tends to remain stable for each growing condition (for example, in arid/rainfed conditions this ratio is lower than in wet/irrigated ones). This means that if, for some reason (damage of the aerial part by freezing temperature, heavy disease attacks, etc.) the canopy is reduced, the trees invest their reserves to promote the regrowth of the canopy and restore the canopy-to-root ratio. This is also the case when roots are damaged, i.e. by tilling too deeply; in this case, the trees invest their reserves to promote the regrowth of the roots and restore the canopy-to-root ratio. Likewise, if pruning is too intense, trees will use their reserves to promote the regrowth of the canopy. When the trees have to restore the right volume of canopy, they produce suckers which are vigorous shoots enabling rapid regrowth since they have only vegetative buds. Therefore, pruning intensity should be matched to the specific pedo-climatic and cultivation conditions in order to maintain the canopy-to-root balance, which means avoiding imbalance between canopy and roots, resulting in the use of resources for the growth of canopy or roots at the expense of reproductive activity (i.e. fruit production). Only in particular situations, such as when rejuvenation or reshaping the canopy are needed, an imbalance caused by very intense pruning can be useful because it stimulates a great emission of vigorous shoots (suckers), which will then be used to rejuvenate/reshape the canopy. In this case, depending on the intensity of the operation, two–four years will be necessary to recover a suitable vegetative/reproductive ratio.

6.4.4.3 Physiological stages during the life cycle of the olive tree

The olive tree has a very long life cycle, even hundreds of years. In nature, plants originating from seed pass through a juvenile stage that can last several years. This stage is characterized by the lack of flowering and by wild-like morphological characteristics (vigorous upright shoots, irregular growth and internode length, three to four leaves per node, small, thick and dark-green leaves, etc.). During this stage the plant invests energy in root development and canopy growth aimed at occupying the surrounding space as quickly as possible, because larger size increases survival. As elongation and branching of the roots and shoots continue, the plant enters the adult stage, which, from a physiological point of view, is the stage of reproduction.

Commercial trees used for establishing new orchards are propagated from mature tissues to avoid the juvenile stage and anticipate the onset of fruit production. In adult trees, juvenility does not completely disappear but persists in the oldest portions of the canopy and especially the stump.

In fact, in case of coppicing at the base of the trunk, resprouting from latent or adventitious buds will produce juvenile shoots (suckers). This is functional, because in these conditions the most important aim of the tree is to regrow the canopy in the shortest possible time, and juvenile vegetation, which does not produce inflorescences (only vegetative buds), makes this possible.

In adult trees it is necessary to balance the vegetative growth (of the canopy and the root system) with fruiting, which is the strongest sink for carbohydrates. When reproductive activity is too high (maximum fruit load) compared to vegetative activity (formation and development of shoots), the growth of new shoots, which is the basis of the following year's yield, and flower differentiation are reduced, affecting fruit production in the following year (alternate bearing). On the other hand, when fruit production is reduced there is excessive vegetative activity. Consequently, in order to maintain constant yields over the years, it is necessary to balance these two activities through a correct application of cultivation practices, *in primis* pruning.

Over the years the tree ages (senescence stage) showing a general reduction in vegetative and reproductive activity, but especially in the vegetative one. This process can be counteracted by pruning a bit more intensely to stimulate vegetative growth. If the trees have accumulated a lot of wood and have very low vegetative activity, very intense extraordinary pruning can be executed to reduce wood, eliminate decayed portions and stimulate the emission of new vigorous shoots (suckers) to rejuvenate the canopy (rejuvenating pruning).

6.4.4.4 The importance of light in the canopy

Regular distribution of light inside the canopy is important to prevent consistent portions from being constantly shaded and hence losing their functionality, leading towards progressive aging (arrest of renewal, absence of fruits, leaf drop, etc.). Heavily shaded leaves are a cost for the plant as they consume more than they produce and, therefore, are eliminated over time (phylloptosis). Well-lighted exposed leaves, on the contrary, have a high net photosynthetic rate and produce carbohydrates that are translocated to all the organs of the trees (fruits, shoots, roots, etc.).

Good light distribution inside the canopy also affects the induction and differentiation of flower buds; in fact, if light availability is constantly below 30% of full sunlight, no flower formation will occur. Furthermore, the fruit located in poorly lighted portions of the canopy shows lesser growth and oil accumulation and quality of the extracted oil is lower.

Pruning prevents branches from excessive shading, promotes their distancing and favors optimal arrangement within the space. Homogeneous light distribution in the internal and basal portion of the canopy enables foliage photosynthetic activity to be maintained. Moreover, lighting promotes the formation of adventitious buds on primary and secondary branches, which can be helpful for the emission of new shoots for vegetative renewal.

6.4.4.5 Canopy architecture and organization

The olive canopy is composed of a woody scaffold bearing vegetative foliage with fruit-bearing shoots. The primary branches are inserted on the main trunk(s) or central axis/leader and increasing orders of branches originate from the previous ones.

There is a hierarchy in the arrangement of the branches which is repeated for every branch order and at every level of the canopy. The terminal portion of the main axis of trees trained to central axis or of the main branches of trees trained to vase has the advantage of attracting more nutrients compared to lower ones and exerts a control over the formation and inclination of the lateral branches (epinasty). The degree of this control is expressed with an intensity positively correlated to the order of the branch and to the height of insertion and negatively to the inclination of the branch itself: the more vertical, the greater vigor and influence on laterals (apical dominance); the more inclination, the lower is the vigor and effect on laterals (and greater is the reproductive aptitude).

If the terminal portion of the primary branches is left unpruned, growth concentrates in the upper part of the canopy and rapid self-shading and physiological decline (senescence) takes place in the lower portions of the canopy, even in young olive trees, especially if planting distances are close. In other words, too many shoots at the top cause excessive nutrient attraction and vegetative growth in this area to the detriment of the basal portions. Thinning the top portions is therefore required to reduce the sink of attraction for nutrients and promote development of the lower parts of the canopy. On the contrary, extreme shoot thinning leads to vigorous vegetative re-sprouting in the apical portion that has been cut or thinned excessively. Therefore, in practice, vegetation in the highest parts of the main axis or main branches has to be thinned, but should remain globose, i.e. not too thin. Accordingly, in hedgerow systems the apical portion of the hedge must not be too dense nor too thin. In general, the secondary branches must have a wider angle of insertion and a shorter length than the primary ones.

6.4.4.6 Fruiting branches and reproductive structures

The fruiting branches are usually bent because of the weight of the vegetative growth and the fruit production. Figure 6.4.17 shows the following types of shoots: suckers, vegetative, mixed, fruiting and exhausted. Among them, the main fruiting structure is the mixed shoot which consists of a one-year-old portion having fruits in the current season and a terminal vegetative portion bearing fruits in the following season (Fig. 6.4.18). Since the distinction of the aforementioned structures is not so clear within the foliage of the fruiting branch, it is advisable to observe areas/zones with different functions so that moving from the inner to the outer part of the canopy, the vegetative area (suckers and vegetative shoots, which mainly produce vegetation), the mixed area (vegetative and mixed

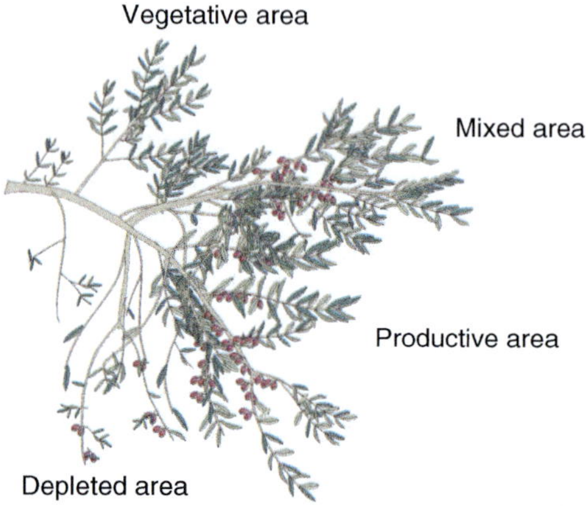

Fig. 6.4.17. A fruiting branch of an adult olive tree, showing different zones: vegetative (suckers, which, if vigorous, are eliminated with pruning; if weak they will drop due to the increasing weight of the new growth and will transform into the mixed zone); mixed (both vegetative and fruiting shoots); productive (both mixed and fruiting shoots); and depleted (exhausted shoots, which have fruited and do not have or have very short new growth which is not enough to ensure a significant production). (Drawing: Francesca Massetani and Enrico Maria Lodolini)

Fig. 6.4.18. Mixed shoots which consist of a one-year-old portion bearing fruits in the current season and a terminal vegetative portion which will bear fruits the following season. (Photo: Franco Famiani)

shoots, in which both vegetation and fruits are produced), the productive area (mixed and fruiting shoots, in which fruits are mainly produced but also new shoots can grow) and the depleted area (prevalence of exhausted shoots that have completed their productive cycle) are recognizable. The different areas have different levels of inclination and branching. Over time, the whole structure increases the downward inclination and the vegetative zone progressively evolves into the mixed, the productive and, finally, the exhausted one (Fig. 6.4.17).

There are significant differences in the structure and temporal evolution of the fruiting branches of olive varieties and therefore pruning must be adjusted accordingly.

6.4.4.7 Effects of branch inclination

As the inclination of lateral branches increases, there is a progressive decrease in the vegetative activity and an increase in the reproductive one (fruiting) (Fig. 6.4.17). This is important in the selection of branches to remove/thin during production pruning.

Inclination is also very important in managing primary branches during both training and production pruning. When trees are trained to vase, the primary branches are inserted in the trunk with an angle that usually varies from 35° to 45° with respect to the vertical. Then, as the branch grows in length/height, this angle can be maintained constant or, more often, as the branch moves away from the trunk the direction of growth is progressively guided towards the vertical. The inclination of the main branches determines important effects on the growth of secondary branches and on the effectiveness of the transmission of vibrations when using trunk shakers for harvesting. If the branch is too open, the growth of very vigorous shoots/branches (suckers) are favored on the upper side, while the shoots/branches on the lower side are weak (Fig. 6.4.19). This phenomenon is known as 'epithony' and must be considered in both training and production pruning. It is important to note that if the inclination of a main branch is wide and maintained throughout its length, it favors the formation of suckers and disadvantages the growth of vegetation in the external part of the canopy, where fruiting should occur. Often, this leads to having good vegetation only on the distal part of the main branches. This makes it important, as branches move away from the trunk, to progressively guide their direction of growth towards the vertical. This, in vase trained trees, reduces the formation of vigorous suckers in the internal part of the canopy and favors the growth of fruiting vegetation in the external part.

Fig. 6.4.19. Main branches that are too open favor the formation of suckers in the internal part of the canopy (upper side of branches) and hamper the growth of vegetation in the external part of the canopy (lower, external side of branches), where fruiting should occur. (Photos: Franco Famiani and Enrico Maria Lodolini)

6.4.5. Main Types of Pruning

In relation to the purpose of pruning, it is possible to distinguish the following types:

1. training pruning (young trees);
2. production pruning (adult trees);
3. rejuvenating/reshaping/regenerating pruning (considered an exceptional intervention).

6.4.5.1 Training pruning

The objectives of training pruning are to build a solid and functional scaffold in the shortest possible time, enabling rapid entry into the adult stage and therefore full production.

Since new plantations are mainly represented by intensive and SHD orchards, indications for correct training pruning are given for these two kinds of plantations. However, the principles and criteria have a general applicability.

During the entire training period, as the main aim is to accelerate tree growth as much as possible, thinning should be reduced to a minimum in order to maximize leaf surface area and, hence, photosynthetic assimilate production. Therefore, cuts must be limited to eliminating vegetative formations, which, if not eliminated, would damage the structure of the tree.

Fig. 6.4.20. A two-year-old tree at planting, which is well structured for training to vase, with a single trunk about 1 m tall and several lateral shoots in the apical portion. (Photo: Franco Famiani)

Training pruning in intensive olive orchards

The free vase is the main training system used in intensive olive orchards. Two-year-old trees are the best for establishing the new plantation (Fig. 6.4.20). They are obtained in the nursery by training the young trees to a single trunk 1–1.2 m in height (minimum 0.8 m), with several lateral shoots in the apical portion. This trunk height will enable a trunk shaker to be used in the future. It is also suitable when harvesting is initially planned to be done manually or with hand-held machines (i.e. in small farms), because, considering the long cycle of the olive orchard, the mechanization of harvesting may become necessary in the future (Fig. 6.4.21).

No pruning is done at the time of planting. Only if there are vigorous branches (which can be recognized because they are relatively long and have a narrow insertion angle and a relatively large diameter) along the trunk is it necessary to remove them (Fig. 6.4.22). The weak ones can be left and will be removed in one to two years, before they reach a size that requires a cut that could affect the normal development of the trunk.

The tree must be secured to a stake with two to three ties using tubular plastic laces or non-rigid material, which must be checked periodically as the tree grows.

Initially, all the lateral shoots/branches above 90–100 cm in height are maintained (supernumerary branches) in order to have

Fig. 6.4.21. Trees in an intensive olive orchard in Algeria with very short trunks, which do not allow the use of a trunk shaker for harvesting. This can also be the result of using one-year-old trees for planting, which, if not trained to obtain a trunk 0.8–1.2 m tall, ramify in the basal portion of the stem. (Photo: Franco Famiani)

Fig. 6.4.22. Vigorous branches along the trunk must be removed when they are young, otherwise they can damage the growth of the trunk. Note the two lateral branches that have negatively affected the growth of the trunk: compare the diameter of the trunk before and after their insertions. They must be removed to have a trunk height at least 0.8 m, better 1.0–1.2 m. The tree before (*left*) and after (*right*) their elimination. (Photos: Franco Famiani)

a high leaf surface area that favors the growth of the young trees. Then the supernumerary branches must be progressively eliminated in three to five years (depending on the growth rate of the trees), in order to obtain the final number of main branches which will form the scaffold of the vase, which depends on the final size of the trees. With distances ranging from 5–7 m between trees, four to five main branches are suggested. The main branches must insert on the trunk with an angle with respect to the vertical not greater than 45° (30–45°) in order to obtain high efficiency with mechanical harvesting with trunk shakers. This inclination is used to ensure the lateral development required in order to occupy the available space for each tree, then the development of the main branches continues in a more or less vertical direction up to the final height of the tree (4–5 m). This structure is very important to ensure good formation of vegetation in the external part of the canopy and to minimize the development of suckers in the internal part of the vase. The following considerations can help to better understand the importance of the inclination of the primary branches. When training the primary branches, the structure that guarantees the best vibration transmission is that of a branch which continues with the same inclination for its entire length, but this tends to cause the apical portion of the branch to grow too externally with the risk of shading the lower portion. Moreover, because of the epithony phenomenon, the emission of suckers is favored in the upper part of the branch and the vigor of the branches developed in the lower part of the branch, which are those that ensure production, is reduced. Therefore, a good compromise is for the main branches to be inserted on the trunk at 30–45° (lower for lower planting distances and *vice versa*) and then, after initial growth at this inclination, train the main branch to an inclination that tends to the vertical. This allows both the development of good external vegetation and a structure that is responsive to trunk shaking.

The direction of growth of the main branches can be guided very easily by widening the angle of growth of excessively vertical ones, which tend to take over the others, and reducing the angle of those that are too weak or too open: by shortening the main branch in correspondence to an outward branching if it is necessary to widen the angle of inclination, or in correspondence to an inward branching if it is necessary to reduce such angle.

As the main branches grow, secondary branches will also develop. Interventions on secondary branches are limited to shortening/eliminating the lowest ones, which cause problems for managing the soil and those which cause excessive shading. The secondary branches must have a wider insertion angle with respect to the vertical than the main ones. This enables a hierarchy between primary and secondary branches to be maintained and to avoid that secondary branches compete with primary ones.

During the entire training period, vigorous suckers or branches which grow in the internal part of the vase must be removed as they tend to grow vertically and vigorously causing excessive shading and enlargement/inclination of lateral branches from which the main ones are chosen and trained (Fig. 6.4.23). Only in some cases, such as in cultivars having an erect vegetative habit, it can be useful to leave some internal vegetation during training because lateral branches tend to grow too erectly.

In some situations, one-year-old trees are used for the plantation. In such cases, trees must be tied to the stake to ensure rapid vertical growth, necessary to reach the height for the development and training of the primary branches (Fig. 6.4.24). In the first year after planting, pruning is limited to eliminating any vigorous shoots on the lower part of the stem, while the weak ones are left. When the main axis exceeds 1.2–1.3 m in height, the apex is cut in order to stimulate the growth of lateral shoots in its apical portions obtaining a tree structure similar to that of two-year-old trees obtained from nurseries. Again, vigorous lateral branches in the middle-lower parts of the trunk must be eliminated (Fig. 6.4.25). Then the training pruning continues as described above.

Fig. 6.4.23. Branches growing in the internal part of the canopy of the young vase must be eliminated as they tend to grow vertically and vigorously causing excessive shading and enlargement/inclination of lateral branches from which the main ones are chosen and trained. The tree before (*left*) and after (*right*) their elimination. (Photos: Franco Famiani)

The use of one-year-old trees is discouraged for establishing intensive olive orchards, because, with respect to two-year-old trees, they prolong the time required to reach the adult stage. Moreover, in small plants, potential attacks by *Prays oleae* or *Palpita unionalis* are much heavier, because the damage they cause at the apex of the main axis of the tree causes the development of a number of new shoots, which make it much more difficult/slower to train a single trunk up to the height of 1–1.2 m. During the training stage, other pests and diseases, such as acarus, pyralid moth (*Euzophera pingüis*), olive knot (*Pseudomonas savastanoi*) and peacock eye spot (*Venturia oleaginea*), must be kept under control. Sometimes, the use of one-year-old trees is the reason for a short trunk, because the trees tend to ramify in the lower portion of the stem (Fig. 6.4.21). Therefore, the use of one-year-old trees requires careful training pruning and good control of biotic adversities.

Training pruning of trees in SHD orchards

Central leader and multi-axis (Smarttree) are the two main training systems. Pruning is essential for managing SHD systems. Training pruning must ensure rapid tree growth and hedgerow formation.

Central leader

To quickly build up the hedgerow, a central vertical axis must be maintained in the young trees. Support structures are used, which are inter-row and header poles and wires (usually two horizontal wires placed at about 0.6 m and 1.6 m from the ground surface), in order to bear the vegetation and hold the drip line, if not buried.

At planting, usually 0.6 m tall olive plantlets are used. They are secured to a 2 m-long cane or iron rod to ensure the formation of the 'leader'. Then, during the first three to four years after planting, pruning

Fig. 6.4.24. One-year-old tree at planting. (Photo: Franco Famiani)

Fig. 6.4.25. In training pruning to vase, vigorous lateral branches in the lower-middle part of the trunk must be eliminated along with vigorous suckers growing in the internal part of the canopy. The tree before (*left*) and after (*right*) their elimination. (Photos: Franco Famiani)

must ensure the growth in height of the central axis and the gradual elimination of the small branches inserted on the stem up to 50–60 cm from the ground; this is necessary to allow soil management and harvesting with the straddle machine. However, the elimination of the small branches on the trunk up to 50–60 cm from the ground must be done gradually, otherwise, if excessive, some suckers can develop at the base of the tree. The growth in height of trees is followed by periodically tying the extension leader to the stake and safeguarding the 'leading function' by eliminating nearby branches that are very vigorous (Figs 6.4.26 and 6.4.27). If the leader is damaged, it must be replaced by a vigorous lower shoot that must be tied to the stake. When approaching the final size of the canopy, the extension shoot can be managed by using heading-back cuts up to 2.5 m in height and with the last flexible portion (50 cm) to prevent it from breaking during over-the-row harvesting.

The growth of lateral branches inserted on the main axis must also be controlled. They are left to grow freely, but with care to eliminate or shorten excessively large and long branches in the inter-row direction, especially if close to the top. This usually encourages the stronger branches to be directed along the row; also because, when young, they can be fastened to the wires of the support structure. In some cases, especially when relatively larger distances along the rows are used, two to four main branches in the direction of the row may be formed. With these interventions, within three to five years, it is possible to transit from a conical canopy of the single tree to the adult continuous hedgerow, which is a vertical vegetation wall 1.5–2.0 m wide and 2.5–3.0 m high.

Fig. 6.4.26. Trees trained to central leader during the first period of training in a SHD orchard in central Italy. (Photo: Franco Famiani)

Fig. 6.4.27. Trees trained to central leader in a SHD orchard in Spain: the growth in tree height is stimulated by periodically tying the extension leader to the stake and eliminating nearby branches that could compete with the extension leader. (Photo: Franco Famiani)

Multi-axis

This is a new-generation SHD system, having a simplified, or even no, support structure. The hedgerow is obtained using pre-prepared plantlets in the nursery with the absence of a predominant central axis (Fig. 6.4.28). Once in the field, the trees immediately take on the shape of a globe/fan (Fig. 6.4.29). According to some reports, pruning can be carried out exclusively mechanically, by cutting the canopy at increasing widths. Topping during the training phase is required only with vigorous trees. Operating in this way, the canopies of the trees assume a flattened globe conformation and within two to three years after planting, the canopies close the space along the row. Ultimately, with this system, the crowns of the trees are widened along the row from the beginning of training and then the growth of all the vegetation is promoted upwards, until reaching the final height of 2.5–3.0 m. At present, very little information/knowledge is available on this system, especially on the effectiveness of complete mechanization of pruning also during the training period. Therefore, further evaluations are required to give a final judgment, also in relation to the cultivars which are used and the pedo-climatic and cultivation conditions.

Use of deflowering to speed up the initial growth of young trees

Deflowering can be considered a pruning operation. In young olive trees, applications of naphthaleneacetic acid (NAA) at concentrations of 160 ppm or 250 ppm at full bloom prevent fruit set and production and, as a result, an increase in vegetative growth (greater shoot elongation and trunk growth) (Famiani *et al.*, 2022). This can be exploited

Fig. 6.4.28. Second-generation super-high-density olive orchard in central Italy, orchard in central Italy, without a support structure and with young multi-axis trees pre-prepared in the nursery. (Photo: Franco Famiani)

Fig. 6.4.29. Second-generation super-high-density olive orchard in central Italy, without a support structure and with multi-axis trees which have taken on the shape of small globe/fan. (Photo: Franco Famiani)

as a powerful technique to accelerate the growth of young olive trees in both intensive (in the first two to four years after planting) and SHD (first and possibly second year after planting) olive orchards.

6.4.5.2 Production pruning

In all olive orchards (traditional, intensive, SHD), the main objectives of production pruning are to:

- maintain the training pruning shape and the volume reached at the end of training pruning in the space available for each tree;
- maintain/improve light penetration and aeration of the canopy (also for reducing susceptibility to diseases and pests);
- reach/maintain a good balance between vegetative and reproductive activities in order to obtain an abundant, high-quality and steady production;
- facilitate the execution of cultural practices, mainly pruning and harvesting;
- remove parts attacked by parasites.

In summary, the main objectives of production pruning are to guarantee a high and constant production (according to the potential in a given environment and applicable cultural practices), a high efficiency of the chosen harvesting system and healthy trees. All this is obtained by specifying and maintaining the right volume and density of the canopies, maintaining the right structure of the trees and minimizing the presence/spread of diseases and pests.

Both reduced and excessive volume has negative effects on potential yield. The most important environmental factors which affect the potential canopy volume per hectare are water availability and soil fertility. Reduced canopy volume does not enable the maximum potential yield to be reached, because trees are not able to maximize the interception of available light. On the contrary, excessive volume causes shading problems and rapid consumption of water in the soil. The higher the leaf surface area, the greater is the consumption of water through transpiration. The result is shorter growth of shoots, smaller fruits and increased fruit drop in summer.

In rainfed conditions, with rainfall of 150–250 mm/year, the volume can range from 2000 m^3/ha to 3000 m^3/ha; with rainfall of 500–600 mm/year, the volume can reach 8000–10,000 m^3/ha; with rainfall of 800–1000 mm/year, it can reach 11,000–12,000 m^3/ha. In irrigated orchards, the volume can reach 13,000–15,000 m^3/ha.

At parity of volume, the layout of trees in the orchard and the structure/shape/density of the canopies are also important for lighting of leaves and, as a result, for maximizing assimilate production. Very few data are available to indicate the optimal vegetation concentration. In central Italy, in trees trained to vase, an optimal vegetation density corresponds to 2 m^2 of leaves/m^3 of canopy and a leaf area index (LAI) of 5–6. The best structure/shape is that which maximizes, at parity of volume, the external surface of the canopy/volume ratio. In this regard, in large trees, a lobular shape can help (Fig. 6.4.30). At parity of volume, it is also important to maximize the leaf-to-wood ratio of the canopy.

In practice, if the canopy volume/ha considered optimal is 10,000–12,000 m^3/ha, the canopy volume/tree is 36–43 m^3 if trees are spaced 6 m × 6 m; and 25–30 m^3 if trees are spaced 5 m × 5 m.

If in the environment and cultural conditions considered, research activity has not quantified the best volume and density of vegetation to be maintained with pruning, they can be established at the farm level. To find the best values it is useful to refer to the volumes suggested above and to the volume and vegetation density normally used in the area considered. Then, higher and lower volumes and thinning intensities (vegetation density) can be applied with pruning and compared, and, on the basis of the results, progressive adjustments can be made in order to establish the best values for the conditions considered. The volume and thinning intensity (intensity of pruning) are correct when the production is high, with good development of new shoots (20–40 cm long) and a limited number of suckers (Fig. 6.4.31). If the production is high (with

Fig. 6.4.30. A large tree with a lobular-shaped canopy, which maximizes the ratio between the external surface of the canopy and its volume. (Photo: Franco Famiani)

Fig. 6.4.31. Examples of different pruning intensities. The canopy volume and pruning intensity are correct when the production is high, with good development of new shoots and a limited number of suckers. (Photos: Franco Famiani)

relatively small fruits and reduced oil content) and the new shoots (constituting the basis of production for the following year) are very short, the volume could be too large and/or the intensity of pruning could be too low. If the production is relatively low, the shoots are long, and numerous suckers have grown, the volume could be too small and/or the intensity of pruning could be excessive. In this regard, the excessive lowering of the canopy by topping the main branches is one of the main causes that lead to an excessive reduction in the volume of the canopy (Fig. 6.4.32). Obviously, these evaluations should also consider other factors that may affect the vegetative and productive performances of the trees (i.e. irrigation, fertilization).

Fig. 6.4.32. Example where the height of the tree, and consequently the volume of the canopy, was excessively reduced. Note that the upper part of the canopy did not bear fruit, whereas, due to excessive pruning, numerous suckers have been produced. An excessive reduction in the height of the canopy is a common error during pruning. (Photo: Franco Famiani)

Pruning intensity must be balanced to reduce/avoid alternate bearing which can be started by favorable conditions for a high yield (high flowering and fruit set) or unfavorable conditions (very high temperatures or rain during flowering), which strongly reduce fruit set. It is well known that very high yields lead to reduction in the growth of new shoots and induction/differentiation of floral buds, affecting the potential yield in the successive year. Therefore, in years when a very high yield is expected, pruning should be more intense in order to reduce the number of floral buds. On the contrary, in years when a low yield is expected, pruning should be very light. Only in the case where trees have been greatly depleted by a very high yield in the previous season, which besides significantly reducing floral induction/differentiation also greatly reduces the accumulation of reserves, intense pruning may be useful in order to concentrate the few reserves over a lower number of vegetative growth points and, in this way, progressively, recover a balanced situation.

In intensive olive orchards, with trees trained according to the vase system, for mechanized harvesting with a trunk shaker, the secondary branches inserted on the main branches must be numerous, short and without long pendant branches (compact canopy). Regarding pendants, the upper secondary branches, being well lit, lengthen and can cover lower ones, which, remaining in the shade, weaken and defoliate and hence could be removed (Fig. 6.4.33). However, this would be a mistake because it would create a structure that would no longer be responsive to the shaker. Therefore, during the adult stage, it is important to keep the development of the secondary branches under control, making shortening cuts if they develop pendants. In the lowest portions of the canopy numerous pendants develop, and must be removed, especially if an interceptor frame is used, because if they come in contact with it, they do not respond to the vibrations of the shaker. This results in a reduction in the overall volume of the canopy and therefore a proportional decrease in production potential. To avoid this, it is necessary to replace the pendants eliminated at the bottom with greater development in the height of the trees. In practice,

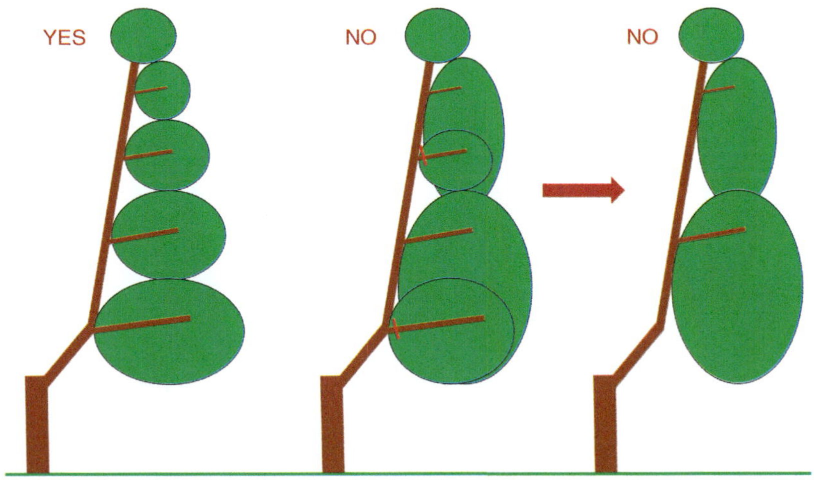

Fig. 6.4.33. The upper secondary branches, being well lit, lengthen and can cover the lower ones, which, due to shading, weaken and defoliate. As a result of this, they could be removed, which would be a mistake because this would create a structure that would no longer be responsive to the shaker. During the adult stage, it is important to keep the development of the secondary branches under control, making shortening cuts if pendants develop. (Ri-elaboration from Famiani, 2017)

vegetation is removed at the bottom and added to the top, where it is much more responsive to the action of the shaker. A good volume is obtained by avoiding excessive depression of the vertical development of the trees: 4–5 m in height is a good reference for a density of 300–400 trees/ha. In fact, these values allow good production and adequate transmission of vibrations up to the highest parts of the canopy. The apical portions of the branches must not be too bare but globose, in order to ensure good production and to better dampen the vibrations transmitted by the shaker, which, in the upper parts of the canopy, if too bare, can cause defoliation and breakages. The secondary branches, as well as being relatively short, must, as far as possible, be linear, that is without goosenecks and/or sharp changes in direction that cause a reduction in vibration transmission.

Efficiency of trunk shakers is significantly influenced by the unit weight of the fruits and their resistance to detachment. Good harvesting yields (> 85% = percentage of detached fruits with respect to the total present on the canopy) are obtained when the ratio between the resistance to detachment of the olives expressed in N and the weight of the fruit expressed in g are equal to or less than 2.3 N/g. This parameter depends on the cultivar (those with relatively large fruits and relatively low detachment resistance are easier to harvest) and the time of harvest (as ripening proceeds, resistance to detachment decreases). Therefore, pruning to develop a structure responsive to the use of the shaker is especially important when the cultivar has small fruits and high detachment resistance and early harvesting is done.

If harvesting is done with hand-held devices, it is especially important to control the height of the trees with pruning. In fact, considering the length of the poles on which these devices are mounted, the height of the trees should not exceed 4.0–4.5 m. Therefore, if trees are taller, it is necessary to lower them with pruning. However, in order to have a good vegetative-productive balance and therefore a good and regular production, it is essential to maintain an adequate canopy volume, so if the height is reduced, the vegetation in the lower portion of the canopy must be increased. If the planting distances allow it, the secondary branches can be extended a little more in order to increase the diameter of the canopy and therefore its volume. In practice, if

Fig. 6.4.34. Example of a tree pruned according to the vase system. Compare with Fig. 6.4.1, which is the same tree before pruning. (Photo: Franco Famiani)

vegetation is eliminated at the top, it will be necessary to add it laterally and/or at the bottom, where harvesting with hand-held machines is easier. However, as vegetation at the bottom is concerned, it is also important to leave space for soil management equipment.

Pruning also has a positive effect on reducing pests and diseases, by improving lighting and aeration of the canopy. In this way it reduces the susceptibility of trees to peacock eye infections and black scale attacks (increased heat in open canopies also causes higher insect mortality). Moreover, by removing the wood infected by olive knot, inoculum of this disease is reduced. In case of severe infections of olive knot (when most of the vegetation has cancers produced by this pathogen), removal of diseased parts with pruning must be done progressively (in two to three years) in order to avoid very great reductions in leaf surface area, which would weaken the plant. In addition, treatments with copper-based compounds should be carried out to avoid further spread of the pathogen.

During production pruning, excessive accumulation of wood must be avoided by ensuring a good turnover of secondary branches by replacing the oldest ones with new ones. This contributes to maintaining a high leaf-to-wood ratio. In very old trees in traditional orchards, restructuring pruning, which also involves the replacement/shortening of primary branches grown excessively in diameter and height, may be necessary in order to restore a good leaf-to-wood ratio.

Technique for production pruning

In all the training systems, production pruning must keep the trees in the available space by using heading back cuts to maintain the height and lateral size of the trees and, as a result, the right volume of the canopy. Moreover, the vegetation density must be adjusted for good lighting/aeration of the canopy and emission of a number of floral buds able to ensure an abundant production and, at the same time, optimal development of shoots, which will be the basis of the production the following year. In a balanced tree, the growth of vigorous suckers is absent or very limited; however, if present, they must be eliminated. The exhausted vegetation should be removed.

Herewith, indications are given for production pruning of trees trained according to the vase system in intensive orchards and the hedgerow of SHD olive orchards. The interventions are indicated in the sequence that enables the easiest and fastest execution.

Intensive orchards – vase system

- Elimination of vigorous suckers on the stump and on the upper part of the main branches in the internal part of the vase. It is very important to note that in balanced trees these formations should be absent or very limited. Often, if present in large numbers, it is the result of excessive pruning intensity the previous year.
- Execution of back cuts (or return cuts) on the top of main branches to maintain the right height of the trees and to avoid that the apical portion of the main branches is larger than the lower part of the canopy, because this would cause shading of the latter as well as mutual shading between trees. The return cuts can be made in correspondence to a lateral branch inserted towards the outside or the inside depending on which enables the best structure to be maintained: towards the outside if the primary branch is too close and *vice versa*. If one of the primary branches is too short/weak, its growth must be favored by deviating its top on a vertically growing secondary branch.

The vegetation left on the top part of the canopy must be less than that in the lower part of the canopy, otherwise there will be excessive vegetative growth on the top at the expense of the lower part, which would also be shaded by the upper canopy. However, excessive thinning of the top which is well lighted and is potentially very productive (heavier fruits with a high content of oil of high quality) should be avoided.

- Elimination or shortening (return cuts) of lateral secondary branches that extend excessively, shading the lower branches.
- Thinning of secondary/tertiary branches. This operation must be done by looking at the density of vegetation, and not the number of branches in a specific area, and interventions should be made only if there is excessive vegetation, caused by not well spaced/overlapping branches. In order to speed up pruning, cuts are made to remove a part of the branch or the whole branch. Never single shoots.

The length of the released/pruned branches must be shorter on the top and progressively longer towards the bottom of the canopy. This is very important in intensive orchards, where trees are close, to ensure good lighting of the whole canopy, and particularly of the lower part.

- Exhausted fruiting branches are present in greater quantity in the basal part of the crown and recognizable because they are defoliated and with few and short shoots. The elimination of all exhausted vegetation is usually very time-consuming. Therefore, in order to speed up this operation, and considering that part of the exhausted vegetation is eliminated with thinning, this operation is limited to the most conspicuous parts (with very few cuts), in the basal part of the canopy. In this regard, it is useful to consider that exhausted/dead vegetation falls down during harvesting.

In Figures 6.4.34, 6.4.35 and 6.4.36, examples of good production pruning of trees trained to vase or globe are reported. When carrying out production pruning, it is also possible to correct possible structural defects (elimination of supernumerary

Fig. 6.4.35. Example of a good structure of trees trained to vase (main branches not too open) and of good pruning according to the vase system: vegetation left on the top part of the canopy must be less than in the basal portion, but excessive thinning must be avoided. The secondary branches must be progressively shorter from the base to the top of the main branches. Intensity of pruning must be adjusted according to the environmental conditions and cultural practices used, especially irrigation. (Photo: Franco Famiani)

Fig. 6.4.36. Example of production pruning on a tree trained to globe in Lebanon. This training system differs from the vase because secondary branches are also trained in the internal top part of the canopy, which assumes a spherical shape. In globe training, it is important to thin vegetation in the upper portion of the canopy, for light penetration. Moreover, excessively long secondary branches on the top lateral portion of the canopy must be avoided as they would shade the lower ones. (Photos: Franco Famiani).

branches, which cause excessive densification of the vegetation and accumulation of wood, and deformed ones) and to eliminate branches damaged by environmental factors and diseases, such as freezing temperatures and hail, which also favor olive knot attacks.

In addition, if numerous suckers are present, this may also be due to some structural defects, such as excessive opening of the main branches and/or excessive planting density, which favors the growth of suckers in the internal (empty) part of the vase, which in this situation is the most lighted. To mitigate this, it is important to try to bring the main branches back vertically with return cuts, encouraging illumination of the lower external parts of the canopy and avoiding excessive protrusion of the upper ones. In extreme situations, structural correction (excessively opened branches) could require reshaping pruning. The emission of suckers is also favored by defoliation caused by disease or pest attacks. In this case, improvement in the control of such adversities must be ensured.

SHD orchards

Once the space between trees along the row has been closed and full fruit production has been reached, the main objectives of pruning are: to maintain the size of the canopy within the limits of the equipment used for harvesting (up to 1.5–2.0 m width and 2.5–3.0 m tall for over-the-row machines); to promote the development of a large number of well-lit and aerated productive branches directly from the primary branches; and to ensure adequate vegetative renewal limiting the biomass removed. Correct canopy management is crucial in this type of planting system in order to avoid excessive shading in the basal portions of the canopy and to maintain an adequate vegetative-reproductive balance, especially when the canopy volume reaches 10,000 m^3 per hectare or more.

Considering these objectives, pruning must be selective and simplified by quickly removing branches that are too rigid (diameter > 3 cm) and protrude from the size limits of the harvesting machine (both laterally and in the apical portion of the canopy), as well as those that are damaged (e.g. broken during harvesting) or that are excessively shaded or that have exhausted their vegetative-reproductive cycle.

The intensity and time of pruning play a pivotal role in vegetative-reproductive control. Too intense interventions carried out at the end of winter, besides promoting the growth of suckers, greatly stimulate vegetative growth in the apical portion of the canopy (progressive upward displacement of the growth) and in the lateral ones distancing the productive branches from the permanent structure with a consequent lateral expansion of the canopy in the inter-row (greater susceptibility to breakage during harvesting). The vigor and architecture of the cultivar is also important in selecting the correct pruning intensity: in weak cultivars having high branching tendency, pruning can be of greater intensity. On the other hand, in medium-vigor cultivars with little branching aptitude, pruning should be carried out more frequently and with reduced intensity to avoid intense vegetative resprouting. As for other fruit tree species trained in high-density or SHD systems, it is preferable to envisage some pruning operations of reduced intensity during the year with specific objectives: to eliminate branches damaged by mechanical harvesting and stimulate the vegetative renewal of productive branches in the lateral portion at vegetative restart (late winter); and to limit the development in height of the upper portion of the canopy (also by executing mechanical topping) and to remove undesirable suckers in summer. In this way, it is possible to stimulate the vegetative growth on the mixed and productive (weaker) shoots and contain the development of the canopy within the limits of the straddle machines used for harvesting.

6.4.5.3 Rejuvenating/reshaping/regenerating pruning

Before proceeding with these kinds of pruning, it is important to evaluate if it would be better to replace the orchards with new ones. This is especially the case when the

Box 1: Olive tree pruning in arid conditions

In very arid climates (rainfall less than 300 mm/year), with irrigation, pruning is done according to the criteria explained in the sections on training and production pruning. In these conditions, regular/annual pruning is necessary because trees grow abundantly, as a result of the long vegetative season that usually characterizes such environments.

If olives are rainfed, and this is the case with a very large part of olive growing in the southern part of the Mediterranean area, often traditional planting and pruning techniques are used; that is, most of the olive trees are established using stumps (protuberances from the base of senescent olive trees). This practice is still adopted and preferred by farmers rather than using plants from semi-woody cuttings, despite the higher risk of disease transmission, uncertainty about varietal authenticity and late fruiting. They justify this choice by the better resistance to extreme conditions due to the lack of rainfall.

The trees are fairly widely spaced, reaching, in some situations (for example in southern Tunisia), distances of 24 m × 24 m. The trees are quite large with an extensive root system, exploiting all the soil moisture. The canopies are trained with vegetation also in the central part of the canopy to protect the woody parts of branches from sun damage, producing a globe-like appearance.

The stumps are taken from the underground part of old, senescent trees destined to be uprooted. This underground part is voluminous and provides an important amount of stumps whose number depends on the unit weight which can reach 4–5 kg. The largest are generally used in the areas of extreme aridity. Those with low volume and weight (1–2 kg) are reserved for plots where water is available for watering during the first two years.

Training pruning

The stumps are covered with soil. Afterwards, latent buds produce a dense vegetative mass forming a bush (even up to ten shoots). These shoots are maintained during the first year and thinning is done the second year to keep only three to four of the best-formed and most vigorous ones. In the third year, only one is selected by pinching the others at 50 cm to favor the retained trunk (these will gradually be eliminated later). It is advisable to keep the 'branching sleeve' around it to protect it from wind and sand. The following year, new branching is formed and any competing vegetation is eliminated. The final height of the trunk from the ground should be 0.8–1 m. Branches are selected that can constitute the four to six main branches. Those not well positioned and that tend to overgrow are pinched off. It is recommended to thin poor vegetation gradually, because the growth in arid conditions is already slow and intense reduction of foliage could accentuate the slow growth. The entry into production of plants is usually slow because of the difficult environmental conditions. The use of larger stumps reduces this time with respect to the use of small ones. The speed of growth is also affected by the fertility and care of the ground. It should be noted that during the first years of establishment, the plants may have small leaves and very short internodes resembling the juvenile form of seedlings.

Production pruning

In arid olive orchards, alternate bearing is frequent because yield depends on rainfall and its distribution. The decision of the final canopy size of the trees depends on the availability of water ensured by rainfall: the lower the availability of water, the lower the canopy volume per hectare. Therefore, pruning must ensure good availability of water to trees, adjusting, with time, the volume and density of vegetation in relation to the amount of available water. Likewise, soil type and quality are important. Pruning should be lighter in relatively fertile soils and heavier in poor soils. A balance is reached when the volume and density applied with pruning enable alternate bearing to be kept to a minimum. However, the behavior of trees greatly depends on the seasonal amount and distribution of rainfall of the year considered, which can be very different from year to year. In practice, as in rainfed arid environments, growth is limited, so pruning can be done every two years, usually, after high production. All the vegetation must be thinned and vegetation in the basal parts of the secondary branches should be maintained. The interior of the canopy is cleared to allow good ventilation without exposing woody parts to sunburning. Pruning is done in December and January for old, weakened and productive trees, in February for young trees and in March/April for vigorous ones.

density of the orchards is low (e.g. too sparse) and if the trees are irregularly spaced and in a poor state of health. In these situations, especially if the environment is ideal for olive production and the land is flat or has a slope that would make it possible for mechanical harvesting, replanting might be the best option. If the hectarage is large, it could be done gradually in order to avoid interruption in the production.

There are situations where environmental factors which are not favorable (low rainfall and/or depth and fertility of the soil) as well as agronomic limitations (i.e. unavailability of water for irrigation) could make it preferable to maintain the orchard. Furthermore, there are circumstances where traditional olive orchards have landscape/historical importance and hence it would be important to maintain them. In addition, in some cases, maintaining at least part of historical/traditional orchards can contribute, with specifically developed marketing strategies, to add value to the production (in terms of selling price). In these cases, rejuvenating/reshaping pruning of trees makes sense and could be useful to improve their productive performance and mechanization, especially for harvesting.

Herewith, indications are given for carrying out these kinds of pruning which, for their nature, are to be considered as an extraordinary kind of intervention.

Rejuvenating pruning

Olive tree aging is accompanied by a decrease in the leaf-to-wood ratio as wood accumulates as the tree ages. Often, this old wood is attacked by pathogens which cause decay. This can bring about progressive weakening of the trees with a significant reduction in the formation of new shoots and, as a consequence, of yield, which, often, also becomes alternant. Usually trees become very large, with shading of the lower portions and concentration of fruiting on the top which makes harvesting more difficult.

There are latent buds in the wood and olive also has ovules. Both of these organs, together with the formation of adventitious buds, enable regeneration, if the canopy is damaged or strongly reduced. The ovules, mainly at the base of the trunk, are also able to produce new roots. Pruning can greatly stimulate the growth of vigorous shoots (suckers) from latent buds and ovules and stimulate the formation of adventitious buds and their development into vigorous shoots. Removing large parts of the canopy with pruning reduces the canopy-to-root ratio, creating vigorous canopy regrowth (suckers). Rejuvenating pruning exploits all these aspects.

Three main situations can usually be distinguished: (i) trees where considerable wood has accumulated, but all parts, including the lower ones, are vital; (ii) trees where the canopy has got out of hand in height and the lower secondary branches are no longer vital; (iii) trees with significant decay on both the main branches and the trunk.

In the first case, skeletonizing pruning can be done. This maintains the main structure but reduces the length of both primary and secondary branches in such a way that with the regrowth of a couple of years, the trees reach the desired size with renewed vegetation. For a good structure, the secondary branches should be progressively shortened from the bottom to the top and the highest could also be removed to promote the formation of new ones in the apical part of the canopy.

In the second case, if the canopy has become very large with excessive wood (large diameter and very tall primary branches) and the lower secondary branches are devitalized, rejuvenation must be done by drastically heading back the primary branches, up to about 0.5–1.0 m from their insertion on the trunk or, in the worst case, up to the trunk (dehorning pruning). Maintaining a portion of the main branches greatly accelerates restoration of the new canopy (Fig. 6.4.37).

Finally, if the entire structure of the tree is damaged, the whole aerial part is eliminated by coppicing the tree. In this case, trees are rebuilt by using the very vigorous suckers produced by the stumps, where numerous ovules are present. Coppicing should be done as close as possible to the

Fig. 6.4.37. An example of regenerating pruning after damages caused by frosting temperatures, by heading back the primary branches up to 0.5–1.0 m from their insertion on the trunk. Elimination of suckers growing on the stump one year after restructuring pruning. (Photos: Franco Famiani)

ground. Removing part of the soil near the stump and cutting the trunk about 10 cm below the surface of the soil produces excellent suckers, which will surely develop their own roots, but it is expensive. In grafted plants, the cut must be made above the graft point. In all cases, the cut must be as low as possible so that the developing suckers will come in contact with the soil. The cuts must be smooth and slightly inclined to avoid water stagnation.

Rejuvenating pruning must be carried out during the rest period, before the trees resume their growth. In this way all the reserves are used to support the regrowth stimulated by rejuvenating pruning. In the first two to four years after pruning, most of the vegetation produced must be left to restore the canopy-to-root ratio. Interventions must be limited to removing the suckers grown on the stump and those very vigorous located in the inner part of the canopy, which, if not eliminated, could slow down the regrowth of vegetation in the external parts of the canopy. Moreover, when a large portion of the main branches has been removed (second case described above), the vigorous suckers growing in the central part of the canopy could cause damage to the structure of the rejuvenated trees. In fact, these suckers growing vertically and very vigorously, shade the lower vegetation, causing excessive enlargement (inclination) of the lateral ramifications that may be used as main branches of the new canopy. In these trees, initially, supernumerary lateral branches are maintained, but then, progressively, they must be thinned up to arriving at the final number of main branches. This operation must be done at the right time, otherwise, if delayed excessively, the supernumerary branches grow too close to each other and do not branch properly in the lower portion. Usually, one year after rejuvenating pruning only suckers grown on the stump are eliminated (Fig. 6.4.37), while the elimination of suckers grown in the central part of the canopy and thinning of lower external ramification are started from the second year.

In the case of coppicing (third case described above), during the first year, suckers do not need to be thinned. Then, if the objective is to obtain a bush vase, thinning of suckers starts in the second or third year to get the two to three main axes of the tree in the fourth to fifth year. If the objective is a single-trunk tree (usually trained to vase), thinning is started the second year to obtain a single axis in the third to the fifth year. The single trunk is more difficult to obtain. Indeed, the developing suckers compete with each other and reciprocally affect their direction of growth. Therefore, in order to obtain one single trunk, which must be vertical, an internal sucker must be selected, creating space around it, and maintaining the nearby suckers in such a way that they do not alter the vertical growth of the selected one (Fig. 6.4.38). The sucker must develop out of the lowest part of the stump, close to the soil, so that as it grows it comes in contact with the soil and develops its own root system.

In Spain, for rejuvenating trees, a method has been proposed which consists in a periodical and cyclic renovation of the main branches (Garcia-Ortiz *et al.*, 2010).

Sometimes, a relatively light rejuvenating pruning can also be useful in intensive olive orchards to re-establish the structure of the main and secondary branches (by shortening them and eliminating long pendants) in order to obtain the best efficiency for the use of trunk shakers for harvesting.

Recovery of abandoned trees

Similar interventions, especially those of the first two cases described for rejuvenating pruning, can also be used for trees that have been abandoned for some years and that are intended to be brought back into production. In this case, usually, there are vigorous suckers growing inside the central part of the trees that have produced a tall and round-shaped canopy (like a globe) (Fig. 6.4.39). Therefore, in addition to the interventions described above, it is very important to remove these suckers (Fig. 6.4.40).

Fig. 6.4.38. An example of regenerating pruning after damages caused by frosting temperatures, by coppicing and subsequent management of sprouted suckers to obtain a tree with one single trunk. (Photo: Franco Famiani)

Fig. 6.4.39. An abandoned tree with vigorous suckers growing in the central part of the canopy which has assumed a globe shape. (Photo: Franco Famiani)

Fig. 6.4.40. An example of restructuring pruning of an abandoned tree according to the vase shape: before (*left*) and after (*right*) pruning. (Photos: Franco Famiani)

Reshaping pruning

In several cases, reshaping is done to adapt trees to mechanical harvesting. For example, trees trained to bush vase are converted into a single trunk vase to be able to use trunk shakers for harvesting. The best main axis is selected, that is, the most vertical and healthy one, whereas the others are removed, sometimes together with their stump and roots, especially if decayed (Fig. 6.4.41). In this case, that is, when only a single trunk is kept, it is also possible to combine an inverted umbrella interceptor with the shaker. In some situations, especially when the distances between the trees are large, two main axes can be kept at a distance that allows the shaker to be hooked up and, in this case, the harvested fruits are collected with nets spread on the ground.

Another situation is trees with long and very open main branches, which have vegetation concentrated in the apical part with a significant amount of pendants and, as a result, do not respond well to trunk shakers. In these trees, the main branches should be shortened and the regrowth trained to obtain more vertical main branches, with numerous and relatively short secondary branches.

A further case is when the main branches are inserted on the trunk at an excessive height, creating an excessively tall canopy and formation of pendants. In this case, the trunk can be shortened, and a new canopy can be formed using the resulting vigorous regrowth. To obtain a trunk 1.0–1.2 m high, which is suitable for a trunk shaker, it must be shortened to about 1.5 m from the ground, because the useful regrowth to make the new branches is not that which

Fig. 6.4.41. An example of reshaping pruning of a multi-trunk bush vase to a single trunk vase. (Photo: Franco Famiani)

grows close to the cut, but the lower ones, which have a more robust insertion as a result of the larger angle of insertion.

In Italy, a particular case of reshaping is the conversion of monocone to vase, which has been done in a number of orchards. It is done by cutting the main axis of monocone at a height which allows conversion to a vase with a trunk 1.0–1.2 m high. This is obtained by cutting the main axis in correspondence of lateral branches, which can then be trained as main branches of the vase, or at a height of about 1.5 m from the ground if there are no lateral branches that can be used.

Pruning of trees damaged by freezing temperatures

Temperatures below −6/−11°C can damage olive trees. Depending on cultivar sensitivity and the lowest reached temperature and the duration, freezing temperatures can cause the following progressive damage: death of buds, dropping of leaves, splitting of bark of shoots, tertiary/secondary branches, primary branches and trunks. Therefore, damage can potentially affect the entire aerial part of the trees.

These trees must be pruned after regrowth has started when latent and adventitious buds give an indication of the extent of damage. In trees that have suffered a high level of defoliation and there is regrowth even in shoots, pruning must be done at a higher intensity than that normally applied to undamaged trees, promoting a significant growth of new shoots. When most one-year-old shoots are damaged and regrowth occurs on two-/three-year-old wood, pruning must be carried out by removing dead parts and those with stunted vegetation (with few and weak shoots). If tertiary/secondary branches are damaged, they must be shortened until reaching parts where there is a good emission of new shoots. If the damage reaches the primary branches, it is necessary to shorten them until reaching the completely intact (not damaged) portion. When action is taken on the main branches, very often,

Box 2: What about rejuvenating pruning of SHD olive orchards?

In SHD olive orchards at the end of their economical lifespan (15–25 years after planting, depending on the climatic conditions and the cultivation practices), it is important to accurately evaluate the benefit of replacing the trees with new ones or applying a major rejuvenation pruning to the canopy. In case of replanting, if the hectarage is large, it could be done gradually in order to maintain a significant production of the whole orchard. In case of major rejuvenation pruning, it can be done when the canopy is excessively old and has a very high wood-to-leaf ratio, indicating that the productive cycle is completed. Two solutions are proposed: (i) coppicing and selecting the best vertical sucker originating from the stump (Fig. 6.4.42); (ii) removing the whole canopy by heading the trunk at about 1.5 m from the ground and leaving some small branches with leaves (Fig. 6.4.43). Both methods rely on the tendency of the tree to re-equilibrate the canopy-to-root ratio in the shortest possible time using latent or adventitious buds. The recovery time to rebuild the hedgerow and restore full fruit production depends on the climatic conditions (length of the growing season) and the cultivation practices, but should be no longer than two to three years. Rejuvenation pruning must be performed before vegetative restart (end of winter). The timing for removing suckers and thinning the supernumerary branches emitted from the headed trunk is crucial during the recovery period, as well as the management of nutrition, irrigation and pest and disease control. Some doubts have been expressed regarding both strategies: lower mechanical stability of the new central leader in the coppiced trees and excessive vegetative development in the headed trunk, leading to rapid self-shading of the inner portion of the canopy, causing premature aging. No scientific studies have been reported on the pros and cons of both rejuvenation pruning methods, but it is very clear that it is an extraordinary intervention that should be planned in long-term perennial cycles together with the possibility of compensating for the lack of production of the trees being rejuvenated with the yield of the other trees in the orchard.

Fig. 6.4.42. Example of coppicing of trees in an SHD orchard in Chile and selection of the best vertical sucker originating from the stump. (Photo: Enrico Maria Lodolini)

it is advisable to shorten them up to about 0.5–1 m from the insertion point in order to renew the whole canopy (Fig. 6.4.37). If the damage involves the whole aerial part, coppicing must be done, as described above. In case of evident damage to the whole aerial part (e.g. large cracks in the trunk and main branches), coppicing could be done before vegetative restart, as the presence of developing suckers makes the execution of coppicing more difficult.

6.4.6 Pruning Frequency

6.4.6.1 Annual pruning

It maintains the chosen/right volume and density of vegetation and thus the lighting and aeration of the canopy, with consequent high efficiency of the canopy and more stable production (lower alternate bearing) and reduced susceptibility to diseases (i.e. peacock eye spot). Moreover, it facilitates mechanical harvesting, because the best structure of the trees is maintained. Finally, it reduces large cuts, resulting in healthier trees and facilitates the agronomic use of pruned materials (smaller pieces are easy to grind and leave on the ground).

6.4.6.2 Biennial pruning

Pruning is more intense because it is executed every two years. Yield is lower in the year of pruning and higher the following one. Vigorous suckers growing inside the canopy should be removed in the year when pruning is not performed.

6.4.6.3 Pluriannual pruning

It is executed every three to five years and, for this reason, is very intense (Fig. 6.4.44).

Fig. 6.4.43. Example of removing the whole canopy by heading the trunk at about 1.5 m from the ground and leaving some small branches with leaves in a SHD orchard in Chile. (Photo: Enrico Maria Lodolini)

Fig. 6.4.44. An example of annual (*left*), biennial (*center*) and triennial (*right*) pruning. The pruning intensity is progressively greater from annual to triennial. (Photos: Franco Famiani)

The yield is very low in the year of pruning and then progressively increases as the canopy volume increases, up to the year of the following pruning. It involves large cuts. It can make the use of harvesting machines less efficient. In the last years before pruning, the trees are more susceptible to peacock eye spot attacks.

Considering pruning frequency, it is important to distinguish intensive/SHD orchards from traditional ones. Moreover, the canopy volume per hectare and the harvesting system need to be considered. Intensive/SHD orchards have close planting distances and large canopy volumes per hectare and this requires annual pruning. In these conditions delayed pruning favors excessive canopy growth and yield toward the top (escape of vegetative growth and yield toward the top) and the consequent progressive reciprocal shading of trees resulting in reduced vegetative growth and production in the lower portions of the canopies. Moreover, trees trained to vase (intensive orchards) and harvested with trunk shakers require a rigid structure and suitable size/volume and trees trained to hedgerows (SHD orchards) harvested with straddle machines require not too long lateral rigid branches and suitable size/volume. These conditions are better ensured with annual pruning.

In traditional orchards with trees widely spaced, there is more flexibility because there is no risk of reciprocal tree shading. In these conditions, in Spain, in a long-term experiment, triennial and quadrennial pruning gave better yields than biennial one. The higher yields were related to higher average canopy volumes, even though the yield efficiency (yield/m^3 canopy) was lower (Garcia-Ortiz *et al.*, 2010). In Italy, also in an intensive orchard of rainfed relatively young trees (ten years old), high yields were obtained with less frequent pruning (biennial and triennial vs. annual) and light-intensity interventions. Also in this case the higher yields were related to higher canopy volumes, and in all the treatments similar yield efficiencies were obtained (indicating that the higher yields were due to the higher canopy volumes) (Tombesi *et al.*, 2008). The results seem to indicate that in order to maximize yield it is important to establish the average volume that corresponds to the potential given by water availability, soil fertility and cultural practices. In other words, it seems that with annual pruning, the intensity is often excessive with the consequent unnecessary reduction of the canopy volume. Indeed, in another work carried out in Italy, it was shown that if the different pruning frequencies (annual, biennial and triennial) were applied maintaining similar average canopy volumes of the trees, similar average yields were obtained, but they were more constant in annually pruned trees. Moreover, the annually pruned trees were, on average, more responsive to mechanical harvesting (Farinelli *et al.*, 2014).

Pruning every three to five years entails larger cuts. These favor decay and quicker aging of trees. Moreover, if the pruned material is ground and left in the soil, the larger size of the pruned material demands the use of more powerful machines.

In conclusion, annual pruning is the most advisable in intensive and SHD orchards as it allows the full productive potential of the orchard to be exploited if the volume and density of the canopy are right. Moreover, the trees maintain a better structure and a more constant canopy volume, making mechanical harvesting easier.

Pluriannual pruning can be used in some particular situations, such as in environments where tree growth is greatly limited by environmental factors such as aridity and high temperatures, and in large trees, which, being widely spaced, are more flexible, especially if it is necessary to use platforms (if trees are very high) and/or available manpower is limited.

In most of the studies carried out on pruning frequency, the average annual time to prune a tree was less with less frequent pruning. This means less manpower. However, this does not necessarily mean a lower cost, because larger cuts require more expensive devices and the larger size of the pruned material makes chopping (in case of incorporation into the soil) more difficult. In addition, in an intensive orchard, with simplified pruning, and similar average canopy volumes, with all pruning frequencies, the average time (minutes/tree/year) of annual pruning tended to be similar to that of biennial and triennial pruning, suggesting that when pruning is executed with a simplified approach and the average annual material removed with pruning (kg/tree/year) is similar, the differences in the average times for pruning with the different frequencies becomes very little (Farinelli *et al.*, 2014).

6.4.7. Time of Pruning

Production pruning is normally carried out during vegetative rest, but can vary according to climatic conditions. In mild areas, pruning can be done from after harvesting to the end of winter. In cold areas, where freezing temperatures may occur, it is preferable to wait until the end of winter/beginning of spring. This is because pruning stimulates the restart of physiological activity, making the trees more sensitive to low temperatures.

Early pruning increases the vigor of new vegetation, whereas late pruning reduces it. Olive trees accumulate reserves in their main structures (roots, trunk, main branches). If pruning is done before translocation to the growing sinks, all the reserves will concentrate in those released with pruning, resulting in more vigorous growth and greater stimulation of shoot growth in the canopy. If pruning is done after translocation has started, part of the reserves will be lost with the vegetation removed and this will cause a slowing down of the growth processes. This means that if tree growth is to be favored, early pruning is better, whereas if the trees are very vigorous, late pruning can contribute to reducing vegetative activity and promoting reproduction.

As far as late pruning is concerned, in the relatively cold area of central Italy, a study, which compared early or late spring pruning showed better results for early pruning in terms of both higher stability in yield and better control of the vegetative growth of trees (Lodolini *et al.*, 2019).

Summer pruning can be utilized to eliminate vigorous suckers in the internal parts of the canopies. This is useful if there are numerous vigorous suckers, which cause considerable shading, and in rainfed conditions their elimination reduces the leaf surface area and, hence, the water requirements of the trees (becoming an intervention useful in case of shortage of water). This could be the case in the first year/s following biennial and pluriannual pruning (Cinosi *et al.*, 2023). Otherwise, as their main growth stops in summer, they can be left because, in the latter part of the vegetative season, they could contribute to the production of assimilates, which would then be available for the trees and also for accumulating reserves. They will then be removed with winter pruning. Excessive sucker growth is not normal, because if the previous pruning was carried out at the right intensity, no suckers, or very few, should grow. Therefore, a high number of suckers is usually the result of intense biennial or pluriannual pruning or intense pruning used to reinvigorate weak trees, rejuvenate old trees, reshape canopies, or regenerate damaged trees (i.e. by freezing temperatures). If the high presence of vigorous suckers is due to rejuvenating/reshaping/restructuring pruning, which greatly reduces the canopies with a consequent reduction in the canopy-to-root ratio, the suckers must be eliminated gradually (in order not to reduce the ratio too much, causing a slow recovery of the trees), only eliminating those which, if not eliminated, would damage the structure of the trees.

In cold areas, intense summer pruning must be evaluated carefully, because some regrowth could be stimulated which could then be damaged by winter temperatures. However, only very intense pruning can cause this problem.

In general, it is never recommended to prune twice a year, because the sum of the two is more expensive than a single pruning. This means that if suckers are removed in summer this should be the only pruning (only in particular situations, such as the occurrence of excessive shading or conditions of lack of water, a double intervention could be considered). Otherwise, if the trees, besides eliminating suckers, require further interventions in the rest of the canopy, it is recommended to remove the suckers during the normal pruning during the rest period of the trees.

6.4.8 How to Prune

6.4.8.1 Manual pruning

Traditional manual pruning using ladders must be abandoned because it requires a lot of manpower and falling from ladders is an unacceptable risk.

6.4.8.2 Facilitated pruning

To reduce the risk of pruner injuries and accelerate the pruning process, it is crucial to use tools that enable all work to be done from the ground. These tools include pole/telescoping pruning saws and shears that do not require ladders and can reach a maximum height of around 6 m. Nevertheless, pruners can work comfortably up to 4.5 m. These tools are manually operated. Moreover, pneumatic and electrical tools are available. In all cases professional equipment must be used, that is they must have a high cutting capacity and be light and robust. Professional tools mounted on poles operated manually can be used efficiently in orchards for annual pruning and with small diameter cuts. Usually, in this situation, in order to speed up the execution of pruning only professional saws mounted on sticks/poles are used for all the pruning interventions (Fig. 6.4.45).

Pneumatic or electrical tools, such as, shears and chainsaws mounted, or not, on poles can be used when numerous branches with a diameter > 3–4 cm are to be cut (Figs 6.4.46 and 6.4.47). Cuts up to 3–6 cm can be made with shears (the larger is the potential cut the heavier is the weight of the tool), whereas chainsaws are used for cuts up to 15–20 cm or more. The weight of these tools depends on the cutting devices and the length of the poles and the material they are made of (i.e. aluminum, fiberglass): hand shears 0.5–0.8 kg, with the poles up to 2–4.5 kg; hand chainsaw about 2 kg, with the pole up to 3–3.5 kg. There are also chainsaws mounted on poles and powered by a gas-powered engine mounted at the bottom of the pole (weight 7–9 kg). Gas-powered chainsaws vary in weight according to size. The weight with a blade from 25 cm to 35 cm can range from 3 kg to 6 kg. These tools are very useful for low-frequency pruning (every two to three years or more, and so cuts are large) and for rejuvenating/reshaping/

Fig. 6.4.45. Pruning from the ground using a professional saw mounted on a pole. (Photo: Franco Famiani)

Fig. 6.4.46. Pruning from the ground using a pneumatic chainsaw mounted on a pole. (Photo: Franco Famiani)

regenerating pruning. Chainsaws mounted on poles can be used with high proficiency. In these cases, the work yard could be made up of operators with chainsaws on poles for the main larger cuts and workers with a normal saw on a pole for refined, smaller pruning.

For selective pruning in SHD olive orchards (i.e. elimination of branches with a diameter > 3–4 cm, growing toward the inter-row), small hand-held chainsaws are often used (electrical or gas-powered).

For training pruning, electrical shears are appropriate.

Pneumatic tools are powered by a compressor that can be wheeled, self-propelled or towed by a tractor, depending on its size. The compressor must ensure the required flux of compressed air, which is about 80–100 l/min for a shear, and 400 l/min for a chainsaw, at the operating pressure of 10 bar for both. Therefore, the size of the compressor must be estimated on the basis of tools that are to be used. For example, if two pneumatic chainsaws are used, the compressor must ensure at least a flux of 800 l/min of compressed air at 10 bar. It is always recommended to have a compressor able to ensure a flux higher than what is normally required, ensuring the best functionality and, if the work yard is increased, it can be easily done, ensuring flexibility that is always beneficial at the farm level. According

Fig. 6.4.47. Pruning from the ground using an electrical chainsaw mounted on a pole. (Photo: Franco Famiani)

to size, electrical tools are powered by 12 V or 24 V batteries, which vary from 1 kg to 5.5 kg. The largest are for high power tools like chainsaws. Batteries can be carried directly by operators, by attaching them to a belt (lighter batteries used for shears) or to an appropriate vest or a particular backpack (heavier ones). Usually, compressors are operated by a gas engine (self-propelled or that of the tractor). The pneumatic tools are connected to the compressor with specific tubes. Therefore, the operators must pull the tube when they move between different trees or positions and this requires additional care/work. Batteries can require up to 7 h for recharge. Battery-powered electrical tools worn directly by operators make movement easy and they are not noisy, even though they cause some fatigue due to the weight of the battery. Gas-powered chainsaws are heavy, noisy and also emit exhaust gas, which can create problems for the operator.

For tall trees, self-propelled or towed lifting platforms can be used (Fig. 6.4.48). They require flat land or very low slopes.

In adult trees, facilitated pruning with respect to traditional pruning enables work time to be reduced up to 50–60%.

Fig. 6.4.48. Self-propelled lifting platform used to prune tall trees. (Photo: Franco Famiani)

To optimize operators' work in the pruning yard, it is very important to use the right tool for the type of work to be done. Heavy tools should be avoided when light ones are enough for the job, as this would cause an unjustified increase in operator fatigue, with a consequent reduction in work efficiency, and, even more importantly, a reduction in the level of safety. It is known that fatigue leads to the occurrence of accidents. Therefore, tools must always be selected according to the work to be done. When using both heavy and light equipment, it is beneficial to rotate operators.

6.4.8.3 Mechanical pruning

Mechanical pruning is done by using pruners with toothed blades (mowing bars), which generally can cut branches up to of 2–5 cm in diameter, making them suitable for light pruning. Then there are pruners with circular saws mounted on a bar which can make cuts up to 15 cm in diameter (Fig. 6.4.49). Circular saws can be replaced by discs with radially inserted knives which, compared to circular saws, have a lower cutting capacity (the maximum diameter of cuts is lower) but make higher-quality cuts on small-diameter branches. Pruners with circular saws are the most widespread. Pruners are generally towed by tractors, but self-propelled ones are also available (Fig. 6.4.50).

With respect to the ground, mechanical pruners make perpendicular or slightly inclined cuts (hedging) and parallel cuts on the top (topping) or at the bottom of the canopy (skirting or trimming). These cuts enable the tree height, lateral expansion of the

Fig. 6.4.49. Pruner with circular saws mounted on an articulated bar on a tractor. (Photo: Franco Famiani)

Fig. 6.4.50. Self-propelled pruner with circular saws mounted on an articulated arm. (Photo: Pierluigi Pierantozzi)

canopy in the inter-row and excessively low vegetation to be contained. Generally, the maximum height at which they can operate is around 6 m.

Unlike manual or facilitated pruning, selective interventions cannot be made with mechanical pruning. Indeed, it removes the vegetative and productive portions indiscriminately. Mechanical pruning cannot operate in the innermost part of the canopy.

Mechanical pruning is based on the choice of the depth of the cuts, the parts of the canopy to be pruned (e.g. hedging on one side or both sides, only topping, hedging + topping, etc.) and the frequency, which depends on the first two choices (Fig. 6.4.51). Frequency is higher (light pruning) if the depth of the cuts is low and/or if only a portion of the canopy is pruned. Frequency is lower in the case of deep cuts (intense pruning) and/or interventions which involve the entire crown.

With regard to the depth of the cut, it is important to be aware that the external portion of the canopy is that grown in the last growing season and therefore it is the one that potentially produces. Therefore, if the cut is very superficial, a large part of this vegetation is kept, which, therefore, will allow a good production in the same year in which the pruning was performed. By increasing the depth of the cut, there will be a progressive reduction up to eliminating (with very deep cut) the fruit production in the year of pruning and a progressive increase in vegetative activity with the formation of new vegetation, which will be able to produce in the following years. With an increase in the depth of cutting, from 25 cm to 75 cm from the external edge of the canopy, and thus of the age and diameter of the branches that are cut, there will be a greater number of vigorous suckers, with a low level of return flowering even three full years after pruning (Albarracín *et al.*, 2018). Regrowth is particularly vigorous in areas with warm climates (long vegetative season) (Cherbiy-Hoffmann *et al.*, 2012).

The time of pruning also affects the vigor of regrowth: delaying pruning close to flowering or to early summer (at the fruit pit hardening stage) reduces regrowth (Albarracín *et al.*, 2017; Tombesi and Farinelli, 2018; Lodolini *et al.*, 2023). The reproductive growth in the first part of the season limits vegetative growth.

In trees that are repeatedly pruned mechanically, a layer of dense vegetation develops in the outer part of the canopy ('hedge effect'), which is not always productive and hinders the penetration of light in the internal and lower parts of the canopy.

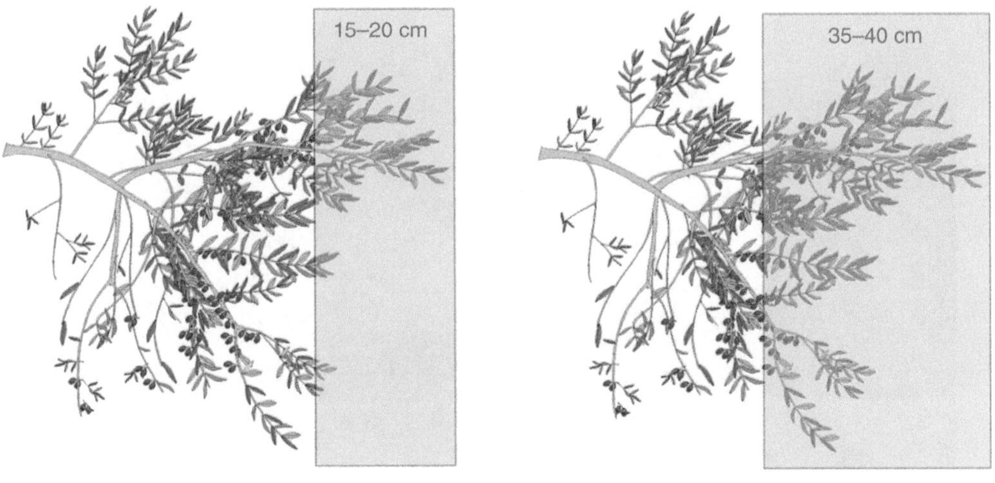

Fig. 6.4.51. Scheme showing the vegetation that is eliminated in relation to the depth of cutting by mechanical prunings. (Drawings: Francesca Massetani and Enrico Maria Lodolini)

Moreover, pruners can cause some abrasions and wounds on the branches and some can be poorly cut. In addition, suckers growing in the central part of the canopy are not removed and this promotes excessive accumulation of wood and premature aging of the canopy, especially in the internal parts. Finally, in varieties sensitive to olive knot, damage to vegetation can enhance the spread of this disease. To reduce these problems, it is necessary to alternate/supplement mechanical pruning with manual/facilitated pruning in order to eliminate the damaged and aged parts and the suckers growing inside the canopy. This creates openings in the canopy with consequent improvement of lighting, which, if not enough, can be further improved by some additional thinning of the external vegetation which causes the 'hedge effect'.

Supplementing with manual/facilitated interventions enables mechanical pruning to be performed even in trees trained with shapes, such as vase, which have parts (i.e. the center) that cannot be reached by pruners (Fig. 6.4.52).

A long-term experiment (19 years) was carried out in Spain in a traditional three-trunk olive orchard (25-year-old trees at the beginning of the experiment) with supplementary irrigation. There were no significant differences in the total yield between trees traditionally pruned and those mechanically pruned (on which manual interventions were carried out every three to four years to remove damaged and dead wood and large suckers) (Garcia-Ortiz *et al.*, 2010).

In Italy, in trials carried out in olive orchards of trees trained to monocone (Fontanazza *et al.*, 1998; Lodolini *et al.*, 2006a, b) and an olive orchard of young trees trained to vase (Giametta and Zimbalatti, 1997), higher yields were obtained with mechanical pruning. These results are surprising and likely due to manual pruning often being done with excessive intensity. Some preliminary good results were also obtained with 15-year-old trees trained to vase on which different mechanical interventions were applied (Farinelli *et al.*, 2011).

In Portugal, a long-term study, carried out in an orchard with trees spaced at m 7 × 3.5, led to the conclusion that mechanical pruning tended to not influence olive yield negatively, except when executed with high frequency (Dias *et al.*, 2022). Moreover, it

Fig. 6.4.52. Effects of topping on a tree trained to vase. (Photo: Enrico Maria Lodolini)

was concluded that complementary manual pruning interventions only make sense for eliminating excess wood accumulated over relatively long periods of time and, therefore, should be carried out sporadically (Dias *et al.*, 2022).

A trial carried out in a SHD olive orchard (1678 trees/ha) in Italy showed that in adult trees (eight years old), canopy restraining of low vigorous cultivars, such as 'Arbequina', 'Arbosana' and 'Koroneiki', could be done with mechanical pruning and manual pruning to cut off branches with a cross-section over 3–4 cm oriented perpendicularly with respect to the direction of the row (Vivaldi *et al.*, 2015). On the other hand, difficulties were encountered with medium-high vigorous cultivars. In another study in a SHD olive orchard, splitting of the pruning period in winter (selective lateral thinning) and summer (mechanical topping) gave the best productive results (Lodolini *et al.*, 2023). Finally, with respect to mechanical pruning, good results were obtained with selective pruning, based on the elimination/shortening of transverse branches of large diameter (rigid), which could cause problems for the straddle machine, and the maintenance of flexible vegetation in the external parts of the canopy. In this way, it was possible to maintain a higher canopy volume, which gave a higher yield than with mechanical pruning, and, at the same time, avoid damages to the harvester or trees during harvesting (Tombesi and Farinelli, 2014, 2018). Obviously, selective pruning is more time-consuming requiring about 25 h/ha of work, whereas mechanical pruning requires 2–6 h/ha (Tombesi and Farinelli, 2014, 2018).

As said, mechanical pruning is not selective and with hedging there is an indiscriminate suppression of both fruiting shoots and exhausted parts of the plant. To make mechanical pruning selective, an innovative device attached to the pruner was developed. It is able to produce a dynamic air fluid action with air jets able to push the young and flexible shoots upwards and downwards, whereas the more rigid vegetation does not move. In this way flexible/productive shoots are maintained, and the aged/rigid parts are cut; preliminary results obtained with this device seem promising (Ottanelli *et al.*, 2019).

The effects of mechanical pruning on the efficiency of olive harvesting still need further evaluation. With trunk shakers, contradictory results have been obtained: in some cases, there were no negative effects, whereas in others there was a reduction in harvested fruit (Dias *et al.*, 2008; Farinelli *et al.*, 2011). For straddle harvesters used for SHD orchards, the canopy size is especially important, also considering the regrowth, within a size that can be harvested without causing damage to both the trees and the machine. The situation is similar in intensive orchards harvested with large straddle machines, like the Colossus, especially used in the southern hemisphere. Mechanical pruning is usually efficient in controlling the size of the canopies with positive effects on the use of canopy shakers (i.e. straddle harvesters, including Colossus). In intensive orchards, harvested with trunk shakers, besides maintaining a correct canopy volume, the structure of the trees is also important. Indeed, for suitable mechanical harvesting with trunk shakers, the harvesting yield (percentage of detached fruits) should be ≥ 80–85%, and if the structure of the trees progressively worsens with time, there will be a similar reduction in harvesting yield, especially in early harvestings and with small fruit cultivars.

At the present time, on the basis of available information, it seems clear that for each situation (orchard type and regrowth potential of trees) mechanical pruning must be done at a frequency that enables a yield similar to normally pruned trees to be obtained. Moreover, it is usually necessary to supplement mechanical pruning with facilitated pruning to remove damaged and aged parts and suckers inside the canopy. On trees with a three-dimension (3D) canopy, this could mean mechanical pruning every four to five years and manual intervention in between.

A particular case is 3D orchards (trees trained to vase) with canopies forming a continuous wide hedgerow (for more details see chapter 5.1). This is a a quite common situation in countries of the southern hemisphere where the cultivation of olive has spread in the last 20–30 years and, often, is characterized by large olive farms (see Box 3). Due to the size of farms and lack of skilled manpower to execute pruning, the mechanization of this practice is an important option. In this situation, in addition to the indications given above for the mechanical pruning of 3D trees, another possibility is to do the topping and hedging in alternate years (biennial mechanical pruning) (Fig. 6.4.53). Hedging is done slightly inclined in order to cut more on the top part of the canopy (shortening possible highly located pendants). In order to maintain a quite constant annual average yield, olive orchards could be divided into two sections to be pruned in alternate years. In addition, in some cases, annual interventions on both sides at low intensity or, alternatively, on one year one side and the following year the other side at higher intensity are considered (Box 3).

In two-dimensional canopies (super-intensive and high-density orchards with flattened canopies, including palmette), mechanical pruning could be done with pluriannual cycles that could be: (i) topping + hedging, topping, topping + hedging, topping (Vivaldi *et al.*, 2015); or (ii) topping + hedging on one side, topping, topping + hedging on the other side, topping (Rius and Lacarte, 2010). In addition, another possibility is topping every year and hedging the two sides of the row in alternate years (one year one side, the following year the other side) (Figs 6.4.54 and 6.4.55). In all cases, in order to reduce the vegetative growth of the apical portions of the hedgerows, topping is usually done in summer (Fig. 6.4.56). Moreover, a rapid passage of manual/facilitated pruning is suggested to refine the pruning by cutting off or shortening large-diameter transverse branches. A combination of mechanical pruning to control tree height (topping) and facilitated pruning to control the enlargement of the lateral portions of the canopy by selective thinning of the vegetation could also be a good option (Lodolini *et al.*, 2023), especially in small- medium-size farms.

Mechanical pruning is more difficult in 3D canopies, whereas it is easier in 2D canopies. The schemes indicated (cycles of interventions) are only indicative. This means that in each situation, even though the suggested scheme could be used as a starting point, the cycle to be used must be optimized taking into account the specific situation of the orchard.

Fig. 6.4.53. Execution of mechanical hedging and topping in a wide-hedge olive orchard in Australia. (Photo: Mauro Ruffolo)

Fig. 6.4.54. SHD orchard in Tunisia pruned mechanically: row where both topping and hedging were executed (*left*) and row which is still unpruned (*right*). (Photo: Franco Famiani)

Fig. 6.4.55. Execution of hedging in a SHD orchard in Chile. (Photo: Enrico Maria Lodolini)

Fig. 6.4.56. SHD orchard in Italy where topping was executed in summer. Note the limited vegetative regrowth at the end of the vegetative season. (Photo: Enrico Maria Lodolini)

Mechanical pruning is particularly interesting for very large farms, where the demand for manpower for pruning is very high. Indeed, mechanization enables a significant reduction in labor and time needed for pruning and this is the main reason why considerable effort is being made to mechanize this practice. With medium-sized trees this reduction can reach values of 96–98% (Giametta and Zimbalatti, 1997; Lodolini *et al.*, 2006a,b; Farinelli *et al.*, 2011). However, if mechanical pruning is supplemented with manual/facilitated pruning, the reduction in manpower is lower and, depending on how these additional interventions are performed, can range from 30% to 80% (Giametta and Zimbalatti, 1997; Fontanazza *et al.*, 1998; Lodolini *et al.*, 2006a,b; Farinelli *et al.*, 2011).

In SHD orchards, mechanization enables a 75–90% reduction in manpower with respect to facilitated pruning (Tombesi and Farinelli, 2018). This percentage is reduced to about 50% if supplemented with facilitated interventions.

In terms of costs, mechanization of pruning gives a reduction due to the lower need for manpower, but the cost related to the purchase, maintenance and use (which implies the utilization of a tractor) of the machinery must be considered.

6.4.9 Pruning Strategies for the Production of Table Olives

In the cultivation for table olive production, large-sized fruit is a very important target. The number of olives per tree is negatively correlated with their weight, so pruning is often executed at a relatively high intensity in order to reduce the number of flower buds and, consequently, the number of fruits. Although large drupes are obtained, yield is reduced and, often, there is an imbalance between vegetative and reproductive activities. In Italy, in Sicily, if trees are heavily loaded, summer pruning is done to reduce the load. However, in some cases, repeated intense pruning over a long time can weaken the trees (Barone *et al.*, 2020).

In brief, in table olive trees, it is important that the number of fruits is not excessive because this results in small-sized fruits, while the value of the production mainly depends on fruit size. At the same time, if the number of fruits is too low there is a loss of value due to low production. This means that in table olives it is much more important to control the fruit load than in oil-olive trees. Furthermore, intense pruning to have large-sized fruits often results in situations of vegetative imbalance. This is why, in addition to normal pruning, fruit thinning is also used as a tool to control tree fruiting. This technique is very powerful because by decreasing the number of fruits without altering the number of leaves, an increase in the leaf-to-fruit ratio is obtained, which promotes greater growth of not-thinned olives and ensures the growth of new shoots which will produce the following year.

Several studies have been carried out on the use of chemical thinning to regulate the number of fruits in table olive trees. At the present time, naphthalene acetic acid (NAA) has been the most reliable in several

Box 3: Olive pruning in the southern hemisphere: the case of Argentina

In the last 20–30 years, olive cultivation has been extended to countries outside the Mediterranean basin. These new areas often have a longer vegetative season due to warmer average temperatures throughout the year, shorter winters and rainfall concentrated in fewer months compared to the Mediterranean.

Most of these countries are located in the southern hemisphere and among them the biggest olive surface area is located in Argentina (around 80,000 ha). Here, the majority of olive orchards are large commercial orchards (> 100 ha) with 400–800 trees/ha and drip irrigation, although there are also small-scale farms (< 10 ha) with 100–300 trees/ha and with traditional management.

In general, a vase-shaped training system is used for both table olives and oil production on small and medium-sized farms, while hedgerow systems are more common for oil production in larger-sized operations. Mechanical harvest is performed using trunk vibro-shakers and huge over-the-row harvesting machines (like Colossus) for vase-shaped trees and hedgerows, respectively. Pruning is usually manual for the vase training system, while it is generally mechanical, or mechanical combined with manual for hedgerow plantations. Pruning is normally carried out during winter. In the case of mechanical pruning, topping is usually done in the summer to adjust the height of the hedgerow to the harvester and to prevent excessive vegetative growth in the upper portion of the canopy. However, if topping is always carried out at the same height, dense vegetation develops, inhibiting light penetration into the inner part of the canopy, leading to acceleration of the aging process with negative effects on fruit production. To solve this problem, topping at 30–35 cm below the normal cutting height is necessary every three to four years.

In recent years, due to the large size of the farms, the lack of skilled workers and the high cost of cultivation management, a large number of companies have opted for mechanical pruning (Fig. 6.4.57). Several types of pruning machines are used, and some locally manufactured. There are knife or disc

Fig. 6.4.57. Wide hedgerows pruned mechanically in Argentina. (Photo: Pierluigi Pierantozzi)

Continued

Box 3: Continued.

trimmers/cutters that can be front/rear mounted on the tractor or self-propelled pruners and even self-propelled pruners with spraying machines to apply fungicides/bactericides immediately after cutting.

Mechanical pruning (topping and hedging) is generally done annually. Pruning can be carried out on both sides of the olive tree with light intensity or alternately, i.e. one year one side and the following year the other side with a higher intensity. Some studies have shown that high-intensity winter pruning is associated with excessive vegetative growth. In addition, greater pruning intensity may favor the formation of suckers, which may cause a loss of potentially productive buds the following year (Albarracín, 2018). Moderate pruning, carried out annually, avoids intense pruning, which, in addition to the aforementioned, is generally associated with the tendency for alternate bearing.

Manual pruning can be used to supplement mechanical pruning. It can be performed quickly after mechanical pruning to remove poorly positioned or woody branches or it can be postponed to after about four months to eliminate suckers resulting from cuts from the mechanical pruner, especially with intense interventions. The latter, although it has given excellent results in terms of growth reduction, return flowering and fruit yield, is not a common practice because it is time-consuming in large orchards. Mechanical tip heading, which is an easier practice to implement, was not effective in controlling shoot growth or in promoting return flowering on new shoots (Albarracín *et al.*, 2019). In hedgerow olive orchards, in both high- or super-high-density plantations located in the southern hemisphere, control of vegetative growth is crucial. To control vigor, the combination of pruning with cultural practices, such as deficit-irrigation strategies, varietal selection and suitable fertilization, is essential (Pierantozzi *et al.*, 2020).

trials (Martin *et al.*, 1990; Rallo and Barranco, 1986; Pastor *et al.*, 1992; Krueger *et al.*, 2005; Barone *et al.*, 2014). The correct choice of timing of application and NAA concentrations are crucial. Timing should be between 12 and 18 days after full bloom (80–90% of flowers opened) when fruits have a transverse diameter of 3–5 mm, while the concentration should be around 150 ppm (Krueger *et al.*, 2005). It is suggested that the dosage is calculated by considering 10 ppm for each day from full bloom to treatment: 120 ppm if applied 12 days after full bloom and 180 ppm if applied 18 days after full bloom (Krueger *et al.*, 2005). Earlier or later treatments have a stronger and reduced effect, respectively. However, in a study in Italy, good results were obtained with application within 12 days after full bloom with 200 ppm NAA (Barone *et al.*, 2014). If the weather is cool, fruit growth slows down and, hence, fruit diameter is a better reference than the number of days after flowering for determining timing. The effectiveness of treatment is also affected by the climatic conditions. In particular, during the few days following treatment, cool temperatures reduce its effect while high temperatures increase it. As well as the use of very high concentrations, treatments when the temperature is higher than 38°C can cause injuries to the new vegetative growth. Chemical thinning is done if fruit set appears excessive. Obviously, chemical thinning cannot be done in organic olive growing.

Trials have shown that the effects of chemical thinning depend on cultivars. It was very effective in the cultivars ‘Manzanilla’ and ‘Hojiblanca’ and less effective in ‘Gordal Sevillana’ (Pastor *et al.*, 1992). In another study, good results were obtained in the cultivars ‘Conservolia’, ‘Bella di Spagna’, ‘Gordal’, ‘Manzanilla’ and Tonda Iblea’, whereas unsatisfactory results were obtained in ‘Moresca’, ‘Picholine’ and ‘Nocellara del Belice’ (Barone *et al.*, 2020).

Due to the variability of the results obtained with different cultivars and environments, the use of chemical thinning must be evaluated at the farm level. Tests should be carried out to optimize timing of application and NAA concentration in relation to the cultivar and environmental conditions.

6.4.10 Use of Pruned Material

The pruning material, which is organic matter, can be used as a byproduct. It is highly discouraged to burn the pruning material in the field, because of the work and costs associated with this operation, the environmental risk of possible fires and the loss of organic matter that it implies. It is highly recommended to shred the pruning material and

Fig. 6.4.58. The pruning material can be shredded and left on the ground for mulching and fertilizing. (Photo: Enrico Maria Lodolini)

leave it on the ground for mulching and fertilizing (Fig. 6.4.58). This is particularly easy in orchards with soil managed with green cover, where shredding of the pruning material can be done together with that of the yard. A part of the carbon from the material added to the soil is re-emitted as a result of respiration, but a part remains in the soil, increasing the amount of carbon stored in the soil, with positive agronomic and environmental repercussions. Pruning material can also be composted and then used as a soil conditioner.

If there are trees infected by *verticillium*, the pruning material must be removed from the orchard.

An alternative use of the pruning material is for energy production. For this, it is collected and chipped or baled (Fig. 6.4.59). If it replaces the use of fossil fuels, there is a reduction in CO_2 emissions in the atmosphere.

Pruning material can also be used as animal feed. Larger pruning material must be separated and the thinner green parts can

Fig. 6.4.59. The pruning material can be baled and used for energy production. (Photo: Franco Famiani)

be baled and used as fodder. However, this kind of use is very minimal.

Other lesser and specialized uses include the use of pruning wood to make furniture or handmade objects.

The amount of material generated by pruning depends on the growing area, the kind of orchard (intensive, SHD, etc.), the cultural practices (especially irrigation), pruning frequency and the kind of pruning (restructuring/rejuvenating pruning produces large amounts of material). In an intensive adult orchard with 280–400 trees/ha, on average, the fresh pruning material (about 50% water) can be 3–6 t/ha/year.

6.4.11. Pruning and Oil Quality

Pruning frequency and intensity do not affect the basic qualitative parameters of the oil, such as free acidity, peroxide value and spectrophotometric absorbances in the ultra-violet, whereas they can have some effects on phenolic and volatile compounds (Farinelli *et al.*, 2014; Cinosi *et al.*, 2022). Pruning affects canopy lighting and light has been reported to have a positive influence on the polyphenolic content and sensorial characteristics of the oil (Proietti *et al.*, 2012; Inglese *et al.*, 2011).

Pruning frequency and intensity result in different canopy volumes and canopy-to-root ratios (Farinelli *et al.*, 2014; Cinosi *et al.*, 2022). In rainfed conditions, this affects the water status of the trees and, consequently, the phenolic and volatile compounds of the oil (Farinelli *et al.*, 2014 Cinosi *et al.*, 2022). A high canopy volume demands greater water consumption and, in rainfed conditions, this often causes water shortage in the summer. This results in higher values of phenolic compounds and lower values of volatile compounds, in particular (E)-2-hexenal and, more in general, C5 and C6 aldehydes (Cinosi *et al.*, 2022, 2023). In practice, pruning can cause effects similar to those obtained with high/low rainfall or irrigation (Inglese *et al.*, 2011). That is, with an increase in water availability there is a reduction in phenolic compounds and an increase in volatile compounds, in particular (E)-2-hexenal.

Annual pruning produces a relatively constant canopy volume and density and, hence, a more constant and balanced oil quality.

6.4.12. Safety Measures for Pruning

Safety in the use of pruning equipment and tools greatly depends on checking the condition of the equipment and its maintenance. In particular, it is important to check the condition of the equipment each time it is used and follow instruction manuals for regular maintenance. It is important to clean, lubricate and sharpen the cutting elements and to check the full efficiency of the safety devices (Famiani *et al.*, 2008b).

Workers must wear personal protective equipment (PPE) to reduce the risk of injury. These consist of: helmet for head protection; visors or goggles for eye protection; high-visibility jacket to be more easily visible to other workers; cut-resistant trousers for when using a chainsaw; gloves that protect against the cold, cuts, vibrations, electric shock; non-slip boots with medium-high legs and reinforced toes that reduce the risk of falling and crushing, if heavy objects fall on the feet; and ear muffs for hearing protection when using noisy tools (e.g. chainsaw) or engines (Famiani *et al.*, 2008b). The PPE to wear depends on the equipment used and is indicated in the instruction manuals of the individual tools and, often, in pictograms shown directly on the equipment.

6.4.13 Pruning and Climate Change

Climate change is producing warm and dry conditions. Pruning, by affecting the leaf surface area of trees, greatly influences water consumption by transpiration. Therefore, it is a practice that can help to adapt/optimize the cultivation to changes produced by climate change.

Moreover, as seen in paragraph 6.4.10, the recycling of pruning residues must be encouraged. By incorporating them into the soil, organic carbon reduces emissions of CO_2 into the atmosphere (Fraga *et al.*, 2021).

References

Albarracín, V. (2018) Efecto del momento e intensidad de poda sobre la demografía de ramas y floración de retorno del olivo (*Olea europaea* L. var Arbequina). PhD thesis, Universidad de Buenos Aires, Argentina.

Albarracín, V., Hall, A.J., Searles, P.S. and Rousseaux, M.C. (2017) Responses of vegetative growth and fruit yield to winter and summer mechanical pruning in olive trees. *Scientia Horticulturae* 225, 185–194.

Albarracín, V., Hall, A.J., Searles, P.S. and Rousseaux, M.C. (2018) Impact of simulated mechanical hedge pruning and wood age on new shoot demography and return flowering in olive trees. *Trees* 32(6), 1767–1777.

Albarracín, V., Hall, A.J., Searles, P.S. and Rousseaux, M.C. (2019) Responses of shoot growth, return flowering, and fruit yield to post-pruning practices and growth regulator application in olive trees. *Scientia Horticulturae* 254, 163–171.

Angeli, L.,Sillari, B. and Cantini, C. (1995) Cespuglio e monocono a confronto. *L'Informatore Agrario* 43, 59–63.

Barone, E., Gullo, G., Zappia, R. and Inglese, P. (1994) Effect of crop load on fruit ripening and olive oil (*Olea europaea* L.) quality. *Journal of Horticultural Science* 69(1), 63–73.

Barone, E., La Mantia, M., Marchese, A. and Marra, F.P. (2014) Improvement in yield and fruit size and quality of the main Italian table olive cultivar 'Nocellara del Belice'. *Scientia Agricola* 71, 52–57.

Barone, E., Briccoli Bati, C. and Di Vaio, C. (2020) Il settore dell'olivicoltura da tavola in Italia: situazione e prospettive. Accademia Nazionale dell'Olivo e dell'Olio, Spoleto (PG), Italy. Available at: www.accademiaolivoeolio.com (accessed 1 June 2023).

Cadahia Cicuendez, P. (1972) *Plantación y poda del Olivo*. Sindicato Nacional del Olivo, Madrid.

Camerini, F., Bartolozzi, F., Vergari, G. and Fontanazza, G. (1999) Analysis of the effects of ten years of mechanical pruning on the yield and certain morphological indexes in an olive orchard. *Acta Horticulturae* 474, 203–207.

Cantini, C. and Panicucci, M. (2002) Managing of a traditional olive orchard by timesaving biennial pruning. *Acta Horticulturae* 586, 361–363.

Cantini, C. and Sillari, B. (1998) Risultati produttivi ed economici di oliveti condotti con diversi sistemi di potatura. *Rivista di Frutticoltura e Ortofloricoltura* 60, 49–54.

Cherbiy-Hoffmann, S.U., Searles, P.S., Hall, A.J. and Rousseaux, M.C. (2012) Influence of light environment on yield determinants and components in large olive hedgerows following mechanical pruning in the subtropics of the Southern Hemisphere. *Scientia Horticulturae* 137, 36–42.

Cinosi, N., Moriconi, F. , Farinelli, D., Marchionni, D., Lodolini, E.M., Rosati, A. and Famiani F. (2023) Effects of summer pruning on the water status and physiology of olive trees and on fruit characteristics and oil quality. *Scientia Horticulturae*.

Cinosi, N., Urbani, S., Marchionni, D., Paoletti, A., Esposto, S. *et al.* (2022) Effects of pruning intensity on water relations and vegetative-reproductive activities of adult olive trees. V Convegno Nazionale dell'Olivo e dell'Olio (abstracts book). SOI, Firenze, Italy.

Civantos, L. and Pastor, M. (1996) Production techniques. In: *World Olive Encyclopedia*. International Olive Oil Council, Madrid, pp. 147–194.

Dias, A.B., Peça, J., Pinheiro, A., Santos, L., Morais, N. and Pereira, A.G. (2008) The influence of mechanical pruning on olive production and shaker efficiency. *Acta Horticulturae* 791, 307–313.

Dias, A.B., Peca, J.O. and Pinheiro, A. (2012) Long-term evaluation of the influence of mechanical pruning on olive growing. *Agronomy Journal* 104, 22–25.

Dias, A.B, Falcão, J., Pinheiro, A. and Peça, J. (2022) Effect of mechanical pruning on olive yield in a high-density olive orchard: an account of 14 years. *Agronomy* 12, 1105.

Famiani, F. (2017) Criteri di potatura per meccanizzare la raccolta. *Olivo e Olio* 2, 30–33.

Famiani, F. (2019) Mantenere le funzioni produttive in oliveti di interesse ambientale e paesaggistico. Proceedings: La multifunzionalità dell'olivicoltura umbra. Spoleto (PG), Italy, 13 December 2019, pp. 15–29.

Famiani F. and Gucci R. (2022) *La raccolta delle olive*. Edagricole – New Business Media, Bologna, Italy.

Famiani, F., Camposeo, S. and Gucci, R. (2021) *Tipologie di impianto per una moderna olivicoltura*. Accademia Nazionale dell'Olivo e dell'Olio, Spoleto (PG), Italy, pp. 77.

Famiani, F., Proietti, P. and Guelfi, P. (2008a) Tecnica di potatura dell'olivo. In: Proietti, P., Famiani, F., Pannelli, G. and Guelfi, P. (eds) *La potatura dell'olivo – manuale per una potatura semplificata ed agevolata*. Agenzia Regionale Umbra per lo Sviluppo e l'Innovazione in Agricoltura, Perugia, Italy, pp. 33–47.

Famiani, F., Proietti, P. and Guelfi, P. (2008b) Potatura dell'olivo: attrezzature, macchine e tipologie di esecuzione. In: Proietti, P., Famiani, F., Pannelli, G. and Guelfi, P. (eds) *La potatura dell'olivo – manuale per una*

potatura semplificata ed agevolata. Agenzia Regionale Umbra per lo Sviluppo e l'Innovazione in Agricoltura, Perugia, Italy, pp. 49–76.
Famiani, F., Proietti, P. and Inglese, P. (2021) *Progettazione e impianto dell'oliveto*. Accademia Nazionale dell'Olivo e dell'Olio, Spoleto (PG), Italy, pp. 75.
Famiani, F., Proietti, P., Lodolini, E.M. and Neri, D. (2009) Gestione della chioma. In: Pisante, M., Inglese, P. and Lercker, G. (eds), *L'ulivo e l'Olio*. Bayer CropScience – Collana Coltura & Cultura, Bologna, Italy, pp. 390–411.
Famiani, F., Hotteit, H. and Dubla, E. (2011) *Improve, Simplify & Facilitate Olive Pruning*. CHIEAM – Agronomic Mediterranean Institute of Bari, Italy. (DVD + illustrative notes, in Arabic and English)
Famiani, F., Cinosi, N., Paoletti, A., Farinelli, D., Rosati, A. and Lodolini, E.M. (2022) Deflowering as a tool to accelerate growth of young trees in both intensive and super-high-density olive orchards. *Agronomy* 12(10), 2319.
Farinelli, D., Baldicchi, A., Frioni, T., Al Hariri, R., Rosati, A., Urbani, S., Servili, M. and Famiani, F. (2014) Effetti della gestione della chioma sull'attività vegeto-produttiva, sulla fisiologia e sulla qualità dell'olio in un oliveto intensivo in Centro Italia. *Acta Italus Hortus* 14, 15.
Farinelli, D., Onorati, L., Ruffolo, M. and Tombesi, A. (2011) Mechanical pruning of adult olive trees and influence on yield and on efficiency of mechanical harvesting. *Acta Horticulturae* 924, 203–209. DOI: 10.17660/ActaHortic.2011.924.25
Ferguson, L., Krueger, W.H., Reyes, H. and Metheney, P. (2002) Effect of mechanical pruning on California black ripe (*Olea europea* L.) cv. 'Manzanillo' table olive yield. *Acta Horticulturae* 586, 281–284.
Ferreira, J. (1979) *Poda*. Explotaciones Olivareras Colaborados. Ministerio de Agricultura, Madrid.
Fodale, A.S., Pannelli, G., Filippucci, B. and Tucci, A. (1986) Prove di potatura agevolata dell'olivo in Sicilia e in Umbria. *Annali Istituto Sperimentale per l'Olivicoltura* 41–47.
Fontanazza, G. (1984) *Allevamento e potatura dell'olivo*. Universale Edagricole, Bologna, Italy.
Fontanazza, G., Camerini, F. and Bartolozzi, F. (1998) Intervento meccanico e manuale nella potatura di produzione. *Olivo e Olio* 1, 27–34.
Fraga, H., Moriondo, M., Leolini, L. and Santos, J.A. (2021) Mediterranean olive orchards under climate change: a review of future impacts and adaptation strategies. *Agronomy* 11, 56.
Garcia-Ortiz, A., Humanes, J., Pastor, M., Morales, J. and Fernandez, A. (2010) Pruning. In: Barranco, D., Fernández-Escobar, R. and Rallo, L. (eds) *Olive Growing*. Ediciones Mundi-Prensa, Madrid, pp. 349–391.
Giametta, G. (2003) Potatura meccanica. In: Fiorino P. (ed.) *Olea – Trattato di Olivicoltura*. Edagricole, Bologna, Italy, pp. 219–221.
Giametta, G. and Zimbalatti, G. (1995) La meccanizzazione della potatura in olivicoltura. In: Proceedings: L'olivicoltura mediterranea: stato e prospettive della coltura e della ricerca, Rende (CS), Italy, pp. 365–373.
Giametta, G. and Zimbalatti, G. (1997) Mechanical pruning in new olive groves. *Journal of Agricultural Engineering Research* 68, 15–20.
Gómez del Campo, M., Morales-Sillero, A., Vita Serman, F., Rousseaux, M.C. and Searles, P.S. (2010) Olive growing in the arid valleys of Northwest Argentina (provinces of Catamarca, La Rioja and San Juan). *Olivae* 114, 23–45.
Gregoriou, C. (2010) Tree training and pruning methods. In: *Good Agricultural Practices for the Near East & North Africa Countries*. FAO, Regional Office for the Near East, Cairo, pp. 109–115.
Gucci, R. and Cantini, C. (2000) *Pruning and Training Systems for Modern Olive Growing*. CSIRO Publishing, Australia.
Gucci, R. and Famiani, F. (2014) Ridurre gli interventi per ottimizzare i costi. *Olivo e Olio* 3, 32–37.
Inglese, P., Famiani, F., Galvano, F., Servili, M., Esposto, S. and Urbani, S. (2011) Factors affecting extra-virgin olive oil composition. *Horticultural Reviews* 38, 83–147.
Krueger, W.H., Maranto, J. and Sibbett, G.S. (2005) Olive fruit thinning. In: Sibbett, G.S., Ferguson, L., Coviello, J.L., Lindstrand, M. (eds) *Olive Production Manual*. University of California, Agriculture and Natural Resources, California, pp. 101–104.
Lo Bianco, R., Proietti, P., Regni, L. and Caruso, T. (2021) Planting systems for modern olive growing: strengths and weaknesses. *Agriculture* 11(6), 494.
Lodolini, E.M., Neri, D., Capogrossi, F. and Capogrossi, C. (2006a) Potatura meccanica dell'olivo: meno costi, più produzione. *L'Informatore agrario* 38, 50–53.
Lodolini, E.M., Neri, D., Capogrossi, F. and Capogrossi, C. (2006b) Preliminary results on olive (*Olea europaea* L.) mechanical hedging. *Proceedings of Olivebioteq 2006*, Vol II. Marsala (TP), Italy, pp. 165–168.
Lodolini, E.M., Endeshaw, T.S., Gangatharan, R., Neri, D. and Santinelli, A. (2011) Olive fruit set in central Italy in response to different pruning systems. *Acta Horticulturae* 924, 195–202.

Lodolini, E.M., Polverigiani, S., Grossetti, D. and Neri, D. (2018a) Pruning management in a high-density olive orchard. *Acta Horticulturae* 1199, 385–390.
Lodolini, E.M., Polverigiani, S., Sirugo, M. and Neri, D. (2018b) Damage to several olive cultivars by two over-the-row harvesters in high-density orchards. *Acta Horticulturae* 1199, 415–419.
Lodolini, E.M., Polverigiani, S., Cioccolanti, T., Santinelli, A. and Neri, D. (2019) Preliminary results about the influence of pruning time and intensity on vegetative growth and fruit yield of a semi-intensive olive orchard. *Journal of Agricultural Science and Technology* 21(4), 969–980.
Lodolini, E.M., Paoletti, A., Nolasco, A., Ceccarelli, D., Rosati, A. *et al.* (2022) Canopy recovery in traditional olive groves. *Acta Horticulturae* 1346, 375–382.
Lodolini, E.M., Polverigiani, S., Giorgi, V., Famiani, F. and Neri, D. (2023) Time and type of pruning affect tree growth and yield in high-density olive orchards. *Scientia Horticulturae* 311, 111831.
Martin, G.C., Nishijima, C. and Rapoport, H.F. (1990) Abscission response of olive flower and fruit population to NAA. XXIII International Horticultural Congress, Firenze, Italy. Vol. 1.
Massenti, R., Ioppolo, A., Veneziani, G., Selvaggini, R., Servili, M., Lo Bianco, R. and Caruso, T. (2022) Low tree vigor, free palmette training form, and high planting density increase olive and oil yield efficiency in dry, sloping areas of Mediterranean regions. *Horticulturae* 8(9), 817.
Metzidakis, I. (2002) Effect of regeneration pruning for the recovery of olive productivity and fruit characteristics in ten olive cultivars. *Acta Horticulturae* 586, 333–336.
Monteleone, J., Zuritz, C.A., Gascón, A.D. and Morábito, J.A. (2013) Efecto del riego y la poda en elcontenido de componentesmenores y ácidos grasos del aceite de oliva virgen varietal Farga. *Grasas y Aceites* 64, 369–377.
Morales, J. (2002) Manejo de la poda del olivar tradicional. *Revista Olivae* 92, 38–43.
Morales, J. (2001) La poda del olivar tradicional. *Revista Oleo* 112–118.
Morales, J., Fernández, A. and Frías, L. (2000) Ensayos de poda en Jaén. *Revista Agricultura* 820, 731–735.
Morettini, A. (1972) *Olivicoltura*. Reda, Rome.
Ortega Nieto, J.M. (1962) *Estudios y experiencias de la poda del olivo*. Ministerio de Agricultura Dirección General de Agricultura, Madrid.
Ortega Nieto, J.M. (1969) *La poda del olivo*. Ministerio de Agricultura Dirección General de Agricultura, Madrid, Spain.
Ottanelli, A., Marone, E. and Fiorino, P. (2019) A new device to improve the mechanical winter pruning in olive trees hedgerows. *Advances in Horticultural Science* 33(1), 113–122.
Pastor, M. and Humanes, J. (1989) Poda del olivo: moderna olivicultura. Ed. Agricola, S.A., Madrid.
Pastor, M., Vega, V. and Humaned, J. (1991) Poda meccánica del olivar en Andalucia. *Máquinas y Tractores Agricolas* (1) 31–40.
Pastor, M., Hermoso, M., Revilla, J., Navarro, C., Morales, J., Vega, V. and Arquero, O. (1992) *Poda de producción del olivar*. Aclareo químico de frutos con ANA. Serie Informaciones Tecnicas 14/92. Servicio de Publicaciones de Consejería de Agricultura y Pesca. Junta de Andalucía.
Peça, J.O., Dias, A.B., Pinheiro, A.C., Santos, L., Morais, N., Pereira, A.G. and Reynolds De Souza, D. (2002) Mechanical pruning of olive trees as an alternative to manual pruning. *Acta Horticulturae* 586, 295–299.
Perez Mohedano, D. (2005) Diferente frequiencias de poda en olivares tradicionales. *Vida Rural* November, 42–44.
Pierantozzi, P., Torres, M., Tivani Keaik, M., Contreras, C., Gentili, L., Parera, C. and Maestri, D. (2020) Spring deficit irrigation in olive (cv. Genovesa) growing under arid continental climate: effects on vegetative growth and productive parameters. *Agricultural Water Management* 238, 106212.
Proietti, P., Palliotti, A., Famiani, F., Preziosi, P. and Antognozzi, E. (1998) Confronto tra le forma di allevamento a monocono e a vaso in diverse cultivar di olivo. *Rivista di Frutticoltura* 7–8d.
Proietti, P., Famiani, F., Nasini, L. and Tombesi, A. (2002) The influence of some agronomic parameters on the efficiency of mechanical harvest on young olive trees. *Acta Horticulturaea* 586, 415–418.
Proietti, P., Famiani, F. and Guelfi, P. (2008a) Principi, basi e scelte per la potatura dell'olivo. In: Proietti, P., Famiani, F., Pannelli, G. and Guelfi, P. (eds) *La potatura dell'olivo – manuale per una potatura semplificata ed agevolata*. Agenzia Regionale Umbra per lo Sviluppo e l'Innovazione in Agricoltura, Perugia, Italy, pp. 7–20.
Proietti, P., Famiani, F. and Guelfi, P. (2008b) Forme di allevamento dell'olivo. Principi, basi e scelte per la potatura dell'olivo. In: Proietti, P., Famiani, F., Pannelli, G. and Guelfi, P. (eds) *La potatura dell'olivo – manuale per una potatura semplificata ed agevolata*. Agenzia Regionale Umbra per lo Sviluppo e l'Innovazione in Agricoltura, Perugia, Italy, pp. 21–31.

Proietti, P., Famiani, F., Guelfi, P. and Pannelli G. (2008c) *La potatura dell'olivo: manuale per una potatura semplificata ed agevolata*. Agenzia Regionale Umbra per lo Sviluppo e l'Innovazione in Agricoltura (ARUSIA), Perugia, Italy.

Proietti, P., Nasini, L., Famiani, F., Guelfi, P. and Standardi, A. (2012) Influence of light availability on fruit and oil characteristics in *Olea europaea* L. *Acta Horticulturae* 949, 243–249

Rallo, L. and Barranco, D. (1986) Influence of the time of application on the response of olive to chemical thinning. *Acta Horticulturae* 179, 709–710.

Rallo, L., Barranco, D., Castro-Garcia, S., Connor, D.J., Gomez del Campo, M. and Rallo, P. (2013) High-density olive plantations. *Horticultural Reviews* 41, 303–384.

Rius, X. and Lacarte, J.M. (2010) Manejo de la plantación. In: La Revolución del Olivar – El Cultivo en Seto. Comgrafic, Barcelona, Spain, pp. 263–287.

Roventini, A. (1936) La ricostituzione olivicola attraverso la potatura. *L'Italia Agricola* 7.

Sibbett, G.S. (2005) Pruning mature bearing olive trees. In: Sibbett, G.S., Ferguson, L., Coviello, J.L. and Lindstrand, M. (eds) *Olive Production Manual*. University of California, Agriculture and Natural Resources, California, pp. 55–59.

Solé Riera, M. and Guiu, M.F. (1996) Poda de producción en arbequina. *Revista Agricultura* 66, 371–374.

Tombesi, A. (1989) Potatura e forme di allevamento nell'olivo. *Rivista di Frutticoltura e Ortofloricoltura* 1.

Tombesi, A. and Correia, J. (2004) Potatura e intercettazione dell'energia radiante nell'olivo. *Rivista di Frutticoltura e Ortofloricoltura* 66, 31–35.

Tombesi, S. and Farinelli, D. (2014) Evaluation of canopy elasticity, light penetration and reciprocal shading for optimal canopy management in high density hedgerow olive orchards. *Acta Horticulturae* 1057, 315–320.

Tombesi, S. and Farinelli, D. (2018) Potatura meccanica di oliveti superintensivi. *Olivo e Olio* 2, 26–30.

Tombesi, A. and Tombesi, S. (2007) Olive pruning and training. In: *Production Techniques in Olive Growing*. International Olive Council, Madrid, pp. 44–81.

Tombesi, A., Boco, M., Pilli, M. and Farinelli, D. (2000) Influenza della potatura sulla raccolta meccanica delle olive con vibratori di tronco. *Rivista di Frutticoltura e Ortofloricoltura* 62, 69–73.

Tombesi, A., Boco, M., Pilli, M., Ruffolo, M. and Farinelli, D. (2008) Intensità e periodicità di potatura dell'olivo, risultati di una prova decennale. Proceedings: Convegno COM.SI.OL – Competitività del sistema olivo in Italia, Spoleto (PG), Italy, pp. 119–126.

Torres, M., Pierantozzi, P., Searles, P., Rousseaux, C., García-Inza, G. *et al.* (2017) Olive cultivation in the Southern hemisphere: flowering, water requirements and oil quality responses to new crop environments. *Frontiers in Plant Science* 8, 1830.

Vitagliano, C. and Bartolini, S. (2003a) Criteri di scelta per le forme di allevamento. In: Fiorino P. (ed.) *Olea – Trattato di Olivicoltura*. Edagricole, Bologna, Italy, pp. 223–233.

Vitagliano, C. and Bartolini, S. (2003b) Potatura. In: Fiorino, P. (ed.) *Olea – Trattato di Olivicoltura*. Edagricole, Bologna, Italy, pp. 211–218.

Vivaldi, G.A., Strippoli, G., Pascuzzi, S., Stellacci, A.M. and Camposeo, S. (2015) Olive genotypes cultivated in an adult high-density orchard respond differently to canopy restraining by mechanical and manual pruning. *Scientia Horticulturae* 192, 391–399.

6.5 Olive Harvesting

S. Castro-Garcia[1], D. Farinelli[2], L. Ferguson[3], J.A. Gil-Ribes[1] and F. Famiani[2]
[1]*Departamento de Ingeniería Rural, Construcciones Civiles y Proyectos de Ingeniería, Universidad de Córdoba, Spain;* [2]*Dipartimento di Scienze Agrarie, Alimentari e Ambientali, Università degli Studi di Perugia, Italy;* [3]*Department of Plant Sciences, University of California Davis, California, USA*

6.5.1 Introduction

The choice of the period of time to carry out the olive harvest and the choice of the method for harvesting greatly influence the quantity and quality of the product and its cost. Lack of hand laborers in olive-producing countries and the high cost of manual harvesting, especially in Europe and the West, are the main reasons why mechanical harvesting of olives is needed. Furthermore, it also reduces labor fatigue in workers. In addition, mechanization enables harvesting to be concentrated in such a way that the farm can more easily pursue its quantitative/qualitative objectives.

Hand harvesting is still an option on small-scale farms (< 1 ha) and in countries where the cost of manpower is low.

The choice of harvesting time and the physiological, mechanical and logistical aspects of mechanical harvesting of both oil and table olives are discussed in this chapter.

6.5.2 Harvesting Time

After cultivar, harvesting timing greatly affects both table olive and oil quantity and quality, and harvester efficiency (Famiani *et al.*, 2002, 2016; Famiani and Gucci, 2022). The force required to detach an unripe olive is greater than that for ripe olives. Therefore, harvesting earlier reduces yield and efficiency. Harvesting later, as ripening proceeds, is easier and more efficient but quality may be reduced (Famiani and Gucci, 2022).

6.5.2.1 Oil synthesis and accumulation

During initial olive fruit growth, oil accumulation is slow. Oil starts accumulating in the pulp after kernel hardening (Famiani *et al.*, 1999; Farinelli *et al.*, 2004; Beltran *et al.*, 2017). During the hottest summer months, oil increases slowly, then at the end of summer/beginning of autumn (September and October in the northern hemisphere), oil accumulates rapidly. Thereafter, the rate of accumulation in most cultivars gradually decreases and then ceases between mid-November and mid-December (northern hemisphere). Most of the oil (96–98%) is in the pulp, with only 2–4% in the seed.

6.5.2.2 Fruit ripening

During ripening, fruits turn from green to colored, through a progressive pigmentation, first of the skin and then of the pulp (from the outside to the inside). Cultivars differ,

DOI: 10.1079/9781789247350.0020

but, generally, in those with gradual color development the oil percentage maximizes when the surface is fully colored: Jaén index of 4 (Famiani *et al.*, 1999, 2004a; Beltran *et al.*, 2017). Varieties that color early achieve maximum oil concentration at higher color values (Jaén index > 4) (Famiani *et al.*, 1999, 2004b). Pulp consistency decreases during ripening. This is greatly influenced by cultivar as different varieties have different cell wall thicknesses and hence different cell wall degradation rates (Lavee, 1996). Simultaneously, fruit weight increases, increasing harvester efficiency. This is an advantage for table olives that are irrigated as close to harvest as possible. However, for oil olives in the final stage of ripening, fruit fresh weight might decrease due to water loss. Oil accumulates, increasing the oil percentage, which, depending on the cultivar, may vary from 10% to 30% of the fruit fresh weight. Evaluating oil percentage as a function of dry weight gives a more accurate estimate of oil content than that based on fresh weight as water weight is eliminated. Indeed, at late harvest, high percentages of oil content expressed on the fresh weight basis are often due to dehydration of the fruit.

In unripe fruit, the fruit detachment force (FDF) can be as high as 6 N, decreasing, depending upon cultivar and climate, with ripening. The rate of FDF development is a major factor in determining harvest time. Pre-harvest fruit drop begins at ~ 4-4.5 N and can be severe at ~ 3 N (Tombesi *et al.*, 2021). Fruit drop is exacerbated by strong winds, rain and parasitic attacks, such as the olive fly. Because fallen olives are generally not harvested, the amount of dropped olives can cause a reduction in the oil obtained. Usually, when the drop exceeds 10% of the olives on the tree, the amount of oil is reduced from those remaining on the canopy, becoming economically unacceptable.

6.5.2.3 Effect of ripening on the olive and on oil quality

With ripening, oil chlorophyll and carotenoid content decrease. Simultaneously, oleic acid increases and palmitic acid decreases, increasing the ratio of unsaturated to saturated fatty acids (Servili *et al.*, 1990; Baccouri *et al.*, 2008; Inglese *et al.*, 2011). Early in ripening, when only the skin is pigmented, volatile aldehydes, particularly trans-2-hexenal responsible for the 'herbaceous' smell, and saturated and unsaturated alcohols at C5 and C6, esters and phenolics, responsible for the bitter and pungent qualities, are highest, decreasing thereafter (García *et al.*, 1996; Angerosa *et al.* 2004). Tocopherols (antioxidants), mainly α-tocopherol, also tend to decrease during ripening (Di Matteo *et al.*, 1992; Baccouri *et al.*, 2008). Oils extracted from olives at a very advanced stage of ripening (overripe), as a result of significant cell wall breakdown can have increased acidity and oxidation (free acidity and peroxide value), because the fruit is more susceptible to rupture of cellular compartmentation during harvesting and subsequent handling of the olives before milling, with consequent activation of oxidative processes and hydrolytic alteration (Inglese *et al.*, 2011).

6.5.2.4 Determining harvest date

Harvesting date greatly affects oil quantity, quality and type. All oil types have their own market sector; thus, harvest date can be determined by the target market (Famiani and Gucci, 2022).

During ripening, extractable oil increases with olive fruit growth and oil accumulation within it but, simultaneously, decreases due to fruit drop. Therefore, early harvesting avoids fruit drop but does not take full advantage of potential fruit growth and oil accumulation. Late harvesting may decrease oil quantity as fruit growth and oil accumulation may not compensate for fruit drop losses. Over 10% fruit drop generally decreases the total oil quantity.

Very early harvesting, olives at 1–2 Jaén index, produces a 'new' oil with a characteristic green color and intense pungency and bitterness. This early harvested oil is marketed as a high-quality and health value 'first fruit' oil with high volatile and phenolic contents. It is expensive due to the low oil content and difficult harvesting.

Box 6.5.1: Ripening indices

In order to establish the harvesting time according to the farm's objective, reference is made to the variation of certain fruit properties, which are correlated to the quantity and/or quality of the oil. These properties are referred to as harvest indices. To use these indices, a 1–2 kg sample of fruit from the upper, lower, outside and inside canopy of at least ten different trees is collected as ripening proceeds.

Pigmentation index

This index is based on a color scale ranging from 0 to 7: Jaén index. It gives useful information about both the quantity and quality of the oil produced. The index is calculated using the following formula:

$$\text{Pigmentation index} = \frac{\sum_{i=0}^{i=7}(i \cdot n_i)}{N}$$

where: n_i = number of olives; $i = 0$ for completely green olives; $i = 1$ for olives with color change from green to light-green-yellow; $i = 2$ for olives with pigmentation on less than 50% of the skin surface; $i = 3$ for olives with pigmentation on more than 50% of the skin surface; $i = 4$ for olives with pigmentation on 100% of the skin surface; $i = 5$ for olives with pigmentation on less than 50% of the pulp; $i = 6$ for olives with pigmentation on more than 50% of the pulp; $i = 7$ for olives with pigmentation on 100% of the pulp; N = total number of olives in the sample.

It is determined in a subsample of at least 100 fruits.

Pulp consistency

It is measured with a penetrometer with a 1–2 mm diameter tip that is inserted into the pulp of the olive. Fruit susceptible to mechanical damage that will degrade oil quality, especially if the olives are not processed immediately after harvesting, will have penetrometer values below 250 g with a 1 mm tip or 500 g with a 1.5 mm tip.

Oil content

Oil content by percentage can be determined by solvent extraction, Soxhlet extraction or nuclear magnetic resonance (NMR) and near infrared (NIR) imaging.

Fruit detachment force

FDF, which is the force needed to detach the fruit, is measured with a dynamometer, or pull-force gauge, in Newtons (N). Typically, 120–150 fruits from at least three sets of three trees are sampled. When FDF values are ~ 5 N for table olives and ~ 4 N for oil olives, good detachment is obtained with mechanical harvesting.

Relationship between fruit detachment force and fruit fresh weight

When the FDF/fruit weight (N/g) is less than 2.3 N/g, mechanical harvesting is more efficient. Generally, harvest yield (percentage of detached fruit) over 85% with trunk shakers and 90–95% with canopy contact harvesting and harvest aides can be obtained (Famiani *et al.*, 2010, 2014; Farinelli *et al.*, 2012c). Moreover, a high harvest work productivity (kg of olive harvested/hour/worker) is obtained.

Pre-harvest fruit drop

At the beginning of ripening, on at least 8–12 trees representative of the olive orchard, at least 8–12 branches (one branch/tree), oriented in different directions and 1.5 m and 2 m above the ground, are selected and labeled. At the time of labeling, and then periodically, the fruits on each branch are counted, and by difference from one count to the next, the percentage of fallen fruits is estimated (Famiani and Gucci, 2022).

Non-destructive ripening indices

Studies are underway to identify indices that take into account several maturation parameters at the same time and are measurable with non-destructive tools/techniques. Promising results have been obtained using near infrared (NIR) technology (Cinosi *et al.*, 2023 and references in it).

Typical oils of high quality and/or high health value are characterized by a high content of volatile and phenolic substances and are obtained by harvesting early. POD (Protected Origin Denomination) and PGI (Protected Geographic Indication) oils also belong to this category. These types of oils are obtained by harvesting the fruit when only a part of the skin is pigmented (Jaén index > 2 < 4). The oils are usually green and have strong sensorial notes of green fruit, bitterness and pungency. These sensations tend to diminish when the index is around 4, the level of coloration at which the maximum quantity of oil in the fruit is usually reached. Then, as the pulp becomes pigmented (Jaén index > 4), the oil progressively takes on a lighter color (from yellowish-green to yellow) and softens, that is there is a reduction in the intensity of bitterness, pungency and fruitiness, with the latter going from 'green' to 'mature'. At the same time, fruit fall increases progressively. Sweet oils also have their own market segment: they are suitable for seasoning foods with delicate flavors and are chosen by those who prefer neutral seasonings. Thus, there is no single quality reference and the choice of the type of oil to be obtained depends on the market segment that is chosen as a reference.

Since color is one of the main harvest indices, it is important to keep in mind that while the majority of cultivars change color gradually, there are some cultivars that change color very slowly and remain green or partially pigmented even in the latter part of the harvest season, while others change color early and quickly. Usually, the first ones, especially if, as often happens, the pulp softens more slowly, the quantity and quality of the oil can be more easily balanced, because they can be harvested later. On the other hand, cultivars with fruit that changes color and softens early, should be harvested early, in order to obtain an oil with marked sensations of bitterness, pungency and green fruit.

Finally, local pests and diseases and climate can dictate harvest time. The threats of late olive fruit fly (*Bactrocera oleae*) attacks, especially in organic olive orchards, and early frost damage, make early harvesting necessary.

6.5.3 Olive Harvesting Methods

The first factor to be considered in developing mechanical harvesting is FDF, and the factors that comprise FDF: stem (peduncle) length, fruit weight and the abscission zone. Stem length is cultivar-dependent. Short-stemmed cultivars are easier to harvest because they have better vibration transmission (Tsatsarelis, 1987).

Fruit weight is also cultivar-dependent but can be improved with cultural practices, particularly irrigation. Larger fruits are easier to harvest and the first to fall with harvesting (Kouraba *et al.*, 2004).

The abscission zone changes with variety and maturity. In early-ripening varieties the abscission zone is between the peduncle-shoot and pedicel-rachis. In the late-maturity ones the abscission zone is between the pedicel-rachis (Castillo-Llanque and Rapoport, 2008). The peduncle-shoot abscission zone includes the stem and leaves as debris. This results in loss of transport capacity, and in the case of liquid transport for table olives, the debris could cause problems in loading and unloading the tanks with liquid transfer pumps.

6.5.3.1 Hand harvesting

In the past, oil olives were frequently harvested by hand from the ground after dropping naturally (Çakmak *et al.*, 2011). This resulted in decreased yields and quality of the product. Currently, manual harvesting is done from the trees. It is done early to avoid fruit drop, wind and rain that exacerbate drop, insect pests, and to improve processed fruit and oil quality. Ladders are used to access the top of the canopy.

Manual harvesting is done with a downward 'milking' motion, pulling the fruit with the worker's fingers into a bucket worn by the picker, or onto the ground. To facilitate hand harvesting, wood, metal or, mainly, plastic combs are used. Combs exert a 'raking' motion and are particularly useful for the detachment of small-sized fruit. Harvesting with poles (wooden, but also fiberglass) detaches the fruit by beating the branches, but is less effective in small fruit varieties with high FDF values and causes tree damage. In the worst case, damage makes trees more susceptible to olive knot bacteria, *Pseudomonas savastanoi*, and to alternate bearing, if a considerable number of current year's apical shoots are broken or damaged. Pole beating can also be combined with trunk shakers or branch-shaking hooks (*aide devise*) to enhance harvest efficiency (Zion *et al.*, 2011). However, poles are increasingly being replaced with hand-held branch or comb shakers. Manual harvesting methods are used primarily on traditional, less productive, smaller farms. For table olives, woven cotton gloves reduce fruit damage during harvesting.

In traditional orchards, manual harvesting efficiency, mainly depending on fruit weight, load/tree and canopy size and height, ranges from 5 to 30 kg/hour/worker (Lavee, 2010; Rallo *et al.*, 2013; Morales-Sillero *et al.*, 2014). Although manual harvesting capacity is low, the percentage of harvested fruits is very high (94–96%).

6.5.3.2 Mechanical harvesting aides

Mechanical harvesting aides are small devises carried by individual workers. They require little tree adaptation and are inexpensive and therefore common, mostly in olive oil harvesting.

Hand-held harvest aides can be autonomous, require a tractor or be self-propelled units. The most common are autonomous hand-held. These gasoline, pneumatic or electrical units are mounted on lightweight, aluminum or carbon-fiber poles, weigh 2–14 kg (the heaviest are those powered by an endothermic gasoline engine) and can reach 4–5 m in height. At the tip of the pole is a branch hook for shaking, combs or beaters. The worker selects the frequency and duration, and specific canopy area, but not amplitude.

The hand-held branch shakers produce a unidirectional vibration with a crankshaft-rod device (Fig. 6.5.1). They are usually powered by a 2-stroke 1 to 3 kW engine. The high and variable rotation speed of the engine is converted into a unidirectional movement on the branch. The branch and the pole are connected with a hook, covered with rubber to avoid bark damage. Branches up to 6 cm in diameter can be grabbed with the hook. These devices can develop a vibration

Fig. 6.5.1. Hand-held branch shaker powered by two-stroke engine combined with manual rod harvesting. (Photo: Sergio Castro-Garcia)

frequency from 12 to 25 Hz, depending on the engine and the branch diameter. The vibration transmitted to the branch has a high amplitude, 6-8 cm, able to produce a branch acceleration with average values of 430 m/s^2 (Kouraba *et al.*, 2004). Several short vibrations can be used on the same branch for a total vibration time of 10 and 15 s per position. The final harvest efficiency, even at high vibration frequencies is greatly influenced by the FDF of the olive cultivar, fruit weight, and branch diameter, length, and the angle of insertion with respect to the vertical.

The hand-held canopy combs or beaters apply an alternating vibration into the canopy's small branches and leaves with penetrating rods (Fig. 6.5.2). The devices have carbon fiber or other lightweight, synthetic, 30–40 cm rods that have a vibration frequency of 6 Hz and 22 Hz. The rod movement can be alternate, rotating or a combination. In most cases they are pneumatically or electrically powered. The former are powered by a compressor that can be wheeled, self-propelled or towed by a tractor. The latter can be powered by 12 V or 24 V batteries (Calvo *et al.*, 2018). Their harvesting efficiency is very high due to acceleration rates, over 600 m/s^2, as well as the possibility of the operator controlling the duration and canopy area selection. In early harvesting, when the fruit is harder to detach, the longer duration of their action can cause considerable breaking of shoots and branches, decreasing the following season's yield and increasing susceptibility to olive knot *Pseudomonas savastanoi*. These combs and beaters can also be used simultaneously with trunk shakers to improve harvester efficiency, especially in early harvesting and in trees which do not have a good structure for the use of the shaker.

Factors affecting harvesting efficiency with mechanical harvesting aides

As olives mature, they are easier to harvest as detachment force is reduced (Farinelli *et al.*, 2012c). Therefore, at an early stage, harvesting time increases and, consequently, labor productivity is lower (kg of olives harvested/hour/worker) (Famiani *et al.*, 2008). Cultivar also greatly influences harvesting efficiency: difficult cultivars (small fruit and high detachment force), early in the season, require up to 50% more time than easier cultivars, while later in the season these differences lessen (Famiani *et al.*, 2008). Early in the season, and with

Fig. 6.5.2. Hand-held canopy beaters powered by electric battery. (Photo: Sergio Castro-Garcia)

difficult cultivars, the persistence with hand-held canopy shakers to detach the fruit leads to greater damage to the vegetation. Early in the season, branch shaking with hooks commonly produces low harvest efficiencies of 50–75%, increasing to > 90–95% later in the season (Benther and Rouina, 2002; Parlati *et al.*, 2006; Famiani *et al.*, 2010). With hand-held devices, the best ratio of detachment force to fruit weight is less than 2.5 N/g (Famiani *et al.*, 2010). The same factors that prolong the shaking time needed to detach the fruits can increase shoot and bark damage.

These aides can be used with all types of training and orchard conformations, as long as tree height does not exceed 4.5 m. The reduction in height with pruning can be compensated by promoting vegetation in the lower canopy in order to maintain a good volume for production (Famiani, 2017). Annual pruning produces the best results. Less-frequent pruning promotes growth and concentration of fruits in the top of the canopy, which decreases yield and worker productivity due to poor fruit accessibility.

When harvesting manually with hand-held aides, nets can increase worker productivity by a factor of 2–3 and even up to 4–5. The highest increases are achieved with heavily cropped trees and a cultivar with a low FDF late in the season (Famiani *et al.*, 2004a; Ferguson, 2006). Under these conditions a skilled worker's productivity usually ranges from 20 to 40 kg/hour (Famiani and Gucci, 2022), with peaks up to 60–65 kg/hour (Deboli *et al.*, 2016; Lavee, 2010).

Harvesting procedures

Hand-held harvesting aides are usually combined with on-ground collection nets, that require crews and coordination. Generally, early harvesting and small varieties with high FDF require more equipment operators and later harvesting of easily detached larger fruits require more net crews (Famiani and Gucci, 2022). The best nets are rectangular and placed in the tree row adjacent to the trees. Efficiency is higher with longer nets, 25–30 m under 3–5 trees for 2–5 hand-held shaker operators and the same number for the nets. As seen above, this ratio can vary as the harvest season progresses as well as the fruit characteristics and tree load.

Net management can be mechanized with nets rolled up in 50–100 m-long reels, towed by tractors, spread in the tree row to collect the fruit, then retracted, emptied and rewound. Tractor-pulled or self-propelled inverted umbrella collection nets, that clamp under individual trees for single-tree harvests, are available but are only used on small farms as they are emptied after each tree, making the work slow and inefficient.

6.5.3.3 Trunk shakers

Trunk shakers are attached to the tree trunk, or a main branch in larger trees, and the vibration transmitted throughout the tree detaches the fruit.

There are multiple similar trunk shakers produced by a number of manufacturers. Breakdowns halt harvesting, and workers require rapid field repair. Therefore, in addition to choosing a specific trunk shaker for power and reliability, it is also essential to deal with a local and responsible supplier.

Trunk shakers can be self-propelled (Fig. 6.5.3) or powered by tractors or other multi-use vehicles, such as telescopic handlers (Fig. 6.5.4). Essentially, a trunk shaker is a machine that converts the hydraulic power, oil pressure and flow, supplied by the power unit, into mechanical power through the rotation of one or two eccentric masses located in the vibration head, which is attached to the tree trunk. The main characteristics of trunk shakers also determine their advantages and disadvantages for mechanical olive harvesting:

1. The vibration is applied to the entire tree. This requires high levels of power, especially for tall trees with large trunks. Older tractors are often unable to provide the necessary power.
2. The detached fruit falls within a few seconds, increasing the working capacity of the trunk shaker, but making it more difficult to manage the detached fruit. In cultivars with staggered ripening, results are more variable.

Fig. 6.5.3. Self-propelled trunk shaker. (Photo: Sergio Castro-Garcia)

3. Trees must be adapted for efficient trunk-shaker harvesting. Results are best with single-trunked trees with three to five upright scaffolds above 45° forming an open vase.

The main trunk shaker components are listed below:

- hydraulic power generating system;
- vibration head and positioning system;
- vibration system;
- trunk clamping system.

HYDRAULIC POWER GENERATING SYSTEM. Trunk shakers require high power for adequate operation. A schematic of the machine's hydraulic system is shown in Fig. 6.5.5. If the trunk shaker is driven by a tractor, the hydraulic power is generated via a power take-off (PTO) connection. Usually, between the PTO of the tractor and the hydraulic pump of the shaker there is a gear system to adjust the rotation speed. It is always advisable to follow the manufacturer's specifications to operate the machine according to the PTO (540 or 1000 rev/min). However, if the trunk shaker is self-propelled, the hydraulic power is generated by the machine's engine via one or more hydraulic pumps. As with any machine, work should start when the transmission oil is at the right temperature; for this reason, the tanks are usually equipped with thermometers or temperature sensors.

Trunk shaker hydraulic systems can be classified as either an open-center or closed-center system. Trunk shakers with an open-center hydraulic system are the most widely used. This system is suited to the operation of trunk shakers as they are operated intermittently and have a limited number of actuators (5–8 hydraulic cylinders and 1 or 2 motors) but have a higher number of hydraulic distributors than a closed-center system. The tractor PTO drives one or more hydraulic pumps with constant displacement, usually rotary vane pumps. In an open-center system, the tractor PTO drives

Fig. 6.5.4. Trunk shaker powered by a self-propelled unit. (Photo: Sergio Castro-Garcia)

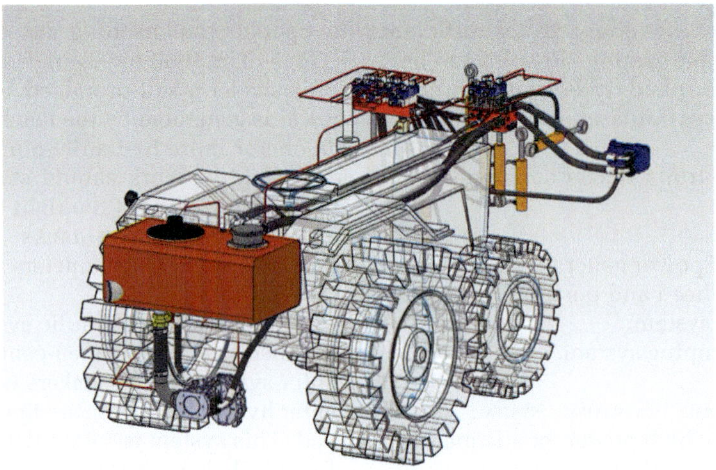

Fig. 6.5.5. Schematic of the hydraulic system of a trunk shaker. (Source: www.hydba.com)

the hydraulic pump, producing a constant oil flow to the hydraulic circuit, but with variable oil pressure depending on the operation of the different actuators. In this system, the machine can have two different displacement pumps. One pump with a smaller displacement dedicated to the operation of the different actuators (double-acting hydraulic cylinders). This gives the operator good control and operating precision when clamping the trunk. Another pump driven in parallel, but with a larger displacement, is used to generate the vibration by powering a hydraulic motor that turns one or two unbalanced masses. In some models, during the vibration process, the oil flow of both pumps can be combined to move the eccentric mass at a higher speed (higher vibration frequency), but they do so at the same maximum pressure. The maximum oil pressure of this system is usually less than 200–220 bar, with flow rates close to 180–220 L/min. These values indicate that the tractor should reach a power of 60–80 kW at the PTO. These values could be lower if the trees have small canopies. In open-center systems, the transmission oil is always in circulation, returning to the tank if the machine is not operated. To reduce the oil temperature, these machines usually have large oil tanks and their own cooling systems. Although the open-center system is simple, it usually does not allow the operator to change the vibration frequency, i.e. the oil flow of the pump or the rotation speed of the unbalanced mass. If the operator chooses to reduce the engine speed, and therefore the PTO speed, the vibration frequency can be reduced, but this will also reduce the available power that the tractor can provide. However, the pump control can be equipped with a load-sensing system to change its flow rate, achieving a proportional system. In both open- and closed-center circuit systems, reversible motors are usually mounted enabling the unbalanced mass to rotate in both directions. To date, there are no studies that demonstrate the advantages of vibrating in one direction and then the other. However, there is evidence that two vibrations of short duration are more effective than one vibration in harvesting efficiency, especially early in the harvest season.

Trunk shakers with a closed-center hydraulic system enable the pump displacement to be adjusted according to the requirements of the machine. Usually, this system has a variable-flow pump with axial pistons that can produce high pressures (400–500 bar), less common in agricultural machinery and more common in specialized machines such as excavators, bulldozers etc. In these, the pump changes its capacity according to the system load, providing a variable flow rate but keeping the pressure in a narrow range. The flow rate of the pump is adjustable to the diameter of the hoses or actuators. This pump allows its displacement to be reduced so that the hydraulic actuators are proportional to the operator's control. However, the high pressure of the pump must be related to the motor characteristics, which, if they are the vane type, cannot exceed 200 bar when using an axial-piston hydraulic motor. During vibration, the displacement of the pump is increased to provide sufficient flow for the motor to reach the required vibration frequency. In this closed-center hydraulic system, the control of the vibration frequency can be done easily without reducing the power reserve of the tractor to operate the machine.

In summary, trunk shakers usually use the open-center hydraulic system because of the simplicity of the configuration of the component, and because it is well adapted to the way they work, with intermittent stops and with a reduced number of hydraulic actuators, i.e. hydraulic motors and cylinders. In general, the open-center system is very reliable, and failures are usually less costly than in closed-center systems. The closed-center systems are more flexible in the different operations than open-center systems, i.e. they only use the required flow at the required pressure, which allows them to work at maximum efficiency and save fuel. This system is highly recommended if the collection system has a large number of actuators, or if robust and fast actuation is required. Nowadays, many manufacturers choose an intermediate solution such as load-sensing systems, which offer a suitable solution for trunk shakers working under demanding operating conditions.

VIBRATION HEAD AND POSITIONING SYSTEM. Trunk shakers can be towed by a tractor, or an all-purpose vehicle, or they can be self-propelled. The main advantage of a tractor is its utility. However, tractors are less maneuverable and have poor operator line-of-sight. Multipurpose machinery, such as telescopic handlers or small excavators, have good operator line-of-sight, high-performance oil hydraulic systems, closed-center hydraulics, good stability and good maneuverability provided by the wheel motors, rubber chain drive or four-wheel steering. Self-propelled trunk shakers have hydrostatic transmissions enabling a high harvesting capacity of 30–40 trees/h. Their high maneuverability reduces operating time and their low center of gravity facilitates operator visibility and stability.

The first step in harvesting is the operator directing the shaker head to clamp the trunk. How the head is mounted on the harvesting unit depends on the orchard design. The head can be mounted in front of, lateral to or behind the drive unit (Fig. 6.5.6). The head can be attached to a front-loader shovel, on a pantograph arm or mounted on a pendulum to harvest multi-trunked trees. The head support system is equipped with anti-vibration systems (silent blocks), chains or slings to prevent the vibration from being transmitted to the harvester. The head support system must isolate the vibration, be able to get under the tree canopy and provide stability while shaking.

Mounting trunk shakers on a front-loader shovel is quick and easy, producing a versatile harvester. However, the tractor must be highly maneuverable to grab the trunks, the working capacity (trees/hour) is low and soil compaction during rainy seasons is common. The side-by-side lateral trunk shakers are well suited to single-trunked trees in orchard rows (Fig. 6.5.7); there is no need to maneuver the machine to attach the shaking head as the positioning system facilitates correction. Rear-mounted trunk shakers are smaller machines which improve operator visibility but they have poor stability and maneuverability. Pendulum-mounted trunk shakers have a small vibrating head, suspended from a sling on an articulated

Fig. 6.5.6. Trunk shaker mounted on the rear of a tractor. (Photo: Sergio Castro-Garcia)

Fig. 6.5.7. Side-by-side lateral trunk shaker. (Photo: Sergio Castro-Garcia)

arm, which can be placed inside a tree with several trunks.

Self-propelled trunk shakers can drop the fruit on the ground, deposit it in nets or convey it into catch frames with cleaning and unloading systems. Dropping the fruit on the ground for collection or using ground nets requires a rather large number of workers.

Trunk shaker heads must grip the trunk correctly to avoid bark damage. While not all trunks are totally vertical, the shaker head must be perpendicular to the trunk. Most vibration heads have at least 6 degrees of freedom: forward and backward, up and down movements, right or left turns, and roll, pitch and yaw to facilitate head positioning. However, the operator has poor visibility. A worker on the ground, a camera, or automation can improve the driver's ability and decrease the time to position the shaker head (Colmenero-Martinez *et al.*, 2018). Improper tree pruning (very low branches) and the presence of weeds limit automation.

VIBRATION SYSTEM. A trunk shaker's vibration is generated by the rotation of one or two eccentric masses within the shaker head. This rotation is expressed as the number of cycles per unit of time, Hz or cycles/s, or as the rotation speed in cycles or revolutions per minute (rpm). Trunk shakers are classified by the vibration path, number of eccentric masses and direction of rotation speed, as orbital, multidirectional and unidirectional models. Figure 6.5.8 shows an example of each vibration path.

Orbital trunk shakers are now preferred due to their simplicity and higher vibration frequencies. Multidirectional shakers are driven by two motors or a crossbelt, with unbalanced masses ranging from 20 kg to 40 kg to generate a multidirectional vibration between 18 Hz and 22 Hz. Previously, orbital trunk shakers were driven by low-powered tractors, used in traditional trees with large canopies, and had low harvesting efficiencies. Now, orbital shakers on more powerful tractors are more efficient, because they can operate at higher frequencies, 28–30 Hz, which is very efficient in intensive orchards with medium-sized trees (Bentaher *et al.*, 2013; Castro-Garcia *et al.*, 2008).

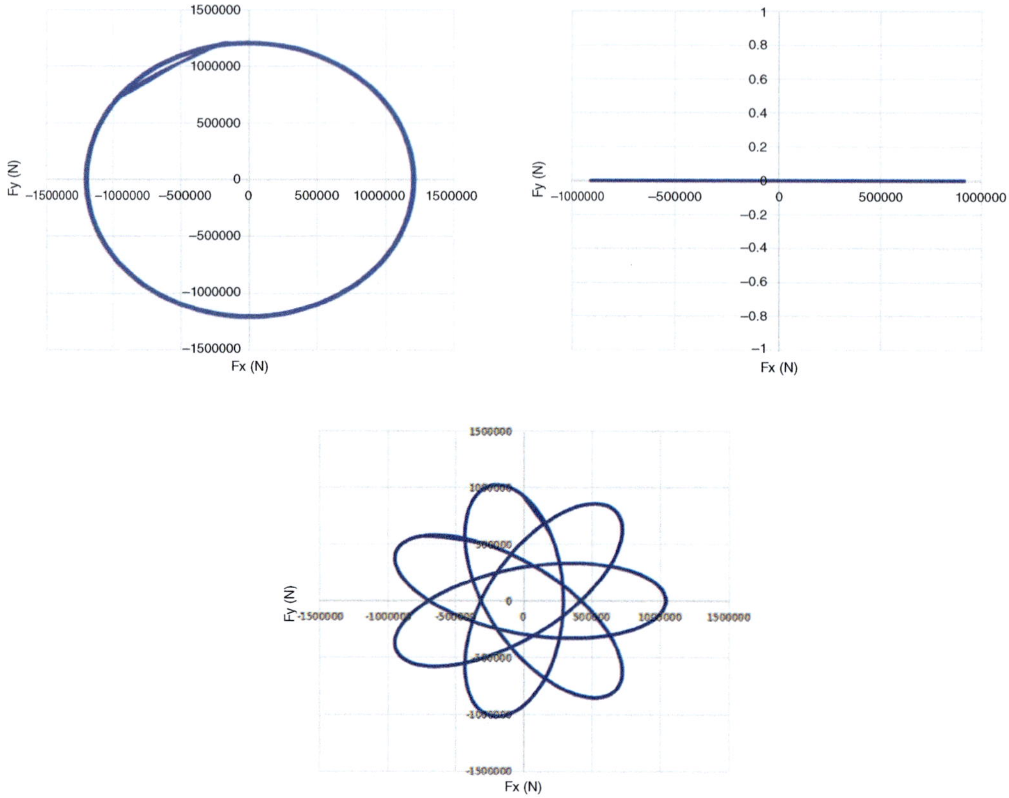

Fig. 6.5.8. Theoretical resultant force (N) in the plane of the vibration head. *Left*: orbital vibration for a mass of 40 kg, eccentricity 10 cm, frequency 28 Hz. *Centre*: multidirectional vibration, for two masses of 30 kg, eccentricity 10 cm, phase π/2 for frequencies of 24 and 18 Hz. *Right*: unidirectional vibration for two 15 kg masses, eccentricity 10 cm, frequency 28 Hz.

Commercial olive trunk shakers vary in their operating ranges. A frequency of 22 Hz and 32 Hz is needed to detach table olives (cv. 'Manzanilla'), which has a high FDF. Orbital shakers producing a 110–280 m/s^2 acceleration in the tree trunk using 15–48 kW of power from a 68–95 kW tractor can average 74% harvesting efficiency (60–87%) (Castro-Garcia *et al.*, 2015). To achieve > 85% harvesting efficiency, an acceleration higher than 180 m/s^2 and a vibration frequency of 28.1 is recommended. For oil olives ('Frantoio', 'Leccino' and 'Cima di Melfi' varieties), 83–95% harvesting efficiencies were obtained with frequencies of 23–27 Hz and an acceleration of 70–99 m/s^2. High harvesting efficiency (97%) was achieved with 'Ortice' and 'Ortolana' varieties with large fruit and an FDF of 3.1–3.8 N (Di Vaio *et al.*, 2012).

The harvester's trunk-clamping system is fundamental for successful vibration transmission. The vibration parameters, mass and eccentricity cannot be modified under field conditions, whereas the rotation speed of the eccentric mass and the harvester clamping can be adjusted according to the operation conditions. The operator can change the trunk-clamping height. Higher clamping heights enable higher vibration frequencies because there is less resistance of the tree to being moved and the amplitude of movement is greater. A high clamping height is useful when the tractor has limited power, the vibration head is small, and the tree is large. Higher clamping increases the vibration amplitude due to the greater trunk elasticity and therefore requires less power (Horvath and Sitkei, 2001). High clamping is not recommended if there is a risk of bark damage. A low clamping height closer to the ground will transmit more vibration to the secondary branches and encounter greater soil resistance. Approximately 15% of the vibration, depending upon soil compaction, is transmitted to the roots (Sola-Guirado *et al.*, 2018). There is no evidence that trunk shaking affects the roots or decreases orchard longevity.

Harvester vibration time is fundamental to harvest efficiency. Olive fruits require longer shaking times than oranges, almonds, pistachios or walnuts. During the harvesting season, operators adjust shaking time according to fruit ripeness. Early in the season the shaking time is longer, and often repeated. For example, early in the season, 'Hojiblanca' table olives need 14 s for 90% harvesting efficiency, but late in the season the 'Picual' oil olive requires 2.1 s (Blanco-Roldán *et al.*, 2009). Two short vs. one long trunk shaking is more effective for early and mid-season harvesting, improving efficiency by 6–9% (Tombesi *et al.*, 2017). Shaking times that are too long damage the tree, remove the following year's fruit wood and reduce equipment life.

TRUNK CLAMPING SYSTEM. The shaker head should clamp perpendicularly to the trunk to avoid, as much as possible, the vertical vibrations, which do not detach fruit and damage the bark (Abdel-Fattah *et al.*, 2003). Some shaker heads can adjust the angle to clamp perpendicularly to the trunks when they are not vertical, i.e. in multi-trunked trees.

How the shaker head clamps the tree varies. The most versatile shaker head has two arms (like a scissor), designed to clamp opposite sides of the trunk (Fig. 6.5.9), meaning that they must be large enough to grasp the trunk. To distribute the force efficiently, the arms should be at equal height and remain parallel when the clamp is completed. The size of the head must be related to the diameter of the trunks to be harvested. Most of the shaking force is delivered between the two points of harvester–trunk contact. This two-arm scissor type harvest head is best for front or lateral trunk shakers. Three-point harvester clamping configurations are typically used on smaller shaking heads. They enable better tree-head contact, good vibration transmission and reduce the risk of bark damage. They are used primarily on trees with small trunks susceptible to bark damage and varieties difficult to harvest because of their low fruit weight, such as 'Arbequina'. The length of the arms must fit the diameter of the trunk. In fact, if the size of the vibrating head is too large, it may not enable firm clamping and slippage may occur. If the head is too small, it may produce tangential stresses that damage the bark. Their lack of versatility, where vibration head size must be in relation to trunk diameter, limits their use. Finally, the shaker can also be equipped with a hook-shaped clamping system for unidirectional shaking. Its use is limited to the shaking of large branches or small trunks.

Fig. 6.5.9. Trunk shaker clamping system with two arms or scissor-type. (Photo: Sergio Castro-Garcia)

Trunk shakers must not damage the tree's bark. The probability of bark damage can be minimized by selecting the correct shaker head padding and using correct clamping procedures and pressures. The padding material is rubber with a hardness ranging from 40 ShA to 70 ShA. An incorrect rubber grip system can reduce shaking transmission by > 50% (Leone *et al.*, 2015). Most commercial trunk shakers incorporate a hydraulic system that ensures that the clamping pressure is not reduced during the vibration process and provides a rigid connection. The gripping system can produce trunk-compression forces greater than 40 kN and requires a large area of contact for the trunk-clamping pressure. Bark damage results if shaking begins before the shaker head is firmly clamped on the trunk, or if the shaker head moves tangentially during shaking, producing shear pressure that sloughs off the bark.

For early harvesting, particularly with table olives, which are irrigated as late as possible to achieve maximum size, soft elastic padding, 40–50 ShA, which has greater deformation, distributes the stress over a wider area. However, softer padding decreases shake transmission to the tree. During early harvests, this can greatly reduce harvesting efficiency, especially when combined with high FDFs. Harder padding, 60–70 ShA, has less deformation, and improved vibration transmission (Bernardi *et al.*, 2018). Hard padding is recommended when there is no risk of bark damage. Padding can be protected by additional layers of elastic or textiles to prevent wear and tear. These layers reduce bark damage, as they reduce the relative displacement between the shaker head and the trunk.

For table olives, additional textile, intermediate non-adherent materials, or lubricants are used between the layers. However, mineral oil should not be used as it degrades the natural rubber padding and could contact the detached fruit. The layers should be secured at the top of the shaker arms leaving the lower part free to move. This could reduce the shear stress that damages the bark if there is displacement between the shaker and the trunk. Some trunk shakers automatically lubricate the layers during the operation. The intensive use of vibration, especially in intensive olive orchards may cause overheating. Therefore, some shaker head paddings have cooling systems to prevent deterioration. During early harvesting with high temperatures in high-density orchards with machines regulated with low clamping pressures, the padding overheats and deteriorates quickly.

Trunk shaker performance is also affected by crop-production practices, particularly irrigation management. Irrigation affects the mechanical properties of the trunk's cambium and therefore bark damage. The cambium is more susceptible to damage seasonally, being more easily damaged in spring than winter, but time of day, morning vs. evening, is unimportant. Starting with 50% deficit irrigation three to seven days before harvest reduces the probability of bark damage by almost half (Vega Macías *et al.*, 2014).

Trunk shaking is faster with a smooth (no irregularities or knots), vertical trunk without suckers and low branches that obscure the operator's vision and increase clamping time. Removing suckers and low branches increases harvester efficiency by 6–9% (Tombesi *et al.*, 2017).

Factors affecting harvesting efficiency with trunk shakers

Cultivar, olive orchard and tree characteristics, crop load and pruning, irrigation and soil management all affect trunk shaker efficiency (Famiani *et al.*, 2004a; Farinelli *et al.*, 2012 a,b; Famiani and Gucci, 2022).

Cultivars with an upright growth habit shake more efficiently than varieties with pendulous, semi-drooping canopies. Fruit weight is a function of variety. Small fruits weighing 1–1.5 g with a high fruit detachment force (FDF) are the most difficult to harvest. Harvesting efficiency is best if the fruit is over 2 g.

FDF is a function of cultivar and decreases during ripening in all cultivars. Varieties with simultaneous ripening and short peduncles are more suitable for mechanical harvesting than those with marked staggered ripening (Castillo-Ruiz *et al.*, 2018). When the FDF(N)-to-fruit weight (g) ratio falls below 2.3 N/g, and trees are well structured, harvest efficiency increases to values ≥ 85% (Farinelli *et al.*, 2012c). Above 3 N/g harvest efficiency is considerably lower.

Trunk shakers are most efficient in orchards with regular rows with at least 5 m between and within the rows, especially if an inverted umbrella catch frame is used with the shaker. Distances within the row can be shorter if the fruit is caught in nets spread on the ground or if using a side-by-side shaker.

Trunk shakers can be used with vase, monocone, globe and, more in general, all single-trunk training systems. However, all must have a single main trunk with no lateral branching below 1–1.2 m (minimum 0.8 m), especially when using an inverted umbrella. If the shaker lacks a catch frame, the trunk can be 0.6 m tall. Trunk shakers are most efficient with trees trained to an upright vase. The upper canopy should not be too bare to ensure good production and reduce dampening of vegetation, such as the breaking of branches and defoliation (Famiani, 2017).

Trunk shakers are most efficient with trunk diameters between 15 cm and 40 cm and canopy volumes up to 30–50 m^3. Individual branch shaking can be done in trees with larger trunks and canopies. The secondary branches must be numerous and short. Mature trees should be pruned annually to maintain short secondary branches without significant pendulous vegetation. If the trunk-shaker has an inverted umbrella catch frame, the lower canopy must be pruned to avoid contact with the catch frame which will dampen the shake and lower harvester efficiency. However, eliminating the lower canopy reduces the canopy volume, lowering yield. This loss can be compensated for by increasing the height of the canopy (Famiani, 2017).

The trunk shaker is difficult to use on the bush-vase system due to their multiple trunks. Restructuring can be done to produce a single trunk per tree, but the cost-effectiveness of this operation must be evaluated.

Annual pruning is the best way to maintain trees with the best structure for the use of trunk shakers. Plurennial pruning can bring about excessive development of pendulous vegetation and suckers inside the canopy, which reduce trunk shaker efficacy (Famiani, 2017; Tombesi *et al.*, 2017).

HARVESTING PROCEDURES. Fully mechanized harvesting is the most efficient. A trunk shaker with an inverted umbrella collecting frame requires two workers; one operates the trunk shaker head and one directs the other in clamping the trunk and drives the tractor with the trailer for the harvested olives (Fig. 6.5.10). The harvesting work productivity ranges from 100 to 400 kg/hour/worker, depending on the characteristics of the olive orchard and, especially, on the olive load of the trees (Famiani *et al.*, 2004a; Di Vaio *et al.*, 2012).

Trunk shakers can also be combined with a catch frame towed by another tractor (Fig. 6.5.11). There is a conveyor belt in the center and a roll of nets on either side, which are unwound and rewound automatically (Fig. 6.5.11) or, depending on the model, with the help of four or more workers. After shaking, the olives are conveyed and dumped into a crate on the frame and the nets are rewound. This system is suitable for flat or gently sloping olive orchards with regularly spaced rows of trees and when the trees have a trunk less than 0.60 m in height. In this case, the trunk shaker cannot be combined with an inverted umbrella collector frame. The size of the equipment requires wide headlands for maneuvering at the ends of the rows. Productivity ranges from 100 to 200 kg/hour/worker (Giametta, 2003; Famiani *et al.*, 2004a; Famiani and Gucci, 2022).

Trunk shakers can also be combined with nets spread on the ground manually by 4–12-worker teams. The number of people working the nets depends on the speed of the shaker. The shaker operator should not have to wait for the nets to be spread (Famiani and Gucci, 2022). Productivity is 40–280 kg of olives/hour/worker (Famiani *et al.*, 2004b; Lavee, 2010; Zion *et al.*, 2011).

Shaking a tree trunk ranges from 1.5 to 3 minutes (160–320 trees in an eight-hour day), depending on the terrain, slope, orchard design and tree structure. Time is reduced with an inverted umbrella vs. ground nets.

In all cases, the harvesting work productivity (kg of olives harvested/h/operator) increases with increasing fruit load per tree. This is because the time to harvest one tree is the same regardless of the amount of fruit on the canopy.

Simultaneous use of hand-held equipment can complement trunk shaking for early harvesting of small-fruited varieties with high FDF and/or when the trees have structural defects, such as secondary branches in the lower half of the crown that form long pendants (2 m or more). The cost-effectiveness

Fig. 6.5.10. Trunk shaker with an inverted umbrella collecting frame. (Photo: Sergio Castro-Garcia)

of this operation depends on the amount of residual fruit and the selling price of the olives or oil (advantage increases as the price of such products increases) (Famiani and Gucci, 2022). Another possibility is to pass again with the machine at a later time. This choice is worthwhile if the production per tree is high, thus giving good harvest productivity in both passes. In other words, it is economical if the additional yield justifies the cost and the second harvest decreases alternate bearing.

In olive orchards with 7 m between rows and short distances between trees within the rows (3–4 m), side-by-side trunk shakers with integrated catch frames and bin carriers can be used for completely mechanized harvesting. The worksite consists of two workers who drive the two units plus another person for handling the crates. These machines require flat or gently sloping terrain, and their main characteristic is speed. So far this type of machine and worksite has been used mainly in large southern hemisphere plantations with very high output per hectare. In trials conducted in Australia in olive orchards of the cultivars 'Barnea' and 'Frantoio', spaced 7 m × 4 m (357 plants/ha), harvesting required 2–4 hours/ha depending on cultivar, time of season, and fruit load (Ravetti and Robb, 2010). These times are comparable to those of continuous harvesting with a straddle harvester. However, the percentage of detached fruit compared to the total amount was not always good. In fact, in early season 'Frantoio', which is more difficult to harvest mechanically than 'Barnea', only 60% of the olives were harvested (Ravetti and Robb, 2010).

In traditional orchards of very large trees, tractor-mounted trunk shakers can be clapped to the main branches. In trees with a good load, with nets under the trees to collect the fruit, 85–103 kg of olives/hour/worker were harvested (Famiani *et al.*,

Fig. 6.5.11. Trunk shaker combined with a catch frame towed by another tractor. (Photo: Franco Famiani)

2014). Usually in very large trees, two to three passes (about every two weeks) are required to complete the harvesting.

6.5.3.4 Canopy shakers

Canopy shakers directly contact the tree canopy with shaking rods that hit the canopy to detach the fruit (Afsah-Hejri *et al.*, 2022). Canopy shaker systems were developed in the 1970s for harvesting juice oranges (Whitney *et al.*, 1974) and are commercially available but not commonly used (Roka *et al.*, 2014; Gil-Ribes *et al.*, 2014). Canopy shakers are currently used for blueberry (Yu *et al.*, 2014) and coffee harvesting (Ferreira Júnior *et al.*, 2016).

There are different kinds of canopy shakers. Self-propelled or tractor-towed canopy shakers for olive are commercially available. Most of them were developed for intensive and super-high-density orchards that have a vertical fruiting wall facing the shaker (Rallo *et al.*, 2013). These canopy shakers can harvest one side of a tree row, or, if operated in pairs, both sides of a tree row, or work as a straddle harvester (Figs 6.5.12 and 6.5.13). In addition, some canopy shakers were developed for the harvesting of traditional orchards, especially those made up of large-sized trees (Fig. 6.5.14). The straddle harvester used in narrow-hedgerow (super-high-density) olive orchards, due to their widespread use and peculiarity, will be treated in a separate section (Fig.6.5.15).

VIBRATION SYSTEM. Canopy shakers detach the fruit through the action of rods, which are mounted radially on a drum or drive shaft or in an oscillating frame with a wide amplitude. The rods penetrate into the tree canopy, moving it with a high amplitude and low-frequency oscillatory motion, strong enough to detach fruit but not enough to cause significant damage to the tree. As the harvester moves down a tree row, the drum's rotation, free or controlled,

Fig. 6.5.12. Self-propelled straddler canopy shaker harvesting an intensive olive orchard (Macteq). (Photo Pierluigi Pierantozzi)

Fig. 6.5.13. Lateral self-propelled olive canopy harvesters (Athena). (Photo: Mario Beltrami)

allows the rods to penetrate the canopy without tangling or breaking the branches (Fig. 6.5.12). Fruit is detached when the oscillating rods contact the branches.

The rods can be nylon, plastic, fiber glass or metal and have a variable length depending on the kind of machine. The operating frequency is usually between 3 Hz and 9 Hz, and the amplitude at the base of the rod is 5–20 cm. The rods may hit the fruits, producing unacceptable bruising for green processed table olives. Less than 20%

Fig. 6.5.14. Canopy shaker mounted on a tractor for harvesting (Oli-Picker). (Photo Franco Famiani)

Fig. 6.5.15. Straddle harvester for narrow-hedgerow orchards. (Photo: Sergio Castro-Garcia)

of harvested table olives and 35% of oil olives were estimated to be bruised during canopy harvesting (Castro-Garcia *et al.*, 2009; Sola-Guirado *et al.*, 2016). Field observations indicate that tree damage with canopy shakers is limited with correct tree training: shaker damage was one-third of that caused by hand harvesting with poles and hand-held equipment (Sola-Guirado *et al.*, 2014). When the rods strike branches, the detached olives fall vertically downward.

When canopy shakers have the correct amplitude, frequency and vibration time, harvester efficiency can achieve 90% where the rods contact the canopy (Bernardi *et al.*, 2016; Ravetti and Robb, 2010; Sola-Guirado *et al.*, 2016). Increasing harvesting efficiency and decreasing fruit damage is the goal (Ferguson *et al.*, 2010). Castro-Garcia *et al.* (2009) reported that canopy shaker damage is triple that of hand harvesting for green processed table olives. In contrast, Ferguson and Castro-Garcia (2014) reported black processed olives had no significant differences in damage in comparing hand and canopy-shaker harvested 'Manzanillo' olives before or after processing. If olive processing is carried out soon after harvesting, high-quality oils are obtained (Famiani *et al.*, 2014, 2020).

One version of these machines consists of a tractor fitted with a movable arm with an oscillating beater, which detaches the olives by beating the vegetation (Fig. 6.5.14). The machine can be operated by a worker on a rotating workstation (turret) or from the ground by remote control. The total weight (base + arm + beater) of the harvester is about 800–850 kg and the minimum power to operate it is 37–45 kW (50–60 HP). The working height is 8.5–10.4 m depending on the model; the length of the beater can vary from 0.8 m to 1.4 m and the spokes can be rigid or flexible. The oscillating beater can also be mounted on arms of more or less large excavators, including crawlers and rotating platforms. The great flexibility and ease of movement of these machines enables use in olive orchards with high structural limitations (e.g. terraced olive orchards).

CATCH FRAME. Fruit is detached by the direct action of the shaking rods. Transmission of vibrations to other canopy areas is low and does not detach fruit. The detached fruits drop vertically into a catch frame. If fruits are hit by the rods or branches, they may not fall vertically into the catch frame, decreasing canopy harvester efficiency slightly.

Canopy shakers designed to harvest trees which form a hedge include a fruit catch frame, and this enables a row to be harvested without stopping, greatly increasing the harvesting capacity (Figs. 6.5.12 and 6.5.13). In this case, once caught in the catch frame the fruit is transported by conveyor belts, buckets or pneumatic systems, cleaned of twigs and leaves by mechanical de-twiggers or horizontal forced air passed across the olives as they are dropped into hoppers, bins or trailers. These machines can be equipped to have continuous unloading, which requires a tractor and another vehicle to travel parallel to the harvester in an adjacent row (Fig. 6.5.14). This option is particularly useful in the case of high production and long tree rows.

When using small canopy shakers, the fruit drops onto nets spread on the ground under the trees (Fig. 6.5.12) or onto an inverted umbrella catching frame towed by another tractor.

Factors affecting harvesting efficiency with canopy shakers and harvesting procedures

The fruit load positively influences work productivity, although at very high fruit load, because of the lower fruit weight and slower ripening, harvesting efficiency tends to be lower. Also, with these machines, the ratio of fruit detachment force/unit weight of olives less than 2.5 N/g gives high values for yield and work productivity (Famiani *et al.*, 2014). When fruit detachment is easier, less damage is also caused to the trees. In fact, high harvesting efficiency with canopy shakers requires both simultaneous adaptation of the harvester and the orchard. Basically, these machines can work in intensive orchards, where trees are planted in rows, or on large, isolated trees, where trunk shakers have limited operation.

With all types of canopy shakers, more than 90% efficiency can be obtained by prolonging the working time until the fruit is completely detached. However, longer working time reduces work productivity and increases branch damage. As a general rule, it is best to avoid prolonging the working time for a slight increase in harvested fruit.

In intensive olive orchards, the trees should be pruned to form a continuous wall with minimal gaps. Fruiting branches should be oriented towards the outer zone and the main or woody branches oriented in the main direction of the hedge. In addition, lower branches must be removed to allow the catch frame to enter. Depending on the canopy shaker model, the hedges can be up to 5 m high and 2–4 m wide. It is recommended that branches be no lower than 0.5–1 m from the ground (Ferguson *et al.*, 2010; Dias *et al.*, 2020). The straddle machine currently available is the Colossus (Maqtec Inc.), which has a 4 m-wide tunnel and is up to 4.5 m high (Fig. 6.5.12). Machines for continuous harvesting, especially the straddle kind, require wide service roads to allow machine turnaround and a distance between rows not less than 6 m. This is especially true for the Colossus, which is very heavy and large (8.5 m long, 6.8 m wide and 5.85 m high).

In an olive orchard of the cultivars 'Barnea' and 'Frantoio', with trees spaced 7 m x 4 m, the straddle harvester (Colossus) gave harvesting efficiencies of 86–91% and 91–96% in early and late harvestings, respectively; there were also cultivar and load effects, as slightly better results were obtained with 'Barnea' and lower efficiencies when trees had a high load (Ravetti and Robb, 2010). The working speed is high considering that the machine harvested 64–182 trees/h, corresponding to a harvesting time of 2–5.5 hours/ha (357 trees/ha) (Ravetti and Robb, 2010). This machine requires that the size of the trees be maintained within the dimensions of the harvesting tunnel. This is not always easy with vigorous trees.

In a preliminary trial on early harvesting of trees of the cultivars 'Frantoio' and 'Leccino' (spaced 6 m × 3 m), a continuous lateral canopy shaker gave a harvesting efficiency of 80–85% (Famiani and Gucci, 2022). Considering that such trees were not originally trained according to a hedge system, it is likely that improvements in the harvesting efficiency can be obtained where trees are trained from the beginning of their development according to the requirements of the machine. Moreover, harvest efficiency should be enhanced when two shakers are used in parallel simultaneously, shaking the canopy from both sides of the tree, determining an amplification of shaking (Fig. 6.5.13). Machines that have tracks instead of wheels cause less soil compaction, because their weight is distributed over a larger surface area. In an intensive table olive orchard with trees pruned for mechanized harvesting with a canopy shaker, an average harvesting efficiency of 92% was achieved (Ferguson and Castro-Garcia, 2014).

Canopy shakers for hedge orchards have a ground speed of 0.5–2 km/h. Ground speed is determined by the need for vibration time. Canopy shakers require hedgerows designed to facilitate turning around at the end of the rows, at least 6–8 m.

Due to the high price of these machines, it is necessary to use them in large areas for continuous harvesting.

Some canopy shakers (Fig. 6.5.14) can be used in traditional irregularly and widely spaced olive orchards with trees that have several, or large, trunks and large canopies. The time required for harvesting is dependent on tree/canopy size: 15–60 minutes/tree for canopy volumes from 150 m^3 to 300 m^3 and up to 8–10 m tall. Larger rods will reduce harvesting time as they can access more canopy with each positioning. In smaller trees, efficiency can be > 90% because the fruit is visible, allowing shaking until most of the fruit is harvested (Famiani *et al.*, 2014). In larger trees visibility is less and efficiency can drop up to 75% (Giametta, 1999; Caricato, 2001; Famiani *et al.*, 2014). Worker productivity varies greatly, 40–145 kg/hour/worker; the lowest is on terraced hillsides (Caricato, 2001; Fiorino *et al.*, 2008; Famiani *et al.*, 2014). Using a tractor-mounted mechanized net winder with a roll of nets unwound as the tractor moves along the row can improve productivity by eliminating hand labor. Alternatively, the combination of an inverted umbrella catcher with a canopy shaker can increase productivity by 35% with respect to using nets.

Experimental prototypes able to harvest large, widely spaced trees, by circling the tree, have also been designed (Sola-Guirado *et al.*, 2016), with interesting results. This canopy shaker has a vibration frequency of 4–5 Hz, an amplitude of 0.12–0.17 m and a forward speed of 0.5–1 km/h, reaching an average harvesting efficiency of 77.3%. It was pointed out that the density of rods was a determining factor in the breakage of branches and shoots and that adaptation of the trees, especially with pruning, was required to improve the contact between the machine and the tree. Elimination of the inner canopy branches where the rods cannot reach the fruit improved the results, without significantly affecting the subsequent yields (Castillo-Ruiz *et al.*, 2017). Another experimental prototype canopy shaker produced 92% harvesting efficiency in a 'Manzanillo' table olive trained to a moderate density (490 trees/ha hedgerow) with mechanical and hand pruning to 3.7 m high, 1.85 m wide and trunk clearance to 1.0 m (Ferguson and Castro-Garcia, 2014).

6.5.3.5 Straddle harvesters for narrow-hedgerow orchards

Super-high-density (SHD) orchards have been expanding rapidly, thanks to the use of low-vigor varieties that are suitable for narrow-hedgerow orchards, mechanical pruning and efficient canopy harvesters. This is a good example of the simultaneous adaptation of the tree, production practices and harvester (Connor *et al.*, 2014). Vineyard harvesters have been used for these olive varieties, which have been selected or bred for less vigor. Hence, they can be trained into a narrow hedge that is maintained with a combination of mechanical and hand pruning.

The self-propelled (Fig. 6.5.15) or tractor-drawn straddle harvesters used in SHD narrow-hedgerow orchards are based on a canopy shaker system. The self-propelled, versus tractor-drawn, harvesters have the highest working capacity. They have 100–130 kW diesel engines and hydrostatic control systems. They are equipped with wheel motors and are highly maneuverable. An all-wheel leveling system that maintains the harvesting head to remain vertical enables operation on slopes up to 20–30%. However, harvester stability on hillsides, where the hedgerows are perpendicular to the slope is a problem, because of slippage that can cause tree damage and a risk of rolling over.

The straddle harvesting head is a tunnel lined with horizontal arched bow-rods on each side, 0.1–0.15 m apart vertically, and alternately on opposite tunnel sides. This arrangement, and their synchronized oscillating movement during harvest, prohibits the bow-rods from striking one another. The horizontal bow-rods deliver a perpendicular vibration to the hedgerow that is three times greater than the vibration in the line of the hedge (Sola-Guirado *et al.*, 2019). The treated plastic or nylon bow-rods are subject to intense vibration and high impact pressure and, as a result of this, especially when working at high temperatures, can break during harvest, decreasing harvester working capacity.

The new straddle machines designed for the harvesting of olive trees, both self-propelled and tractor-towed, have a large tunnel. They can harvest trees with a total above-ground height of 4–4.5 m (canopy 3.3–3.8 m) and a canopy width of 2–2.5 m. Trees taller than the straddler tunnel do not create a problem if the vegetation is flexible (Tombesi and Farinelli, 2017). Recently, even larger straddle machines have been designed.

The canopies are guided into the harvesting head by deflectors at the harvester head tunnel entrance, helping to minimize branch damage. Upon entering the tunnel the arched bow-rods progressively compress the canopy from 60–80 cm to 20–30 cm wide. The vibration produced by the bow-rods oscillating against the canopy detaches the fruit within the canopy harvesting head.

Harvest efficiency is improved with the correct combination of hedgerow and harvester head working height, ground speed and vibration frequency. If the hedgerow is too high, the lower canopy, below the harvesting head, will not be harvested. The minimum height of the harvester head is determined by the lower canopy and if the hedgerow is on a ridge. The harvester ground speed is 2–3 km/h depending on the canopy volume and production. When the canopy is large and heavily cropped, the ground speed is decreased to increase harvesting efficiency. When the orchard is young with little crop, the ground speed can be increased up to 3 km/h. The adjustable vibration frequency ranges from 6.5 Hz to 8 Hz (380–480 cycles/min). Using these parameters at 2 km/h, the bow-rods impact each tree 2000 times, with an average peak acceleration of two to six times that produced by trunk shakers (Sola-Guirado *et al.*, 2019).

Straddle harvesters, particularly at the tunnel entrance, detach and lose a small percentage of fruit and some fruit also remains on the tree, both decreasing final harvest efficiency. Attempting to achieve higher efficiencies with slower ground speeds, or higher acceleration can result in tree damage.

As all canopy shaker harvesters, an immediate post-harvest prophylactic copper treatment is recommended to prevent olive knot bacteria, *Pseudomonas savastanoi*, from entering the wounds and leaf injuries caused by the machine (Tous *et al.*, 2014).

When detached fruit falls to the bottom of the harvester, unless deflected by hitting a rod or branch, it drops onto a conveyer belt at the bottom of the harvester. The fruit is prevented from falling to the ground by a system of flexible mechanical 'petals' that close the gap around the tree trunk. The conveyer belt moves the fruit to a collection system, usually baskets, that lift the fruit up to be transferred into two hoppers or collection trailers moving alongside the adjacent row. There are two hoppers to maintain the harvester's balance. Each hopper has a volume of approx. 1 m^3. The hoppers limit the length of the tree row as these harvesters cannot go backwards or forwards without harvesting the tree. For this reason, larger orchards use continuously moving bin collectors driving alongside the adjacent row; as the bin is filled, it moves forward to be replaced by another coming from behind. With both fruit collection systems, as the fruit is discharged, an 'air leg', a fan producing a strong horizontal steam of air perpendicular to the flow of discharged fruit, blows out the lighter leaves and shoots. Some models also have horizontal rods, 'de-twiggers', at the end of the harvesting head to remove the larger-diameter branches broken off during harvesting before the fruit is moved to the top of the harvester.

In highly productive orchards, the high working capacity of straddle harvesters and the reduced labor requirements decrease olive oil harvesting costs by 25–33% compared to trunk shakers (Zion *et al.*, 2011). In Spain, it was estimated that costs were 0.06 €/kg for SHD with straddle harvesters, compared to 0.09 €/kg and 0.12 €/kg for irrigated and rainfed intensive orchards harvested with a trunk shaker.

Currently, SHD oil olive orchards are harvested early in the season for higher oil quality and to avoid the rains that limit or stop harvesting and/or increase fruit drop. Continuing harvesting through the night to reduce the impact of high temperatures on oil quality and maximize the use of expensive machines has also been tried. However, the harm to night-roosting birds resulted in regulations against night harvesting (Da Silva and Mata, 2019).

The degree of fruit damage caused by the straddle harvester depends on tree age. Damage is limited in young trees with flexible branches (Camposeo *et al.*, 2010). As the orchard ages the rigid structure increases in size, increasing potential tree damage, particularly in more vigorous varieties with less-supple branches. In Spain, six years after planting, up to 30% of the branches in the cultivars 'Koroneiki' and 'FS-17' were damaged compared to 10% in 'Arbequina' (Tous *et al.*, 2010). Correct training and annual pruning can significantly reduce harvester damage; rigid branches greater than 4 cm in diameter extending into the orchard alley must be removed by hand pruning or shortened with mechanical pruning. The former is preferred.

Straddle harvesters can be equipped with yield monitors. Potentially, yield monitors can generate information for management decisions, which will facilitate precision olive farming.

Factors affecting the harvesting efficiency of straddle harvesters

The main limitation to using straddle harvesters is the size of the tree's rigid structure; the tree must fit into the harvest tunnel without harming the tree or harvester. New available machines, having relatively larger tunnels, are reducing this limitation.

Crop load, where fruit is present in the canopy and individual fruit weight and maturity, which dictate the FDF:fruit weight ratio, all affect harvester efficiency. Generally, straddle harvesters have very high harvesting efficiencies, over 95% (Farinelli and Tombesi, 2015) or higher with large fruit (Jiménez *et al.*, 2017). Harvesting efficiency is reduced with increased canopy volumes and higher FDF:fruit weight ratios (Farinelli and Tombesi, 2015). High harvesting efficiencies are possible with FDF:fruit weight ratios below 2.7 N/g (Camposeo *et al.*, 2013). In the worst conditions, small fruit and an early harvest, harvest efficiency is around 85%, but approaches 100% in the best situations (Camposeo *et al.*, 2010; Camposeo, 2017; Giametta and Bernardi, 2010). Harvest efficiencies above 95% are consistently achievable in well-managed olive orchards when the FDF:fruit weight ratio is about 2 N/g (Camposeo *et al.*, 2010; Camposeo, 2017).

Straddle harvesters have a high working capacity, 1.7–3 h/ha at 1–3 km/ h (Camposeo *et al.*, 2010; Morales-Sillero *et al.*, 2014). After tree size, slope and row length are the primary factors affecting ground speed. Although self-propelled machines have self-leveling systems, transverse and longitudinal slopes should not exceed 20%. Efficiency increases greatly with long rows as turning time at the headlands constitutes the second greatest portion, 25%, of the straddle harvester's non-operating time. Unloading fruit from the hoppers is the highest portion (75%) of a straddle harvester's downtime (Giametta and Bernardi, 2010). Therefore, downtime can be slightly mitigated with orchard design, also ensuring a road system to facilitate harvester transport (Giametta and Bernardi, 2010).

The speed of the machine also depends on the size of the trees. Generally, harvesting time is faster when the trees are younger and slower as the trees get older. Harvesting time also increases as production increases (Giametta and Bernardi, 2010).

6.5.4 Fruit Cleaning, Transport and Storage

6.5.4.1 Cleaning

Harvester efficiency is defined as the percentage of fruit on the tree that is collected in the mechanical harvester fruit receiver or the inverted umbrella and conveyed into the bins or bank out wagons attached to the trunk shakers or the unattached continuous loading hopper trailers that travel alongside in the adjacent tree row, described above. For reasons of efficiency and potential contamination that would affect oil quality, and cost, fruit not collected by the harvester collection system is not collected from the ground.

Trees not trained and pruned for mechanical harvesting, or those with a high olive knot infestation, are particularly prone to twig and leaf detachment during harvest. And, though minor, the leaves and twigs can scratch the olive fruits, accelerating the natural fruit senescence that degrades both table and olive oil quality. Therefore, most mechanical harvesters have modifications to remove the leaves and twigs from the harvested olives. Trunk shaking harvesters, except inverted umbrella types, have rotating twig removers, a 'de-twigger', at the far end of the horizontal conveying systems, then an 'air leg', a fan blowing horizontally across and perpendicular to the olives to remove the leaves as the olives are dumped into bins or bank out wagons. Continuous canopy contact harvesters for SHD and HD orchards also have fans blowing horizontally and perpendicular to the flow of olives being dumped into the hopper trailers. Removing twigs and leaves at harvest collection is more efficient than doing it at the processing plant.

6.5.4.2 Transport and storage

The time elapsed, and temperatures, between harvest and processing are among the critical factors affecting table and olive oil quality. Ideally, olives should be processed within 12 hours, maximum 24 hours, from harvest, especially when straddle harvesters are used (Famiani *et al.*, 2020). Meanwhile, they should be kept as cool as possible, not more than 15–20°C during post-harvest transport and before processing. Until recently, temperature at harvest and during transport was not a concern. However, as climatic conditions have changed, warm ambient temperatures during a September harvest can be as high as 30°C. These temperatures limit the starting date for early-season harvest varieties like 'Arbequina', 'Arbosana' and 'Koroneiki' as the oil quality will be degraded by the extreme temperatures during transport. Also, harvested olives left in the sun burn within an hour.

Depending on the harvest method and/or orchard size, table and oil olives can be transported to the mill in small 25 kg perforated crates, standard orchard 300–500 kg bins or large capacity 5000–10,000 kg trailers. Transport of actively respiring fruit harvested at temperatures above 20°C, particularly if in 500 kg field bins or large hopper trailers with low surface-to-volume ratios, can result in asphyxia, anaerobic fermentation and high levels of alcoholic esters; all produce 'off flavors' in table and oil olives. Also, field and transport temperatures above 30°C may prevent extraction of high-quality oil. Malaxation is most efficient between 25°C and 30°C. If olives enter the plant at 25°C and crushing raises the olive paste temperature 5°C, oil extraction may increase significantly, and the quality of the oil may be reduced.

Thus far, temperature control during transport has not been examined. Harvesting at night was tried but was soon abandoned due to the harmful effects on bird life (Da Silva and Mata, 2019). However, controlling temperatures during transport, sampling internal fruit-carrier temperatures upon arrival at the processing plant and pre-cooling fruit before processing may become part of the harvest and transport chain to ensure high table olive and oil quality and optimal oil extraction, as well as determining payment to the producer.

Storage

If harvesting and processing capacity are not synchronized, the olives will need to be stored. Short-term storage, not exceeding 12–18 hours, can be considered (Famiani *et al.*, 2020; Plasquy *et al.*, 2021). Prolonged storage before processing degrades the quality of the oil. If necessary, storage is usually done in the same crates, bins or hopper trailers used to transport the fruit or in specifically designed, clean stainless steel fruit hoppers in coolers. The olives are kept as cool as possible, 4–6°C, but below this temperature, olives sustain chilling injury (Plasquy *et al.*, 2021).

6.5.5 Harvesting Systems on Fruit Integrity

Harvesting causes more or less damage to the olives and consequently to the extracted oil (Famiani *et al.*, 2020). To minimize damage, utmost care must be exerted in all phases, from fruit detachment and catching to placing the olives in crates, bins and trailers and transporting them to the farm/mill. Recently, four different harvesting methods were evaluated on the 'Frantoio' and 'Arbequina' varieties: by hand, by hand with plastic rakes or pneumatic combs, trunk shaker with catch frame, and straddle harvester. Gentle harvesting by hand produced the least damage and straddle harvesting produced the most damage (Famiani *et al.*, 2020). As damage to the fruit increases, the total polyphenols and the secoiridoids in the oil decrease proportionally. Moreover, damaged olives produce oils with lower amounts of aldehydes and esters and higher amounts of alcohols and ketones.

The method of harvesting and storage time are vital in producing the degree of damage to the fruit and consequently the degree of oil quality. Harvesting drupe by drupe in order to minimize damage is not feasible, either for the time needed to do it or for the cost, but it is very useful for comparing intact fruit with mechanically harvested fruit in order to evaluate the impact on oil quality.

It is clear that the higher the level of mechanization the more necessary it is to reduce the storage time of the olives to avoid quality degradation of the oil. With mechanical harvesting it is necessary to control the post-harvest phase, avoiding thermal abuse to the olives and decreasing transport time to processing at the mill. It is very important to bear in mind that the negative effects of mechanization are generally mitigated/compensated by the increase in harvesting productivity, which makes it possible to concentrate harvesting in the period that is optimal for the production objective (quantity and quality) and to avoid or reduce the need for storage (Famiani *et al.*, 2020).

Unlike the modern oil olive industries, the table olive industries have not adapted mechanical harvesting. Spain and California, and even more so the less-developed countries, still hand harvest table olives. The major Spanish table olive cultivars are the 'Hojiblanca' and 'Manzanilla de Sevilla'. The major California table olive cultivar is 'Manzanillo'. Spain processes most of its olives into the 'Spanish'-style green olive. California processes their olives into the 'California Black Ripe' olive. Ferguson *et al.* (2010) and Ferguson and Castro-Garcia (2014) experimentally demonstrated the successful use of both trunk shaking and canopy contact harvesters to be efficient and economical in harvesting table olives. Neither a sensory nor a consumer panel could distinguish mechanically harvested olives from hand-harvested olives after processing. However, the Spanish table olive industry is still struggling with mitigating the bruising produced by mechanical harvesting. The oxidation in processing that produces a 'Black Ripe' table olive masks the superficial bruises that do not affect fruit eating quality. When processed into a green 'Spanish'-style olive, the bruises are visible and significantly decrease quality (Jiménez *et al.*, 2017).

Both the trunk shakers and the canopy shakers are suitable machines for table olive harvesting (Ferguson and Castro-Garcia, 2014). In multiple trials, both harvesting methods produced fresh olives that received total adjusted prices per ton and canning percentages equal to those of hand-harvested fruit. Similarly, trained sensory and consumer panel evaluations of the processed fruit harvested by both harvesting technologies were also equal to those of hand-harvested fruit.

The experience with table olives in California shows that the harvesting efficiency, both of commercial trunk shakers and a prototype canopy harvester, is increased by simultaneously improving harvester technology and using mechanically pruned orchards and new wide-hedgerow orchards. After six years of trials with mechanical topping and hedging vs. hand-pruned trees, mechanically pruned trees achieved an annual average of 9400 kg/ha, compared to 11,900 kg/ha for hand-pruned trees. Approximately 85% of the olives from the hand-pruned trees were cannable, compared to 86% of the mechanically pruned olives. The harvesting efficiency, in both cases of pruning, was over 80%. Likewise, the pruning treatments did not produce significant differences in terms of the percentage of cannable olives, fruit-size distribution or value per ton. Canopy volume is a key factor in the adaptation of orchards to mechanical harvesting. There is a considerable reduction in canopy volume, as an individual new hedgerow tree canopy has a volume of 9.2 m^3 and can produce an annual average of 8300–9600 kg/ha compared to 9400–11,900 kg/ha for trees with canopy volumes of 35.3 m^3 and 82.4 m^3, respectively. The experience developed indicates that adapting olive orchards with mechanical pruning does not decrease the average annual yield and can produce table olive orchards that can be harvested mechanically at a cost and speed that are competitive with manual harvesting.

6.5.6. Occupational Hazards in Mechanical Olive Harvesting

The occupational hazards associated with olive harvesting are diverse and often require special consideration when using agricultural machinery. Olive harvesting is characterized by a high degree of temporary employment, where workers are often seasonal and sometimes unaware of the risks involved in the operation.

The early harvesting of table olives usually involves high ambient temperatures. Attention must be paid to possible heat stress, which can be reduced by easy access to water and other abundant beverages (avoiding alcohol and caffeine), light meal breaks, the use of hats and sun protection. At the end of the harvesting period, harvesting takes place on shorter-light days, and the starting of the working day is earlier and coincides with lower daytime temperatures. Appropriate warm clothing should be worn to avoid hypothermia.

Workers walking through the orchard risk slipping and falling, especially on sloping and muddy ground, on frosty days or when there are gullies. It is advisable that the paths be free of materials in the work areas. Special attention must be paid to the use of ladders when harvesting table olives, avoiding excessive effort when moving them, tripping or falling. Useful preventive measures are: never climb the tree, check the condition of the ladder, use by only one worker and make sure that it is well supported on the ground.

Olive tree beating, whether manual or mechanical, involves the possibility of the fruit flying off and hitting the eyes or face. It is important to avoid working near workers with poles or machines handling the olives and, if not, to wear protective eyewear.

Handling of detached fruit usually involves moving heavy loads and the use of unsuitable posture, maintained, or repeated over time. Heavy containers should be avoided, loads should never be carried on the head, and lifting should be done with the legs and not with the spinal column, and sudden movements with the back should be avoided.

Olive harvesting is not exempt from electrical hazards. Special care must be taken with aerial power lines crossing the orchard, when operating cranes to load the fruit onto trailers, when tipping trailers or when using poles, especially fiber-glass poles, when a storm is approaching.

The ergonomics of the electric hand-held harvesters have been improved compared to gasoline-engine models. Despite this technical evolution, operators continue to be exposed to both upper-limb muscle-skeletal disorder (ULMSD) and hand-arm vibration (HAV) risks. It is interesting to note that 10% of workers using these machines have traumatic vasospastic disease after 1–4.7 years of using the right hand (Çakmak *et al.*, 2011). The prolonged use of these machines could lead to the hand-arm vibration (HAV) syndrome, which can affect the nervous, musculoskeletal, and vascular peripheral structures of the upper limbs. Furthermore, they are often used in conditions unfavorable for the operator; sloping, uneven ground and at low or high temperatures. It is advisable to select machines that transmit less vibration and noise to the operator and are as light as possible. In addition, reducing working time, as well as rotating their use among operators during the working day decreases exposure time.

The use of fuel in machines or equipment involves a risk. Containers other than those approved for transport should never be used, spillage should be avoided, and the engine should be switched off and cold before refuelling.

Personal protective equipment is essential for the safety of operators and must be maintained in good condition and replaced when deteriorated. Protective equipment includes safety footwear (high boots with reinforced toe cap), protective goggles and gloves (there are also gloves able to mitigate vibrations transmitted by hand-held machines). Gloves must be suitable for the working conditions, from picking green fruit, handling sticks, to refuelling equipment.

The use of agricultural machinery requires additional protective measures. Hearing protection is necessary with some stationary equipment (screens, cleaners) or mobile equipment (tractors, harvesters, blowers), and in case of dust, respiratory protection is necessary, and workers should be protected with masks and goggles. In case of using harvesting machinery together with workers on the ground, it is essential to organize the work in separate lines and to provide workers with high-visibility clothing to reduce the risk of being run over.

The use of harvesting machinery involves the risk of severe accidents for workers. The main risks are usually crushing by overturned machinery (tractors and harvesters), getting caught in transmission parts (tractor power take-off) or by moving parts of the machines (fruit-cleaning systems or mechanical systems for moving nets on the ground) or being run over by the machines. The use of machines for purposes other than those for which they are designed, such as the use of off-road vehicles to pull nets with fruit, or getting on or off a moving machine must be avoided. Caution must be taken to avoid being run over by fast-moving machinery, such as self-propelled shakers, or when backing up.

The most appropriate preventive measures require operators to be trained and experienced in the use of machinery, inspection and maintenance of equipment, especially safety guards on moving parts. The organization of work, the training of workers and the identification of those responsible for operations is essential for reducing risks and avoiding accidents.

References

Abdel-Fattah, H., Shackel, K. and Slaughter, D. (2003) Substantial vertical tree displacements occur during almond shaker harvesting. *Applied Engineering in Agriculture* 19(2), 145–150. https://doi: 10.13031/2013.13103

Afsah-Hejri, L., Homayouni, T., Toudeshki, A., Ehsani, R., Ferguson, L. and Castro-García, S. (2022) Mechanical harvesting of selected temperate and tropical fruit and nut trees. *Horticultural Reviews*,171–242. https://doi.org/10.1002/9781119851981.CH4

Angerosa, F., Servili, M., Selvaggini, R., Taticchi, A., Esposto, S. *et al.* (2004) Volatile compounds in virgin olive oil: occurrence and their relationship with the quality. *Journal of Chromatography A* 1054, 17–31. https://doi.org/10.1016/j.chroma.2004.07.093

Baccouri, O., Guerfel, M., Baccouri, B., Cerretani, L., Bendini, A. *et al.* (2008) Chemical composition and oxidative stability of Tunisian monovarietal virgin olive oils with regard to fruit ripening. *Food Chemistry* 109, 743–754. https://doi.org/10.1016/j.foodchem.2008.01.034

Beltran, G., Uceda, M., Hermoso, M. and Frias, L. (2017) Maduracion. In: *El cultivo del olivo*, 7th edn. Mundi-Prensa, Madrid, pp. 187–212.

Bentaher, H. and Rouina, B.B. (2002) Mechanical harvesting of 'Chemlali de Sfax' olive-trees. *Acta Horticulturae* 586, 365–368. https://doi.org/10.17660/ActaHortic.2002.586.73

Bentaher, H., Haddar, M., Fakhfakh, T. and Mâalej, A. (2013) Finite elements modeling of olive tree mechanical harvesting using different shakers. *Trees* 27, 1537–1545. https://doi.org/10.1007/s00468-013-0902-0

Bernardi, B., Benalia, S., Fazari, A., Zimbalatti, G., Stillitano, T. *et al.* (2016) Mechanical harvesting in traditional olive orchards: Oli-picker case study. *Agronomy Research* 14(3), 683–688.

Bernardi, B., Castro-García, S., Blanco-Roldán, G.L., Benalia, S., Fazari, A. *et al.* (2018) A comparison between two materials for head-clamping pads of olive tree shakers: first trials. *Acta Horticulturae* 1199, 403–408. 10.17660/ActaHortic.2018.1199.64

Blanco-Roldán, G.L., Gil-Ribes, J.A., Kourba, K. and Castro-García, S. (2009) Effects of trunk shaker duration and repetitions on removal efficacy for the harvesting of oil olives. *Applied Engineering in Agriculture* 25(3), 329–334. DOI: 10.13031/2013.26883

Çakmak, B., Saraçoğlu, T., Alayunt, F.N. and Özarslan, C. (2011) Vibration and noise characteristics of flap type olive harvesters. *Applied Ergonomics* 42(3), 397–402. https://doi.org/10.1016/j.apergo.2010.08.015

Calvo, A., Romano, E., Preti, C., Schillaci, G. and Deboli, R. (2018) Upper limb disorders and hand-arm vibration risks with hand-held olive beaters. *International Journal of Industrial Ergonomics* 65, 36–45. https://doi.org/10.1016/j.ergon.2018.01.018

Camposeo, S. (2017) Utilizzo delle scavallatrici per l'altissima densità. *Olivo e Olio* 3, 48–50.

Camposeo, S., Bellomo, F. and Godini, A. (2010) Aspetti quanti-qualitativi della raccolta meccanica in un giovane oliveto superintensivo. *Rivista di Frutticoltura* 6, 76–78.

Camposeo, S., Vivaldi, G.A. and Gattullo, C.E. (2013) Ripening indices and harvesting times of different olive cultivars for continuous harvest. *Scientia Horticulturae* 151, 1–10. https://doi.org/10.1016/j.scienta.2012.12.019

Caricato, F. (2001) Prove di raccolta meccanica su olivi secolari. *L'Informatore Agrario* 42, 89–90.

Castillo-Llanque, F. and Rapoport, H.F. (2008) Indentifying the location of olive fruit abcission. *Scientia Horticulturae* 120(2), 292–295. https://doi.org/10.1016/j.scienta.2008.11.006

Castillo-Ruiz, F.J., Sola-Guirado, R.R., Castro-Garcia, S., Gonzalez-Sanchez, E.J., Colmenero-Martinez, J.T. *et al.* (2017) Pruning systems to adapt traditional olive orchards to new integral harvesters. *Scientia Horticulturae* 220, 122–129. https://doi.org/10.1016/j.scienta.2017.03.043

Castillo-Ruiz, F.J., Tombesi, S. and Farinelli, D. (2018) Tracking olive fruit movement and twisting during the harvesting process using video analysis. *Acta Horticulturae* 1199, 409–414. https://doi.org/10.17660/ActaHortic.2018.1199.65

Castro-Garcia, S., Blanco-Roldan, G.L., Gil-Ribes, J.A. and Aguera-Vega, J. (2008) Dynamic analysis of olive trees in intensive orchards under forced vibration. *Trees – Structure and Function* 22(6), 795–802. https://doi.org/10.1007/s00468-008-0240-9

Castro-Garcia, S., Rosa, U.A., Gliever, C.J., Smith, D., Burns, J.K. *et al.* (2009) Video evaluation of table olive damage during harvest with a canopy shaker. *Horttechnology* 19(2), 260–266. https://doi.org/10.21273/HORTSCI.19.2.260

Castro-Garcia, S., Castillo-Ruiz, F.J., Jimenez-Jimenez, F., Gil-Ribes, J.A. and Blanco-Roldan, G.L. (2015) Suitability of Spanish 'Manzanilla' table olive orchards for trunk shaker harvesting. *Biosystems Engineering*,129, 388–395. https://doi.org/10.1016/j.biosystemseng.2014.11.012

Cinosi, N., Portarena, S., Almadi, L., Berrettini, A., Torres, M., Pierantozzi, P., Villa, F., Galletti, A., Famiani, F. and Farinelli, D. (2023) Use of portable devices and an innovative and non-destructive index for in-field monitoring of olive fruit ripeness. *Agriculture* 13: 194. https://doi.org/10.3390/ agriculture13010194

Colmenero-Martinez, J.T., Blanco-Roldán, G.L., Bayano-Tejero, S., Castillo-Ruiz, F.J., Sola-Guirado, R.R. *et al.* (2018) An automatic trunk-detection system for intensive olive harvesting with trunk shaker. *Biosystems Engineering* 172, 92–101. https://doi.org/10.1016/j.biosystemseng.2018.06.002

Connor, D.J., Gómez-del-Campo, M., Rousseaux, M.C. and Searles, P.S. (2014) Structure, management and productivity of hedgerow olive orchards: a review. *Scientia Horticulturae* 169, 71–93. https://doi.org/10.1016/j.scienta.2014.02.010

Da Silva, L.P. and Mata, V.A. (2019) Stop harvesting olives at night: it kills millions of songbirds. *Nature* 569 (7755), 192.

Deboli, R., Calvo, A. and Preti, C. (2016) Vibration and impulsivity analysis of hand held olive beaters. *Applied Ergonomics* 55, 258-267. https://doi.org/10.1016/j.apergo.2015.11.003

Di Matteo, M., Spagna Musso, S., Grassog, G. and Bufalo, G. (1992) Caratterizzazione agronomica e merceologica in relazione al grado di maturazione delle produzioni di alcune cultivar olearie della provincia di Avellino. *Rivista della Società Italiana di Scienza dell'Alimentazione* 21, 35–36.

Di Vaio, C., Marallo, N., Nocerino, S. and Famiani, F. (2012) Mechanical harvesting of oil olives by trunk shaker with a reversed umbrella interceptor. *Advances in Horticultural Science* 26 (3–4), 1–4. DOI: 10.1400/220111

Dias, A.B., Falcão, J.M., Pinheiro, A. and Peça, J.O. (2020) Evaluation of olive pruning effect on the performance of the row-side continuous canopy shaking harvester in a high density olive orchard. *Frontiers in Plant Science* 10, 1631. https://doi.org/10.3389/fpls.2019.01631

Famiani, F. (2017) Criteri di potatura per meccanizzare la raccolta. *Olivo e Olio* 2, 30–33.

Famiani, F., Alfei, B., D'Ambrosio, A., Rosati, S. and Pannelli, G. (1999) Biologia fiorale e di fruttificazione delle principali varietà locali di olivo in Umbria. In: Atti V Convegno Nazionale 'Biodiversità: biodiversistà e sistemi ecocompatibili', Caserta, Italia, pp. 204–213.

Famiani, F., Proietti, P., Farinelli, D. and Tombesi, A. (2002) Oil quality in relation to olive ripening. *Acta Horticulturae* 586, 671–674. https://doi.org/10.17660/ActaHortic.2002.586.143

Famiani, F., Farinelli, D. and Proietti P. (2004a) Raccolta delle olive. In: *Aggiornamenti sulle Tecniche Colturali in Olivicoltura*. Accademia Nazionale dell'Olivo e dell'Olio, Spoleto (PG), Italia, pp. 123–162.

Famiani, F., Proietti, P., Boco, M., Marucci, B., Alfei, B. *et al.* (2004b) Caratteristiche vegeto-produttive e qualità dell'olio della cultivar di olivo 'Rosciola'. In: Atti convegno nazionale 'Germoplasma Olivicolo e Tipicità dell'Olio', Perugia, Italia, pp, 284–289.

Famiani, F., Giurelli, A., Proietti, P., Nasini, L., Farinelli, D. *et al.* (2008) Sì alla raccolta agevolata in oliveti tradizionali e intensivi. *L'Informatore agrario* 4, 103–107.

Famiani, F., Baldicchi, A., Farinelli, D., Nasini, L. and Proietti, P. (2010) L'efficienza dell'agevolatrice è influenzata dalla cultivar. *Olivo e Olio* 10, 44–51.

Famiani, F., Farinelli, D., Rollo, S., Camposeo, S., Di Vaio, C. *et al.* (2014) Evaluation of different mechanical fruit harvesting systems and oil quality in very large size olive trees. *Spanish Journal of Agricultural Research* 12(4), 960–972. https://doi.org/10.5424/sjar/2014124-5794

Famiani, F., Farinelli, D., Caruso, G. and Gucci, R. (2016) Raccolta agevolata e meccanizzata. *Olivo e Olio* 6, 47–50.
Famiani, F., Farinelli, D., Urbani, S., Al Hariri, R., Paoletti, A. *et al.* (2020) Harvesting system and fruit storage affect basic quality parameters and phenolic and volatile compounds of oils from intensive and super-intensive olive orchards. *Scientia Horticulturae* 263, 109045. https://doi.org/10.1016/j.scienta.2019.109045
Famiani, F. and Gucci, R. (2022) La raccolta delle olive. In: Edagricole – Edizioni Agricole di New Business Media, Boloegna, Italia.
Farinelli, D. and Tombesi, S. (2015) Performance and oil quality of 'Arbequina' and four Italian olive cultivars under super high density hedgerow planting system cultivated in central Italy. *Scientia Horticulturae* 192, 97–107. https://doi.org/10.1016/j.scienta.2015.04.035
Farinelli, D., Boco, M. and Tombesi A. (2004) Influenza dell'andamento stagionale sull'accumulo dell'olio in cultivar di olivo. In: Atti convegno nazionale 'Germoplasma Olivicolo e Tipicità dell'Olio', Perugia, Italia, pp. 165–169.
Farinelli, D., Ruffolo, M., Boco, M. and Tombesi, A. (2012a) Yield efficiency and mechanical harvesting with trunk shaker of some international olive cultivars. *Acta Horticulturae* 949, 379–384. https://doi.org/10.17660/ActaHortic.2012.949.55
Farinelli, D., Scatolini, G., Siena M. and Tombesi, S. (2012b) Efficienza dei vibratori di tronco nella raccolta meccanica delle olive. *Rivista di Frutticoltura* 10, 82–84.
Farinelli, D., Tombesi, S., Famiani, F. and Tombesi, A. (2012c) The fruit detachment force/fruit weight ratio can be used to predict the harvesting yield and the efficiency of trunk shakers on mechanically harvested olives. *Acta Horticulturae* 965, 61–64. https://doi.org/10.17660/ActaHortic.2012.965.5
Ferguson, L. (2006) Trends in olive fruit handling previous to its industrial transformation. *Grasas y aceites* 57(1), 9–15. DOI: https://doi.org/10.3989/gya.2006.v57.i1.17
Ferguson, L. and Garcia, S.C. (2014) Transformation of an ancient crop: preparing California 'Manzanillo' table olives for mechanical harvesting. *HortTechnology* 24(3), 274–280. https://doi.org/10.21273/HORTTECH.24.3.274
Ferguson, L., Rosa, U.A., Castro-Garcia, S., Lee, S.M., Guinard, J.X. *et al.* (2010) Mechanical harvesting of California table and oil olives. *Advances in Horticultural Science* 24(1), 53–63. DOI: 10.1400/132343
Ferreira Júnior, L.G., Da Silva, F.M., Ferreira, D.D. and Sales, R.S. (2016) Recommendation for mechanical harvesting of coffee based on vibration behavior settings rods harvesters. *Ciencia Rural* 46(2), 273–278. DOI:10.1590/0103-8478cr20141679
Fiorino, P., Marone, E. and Ottanelli, A. (2008) Cultivar, produttività e sistemi di raccolta meccanica: un quadriennio di confronti. In: Atti convegno 'COM.SI.OL – Competitività del Sistema Olivo in Italia', Spoleto (PG), Italia, pp. 5–16.
García, J.M., Seller, S. and Pérez-Camino, C. (1996) Influence of fruit ripening on olive oil quality. *Journal of Agricultural and Food Chemistry* 44, 3516–3520. https://doi.org/10.1021/jf950585u
Giametta, F. (1999) Macchine innovative per olivi tradizionali. *Terra e Vita* 17, 76–80.
Giametta G. (2003) La raccolta delle olive. In: *Olea – Trattato di Olivicoltura*. A cura di P. Fiorino. Edagricole, Bologna, Italia, pp. 235–261.
Giametta, G. and Bernardi, B. (2010) Olive grove equipment technology. Straddling trees: mechanized olive harvests. *Advances in Horticultural Science* 24(1), 64–70. https://www.jstor.org/stable/42882755
Gil-Ribes, J.A., Ferguson, L., Castro-Garcia, S. and Blanco-Rodán, G.L. (2014) How agricultural engineers develop mechanical harvesters: the university perspective. *HortTechnology* 24(3), 270–273. https://doi.org/10.21273/HORTTECH.24.3.270
Horvath, E. and Sitkei, G. (2001) Energy consumption of selected tree shakers under different operational conditions. *Journal of Agricultural Engineering Research* 80(2), 191–199. https://doi.org/10.1006/jaer.2001.0727
Inglese, P., Famiani, F., Galvano, F., Servili, M., Esposto, S. *et al.* (2011) Factors affecting extra-virgin olive oil composition. *Horticultural Reviews* 38, 83–147.
Jiménez, M.R., Casanova, L., Suárez, M.P., Rallo, P. and Morales-Sillero, A. (2017) Internal fruit damage in table olive cultivars under superhigh-density hedgerows. *Postharvest Biology and Technology* 132, 130–137. https://doi.org/10.1016/J.POSTHARVBIO.2017.06.003
Kouraba, K., Blanco-Roldán, G.L., de Jaime-Revuelta, M.A. and Barranco-Navero, D. (2004) Suitability of olive varieties for mechanical shaker harvesting. *Olivae* 101, 39–43.
Lavee, S. (1996) Olive biology and physiology. In: *World Olive Encyclopedia*. International Olive Oil Council, Barcelona, Spain, pp. 59–110.
Lavee, S. (2010) Integrated mechanical, chemical and horticultural methodologies for harvesting of oil olives and the potential interaction with different growing systems: a general review. *Advances in Horticultural Science* 24(1), 5–15. DOI: 10.1400/132337

Leone, A., Romaniello, R., Tamborrino, A., Catalano, P. and Peri, G. (2015) Identification of vibration frequency, acceleration, and duration for efficient olive harvesting using a trunk shaker. *Transactions of the ASABE* 58(1), 19–26. DOI: 10.13031/trans.58.10608

Morales-Sillero, A., Rallo, P., Jiménez, M.R., Casanova, L. and Suárez, M.P. (2014) Suitability of two table olive cultivars ('Manzanilla de Sevilla' and 'Manzanilla Cacereña') for mechanical harvesting in super-high-density hedgerows. *HortScience* 49(8), 1028–1033. https://doi.org/10.21273/HORTSCI.49.8.1028

Parlati, M.V., Ripa V., Padula, G., De Rose, F. and Iannotta, N. (2006) Efficienza di alcune macchine per la raccollta impiegate su diverse cultivar di olivo dell'Italia centrale. In: Atti convegno nazionale 'Maturazione e raccolta delle olive: strategia e tecnologie per aumentare la competitività in olivicoltura'. Alanno (PE), Italia, pp. 137–141.

Plasquy, E., García Martos, J.M., Florido Fernández, M.D.C., Sola-Guirado, R.R. and García Martín, J.F. (2021) Cold storage and temperature management of olive fruit: the impact on fruit physiology and olive oil quality: a review. *Processes* 9, 1543. https://doi.org/10.3390/PR9091543

Rallo, L., Barranco, D., Castro-García, S., Connor, D.J., del Campo, M.G. *et al.* (2013) High-density olive plantations. *Horticultural Reviews* 41, 303–383. https://doi.org/10.1002/9781118707418.ch07

Ravetti, L. and Robb, S. (2010) Continuous mechanical harvesting in modern Australian olive growing systems. *Advances in Horticultural Science* 24(1), 71–77. http://www.jstor.org/stable/42882756

Roka, F.M., Ehsani, R.J., Futch, S.H. and Hyman, B.R. (2014) Citrus mechanical harvesting systems – continuous canopy shakers. University of Florida, Food and Economic Resources Department, UF/IFAS Extension.

Servili, M., Montedoro, G.F., Pannelli, G. and Famiani, F. (1990) Influenza delle variabili pedoclimatiche, tecnologiche e varietali sulla qualità degli oli vergini di oliva. In: Atti convegno 'Problematiche qualitative dell'olio di oliva', Sassari, Italia, pp. 231–245.

Sola-Guirado, R.R., Castro-García, S., Blanco-Roldán, G.L., Jiménez-Jiménez, F., Castillo-Ruiz, F.J. *et al.* (2014) Traditional olive tree response to oil olive harvesting technologies. *Biosystems Engineering* 118, 186–193. https://doi.org/10.1016/j.biosystemseng.2013.12.007

Sola-Guirado, R.R., Jimenez-Jimenez, F., Blanco-Roldan, G.L., Castro-Garcia, S., Castillo-Ruiz, F.J. *et al.* (2016) Vibration parameters assessment to develop a continuous lateral canopy shaker for mechanical harvesting of traditional olive trees. *Spanish Journal of Agricultural Research* 14(2). https://doi.org/10.5424/sjar/2016142-7909

Sola-Guirado, R.R., Bernardi, B., Castro-García, S., Blanco-Roldán, G.L., Benalia, S. *et al.* (2018) Assessment of aerial and underground vibration transmission in mechanically trunk shaken olive trees. *Journal of Agricultural Engineering* 49(3), 191–197.

Sola-Guirado, R.R., Aragon-Rodriguez, F., Castro-Garcia, S. and Gil-Ribes, J. (2019) The vibration behaviour of hedgerow olive trees in response to mechanical harvesting with straddle harvester. *Biosystems Engineering* 184, 81–89. https://doi.org/10.1016/j.biosystemseng.2019.06.009

Tombesi, A., Tombesi, S. and Gucci, R. (2021) Sviluppo del frutto. Collana divulgativa dell'Accademia, Accademia Nazionale dell'Olivo e dell'Olio, vol. II. Pubblicazione on line: www.accademiaolivoeolio.com (accessed 21 May 2023).

Tombesi, S. and Farinelli, D. (2017) Canopy management in super high-density olive orchards: relationship between canopy light penetration, canopy size and productivity, *Acta Horticulturae* 1177, 87–91.

Tombesi, S., Poni, S., Palliotti, A. and Farinelli, D. (2017) Mechanical vibration transmission and harvesting effectiveness is affected by the presence of branch suckers in olive trees. *Biosystems Engineering* 158, 1–9. https://doi.org/10.1016/J.BIOSYSTEMSENG.2017.03.010

Tous, J., Romero, A. and Hermoso, J.F. (2010) New trends in olive orchard design for continuous mechanical harvesting. *Advances in Horticultural Sciences* 24(1), 43–52. DOI: 10.1400/132342

Tous, J., Romero, A., Hermoso, J.F., Msallem, M. and Larbi, A. (2014) Olive orchard design and mechanization: present and future, *Acta Horticulturae* 1057, 231–246. 10.17660/ActaHortic.2014.1057.27

Tsatsarelis, C.A. (1987) Vibratory olive harvesting: the response of the fruit-stem system to fruit removing actions. *Journal of Agricultural Engineering Research* 38(2), 77–90. https://doi.org/10.1016/0021-8634(87)90120-X

Vega Macías, V., Orgaz Rosúa, F., Garay, J., Hidalgo Moya, J.C., Hidalgo Moya, J.J. *et al.* (2014) Influencia del estado hídrico del olivo en el descortezado del tronco en la recolección. *Vida rural* 375, 36–44.

Whitney, J.D., Hedden, S.L. and Sumner, H.R. (1974) Harvesting 'Valencia' oranges with a vertical foliage shaker. *Citrus Ind* 55(11), 15–18.

Yu, P., Li, C., Takeda, F., Krewer, G., Rains, G. and Hamrita, T. (2014) Measurement of mechanical impacts created by rotary, slapper, and sway blueberry mechanical harvesters. *Computers and Electronics in Agriculture* 101, 84–92. https://doi.org/10.1016/j.compag.2013.12.001

Zion, B., Bechar, A., Regev, R., Shamir, N., Weissblum, A., *et al.* (2011) Mechanical harvesting of olives: an operations study. *Israel Journal of Plant Sciences* 59(1), 71–84. DOI: 10.1560/IJPS.59.1.71

7.1 Responses to Abiotic Stresses

L. Sebastiani[1] and R. Gucci[2]
[1]*Crop Science Research Center, Sant'Anna University, Pisa, Italy;*
[2]*Department of Agriculture, Food and Environment, University of Pisa, Italy*

7.1.1 Adaptation to Mediterranean Climates

Plants are living organisms that closely interact with the environment in which they live and for this reason they have developed complex and different mechanisms that allow them to survive and grow under a multitude of abiotic (non-living) chemical and physical factors. Abiotic factors include water availability, temperature, mineral nutrients and trace elements, salinity and many others related to the atmospheric, soil and water dynamic processes which vary in time and space. When abiotic factors are outside the optimum range for growth, plants experience negative physiological consequences. Adaptation to the environment involves genetic changes in the population that the natural selection fixes and maintains generation after generation. The evolution of adaptive mechanisms to specific environments frequently involves processes that allow avoidance of the damaging effects (i.e. active growth cycle under favorable temperature and water availability period, or avoid the uptake of toxic substances). On the other hand, there are responses that do not require genetic modifications and that each plant uses to react to the frequent day-by-day or seasonal changes in environmental conditions. These responses can alter plant physiology and morphology and allow them to better survive to the new environmental circumstances; and if this happens by repeated exposure to stressful conditions, this process is known as acclimation (Fig. 7.1.1). Plant responses to the environment are known as phenotypic plasticity, that consist in non-permanent changes in physiology and morphology that are reversible if the factors responsible for them change. The combination of adaptation and acclimation processes contribute to the overall tolerance to abiotic stress for individual species.

Olive trees are adapted to Mediterranean climatic conditions: mild, wet winters with temperatures that drop below 10°C but rarely below 0°C, and warm, dry summers. More recently, olive orchards were planted in different warm-temperate regions of the world such as North and South America, Australia, New Zealand and South Africa, and even in the monsoon systems of China and India. Both in Mediterranean and in the new cultivation areas, when weather conditions become more extreme (drought, high or low temperatures) or soil conditions are not optimal for growth (salinity, low oxygen, nutrient deficiencies), olive plants are subjected to abiotic stresses.

DOI: 10.1079/9781789247350.0021

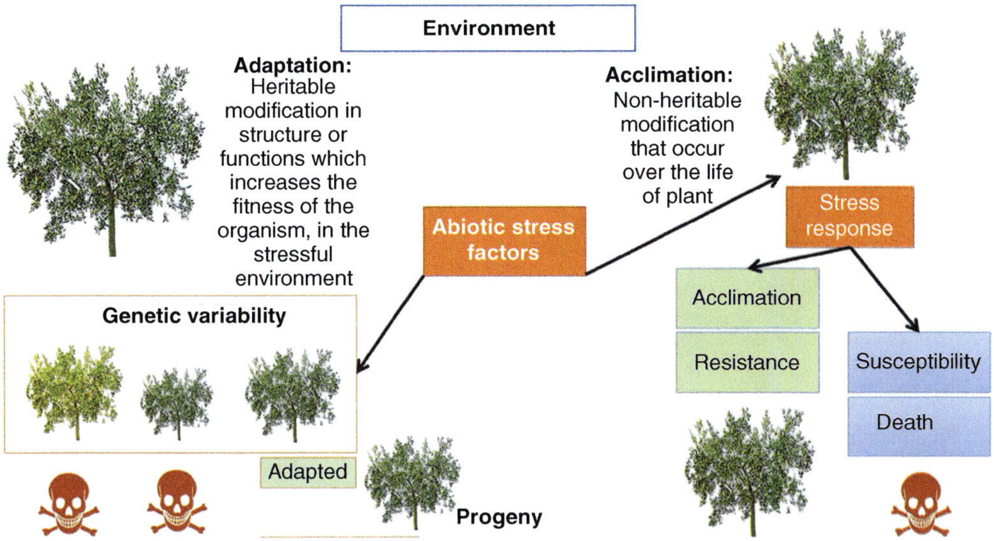

Fig. 7.1.1. Schematic representation of the adaptation and acclimation mechanisms in plants. The combination of adaptation and acclimation contribute to the overall tolerance to abiotic stress of a species.

Damages due to abiotic stresses are seldom readily identified in olive orchards since plants are often grown in non-specialized planting systems that are managed with limited cultural practices. However, olive cultivation has now been modified from traditional low-density and low-input to high-density and high-input growing systems due to the renewed interest in extra-virgin olive oil for its beneficial health effects. Under the new cultivation systems, the effect of abiotic stresses on olive might have a strong impact on both yield and quality. The aim of this chapter is to focus on the physiological responses of olive trees to drought, salinity and temperature stress, while for other abiotic stresses the reader can refer to the existing literature.

7.1.2 Water Deficit

Water deficit strongly affects plant growth and modifies plant anatomy, morphology and physiology. In general, plants on dry locations have smaller organ size since growth is highly correlated with water availability.

Under water deficit, cellular dehydration occurs with multiple effects on physiological processes. One primary effect is the reduction in cell turgor potential (Ψ_p) and cell volume since the Ψ_w of the apoplast becomes more negative than that of the symplast and water exits the cell. After turgor loss and cell dehydration the molecules inside the cell become more concentrated and might also become cytotoxic, increasing the cell osmotic potential ($\Psi\pi$). Hydraulic resistance will increase as a consequence of water deficit. Root shrinkage determines fine root detachment from the soil particles and root hair damage increases the hydraulic resistance, cavitation (breakage of the water columns under tension within the xylem), reduced leaf expansion, stomatal closure, photosynthesis inhibition, leaf abscission, altered carbon partitioning, membrane and protein destabilization, reactive oxygen species production and cell death (Fig. 7.1.2).

The olive has several adaptive traits to water scarcity that help the plant overcome some of the negative impacts of drought. In Table 7.1.1, the most relevant characteristics for root, stem and leaf are listed.

Stems play a main role in water conduction to other parts of the plant and its

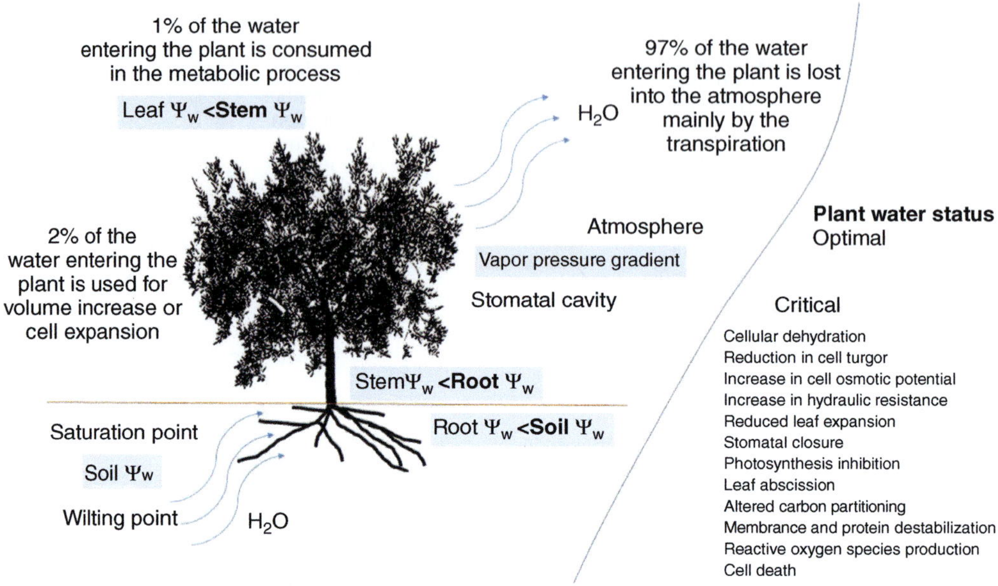

Fig. 7.1.2. Main driving forces for water flow from the soil through the plant to the atmosphere (*left*) and physiological and biochemical changes caused by critical plant water status.

Table 7.1.1. Olive adaptive traits to drought.

Organ	Trait	Response to drought
Root	Capacity to explore the topsoil layers	Advantage from scarce and intermittent rainfall events
	Active roots close to the trunk	Increases the water-absorbing efficiency from rainfall running down the stem
Stem	High proportion of small-diameter vessels uniformly distributed in the annual growth ring	Lower hydraulic conductivity and less susceptibility to embolism
Leaf	Thick cuticle and stomata located on the abaxial side of the leaf	Barrier to water loss
	Dense and multi-layered *indumentum* of peltate scales	Microenvironmental conditions favorable to gas exchange in dry conditions
	Small epidermal cells and higher number of epidermal cells per leaf surface	Resistant to collapse, limit water loss

xylem characteristics are relevant. In olive stems, xylem is diffuse and porous, with small-diameter vessels uniformly distributed in the annual growth ring, rich in fibers and with scarce parenchymatic cells (Salleo *et al.*, 1985; Rossi *et al.*, 2013). Lo Gullo and Salleo (1990) showed that the diameter of the xylem vessels in nodal and inter-nodal portions of a one-year-old stem is less than 40 μm, and as much as 90% of the vessels have a diameter of less than 20 μm, which makes olive adapt to an environment with limited water availability (Fig. 7.1.3). Under drought stress, phenotypic plasticity of olive stem anatomy is observed and the frequency of vessel diameter classes in stem rings changes. Rossi *et al.* (2013) reported that irrigated trees have fewer vessels (< 20 μm diameter) than rainfed trees (4.3–8.3% and 8.3–15%, respectively), while the diameter classes higher than 20 μm were, on average, more abundant in irrigated than in rainfed trees. This shift in vessel characteristics could be explained by the close relationship between vessel diameters and hydraulic conductivity. Smaller vessels have a lower hydraulic conductivity than larger ones. This property and other anatomical characteristics make smaller vessels less susceptible to the embolism phenomena and consequent cavitation that could destroy the xylem functionality under severe drought. Moreover, the xylem vessel functionality can also be affected by the aspiration of air into the transpiration stream through the pit membrane of adjacent embolized vessels, a phenomenon known as air-seeding. When gas bubbles form, they could rapidly spread in a large portion of the plant hydraulic system provoking a generalized hydraulic failure in the water-transport mechanism. For this reason, under drought conditions, the olive produces a higher number of small vessels thus reducing the probability of embolism occurring in the xylem and likely increase its resistance to water-deficit stress. Salleo and Nardini (1999) estimated that only 5% of the xylem vessels of the stem undergo cavitation when the leaf Ψ_w approaches –3.5 MPa.

Leaves are the organs responsible for the large part of water loss. A common trait for the species adapted to xeric habitats is the presence of thick and tough leaves (Larcher, 1995). Other useful traits comprise a reduction of the leaf size, leaf rolling, dense leaf pubescence, deeply developing stomata, accumulation of mucilage and other secondary metabolites in the mesophyll, and increase of mesophyll compactness (Bosabalidis and Kofidis, 2002). Olive leaves are evergreen and long living (2–3 years) with many sclerophylly traits. These traits include a thick cuticle and stomata located on the abaxial side of the leaf. Moreover, in olive leaf, the stomata are not directly exposed to the atmosphere, but are protected by a dense and multi-layered *indumentum* of peltate scales (Marchi *et al.*, 2005, 2007). This *indumentum* creates microenvironmental conditions that are favorable to gas exchange in dry conditions and limit water loss (Besnard *et al.*, 2009) and when the water stress increases the leaf blade folds and the leaf petiole lowers its insertion angle onto the stem. These processes diminish the leaf surface directly hit by the sun and expose the abaxial surface, rich in peltate scales, to sunlight, thus increasing the leaf reflectance of the incident light while diminishing its transmittance (Baldini *et al.*, 1997) (Fig. 7.1.4). Under drought conditions, olive leaves undergo an acclimation process leading to anatomical alterations that increase the sclerophylly characteristics and significantly decrease water consumption. Such changes, principally, comprise an increase of the cuticle thickness and the density of stomata and of the peltate scales. A significant reduction of the cell size of the epidermal and mesophyll cells is also observable (Bosabalidis and Kofidis, 2002). The small size and the straight (not sinuous) walls of the epidermal cells contribute to a significant resistance against cells collapsing under drought (Oertli *et al.*, 1990). Moreover, small epidermal cells are at least 20 times more resistant to collapse than large ones and a higher number of epidermal cells per leaf surface unit limit water loss through cuticular transpiration. Regarding phenotypic plasticity, differences among cultivars exist. Bacelar *et al.* (2004) found an enhancement in sclerophylly in 'Manzanilla', as measured by

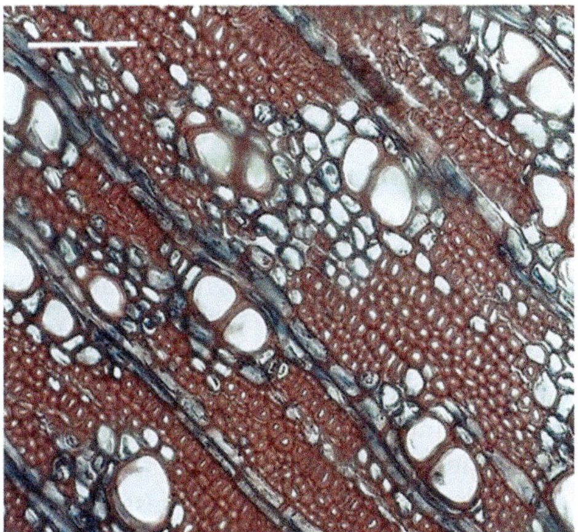

Fig. 7.1.3. Olive xylem tissue showing xylem vessels (large red), fibers (small red), axial- and ray-parenchyma cells (blue). White bar = 100 µm. (Photo: L. Sebastiani)

Fig. 7.1.4. (A) water stress (+) lowers the insertion angle of the leaf petiole onto the stem and (B) increases the leaf blade folding. (C) dense and multilayered *indumentum* of peltate scales on the abaxial surface of the leaf. (D) peltate scales artificially removed from the leaf exposed the leaf surface where stomata are localized. (Photos: L. Sebastiani)

the production of more parenchyma tissues and increased protective structures like the upper cuticle and the upper and lower epidermis. In other cultivars, like 'Cobrançosa', water loss was prevented through high-density foliar tissue and thicker cuticle and trichome layers.

These adapting traits and acclimation processes make the olive tree a species with a high resistance to water scarcity. However, in cultivation we are interested in growing olive plants under optimal water conditions. Bongi and Palliotti (1994) estimated that for each gram of olive fruit dry matter, approximately 315 g of water are absorbed by the roots, transported towards the stems and the leaves and then lost into the atmosphere. To ensure this continuous and intense water flow, the maintenance of hydraulic functionality is crucial. At the beginning of drought, the olive tree could finely regulate water loss while continuing to maintain water uptake. This capacity enables the olive to preserve the cell turgor whereas stress-induced dehydration phenomena increase (Connor and Fereres, 2005). The adaptive traits and the phenotypic plasticity for resistance to drought stress make visible symptoms appear late in field-grown trees even if growth and yield have been already negatively affected by water scarcity. The most common symptoms include early senescence and leaf drop but the progression of symptoms, as reported in several controlled experimental trials, first starts with leaf blade folding and changes in leaf angle and then ends with leaf shedding. In the early stages, visible symptoms are mainly localized in the youngest leaves and occur during the warmest period of the day. Loss of turgor determines a decrease in transpiration that is negative for plant productivity, but it has a positive feedback since it limits irreversible damages to the cells due to temperature rise and photoinhibition mechanisms (Schwabe and Lionakis, 1996). As water stress progresses further, other symptoms appear and they include chlorosis, bronzing, and blade folding of older leaves (Gucci *et al.*, 2003), shoot wilting and shriveled fruits.

The olive has the capacity to activate tolerance mechanisms to drought stress triggering an array of physiological responses. The aim is to achieve a physiological range for the tissue/cell water status that is still acceptable for the metabolic processes (Lo Gullo and Salleo, 1988). The first mechanism is to establish a high-water potential gradient between leaves and roots, and, consequently, maintain the capacity to extract water from soil when water content decreases. Experiments showed that the olive has the capacity to extract water up to a soil water potential (ψ_{soil}) of –2.5/–3.5 MPa (Lo Gullo and Salleo, 1988; Dichio *et al.*, 2003, 2005) before wilting symptoms become visible. Some studies suggest that olive plants might achieve Ψ_w of –6 MPa or even –8 MPa and tolerate severe leaf dehydration (water loss of 40%) while retaining the full rehydration capacity when stress conditions disappear (Rhizopoulou *et al.*, 1991).

Plant capacity to withstand dehydration is mainly achieved through turgor maintenance and protection of cellular functions by specific solutes. These effects are driven by osmotic adjustment and accumulation of cellular compatible solutes (Blum, 2017). In olive, during the early stages of water deficit (when the water absorbed by the roots is no longer sufficient to replenish the losses due to the transpiration from leaves), both the Ψ_w and the Ψ_p decrease rapidly (Lo Gullo and Salleo, 1988). These drought-induced physiological changes inhibit cell division and expansion in the growing organs, and the water lost from the tissues triggers a reduction in $\Psi\pi$. Change in $\Psi\pi$ can be due to passive (dehydration itself that increases the solute concentration into the cell) and to active mechanisms that generate an accumulation of intra-cellular solutes. Olives achieve a high level of osmotic adjustment and synthetize and accumulate osmotically active and metabolically compatible solutes. Active mechanisms of osmotic regulation in response to water-deficit stress enable plants to tolerate short- or long-term water-deficit stress. The main molecules that plant cells accumulate in this process are soluble carbohydrates, inorganic cations, and organic acids or amino acids (Ingram and Bartels, 1996). Osmotic adjustments have long been known to sustain photosynthesis via turgor maintenance as stress develops (Mervyn and Ludlow, 1987) and contribute to the plant's ability to extract water when the Ψ_w of the soil lowers as a consequence of water scarcity; in many crops, a positive and significant association between osmotic regulation and yield is frequent (Blum, 2017). In olive plants at a pre-dawn Ψ_w leaves of –5.2 MPa, Xiloyannis *et al.* (1999) detected that roots of 4–5 mm undergo an osmotic adjustment around 1.67 MPa and roots of $\leq$ 1 mm to 0.2 MPa. The main molecules involved in osmotic adjustments are mannitol, glucose and organic acids. Dichio *et al.* (2005) reported an osmotic adjustment ranging from 2.4 MPa to 3.8 MPa in severely stressed trees and estimated that the active osmotic adjustment due to mannitol was 0.45–0.8 MPa for leaves and 0.75–1.42 MPa for roots. The osmotic adjustments in roots are responsible for the maintenance of the cell turgor required to avoid the detachment of fine roots from soil particles and for the set-up of a root-to-soil Ψ_w gradient that is necessary for the conservation of a water flow under drought.

Unlike most other species, g_s in olive declines gradually as water-deficit stress increases causing a gradual reduction in A_n and only when stem Ψ_w becomes very low < –4 MPa, A_n decreases much faster than g_s (Moriana *et al.*, 2002). This happens when the water deficit begins drought stress and causes the closure of stomata and the consequent reduction in g_s and A_n. With the progression of water deficit, the negative effects also affect the non-stomatal components, such as those related to g_m or to the light-dependent reactions (i.e. the photosystem II) (Angelopoulos *et al.*, 1996). Changes in g_m depend on many factors, such as leaf temperature and air vapor pressure deficit. Regarding leaf temperature, Diaz-Espejo *et al.* (2007) measured values for g_m of 0.224 mol m^{-2} s^{-1} at 29.6°C, and of 0.14 mol m^{-2} s^{-1} at 40°C. Therefore, in summer, g_m can decrease both for drought and for the high temperatures that are very common in the Mediterranean region. Anatomical investigations showed that g_m correlates with the surface of the chloroplast and the mesophyll cell-wall thickness (Tomás *et al.*, 2013). Moreover, molecular players, such as aquaporins and carbonic anhydrase, have been found to be actively involved in the regulation of both g_s and g_m in olive. Perez-Martin *et al.* (2014) showed, in a water stress and recovery experiment, that leaf gas exchange, chlorophyll fluorescence and plant water status have some correlations with the gene expression of *OePIP1.1* and *OePIP2.1* aquaporins as well as with stromal carbonic anhydrase. Using modeling approaches, the authors suggest that both *OePIP1.1* and *OePIP2.1* expression could explain most of the g_s and g_m variations under drought, while the stromal carbonic anhydrase had a small effect on g_m (Fig. 7.1.5).

Under severe water-limiting conditions, olive leaf could approach the point of turgor loss (daily minimum Ψ_w of the stem lower than –3.5 MPa). In this condition, g_s and transpiration are very low and the RWC of the leaf approaches 75% (Lo Gullo and Salleo, 1988; Giorio *et al.*, 1999). In these circumstances, the recovery of the water

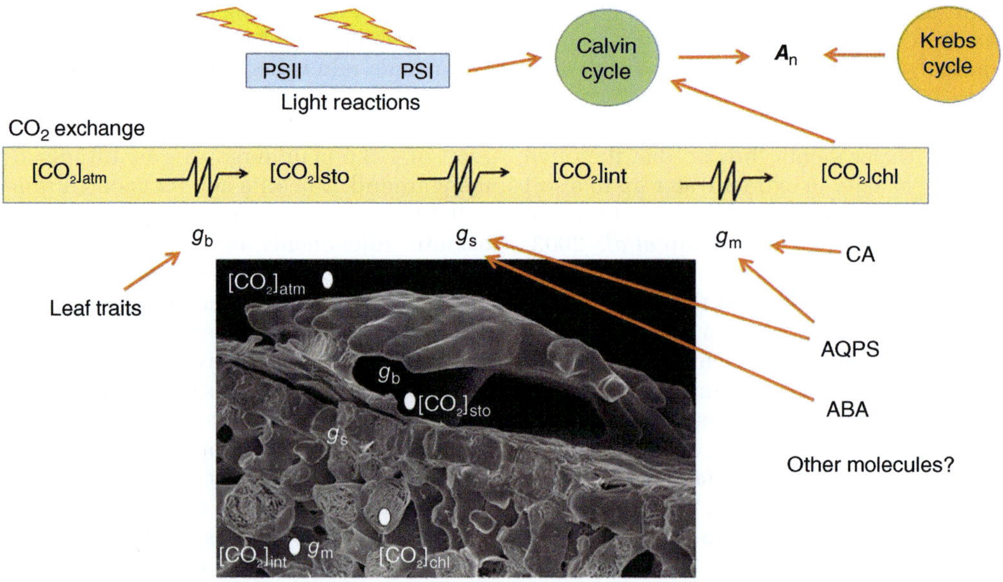

Fig. 7.1.5. Possible mechanisms that lead to a decrease in net CO_2 assimilation rate (A_n) under drought stress. CO_2 exchange pathway is regulated by boundary layer conductance (g_b), stomatal conductance (g_s) and mesophyll conductance (g_m) that under water-limiting conditions can change due to morphological or biochemical modifications. Light reactions can be affected under severe drought. Photosynthetic and respiration cycles are the final actors determining A_n values. PSII, photosystem II; PSI photosystem II, ABA, Abscisic Acid, CA. carbonic anhydrase; AQPs aquaporins; $[CO_2]_{atm}$, CO_2 concentration in the atmosphere; $[CO_2]_{sto}$, CO_2 concentration close to the stomata; $[CO_2]_{int}$, CO_2 concentration in the leaf internal air space; $[CO_2]_{chl}$, CO_2 concentration close to the chloroplast; the white dot on the leaf section shows where the CO_2 concentration is measured.

transpired by the leaf is not efficient due to the failure of the hydraulic conductivity of the xylem. In young shoots (current year), a loss of hydraulic conductivity near 50% at Ψ_w of −5 MPa in the xylem has been observed (Torres-Ruiz *et al.*, 2014). However, in stem, the reduction in hydraulic conductivity due to cavitation is low (25–30%) and seems to be caused mainly by the large-diameter vessels, which are the most vulnerable to the cavitation process but they are also the least represented as discussed before in this chapter (Torres-Ruiz *et al.*, 2013).

The anatomical variables responsible for the xylem vulnerability to embolism are not very clear yet. Some authors suggest that the diameter of the inter-vessel pit membrane pore is more relevant than the diameter of the vessel itself (Tyree and Sperry, 1988). If the pit area hypothesis (Wheeler *et al.*, 2005) is true, then the loss of hydraulic conductivity is mainly dependent on the largest pit pore of a vessel. Scientific evidence shows that under tension, the water present in the vessels contains a large number of nanometric bubbles and that these bubbles correlate with the structure of the pit membrane, rather than with the pore diameter (Jansen *et al.*, 2009). When bubbles are formed, they can explode, leading to embolism, or undergo a repair mechanism (Brodersen *et al.*, 2013).

Drought conditions are not permanent during the growing season and plants are normally subjected to water scarcity followed by re-watering processes. Recovery after drought is preceded by a period of inertia in leaf activity, such that the full recovery of the leaf functions during the process of rehydration normally takes a few days. Short-term studies on water-use dynamics in olive after re-watering by heat-pulse measurements of the sap flux (Moreno *et al.*, 1996) showed differences between irrigated and non-irrigated trees. Up to three days after irrigation, the irrigated plants maintained a transpiration rate of 1.65 $mm^3\ mm^{-2}\ d^{-1}$ on a leaf basis. After this phase, the rate of water use declined and transpiration fell. The sap flow in the near-surface roots dropped concomitantly. In non-irrigated trees, irrigation raised the transpiration rate to 1.12 $mm^3\ mm^{-2}\ d^{-1}$ only and leaf-water potential did not recover. These data show that after a re-watering following a drought stress, olives adopt a water-saving strategy.

The phytohormone abscisic acid (ABA) could have a major role as an endogenous messenger in the root-to-shoot signaling mechanism under drought stress. ABA accumulation induces stomatal closure (Christmann *et al.*, 2006) and long-lasting effects on plant hydraulic efficiency via the activity of aquaporins (AQPs) contributing to the preservation of plant water status (Parent *et al.*, 2009). A significant correlation between ABA and Ψ_w of the leaves was measured in olive (Kitsaki and Drossopoulos, 2005) as well as cultivar-specific differences that reflected the degree of stress tolerance: drought-tolerant 'Chemlali' cultivar accumulated lower levels of ABA in its leaves than the drought-sensitive 'Chetoui' (Guerfel *et al.* 2009).

Water movement across membranes is facilitated by AQPs and these proteins are present in all living organisms (Preston *et al.*, 1992). Under water scarcity, higher AQP expression can lead to an increase in water permeability (Yamada *et al.*, 1997) while a decrease in AQP expression can ensure a better conservation of water (Smart *et al.*, 2001). In plants, these two mechanisms seem to work simultaneously to maintain adequate water status in the tissues. In olive, the discovery and characterization of AQP have been done by Secchi *et al.* (2007). These authors found plasma membrane (*OePIP1.1*, *OePIP2.1*) and tonoplast (*OeTIP1.1*) putative AQPs and showed that during water deficit the expression levels of AQP genes decreased in all organs, thus assuming a role in water conservation. AQPs, together with carbonic anhydrase, could also serve for acclimation strategies modulating the internal conductance to CO_2 under water stress (Perez-Martin *et al.*, 2014).

Drought stress is usually associated with high temperature and irradiation. These additional stress factors cause photoinhibition, photooxidation and photorespiration (Osmond *et al.*, 1997) and boost the production of reactive oxygen species (ROS). ROS are harmful molecules also acting as cellular stress-response signals, and for these reasons their concentration in cells and tissues must be under control (Noctor, 2006). ROS homeostasis in plant cells is regulated non-enzymatically (α-tocopherol, β-carotene, phenolic compounds, ascorbate and glutathione) and enzymatically (superoxide dismutase, catalase, peroxidase, among others). In olive, the activities of antioxidant enzymes have been investigated in 'Coratina' cultivar under water deficit and after the recovery, and, in general, the activity of antioxidant enzymes in leaves and roots was reduced during recovery (Sofo *et al.*, 2004). Moreover, as shown for other species, increased levels of proline were also found since proline accumulation increases the osmotic potential of the cells and high proline contents are compatible with cellular functions.

The phenotypic plasticity responses activated by water deficit result in high energy expenditure, which has adverse effects on growth and production in both current and subsequent years. The effects of water scarcity on olive production largely depend on the biological phase during which water stress arises. Relevant damages can be observed if water scarcity occurs during flowering, beginning of pit hardening, and beginning of fruit ripening. At flowering, abnormalities in flower formation, and reduction in flower and fruit number have been observed. In the early stages of fruit development and ripening, an increase in fruit drop and a reduction in fruit size at ripening are the consequences of the stress (Inglese *et al.*, 1996). Fruit growth under drought conditions leads to a decrease in fruit fresh weight and volume affecting the mesocarp cell size, rather than the cell number (Costagli *et al.*, 2003). Endocarp area is modified by water stress since its growth reached only 40% of that expected under optimal water supply (Rapoport *et al.*, 2004). Field experiments proved that water supplies amounting to 33% only of the crop evapotranspiration (ET_c) during the dry season could be enough to increase yield significantly in comparison to non-irrigated plants. Both a larger number of fruits per plant and a higher weight per fruit (d'Andria and Morelli, 2000) determine this positive effect. Other field experiments with olive trees irrigated from the beginning of pit hardening to early fruit veraison with 100%, 66% or 33% of ET_c showed that yield differences are mainly associated to the mean fruit weight, and that the negative effects were more intense at flowering or at the beginning of fruit growth.

Water deficit can have an impact on oil quality (Caruso *et al.*, 2014; Gómez-del-Campo *et al.*, 2014). In temperate environments, the acidity and fatty acid composition of the oil produced under dry-farming conditions does not differ from that under irrigation conditions, while it has a higher concentration of polyphenols (Caruso *et al.*, 2014; Gucci *et al.*, 2019; Patumi *et al.*, 2002; d'Andria *et al.*, 2004, 2009; Servili *et al.*, 2007). In arid climates, water deficit can lead to significant changes in fatty acid composition (Lavee and Schachtel, 1999). As a general rule, the quantity of olive oil per tree or per unit area decreases under conditions of water scarcity and this effect gets stronger as the climate becomes dryer (Grattan *et al.*, 2006; Tognetti *et al.*, 2006).

7.1.3 Salinity

Soil salinity can reduce crop production in arid and semi-arid regions, since cultivated species have evolved and been cultivated on arable land and, therefore, most of them are salt-sensitive. These plants are classified as 'glycophytes' in comparison to 'halophytes'. Halophytes can complete their life cycle in a salt ion-rich environment where many of the salt-sensitive species will die (Panta *et al.* 2014). High concentration of salts in the soil may be due to the geological origin of soils, the high evaporative demand of the environment and the poor quality of irrigation water. Salinity problems are becoming more extensive due to unsustainable irrigation practices and pressures for producing in marginal land. In the last few decades, the increase in agricultural production and competition for water with the civil and industrial sector have led to the use of poor-quality water rich in ions that increase the risk of soil salinity. Sodium (Na^+) and chlorine (Cl^-) ions are usually responsible for salinization, whereas other ions like calcium (Ca^{2+}) and sulfate (SO_4^{2-}) are more rarely involved. In coastal areas of the Mediterranean basin, both the presence of numerous artesian wells and the high evapotranspiration during summer often cause temporary infiltration of seawater into freshwater aquifers, exposing the nearest crop-growing areas to an increasing risk of salt stress.

The costs of soil salinity for farmers are relevant and the economic impacts in terms of reduced yield are relevant. Among fruit trees, olive is classified as a species moderately resistant to salinity (FAO, 1985; Rugini and Fedeli, 1990; Gucci and Tattini, 1997). A reduction in olive production of only 10% at the electrical conductivity of a saturated-paste extract (EC_e) of 4–6 dS m^{-1} (6–8 dS m^{-1} in calcium-rich soils) has been measured (Bernstein, 1965). According to Maas and Hoffman (1977), a mathematical relationship between yield and EC_e can be established in olive: yield reduces linearly as salinity increases beyond a 3–4 dS m^{-1} EC_e threshold. Olive capability to tolerate high values of EC_e depends on the presence in the irrigation water of low percentages of NaCl or Na_2SO_4 (Bartolini *et al.*, 1991). Young plants significantly reduce their growth when exposed to NaCl concentrations in the range of 40–100 mM, whereas values higher than 100 mM impact plant survival. Rugini and Fedeli (1990) estimated around 137 mM as the maximum value of NaCl concentration that the olive tree can tolerate, and Gucci and Tattini (1997) around 30 mM the value of NaCl concentration at which the reductions in production were measurable. Factors such as cultivar, age of the plant and agronomic and environmental variables modify the response of olive to salinity (Chartzoulakis, 2005). Differences in salt tolerance between 'Frantoio' (resistant) and 'Leccino' (sensitive) are well known (Tattini *et al.*, 1992), as well as for the resistant 'Kalamata' in comparison to the less-resistant 'Mastoidis' and 'Amphissis' (Chartzoulakis *et al.*, 2002).

Symptoms of salt stress comprise chlorosis and necrosis of leaves, desiccation of flowers, abortion of ovary and desiccation of the apex of the leaves. Necrosis of stem tip and root and abscission of leaves appeared when the stress conditions become intense (Gucci *et al.*, 2003). Salinity reduces olive growth, with stressed plants showing short internodes and small leaves, a response typical of glycophyte species (Munns and Termaat, 1986). Through leaf abscission, the olive plant reduces its transpiration and eliminates toxic ions that accumulate in the older leaves (Bongi and Loreto, 1989; Loupassaki *et al.*, 2002). The salinity impact on roots is negative, but the canopy-to-root ratio decreases, proving that injury of the aerial part is stronger (Therios and Misopolinos, 1988; Bongi and Loreto, 1989; Gucci and Tattini, 1997). Negative impacts of this stress are also: (a) reduced viability and germination of pollen; (b) low number of flowers per inflorescence; (c) decrease in fruit set; (d) smaller fruit size (Gucci *et al.*, 2003). Despite that fruit weight is reduced under salinity, the oil content is not modified (Chartzoulakis, 2005) and fruit yield is more affected than oil yield (Gucci *et al.*, 2003).

Two main mechanisms are responsible for the negative effects of salinity on the plant: the osmotic effect of the salt in the soil that caused a decrease in the Ψ_w and the toxic effect of the salt (Na^+ and Cl^-) within the cells (Marschner, 1995; Munns and Tester, 2008) (Fig. 7.1.6). The response time is rapid to the increase in external Ψ_w and slower to the accumulation of Na^+. The osmotic phase starts immediately after the salt concentration around the roots increases to a threshold level and the visible symptom is a decrease in the rate of shoot growth. The ion-specific phase starts when salt accumulates to toxic concentrations in the old leaves. Old leaves do not expand and the salts reaching them are not diluted, so they die. On the contrary, young leaves continue to grow and dilute the salt uptake. When the rate at which old leaves die overcomes the capacity of the plant to produce new leaves, the whole plant photosynthetic capacity is not able to sustain growth anymore (Munns

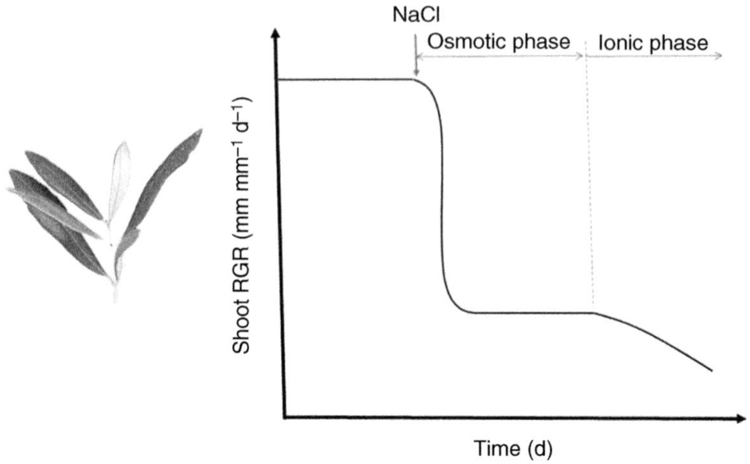

Fig. 7.1.6. Salinity stress occurs in two phases. The rapid response to the increase in external osmotic pressure (osmotic phase) and the slower response due to the accumulation of Na^+ in leaves (ionic phase). The solid line represents the change in the Shoot Relative Growth Rate before and after the addition of NaCl. (Redrawn from Munns and Tester, 2008)

and Tester, 2008). According to Munns and Tester (2008), the mechanisms of salinity tolerance fall into three categories: tolerance to osmotic stress; Na^+ exclusion; and tissue tolerance (Table 7.1.2).

In olive, salt tolerance has been proven to be associated with the ability to retain Na^+ and Cl^- in the root, limiting the translocation into the xylem, and the transport and accumulation in buds and leaves (Gucci and Tattini, 1997; Chartzoulakis, 2005). Tolerant cultivars limit salt translocation to the aerial part (Tattini *et al.*, 1995; Chartzoulakis, 2005) and show a gradient of Na^+ and Cl^- that decreases from the base to the apex of the shoot. 'Leccino' (less-tolerant) cultivar has a lower exclusion of Na^+ from the shoot than 'Frantoio' (tolerant) cultivar at 120 mM NaCl (Gucci and Tattini, 1997; Gucci *et al.*, 2003). Rossi *et al.* (2015) investigated the role of the root anatomical modifications on the uptake and localization of Na^+. Data showed that endodermis apoplastic barriers were formed significantly closer to the root tip in 'Leccino' than in 'Frantoio' and that Na^+ has a concentration gradient in the root that decreases from the exodermis to the stele tissues. Overall, these processes in the roots seem to play a role in reducing Na^+ fluxes into the root and consequently from the root to the shoot. Moreover, recent data (Sodini *et al.*, 2022) showed that 'Frantoio' capability to mobilize elements, especially Ca^{2+}, together with the overexpression of key genes for sodium management, could be a crucial step for salt tolerance in this cultivar.

A side-effect of Na^+ and Cl^- accumulation is the imbalance in the homeostasis of essential elements. Calcium has an important role in reducing Na^+ toxicity (Melgar *et al.*, 2006) and a high Ca^{2+}/Na^+ ratio is important to reduce the negative effects of Na^+ on the depolarization and selectivity of the olive plasma membranes (Rinaldelli and Mancuso, 1996). Under salinity, K^+ undergoes a reduction in many glycophyte species (Greenway and Munns, 1980), and in olive this has been observed both in roots and old leaves (Bartolini *et al.*, 1991; Tattini *et al.*, 1995; Chartzoulakis *et al.*, 2002). In young and growing organs, an appropriate K^+/Na^+ ratio is necessary to maintain a physiological ionic balance (Chartzoulakis, 2005). Salinity can also interfere with nitrogen uptake via reductions in the activity of the enzyme nitrate reductase (Tabatabaei, 2006).

Salinity alters the plant's water relations. At the initial stages of salinity stress, the Ψ_w of the soil solution becomes more

Table 7.1.2. The three components of salinity tolerance organized by plant processes. (Redrawn from Munns and Tester, 2008)

	Mechanisms of salinity tolerance		
Physiological process	Osmotic stress	Na^+ exclusion	Tissue tolerance
Sensing and signaling (roots)	Adjustment in long-distance signaling	Regulation of net ion transport to shoot	Control of vacuolar loading
Shoot growth	Decreased inhibition of cell expansion and lateral bud development	-	Delay in senescence of old leaves (carbon source)
Photosynthesis	Decreased stomatal closure	Avoidance of ion toxicity in chloroplasts	Delay in ion toxicity in chloroplasts
Na^+ in shoots	Increased osmotic adjustment	Reduced Na+ long-distance transport	Reduced energy spent on Na^+ exclusion
Na^+ in vacuoles	Increased osmotic adjustment	Increased Na+ sequestration into root vacuoles	Increased Na+ sequestration into leaf vacuoles
Accumulation of organic solutes	Increased osmotic adjustment	Regulation of transport processes to reduce Na^+ accumulation	Accumulation of compatible solutes in cytoplasm

negative owing to the presence of osmotic salts. In these conditions, roots have more difficulties taking up water (Therios and Misopolinos, 1988), the cell-water status is affected and transpiration decreases (Gucci *et al.*, 1997). 'Frantoio' and 'Leccino' plants exposed to 100 and 200 mM NaCl, for example, showed a more negative Ψ_w at predawn (mainly caused by a reduction of the $\Psi\pi$) and a significant reduction in RWC (Gucci *et al.*, 1997). Since the reduction in Ψ_w of the leaf is proportional to the salt concentration of the external solution, olives maintain a stable Ψ_w gradient between saline soil and plant cells. The Ψ_p of the cells can be maintained if the salt concentration does not become too high or the stress duration is not too long, and olive cells accumulate glucose and mannitol as osmotically active molecules (Tattini *et al.*, 1996) that provide 25–30% of the total osmotic effect, while the inorganic solutes are responsible for the other part (Gucci *et al.*, 1998). The thick cuticle and tightly packed mesophyll cells of the olive leaves undergo morphological changes under salinity (Bongi and Loreto, 1989) and photosynthetic activity decreases depending on salt concentration and genotype tolerance (Bongi and Loreto, 1989; Tattini *et al.*, 1995; Chartzoulakis *et al.*, 2002). Stomatal limitations to photosynthesis seem to be prevalent during the early stages of salt stress but some authors (Loreto *et al.*, 2003) showed that both g_s and g_m have a role under salinity.

Although we are aware that salinity negatively impacts olive yield, the effect of mild levels of salinity on fruit physiology is still poorly studied. Moretti *et al.* (2018) studied the effect of salinity on fruit physiology of the salt-sensitive genotype 'Leccino', which translocates sodium from roots into the shoots. Fruits of plants exposed at 60 mM of NaCl during the phenological stages including between 'beginning of pit hardening' – 'veraison' – showed Na accumulation in flesh depending on the ripening stage without negative effects on fruit yield parameters. On the contrary, total phenols increase (58%) and the salt treatment induces a thickening of cuticle, epidermis, hypodermis and outer mesocarp cells, that could be considered useful traits against other stress. More recently (Moretti *et al.*, 2019), the effect of slightly higher salinity (80 mM NaCl) on fatty acid composition and desaturase gene expression in fruit mesocarp of the same cultivar proved that salinity accelerates the ripening process, increases fruit number and decreases total fatty acid content in fruits with 100% purple skin; while oleic acid increases and linoleic acid decreases in fruits with ≤ 50% purple skin (Fig. 7.1.7). These changes could be explained by the synergic regulation of genes involved in olive fruit fatty acid biosynthetic pathways and have, as a consequence, an increase in the oleic/linoleic (41%) and a decrease in the polyunsaturated/monounsaturated ratios (30%) proving that a mild level of salinity could have positive effects on quality.

7.1.4 Temperature

Climate change and the progressive expansion of olive cultivation areas have increased the likelihood of both high- and low-temperature stress. The olive trees have some degree of resistance to high temperatures, but poorly tolerate low temperatures.

Olive trees mainly grow in the Mediterranean region where they are often exposed to high temperatures, severe water shortages and high light intensity. Under these conditions, stem sunburn and leaf chlorosis may become evident. In order to control heat balance, the olive trees have developed small leaves, stomata that remain partially open under drought, a thick *indumentum* of trichomes on the abaxial side that filter out UV light and reflect infrared radiation and a mechanism to adjust the leaf-insertion angle on the stem (Baldini *et al.*, 1997; Liakoura *et al.*, 1997). Photosynthesis is sensitive to high temperatures, due to the heat sensitivity of thylakoid membranes (Berry and Bjorkman, 1980). The critical threshold for chlorophyll fluorescence is at 46–47°C, since at 48–49°C leaf necrosis starts appearing (Gucci *et al.*, 2003). Exposure to 38°C and high PPF for 3 h decreases apparent quantum yield by 80%, although it recovers within 4 h from returning to 26°C (Bongi and Long, 1987).

Cultivars from northern areas are less tolerant to high temperatures than those from southern areas (Mancuso and Azzarello, 2002) and a seasonal dynamic in heat tolerance has also been observed (Kappen, 1981); in the summer there is an increase of 3–4°C in tolerance with respect to winter. Flowering, fruit set and growth are sensitive to high temperature, especially when higher than 35°C. Fruit shriveling, necrosis and abscission are caused by the increase in tissue temperature during the first few weeks of fruitlets growth, and during fruit development high temperatures decrease the oleic acid concentration and modify the fatty acid composition (Garcia-Inza *et al.*, 2014).

In the Mediterranean climate, winter is usually characterized by mild temperatures, but unfortunately, low-temperature damage to olive plants has been observed at intervals of 25–40 years. Low-temperature stress could be determined by temperatures between

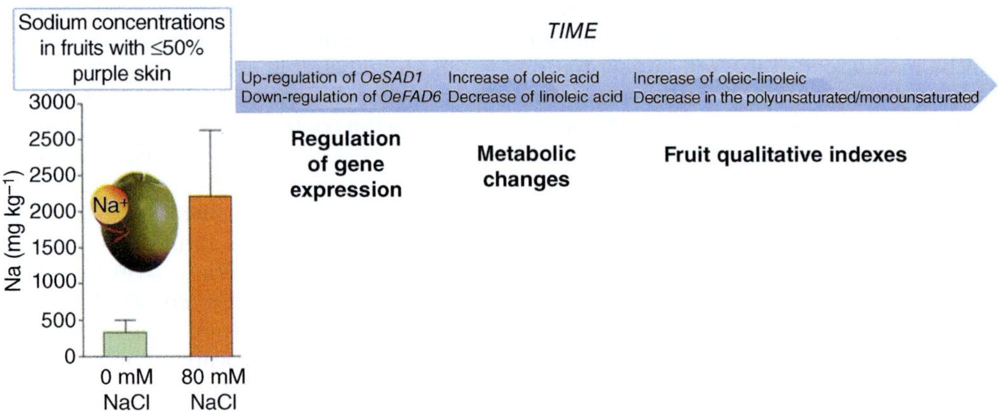

Fig. 7.1.7. Mild level of salinity stress induces a synergic regulation of genes involved in olive fruit fatty acid biosynthetic pathways and an increase in the oleic/linoleic, while decreasing the polyunsaturated/monounsaturated ratios, having a positive effect on quality. (Modified from Moretti *et al.*, 2019).

10°C and 15°C and 0°C (chilling stress) or below 0°C (freezing stress). Chilling stress in olive causes a slowdown of the metabolic processes since the optimal temperatures for metabolism are in the range of 20–30°C (Rinaldelli and Mancuso, 1994). Reduction in enzymatic activity, water and nutrient uptake and respiration and photosynthetic efficiency are the main targets that determine growth reductions in this stress condition (Mancuso, 2000). Freezing stress is more dangerous and could easily determine olive plant death. Genotype, phenological stage, nutritional and health status, age and type of organ are the key plant factors in freezing tolerance. Environmental conditions, like humidity, wind direction and velocity, temperature drop rates and freezing duration, have also a relevant role in freezing-stress intensity. Freezing damage is visible as shoot-tip desiccation, leaf drop, bark longitudinal cracks and sapwood split. Freezing temperatures cause ice crystal formation both intra-cellular and extracellular. The intra-cellular ice physically destroys the cell membranes and organelles, while the extracellular ice (usually formed before intra-cellular ice from the extracellular water) may not be responsible for immediate cell damage (if the cells are acclimated to freezing) and cause cellular dehydration. This happens since ice formation decreases the Ψ_w in the apoplastic space determining a Ψ_w gradient from the simplast to the apoplast that is responsible for gradual release of water from the cell (exosmosis). In these conditions an adequate permeability to water of the cell membrane delays the intra-cellular freezing events (Levitt, 1980), since the intra-cellular solute concentration increases and this gradually lowers the intra-cellular freezing temperature. The accumulation of compatible solutes into the cell reduces further the risk of intra-cellular freezing; however, when the temperature drops below a threshold, the intra-cellular ice is formed from nucleation points into the cytoplasmic solution and this process is strongly influenced by the cell acclimatization to freezing (Levitt, 1980). In olive, Fiorino and Mancuso (2000) reported the presence of a super-cooling mechanism that allows conservation of water in the liquid state well below the freezing temperatures (down to –7/–18°C). In olive, the freezing temperature values vary between organs, with the following descending order of sensitivity: fruits, roots, leaves, branches and buds (Gucci *et al.*, 2003). Moreover, olive cultivars showed a different tolerance to freezing (La Porta *et al.*, 1994; Bartolozzi and Fontanazza, 1999) and LT_{50} (lethal temperature at which 50% of damage occurs) values range from –12.0 in 'Coratina', –12.8 in 'Frantoio' and –15.6 in 'Leccino' up to –18.3°C in 'Ascolana' (Mancuso, 2000). Freezing tolerance occurs only in genetically competent species, through a mechanism known as acclimation. Acclimation to freezing follows a complex series of physiological events mainly activated by temperature and photoperiodic stimuli (Sakai and Larcher, 1987). In the northern hemisphere, acclimation starts in autumn, when temperatures drop below 10–15°C and the day length shortens. Since olive is an evergreen species and lacks true dormancy, once acclimated it can easily lose acclimation if mild temperatures persist in winter. A partial loss of acclimation potential has been observed after six days when temperatures reach values above 16°C (Gucci *et al.*, 2003). In olive fruits, the cold sensitivity was high in the epi- and mesocarp cells before oil body formation and decreased during oil biogenesis (Matteucci *et al.*, 2011).

7.1.5 Conclusion

Olive physiology and biochemistry during abiotic stresses have been extensively investigated and studies disclosed some of the tolerance mechanisms and highlighted wide genotypic variability. Data on olive performance in the new growing areas and those on the effects of high spring temperature and low and high rain on olive productivity are still missing. Studies on olive responses to multiple environmental stresses are required (Fernández, 2014; Hernández *et al.*, 2021) to cope with the climate change scenario that predicts rising temperatures, increase in precipitation abnormalities, freezing temperatures and salinization.

References

Angelopoulos, K., Dichio, B. and Xiloyannis, C. (1996) Inhibition of photosynthesis in olive trees (*Olea europea* L.) during water stress and rewatering. *Journal of Experimental Botany* 301, 1093–1100.

Bacelar, E.A., Correia, C.M., Moutinho-Pereira J.M., Gonccalves, B.C., Lopes, J.I. and Torres-Pereira, J.M.G. (2004) Sclerophylly and leaf anatomical traits of five field-grown olive cultivars growing under drought conditions. *Tree Physiology* 24, 233–239.

Baldini, E., Facini, O., Nerozzi, F., Rossi, F. and Rotondi, A. (1997) Leaf characteristics and optical properties of different woody species. *Trees* 12, 73–81.

Bartolini, G., Mazuelos, C. and Troncoso, A. (1991) Influence of Na_2SO_4 and NaCl salts on survival, growth and mineral composition of young olive plants in inert sand culture. *Advances in Horticultural Science* 5, 73–79.

Bartolozzi, F. and Fontanazza, G. (1999) Assessment of frost tolerance in olive (*Olea europaea* L.). *Scientia Horticulturae* 81, 309–319.

Bernstein, L. (1965) *Salt Tolerance of Fruit Crops*. USDA Agriculture Research Bulletin.

Berry, J. and Bjorkman, O. (1980) Photosynthetic response and adaptation to temperature in higher plants. *Annual Review Plant Physiology* 31, 491–543.

Besnard, G., Rubio de Casas, R., Christin, P.A. and Vargas, P. (2009) Phylogenetics of *Olea* (Oleaceae) based on plastid and nuclear ribosomal DNA sequences: tertiary climatic shifts and lineage differentiation times. *Annals of Botany* 104, 143–160.

Blum, A. (2017) Osmotic adjustment is a prime drought stress adaptive engine in support of plant production. *Plant Cell Environment* 40, 4–10.

Bongi, G. and Long, S.P. (1987) Light-dependent damage to photosynthesis in olive leaves during chilling and high temperature stress. *Plant, Cell and Environment* 10, 241–249.

Bongi, G. and Loreto, F. (1989) Gas-exchange properties of salt-stressed olive (*Olea europaea* L.) leaves. *Plant Physiology* 90, 1408–1416.

Bongi, G. and Palliotti, A. (1994) Olive. In: Schaffere, B. and Andersen, P.C. (eds) *Handbook of Environmental Physiology of Fruit Crops. Vol. 1: Temperate Crops*. CRC Press, Boca Raton, Florida, pp. 165–187.

Bosabalidis, A.M. and Kofidis, G. (2002) Comparative effects of drought stress on leaf anatomy of two olive cultivars. *Plant Science* 163, 375–379.

Brodersen, C.R., McElrone, A.J., Choat, B., Lee, E.F., Shackel, K.A. and Matthews, M.A. (2013) *In vivo* visualizations of drought-induced embolism spread in *Vitis vinifera*. *Plant Physiology* 161, 1820–1829.

Caruso, G., Gucci, R., Urbani, S., Esposto, S., Taticchi, A. *et al.* (2014) *Effect of different irrigation volumes during fruit development on quality of virgin olive oil of cv.* Frantoio. *Agricultural Water Management* 134, 94–103.

Chartzoulakis, K. (2005) Salinity and olive: growth, salt tolerance, photosynthesis and yield. *Agricultural Water Management* 78, 108–121.

Chartzoulakis, K., Loupassaki, M., Bertaki, M. and Androulakis, I. (2002) Effects of NaCl salinity on growth, ion content and CO_2 assimilation rate of six olive cultivars. *Scientia Horticulturae* 96, 235–247.

Christmann, A., Moes, D., Himmelbach, A., Yang, Y. and Grill, E. (2006) Integration of abscisic acid signalling into plant responses. *Plant Biology* 8, 314–325.

Connor, D.J. and Fereres, E. (2005) The physiology of adaptation and yield expression in olive. *Horticultural Review* 34, 155–229.

Costagli, G., Gucci, R. and Rapoport, H.F. (2003) Growth and development of fruits of olive 'Frantoio' under irrigated and rainfed conditions. *Journal of Horticultural Science and Biotechnology* 78, 119–124.

d'Andria, R., Lavini, A., Morelli, G., Patumi, M., Terenziani, S. *et al.* (2004) Effect of water regime on five pickling and double aptitude olive cultivars (*Olea europaea* L.). *Journal of Horticultural Science and Biotechnology* 78, 15–23.

d'Andria, R., Lavini, A., Morelli, G., Sebastiani, L. and Tognetti, R. (2009) Physiological and productive responses of *Olea europaea* L. cultivars Frantoio and Leccino to a regulated deficit irrigation regime. *Plant Biosystems* 143, 222–231.

d'Andria, R. and Morelli, G. (2000) Irrigation regime affects yield and oil quality of olive trees. *Acta Horticulturae* 586, 273–276.

Diaz-Espejo, A., Nicolás, E. and Fernández, J.E. (2007) Seasonal evolution of diffusional limitations and photosynthetic capacity in olive under drought. *Plant Cell Environment* 30, 922–933.

Dichio, B., Xiloyannis, C., Angelopoulos, K., Nuzzo, V., Bufo, S.A. and Celano, G. (2003) Drought-induced variations of water relations parameters in *Olea europaea*. *Plant and Soil* 257, 381–389.

Dichio, B., Xiloyannis, C., Sofo, A. and Montanaro, G. (2005) Osmotic regulation in leaves and roots of olive trees during a water deficit and rewatering. *Tree Physiology* 26, 179–185.
FAO (1985) Water quality for agriculture. FAO Irrigation and Drainage paper 29. Rome.
Fernández, J.E. (2014) Understanding olive adaptation to abiotic stresses as a tool to increase crop performance. *Environmental and Experimental Botany* 103, 158–179.
Fiorino, P. and Mancuso, S. (2000) Differential thermal analysis, deep supercooling and cell viability in organs of *Olea europaea* at subzero temperatures. *Advances in Horticultural Science* 14, 23–27.
García-Inza, G.P., Castro, D.N., Hall, A.J. and Rousseaux M.C. (2014) Responses to temperature of fruit dry weight, oil concentration, and oil fatty acid composition in olive (*Olea europaea* L. var. 'Arauco'). *European Journal of Agronomy* 54, 107–115.
Giorio, P., Sorrentino, G. and d'Andria, R. (1999) Stomatal behaviour, leaf water status and photosynthetic response in field-grown olive trees under water deficit. *Environmental and Experimental Botany* 42, 95–104.
Gómez-del-Campo, M., Pérez-Expósito, M.Á., Hammami, S.B.M., Centeno, A. *et al.* (2014) Effect of varied summer deficit irrigation on components of olive fruit growth and development. *Agricultural Water Management* 137, 84–91.
Grattan, S.R., Berenguer, M.J., Connell, J.H., Polito, V.S., Vossen, P.M. *et al.* (2006) Olive oil production as influenced by different quantities of applied water. *Agricultural Water Management* 85, 133–140.
Greenway, H. and Munns, R. (1980) Mechanisms of salt tolerance in non-halophytes. *Annual Review of Plant Physiology* 31, 149–190.
Gucci, R. and Tattini, M. (1997) Salinity tolerance in olive. *Horticultural Reviews* 21, 177–214.
Gucci, R., Lombardini, L. and Tattini, M. (1997) Analysis of leaf water relations of two olive (*Olea europaea* L.) cultivars differing in tolerance to salinity. *Tree Physiology* 17, 13–21.
Gucci, R., Moing, A., Gravano, E. and Gaudillère, J.P. (1998) Partitioning of photosynthetic carbohydrates in leaves of salt-stressed olive plants. *Australian Journal of Plant Physiology* 25, 571–579.
Gucci, R., Mancuso, S. and Sebastiani, L. (2003) Resistenza agli stress ambientali. In: Fiorino, P. (ed.) *Olea – Trattato di Olivicoltura*. Edagricole. Edizioni Agricole de Il Sole 24 ORE Edagricole Srl pp 91–111.
Gucci, R., Caruso, G., Gennai, C., Esposto, S. *et al.* (2019) Fruit growth, yield and oil quality changes induced by deficit irrigation at different stages of olive fruit development. *Agricultural Water Management* 212, 88–98.
Guerfel, M., Beis, A., Zotos, T., Boujnah, D., Zarrouk, M. and Patakas, A. (2009) Differences in abscisic acid concentration in roots and leaves of two young olive (*Olea europaea* L.) cultivars in response to water deficit. *Acta Physiologiae Plantarum* 31, 825–831.
Hernández, M.L., Moretti, S., Sicardo, M.D., García, ., Pérez, A. *et al.* (2021) Distinct physiological roles of three phospholipid:diacylglycerol acyltransferase genes in olive fruit with respect to oil accumulation and the response to abiotic stress. *Frontiers in Plant Science* 12, 751959.
Inglese, P., Barone, E. and Gullo, G. (1996) The effect of complementary irrigation on fruit growth, ripening pattern and oil characteristics of olive (*Olea europaea* L.) cv. Carolea. *Journal of Horticultural Science and Biotechnology* 71, 257–263.
Ingram, J. and Bartels, D. (1996) The molecular basis of dehydration tolerance in plants. *Annual Review of Plant Physiology and Plant Molecular Biology* 47, 377–403.
Jansen, S., Choat, B. and Pletsers, A. (2009) Morphological variation of intervessel pit membranes and implications to xylem function in angiosperms. *American Journal of Botany* 96, 409–419.
Kappen, L. (1981) Ecological significance of resistance to high temperatures. In: Lange, O.L., Nobel, P.S., Osmond, C.B. and Ziegler, H. (eds) *Encyclopedia of Plant Physiology*. Springer, Berlin, Vol. 12A, pp. 439–474.
Kitsaki, C.K. and Drossopoulos, J.B. (2005) Environmental effect on ABA concentration and water potential in olive leaves (*Olea europaea* L. cv. 'Koroneiki') under non-irrigated field conditions. *Environmental and Experimental Botany* 54, 77–89.
La Porta, N., Zacchini, M., Bartolini, S., Viti, R. and Roselli, G. (1994) The frost hardiness of some clones of olive cv Leccino. *Journal of Horticultural Science and Biotechnology* 69, 433–435.
Larcher, W. (1995) *Physiological Plant Ecology*, 3rd edn. Springer, Berlin.
Lavee, S. and Schachtel, J. (1999) Interaction of cultivar rootstock and water availability on olive tree performance and fruit production. *Acta Horticulturae* 474, 399–401.
Levitt, J. (1980) *Responses of Plants to Environmental Stresses, Vol I. Chilling, freezing and high temperature stresses*. Academic Press, New York.
Liakoura, V., Stefanou, M., Manetas, Y., Cholevas, C. *et al.* (1997) Trichome density and its UV-B protective potential are affected by shading and leaf position on the canopy. *Environmental and Experimental Botany* 38, 223–229.

Lo Gullo, M.A. and Salleo, S. (1988) Different strategies of drought-resistance in three Mediterranean sclerophyllous trees growing in the same environmental conditions. *New Phytologist* 108, 267–276.

Lo Gullo, M.A. and Salleo, S. (1990) Wood anatomy of some trees with diffuse- and ring-porous wood: some functional and ecological interpretations. *Giornale Botanico Italiano* 124, 601–613.

Loreto, F., Centritto, M. and Chartzoulakis, K. (2003) Photosynthetic limitations in olive cultivars with different sensitivity to salt stress. *Plant, Cell and Environment* 26, 595–601.

Loupassaki, M.H., Chartzoulakis, K., Digalaki, N. and Androulakis, I.I. (2002) Effects of salt stress on concentration of nitrogen, phosphorus, potassium, calcium, magnesium and sodium in leaves, shoots and roots of six olive cultivars. *Journal of Plant Nutrition* 25, 2457–2482.

Maas, E.V. and Hoffman, G.J. (1977) Crop salt tolerance-current assessment. *Journal of the Irrigation and Drainage Division* 103, 115–134.

Mancuso, S. (2000) Electrical resistance changes during exposure to low temperature measure chilling and freezing tolerance in olive tree (*Olea europaea* L.) plants. *Plant, Cell and Environment* 23, 221–229.

Mancuso, S. and Azzarello, E. (2002) Heat tolerance in olive. *Advances in Horticultural Science* 16, 125–130.

Marchi, S., Sebastiani, L. and Gucci, R. (2005) Changes in sink-source relationships during shoot development in olive. *Journal of the American Society for Horticultural Science* 130, 631–637.

Marchi, S., Guidotti, D., Sebastiani, L. and Tognetti, A. (2007) Changes in assimilation capacity during leaf development in broadleaved *Prunus persica* and sclerophyllous *Olea europaea*. *Journal of Horticultural Science and Biotechnology* 82, 69–78.

Marschner, H. (1995) *Mineral Nutrition of Higher Plants*, 2nd edn. Academic Press, San Diego, California.

Matteucci, M., D'Angeli, S., Errico, S., Lamanna, R., Perrotta, G. and Altamurra, M.M. (2011) Cold affects the transcription of fatty acid desaturases and oil quality in the fruit of *Olea europaea* L. genotypes with different cold hardiness. *Journal of Experimental Botany* 62, 3403–3420.

Melgar, J.C., Benlloch, M. and Fernández-Escobar, R. (2006) Calcium increases sodium exclusion in olive plants. *Scientia Horticulturae* 109, 303–305.

Mervyn, M. and Ludlow, M.M. (1987) Contribution of osmotic adjustment to the maintenance of photosynthesis during water stress. In: Biggins, J. (ed.) *Progress in Photosynthesis Research* Springer, New York, pp. 161–168.

Moreno, F., Fernandez, J.E., Clothier, B.E. and Green, S.R. (1996) Transpiration and root water uptake by olive trees. *Plant Soil* 184, 85–96.

Moretti, S., Francini, A., Minnocci, A, and Sebastiani, L. (2018) Does salinity modify anatomy and biochemistry of *Olea europaea* L. fruit during ripening? *Scientia Horticulturae* 228, 33–40.

Moretti, S., Francini, A., Hernández, M.L., Martínez-Rivas, J.M. and Sebastiani, L. (2019) Effect of saline irrigation on physiological traits, fatty acid composition and desaturase genes expression in olive fruit mesocarp. *Plant Physiology and Biochemistry* 141, 423–430.

Moriana, A., Villalobos, F.J. and Ferreres, E. (2002) Stomatal and photosynthetic responses of olive (*Olea europaea* L.) leaves to water deficits. *Plant Cell and Environment* 25, 395–405.

Munns, R. and Termaat, A. (1986) Whole plant responses to salinity. *Australian Journal of Plant Physiology* 13, 143–160.

Munns, R. and Tester, M. (2008) Mechanisms of salinity tolerance. *Annual Review of Plant Biology* 59, 651–681.

Noctor, G. (2006) Metabolic signalling in defence and stress: the central roles of soluble redox couples. *Plant Cell and Environment* 29, 409–425.

Oertli, J.J., Lips, S.H. and Agami, M. (1990) The strength of sclerophyllous cells to resist collapse due to negative turgor pressure. *Acta Oecologica* 11, 281–289.

Osmond, B., Badger, M., Maxwell, K., Björkman, O. and Leegood, R. *et al.* (1997) Too many photons: photorespiration, photoinhibition and photooxidation. *Trends in Plant Science* 2, 119–121.

Parent, B., Hachez, C., Redondo, E., Simonneau, T., Chaumont, F. and Tardieu, F. (2009) Drought and abscisic acid effects on aquaporin content translate into changes in hydraulic conductivity and leaf growth rate: a trans-scale approach. *Plant Physiology* 149, 2000–2012.

Patumi, M., d'Andria, R., Marsilio, V., Fontanazza, G., Morelli, G. *et al.* (2002) Olive and olive oil quality after intensive monocone olive growing (*Olea europaea* L., cv. Kalamata) in different irrigation regimes. *Food Chemistry* 77, 27–34.

Panta, S., Flowers, T., Lane, P., Doyle, R. *et al.* (2014) Halophyte agriculture: success stories. *Environmental and Experimental Botany* 107, 71–83.

Perez-Martin, A., Michelazzo, C., Torres-Ruiz, J.M., Flexas, J., Fernández, J.E. *et al.* (2014) Regulation of photosynthesis and stomatal and mesophyll conductance under water stress and recovery in olive trees:

correlation with gene expression of carbonic anhydrase and aquaporins. *Journal of Experimental Botany* 65, 3143–3156.
Preston, G.M., Carroll, T.P., Guggino, W.B. and Agre, P. (1992) Appearance of water channels in *Xenopus oocytes* expressing red cell CHIP28 protein. *Science* 256, 385–387.
Rapoport, H.F., Costagli, G. and Gucci, R. (2004) The effect of water deficit during early fruit development on olive fruit morphogenesis. *Journal of the American Society for Horticultural Science* 129, 121–127.
Rhizopoulou, S., Meletiou-Christou, M.S. and Diamantoglou, S. (1991) Water relations for sun and shade leaves of four Mediterranean evergreen sclerophylls. *Journal of Experimental Botany* 42, 627–635.
Rinaldelli, E. and Mancuso, S. (1994) Cell transmembrane electropotentials in adventitious roots of *Olea europaea* L. cv. Frantoio as related to temperature, respiration, external potassium, anoxia, and 2,4-dinitrophenol treatments. *Advances in Horticultural Science* 8, 229–234.
Rinaldelli, E. and Mancuso, S. (1996) Response of young mycorrhizal and non-mycorrhizal plants of olive tree (*Olea europaea* L.) to saline conditions. I. Short-term electro-physiological and long-term vegetative salt effects. *Advances in Horticultural Science* 10, 126–134.
Rossi, L., Sebastiani. L., Tognetti, R., d'Andria, R., Morelli, G. and Cherubini, P. (2013) Tree-ring wood anatomy and stable isotopes show structural and functional adjustments in olive trees under different water availability. *Plant and Soil* 372, 567–579.
Rossi, L., Francini, A., Minnocci, A. and Sebastiani, L. (2015) Salt stress modifies apoplastic barriers in olive (*Olea europaea* L.): a comparison between a salt-tolerant and a salt-sensitive cultivar. *Scientia Horticulturae* 192, 38–46.
Rugini, E. and Fedeli, E. (1990) Olive as an oilseed crop. In: Bajaj, Y.P.S. (ed.) *Biotechnology in Agriculture and Forestry: Legumes and Oil-seed Crops*. Springer-Verlag, Berlin, pp. 593–641.
Sakai, A and Larcher, W. (1987) *Frost survival of plants: responses and adaptation to freezing stress. Ecological Studies* 62, Springer-Verlag, Berlin.
Salleo, S. and Nardini, A. (1999) Ecofisiologia di *Olea oleaster* Hoffmgg et Link: verso un modello predittivo dell'adattamento all'aridità. *Olivo Olio* 2, 70–79.
Salleo, S., Lo Gullo, M.A. and Olivieri, F. (1985) Hydraulic parameter measured in 1-year old twigs of some Mediterranean species with diffuse-porous wood: changes in hydraulic conductivity and their possible functional significance. *Journal of Experimental Botany* 36, 1–11.
Schwabe, W.W. and Lionakis, S.M. (1996) Leaf attitude in olive in relation to drought resistance. *The Journal of Horticultural Science* 71, 157–166.
Secchi, F., Lovisolo, C., Uehlein, N., Kaldenhoff, R. and Schubert, A. (2007) Isolation and functional characterization of three aquaporins from olive (*Olea europaea* L.). *Planta* 225, 381–392.
Servili, M., Esposto, S., Lodolini, E.M., Selvaggini, R., Taticchi, A. *et al.* (2007) Irrigation effects on quality, phenolic composition and selected volatiles of virgin olive oil cv Leccino. *Journal of Agricultural and Food Chemistry* 55, 6609–6618.
Smart, L.B., Moskal, W.A., Cameron, K.D. and Bennett, A.B. (2001) *MIP* genes are down-regulated under drought stress in *Nicotina glauca*. *Plant and Cell Physiology* 42, 686–693.
Sodini, M., Astolfi, S., Francini, A. and Sebastiani, L. (2022) Multiple linear regression and linear mixed models identify novel traits of salinity tolerance in *Olea europaea* L. *Tree Physiology* 42, 1029–1042.
Sofo, A., Dichio, B., Xiloyannis, C. and Masia, A. (2004) Effects of different irradiance levels on some antioxidant enzymes and on malondialdehyde content during rewatering in olive tree. *Plant Science* 166, 293–302.
Tabatabaei, S.J. (2006) Effects of salinity and N on the growth, photosynthesis and N status of olive (*Olea europaea* L.) trees. *Scientia Horticulturae* 108, 432–438.
Tattini, M., Bertoni, P. and Caselli, S. (1992) Genotypic responses of olive plants to sodium chloride. *Journal of Plant Nutrition* 15, 1467–1485.
Tattini, M., Gucci, R., Coradeschi, M.A., Ponzio, C. and Everard, J.D. (1995) Growth, gas exchange and ion content in *Olea europaea* plants during salinity and subsequent relief. *Physiologia Plantarum* 95, 203–210.
Tattini, M., Gucci, R., Romani, A., Baldi, A. and Everard, J.D. (1996) Changes in non-structural carbohydrates in olive (*Olea europaea*) leaves during root zone salinity stress. *Physiologia Plantarum* 98,117–124.
Therios, I.N. and Misopolinos, N.D. (1988) Genotypic response to sodium chloride salinity of four major olive cultivars (*Olea europaea* L.). *Plant and Soil* 106, 105–111.
Tognetti, R., d'Andria, R., Lavini, A. and Morelli, G. (2006) The effect of deficit irrigation on crop yield and vegetative development of *Olea europaea* L. (cvs. Frantoio and Leccino). *European Journal of Agronomy* 25, 356–364.

Tomás, M., Flexas, J. Copolovici, L., Galmés, J. *et al.* (2013) Importance of leaf anatomy in determining mesophyll diffusion conductance to CO_2 across species: quantitative limitations and scaling up by models. *Journal of Experimental Botany* 64, 2269–2281.

Torres-Ruiz, J.M., Diaz-Espejo, A., Perez-Martin, A. and Hernandez-Santana, V. (2013) Loss of hydraulic functioning at leaf, stem and root level and its role in the stomatal behaviour during drought in olive trees. *Acta Horticulturae* 991, 333–339.

Torres-Ruiz, J.M., Cochard, H., Mayr, S., Beikircher, B., Diaz-Espejo, A. *et al.* (2014) Vulnerability to cavitation in *Olea europaea* current-year shoots: further evidence of an open-vessel artifact associated with centrifuge and air-injection techniques. *Physiologia Plantarum* 152, 465–474.

Tyree, M.T. and Sperry. J.S. (1988) Do woody plants operate near the point of catastrophic xylem dysfunction caused by dynamic water stress? Answers from a model. *Plant Physiology* 88, 574–580.

Wheeler, J.K., Sperry, J.S., Hacke, U.G. and Hoang, N. (2005) Inter-vessel pitting and cavitation in woody Rosaceae and other vesselled plants: a basis for a safety versus efficiency trade-off in xylem transport. *Plant, Cell and Environment* 28, 800–812.

Xiloyannis, C., Dichio, B., Nuzzo, V. and Celano, G. (1999) Defence strategies of olive against water stress. *Acta Horticulturae* 474, 423–426.

Yamada, S., Komori, T., Myers, P.N., Kuwata, S., Kubo, T. and Imaseki, H. (1997) Expression of plasma membrane water channel genes under water stress in *Nicotiana excelsior*. *Plant and Cell Physiology* 38, 1226–1231.

7.2 Pest Management in Olive Orchards

R. Petacchi[1], M. Picchi[1], A. Canale[2], G. Benelli[2], V. Zeni[2], P. Sacchetti[3] and A. Belcari[3]
[1]*Institute of Life Sciences, Sant'Anna School of Advanced Studies, Pisa, Italy;*
[2]*Department of Agriculture, Food and Environment, University of Pisa, Italy;*
[3]*Department of Agriculture, Food, Environment and Forestry, University of Florence, Italy*

This chapter refers to traditional olive production in southern Europe, in particular in Italy. Therefore, it does not refer to intensive and super-intensive cultivation systems that need specificpest management strategies.

7.2.1 Key Pest: *Bactrocera oleae*

7.2.1.1 Introduction

The olive fruit fly, *Bactrocera oleae* (Rossi) (Diptera: Tephritidae) is the key pest of the olive grove agroecosystem worldwide. The larvae are monophagous, feeding exclusively on *Olea* species, including cultivated and wild *O. europaea* (Daane and Johnson, 2010); its distribution is therefore confined to regions where olive trees are cultivated. Currently, it is spread in the Mediterranean basin, South and Central Africa, Canary Islands, the Near and Middle East, California, Hawaii, and Central America (EPPO, 2020). Despite its abundance in the Mediterranean basin, *B. oleae* most likely originated from sub-Saharan Africa, from which domesticated cultivars derive. The recent invasion of *B. oleae* in the American region was likely due to an unintended spread of individuals from the Mediterranean area (Nardi *et al.*, 2005).

Adult females can cause direct damage by laying eggs under the skin of olives; the eggs hatch into larvae that feed in the pulp of the olive. Infested fruits quickly lose their quality becoming inedible with a subsequent reduction in oil production (Malheiro *et al.*, 2015). In addition, changes in the fatty acid profile (i.e. increased percentage of fatty acids (FA), peroxide value (PV), K232, and K270) are observed, causing a lowering of the olive oil quality (Gucci *et al.*, 2012). Economic losses affect all the steps of olive production; olive farmers must deal with productivity losses due to fruit drop, rot fruits, and lower-income in table olives and olive oil production; industrials are affected due to low quality of fruits for table olives and olive oil extraction, leading to a consequent lower income. At the end, infested table olives are rejected by the consumer, and olive oil quality is reduced drastically, influencing also its nutritional and beneficial properties (Malheiro *et al.*, 2015).

To reduce the *B. oleae* economic impact worldwide, chemical control (i.e. cover or bait sprays) based on the use of conventional insecticides (i.e. organophosphate or pyrethroids), biotechnical tools (e.g. sterile insect technique (SIT) and male annihilation technique (MAT)), and biological control have been used as tools in conventional,

DOI: 10.1079/9781789247350.0022

integrated pest management (IPM) and biocontrol programs in the last decades (Daane and Johnson, 2010; Canale and Benelli, 2012). Of note, research carried out on *B. oleae* courtship and mating behavior could help to unravel mate dynamics to develop proper behavior-based control strategies.

7.2.1.2 Biology

Bactrocera oleae is a polyvoltine species; this pest completes the first annual generation in springtime (Marchini *et al.*, 2017); later, from July to November, three to five generations per year are expected in the Mediterranean region (Katsoyannos, 1992) (Fig. 7.2.1). Its phenology varies according to abiotic factors, such as temperature, humidity, microclimate conditions, altitude, latitude, and the availability and quality of drupes (Delrio and Lentini, 2016; Daane and Johnson, 2010). Usually, *B. oleae* is best adapted to the autumn period, when olive drupes are in optimal conditions for larval growth (Tzanakakis, 2006). Initial infestation can have different timing depending on the ovarian maturation of females which is temperature-dependent, stopping at 12°C and decreasing at 29°C (Daane and Johnson, 2010). Females lay eggs underneath the epidermis of olive fruits, usually one egg per drupe (Preu *et al.*, 2020); on average, a female can oviposit four to 19 eggs per day and a total of 250–300 eggs in its lifetime (Daane and Johnson, 2010). Relying on visual and olfactory cues for locating suitable hosts, females usually choose unripe (Malheiro *et al.*, 2019) and low-infested fruits (Giunti *et al.*, 2020). This phase is often characterized by female–female aggressive interactions to gain and defend the single oviposition site (Benelli, 2014). After the oviposition, a deterrent display occurs, with the female sucking up and spreading the exudate from the oviposition puncture on the drupe surface (Benelli *et al.*, 2014).

Bactrocera oleae sexual communication and aggressive behavior are detailed below in section 7.2.1.3. The egg incubation period generally does not exceed two to three days at 28–30°C, but in winter it can last up to three weeks. The lower thermal threshold ranges from 7.5°C to 10°C (Tsisipis, 1977). The newly born larva starts feeding on the pulp of olives; since they are monophagous, their development is strictly related to the presence of *Olea* fruits. Once larval stages are completed, the third instar larva opens an exit hole on the drupe, and usually pupates outside the fruit, in the soil (Daane and Johnson, 2010), although if the climate conditions are favorable, *B. oleae* larvae can also pupate inside the fruit. At optimal temperatures, the immature stages (i.e. egg, three larval instars, and pupae) can last from 16 to 20 days (Preu *et al.*, 2020).

The *B. oleae* life cycle is summarized in Fig. 7.2.1. Unlike larvae, adults feed on multiple food sources, including floral nectar and pollen, insect honeydew, and even bird feces and yeasts (Daane and Johnson, 2010; Malheiro *et al.*, 2015; Preu *et al.*, 2020). Even if adults have a rather wide range of food sources, their diet can be poor in amino acids, essential for ovarian maturation and other physiological activities (Ben-Yosef *et al.*, 2010). Thanks to the gut symbiotic bacterium *Candidatus* Erwinia dacicola, present in all life stages of *B. oleae* (Estes *et al.*, 2012), the olive fruit fly adults can exploit non-essential amino acids as a source of nitrogen (Ben-Yosef *et al.*, 2010). This bacterium is vertically transferred to the progeny (Bigiotti *et al.*, 2020), allowing the insect to cope with olive-plant-produced compounds, such as oleuropein, a toxic phenolic glycoside (Ben-Yosef *et al.*, 2014).

7.2.1.3 Aggressive behavior and sexual communication in *B. oleae*

After emergence, adults (Fig. 7.2.2) complete the gonad maturation reaching sexual maturity within three to eight days (Canale *et al.*, 2012). Before courtship events occur, the flies start fast and whirling flights, forming non-compact swarms mainly composed of males (Benelli *et al.*, 2012). Each swarm establishes itself on the windward side of an olive plant (lekking site). Within each

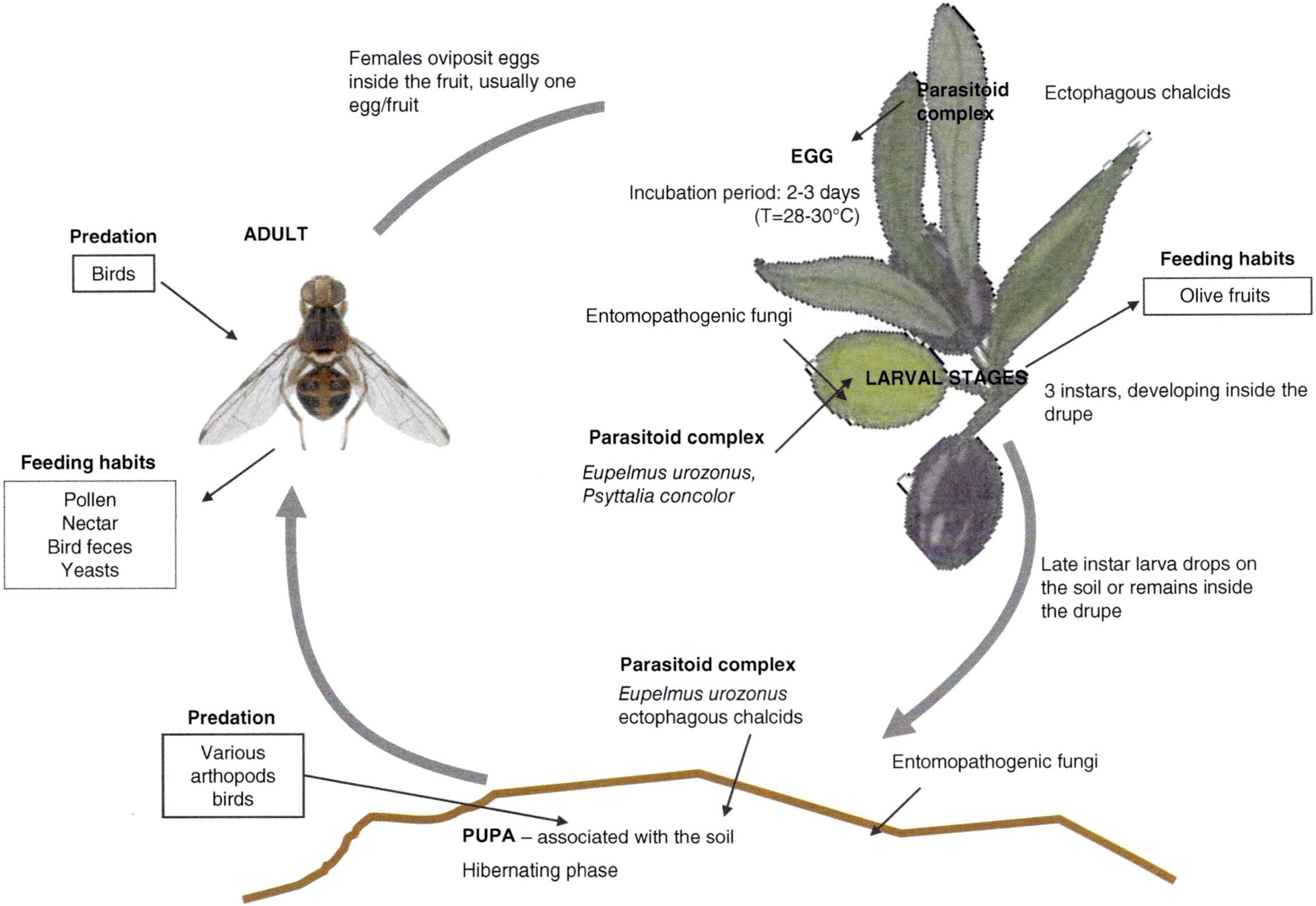

Fig. 7.2.1. The olive fruit fly, *Bactrocera oleae* (Rossi) (Diptera: Tephritidae): life cycle, main ecological traits and natural enemies regulating its populations. (From University of Pisa)

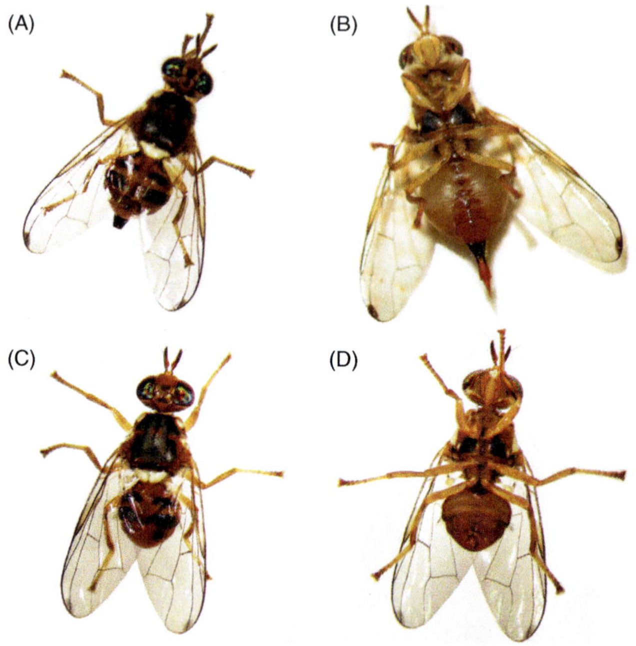

Fig. 7.2.2. *B. oleae* female (*top*) and male (*bottom*) total views. Dorsal (A and C) and ventral (B and D) views, respectively. Ventral view of female highlights the ovipositor partially everted and abdomen with eggs. (BioLabs, Sant'Anna School of Pisa)

group, males exhibit aggressive behavior to defend a small territory (typically a leaf or an olive young twig) against other males (Benelli, 2014). Aggressive displays characterizing male–male interactions are reciprocal wing wavings, fast head-butting acts, and boxing with forelegs (Benelli, 2014, 2015). Benelli *et al.* (2015) showed that prior residence at a given site enhances the male's chances to win. Then, the males release olfactory cues attracting females to lekking sites (Landolt, 1997; Benelli *et al.*, 2014).

In *B. oleae*, courtship and mating are routed by acoustic, physical and olfactory cues (Mazomenos, 1989; Benelli, 2014). Olfactory cues are sex pheromones, i.e. chemical messengers with intra-specific action, whose purpose is to attract the conspecific individual of the opposite sex for mating. At variance with other Tephritidae species, both sexes of *B. oleae* can produce sex pheromones (Carpita *et al.*, 2012; Canale *et al.*, 2013). The production of sex pheromones is age-related, reaching a maximum secretion after 21 days from fly emergence (Canale *et al.*, 2013, 2015; Fusini *et al.*, 2018). Specific semiochemicals produced by *B. oleae* female have been deeply investigated over the last decades. First evidence reported that virgin females produce a multicomponent sex pheromone containing 1,7–dioxaspiro [5.5] undecane (DSU, commonly known as olean) and methyl dodecanoate, both produced in rectal glands, and alpha-pinene and nonanal (Baker *et al.*, 1980; Mazomenos and Haniotakis, 1981, 1985). DSU, which is the most abundant component of the mix, elicits positive chemotaxis in males (Mazomenos and Haniotakis, 1981, 1985). Recent studies stressed the presence of at least nine sex-specific compounds produced in the rectal glands of virgin females, plus another compound detected in both sexes [ethyl(*Z*)-9-octadecenoate] (Canale *et al.*, 2015). Interestingly, three of the compounds found in the female rectal glands have never been detected in other Tephritidae species, i.e. ethyl decanoate, n-butyl dodecanoate and ethyl(*Z*)-9-octodecenoate. The overall quantities produced by 21-day-old females are lower if compared to DSU production in females of the same age (Canale *et al.*, 2015). Among these compounds, two molecules evoke attraction towards conspecific flies, i.e. methyl hexadecanoate (C16Me) and ethyl decanoate (C10Et). The first one achieves dose-dependent attraction towards males (Canale *et al.*, 2015; Benelli, 2014; Benelli *et al.*, 2014), while C10Et is attractive to females. The chemoecological role of this molecule is still unclear; C10Et may be involved in female–female aggression dynamics leading to maintaining single oviposition sites (Canale *et al.*, 2015). As discussed above, female–female aggressive interaction is not uncommon among tephritids, representing a possible way for reducing larval competition for food, which can be detrimental to *B. oleae* progeny if more than one oviposition per fruit occurs (Benelli, 2014). Recently, a study conducted by Fusini *et al.* (2018) identified 11 new sex-specific components in female rectal glands: four generic saturated butyl esters; two hexadecenoatemethyl esters, two hexadecenoate ethyl esters, two octadecenoatemethyl esters and one octadecenoate ethyl ester. Their abundance is also age-related. The behavioral role of these molecules still needs to be investigated.

Unlike female-borne compounds, male semiochemicals have been investigated only recently. Of note, *B. oleae* males can produce DSU in rectal glands from first day of emergence, reaching its maximum with the sexual maturation, then stops its biosynthesis by the eleventh day of life (Canale *et al.*, 2013). The role of DSU is still unclear; it cannot be considered as a case of chemical mimicry, since mating performance of young males is not higher than adult ones (Benelli *et al.*, 2013). However, young males may indirectly benefit from DSU production by distracting competitors away from females. After reaching the sexual maturation, *B. oleae* males start producing a sex-specific pheromone, (*Z*)-9-tricosene, also known as muscalure, which evokes attraction to females during the close-range phase (Carpita *et al.*, 2012; Canale *et al.*, 2013). The production and storage of (*Z*)-9-tricosene are borne by the glands associated with rectal ampulla (Carpita *et al.*, 2012). However, evidence suggests that also the urotergal glands (i.e. two visible glandular areas in the fifth urotergite of both sexes) could play a role in the olfactory communication during courtship and mating of *B. oleae*. A study conducted by Canale *et al.* (2012) provided that the presence of (*Z*)-9-tricosene on the urotergal glands and the trends of its amount were directly correlated to its production in the rectal ampulla. The presence of muscalure can be explained because of the transport of sex pheromone from the rectal ampulla to the urotergal glands by a peculiar leg-rubbing behavior performed during courtship (Canale *et al.*, 2013). Leg rubbing is simultaneously performed with wing vibrations and stridulation, two displays that are functional to achieve high male mating success and may help in the dissemination of the sex pheromone (Benelli *et al.*, 2012). *B. oleae* males vibrate their wings and screech by rubbing the microtrichia against the abdominal pecten, located at the apex of the third tergite (Feron and Andrieu, 1962). According to Feron and Andrieu (1962), male wing vibration is a courtship cue that may serve to test female willpower. Further studies showed that the female mating decision is affected by the quality of male wing vibrations (Benelli *et al.*, 2012). Male wing vibration usually continues during the copulation-attempt phase. In this last phase, the male jumps on the female and tries to align the body along her axis (Benelli *et al.*, 2012). If females are receptive, they remain motionless allowing copulation. On the other hand, non-receptive females do not remain still, moving away from the undesired partner. During copulation, the male and female remain still, with periodic male leg-rubbing acts. At the end of the mating, cleaning behavior of the body is noted in both sexes. Males do not show preferences between virgin or already-mated females (Benelli *et al.*, 2012). In *B. oleae*, as for other Tephritidae species, mating occurs usually at dusk, regulated by a circadian clock (Bertolini *et al.*, 2018). The diurnal emission of female sex pheromone increases just before or after twilight (Levi-Zada *et al.*, 2012). In this time window, male sexual activity should be synchronized with female receptivity (Bertolini *et al.*, 2018). According to observations made by Loher and Zervas (1979), sexual communication and mating occur during the last three or hours of the photophase, decreasing as it gets dark. Long-range attraction stops with the beginning of the night, though short-range walking is still possible (Loher and Zervas, 1979).

Basic knowledge about *B. oleae* sexual communication can contribute to the development of novel tools to be adopted in IPM programs. Understanding the sexual selection process in olive fruit fly may be also useful for the implementation of the Sterile Insect Technique (SIT). Low mating ability of sterile males is a concern in species like *B. oleae* that has complex mating systems for the presence of leks and choosy females (Lance and McInnis, 2005). In addition, semiochemicals that influence courtship and mating behavior might be employed to improve IPM programs by reducing longevity, fecundity and adult dispersal.

7.2.1.4 Population dynamics

The fluctuations of insect populations depend on abiotic factors (e.g. temperature, soil physico-chemical properties, and water availability), topography, host phenology, pest-host synchrony and density dependence dynamics (i.e. previous population size) (Carriere *et al.*, 2006; Ordano *et al.*, 2015). In addition, global warming may further modify the distribution of a pest (Hodgson *et al.*, 2011; Marchi *et al.*, 2016).

The role of temperatures for *B. oleae* is of relevance in all the developmental phases of the pest throughout all seasons, including overwintering and spring flights (Marchi *et al.*, 2016). It has been suggested that higher temperatures may have a great impact on crop phenology and host–pest interaction in the Mediterranean region (Ponti *et al.*, 2014), and that climate warming may boost the spread of *B. oleae* population to cooler areas, while its fitness will be reduced in relatively hot areas (Marchi *et al.*, 2016). In the Mediterranean region, the size of *B. oleae* population increases after blooming and pit-hardening, reaching the maximum in late summer and fall and then decreasing in winter (Tznakakis, 2006). Temperatures above 31°C increase mortality, while the reproductive activity is reduced, and reproductive dormancy is induced (Gutierrez *et al.*, 2009; Ordano *et al.*, 2015). The 50% of mortality of immature stages is observed at 30°C (Preu *et al.*, 2020). The optimum for physiological activities of *B. oleae* ranges from 25°C to 28°C; below 23°C, general activities and reproduction decrease, stopping at 17°C. Overwintering mortality is a key factor for *B. oleae* population dynamics (Orsini *et al.*, 2007). During autumn, the larvae exit the fruit and pupate in the soil. Pupal mortality is influenced by low temperature and high soil moisture (Neuenschwander *et al.*, 1981); however, it may also depend on predation by ants and other ground-dwelling predators (see details below in section 7.2.1.10) and food-searching birds, like the blackbird (*Turdus merula*), the robin (*Erithacus rubecula*) or the starling (*Sturnus vulgaris*) (Neuenschwander *et al.*, 1981, 1983; Orsini *et al.*, 2007; Gonçalves *et al.*, 2012a). *Bactrocera oleae* population dynamic varies also according to the seasoning. The spatial patterns of *B. oleae* population shift from high elevations in summer to lower altitudes in late summer (Castrignanò *et al.*, 2012), suggesting a strong influence of the elevation on *B. oleae* population density.

In conclusion, olive fruit fly populations are expected to fluctuate and be highly seasonal in cold areas (e.g. mild summer and cold winter and nights) with higher reproductive and flying activities during summer and autumn. In coastline and lowland areas, characterized by harsh summer and mild winter, highly reproductive and flying activities can be observed during autumn, winter and spring. In these areas, *B. oleae* reduces its activities during summer (diapause) (Ordano *et al.*, 2015).

7.2.1.5 Integrated Pest Management

The olive tree is one of the most important crops in the Mediterranean area both to produce olive oil and table olives and being an essential component of our landscape. Further, it is really a complex and balanced agroecosystem from a biodiversity point of view as more than 100 species of animals are known to live at the expense of the plant (Tzanakakis, 2006). Many insects and arachnids live within the system establishing various types of symbiosis with the animals that live on the olive tree as well as with organisms resident on spontaneous plants growing in the agroecosystem (Tzanakakis, 2006; Picchi *et al.*, 2016; Ortega and Pascual, 2014). This stability and complexity allowed the reduction of the number of pestiferous species in the system thanks also to the progressive reduction of chemical treatments occurring in the last 30 years. As a matter of fact, only the olive fruit fly, *B. oleae*, can be termed 'key species', since it is always present in the system with such population density that it frequently causes severe damage to the olive oil production from a quantitative and qualitative point of view (Gucci *et al.*, 2012). Therefore, it is very important to set appropriate control strategies for the crop with the utmost attention aimed at avoiding undesirable side-effects in the delicate trophic chains and, in addition, to follow the economic thresholds established for some pests of the olive crop. Finally, in view of the recent EU regulation (art. 1 of Commission Implementing Regulation (EU) 2019/1090 of 26 June 2019 concerning the non-renewal of approval of the active substance dimethoate), today's challenge is to develop new control strategies against the *B. oleae* adult populations as well as to improve any other environmentally friendly control method.

At present, control strategies against *B. oleae* rely mainly on the use of chemicals and biotechnical methods, while the biological control has not given satisfactory results even if there were several attempts in the past, mainly concerning the introduction of exotic parasitoids (Silvestri *et al.*, 1908) or mass releases of natural enemies of the fly (Monastero, 1968; Kapatos *et al.*, 1977; Delrio *et al.*, 2005). As already underlined, a quite high number of natural enemies of the olive fruit fly live in the system (see also section 7.2.1.10) which do not affect, unfortunately, as far as we know at present, the population dynamics of the fly.

Chemical compounds employed in controlling the *B. oleae* can be applied for killing immatures inside the olives (cover sprays) or against the adult populations (adulticide treatments). It is obvious that both these two strategies are not 'environmentally friendly' and give rise to negative side-effects within the system.

7.2.1.6 Chemical control: cover sprays

After dimethoate has been banned in EU countries, this kind of strategy against the immatures of the *B. oleae* almost completely loses its importance, also in consideration of the fact that the two active ingredients admitted (in Italy) in conventional olive growing such as acetamiprid and phosmet do not show the same efficacy of the first organophosphate. However, since this strategy is still allowed, it is necessary to follow the most appropriate sampling methods to accurately establish the treatment time in compliance with the economic action thresholds. Just this advice could allow reduction of the number of cover sprays that very often are applied in excess.

The monitoring of the pre-imaginal population therefore starts from the taking of a sample of at least 100–200 olives per sample area (100–200 plants). The olives must be picked randomly, at the rate of no more than one or two drupes per plant, from the different parts of the canopy (Chesi and Quaglia, 1982).

The sampled olives must then be examined to assess the level and type of infestation. Such analysis should be carried out by qualified personnel in a special facility using a good lens suitable for the purpose or, better, using a stereoscopic microscope. Usually, this activity is carried out by technicians from professional organizations, producer associations, oil mills etc. In this case, it is worth underlining how adequate professional growth in this area by the farmers themselves could be of great support for the rationalization of control measures against the *B. oleae*. The infestation should be considered: 'active' (eggs, first larval instars) and 'damageous' (third instar larva, pupa, drupes with exit hole). Of course, only youngest immatures could be killed by cytotropic insecticides, so the most important thing is to proceed with the sampling weekly until the action threshold is reached. In Italy, the economic threshold of intervention has been estimated on values ranging from 10% to 15% of active infestation.

In a modern concept of IPM, cover sprays should be abandoned since their impact on the olive system is remarkable (Haniotakis, 2005). Besides, in the light of recent studies that proved an important complexity present in this agroecosystem, this kind of strategy can give rise to significant negative effects on arthropods living on the plant and in the ground. Further drawbacks of the cover sprays that should be taken into consideration are the presence of residuals in the olive oil (Rastrelli *et al.*, 2002) as well as the complex problems related to the *B. oleae* resistance, induced by the long-lasting application of systemic and/or cytotropic insecticides used as cover treatments in wide olive crop areas (Vontas *et al.*, 2002; Skouras *et al.*, 2007).

7.2.1.7 Chemical control: adulticide treatments

Preventive control methods are aimed at reducing the adult population to minimize the oviposition. This goal can be obtained by killing the highest number of adult flies, traditionally using chemical insecticides, or preventing egg laying, using physical methods. Until some years ago, adulticide treatments were based mainly on the use of poisoned bait sprays. Baits normally consist of protein hydrolysate (attractants) mixed with organophosphates, pyrethroids or neoticotinoids as active ingredients. Flies, especially immature females that require protein-rich meal for their oogenesis, are attracted by odour sources emitted from decaying proteins and feed on these attractants. The poisoned bait spray residue on the foliage is dissolved by the insect's saliva and ingested by the flies, killing them (Sarwar, 2015).

Sprays can be performed by ground or aerial applications; it is obvious that aerial applications are not so environmentally friendly due to: (i) drift (Woods *et al.*, 2001); (ii) side-effects on many animal species living in the treated areas. Just think of the enormous impact caused by aerial treatments against tsetse fly on aquatic vertebrates and invertebrates and on birds in Botswana (Kurugundla *et al.*, 2012). Very often, toxic baits are applied by a ground spot spraying technique that reduces the impact of side-effects. Protein bait sprays are less harmful to beneficial insects, so making them suitable for their use in IPM programs (Sarwar, 2015).

Bait applications against *B. oleae* started very early in Greece both with ground and aerial applications (Orphanidis *et al.*, 1971). In Italy, bait sprays have been mainly carried out in the 1980s and 1990s; after that the use of cytotropic insecticides (dimethoate) against the immatures inside the drupes prevailed in most of the Italian olive-growing areas (Delrio, 1989).

In the late 1990s in the USA, a new bait-insecticide based on the active ingredient spinosad was proved to be effective against the medfly *Ceratitis capitata* (Vargas *et al.*, 2002). Spinosad, a toxin derived from a soil bacterium *Saccharopolyspora spinosa*, showed a low mammalian toxicity and reduced environmental impact on parasitoids (Stark *et al.*, 2004). In the Hawaiian islands, spinosad showed an adequate replacement for cover sprays based on the organophosphates in the control programs of *Bactrocera cucurbitae* and *B. dorsalis* (Prokopy *et al.*, 2003; Piñero *et al.*, 2009). Concerning the *B. oleae*, first attempts to control this pest with Spinosad-based baits date back to 2000 in Egypt (Moustafa , 2009) while the efficacy of this product towards *B. oleae* and the impact on non-target arthropods was evaluated first in Portugal (Gonçalves *et al.*, 2012b). In Italy, the selectivity of GF-120 (Spinosad) towards many pollinators has been shown (Infusino *et al.*, 2014) and in Greece the same formulation proved to be extremely selective towards *Chrysopa* sp., Coccinellidae predators and Hymenopteran parasitoids (Vergoulas *et al.*, 2007). Anyway, other studies carried out in Spanish olive crops demonstrated some effects on the canopy non-target arthropod fauna, especially towards predators such as *Orius* spp. and Hymenoptera Aphelinidae (Pascual *et al.*, 2014).

Although some works have shown the product's less-than-complete selectivity, Spinosad sprays are also allowed in organic olive crops being the insecticide of microbial origin and regulated by European legislation on its use in organic farming. In Tuscany, not more than eight treatments against *B. oleae* are allowed in the olive crops.

7.2.1.8 Physical methods

Particle films technology (PFT) was developed at the end of the 1990s to set up a sprayable means as effective as reflective mulches in repelling insects. Very soon, PFT was demonstrated to be a successful control method against a variety of pests, such as psyllids, moths and weevils (Glenn *et al.*, 1999; Glenn and Puterka, 2005), although it provides a variety of useful effects to horticultural crops (Sharma *et al.*, 2015). This physical control method consists in spraying a chemically inert, water dispersible, finely powdered kaolin, that, after drying, forms a particle film on the overall plant surface making it unfavorable to small insects moving over plant organs (Sharma *et al.*, 2015). Besides physical repellency, particle films exert multiple modes of action on insects, interfering with their behavior and biology (hindered host plant selection, feeding activity, egg laying, settling), causing also direct toxicity (Puterka *et al.*, 2005; Sharma *et al.*, 2015).

The first documented application of PFT to control *B. oleae* occurred in Syria, produced a significant reduction in olive infestation and a long-lasting fruit protection proving that the technique was comparable to synthetic insecticides and feasible for olive fruit fly management (Saour and Makee, 2004). Since then, this control method has been evaluated mainly for applications in organic olive orchards in Italy (Caleca and Rizzo, 2007; Pennino *et al.*, 2006; Caleca *et al.*, 2010) and Spain (Pascual *et al.*, 2010) but it spread also in other olive-growing countries such as Iran (Mozhdehi and Kayhanian, 2014), reducing significantly fruit infestation, sometimes more effectively than dimethoate baits, also in high-pest-density conditions (González-Núñez *et al.*, 2020). Specific effects of kaolin on *Ceratitis capitata* (Wiedemann) females consisting in reduced adhesion to the fruit surface have been recently disclosed (Salerno *et al.*, 2020): likely the same mechanism of action may account for the successful achievements of PFT on *B. oleae* control.

Since inert dusts are known to be detrimental to natural enemies, field trials and lab experiments have been implemented in different countries to evaluate possible negative drawbacks of kaolin-based products on several predators and parasitoids dwelling on olive tree agroecosystem. Lab tests showed that kaolin was not toxic to *Psyttalia concolor* (Bengochea *et al.*, 2014a), *Chrysoperla carnea* (Stephens) (Porcel *et al.*, 2011; Bengochea *et al.*, 2014b) and other natural enemies, except *Anthocoris nemoralis* (Bengochea *et al.*, 2013), although sublethal effects were evidenced and a general repellency was observed. Conversely, studies carried out in the field revealed, on average, a detrimental effect of the PFT on arthropod biodiversity with reduction of some species or groups such as spiders (Scalercio *et al.*, 2010) and some coccinellids (Pascual *et al.*, 2010; González-Núñez *et al.*, 2020).

Nowadays the use of particle films formed by processed kaolin-based products is one of the most profitable methods among non-insecticidal alternatives for controlling *B. oleae* infestations, especially in organically managed orchards. Applications of this physical method proved that OFF adults can be easily affected by alteration of the fruit surface or, more generally, by the surface's characteristics. As a matter of fact, studies on olive epicuticular waxes of different cultivars pointed out that the OFF attachment to the olive surface varies depending on the structure of the waxy coverage of the fruit (Rebora *et al.*, 2020). Hopefully, these findings could be integrated with the development of PFT to better understand the numerous mechanisms of action and enhance the effectiveness of this method against the adults of *B. oleae*.

7.2.1.9 Biological control

Biological control as a method to manage olive fruit fly populations has long been believed to be an untapped but potentially effective strategy, worthy to be further investigated. In Mediterranean countries, since the control action of native natural enemies is ineffective in reducing olive crop infestation in cultivated olive orchards, biological control has been attempted during the whole of the 20th century, following principles of both classical and augmentative biological control. First attempts were undertaken at the beginning of the century when Silvestri, to increase the *B. oleae* natural enemy complex, introduced repeatedly to Italy parasitoids native to Africa and other continents, since the *B. oleae* was considered indigenous to this continent (oppure: he was convinced olive fly was not native to Europe) (Silvestri *et al.*, 1908). After the discovery of *Psyttalia concolor* (Szépligeti) in Tunisia, the research on biological control focussed on this endophagous, larval-pupal parasitoid: it was mass-reared on medfly and large-scale field releases were performed in Sicily producing satisfactory reduction on olive infestation levels (Monastero and Delanoue, 1966; Monastero, 1968). Afterwards, introduction programs based on field releases of *P. concolor* were conducted in other Italian regions as well as in Mediterranean countries testing also different procedures such as anticipated releases against spring populations performed in Greece during the 1970s (Kapatos *et al.*, 1977; Liaropoulos *et al.*, 1977).

For extensive (exhaustive) documentations on the use of *P. concolor* in biological control, see Wharton and Yoder (1967) at: http://paroffit.org/public/public_content/show/13309?content_template_id=54 (accessed 28/5/23). Because of the repeated introductions, the parasitoid has established in warmer locations of southern Mediterranean countries (islands, some coastal areas, south Italy and Greece) and the impact on *B. oleae* populations developed in cultivated olives is highly variable, although assessed only in a few countries. For instance, in Sardinia, in 1997–2001, the parasitism rate swung from 0% to 100% (Delrio *et al.*, 2005); in Balearic Islands, in organic orchards, an average parasitism rate of 23%, varying with orchards and sampling time, was recorded (Miranda *et al.*, 2008), while in Sicily in 2010 it reached approximately 33% of the pupae (Caleca *et al.*, 2017) and fluctuated from 0% to 55% in Greece, depending on olive tree cultivar and on the year (Garantonakis *et al.*, 2017). However, such variable parasitism rates are not effective in preventing damage and economic losses. In general, all these experiences proved that *P. concolor* can be used effectively against the *B. oleae* only with inundative releases aimed at producing short-term, temporary increase in parasitism rate with subsequent reduction of the olive infestation.

More recently, other allochthonous parasitoids have been considered for classical biological control in Mediterranean countries, such as *Diachasmimorpha longicaudata* (Ashmead), a larval-pupal parasitoid released occasionally in different countries (Jiménez and Castillo, 1992), *Fopius arisanus* (Sonan) that is an oo-pupal parasitoid evaluated in Italy (Calvitti *et al.*, 2002) and *Psyttalia lounsburyi* (Silvestri) studied in France (Malausa *et al.*, 2010).

After the accidental introduction of *B. oleae* in California, renewed interest on biological control emerged, promoting also expeditions to Africa and Asia aimed at searching and evaluating new or known but neglected parasitoid species (Daane and Johnson, 2010). More attention was devoted to species from sub-Saharan Africa where parasitism rates are known to be usually higher than in other countries (Mkize *et al.*, 2008). The extensive researches carried out on resident and imported parasitoids for their evaluation as potential biological control agents against the *B. oleae* have been reviewed by Daane and Johnson (2010) who concluded, once again, that these parasitoids 'do not provide reliable control of olive fruit fly', endorsing as the main reason that the *B. oleae* prefers larger fruit for oviposition while parasitoids are hindered, to a different extent, from reaching *B. oleae* larvae developing inside large olives. For future perspectives of biological control of *B. oleae* the same authors (Daane and Johnson, 2010) suggest also to investigate further *D. longicaudata* and *D. kraussii* Fullaway, two parasitoids with longer ovipositor but causing concern for possible side-effects on non-target species, and to study the effects of cues released from olives on parasitoid's behavior which was recently undertaken (Sacchetti *et al.*, 2019).

7.2.1.10 Conservation biological control

Conservation biological control (CBC) is a salient point in olive orchard management strategies converging the need to suppress pests and the search for eco-friendly IPM solutions in light of sustainability of agriculture as one of the main objectives of the European strategies of the Green Deal and the Sustainable Development Goals of the United Nations Agenda 2030. Agriculture should find an alternative and innovative solution to pest management, and biological control is better applied in perennials that suffer the least disruption and human intervention than annual crops.

Olive, with its characteristics, is, potentially, a good candidate for reducing chemical inputs. Such a complex agroecosystem (Arambourg, 1986) has both traits of crop and semi-natural habitats: the management and pesticide application are limited, with no-tillage, maintaining spontaneous vegetation at the ground. Besides, olive trees have higher stability for years, structurally and vegetationally more diverse than annual crops and thus providing a high diversity of niches such as drywalls and stones on the ground. Moreover, olive does not suffer heavy periodical disruption action (except for the harvest in autumn) that could affect the indigenous community of invertebrates that consist of potential natural enemies of the olive fruit fly: from various researches, carabids, staphylinids and spiders can be considered good-candidate taxa for biological control programs in olive orchards, even with limited success against the olive fruit fly up to now. Further agroecological studies are needed to investigate how habitat management can affect the survival of natural enemies in the agroecosystem and how to conserve and enhance their presence within the cropped field, because a good CBC strategy does not necessarily need to add new elements, but rather to conserve and improve the already existing ones.

Since the life cycle of the olive fruit fly is synchronized with fruit production (Tzanakakis, 2006), control strategies should be considered for the timing of the specific stage of the pest. In late spring and during summer, adults find their mate flying in the olive orchard and females oviposit in olive fruits they carefully chose by visual and tactile inspection. Usually, the life cycle hosted by olive trees consists of three to four generations lasting until autumn. During this stage, they are exposed to flying natural enemies or to predators living in the tree canopies, whereas before the overwintering phase, most larvae leave the fruit to pupate in the soil and, consequently, are subjected to ground-dwelling natural enemies.

The eggs are usually well protected from natural enemies inside the drupe, except for the effect of *Lasioptera berlesiana* Paoli, 1907 (Diptera: Cecidomyiidae), an occasional parasitoid that can recognize the stings made by the olive fruit fly soon after the attack on drupes, and lay eggs near the olive fruit fly. The olive fruit midge may occasionally transmit the fungus *Camarosporium dalmaticum* (Iannotta *et al.*, 2012) causing pre-harvest fruit drop.

The growth of the larva takes place inside the drupes and, at this stage of development, the main biotic enemies consist of the parasitoid assemblage composed of four endemic solitary wasp species that are reported in the northwestern Mediterranean basin: *Eurytoma martellii* Domenichini,

1960 (Hymenoptera: Chalcidoidea, Eurytomidae); *Eupelmus urozonus* Dalman, 1820 (Eupelmidae); *Pnigalio agraules* (Walker, 1839) (Eulophidae); and *Cyrtoptyx latipes* (Rondani, 1874). (Pteromalidae) and one more species, the koinobiont endoparasitoid *P. concolor*, has been used for biological control of the olive fruit fly by inundative and propagative releases (see details in previous paragraph; Russo, 1967; Arambourg, 1986; Canale and Loni, 2006).

E. martellii is an idiobiont third-instar larval ectoparasitoid of *B. oleae*, on which it carries out several generations per year but it is mainly active during summer (Arambourg, 1986); its ecology is not well known (Boccaccio and Petacchi, 2009), and, to date, no alternative hosts have been reported.

E. urozonus (Fig. 7.2.3) is a generalist ectoparasitoid, multivoltine and idiobiont, living at the expense of the third instar larvae and pupae of the olive fruit fly, where it can also behave as a hyperparasitoid or even as self-parasitoid (Russo, 1967). Besides, in olive orchards, it could also attack *Prays oleae* (Alissandrakis *et al.*, 2018) as *P. agraules*, which is an idiobiont, generalist ectoparasitoid with several generations per year, which attacks the latest instar of larvae of the olive fruit fly. They were both dominant in Tuscan olive groves, fostered by surrounding semi-natural habitat that hosts alternative prey (Boccaccio and Petacchi, 2006). In addition, augmentative release of *E.urozonus* has been attempted, showing the inadequacy of this species to be used as biological control agent since it can perform as hyperparasitoid after being released (Delrio, 2010).

C. latipes is a generalist parasitoid and it has been reported in Italy, but with a patchy distribution and considered rare in all localities in which it has been found. After the larval stage, the pupa could be found inside the drupe or the larva could drop off to the soil, to reach a suitable place to overwinter. The larva easily moves in the soil with its high capacity of dispersal (Yokoyama, 2012) and burrows in to the first 4 cm of topsoil in a few minutes (Cavalloro and Delrio, 1975).

In the soil, the snappy larva and the hidden pupae are exposed to generalist

Fig. 7.2.3. *Eupelmus urozonus* Dalman: (A) dissection of olive fruit with evidence of larval stage after *B.oleae* predation; (B) pupal stage; (C) particular of exit hole on olive surface; (D) adult female with characteristic ovipositor. (Photos: BioLabs, Sant'Anna School of Pisa)

ground-dwelling predators such as beetles, especially individuals of the family Carabidae and Staphylinidae, ants, spiders, opiliones, centipedes, earwigs, and chilopods (Cavalloro and Delrio, 1975; Neuenschwander *et al.*, 1983; Orsini *et al.*, 2007; Dinis *et al.*, 2016) besides pathogens and mold that could increase mortality. Ants are ubiquitous in the canopy and dominate all other invertebrate groups.

In the Monte Pisano, field experiments took place in 2014 and 2015 with exclusion cage suggesting that the effect of generalist arthropods on pupae exposed are higher in organic olive orchards (Picchi *et al.*, 2017a). Moreover, the effect is more dependent on local conditions rather than the composition of nearby habitats. Also, Albertini *et al.* (2018a) stressed the potential role of some carabid predator species in olive fruit fly pest control, having a peak of activity in October and November, when pest pupae are abundant in the soil.

Dinis *et al.* (2016) sampled soil arthropods in olive orchards from January to May and found that they had a strong impact on the decline of olive fruit fly pupae in the soil. Moreover, the analysis of the gut content from the field sample confirms the relevant role of rove beetles and some species of carabid against the overwintering pupa (Albertini *et al.*, 2018a, b). Also, the ground cover increases spider abundance (Paredes *et al.*, 2013) but no studies have been found on an effective contribution to pupae control. To our knowledge, the early impact of natural enemies on overwintering pupae of olive fruit fly is still not known.

In spring, the adult breaks out of the pupal cocoon and emerges from the surface to search for food and mate. After reproduction, they fly to seek for the best olive available and oviposit the eggs. In this stage, the adult spends a lot of time in canopies of olives. At this point, spiders are prepared to catch flies in their spider net or by ambushes, depending on the hunting strategy they use. Spiders are the most abundant predators in olive canopies (Cárdenas *et al.*, 2015; Benhadi Marín *et al.*, 2017) and many variables at different scales could influence their abundance and richness in crops, such as the type of the nearby habitats, or the landscape configuration and composition (Duelli and Obrist, 1998) or by local variables such as management (Bengtsson *et al.*, 2005) or by the abundance of stones in the ground (Benhadi-Marín *et al.*, 2018). Moreover, the availability of food in an area is a relevant factor to assemble the spiderweb (Marc *et al.*, 1999). Despite recent studies, their contribution to the control of pest, and, in the given case, against olive fruit fly is quite underestimated.

Recent studies in the Monte Pisano added data in ecology and landscape dependency of the richness and abundance of spiders in olive (Picchi *et al.*, 2016, 2020; Picchi, 2020). From the spider perspective, the olive grove is a mesophile habitat with species typical of drier habitat. The case study of 2016 recommends to further deepen the contribution of spiders: the negative correlation between the activity density of the olive fruit fly and the abundance of web builders suggest that spiders affect the distribution of the olive fruit fly. Olive trees with a higher density of Linyphiidae had a reduced density of flies searching for the substrate and it could be linked to the intimidation effect that webs have on olive flies: in fact, dipterans integrate olfactory stimuli with visual cues to locate food (Wall and Fisher, 2001) and spider silk is a warning signal of the presence of predators in canopies that may deter herbivores (Rypstra and Buddle, 2013), especially if there are dense webs as made by Linyphiidae spiders (Segoli *et al.*, 2004). Thus, we need to define if canopies that contain high amounts of spiders have also a reduced olive infestation.

7.2.1.11 Microbial insecticides

Microbial insecticides selectively toxic to *B. oleae* have been developed to a certain extent. Different species and strains of *Bacillus* spp. have been tested in laboratory and field experiments (Karamanlidou *et al.*, 1991; Alberola *et al.*, 1999; Navrozidis *et al.*, 2000; Ruiu *et al.*, 2007, Mostakim *et al.*, 2012) with promising results but they were not so effective as to receive interest from the pesticide industry. Among different entomopathogenic fungi, *Beauveria bassiana* strain ATCC 74040 was effective in reducing olive infestation and it was registered as a commercial insecticide to control *B. oleae* adults (Konstantopoulou and Mazomenos, 2005; Benuzzi *et al.*, 2007) while other soil fungi like *Metarhizium brunneum* have been evaluated for soil treatment against *B. oleae* pupae with evident reduction on the emerged adult population (Yousef *et al.*, 2013). In consideration of the high biodiversity of the olive agroecosystem, in our opinion entomopathogenic fungi should be used after close evaluation and when climatic conditions are favorable.

7.2.1.12 Symbiosis in the olive fruit fly: new perspectives in IPM

The olive fly establishes two kinds of relationship with microorganisms; the first is due to the microbiome of the olive phylloplane, the second with a not-culturable bacterium living in the alimentary canal of the fly: *Candidatus* Erwinia dacicola (Capuzzo *et al.*, 2005; Sacchetti *et al.*, 2008; Fig. 7.2.4). In field, *B. oleae* adults, especially immature females, are continuously searching for protein to satisfy their need for oogenesis. Phylloplane bacteria are an important protein source, and thanks to their specialized mouthparts flies are able to hold bacteria in their saliva and suck through the food canal a huge amount of these microorganisms (Sacchetti *et al.*, 2008). In addition, the presence, nearly

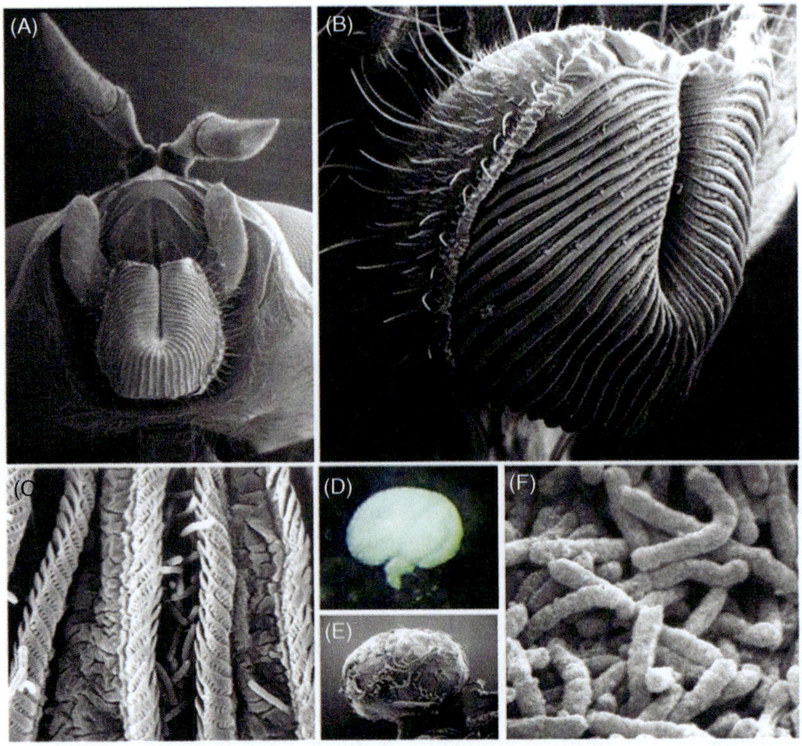

Fig. 7.2.4. Morphological structures of the olive fruit fly involved in the relationships with bacteria: (A) head with mouthparts; (B) micrograph of the labella with pseudotracheae; (C) high magnification of pseudotracheae showing the *lumen* filled with acquired bacteria; (D) milky oesophageal bulb full of *Ca.* Erwinia dacicola colonies; (E) micrograph of the oesophageal bulb; (F) biofilm of *Ca.* Erwinia dacicola within the oesophageal bulb. (Photos: Patrizia Sacchetti & Antonio Belcari, University of Florence)

constant, of *Ca.* E. dacicola in some specialized tracts of the gut (Estes *et al.*, 2012) increases fecundity of females fed with different nitrogen sources such as non-essential amino acids, urea and bird droppings (Ben-Yosef *et al.*, 2014).

Females can transmit this symbiont to their progeny via eggs that are smeared with bacteria in the last tract of the reproductive system, which is in common with the alimentary canal (Mazzini and Vita, 1983; Sacchetti *et al.*, 2008; Bigiotti *et al.*, 2019). Bacteria are then ingested by the hatching larva and become an intra-cellular life in the gastric ceca (Estes *et al.*, 2009). Their presence allows larval development in green drupes, counteracting the action of oleuropein (Ben-Yosef *et al.*, 2015).

Basic knowledge on the olive fly symbiosis allows us to understand that symbioses are very important for the fly fitness and survival. For this reason, the hypothesis that interrupting the relationships between the *B. oleae*, phylloplane microorganisms and *Ca.* E. dacicola was tested planning field and laboratory trials (Fig. 7.2.5). Firstly, treatments against the *B. oleae* performed with copper products showed a reduction of the infestation in field, but it was not possible to ascertain whether the reduction was due to copper as bactericidal or deterrent product (Belcari *et al.*, 2003). Only when the complexity of this symbiosis began to be clarified was it possible to understand, also through laboratory experiments, that copper interferes with the symbiosis of the *B. oleae*, killing the symbiont in the gut or strongly decreasing its presence (Sacchetti *et al.*, 2008, 2017; Bigiotti *et al.*, 2019). Therefore, in some extensive olive orchards treated with copper products, *B. oleae* searching for phylloplane bacteria, fed on the leaves, surface and suck the copper that later acts as a killing agent of the symbiont. Of course, copper being a heavy metal is not so environmentally friendly, but the door is opened

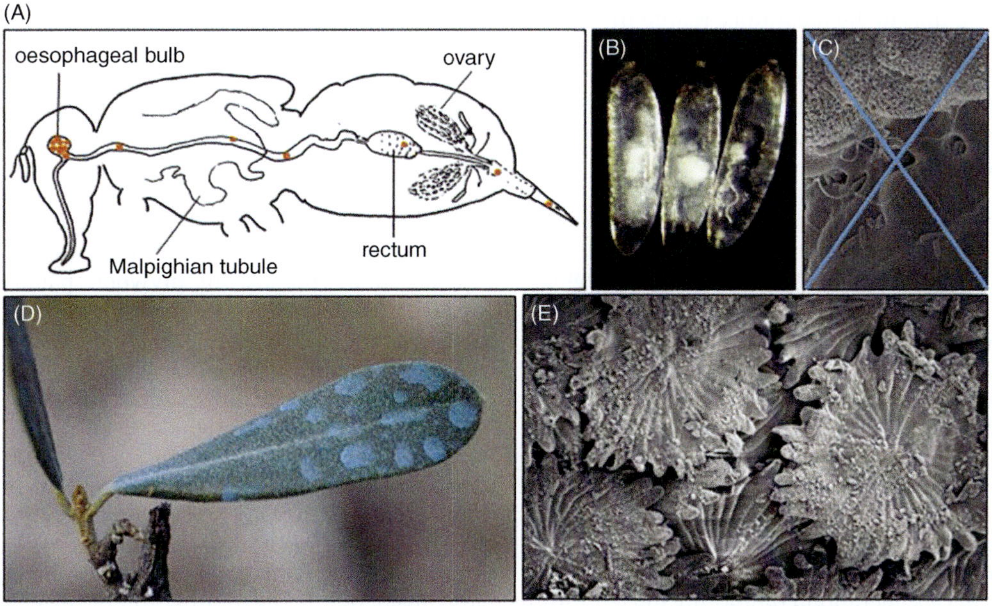

Fig. 7.2.5. Symbiosis disruption in the olive fruit fly. Olive fruit fly feeds on phylloplane treated with copper products or other bactericidal compounds and loses the symbiont. (A) vertical transmission of the symbiont *Ca.* Erwinia dacicola; (B) eggs; (C) micrograph of the aeropylar area with symbiont colonies; (D) olive leaf treated with copper products; (E) Micrograph of trichomes on the abaxial surface of an olive leaf treated with copper products covered with copper residues. (Photos: Patrizia Sacchetti & Antonio Belcari, University of Florence)

and now the research should be addressed towards other substances that could play a symbionticide activity such as propolis (Bigiotti *et al.*, 2019, 2020) or natural compounds with proven bactericide action (Schrader *et al.*, 2010; Khater, 2012; Sinno *et al.*, 2020).

Survival of *Ca.* Erwinia dacicola, like most microorganisms, is strongly affected by substances such as antibiotics, fungicides and preservatives. That is a reason why in small- and in large-scale rearings, the *B. oleae* do not show the symbiont and lose their fitness (Estes *et al.*, 2012; Sacchetti *et al.*, 2019). SIT requires for its application in the field that the insect can be reared in large quantities and the colony should maintain a constant fitness. Indeed, lab-released sterile males should compete with wild males for mating purposes (Estes *et al.*, 2012). It has been demonstrated that a small-scale rearing of the *B. oleae* can be maintained without adding antibiotics in the artificial diet (Dimou *et al.*, 2010); besides, the symbiont *Ca.* E. dacicola can be transferred from the wild flies to the laboratory-reared flies (Bigiotti *et al.*, 2019). This latest discovery paves the way for future improvements of mass-rearing procedures of *B. oleae*, which are mandatory for future SIT programs at area-wide scale.

7.2.1.13 Lure-and-kill and mass trapping

Among different pest-control methods, the lure-and-kill (L&K) and mass trapping (MT) techniques provide the most important ecological benefit since they do not input chemical residues or inert material (like kaolin) into the environment. Only small amounts of semiochemicals are released into the air surrounding the device to modify the behavior of the target insect pests. The L&K technique represents the evolution of MT, although the two terms sometimes are used incorrectly as synonyms: MT consists in trapping the highest number of insects attracted by different kinds of stimuli and lures towards a releasing source and catching them in a device using glue or water. In lure-and-kill, instead of being restrained in a trap, insects get in contact with the device impregnated with contact insecticide and are killed (Jones, 1998). A detailed history of the techniques based on attracting and/or trapping designed for the control of tephritids has been reviewed by Navarro-Llopis and Vacas (2014) giving plentiful accounts for the main achievements on traps (shape and color), attractants and interactions between different stimuli. Concerning *B. oleae*, although the use of trapping as a means to reduce the presence of *B. oleae* adults in olive orchards was known even before synthetic insecticides were issued in the market, pioneering researches on the use of attractants started at the end of the 1970s and were boosted after the identification of the sexual pheromone emitted by *B. oleae* females that allowed the synthetic production of the main pheromone component (olean) (Baker *et al.*, 1980). An intensive research activity flourished, mainly in Greece, promoting laboratory and in-field trials that provided knowledge on olive fly biology, food attractants and trap designs that, together with new findings on pheromone, paved the way for succeeding applications in trapping techniques (Nestel *et al.*, 2016). First field trials relied on the use of yellow sticky traps which soon showed the unfavorable limitation of catching also beneficial insects (Neuenschwander, 1982). Shortly, a more convenient system was set up, based on the use of plywood panels soaked in deltamethrin and baited with two dispensers, one containing the *B. oleae* sex pheromone, attracting males, and another with a food attractant, ammonium bicarbonate, luring females (Haniotakis *et al.*, 1986). The L&K technique devised during the 1980s was applied in different olive ecosystems and habitats, in comparison with other control methods (Broumas *et al.*, 1985). In general, all the field trials proved the main advantage, that is the elimination of insecticide sprays, showing weaknesses as well as effectiveness recorded over years with *B. oleae* in high-density or small treated areas (Haniotakis *et al.*, 1991). The L&K technique was further modified and underwent improvements, resulting in a commercial device with a killing capacity that has been a widely used method in Greece (Broumas *et al.*, 2002; Mazomenos *et al.*, 2002; Ragoussis, 2005). Consequently, the technique was tested and applied also in other Mediterranean countries (Hepdurgun *et al.*, 2009; Tsolakis *et al.*, 2011) with partial success, due mainly to the limited extent of the treated orchards. So far, the only area-wide example of the L&K outside Greece has been in Italy, where this control method was applied for three years on a surface of 1250 ha in comparison with chemical sprays (Petacchi *et al.*, 2003). Besides being able to reduce *B. oleae* infestation on nearly all the treated orchards, the study provides practical advice for the deployment of the device (one device/tree, half of the devices placed at the end of the summer) and the 'suitability threshold index', a measure to predict whether an area is appropriate for L&K, avoiding risk of failure: '...an area is suitable for MT when active infestation does not exceed 30% on the majority of farms within the third week of October of every year' (Petacchi *et al.*, 2003). It is clear that the L&K technique, if applied properly, is a reliable control method, with undeniable environmental advantages and high potential in the long term for pest management (El-Sayed *et al.*, 2009) but it is still inconvenient for farmers in many countries without public financial assistance (Nestel *et al.*, 2016). For this reason, research aimed at designing and evaluating cheaper devices that can be used for L&K (Noce *et al.*, 2009; Yokoyama, 2014) or for MT (Ros *et al.*, 2009; Yasin *et al.*, 2014) have been developed, to make these techniques more affordable and extensively applied. In fact, it is worth underlining that both L&K and MT are tools that can be easily and successfully combined with other sustainable control methods such as augmentative (Liaropoulos *et al.*, 2005; Hepdurgun *et al.*, 2009) and conservative biological control.

7.2.1.14 Area-wide management strategies

This short review on the control methods against *B. oleae* should not lead us to think that one method or the other can be used separately, but that they have to be combined together or used in succession. Any adopted strategy must be carried out in large olive-growing areas, the so-called area-wide control strategy. In the Hawaiian islands, the AWPM (area-wide pest management) of fruit flies of economic importance showed, as a primary goal, the development of an integrated biologically based pest management approach that will result in area-wide suppression and control of fruit flies throughout selected agricultural areas of Hawaii. The IPM program had to integrate two or more technologies into a comprehensive package that is economically viable, environmentally acceptable and sustainable (Vargas *et al.*, 2007). For example, for the *B.oleae* control, at present, we could give some suggestions in planning integrated strategies such as: (i) environment conservation in terms of reduction or elimination of insecticides, herbicides and chemicals that, in general, affect negatively the system; (ii) adult and immature stages monitoring with different kind of traps and checking the infestation on fruits; and (iii) applying different control strategies against the adult population in relation to the environment and on the climatic trend, which is of a primary importance for the development of the olive fly populations (Wang *et al.*, 2009; Johnson *et al.*, 2011). For example, olive growers can apply kaolin (or kaolin + copper) before *B. oleae* starts to oviposit inside the drupes (appropriate time – olive pit hardening) and then check weekly the population trend both via traps and olive samples. If there is infestation growth, before applying any other control methods, it is important to check the weather forecast since the infestation could break down with high temperatures together with low levels of relative humidity. If no growth, we could decide to apply Spinosad bait and check again, as the population should reduce with this kind of treatment. We could also make an attempt to improve the number of traps for mass-trapping purposes or make cover sprays in order to reduce the adult and pre-imaginal population. These examples just underline the importance of using different control strategies. Of course, to achieve the best results, control procedures might be applied on a large scale according to the AWPM philosophy.

7.2.1.15 *B. oleae* control: information and communication technology approach in area-wide pest management

As already mentioned, the concept of area-wide pest management of agricultural pests is based on a set of principles that are somewhat different from those of traditional integrated pest management (IPM). APM has evolved as a component of IPM and is currently viewed as an effective method to manage pests of economic importance using an organized and coordinated attack on pest populations over large areas (multi-field or farm).

APM is most effective when conducted against a single or small groups of pests over large geographical areas that are delineated by biological criteria associated with pest colonization and dispersal potential. Additionally, AWPM should be coordinated by organizations rather than individuals and should focus on reducing and maintaining a pest population below an economically damaging level. Monitoring network systems (Damos, 2015) also in olive growing is based on many fields and automated weather stations spread over a large area. Data from stations are used in forecasting models that seek to offer early warnings of pest population dynamics and infestations. In olive growing APM, the use of a web-based database to manage data from the monitoring network could produce benefits in the communications among different subjects geographically dispersed (Splike and Köstler, 1999). APM data management is closely related to internet-based information delivery systems. Many projects showed the possibility of using the internet to disseminate information for plant protection (Thysen *et al.*, 1999) and researchers agree that computer-aided forecasting and related decision support systems (Tomassone *et al.*, 2006) make pest control more sustainable by avoiding unwanted consequences of pesticide applications (Damos, 2015).

About *B. oleae* control, recently, in Italy, a system was developed (Petacchi *et al.*, 2001) that allows users to insert, share, elaborate and distribute data collected through monitoring networks. The system aims to manage area-wide monitoring networks on olive fruit fly and the specific objectives are: (i) to facilitate the project coordination at regional and farmer-organization level and the communications among partners; (ii) to standardize the service using the same data categories and similar working protocols; (iii) to distribute data and elaboration to the final user; (iv) to provide further tools to support decisions and evaluation in IPM during the growing season; and (v) to create a large dataset for applied ecology studies, to allow analysis and comparison of biological phenomena and the efficacy assessment of IPM strategies.

Web-based decision support systems provide user-interactive content and real-time pest forecasting facilitated by the use of climatic-driven models. For example, in central Italy (Tuscany region), *B. oleae* control by means of web-based decision support systems (http://agroambiente.info.regione.toscana.it/agro18/ – accessed 28/5/23) currently works with the participatory process where different olive-growing actors have been involved and play a role: regional administrators for project coordination and controls both technically and administratively; producer organizations to have real-time information and to perform effective pest control; technicians to produce a user-friendly interface and good data elaboration to support their technical advice; and researchers to obtain good-quality datasets for scientific investigations.

7.2.2 Minor Pests

7.2.2.1 *Prays oleae* (Bernard, 1788)

Biology and population dynamics

Prays oleae (Bernard, 1788) (Lepidoptera: Praydidae), also known as the olive moth, is a lepidopteran pest of economic importance in olive groves in the Mediterranean basin,

Black Sea, Middle East and the Canary Islands (Tzanakakis, 2006) causing quanti-qualitative losses. *Prays oleae* is a monophagous species (Gonzalez *et al.*, 2015) completing three generations per year (Fig. 7.2.6), each one developing on different organs of the olive tree (i.e. leaves, flowers and fruits). Larval trophic activity can cause relevant damage to olive trees, as well as the premature fall of young olives (Tzanakakis, 2006). First-generation larvae, the anthophagous (Ant) ones, consume the flower buds and flowers during the springtime. In the summer, second-generation larvae, carpophagous (Car) ones, attack small fruits (~5 mm in diameter), digging tunnels inside the drupe up to the endocarp, causing early drop. After completing larval development, they emerge from the fruits to pupate on branches, causing a second fruit dropping. Third-generation larvae, phyllophagous (Phy) ones, mine into the leaves where they overwinter and complete the development at the beginning of the spring (Tzanakakis, 2006). Bento *et al.* (2001) estimated losses up to 368 €/ha in the anthophagous generation and 535 €/ha in the carpophagous generation. The reproduction synchronism of *P. oleae* with its host plant is crucial for its survival and fitness (Gonzalez *et al.*, 2015). The adults have crepuscular habits, staying among the vegetation during the day, and are active at temperatures of at least 12°C (Tremblay, 1986; Delrio and Lentini, 2003). Mating takes place immediately after adult emergence. Olfactory cues (i.e. sex pheromones) are essential in this phase. The main component of the female sex pheromone has been identified by Campion *et al.* (1979) as an aldehyde, i.e. (*Z*)-7-tetradecenal, with a wide range of action. The oviposition occurs immediately after mating. A female can lay about 200 eggs over her lifetime, and is able to oviposit for about a month, with the maximum intensity in the first week (Tremblay, 1986; Delrio and Lentini, 2003). The Phy generation begins at the end of the autumn, resulting from the oviposition of the

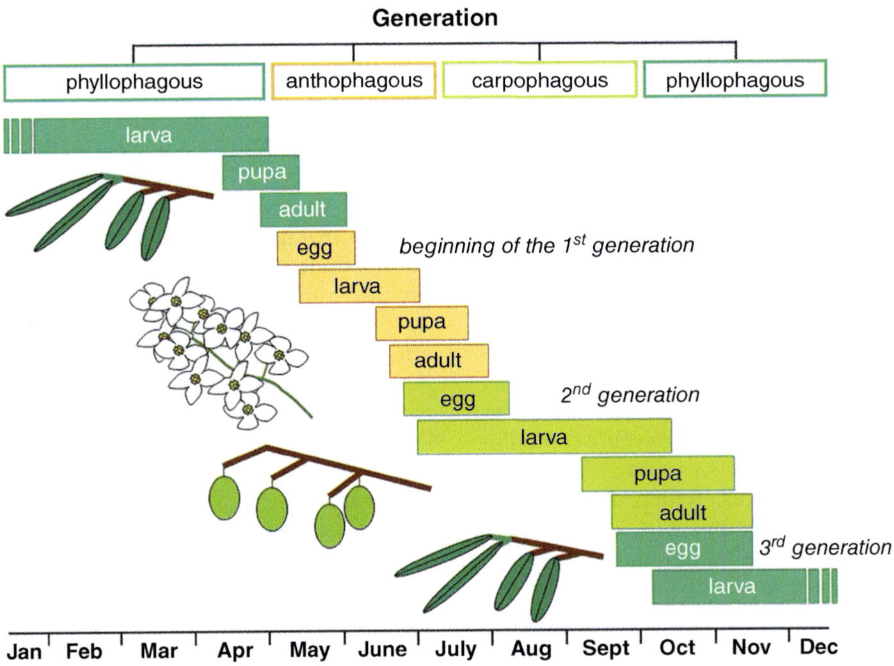

Fig. 7.2.6. Life cycle of the olive moth, *Prays oleae*, showing the three generations developing on flowers, fruits and leaves. (Photo: Patrizia Sacchetti & Antonio Belcari, University of Florence)

Car-generation females. Its development period is the longest, and generally ranges from September to the beginning of April. In the middle of winter, only serpentine mines can be found on the leaves, then larvae dig different-shaped mines after each molt. Larval development ends in the spring (March/April), followed by chrysalis stage. Adults of the Phy generation mate, then females lay eggs, usually one per bud, from April to June (Delrio and Lentini, 2003). The Ant generation larvae are present from May to the end of June, feeding first on buds, then on flowers (Fig. 7.2.7b). Their development is completed in about a month, followed by chrysalis, lasting 1–2 weeks. Adults emerge from the end of May to the first half of July, with the greatest presence in June. Once mating occurs, the females lay their eggs on the calyx of the small drupes. In general, the oviposition is earlier on table olives, later in oil ones; the drupes affected by the oviposition are mostly the size of a peppercorn (Delrio and Lentini, 2003). Typically, with low and moderate infestations, only one egg is laid for each olive. The Ant generation is followed by the Car one. Car larvae penetrate the olive directly through the bottom side of the egg. An early drop of olives occurs during the penetration phase. The penetration of the larvae stops when the endocarp is lignified, which, on average, occurs in the first half of August in central Italy. Rarely, before the hardening of the endocarp, the larva can also attack the growing pulp causing desiccation and summer drop. After lignification, it is possible to detect the attack by opening the core with nippers: the seed appears eroded, with darkened or necrotized areas. Larval development is completed – on table olives – from the second half of August. From this period onwards, the larva exits causing the olives to fall. Damaged olives, which do not detach when the larva comes out, are, however, destined to fall before ripening due to the low resistance at the insertion of the peduncle on the drupe (Fig. 7.2.7c). The pupation, in the ground or on the tree, lasts about two weeks (Tremblay, 1986; Delrio and Lentini, 2003). Every year, damage can vary depending on climatic conditions. Eggs and newly hatched larvae are sensitive to low relative humidity (R.H.) and high temperatures. Eggs can dry out with an R.H. lower than 60%, while the upper temperature threshold is 30°C for the larvae. In addition, olive moth population dynamics are correlated to host-plant characteristics and the occurrence of natural antagonists (Tzanakakis, 2006).

Management of P. oleae

The main tools to control this moth rely on IPM and biocontrol. Usually, *Prays oleae* attacks are not so severe as to justify conventional chemical control; indeed, all interventions are usually aimed to control the Car generation, leaving Phy and Ant generations to natural control dynamics. Given the mechanism of this infestation, the active ingredients must be endotherapeutic (systemic or cytotropic); to penetrate the drupe and reach the larva, organophosphates can be adopted (Lopez-Villalta, 1999). Chemical treatments increase potential risk of adverse ecological, toxicological and economic effects (Youssef *et al.*, 2004). The application of *Bacillus thuringiensis* serovar. *kurstaki* is a possible alternative, but it is effective only on external feeders (Phy generation). Car generation larvae cannot be controlled through this method (Herz and Hassan, 2006). Biotechnological methods (e.g. mating disruption (MD)) have been used to monitor and control this pest for decades. Pest control using sex pheromone-mediated mating disruption has proved successful for several moth pests (Murray and Alston, 2020; Lucchi *et al.*, 2018). MD has no adverse effects on non-target organisms, reducing mating rates by dispersing (*Z*)-7-tetradecenal throughout the cropping system (Campion *et al.*, 1979). Recently, effective MD formulation has been developed. For example, Mazomenos *et al.* (1999) tested two different formulations of the *P. oleae* sex pheromone, comparing MD to plots treated with *B. thuringiensis*, synthetic insecticides or untreated. Two different doses/ha of the sex pheromone (i.e. 40 g/ha and 50 g/ha) achieve a better control over other control programs, offering an effective and eco-friendly tool to manage *P. oleae*. Kikionis *et al.* (2017) make

Fig. 7.2.7. *Prays oleae*. (A) adult; (B) olive blossom damaged by an antophagous larva; (C) olives with characteristic exit holes produced by fully grown larvae; (D) fully grown larva just escaped from an olive fruit. (Photos: Patrizia Sacchetti & Antonio Belcari, University of Florence)

a further step towards an eco-friendly tool using micro-/nanofibers as a biodegradable slow-release system for the *P. oleae* pheromone. The latter (5%, 10% and 20% w/w) has been incorporated in electrospinning polymer blends, allowing the production of fiber mats with variable loading levels and release rates. In all cases the release of pheromones has been ensured for more than 16 weeks. Furthermore, (*Z*)-7-tetradecenal has been used to monitor the populations of this moth; earlier, Ramos *et al.* (1989) proposed pheromone-baited traps to monitor *P. oleae* populations, allowing an effective timing of control measures.

IPM programs to manage *P. oleae* also rely on the adoption of biological control agents to maintain the pest under the economic threshold level. Bento (1999) studied the effectiveness of biocontrol using chrysopids, predators of eggs and larval stages of *P. oleae* in Portuguese olive groves. In natural settings, the most active predator on eggs and larvae of *P. oleae* is *C. carnea*. However, to increase its efficiency as a biocontrol agent, higher amounts of *C. carnea* larvae per tree should be released. A significant reduction of the pest population has been obtained after releasing 90 larvae per tree, while the *P. oleae* population has been almost halved by releasing 360 larvae of *C. carnea* per tree (Bento, 1999).

In research programs for the biological control of *P. oleae*, egg parasitoids of the genus *Trichogramma* are also proved to be effective (Bento, 1999). Hegazi *et al.* (2007a) highlighted that indigenous *Trichogramma* species (i.e. *T. bourarachae*, *T. cordubensis* and *T. euproctidis*) accomplished higher egg parasitism if compared to a common commercially available species (*T. evanescens*) in Egyptian olive groves. In this study, the releases of *T. evanescens* resulted in moderate parasitism ranging from 0.2% to 54.2%, while the parasitization rates of indigenous parasitoids have been found higher, up to 91% of eggs parasitized by *T. bourarachae*, especially later in the season (Hegazi *et al.*, 2007a). Lower parasitism rates of *T. evanescens* are a possible consequence of poor adaptation to particular environmental conditions of olive groves (Hegazi *et al.*, 2007a). In addition, in olive groves ecosystems, a rich complex of parasitoids naturally occurs in association with *P. oleae*, leading to high mortality rates and highlighting the importance of these beneficial natural enemies in the control of this pest. A study conducted by Nave *et al.*, (2017) identified 22 parasitoid species in Portuguese olive groves. Among them *Ageniaspis fuscicollis* (Dalman) (Hymenoptera: Encyrtidae), *Chelonus elaeaphilus* Silvestri (Hymenoptera: Braconidae), and *Elasmus flabellatus* (Fonscolombe) (Hymenoptera: Elasmidae) exhibit high parasitization rates, 78.7% against Ant, 50% against Car, and 4–30% versus Ant and Phy generations, respectively. The richness of parasitoids is higher in the Ant generation, corresponding to a generation with greater biodiversity in the sampled community, and lower in the Car generation (Nave *et al.*, 2017).

In conclusion, IPM programs of *P. oleae* might rely on the adoption of sex pheromone-mediated MD, which significantly reduces mating rates. In addition, the parasitoid and predator complex already present in the olive groves is sufficient to control this pest. To maintain or to increase the biodiversity of the natural enemy's complex in the olive orchard, it is mandatory to reduce chemical sprays with synthetic insecticides and to create a suitable ecological infrastructure to provide resources, with special reference to essential food sources, boosting the fitness of biocontrol agents (Benelli *et al.*, 2017). Besides, to improve the effectiveness of some predators, such as *C. carnea*, augmentative biological control strategies may be considered in the case of a too-low density of the natural enemy to control the pest effectively.

7.2.2.2 *Dasineura oleae* (Angelini, 1831)

Biology and population dynamics

Dasineura oleae (Angelini, 1831) is a gall midge bound to olive leaves with a Mediterranean distribution. This species is autochthonous in Italy, first described in Verona in 1831.

It has always been considered as a minor pest in olive groves (Giraldi, 1952; Darvas *et al.*, 2000) but in recent years, several outbreaks have been reported in many countries: from Turkey, Syria, Greece, Montenegro and Palestinian territories (Doğanlar, 2011; Simoglou *et al.*, 2012; Baidaq *et al.*, 2015; Batta, 2019) to Italy in 2012, starting from olive fields near to Lake Garda and in Emilia Romagna, near Rimini (Boselli and Bariselli, 2015). In Tuscany (central Italy), the infestation traced out a patchy distribution in the southern part of the region in the entire province of Grosseto in 2014 (Picchi *et al.*, 2017b; Petacchi *et al.*, 2019; Tondini and Petacchi, 2019). This area began to bear a severe infestation from 2016 when a regional monitoring network was established and is still ongoing.

The effect on plants consists of galls on leaves, stimulated by the larvae feeding on the leaves' mesophyll causing the formation of superficial unilocular extended bulges of 3–5 mm in length (Barnes, 1948).

Usually, the gall density leads to negligible damage, however high density of galls may affect photosynthesis (Huang *et al.*, 2014) and nutrient transport (Motta *et al.*, 2005), which could cause defoliation (Skuhravy *et al.*, 1980). Yield losses could be linked to pest damage (Simoglou *et al.*, 2012) but significant experimental data are still not available except Caselli *et al.* (2020), which suggested that the pest may trigger on-leaf morphology and some physiological activities of olive leaves related to photosynthesis performance.

The life cycle, from first instar larvae to pupae (Fig. 7.2.8), takes place inside the galls (Arambourg, 1986). The female, easily recognizable by a light-red abdomen, lays various eggs in clusters on the surface of young leaves, more attractive and easier to

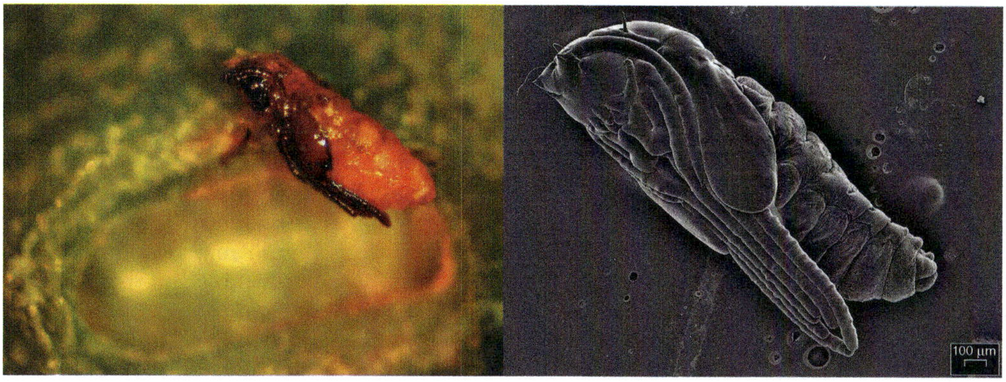

Fig. 7.2.8. *Dasineura oleae* pupal stage: (*left*) removed by gall sectioned; (*right*) LTSEM micrograph. (Photos: BioLabs, Sant'Anna School of Pisa)

digest, which usually contain enough water and nutrient concentrations (Cates, 1980).

The first instar larva feed on the leaves' mesophyll fostering the reaction of the plant. Inside the gall, the development of the pest takes place until the flicker of the adult that, again, searches for a mate to reproduce. The number of annual generations depends on abiotic factors: usually the pest develops only one generation each year, but in suitable climatic conditions, two generations could be observed (Doğanlar *et al.*, 2011; Baidaq *et al.*, 2015) or in coastal areas (Batta, 2019), as observed in Tuscany by Tondini and Petacchi (2019). In fact, it was observed that the first generation was in spring, when new foliage of olive was available, and the second generation was observed in November, during the winter vegetative phase and no diapause larvae were noticed.

Management of D. oleae

Considering that *Dasineura oleae* has always been considered a minor pest and not much is known about its biology, no proper management strategy has been attempted or applied. However, the recent area-wide outbreaks and variability, at the farm scale, of the percentage of infestation and parasitization suggests management based on knowledge of these data. In this regard, it is necessary to increase knowledge about the pest, and one useful approach to deal with it is surveillance through a monitoring network that provides data about the infestation level. The understanding of the population dynamics allows providing guidelines to farmers to help them make informed management decisions, as planned by the IPM strategy (Barzman *et al.*, 2015).

In spring, together with the infestation level, it is feasible to evaluate the parasitism rate, thus enabling the estimation of the effect of the main natural enemies of *D. oleae* and to properly adjust the strategy in the olive field.

Olive field sampling points should be representative of the area and, in the Tuscan case study (Picchi *et al.*, 2020 unpublished results), we started from reports and land use, defining 30 olive fields in which samples were picked. The monitoring network was established considering reports from technicians of extension services and consortia and considering the extension of the area dedicated to olive orchards through the 'Utilised Agricultural Area' (UAA) value for each municipality.

Inside galls, *Dasineura oleae* is well protected from chemicals (Lyrene and Payne, 1996) but they are liable to parasitoids that can reach the pest inside the gall. Parasitism could reduce infestation and could be a key factor in stabilizing the pest population and reducing the risk or, sometimes, even stopping the outbreaks (Rhodes *et al.*, 2014).

Considering the above, the management of *D. oleae* is necessarily based on CBC.

Platygastridae wasps are effective against gall midges and in Tondini and Petacchi (2019), there were found four of 16 species of five families (Platygastridae, Pteromalidae, Eulophidae, Eupelmidae and Torymidae) reported for *D. oleae*. Among these species, *Platygaster oleae* Szelenyi, *Platygaster demades* Walker (Platygastridae), *Eupelmus urozonus* Dalman (Eupelmidae), *Quadrastichus dasineurae* Doğanlar, LaSalle, and Sertkaya et Doğanlar (Eulophidae) have been suggested to be the most effective (Doğanlar *et al.*, 2009; Doğanlar, 2011 Baidaq *et al.*, 2015).

Tondini and Petacchi (2019) found koinobiont endoparasitoids, three generalists – oligophagous or polyphagous – and one, *Platygaster oleae* Szelenyi 1940, specific for *D. oleae*. They all are koinobiont endoparasitoids; *Mesopolobus aspilus* (Walker, 1835) and *Mesopolobus mediterraneus* (Mayr, 1903) of Chalcidoidae family and *Platygaster demades* Walker 1835 – together with *P. oleae* – belong to Platygasteridae.

Indeed, *M. mediterraneus* and *M. aspilus* are polyphagous parasitoids and hyperparasitoids that attack several species belonging to Coleoptera (Vidal, 1997) and Diptera: including numerous Cecidomyiidae (Graham, 1969; Askew *et al.*, 2001; Doğanlar *et al.*, 2011). Moreover, it has been reported that such parasitoids attack some species of Hemiptera (Öncüer, 1991), Hymenoptera (Askew *et al.*, 2001; Gomez *et al.*, 2006) and Lepidoptera (Viggiani, 1967).

These hosts are associated with plants such as *Juniperus* spp. (Askew *et al.*, 2001), *Quercus* spp. (Gomez *et al.*, 2006; Xiao *et al.*, 2016), *Pinus* spp. (Askew *et al.*, 2001), *Cytisus scoparius* and other species (Vidal, 1997). *P. demades* is an oligophagous parasitoid that attacks *D. oleae*, *Dasineura pyri* (Bouché, 1847), *Dasineura mali* (Kieffer, 1904), and *Wachtliella ericina* (Low, 1885), which are respectively associated with *O. europaea*, *Malus domestica* Borkh (1803), *Pyrus communis* subsp. *pyraster* (L.) Ehrh. and *Erica* spp.

The food network established among parasitoids, phytophagous pests, and the corresponding host plants define a complex framework that could depend on the nearby landscape. The configuration and/or the composition of the habitats can enhance predators and parasitoids and, thus, can lead to higher rates of pest suppression (Bengtsson *et al.*, 2005; Bianchi *et al.*, 2006; Gallé *et al.*, 2019; Haan *et al.*, 2020).

Usually, a high proportion of semi-natural habitat benefits natural enemies (Chaplin-Kramer and Kremen, 2012, but see Hawro *et al.*, 2015), providing natural enemies with resources such as alternative hosts, that depend on host plants' composition, and/or overwintering refugia. The abundance of *Erica* sp., the host plants of alternative prey of *P. demades*, seems the main variable to increase parasitism against *D. oleae* in southern Tuscany. Where this species is abundant in adjacent habitats of the crop, the parasitism rate in the olive field increases (Picchi *et al.*, 2020, unpublished results).

The adult spends its life around the canopy of olive, looking for young leaves and mates. To our knowledge, no experimental approaches were made to estimate the effect of predators on this fly, but spiders, abundant in olive canopies (Morris and Campos, 2000; Picchi *et al.*, 2016) could be an interesting group, able to catch flies by ambush and through webs.

7.2.2.3 *Palpita vitrealis* (Rossi, 1794)

Biology and population dynamics

Palpita vitrealis (Rossi, 1784) (Lepidoptera: Pyralidae), also known as jasmine moth, is currently considered a minor pest damaging olive agroecosystems. The jasmine moth is diffused in the Mediterranean region, Asia Minor and North Africa. *O. europea* is not the only host of *P. vitrealis*, the larvae also attack other species belonging to Oleaceae (e.g. *Ligustrum vulgare*, *Jasminum officinale*, *O. europaea*, *Phillyrea media*, *Fraxinus* spp.) and *Arbutus unedo* (Ericacae) (Hegazi *et al.*, 2007b). *Palpita vitrealis* is a multi-voltine species with several overlapping generations per year (Hegazi *et al.*, 2007b). In Italy, four to five generations can occur, while the moth completes ten generations per year in Egypt (Badawi *et al.*, 1976).

Shehata *et al.* (2003) outlined that *P. vitrealis* moth has crepuscular habits and mating can last from one to three hours. Usually, eggs are laid singly, in series or in masses on the lower surface of the leaves; the incubation period is temperature-dependent. Over 35°C, all eggs fail to hatch (Fig. 7.2.9), while the shortest incubation period (three days) is obtained at 30°C, and the longest (112 days) occurs at 15°C. The effect of R.H. seems negligible, unlike temperature (Badawi *et al.*, 1976). Newly hatched larvae feed on the under-leaf part of young tender leaves, while older ones prefer parenchyma. Infested flower buds usually fall before fruit setting. Fruits may also be attacked, and larvae often make irregular holes everywhere in the skin, sometimes leading to fruit fall (Shehata *et al.*, 2003). Larval feeding often occurs in shaded areas, away from direct sunshine or light. During immature stages (six instars), the larva constructs a shelter, joining several leaves or parts of it together with silk threads, before pupating into a brown cocoon (Shehata *et al.*, 2003). Immature stages are shorter as temperature rises, but the development fails at 35°C (Badawi *et al.*, 1976). The overall duration of the last instar (sixth) is always the longest, lasting about twice as long if compared to the duration of the first instar. These stages are also slightly influenced by R.H. The pupation takes place in a silken cocoon; its duration varies according to temperature and R.H., increasing as these two factors rise. Overall, female pupation is shorter than the male one (Badawi *et al.*, 1976). When this stage is completed the moth is ready to emerge (Fig. 7.2.10). All pupae die at 35°C, the optimum for adult emergence is between 25°C and 30°C.

Fig. 7.2.9. *Palpita vitrealis*: larva emerging from the egg. (Photo: BioLabs, Sant'Anna School of Pisa)

Management of P. vitrealis

The adoption of chemical insecticides to control *P. vitrealis* populations is not recommended, since this pest rarely causes serious yield losses. In addition, some chemical treatments (i.e. organophosphates and deltamethrins), useful to manage *B. oleae*, are effective also against *P. unionalis*. Arambourg (1986) recommended the use of *B. thuringiensis* serovar. *kurstaki* to control juvenile stages of *P. vitrealis* larvae, but treatments must be performed early, because late larval stages are less sensitive to *B. thuringensis*.

P. vitrealis complex of naturally occurred antagonists include species belonging to *Apanteles* (Hymenoptera: Braconidae), *Xanthopimpla* (Hymenoptera: Ichneumonidae), *Brachymera* (Diptera: Tachinidae), *Eupeodes* (Diptera: Syrphidae) genera (Nasr *et al.*, 2002; Triggiani, 1971). *Apanteles xanthostigma* is a larval parasitoid of *P. vitrealis*, considered one of the most effective biocontrol agents against this pest (Triggiani, 1971). Among predators, the larvae of *Eupoedes corollae* (Fabricius) (Diptera: Syrphidae) are highly effective against *P. vitrealis*

Fig. 7.2.10. Palpita *vitrealis*: adult female at rest on olive leaf. The wings, with their ochreous-brown color, make it unmistakable. (Photo: BioLabs, Sant'Anna School of Pisa)

larval stages. Further studies are needed to better understand the antagonist complex of *P. vitrealis*, since fragmented data are currently reported.

Another possible way to manage (and monitor) *P. vitrealis* is the use of its sex pheromone. The female sex pheromone of *P. vitrealis* mainly consists of two components, (E)-11-hexadecenal (E11-16: Ald) and (*E*)-11-hexadecenyl acetate (E11-16: Ac) (Mazomenos *et al.*, 1994). An effective way to monitor the moth population is the combination of the binary pheromone blend with the funnel trap (Athanassiou *et al.*, 2004). Also, a study conducted by Hegazi *et al.* (2007b) demonstrates a continuous decrease in egg densities because of MD carried out with the above-mentioned pheromone components against this pest (i.e. E11-16: Ac: E11-16: Ald at the ratio 70:30). In these experiments, the active pheromone dose applied was 80 g/ha, and two to three dispensers/tree were placed at head height in the canopy of olive trees.

However, a major issue in MD of *P. vitrealis* is the immigration of gravid females from the surrounding into MD areas (Mazomenos *et al.*, 1994): this can be avoided by integrating MD with biocontrol methods focussing on eggs and first instar predation or parasitization (Hegazi *et al.*, 2007b).

Further studies should focus on the pheromone delivery methods for MD of multiple pest species (e.g. a dispenser able to release both *P. oleae* and *P. vitrealis* sex pheromones). In addition, little is known about the antagonist complex of the jasmine moth; further studies should investigate antagonist biodiversity correlated to *P. vitrealis* to properly develop biocontrol strategies.

References

Alberola, T.M., Aptosoglou, S., Arsenakis, M., Bel, Y., Delrio, G. *et al.* (1999) Insecticidal activity of strains of *Bacillus thuringiensis* on larvae and adults of *Bactrocera oleae* Gmelin (Dipt. Tephritidae). *Journal of Invertebrate Pathology* 74, 127–136.

Albertini, A., Santos, S.A.P., Martins, F., Pereira, J.A., Lino-Neto, T. *et al.* (2018a) Detection of *Bactrocera oleae* (Diptera: Tephritidae) DNA in the gut of the soil species *Pseudoophonus rufipes* (Coleoptera, Carabidae). *Spanish Journal of Agricultural Research* 16(3), 1–9.

Albertini, A., Marchi, S., Ratti, C., Burgio, G., Petacchi, R. *et al.* (2018b) *Bactrocera oleae* pupae predation by *Ocypus olens* detected by molecular gut content analysis. *BioControl* 63, 227–239.

Alissandrakis, E., Psirofonia, P., Kavallieratos, N.G., Stanković, S.S. and Žikić, V. (2018) Parasitic wasps related to *Prays oleae* (Bernard, 1788) (Lepidoptera, Praydidae) in olive orchards in Greece. *ZooKeys* 773, 143–154.

Arambourg, Y. (1986) *Entomologie oleicole*. Conseil Oleicole International, Madrid.

Askew, R.R., Blasco-Zumeta, J. and Pujade-Villar, J. (2001) Chalcidoidea and Mymarommatoidea (Hymenoptera) of a *Juniperus thurifera* L. forest of Los Monegros region, Zaragosa. *Monografias Sociedad Entomológica Aragonesa* 4, 46.

Athanassiou, C.G., Kavallieratos, N.K. and Mazomenos, B.E. (2004) Effect of trap type, trap color, trapping location, and pheromone dispenser on captures of male *Palpita unionalis* (Lepidoptera, Pyralidae). *Journal of Economic Entomology* 97(2), 321–329.

Badawi, A., Awadallah, A.M. and Foda, S.M. (1976) On the biology of the olive leaf moth *Palpita unionalis* Hb. (Lep., Pyralidae). *Zeitschrift für Angewandte Entomologie* 80, 103–110.

Baidaq, Z.M., Ramadhane, A.M. and Tara, R.A. (2015) Biological synchronization of the endo-parasitoid *Platygaster demades* Walker (Hymenoptera, Platygasteridae) with its host the olive leaf midge *Dasineura oleae* F. Loew (Diptera: Cecidomyiidae). *SSRG International Journal of Agriculture & Environmental Science* 2(3), 1–8.

Baker, R., Herbert, R.H, Howse, P.E., Jones, O.T., Francke, W. *et al.* (1980) Identification and synthesis of the major sex pheromone of the olive fruit fly, *Dacus oleae*. *Journal of Chemical Society, Chemical Communications* 1, 52–54.

Barnes, H.F. (1948) *Gall Midges of Economic Importance. Vol. III. Gall Midges of Fruit.* Wiley.

Barzman, M., Bàrberi, P., Birch, A.N.E, Boonekamp, P., Dachbrodt, S. *et al.* (2015) Eight principles of integrated pest management. *Agronomy for Sustainable Development* 35(4), 1199–1215.

Batta, Y.A. (2019) New findings on infestation and phenology of *Dasineura oleae* Angelini (Diptera: Cecidomyiidae), an emerging pest on olive trees in the Palestinian Territories. *Journal of Plant Diseases and Protection* 126(1), 55–66.
Belcari, A., Sacchetti, P., Marchi, G. and Surico, G. (2003) La mosca delle olive e la simbiosi batterica. *Informatore Fitopatologico* 9, 55–59.
Benelli, G. (2014) Aggressive behavior and territoriality in the olive fruit fly, *Bactrocera oleae* (Rossi) (Diptera: Tephritidae), role of residence and time of day. *Journal of Insect Behavior* 27, 145–161.
Benelli, G. (2015) Aggression in Tephritidae flies: where, when, why? Future directions for research in integrated pest management. *Insects* 6(1), 38–53.
Benelli, G., Canale, A., Bonsignori, G., Ragni, G., Stefanini, C. *et al.* (2012) Male wing vibration in the mating behavior of the olive fruit fly *Bactrocera oleae* (Rossi) (Diptera: Tephritidae). *Journal of Insect Behavior* 25, 590–603.
Benelli, G., Bonsignori, G., Stefanini, C., Raspi, A. and Canale, A. (2013) The production of female sex pheromone in *Bactrocera oleae* (Rossi) young males does not influence their mating chances. *Entomological Science* 16, 47–53.
Benelli, G., Daane, K., Canale, A., Niu, C.-Y., Messing, R. *et al.* (2014) Sexual communication and related behaviours in Tephritidae: current knowledge and potential applications for Integrated Pest Management. *Journal of Pest Science* 87, 385–405.
Benelli, G., Desneux, N., Romano, D., Conte, G., Messing, R.H. *et al.* (2015) Contest experience enhances aggressive behaviour in a fly: when losers learn to win. *Scientific Reports* 5, 9347.
Benelli, G., Giunti, G., Tena, A., Desneux, N., Caselli, A. *et al.* (2017) The impact of adult diet on parasitoid reproductive performance. *Journal of Pest Science*, 90(3), 807–823.
Bengochea, P., Amor, F., Saelices, R., Hernando, S., Budia, F. *et al.* (2013) Kaolin and copper-based products applications: ecotoxicology on four natural enemies. *Chemosphere* 91, 1189–1195.
Bengochea, P., Budia, F., Viñuela, E. and Medina, P. (2014a) Are kaolin and copper treatments safe to the olive fruit fly parasitoid *Psyttalia concolor*? *Journal of Pest Science* 87, 351–359.
Bengochea, P., Saelices, R., Amor, F., Adán, Á., Budia, F. *et al.* (2014b) Non-target effects of kaolin and coppers applied on olive trees for the predatory lacewing *Chrysoperla carnea*. *Biocontrol Science and Technology* 24, 625–640.
Bengtsson, J., Ahnström, J. and Weibull, A.C. (2005) The effects of organic agriculture on biodiversity and abundance: a meta-analysis. *Journal of Applied Ecology* 42(2), 261–269.
Benhadi-Marín, J., Pereira, J.A., Bento, A., Sousa, P. and Santos, S.A.P. (2017) Biodiversity of spiders in agroecosystems: from community structure to conservation biological control of pest. In: Santos, S.A.P. (ed.) *Natural Enemies: Identification, Protection Strategies and Ecological Impacts*. Nova Science Publishers, New York, pp. 43–109.
Benhadi-Marín, J., Pereira, J.A., Barrientos, J.A., Sousa, J.P. and Santos, S.A.P. (2018) Stones on the ground in olive groves promote the presence of spiders (Araneae). *European Journal of Entomology* 115, 372–379.
Bento, A. (1999) Contribuição para o estabelecimento de um programa de protecção integrada contra a traça-da-oliveira, *Prays oleae* (Bern.) em Trás-os-Montes. *Tese de Doutoramento em Engenharia Agrícola.* UTAD, Vila Real, Portugal.
Bento, A., Torres, L.M. and Lopes, J. (2001) Avaliação de prejuízos causados pela traça da oliveira, *Prays oleae* (Bern.) em Trás-os-Montes. *Revista de Ciências Agrárias* 24, 89–96.
Benuzzi, M., Albonetti, E., Fiorentini, F. and Ladurner, E. (2007) A *Beauveria bassiana*-based bioinsecticide for the microbial control of the olive fly (*Bactrocera oleae*). *IOBC WPRS Bulletin* 30(9), 125.
Ben-Yosef, M., Aharon, Y., Jurkevitch, E. and Yuval, B. (2010) Give us the tools and we will do the job: symbiotic bacteria affect olive fly fitness in a diet-dependent fashion. *Proceedings of the Royal Society B, Biological Sciences* 277, 1545–1552.
Ben-Yosef, M., Pasternak, Z., Jurkevitch, E. and Yuval, B. (2014) Symbiotic bacteria enable olive flies (*Bactrocera oleae*) to exploit intractable sources of nitrogen. *Journal of Evolutionary Biology* 27, 2695–2705.
Ben-Yosef, M., Pasternak, Z., Jurkevitch, E. and Yuval, B. (2015) Symbiotic bacteria enable olive fly larvae to overcome host defences. *Royal Society Open Science* 2, 150170.
Bertolini, E., Kistenpfennig, C., Menegazzi, P., Keller, A., Koukidou, M. *et al.* (2018) The characterization of the circadian clock in the olive fly *Bactrocera oleae* (Diptera: Tephritidae) reveals a *Drosophila*-like organization. *Scientific Reports* 8(1), 816.
Bianchi, F.J.J.A., Booij, C.J.H. and Tscharntke, T. (2006) Sustainable pest regulation in agricultural landscapes: a review on landscape composition, biodiversity and natural pest control. *Proceedings of the Royal Society B, Biological Sciences* 273(1595), 1715–1727.

Bigiotti, G., Pastorelli, R., Belcari, A. and Sacchetti, P. (2019) Symbiosis interruption in the olive fly: effect of copper and propolis on *Candidatus* Erwinia dacicola. *Journal of Applied Entomology* 143, 357–364.
Bigiotti, G., Sacchetti, P., Pastorelli, R., Lauzon, C.R. and Belcari, A. (2020) Bacterial symbiosis in *Bactrocera oleae*: an Achilles' heel for its pest control. *Insect Science* 28(4), 874–884.
Boccaccio, L. and Petacchi, R. (2009) Landscape effects on the complex of *Bactrocera oleae* parasitoids and implications for conservation biological control. *BioControl* 54(5), 607–616.
Boselli, M. and Bariselli, M. (2015) Il pericolo corre sulle foglie degli ulivi. *Avversità* 60–62.
Broumas, T., Liaropoulos, C., Yamvrias, C., Haniotakis, G., Cavallaro, R. and Crovetti, A. (1985) Experiments on the control of the olive fruit fly by mass trapping. In: *Proceedings of the CEC/FAO/IOBC International Joint Meeting, Pisa, 3–6 April 1984*. A.A. Balkema, Rotterdam, The Netherlands, pp. 411–419.
Broumas, T., Haniotakis, G., Liaropoulos, C., Tomazou, T. and Ragoussis, N. (2002) The efficacy of an improved form of the mass-trapping method, for the control of the olive fruit fly, *Bactrocera oleae* (Gmelin) (Dipt., Tephritidae): pilot-scale feasibility studies. *Journal of Applied Entomology* 126(5), 217–223.
Caleca, V. and Rizzo, R. (2007) Tests on the effectiveness of kaolin and copper hydroxide in the control of *Bactrocera oleae* (Gmelin). *IOBC/WPRS Bulletin* 30, 111–117.
Caleca, V., Lo Verde, G., Lo Verde, V., Palumbo Piccionello M. *et al.* (2010) Control of *Bactrocera oleae* and *Ceratitis capitata* in organic orchards: use of clays and copper products. *Acta Horticulturae* 873, 227–233.
Caleca, V., Giacalone, C., Maltese, M. and Tortorici, F. (2017) Contenimento naturale di Bactrocera oleae (Rossi): clima o parassitoidi? Confronto tra Western Cape (SUD AFRICA) e Sicilia. *Atti Accademia Nazionale Italiana di Entomologia* LXIV, 99–105.
Calvitti, M., Antonelli, M., Moretti, R. and Bautista, R.C. (2002) Oviposition response and development of the egg-pupal parasitoid *Fopius arisanus* on *Bactrocera oleae*, a tephritid fruit fly pest of olive in the Mediterranean basin. *Entomologia Experimentalis et Applicata* 102, 65–73.
Campion, D.G., McVeigh, L.J., Polyrakis, J., Michaelakis, S., Stravrakis, G.N. *et al.* (1979) Laboratory and field studies of the female sex pheromone of the olive moth, *Prays oleae*. *Experientia* 35, 1146–1147.
Canale, A. and Benelli, G. (2012) Impact of mass-rearing on the host-seeking behavior and parasitism by the fruit fly parasitoid *Psyttalia concolor* (Szèpligeti) (Hymenoptera, Braconidae). *Journal of Pest Science* 85, 65–74.
Canale, A. and Loni, A. (2006) Host location and acceptance in *Psyttalia concolor*: role of host instar. *Bulletin of Insectology* 59(1), 7–10.
Canale, A., Carpita, A., Conti, B., Canovai, R. and Raspi, A. (2012) Effect of age on 1,7-dioxaspiro-[5.5]-undecane production in both sexes of olive fruit fly, *Bactrocera oleae* (Rossi) (Diptera: Tephritidae). *IOBC/WPRS Bulletin* 72, 219–225.
Canale, A., Germinara, S.G., Carpita, A., Benelli, G., Bonsignori, G. *et al.* (2013) Behavioural and electrophysiological responses of the olive fruit fly, *Bactrocera oleae* (Rossi) (Diptera: Tephritidae) to male- and female-borne sex attractants. *Chemoecology* 23, 155–164.
Canale, A., Benelli, G., Germinara, G.S., Fusini, G., Romano, D. *et al.* (2015) Behavioural and electrophysiological responses to overlooked female pheromone components in the olive fruit fly, *Bactrocera oleae* (Diptera: Tephritidae). *Chemoecology* 25, 147–157.
Capuzzo, C., Firrao, G., Mazzon, L., Squartini, A. and Girolami, V. (2005) '*Candidatus* Erwinia dacicola', a coevolved symbiotic bacterium of the olive fly *Bactrocera oleae* (Gmelin). *International Journal of Systematic and Evolutionary Microbiology* 55, 1641–1647.
Cárdenas, M., Pascual, F., Campos, M. and Peká, R.S. (2015) The spider assemblage of olive groves under three management systems. *Environmental Entomology* 44(3), 509–518.
Carpita, A., Canale, A., Raffaelli, A., Saba, A., Benelli, G. *et al.* (2012) (*Z*)-9-tricosene identified in rectal gland extracts of *Bactrocera oleae* males: first evidence of a male-produced female attractant in olive fruit fly. *Naturwissenschaften* 99, 77–81.
Carriere, Y., Ellsworth, P.C., Dutilleul, P., Ellers-Kirk, C., Barkley, V. *et al.* (2006) A GIS-based approach for areawide pest management: the scales of *Lygus hesperus* movements to cotton from alfalfa, weeds, and cotton. *Entomologia Experimentalis et Applicata* 118, 203–210.
Caselli, A, Francini, A., Minnocci, A. and Petacchi R. (2020) *Dasineura oleae*: morphological and physiological characterization following the midge attack on olive leaves. *Journal of Plant Disease and Protection* 128, 173–182.
Castrignanò, A., Boccaccio, L., Cohen, Y., Nestel, D., Kounatidis, I. *et al.* (2012) Spatio-temporal population dynamics and area-wide delineation of *Bactrocera oleae* monitoring zones using multi-variate geostatistics. *Precision Agriculture* 13, 421–441.
Cates, R.G. (1980) Feeding patterns of monophagous, oligophagous, and polyphagous insect herbivores: the effect of resource abundance and plant chemistry. *Oecologia* 46(1), 22–31.

Cavalloro, R. and Delrio, G. (1975) Osservazioni sulla distribuzione e sopravvivenza delle pupe di *Dacus oleae* Gmelin nel terreno. *Redia* 56, 167–176.

Chaplin-Kramer, R. and Kremen, C. (2012) Pest control experiments show benefits of complexity at landscape and local scales. *Ecological Applications* 22(7), 1936–1948.

Chesi, F. and Quaglia, F. (1982) Ricerche sulle metodologie di campionamento per la valutazione dell'infestazione dacica. Confronto delle varianze in un campione ampio ed in uno ridotto. Studi preliminari in due anni di sperimentazione condotti ad Asciano Pisa (1980–1981). *Frustula Entomologica*, Pisa, n.V(XVIII), 111–116.

Daane, K.M. and Johnson, M.W. (2010) Olive fruit fly: managing an ancient pest in modern times. *Annual Review of Entomology* 55, 151–169.

Damos, P. (2015) Modular structure of web-based decision support systems for integrated pest management: A review. *Agronomy for Sustainable Development* 35, 1347–1372.

Darvas, B., Skuhrava, M. and Andersen, A. (2000) Agricultural dipteran pests of the Palaearctic region. In: Papp, L. and Darvà, B. (eds) *Contributions to a Manual of Palaearctic Diptera. Vol. 1: General and Applied Dipterology*. Science He, Budapest, pp. 565–650.

Delrio, G., (1989) Mass trapping experiments to control the olive fruit fly in Sardinia. In: Cavalloro, R. (ed.) *Fruit Flies of Economic Importance 87*. Proceedings of CEC/IOBC International Symposium. A.A. Balkema, Rotterdam, The Netherlands, pp. 419–425.

Delrio, G. (2010) Biological control of olive pests in the Mediterranean region: integrated protection of olive crops. *IOBC/WPRS Bulletin* 53, 85–92.

Delrio, G. and Lentini, A. (2003) Difesa, fitofagi dell'olivo: Tignola. In: Bandino, G., Dettori, S. (eds) *Manuale di olivicoltura*. Consorzio Interprovinciale per la Frutticoltura Cagliari Oristano Nuoro, Cagliari, Italy, 247–254.

Delrio, G. and Lentini, A. (2016) Dinamica e fattori di regolazione delle popolazioni della mosca delle olive. *Atti Accademia Nazionale Italiana di Entomologia* LXIV, 55–62.

Delrio, G., Lentini, A. and Satta, A. (2005) Biological control of olive fruit fly through inoculative releases of *Opius concolor* Szépl. *IOBC/WPRS Bulletin* 28(9), 53–58.

Dimou, I., Rempoulakis, P. and Economopoulos, A.P. (2010) Olive fruit fly [*Bactrocera* (*Dacus*) *oleae* (Rossi) (Diptera: Tephritidae)] adult rearing diet without antibiotic. *Journal of Applied Entomology* 134(1), 72–79.

Dinis, A.M., Pereira A.J., Pimenta, M.C., Oliveira, J., Benhadi-Marín, J. *et al.* (2016) Suppression of *Bactrocera oleae* (Diptera: Tephritidae) pupae by soil arthropods in the olive grove. *Journal of Applied Entomology* 140, 677–687.

Doğanlar, M. (2011) Parasitoids complex of the olive leaf gall midges, *Dasineura oleae* (Angelini 1831) and *Lasioptera oleicola* Skuhravá, 2011 (Diptera: Cecidomyiidae) in Hatay, Turkey, with descriptions of new genus and species from Tetrastichinae (Hymenoptera, Eulophidae). *Turkiye Entomoloji Dergisi* 35(2), 245–264.

Doğanlar, M., Lasalle, J., Sertkaya, E. and Doğanlar, O. (2009) A new species of *Quadrastichus* Girault, 1913 (Hymenoptera, Eulophidae, Tetrastichinae) from Hatay, *Turkey. Turkish Journal of Zoology* 33(3), 309–314.

Doğanlar, M., Sertkaya, E. and Skuhravá, M. (2011) Pest status of olive leaf gall midge *Dasineura oleae* (Angelini, 1831), description of *Lasioptera oleicola* Skuhravá sp. new (Diptera: Cecidomyiidae) and effectiveness of parasitoids on their populations in Hatay Turkey. *Turkiye Entomoloji Dergisi* 35(2), 265–284.

Duelli, P. and Obrist, M.K. (1998) In search of the best correlates for local organismal biodiversity in cultivated area. *Biodiversity and Conservation* 7(3), 297–309.

El-Sayed, A.M., Suckling, D.M., Byers, J.A., Jang, E.B. and Wearing, C.H. (2009) Potential of 'lure and kill' in long-term pest management and eradication of invasive species. *Journal of Economic Entomology* 102, 815–835.

EPPO (2020) *Bactrocera oleae distribution*. Available at: https//gd.eppo.int/taxon/DACUOL/distribution (accessed 28 September 2020).

Estes, A.M., Hearn, D.J., Bronstein, J.L. and Pierson, E.A. (2009) The olive fly endosymbiont, '*Candidatus* Erwinia dacicola', switches from an intracellular existence to an extracellular existence during host insect development. *Applied and Environmental Microbiology* 75, 7097–7106.

Estes, A.M., Nestel, D., Belcari, A., Jessup, A., Rempoulakis, P. *et al.* (2012) A basis for the renewal of sterile insect technique for the olive fly, *Bactrocera oleae* (Rossi). *Journal of Applied Entomology* 136, 1–16.

Feron, M. and Andrieu, A.J. (1962) Etude des signeaux acoutiques du male dans le comportament sexual des *Dacus oleae* Gmel. (Diptera: Trypetidae). *Annales des Epiphyties* 13, 269–276.

Fusini, G., Barsanti, D., Angelici, G., Casotti, G., Canale, A. *et al.* (2018) Identification and synthesis of new sex-specific components of olive fruit fly (*Bactrocera oleae*) female rectal gland, through original Negishi reactions on supported catalysts. *Tetrahedron* 74(3), 4381–4389.

Gallé, R., Happe, A.K., Baillod, A.B., Tscharntke, T. and Batáry, P. (2019) Landscape configuration, organic management, and within-field position drive functional diversity of spiders and carabids. *Journal of Applied Ecology* 56(1), pp. 63–72.

Garantonakis, N., Varikou, K. and Birouraki, A. (2017) Parasitism of *Psytallia concolor* (Hymenoptera: Braconidae) on *Bactrocera oleae* (Diptera: Tephritidae) infesting different olive varieties. *Phytoparasitica* 45(4), 461–469.

Giraldi, G. (1952) Osservazioni preliminari sulla *Dasineura (Perrisia) oleae* Löw nel Veronese. *Annali della Sperimentazione Agraria* 7, 827–836.

Giunti, G., Campolo, O., Laudani, F., Algeri, G.M. and Palmeri, V. (2020) Olive fruit volatiles route intraspecific interactions and chemotaxis in *Bactrocera oleae* (Rossi) (Diptera: Tephritidae) females. *Scientific Reports* 10, 1666.

Glenn, D.M., Puterka, G.J., Vanderzwet, T., Byers, R.E. and Feldhake, C. (1999) Hydrophobic particle films: a new paradigm for suppression of arthropod pests and plant diseases. *Journal of Economic Entomology* 92, 759–771.

Glenn, D.M. and Puterka, G.J. (2005) Particle films: a new technology for agriculture. In: Janick, J. (ed.) *Horticultural Reviews* 31, 1–44.

Gomez, J.F. *et al.* (2006) Los Chalcidoidea (Hymenoptera) asociados con agallas de Cinípidos (Hymenoptera, Cynipidae) en la Comunidad de Madrid. *Graellsia (número extraordinario)* 62, 293–331.

Gonçalves F.M., Rodrigues M.C., Pereira J.A., Thistlewood, H. and Torres L.M. (2012a) Natural mortality of immature stages of *Bactrocera oleae* (Diptera: Tephritidae) in traditional olive groves from north-eastern Portugal. *Biocontrol Science and Technology* 22, 837–854.

Gonçalves, M.F., Santos, S.A.P. and Torres, L.M. (2012b) Efficacy of spinosad bait sprays to control *Bactrocera oleae* and impact on non-target arthropods. *Phytoparasitica* 40(1), 17–28.

Gonzalez, D., Cabral, J.A., Torres, L. and Santos, M. (2015) A cohort-based modelling approach for managing olive moth *Prays oleae* (Bernard, 1788) populations in olive orchards. *Ecological Modelling* 296, 46–56.

González-Núñez, M., Pascual, S., Cobo, A., Seris, E., Cobos, G., Fernández, C.E. and Sánchez-Ramos, I. (2020) Copper and kaolin sprays as tools for controlling the olive fruit fly. *Entomologia Generalis* 41(1).

Graham, M.W. (1969) The Pteromalidae of north-western Europe (Hymenoptera, Chalcidoidae). *Bulletin of the British Museum (Natural History) Entomology,* Supplement. Vol. 16.

Gucci, R., Caruso, G., Canale, A., Loni, A., Raspi, A. et al (2012) Qualitative changes of olive oils obtained from fruits damaged by *Bactrocera oleae* (Rossi). *HortScience* 47, 301–306.

Gutierrez, A.P., Ponti, L. and Cossu, Q.A. (2009) Effects of climate warming on olive and olive fly (*Bactrocera oleae* (Gmelin)) in California and Italy. *Climatic Change* 95, 195–217.

Haan, N.L., Zhang, Y. and Landis, D.A. (2020) Predicting landscape configuration effects on agricultural pest suppression. *Trends in Ecology and Evolution* 35(2), 175–186.

Haniotakis, G.E. (2005) Olive pest control: present status and prospects. Proceedings of the European Meeting of the IOBS/WPRS Working Group 'Integrated Protection of Olive Crops', Chania, Greece, 29–31 May 2003. *Bulletin OILB/SROP* 28(9), 1–9.

Haniotakis, G.E., Kozyrakis, E. and Bonatsos, C. (1986) Control of the olive fruit fly, *Dacus oleae* Gmel. (Dipt., Tephritidae) by mass trapping: pilot scale feasibility study. *Journal of Applied Entomology* 101, 343–352.

Haniotakis, G., Kozyrakis, M., Fitsakis, T.H. and Antonidakj, A. (1991) An effective mass trapping method for the control of *Dacus oleae* (Diptera: Tephritidae). *Journal of Economic Entomology* 84, 564–569.

Hawro, V., Ceryngier, P., Tscharntke, T., Thies, C., Gagic, V. *et al.* (2015) Landscape complexity is not a major trigger of species richness and food web structure of European cereal aphid parasitoids. *BioControl* 60(4), 451–461.

Hegazi, E., Herz, A., Hassan, S.A., Khafagi, W.E., Agamy, E. *et al.* (2007a) Field efficiency of indigenous egg parasitoids (Hymenoptera, Trichogrammatidae) to control the olive moth (*Prays oleae*, Lepidoptera, Yponomeutidae) and the jasmine moth (*Palpita unionalis*, Lepidoptera, Pyralidae) in an olive plantation in Egypt. *Biological Control* 43(2), 171–187.

Hegazi, E.M., Konstantopoulou, M.A., Milonas, P., Herz, A., Mazomenos, B.E. *et al.* (2007b) Mating disruption of the jasmine moth *Palpita unionalis* (Lepidoptera, Pyralidae) using a two pheromone component blend: a case study over three consecutive olive growing seasons in Egypt. *Crop Protection* 26(6), 837–844.

Hepdurgun, B., Turanli, T. and Zumreoglu, A. (2009) Control of the olive fruit fly, *Bactrocera oleae* (Diptera: Tephritidae) through mass trapping and mass releases of the parasitoid *Psyttalia concolor* (Hymenoptera: Braconidae) reared on irradiated Mediterranean fruit fly. *Biocontrol Science and Technology* 19, 211–224.

Herz, A. and Hassan, S.A. (2006) Are indigenous strains of *Trichogramma* sp. (Hym., Trichogrammatidae) better candidates for biological control of lepidopterous pests of the olive tree? *Biocontrol Science and Technology* 16(8), 841–857.

Hodgson, J.A., Moilanen, A., Wintle, B.A. and Thomas, C.D. (2011) Habitat area, quality and connectivity, striking the balance for efficient conservation. *Journal of Applied Ecology* 48, 148–152.

Huang, M.Y., Chou, H.-M., Chang, Y.-T., Chang, C.-M. *et al.* (2014) The number of cecidomyiid insect galls affects the photosynthesis of *Machilus thunbergii* host leaves. *Journal of Asia-Pacific Entomology* 17(2), 151–154.

Iannotta, N., Belfiore, T., Noce, M.E., Scalercio, S. and Vizzarri, V. (2012) Correlation between *Bactrocera oleae* infestation and *Camarosporium dalmaticum* infection in an olive area of southern Italy. *Acta Horticulturae* 949, 309–316.

Infusino, M., Iannotta, N. and Scalercio, S. (2014) Field selectivity of a spinosad-poisoned bait toward target tephritids and non-target insects. *Bollettino della Società Entomologica Italiana* 146, 11–15.

Jiménez, A. and Castillo, E. (1992) *Biosteres longicaudatus* (Ashmead), un parasitoide de las moscas de las frutas. Su cria y posibilidades de empleo en control biologico. *Boletin de Sanidad Vegetal Plagas* 18, 139–148.

Johnson, M., Wang, X.G., Nadel, H., Opp, S., Lynn-Patterson, K., Stewart-Leslie, J. and Daane, K. (2011) High temperature affects olive fruit fly populations in California's Central Valley. *California Agriculture* 65(1), 29–33.

Jones, O.T. (1998) Practical applications of pheromones and other semiochemicals. In: Howse, P., Stevens, I. and Jones, O.T. (eds) *Insect Pheromones and Their Use in Pest Management.* Chapman & Hall, London, pp. 280–300.

Kapatos, E., Fletcher, B.S, Pappas, S. and Laudeho, Y. (1977) The release of *Opius concolor* and *O. concolor* var. *siculus* [Hym.: Braconidae] against the spring generation of *Dacus oleae* [Dipt.: Trypetidae] on Corfu. *Entomophaga* 22(3), 265–270.

Karamanlidou, G., Lambropoulos, A.F., Koliais, S.I., Manousis, T., Ellar, D. *et al.* (1991) Toxicity of *Bacillus thuringiensis* to laboratory populations of the olive fruit fly (*Dacus oleae*). *Applied and Environmental Microbiology* 57(8), 2277–2282.

Katsoyannos, P. (1992) Olive pests and their control in the Near East. *FAO Plant Production and Protection, Paper 115.* FAO, Rome.

Khater, H.F. (2012) Prospects of botanical biopesticides in insect pest management. *Pharmacologia* 3(12), 641–656.

Kikionis, S., Ioannou, E., Konstantopoulou, M. and Roussis, V. (2017) Electrospun micro/nanofibers as controlled release systems for pheromones of *Bactrocera oleae* and *Prays oleae*. *Journal of Chemical Ecology* 43, 254–262.

Konstantopoulou, M.A. and Mazomenos, B.E. (2005) Evaluation of *Beauveria bassiana* and *B. brongniartii* strains and four wild-type fungal species against adults of *Bactrocera oleae* and *Ceratitis capitata*. *BioControl* 50, 293–305.

Kurugundla, C.N., Kgori, P.M. and Moleele, N. (2012) Management of tsetse fly using insecticides in northern Botswana. In: Perveen, F.K. (ed.) *Insecticides: Pest Engineering.* InTech Open, pp. 449–476.

Lance, D.R. and McInnis, D.O. (2005) Biological basis of the sterile insect technique: principles and practice. In: Dyck, V.A., Hendrichs, J. and Robinson, A.S. (eds) *Area-wide Integrated Pest Management.* Springer, Dordrecht, The Netherlands, pp. 69–94.

Landolt, P.J. (1997) Sex attractant and aggregation pheromones of male phytophagous insects. *American Entomologist* 43(1), 12–22.

Levi-Zada, A., Nestel, D., Fefer, D., Nemni-Lavy, E., Deloya-Kahane, I. *et al.* (2012) Analyzing diurnal and age-related pheromone emission of the olive fruit fly, *Bactrocera oleae* by sequential SPME-GCMS analysis. *Journal of Chemical Ecology* 38, 1036–1041.

Liaropoulos, C., Louskas, C., Canard, M. and Laudeho, Y. (1977) Lachers d'*Opius concolor [Hym.: Braconidae]* dans des populations de printemps de *Dacus oleae [Dipt.: Trypetidae]*: expérimentation en Grèce continentale. *Entomophaga* 22(3), 259–264.

Liaropoulos, C., Mavraganis, V.G., Broumas, T. and Ragoussis, N. (2005) Field tests on the combination of mass trapping with the release parasite *Opius concolor* (Hymenoptera: Braconidae), for the control of the olive fruit fly *Bactrocera oleae* (Diptera: Tephritidae). *IOBC/WPRS Bulletin* 28, 77.

Loher, W. and Zervas, G. (1979) The mating rhythm of the olive fruitfly, *Dacus oleae* Gmelin. *Zeitschrift für Angewandte Entomologie* 88, 425–435.
López-Villalta, C.M. (1999) *Olive Pest and Disease Management*. International Olive Oil Council, Madrid.
Lucchi, A., Sambado, P., Juan Royo, A.B., Bagnoli, B., Conte, G. *et al.* (2018) Disrupting mating of *Lobesia botrana* using sex pheromone aerosol devices. *Environmental Science and Pollution Research* 25, 22196–22204.
Lyrene, P. and Payne, J. (1996) Blueberry gall midge. *Journal of Small Fruit & Viticulture* 3(2–3), 111–124.
Malausa, J.C., Blanchet, A., Bon, M.C., Cheyppe-Buchmann, S., Groussier-Bout, G. *et al.* (2010) Introduction of the African parasitoid *Psyttalia lounsburyi* in south of France for the classical biological control of *Bactrocera oleae*: will hybridization affect establishment and population growth? *IOBC/WPRS Bulletin* 53, 49–55.
Malheiro, R., Casal, S., Baptista, P. and Pereira, J.A. (2015) A review of *Bactrocera oleae* (Rossi) impact in olive products: from the tree to the table. *Trends in Food Science & Technology* 44(2), 226–242.
Malheiro, R., Casal, S., Pinheiro, L., Baptista, P. and Pereira, J.A. (2019) Olive cultivar and maturation process on the oviposition preference of *Bactrocera oleae* (Rossi) (Diptera: Tephritidae). *Bulletin of Entomological Research* 109, 43–53.
Marc, P., Canard, A. and Ysnel, F. (1999) Spiders (Araneae) useful for pest limitation and bioindication. *Agriculture, Ecosystems & Environment* 74(1–3), 229–273.
Marchi, S., Guidotti, D., Ricciolini, M. and Petacchi, R. (2016) Towards understanding temporal and spatial dynamics of *Bactrocera oleae* (Rossi) infestations using decade-long agrometeorological time series. *International Journal of Biometeorology* 60, 1681–1694.
Marchini, D., Petacchi, R. and Marchi, S. (2017) *Bactrocera oleae* reproductive biology: new evidence on wintering wild populations in olive groves of Tuscany (Italy). *Bulletin of Insectology* 70(1), 121–128.
Mazomenos, B.E. (1989) *Dacus oleae*. In: Robinson, A.S. and Hooper, G. (eds) *World Crop Pests – Fruit Flies, Their Biology, Natural Enemies and Control*. Vol. 3A. Elsevier, Rotterdam.
Mazomenos, B.E., and Haniotakis, G.E. (1981) A multicomponent female sex pheromone of *Dacus oleae* (Gmelin): isolation and bioassay. *Journal of Chemical Ecology* 7(2), 137–144.
Mazomenos, B.E., and Haniotakis, G.E. (1985) Male olive fruit fly attraction to the synthetic sex pheromone components in laboratory and field tests. *Journal of Chemical Ecology* 11, 397–406.
Mazomenos, B.E., Raptopoulos, D., Lefkidou, I. and Mazomenos-Pantazi, A. (1994) Female sex pheromone components of jasmine moth *Palpita unionalis* (Lepidoptera, Pyralidae). *Journal of Chemical Ecology* 20, 745–751.
Mazomenos, B.E., Ortiz, A., Mazomenos-Pantazi, A., Stefanou, D., Stravrakis, N. *et al.* (1999) Mating disruption for the control of the olive moth, *Prays oleae* (Bern) (Lep., Yponomeutidae) with the major sex pheromone component. *Journal of Applied Entomology* 123(4), 247–254.
Mazomenos, B.E., Pantazi-Mazomenou, A. and Stefanou, D. (2002) Attract and kill of the olive fruit fly *Bactrocera oleae* in Greece as a part of an integrated control system. *IOBC/WPRS Bulletin* 25, 137–146.
Mazzini, M. and Vita, G. (1983) Identificazione submicroscopica del meccanismo di trasmissione del batterio simbionte in *Dacus oleae* (Gmelin) (Diptera: Trypetidae). *Redia* 64, 277–301.
Miranda, M., Miquel, M., Terrassa, J., Melis, N. and Monerris, M. (2008) Parasitism of *Bactrocera oleae* (Diptera: Tephritidae) by *Psyttalia concolor* (Hymenoptera; Braconidae) in the Balearic Islands (Spain). *Journal of Applied Entomology* 132, 798–805.
Mkize, N., Hoelmer, K.A. and Villet, M.H. (2008) A survey of fruit-feeding insects and their parasitoids occurring on wild olives, *Olea europaea* ssp. *cuspidata*, in the Eastern Cape of South Africa. *Biocontrol Science and Technology* 18(10), 991–1004.
Monastero, S. (1968) Nouvelle experimentation à grande échelle de la lutte biologique contre la mouche de l'olive (*Dacus oleae* Gmel.) au moyen d'*Opius concolor* siculus Mon. en Sicile en 1967. *Entomophaga* 13(3), 251–261.
Monastero, S. and Delanoue, P. (1966) Lutte biologique expérimentale contre la mouche de l'olive (*Dacus oleae* Gmel.) au moyen d'*Opius concolor* Szepl. *siculus* Mon. dans les îles Éoliennes (Sicile) en 1965. *Entomophaga* 11, 411–432.
Morris, T.I. and Campos, M. (2000) The olive branch: a passageway for predators? [Las ramas de olivo: un camino para los depredadores? *Zoologica Baetica* 11, 67–75.
Mostakim, M., Iraqui, M., Benbrahim, K.F., Houari, A., Gounni, A.S. *et al.* (2012) Biocontrol potential of a *Bacillus subtilis* strain against *Bactrocera oleae*. *Annals of Microbiology* 62(1), 211–216.
Motta, L.B., Kraus, J.E., Salatino, A. and Salatino, M.L.F. (2005) Distribution of metabolites in galled and non-galled foliar tissues of *Tibouchina pulchra*. *Biochemical Systematics and Ecology* 33(10), 971–981.
Moustafa, S.A. (2009) Response of the Mediterranean fruit fly, *Ceratitis capitata* (Wied.) and peach fruit fly, *Bactrocera zonata* (Saund.) to some food attractants. *Egyptian Journal of Agricultural Research* 2, 111–118.

Mozhdehi, M.R.A. and Kayhanian, A.A. (2014) Application of detterent compound for control of olive fruit flies *Bactrocera oleae* Gmelin (Diptera: Tephritidae). *Romanian Journal of Plant Protection* 7, 24–30.

Murray, M. and Alston, D.G. (2020) *Codling Moth Mating Disruption (factsheet)*. Utah State University, Utah Pests Extension.

Nardi, F., Carapelli, A., Dallai, R., Roderick, G.K. and Frati, F. (2005) Population structure and colonization history of the olive fly, *Bactrocera oleae* (Diptera: Tephritidae). *Molecular Ecology* 14, 2729–2738.

Nasr, F.N., Abou-Elkhair, S.S., Stefanos, S.S., Youssef, A.A. and Shehata, W.A. (2002) New record of some biological control agents of *Palpita unionalis* Hubn. (Lepidoptera, Pyralidae) and *Prays oleae* Bern. (Lepidoptera, Yponomeutidae) in olive groves in Egypt. *Egyptian Journal of Biological Pest Control* 12(2), 29.

Navarro-Llopis, V. and Vacas, S. (2014) Mass trapping for fruit fly control. In: Shelly, T., Epsky, N., Jang, E.B., Reyes-Flores, J. and Vargas, R. (eds) *Trapping and the Detection, Control, and Regulation of Tephritid Fruit Flies*. Springer, Dordrecht, The Netherlands, pp. 513–555.

Nave, A., Gonçalves, F., Teixeira, R., Amaro Costa, C., Campos, M. *et al.* (2017) Hymenoptera parasitoid complex of *Prays oleae* (Bernard) (Lepidoptera, Praydidae) in Portugal. *Turkish Journal of Zoology* 41, 502–551.

Navrozidis, E.I., Vasara, E., Karamanlidou, G., Salpiggidis, G.K. and Koliais, S.I. (2000) Biological control of *Bactocera oleae* (Diptera: Tephritidae) using a Greek *Bacillus thuringiensis* isolate. *Journal of Economic Entomology* 93, 1657–1661.

Nestel, D., Rempoulakis, P., Yanovski, L., Miranda, M.A. and Papadopoulos, N.T. (2016) The evolution of alternative control strategies in a traditional crop: economy and policy as drivers of olive fly control. In: Horowitz, A.R. and Ishaaya, I. (eds) *Advances in Insect Control and Resistance Management*. Springer, Dordrecht, The Netherlands, pp. 47–76.

Neuenschwander, P. (1982) Beneficial insects caught by yellow traps used in mass-trapping of the olive fly, *Dacus oleae*. *Entomologia Experimentalis et Applicata* 32, 286–296.

Neuenschwander, P., Michelakis, S. and Bigler, F. (1981) Abiotic factors affecting mortality of *Dacus oleae* larvae and pupae in the soil. *Entomologia Experimentalis et Applicata* 30, 1–19.

Neuenschwander, P., Bigler, F., Delucchi, V. and Michelakis, S. (1983) Natural enemies of preimmaginal stages of *Dacus oleae* Gmel. (Dipt., Tephritidae) in Western Crete. I. Bionomics and phenologies. *Bollettino del Laboratorio di Entomologia Agraria Filippo Silvestri* 40, 3–32.

Noce, M.E., Belfiore, T., Scalercio, S., Vizzarri, V. and Iannotta, N. (2009) Efficacy of new mass-trapping devices against *Bactrocera oleae* (Diptera: tephritidae) for minimizing pesticide input in agroecosystems. *Journal of Environmental Science and Health, Part B* 44, 442–448.

Öncüer, C. (1991) *A Catalogue of the Parasites and Predators of Insect Pests of Turkey*. Available at: https://www.semanticscholar.org/paper/A-catalogue-of-the-parasites-and-predators-of-pests-Oncuer/ed9815d3b9d4011ab68ae1922744abaec3b07cfd (accessed 29 May 2023).

Ordano., M., Engelhard, I., Rempoulakis, P., Nemny-Lavy, E., Blum, M. *et al.* (2015) Olive fruit fly (*Bactrocera oleae*) population dynamics in the eastern Mediterranean: influence of exogenous uncertainty on a monophagous frugivorous insect. *PLOS ONE*, 10(5), e0127798.

Orphanidis, P.S., Mourikis, P.A. and Souliotis, M.C. (1971) Note sur la lutte contre *Dacus oleae* (Gmel.) au moyen de pulvérisations d'appâts à bas et ultra-bas volume. *EPPO Bulletin* 1, 57–62.

Orsini, M.A., Daane, K.M., Sime, K.R. and Nelson, E.H. (2007) Mortality of olive fruit fly pupae in California. *Biocontrol Science and Technology* 17, 797–807.

Ortega, M. and Pascual, S. (2014) Spatio-temporal analysis of the relationship between landscape structure and the olive fruit fly *Bactrocera oleae* (Diptera: Tephritidae). *Agricultural and Forest Entomology* 16(1), 14–23.

Paredes, D., Cayuela, L. and Campos, M. (2013) Synergistic effects of ground cover and adjacent vegetation on natural enemies of olive insect pests. *Agriculture, Ecosystems and Environment* 173, 72–80.

Pascual, S., Cobos, G., Seris, E. and González-Núñez, M. (2010) Effects of processed kaolin on pests and non-target arthropods in a Spanish olive grove. *Journal of Pest Science* 83(2), 121–133.

Pascual, S., Cobos, G., Seris, E., Sánchez-Ramos, I. and González-Núñez, M. (2014) Spinosad bait sprays against the olive fruit fly (*Bactrocera oleae* (Rossi)): effect on the canopy non-target arthropod fauna. *International Journal of Pest Management* 60, 258–268.

Pennino, G., Pane, G., Raiti, G, Perri, E., Carovita, M.A. *et al.* (2006) Three years field trials to assess the effect of kaolin made particles and copper on olive-fruit fly (*B. oleae* Gmelin) infestations in Sicily. In: *Proceedings, DCA – Università di Palermo*; Regione Siciliana – Assessorato Agricoltura e Foreste, II, pp. 303–306.

Petacchi, R., Guidotti, D. and Rizzi, I. (2001) Experiences on using web-based data management system for area-wide monitoring network in olive fruit fly IPM. *Proceedings EFITA 2001*, Montpellier, France, 18–20 June, pp. 53–58.

Petacchi, R., Rizzi, I. and Guidotti, D. (2003) The 'lure and kill' technique in *Bactrocera oleae* (Gmel.) control: effectiveness indices and suitability of the technique in area-wide experimental trials. *International Journal of Pest Management* 49, 305–311.

Petacchi, R. Picchi, M.S., Tondini, E., Caselli, A. and Monteforti, G. (2019) Lotta biologica conservativa alla cecidomia dell'olivo. *Informatore Agrario* 44, 43–45.

Picchi, M.S. (2020) Spiders (Araneae) of olive groves and adjacent semi-natural habitats from central Italy. *Arachnologische Mitteilungen* 60(1), 1–11.

Picchi, M.S., Bocci, G., Petacchi, R. and Entling H.M. (2016) Effects of local and landscape factors on spiders and olive fruit flies. *Agriculture, Ecosystems & Environment* 222, 138–147.

Picchi, M.S., Marchi, S., Albertini, A. and Petacchi, R. (2017a) Organic management of olive orchards increases the predation rate of overwintering pupae of *Bactrocera oleae* (Diptera: Tephritidae). *Biological Control* 108, 9–15.

Picchi, M.S., Marchi, S. and Petacchi, R. (2017b) Cecidomia delle foglie dell'olivo, nuovo rischio o vecchio problema? *Informatore Agrario* 16, 49–52.

Picchi, M.S., Bocci, G., Petacchi, R. and Entling, H.M. (2020) Taxonomic and functional differentiation of spiders in habitats in a traditional olive producing landscape in Italy. *European Journal of Entomology* 117, 18–26.

Piñero, J.C., Mau, R.F.L. and Vargas, R.I. (2009) Managing oriental fruit fly (Diptera: Tephritidae) with spinosad-based protein bait sprays and sanitation in papaya orchards in Hawaii. *Journal of Economic Entomology* 102, 1123–1132.

Ponti, L., Gutierrez, A.P., Ruti, P.M. and Dell'Aquila, A. (2014) Fine-scale ecological and economic assessment of climate change on olive in the Mediterranean Basin reveals winners and losers. *PNAS* 111, 5598–5603.

Porcel, M., Cotes, B. and Campos, M. (2011) Biological and behavioral effects of kaolin particle film on larvae and adults of *Chrysoperla carnea* (Neuroptera: Chrysopidae). *Biological Control* 59, 98–105.

Preu, M., Frieß, J.L., Breckling, B. and Schröder, W. (2020) Case study 1: olive fruit fly (*Bactrocera oleae*). In: von Gleich, A. and Schröder, W. (eds) *Gene Drives at Tipping Points*. Springer, Cham, Switzerland, pp. 79–101.

Prokopy, R.J., Miller, N.W., Piñero, J.C., Barry, J.D., Tran, L.C. *et al.* (2003) Effectiveness of GF-120 fruit fly bait spray applied to border area plants for control of Melon flies (Diptera: Tephritidae). *Journal of Economic Entomology* 96, 1485–1493.

Puterka, G.J., Glenn, D.M. and Pluta, R.C. (2005) Action of particle films on the biology and behavior of pear psylla (Homoptera: Psyllidae). *Journal of Economic Entomology* 98, 2079–2088.

Ragoussis, N. (2005) Contribution to the biological olive agriculture: efficient control of the olive fruit fly by the ECO-TRAP®. *IOBC/WPRS Bulletin* 28, 29–35.

Ramos, P., Campos, M., Ramos, J.M. and Jones, O.T. (1989) Nine years of studies on the relationship between captures of male olive moths, *Prays oleae* Bern. (Lepidoptera, Hyponomeutidae) in sex pheromone baited traps and fruit infestation by the subsequent larval generation (1979–1987). *Tropical Pest Management* 35(2), 201–204.

Rastrelli, L., Totaro, K. and De Simone, F. (2002) Determination of organophosphorus pesticide residues in Cilento (Campania, Italy) virgin olive oil by capillary gas chromatography. *Food Chemistry* 79(3), 303–305.

Rebora, M., Salerno, G., Piersanti, S., Gorb, E. and Gorb, S. (2020) Role of fruit epicuticular waxes in preventing *Bactrocera oleae* (Diptera: Tephritidae) attachment in different cultivars of *Olea europaea*. *Insects* 11, 189.

Rhodes, E.M., Benda, N.D. and Liburd, O.E. (2014) Field distribution of *Dasineura oxycoccana* (Diptera: Cecidomyiidae) adults, larvae, pupae, and parasitoids and evaluation of monitoring trap designs in Florida. *Journal of Economic Entomology* 107(1), 310–318.

Ros, J.P., Seris, E., Castillo, E., Cobo, A. and Gonzalez-Nuñez, M. (2009) A new advanced step to make a useful 'Mass Trapping Method' to control the olive fruit fly *Bactrocera oleae* (Rossi): comparative study of a new attractant. *Boletín de Sanidad Vegetal, Plagas* 35, 391–400.

Ruiu, L., Delrio, G., Floris, I., Satta, A. and Solinas, M. (2007) Histopathological observations in the midgut and behaviour of olive fruit fly (*Bactrocera oleae* Gmelin) adults treated with a strain of *Bacillus thuringiensis* Berliner. *IOBC/WPRS Bulletin* 30(9), 67.

Russo, L. (1967) Studio sugli Imenotteri parassiti della Mosca delle olive (*Dacus oleae* Gmel.) in Campania. *Bollettino del Laboratorio di Entomologia Agraria Filippo Silvestri* 25, 177–226.

Rypstra, A.L. and Buddle, C.M. (2013) Spider silk reduces insect herbivory. *Biology Letters* 9(1), 20120948.

Sacchetti, P., Granchietti, A., Landini, S., Viti, C., Giovanetti, L. *et al.* (2008) Relationships between the olive fly and bacteria. *Journal of Applied Entomology* 132, 682–689.
Sacchetti, P., Liscia, A., Pastorelli, R., Bigiotti, G., Guidi, R. *et al.* (2017) Le simbiosi batteriche nella mosca delle olive, *Bactrocera oleae*: dalla ricerca di base allo sviluppo di nuove strategie di controllo. *Atti Accademia Nazionale Italiana di Entomologia* 64, 93–98.
Sacchetti, P., Rosi, M.C., Rispoli, S., Berni, G., Bigiotti, G. *et al.* (2019) Behavioural response of *Diachasmimorpha longicaudata* (Hymenoptera: Braconidae) to cues from infested and non-infested olives. *IOBC/WPRS Bulletin* 141, 183–187.
Salerno, G., Rebora, M., Kovalev, A., Gorb, E. and Gorb, S. (2020) Kaolin nano-powder effect on insect attachment ability. *Journal of Pest Science* 93(1), 315–327.
Saour, G. and Makee, H. (2004) A kaolin-based particle film for suppression of the olive fruit fly *Bactrocera oleae* Gmelin (Dip., Tephritidae) in olive groves. *Journal of Applied Entomology* 128, 28–31.
Sarwar, M. (2015) Attraction of female and male fruit flies (Diptera: Tephritidae) to bait spray applications for reduction of pest populations. *International Journal of Animal Biology* 1, 225–230.
Scalercio, S., Belfiore, T., Noce, M.E., Vizzarri, V. and Iannotta, N. (2010) Impact of kaolin and *Beauveria bassiana* treatments against olive fly on the non-target arthropods of the olive ecosystem. *Acta Horticulturae* 873, 329–336.
Schrader K.K., Andolfi, A., Cantrella, C.L., Cimmino, A., Duke, S.O. *et al.* (2010) A survey of phytotoxic microbial and plant metabolites as potential natural products for pest management. *Chemistry and Biodiversity* 7, 2261–2280.
Segoli, M., Maklakov, A., Gavish, E., Tsurim, I. and Lubin, Y. (2004) The effect of previous foraging success on web-building behaviour in the sheet-web spider, *Frontinellina* cf. *frutetorum* (Araneae Linyphiidae). *Ethology, Ecology & Evolution* 16(4), 291–298.
Sharma, R.R., Reddy, S.V.R. and Datta, S.C. (2015) Particle films and their applications in horticultural crops. *Applied Clay Science* 116, 54–68.
Shehata, W., Abou-Elkhair, S., Youssef, A. and Nasr, F.N. (2003) Biological studies on the olive leaf moth, *Palpita unionalis* Hübner (Lepid., Pyralidae), and the olive moth, *Prays oleae* Bernard (Lepid., Yponomeutidae). *Journal of Pest Science* 76, 155–158.
Silvestri, F., Martelli, G. and Masi, L. (1908) Sugli Imenotteri parassiti ectofagi della mosca delle olive fino ad ora osservati nell'Italia meridionale e sulla loro impotanza nel combattere la mosca delle olive. *Bollettino del Laboratorio di Zoologia generale e agraria della R Scuola superiore d'agricoltura in Portici* 2, 18–82.
Simoglou, K.B., Karataraki, A., Roditakis, N.E. and *Roditakis, E.* (2012) *Euzophera bigella* (Zeller) (Lepidoptera: Pyralidae) and *Dasineura oleae* (F. Low) (Diptera: Cecidomyiidae): emerging olive crop pests in the Mediterranean? *Journal of Pest Science* 85(2), 169–177.
Sinno, M., Bézier, A., Vinale, F., Giron, D., Laudonia, S. *et al.* (2020) Symbiosis disruption in the olive fruit fly, *Bactrocera oleae* (Rossi), as a potential tool for sustainable control. *Pest Management Science* 76(9), 3199–3207.
Skouras, P.J., Margaritopoulos, J.T., Seraphides, N.A., Ioannides, I.M., Kakani, E. *et al.* (2007) Organophosphate resistance in olive fruit fly, *Bactrocera oleae*, populations in Greece and Cyprus. *Pest Management Science* 63, 42–48.
Skuhravy, V., Skuhrava, M. and Brewer, J.W. (1980) Evaluation of plant damage caused by three species of gall midges (Diptera: Cecidomyiidae). *Zeitschrift für Angewandte Entomologie*, 90(1–5), 184–190.
Spilke, J. and Köstler, W. (1999) Utilization of on-line database in regional information networks in agriculture. *Proceedings of Second EFITA Conference*, Bonn, Germany, 27–30 September.
Stark, J.D., Vargas, R.I. and Miller, N.W. (2004) Toxicity of spinosad in protein bait to three economically important tephritid fruit fly species (Diptera: Tephritidae) and their parasitoids (Hymenoptera: Braconidae). *Journal of Economic Entomology* 97, 911–915.
Thysen, I., Boll, P.S. and Jensen, A.L. (1999) *Web-based information chains in agriculture – an example from Pl@nteinfo. Proceedings of Second EFITA Conference*, Bonn, Germany, 27–30 September.
Tomassone, D., Petacchi, R. and Belcari, A. (2006) Modelli matematici e supporti informatici per la gestione della difesa da *Bactrocera oleae* (Rossi), panoramica delle possibilità attuali e prospettive future. *Informatore Fitopatologico* 11, 31–36.
Tondini, E. and Petacchi, R. (2019) First observations on the parasitoid complex and on the biology of *Dasineura oleae* during an outbreak in Tuscany, *Italy. Bulletin of Insectology* 72(1), 93–102.
Tremblay, E. (1986) *Entomologia applicata. Volume II Parte II.* Liguori Editore, Napoli, Italy.
Triggiani, O. (1971) *Margaronia unionalis* Hübner (Lepidoptera, Pyralidae) (olive pyralid). *Entomologica* 7, 29– 47.

Tsitsipis, J.A. (1977) Effect of constant temperatures on eggs of olive fruit fly, *Dacus oleae* (Diptera: Tephritidae). *Annales de Zoologie Ecologie Animale* 9,133–139.

Tsolakis, H., Ragusa, E. and Tarantino, P. (2011) Control of *Bactrocera* oleae by low environmental impact methods: NPC methodology to evaluate the efficacy of lure-and-kill method and copper hydroxide treatments. *Bulletin of Insectology* 64, 1–8.

Tzanakakis, M.E. (2006) *Insects and Mites Feeding on Olive: Distribution, Importance, Habits, Seasonal Development and Dormancy.* Applied Entomology Library vol. 1. Brill, Leiden, The Netherlands.

Vargas, R.I., Miller, N.W. and Prokopy, R.J. (2002) Attraction and feeding responses of Mediterranean fruit fly and a natural enemy to protein baits laced with two novel toxins, Phloxine B and Spinosad. *Entomologia Experimentalis et Applicata* 102, 273–282.

Vargas, R.I., Mau, R.L. and Jang, E.B. (2007) The Hawaii fruit fly area-wide pest management program: accomplishments and future directions. *Proceedings of the Hawaiian Entomological Society* 39, 99–104.

Vergoulas, P.V., Prophetou-Athanassiadou, D., Alimi, E., Ben Salah, H., Mavrotas, C. *et al.* (2007) Bait application effect of Spinosad success tm 0.24CB (GF-120) formulation, on *Bactrocera oleae* Gmel. (Dacuol), and impact on other non target organisms in olive trees. *IOBC/WPRS Bulletin* 30, 131.

Vidal, S. (1997) Determination list of entomophagous insects Nr. 13. *Bulletin. Section Regionale Ouest Palaearctique, Organisation Internationale de Lutte Biologique* 20(2), 22.

Viggiani, G. (1967) Studies on Hymenoptera Chalcidoidea. XI. Chalcidoidea of the Parco Gussone (1st note). *Bollettino del Laboratorio di Entomologia Agraria 'Filippo Silvestri', Portici* 25, 150–162.

Vontas, J.G., Hejazi, M.J., Hawkes, N.J., Cosmidis, N., Loukas, M. *et al.* (2002) Resistance-associated point mutations of organophosphate insensitive acetylcholinesterase, in the olive fruit fly *Bactrocera oleae*. *Insect Molecular Biology* 11, 329–336.

Wall, R. and Fisher, P. (2001) Visual and olfactory cue interaction in resource-location by the blowfly, *Lucilia sericata*. *Physiological Entomology* 26(3), 212–218.

Wang, X.G., Johnson, M.W., Daane, K.M. and Nadel, H. (2009) High summer temperatures affect the survival and reproduction of olive fruit fly (Diptera: Tephritidae). *Environmental Entomology* 38(5), 1496–1504.

Wharton, R.A. and Yoder, M.J. (1967) Parasitoids of fruit-infesting tephritidae. Available at: http://paroffit.org. (accessed September 2020).

Woods, N., Craig, I.P., Dorr, G. and Young, B. (2001) Spray drift of pesticides arising from aerial application in cotton. *Journal of Environmental Quality* 30, 697–701.

Xiao, H. Sun, L., Jiao, T. and Li, Z. (2016) A revision of Chinese species of *Mesopolobus* Westwood (Hymenoptera, Pteromalidae) with descriptions of four new species from China. *Zoological Systematics* 41(1), 64–81.

Yasin, S., Rempoulakis, P., Nemny-Lavy, E., Levi-Zada, A., Tsukada, M., Papadopoulos, N. and Nestel, D. (2014) Assessment of lure and kill and mass-trapping methods against the olive fly, *Bactrocera oleae* (Rossi), in desert-like environments in the Eastern Mediterranean. *Crop Protection* 57, 63–70.

Yousef, M., Lozano-Tovar, M.D., Garrido-Jurado, I. and Quesada-Moraga, E. (2013) Biocontrol of *Bactrocera oleae* (Diptera: Tephritidae) with *Metarhizium brunneum* and its extracts. *Journal of Economic Entomology* 106(3), 1118–1125.

Youssef, A.I., Nasr, F.N., Stefanos, S.S., Elkhair, S.S.A., Shehata, W.A. *et al.* (2004) The side-effects of plant protection products used in olive cultivation on the hymenopterous egg parasitoid *Trichogramma cacoeciae* Marchal. *Journal of Applied Entomology* 128, 593–599.

Yokoyama, V.Y. (2012) Mobility of olive fruit fly (Diptera: Tephritidae) late third instars and teneral adults in test arenas. *Environmental Entomology* 41(5), 1177–1183.

Yokoyama, V.Y. (2014) Response of olive fruit fly (Diptera: Tephritidae) to an attract-and-kill trap in greenhouse cage tests. *Journal of Insect Science* 14.

7.3 Olive Diseases

G.E. Agosteo[1], A. Ragazzi[2], G. Surico[2] and S.O. Cacciola[3]
[1]*Mediterranean University, Department of Agriculture, Reggio Calabria, Italy;* [2]*Dipartimento di Scienze e Tecnologie Agrarie, Alimentari, Ambientali e Forestali, Firenze, Italy;* [3]*Department of Agriculture, Food and Environment, University of Catania, Italy*

This chapter is not intended to be a comprehensive review of diseases affecting olive but addresses only infectious diseases of global significance for the olive industry and, in particular, for the Mediterranean region, which is the major olive-growing area in the world. For each disease, essential information is provided with particular emphasis on the selection of cultivars for disease resistance, which is a fundamental aspect for any integrated and eco-sustainable disease-management strategy. Diseases are not listed in order of relevance but are grouped according to an etiological criterion and the organ of the tree they infect. Particularly detailed information is provided as regards the new disease named 'olive quick decline syndrome' caused by the bacterium *Xylella fastidiosa*, due to its destructive effects. It emerged about ten years ago in a restricted area of southern Italy and the epidemic is inexorably spreading. Moreover, it has been more recently found on olive in South America and a very restricted outbreak has been intercepted in southern France, but promptly eradicated. The pathogen is also present in Spain and Portugal.

7.3.1 Virus Diseases

To date, 15 different viruses from six families and eight genera – *Olive leaf yellowing associated virus* (OLYaV), *Cherry leaf roll virus* (CLRV), *Strawberry latent ring spot virus* (SLRSV), *Arabis mosaic virus* (ArMV), *Olive latent virus-1* (OLV-1), *Cucumber mosaic virus* (CMV), *Olive latent virus-2* (OLV-2), *Olive latent virus-3* (OLV-3), *Olive semilatent virus* (OSLV), *Olive mild mosaic virus* (OMMV), *Olive yellow mottling and decline associated virus* (OYMDaV), *Olive latent ringspot virus* (OLRSV), *Olive vein yellows associated virus* (OVYaV), *Tobacco necrosis virus* (TNV) and *Tobacco mosaic virus* (TMV) – have been isolated from olive trees. Infections are in most cases symptomless and/or reported only in one tree or a very limited number of trees, as in the case of OLRSV, OVYaV and OSLV. Some olive viruses have been found to be associated with particular symptoms, such as deformation of fruits and leaves and various patterns of leaf discoloration, ranging from mild chlorosis to bright yellow mottling, but there is no clear evidence of their etiological involvement in the disease. As a consequence, the identification of olive-infecting viruses is performed by molecular techniques.

Although the intervention of vectors in the transmission of some olive viruses cannot be ruled out, the spread of viruses takes place by means of infected propagation material and virus dissemination occurs prevalently through the trade of

DOI: 10.1079/9781789247350.0023

nursery plants (Martelli, 2011; Çağlayan *et al.*, 2011).

In 1993, Italy was the first country to publish standards for certification of olive propagation material. Certified olive plants should be free from *Pseudomonas savastanoi* pv. *savastanoi*, *Verticillium dahliae* and six viruses (OLV-1, OLV-2, CMV, ArMV, CLRV and SLRSV).

7.3.2 Diseases Caused by Prokaryotes

7.3.2.1 Olive quick decline syndrome

The causal agent

Xylella fastidiosa (Wells *et al.*, 1987) is a Gram-negative gamma-proteobacterium, rod-shaped, aerobic, non-flagellated (but fimbricated), asporigenic bacterium, approximately 0.25–0.35 x 0.9–3.5 µm, in the family *Xanthomonadaceae*. The name *Xylella* derives from the Greek *xylon* meaning piece of wood, while the epithet *fastidiosa* refers to the bacterium which is slow-growing and difficult to culture *in vitro*. In fact, in phytobacteriology, the group of *fastidous bacteria* includes bacteria that can only grow, and with difficulty, on artificial complex substrates whose composition includes carbon and nitrogen sources (peptone, triptone, soitone, yeast extract, amino acids), inorganic salts, iron (ferric chloride heme or ferric pyrophosphate), citrate, succinate, starch and bovine serum albumin, or charcoal. On these substrates, *X. fastidiosa* generally forms small colonies (1 mm in diameter or slightly more) with a convex shape, a smooth or wrinkled surface and an entire or finely wavy margin. Some strain of *Xylella* can also grow on simpler substrates such as potato starch.

Phytosanitary category

X. fastidiosa is one of the most dangerous plant pathogens worldwide, causing a variety of diseases, with huge economic impact for agriculture, landscape and the environment.

The European Plant Protection Organization (EPPO) included, initially, *X. fastidiosa* in the A1 quarantine pest list (organisms that are not present in the region), but after its spread in France and Spain, as well as in Italy, in 2018 it was moved to the A2 list (quarantine organisms that are present in certain areas within the EPPO region) (EFSA, 2018).

Origin and diffusion

X. fastidiosa first appeared on 2- to 30-year-old vine plants around Anaheim, Orange County, California. The disease was first observed in 1882, but the damage did not begin to become apparent and attract the attention of scientists until 1885; it then exploded into a severe form around 1887.

The disease quickly spread to other counties in California and then to 30 other US states and Canada (British Columbia, Ontario, Saskatchewan). Towards the south, *Xylella* is present in Mexico, Central and South America; Costa Rica, Puerto Rico, Paraguay, Brazil, Argentina and Venezuela.

Outside the American continent, *Xylella* has been reported in Italy, France, Spain, Portugal, Lebanon, Israel and Iran. A different species of *Xylella* (*X. taiwanensis*) is present in the island of Taiwan (Su *et al.*, 2016).

The center of origin of *X. fastidiosa* seems to be Central/South America and not, as was always believed, North America. Several experimental data indicate that the *Xylella* population attacking grapevines arrived in California with a coffee plant imported from Costa Rica (Table 7.3.1). Other phytopathogenic variants, including, to some extent, the *pauca* subspecies, would have differentiated from the new North American bacterial populations. Thus, some experimental evidence seems to show that the *multiplex* subspecies, accidentally introduced in South America, transferred part of the genetic material to a local non-pathogenic bacterial population which, as a consequence of this acquisition, became capable of infecting citrus and coffee species and causing 'variegated chlorosis' of citrus and 'leaf blight' of coffee, respectively. The same CoDiRO strain, the one that caused the olive quick decline syndrome in Apulia, Italy, and which possesses the genetic characteristics of the *pauca* subspecies, is said to have been generated in Costa Rica as a result of the exchange of genetic material between an unidentified population of *X. fastidiosa* and a population of *X. fastidiosa* subsp. *pauca*, both from South America. From Costa Rica, the CoDiRO strain would have arrived in Salento around 2008 with an oleander or coffee plant.

Host range and specialization

X. fastidiosa is capable of infecting, to date, 563 plant species from 264 genera in 82 botanical families (EFSA, 2018). Of these 563 species, 312 have been confirmed as host plants by examining, with two different molecular assays, the actual presence of the bacterium in plant tissues. It was then demonstrated that of the 312 species mentioned above, 192 can be infected under natural conditions. However, not all infected host species show the symptoms; indeed, most of them remain asymptomatic for long periods of time during which the colonized plants may act as reservoirs of inoculum. Different levels of susceptibility within and between hosts have also been observed.

All populations were originally collected in a single species, the one that causes the so-called Pierce's disease on vines: *X. fastidiosa*. Subsequent studies have shown that there is a certain host specificity of different populations of *X. fastidiosa* in nature, so that six different subspecies have been differentiated (Table 7.3.1). The population of *X. fastidiosa* identified in Apulia belongs to the subspecies *pauca*. Within each subspecies, individual populations of the bacterium can differ from each other genetically and in terms of the range of plant host.

Xylella *population typing and spread in Europe*

The Multi Locus Sequence Typing (MLST) technique was used for the intra-specific typing of *X. fastidiosa*. It is based on the sequencing of internal segments of seven housekeeping genes (*cysG*, *gltT*, *holC*, *leuA*, *malF*, *nuoL*, *petC*), i.e. constitutive genes in which polymorphism of certain nucleotides induces allelic variations that result in certain phylotypes. On the basis of the allelic variations displayed by specific populations of a given *Xylella* subspecies, various sequence type (ST) profiles have been identified. Each ST is basically assimilated to a strain or isolate of a given subspecies: two strains sharing the same allelic variation in the seven genes considered are assigned to the same ST. The MLST technique has enabled to outline the spread of *Xylella* in the Mediterranean basin. In practice, seven different STs included in three different

Table 7.3.1. Host plants, geographical distribution, first reports and origin of *Xylella fastidiosa* subspecies[a].

Subspecies	Main host(s)	Geographical distribution	First report[c]	Geographical origin[d]
pauca	grapevine	NA, CA, MB	1882–1887	CA
sandyi	numerous[b]	NA, SA, MB	1980	NA
morus	citrus, coffee, olive	SA, CA, MB	1987 (CVC); 1995 (CLS)	SA
tashke	oleander	NA	late 1980s, 20th century	NA
pauca	mulberry	NA	early 1980s, 20th century	NA
sandyi	*Chitalpa tashkentensis*	NA	early 2000s	NA

[a]To date, of the six subspecies, only *fastidiosa* and *multiplex* are recognized as valid names by the International Commission on Taxonomy of Phytopathogenic Bacteria of the ISPP (International Society of Plant Pathology).
[b]almond, oak, elm, oleander, maple, sycamore, *Prunus* spp., grapevine, etc.
[c]CVC, Citrus Variegated Chlorosis; CLS, Coffee Leaf Scorch
[d]CA, Central America; NA, North America; SA, South America; MB, Mediterranean basin

subspecies have been identified, indicating that there have been several introductions of *Xylella fastidiosa* in Europe and, more generally, in the Mediterranean basin and that, so far, there has been a mainly local spread of each introduced ST (Table 7.3.2).

Plant–pathogen interaction

X. fastidiosa multiplies only in the xylem vessels of stem, roots and leaves. The type of symptoms of infected plants are often the expression of an occlusion of the conducting vessels by bacterial masses organized in biofilms as well as by thylls and gums produced by the plant in response to infection. As a result, symptoms from *Xylella* often resemble those caused by permanent water stress. The occlusive action of bacterial biofilms is also exacerbated by the activity of an enzyme complex (LesA) secreted by bacterial cells (Nascimento *et al.*, 2016). In some cases, the infection leads the plant to a rapid death.

In some hosts, the bacterium remains confined to the area of infection, whereas in susceptible plants it spreads along the vessels in both directions until it invades all the organs of the plant. The incubation period of the disease, i.e. the time between infection and the appearance of symptoms, can vary from a few months to more than a year depending on the host plant, its age, vegetative conditions, and also the genotype of the infecting bacterial population.

The bacterium is normally easily acquired by the insect vector; it is immediately transmissible (i.e. there is no latency period between acquisition of the pathogen and the ability to inject it into a healthy plant) and persists indefinitely in the adult insect. Therefore, the bacterium is able to overwinter in its vector if the latter remains obviously viable throughout the winter. Since there is no transovarian transmission, there is also no possibility of overwintering in the insect's eggs. More commonly, the bacterium overwinters in the infected host.

Normally, it is not transmitted (but this possibility is not entirely excluded) through infected seed.

The reproduction of *Xylella* in plants appears to be conditioned by temperature: values between 25°C and 32°C (28°C being the optimal temperature for *in vitro* growth) would be favorable for an epidemic development of the disease; on the contrary, temperatures below 12–17°C and above 34°C could negatively affect the survival of the bacterium in host plants. *X. fastidiosa*'s sensitivity to low winter temperatures partly explains its geographical distribution, which appears to be limited to tropical and subtropical areas. However, *X. fastidiosa* has also been reported in Canada, where temperatures can be quite cold (Goodwin and Zhang, 1997).

Propagation means of the bacterium

A natural means of spreading *X. fastidiosa* is by insect vectors. These insects fly short

Table 7.3.2. Presence of *Xylella fastidiosa* in Europe.

Sequence type (ST) and subspecies	Country and provinces
ST1, *fastidiosa*	Spain (Balearic Islands)
ST6, *multiplex*	France (Corsica, PACA Region*);Spain (Balearic Islands, Province of Alicante, Autonomous Region of Madrid)
ST7, *multiplex*	France (Corsica, PACA region);Portugal (metropolitan area of Porto); Spain (Balearic Islands)
ST53, *pauca*	France (PACA region);Italy (Apulia)
ST80, *pauca*	Spain (Balearic Islands)
ST81, *multiplex*	Spain (Balearic Islands)
ST87, *multiplex*	Italy (Tuscany)

*Provenza-Alpes-Cote d'Azur

distances (up to 100 m), but if pushed by the wind they can travel even longer distances.

There is also a passive possibility of transporting *X. fastidiosa* over short and even long distances: the clothes or other body parts of people to which the vector can attach itself or even mechanical means of transport.

Apart from all these possibilities, the most effective way of transport over any distance is through the movement of infected plants.

Vectors

The bacterium lives in the plant xylem tissue and it is normally spread by insect vectors feeding from the plant xylem. Theoretically, all xylem insects would, until proved otherwise, be able to acquire the bacterium from an infected plant and transmit it to a healthy plant (Purcell, 1989; Almeida, 2016). Transmission is, therefore, completely non-specific. In contrast to the high transmission potential that exists in nature, however, it should be noted that the transmission efficiency varies greatly depending on the species of insect, the host plant and also the genotype of *Xylella* (Redak *et al.*, 2004; Lopes *et al.*, 2010). Thus, it may also be the case that the transmission efficiency is so low as to be negligible for infection purposes.

All vector insects of *X. fastidiosa* belong to the Section Auchenorrhyncha (suborder Homoptera, genus Rhynchus) and are included in the three superfamilies Cicadidae, Cercopoidea and Membracoidea (in general, all xylem-inhabiting and sap-feeding hemipterans). All the species of the first two superfamilies (cicadas and spittoons) have xylem-feeding behavior; however, among the Membracoidea, only the species of the subfamily Cicadellinae, family Cicadellidae (cicadas), are xylem-sap-feeding insects.

In Europe, the vectors of *X. fastidiosa* are *Philaenus spumarius* (Cercopoidea: Aphrophoridae), *Neophilaenus campestris* and *Philaenus italosignus* (Aphrophoridae). The last two vectors can transmit *X. fastidiosa* only. Under experimental conditions (Cavalieri *et al.*, 2018).

The most effective and so far only vector of *X. fastidiosa*, under natural conditions, in European territory, is the meadow spittlebug *P. spumarius* (Saponari *et al.*, 2014; Cornara *et al.*, 2017). The term 'spittlebug' is derived from the characteristic production of a very evident whitish foam on the infested vegetation, similar in appearance to human saliva. The function of this foam is to protect the juvenile stages (three ages of neanid and two of nymph) from the drying action of atmospheric agents and predation by natural enemies.

P. spumarius is a univoltine species (only one cycle per year), highly polyphagous (juveniles can develop on at least 375 plant species, mainly herbaceous dicotyledons), with adults and juveniles having a different host range. The juvenile forms prefer the more tender parts of plants and, in any case, herbaceous plants, in particular of the Asteraceae, Fabaceae and Apiaceae families (Di Serio *et al.*, 2019; Dongiovanni *et al.*, 2018a) that often grow wild in cultivated fields or at their margins, or in natural or semi-natural areas. On the contrary, the adults (about 5 mm long and blackish to light-brown in color) preferentially feed on woody plants. However, both juveniles and adults do not cause significant damage to crops. For this reason, *P. spumarius* has never been considered a damaging insect to the extent that it has been the subject of any kind of control measures (occasional minor damage has been observed on grapevines, alfalfa, strawberries and ornamental and horticultural plants). Today, it is subject to control measure, but only as a vector for *X. fastidiosa*.

In Apulia, the first juveniles (first-age nymphs) generally appear at the end of February/beginning of March (Cornara *et al.*, 2017; Di Serio *et al.*, 2019). At this stage, the meadow froghopper encounters the olive tree and if the olive tree is infected, the insect that stings and feeds on it can acquire the bacterium and then transmit it to healthy plants by feeding on and moving through the crowns of olive trees or other shrub or tree species. Once acquired, the bacterium multiplies in the first part of the digestive tract and persists there throughout the life of the insect.

With the autumn grassy shoots, favored by the first rains in late summer/early autumn, the adults move from tree species back to herbaceous plants.

Adult mating is observed from late August to early September and continues into autumn (Cornara *et al.*, 2018). Overall, each female lays an average of 20 eggs, preferably, so it has been observed, in the space between the leaf sheath and the stem of the stubble. Other oviposition sites are tree bark, vegetation residues or the soil surface itself.

The biological cycle of the meadow froghopper is strongly influenced by climatic conditions. A mild winter, with temperatures above seasonal averages, can induce an earlier hatching of the eggs and, consequently, the entire development cycle of the insect.

Vector–pathogen interaction

Transmission to the xylem by the insect occurs with characteristics of both persistent and non-persistent transmission. Indeed, *X. fastidiosa* vectors are capable of inoculating the pathogen immediately after a feeding sting, without the need for an incubation period (as in non-persistent transmission), infectivity can be maintained throughout life (as in persistent transmission), but not through molting (as in non-persistent transmission). This particular mode of transmission is explained by the fact that, despite the persistent form of the infection, the bacterium is localized only in the first part of the digestive tract (adheres to the walls of the pre-ciliary and ciborium), of ectodermal origin (and therefore subject to renewal after each molt), and therefore does not colonize the entire body of the insect (i.e. it is not circulatory) (Almeida, 2016).

Temperature plays an important role in vector–pathogen interaction: it influences bacterial multiplication in the vector, the establishment of the bacterium in the host plant and the behavior of the vector in its natural environment.

Main hosts and symptoms of Xylella *diseases*

Important hosts of *X. fastidiosa* are vines, almonds and some other stone fruits, several citrus species, coffee, oleander, alfalfa, olives, walnuts, several ornamental and forest species, as well as typical Mediterranean shrubs and weeds.

The symptoms are quite variable. In peach trees the growth of the young branches is stunted, the foliage is denser, the lateral branches become horizontal or curved downwards, the whole crown appears uniform, compact, rounded. Leaves and flowers appear earlier and leaf fall is delayed in autumn. In the case of citrus fruits, the most obvious symptom is internal leaf chlorosis similar to that caused by a zinc deficiency. On many other hosts, the most characteristic symptom is leaf scorch. For example, in grapevines, the first known host of *X. fastidiosa*, the first sign of the disease (Pierce's disease) is a sudden desiccation of a portion of the leaf blade. The desiccated part turns brown while the adjacent tissue turns yellow or red. The symptomatic leaves wither and fall off at the petiole. The shoots often ripen irregularly and may show patches that are still green, alternating with brownish portions of tissue. Vegetative recovery is delayed and stunted and normally the infected plant dies in one or two years.

Symptoms of leaf bronzing are also observed on oleander, oak and other forest species, almond trees, *Polygala myrtifolia*, one of the most susceptible species within the ongoing epidemic in Europe, and also on olive trees, the species chosen from scratch or almost chosen by *X. fastidiosa* to make its appearance in the Mediterranean basin.

The syndrome incited by Xylella fastidiosa *on olive*

THE CAUSAL AGENT OF THE DISEASE. The first indications of a possible interaction between *X. fastidiosa* and olive trees date back to 2003/2004. During these years, researchers at the University of Riverside, California, isolated *X. fastidiosa* from several species of ornamental and fruit-bearing plants (*Hemerocallis* sp., jacaranda, magnolia, fragrant broom, cherry, *Cercis canadensis*, liquidambar, peach, plum, olive, etc.). In particular, the strain obtained from olive trees, designated G12, was assigned to the *multiplex* subspecies (Hernandez-Martinez *et al.*, 2006). A few years later (2008), following reports of increased mortality in olive plants affected by olive leaf scorch disease (OLSD) in the Los Angeles area, an investigation into the association between *X. fastidiosa* and symptoms on olive leaves was carried out. The results of the survey are not very clear, in the sense that it was not established for sure, neither then nor a few years later, whether or not *X. fastidiosa* was responsible for leaf scorch symptoms (Krugner *et al.*, 2010, 2014).

More or less since 2010, the symptoms observed on olive trees in California have also started to be seen in Apulia (south Italy). Initially, the two events were not linked; other possible causes were investigated (Martelli, 2016): toxicity due to groundwater pollution; attacks of 'leprosy', a mycosis caused by *Colletotrichum* spp.; infestations of the lepidopteran beetle *Zeuzera pyrina* (yellow rhodiless); inadequate agricultural practices; root rots. After ruling out these different possibilities, the presence of *X. fastidiosa* DNA was finally detected in late summer 2013 (Saponari *et al.*, 2013). A few months later, the bacterium was isolated in pure culture.

Initially, due to the frequent association on older and highly symptomatic olive trees in Apulia of *X. fastidiosa* with some species of lignicolous fungi (*Phaecremonium* spp., *Phaemoniella* spp.) and tunnels dug by *Zeuzera pyrina* larvae (Nigro *et al.*, 2013), the syndrome was named Olive Quick Decline Complex (CoDiRO). Subsequently, the predominant, if not exclusive, role of *X. fastidiosa* was established and the disease was more simply named Olive Quick Decline Syndrome (OQDS). However, the acronym CoDiRO has not been abandoned, but has remained to indicate the population of *X. fastidiosa* that has attacked the olive tree in Puglia.

SYMPTOMS OF THE DISEASE. The first symptoms of OQDS are the appearance of leaf scorch and dryness of branches and small branches with a patchy distribution in the upper canopy (Fig. 7.3.1). As time passes, the dry spots increase in size and severity and spread to the entire aerial part of the plant, which wilts and dies within two to three years (depending on the cultivar and size of the plant) (Fig. 7.3.2). Tissue desiccation starts at the apex of the leaves and progresses towards the petiole, spreading to the entire leaf blade. Dead leaves remain on the branches from which they detach with the autumn rains. The disease in Apulia affected all olive trees regardless of age, size and cultivar. Its outcomes, however, have been particularly devastating on two specific cultivars: 'Cellina di Nardò' and 'Ogliarola salentina', the two varieties that dominate olive growing in Salento (Martelli, 2016).

THE APULIAN STRAIN OF *XYLELLA FASTIDIOSA*. On the basis of specific molecular evidence, the *X. fastidiosa* population that infected olive trees in Salento, Italy (CoDiRO strain), was assigned to the subspecies *pauca*. This population was found to be genetically identical to a strain obtained from oleander in Costa Rica and assigned the number 53 as 'Sequence Type' (ST53). Following this finding, it was assumed that *Xylella* arrived in Italy from Central America with infected plant material. In artificial inoculation tests, the CoDiRO strain was not found to infect vine or citrus species.

THE DISEASE IN ARGENTINA AND BRAZIL. In late 2013, the bacterium *Xylella fastidiosa* was detected in traditional olive orchards in the provinces of La Rioja, Córdoba and Catamarca, in Argentina. Diseased plants (usually of more than 50 years of age) showed marked decline, desiccated branches and apical necrosis in leaves. Olive cultivar 'Arauco' was the most severely affected, whereas the bacterium was detected only in isolated individuals of the cultivar 'Frantoio'. The bacterium detected in Argentina was identified as *X. fastidiosa* subsp. *pauca* ST 69. This same ST has also been obtained in an analysis of citrus from north-eastern Argentina (Misiones and Corrientes) (Coletta-Filho *et al.*, 2017).

At the end of 2014, typical symptoms of OQDS were also observed on olive trees

Fig. 7.3.1. Alliste (Lecce, Apulia, southern Italy), November 2015. Thousand-year-old olive tree ('Il Gigante di Alliste') with twigs and branches drying out.

in Brazil. Diseased trees were found in olive orchards of Maria da Fé (Minas Gerais) and São Bento do Sapucaí (São Paulo), which are approximately 130 km apart. Laboratory studies (morphology, PCR tests, sequencing, phylogenetic analysis, MLST) assigned the bacterium found in Brazil to *Xylella fastidiosa* subsp. *pauca* STs 16 (the most prevalent), 84, 85 and 86 (three STs of new determination). All together, the results obtained in Brazil showed that genetically diverse strains of *X. fastidiosa* subsp. *pauca* were widely present in olive orchards in south-eastern Brazil.

In late 2019, *X. fastidiosa* subsp. *pauca* ST 53, the same ST present in Apulia, was detected in two olive plants at Mentone and one at Cap d'Antibes. The plants have been immediately uprooted and burned.

VARIETAL SUSCEPTIBILITY In the Salento area, the disease caused by *X. fastidiosa* has had devastating effects, partly because the two prevalent varieties in the region, 'Ogliarola salentina' and 'Cellina di Nardò', have been found to be extremely susceptible to the bacterium. In contrast, plants of another variety, 'Leccino', which is not as widespread as the other two, although infected, usually show only minor symptoms of desiccation or even appear asymptomatic. Appropriate studies on the subject have also shown that in infected 'Leccino' plants, compared to 'Ogliarola salentina', only a low concentration of bacteria is present (Giampetruzzi *et al.*, 2016). This is a sign that the plant has a certain tolerance, if not resistance, to the bacterium so as to limit its multiplication in the penetration sites and

Fig. 7.3.2. Alliste (Lecce, Apulia, southern Italy). The same Olive tree as in Fig. 7.3.1 photographed in November 2017.

its spread to the rest of the plant. Another cultivar that shows appreciable resistance to *X. fastidiosa* (exemplified by a low percentage of infected plants and a very low titer of the bacterium in infected tissues), perhaps even higher than that shown by 'Leccino', is the cv. 'FS-17' (also known as 'Favolosa'), patented as a clonal rootstock, but also used for the production of olives for oil. Therefore, there are important sources of resistance to *Xylella* in the olive tree, which can be used, if not to defeat this bacterium, at least to live with it (Boscia *et al.*, 2017).

Other cultivars from which the bacterium has been isolated include: 'Arauco' and 'Frantoio' in Argentina; and 'Negroa', 'Coratina', 'Rodapés', 'Grappolo', 'Arbequina', 'Arbosana', 'Ascolana' and 'Koroneiki' in Brazil.

Control strategies against Xylella

General measures against *X. fastidiosa* in the European territory are included in the EU Decision 2015/789 of 18 May 2015 as amended by the Commission Implementing Decision (EU) 2017/2352 of 14 December 2017. When the presence of the specified organism (i.e. *X. fastidiosa*) is identified, the Member State shall without delay demarcate an area consisting of an 'infected zone' and a 'buffer zone'. The 'buffer zone' is an area pathogen free at the border of the infected zone, it shall be of a width of at least 5 km surrounding the 'infected zone'.

In Apulia, where the pathogenic action of *X. fastidiosa* has been particularly devasting on olive plants, the 'infected zone' is an area delimited to the north by a line extending continuously from the Adriatic to the Ionian Sea. The 'buffer zone' is represented by a strip of land 10 km wide. Moreover, a 'containment zone' has also been defined, i.e. an area within the 'infected zone', immediately below the 'buffer zone', in which all necessary measures must be implemented to limit the spread of the bacterium.

Control of *X. fastidiosa* can be carried out by different means and using different strategies. The targets of control interventions can also be different: the vector, the bacterium, both the vector and the bacterium, and finally the host plants that act as reservoirs of *X. fastidiosa*. Control can then be carried out according to different criteria depending on the zone where it is necessary to operate: the 'buffer zone', the 'containment zone', the 'infected zone'. In the first two zones, the effort must be to try to stop the advance of the bacterium; in the third, it is simply a matter of finding the best ways to live with the disease while waiting and hoping to develop more definitive control means of *X. fastidiosa*.

The 'Emergency measures for the prevention, control and eradication of *Xylella fastidiosa*', updated on 6 April 2018 (Decree No. 4999 of the Italian Republic) pursue the objectives to reduce the population of *P. spumarius* adults as much as possible, minimizing new infections, and to reduce the sources of inoculum, represented by infected plants, as much as possible.

In the 'buffer zone', within a radius of 100 m around even a single plant found to be infected, appropriate phytosanitary treatments must first be carried out against the vectors of *X. fastidosa* and plants that may harbor the vectors, and then eradication measures must be taken consisting of the immediate removal of: (a) host plants of the bacterium, irrespective of their health status; (b) infected plants; (c) plants showing symptoms.

In the 'containment area', infected plants must be removed, carrying out, prior to the removal, appropriate phytosanitary treatments against vectors and, within a 100 m radius around infected plants, sampling and testing of host plants must be realized at regular intervals and at least twice a year.

In the 'infected area', it is not compulsory to uproot infected plants. Plants seriously affected by the disease and showing a considerable and permanent loss of productivity may be eradicated. Olive trees must be planted in accordance with good agricultural practice and pruned in order to improve the aeration of the tree and its vegetative condition. Vectors of *X. fastidiosa* must be controlled by mechanical operations, to eliminate wild herbaceous plants, and insecticide treatments in the early hours of the morning, when the insects are not very mobile. Dimethoate, which is used in olive groves to control the olive fly (*Bactrocera oleae*) and the olive moth (*Prays oleae*), has low efficacy against *P. spumarius*. Formulations based on buprofenzin, pymetrozine and spirotetramat have also shown negligible efficacy. In contrast, the pyrethroids deltamethrin and lambda-cyhalothrin and the neonicotinoids imidacloprid, thiamethoxam and acetamiprid (and also etofenprox) have been shown to have high knock-down power and a persistence of almost two weeks. Results of some interest were also obtained with calendar applications of kaolin on healthy plants exposed to natural *Xylella* infections and with sweet orange essential oil (which can also be used organically) (Dongiovanni *et al.*, 2018b).

With regard to the pruning of the trees that are the hosts of the bacterium, it is recommended to remove new vegetation and suckers that are softer and more attractive to the vector as often as possible, thus greatly reducing the mutual risk of infection transmission. Tools used for pruning must be disinfected with a 2% solution of sodium hypochlorite or quaternary ammonium salts.

On the other hand, no effective means or control methods of the bacterium are available. A mixture of copper and zinc complexed with citric acid (in the form of hydracide) (Dentamet) has only been shown to reduce the reproduction of the bacterium in the olive tree and to alleviate the symptoms of the disease (Scortichini *et al.*, 2018). Confirmation of Dentamet's use in the field is awaited. Apart from this and other similar attempts to control OQDS, the best way to control this disease is to cultivate olive varieties that are tolerant or resistant to *X. fastidiosa*.

7.3.2.2 Olive knot

Olive knot (OK) disease, caused by the Gram-negative bacterium *Pseudomonas savastanoi* pv. *savastanoi* (*Pss*), is one of the most serious diseases affecting olive trees in most olive-growing areas of the world, in particular the Mediterranean region (Young, 2004). It can substantially reduce yield, and there is evidence that a high incidence of the disease compromises the quality of oil by imparting an off-flavor (Lelliot, 1988). The first documented record of OK dates back to the 4th century BC by the Greek Theophrastus. The name of the disease refers to the tumors or hyperplastic galls (knots) caused by this bacterium on the aerial part of the olive trees, mainly on twigs, shoots, branches and stem, and only occasionally on leaves and fruits (Fig. 7.3.3). OK can be considered a chronic disease as its symptoms may persist for many years on the tree. Galls form prevalently at the point of emergence of new leaves or on leaf scars, pruning cuts, wounds and abrasions. Initially, they are small swellings that develop within a few months into more or less round, irregularly splitted, ochre knots, spongy at first and becoming woody and brown later; on shoots and small branches knots are up to about 2 cm in diameter. On fruit petioles and leaf midribs, small knots or tubercles may develop and cause premature abscission (Lelliot, 1988).

Pss is able to survive as an epiphyte on olive phyllosphere where its populations fluctuate depending on the seasons, with peaks in April and October. The bacterium is a wound-infecting pathogen; it enters the plant through wounds caused by harvesting, pruning, hail, frost and leaf-scars; it needs molecular signals released by the plant as a consequence of wounds to activate the genesis of tumor (Buonaurio *et al.*, 2015). In fresh wounds, *Pss* initially colonizes the tissue around the infection and court by pectolytic and hemicellulolytic enzymes disrupts the surrounding tissues producing lysigenous cavities that are filled by bacterial cells. Alternatively, it may invade the vascular tissue and become systemic (Marchi *et al.*, 2009). It was shown that in artificial inoculations, secondary tumors formed in distant positions from inoculated wounds on the stem, suggesting the migration of the pathogen within the plant (Penyalver *et al.*, 2006). Knots form as a consequence of an increase in size (hypertrophy) and an abnormal division (hyperplasia) of plant cells induced by the production of indol-3-acetic acid (IAA) and cytokinins, which are the main virulence factors of the pathogen (Iacobellis, 2001). Besides the phytohormones, other virulence factors are involved in disease development (Buonaurio *et al.*, 2015). Quorum-sensing intercellular regulation and communication system, mediated by N-acylhomoserinelactones, also plays a major role in *Pss* virulence (Hosni *et al.*, 2011). Inside olive knots, *Pss* constitutes a microbial consortium with other resident endophytic bacteria, including

Fig. 7.3.3. Symptoms of olive knot caused by the bacterium *Psudomonas savastanoi* pv. *savastanoi*.

species of *Pantoea, Clavibacter, Curtobacterium, Enterobacter, Erwinia, Hymenobacter, Kineococcus, Pectobacterium* and *Sphingomonas,* that are not harmful to the plant (Buonaurio *et al.*, 2015). It has been suggested that such microbial consortium may influence the disease process (Mina *et al.*, 2020).

Bacterial cells are abundant in and exude from developing galls and are spread by rain splashes. Incidence of the disease is greatest in windy places, after hailstorms and frost, where pruning is careless, in small fruited cultivars that are harvested by beating trees, and also depends on the susceptibility of cultivars. Table 7.3.3 reports a list of olive cultivars that were rated phenotypically for their susceptibility to OK disease. This list is only indicative and shows many incongruences as in artificial inoculations disease severity and cultivar susceptibility are influenced by the dose of inoculum, the virulence of *Psv* strain and the age of test plants (Penyalver *et al.*, 2006). However, no olive cultivar so far tested can be considered immune or completely resistant.

The management of OK disease is based on preventive measures. In areas at risk in the Mediterranean region, annual pruning, which is also the opportunity to remove shoots and twigs with knots, should be completed by February and the main fertilizer application should be in January/February. Spraying with copper compounds in late autumn and in spring is advisable. These treatments are also effective against olive scab and cercosporiose.

7.3.2.3 Phytoplasma diseases

Phytoplasmas are a large group of wall-less bacteria that constitute a monophyletic clade within the class Mollicutes. Their classification is based on the sequencing of the conserved 16S rRNA gene followed by the restriction fragment length polymorphism analysis (RFLP). They are currently classified within the provisional genus '*Candidatus* Phytoplasma'. Four phytoplasma species have been isolated from olive trees in Italy and Spain.

The phytoplasmas identified in olive trees have been classified as 16S-IB (Aster yellows group), 16S-VA (Elm yellows group) and 16S-XIIA (Stolbur group). Infected plants show various symptoms such as foliar yellowing, proliferation (witches' broom), shortening of the internodes, fasciations, bud failure, floral abortion and hypertrophy of inflorescences. Also mixed infections, with double or multiple phytoplasmas, were detected in olive trees. It is suspected that the cicadellid *Hyalesthes obsoletus* can transmit the phytoplasmas in the field.

Phytoplasmas appear to be ubiquitous in olive plants but a clear association of the disease with one or more *Candidatus* Phytoplasma species has not been established (Pasquini *et al.*, 2000; Caglayan *et al.*, 2011). Consequently, the Olive phytoplasma disease was added to the EPPO Alert List in 1999 and then deleted in 2000.

7.3.3 Fungus and Oomycete Diseases

7.3.3.1 Olive leaf spot (Scab, Peacock spot)

Introduction

Olive leaf spot (OLS), caused by the fungus *Venturia oleaginea* (synonyms *Spilocaea oleagina* and *Fusicladium oleagineum*), is one of the most common and serious diseases of olive groves. Since the teleomorphic stage of *V. oleaginea* is not known, the fungus has been classified for a long time in the mitosporic fungi, ord. Hyphales, fam. Dematiaceae. According to rDNA sequencing analysis, it was more recently placed in the Dothideomycetes class of the Ascomycota phylum and, based on phylogenetic analysis, it was assumed that it is an anamorphic phase of the genus *Venturia*.

In recent years, due to climate change, the incidence and severity of OLS in the Mediterranean region have increased significantly during late winter and early spring, causing very serious damages.

The disease is widespread in all olive-growing areas of the world and is recognized by several names, the most common of which recall the similarity of leaf symptoms

Table 7.3.3. Level of susceptibility of olive cultivars and genotypes to olive knot reported from different countries[a].

Cultivar and susceptibility class[bc]
'Frantoio' (H), 'Ascolana Dura'(H), 'Manzanilla'(H), 'Picholine Marocaine' (M), 'Dahbia' (L), 'Gordal Sevillana' (L) (Benjama *et al.*, 1987)
'Cordovil de Serpa' (H), 'Galega Vulgar' (H), 'Branquita' (M), 'Santulhana' (M), 'Rendonil' (L), 'Cobrancosa' (L) (Marcelo *et al.*, 1999)
'Cellina di Nardò' (H), 'Frantoio' (H), 'FS-17' (H), 'Morcana' (H), 'Nociara' (H), 'Ogliarola' (H), 'Pendolino' (H), 'Carolea' (M), 'Bella di Spagna' (M), 'Cerasella' (M), 'Cima di Melfi' (M), 'Coratina' (M), 'Corniola' (M), 'Dolce Agogia' (M), 'Leucocarpa' (M), 'Maiatica di Ferrandina' (M), 'Noica' (M), 'N3' (M), 'Nolca' (M), 'San Felice' (M) (Sisto *et al.*, 2001)
'Nocellara del Belice' (H), 'Manzanillo'(H), 'Picholine'(H), 'Leccino' (M), 'Carolea'(L), 'Koroneiki'(L), 'Pendolino'(L), 'Coratina'(L) (Young, 2004)
'Arbequina' (H),'Arróniz' (H), 'Nevadillo Blanco de Jaén' (H), 'Pajarero' (H),'Picudo' (H),'Vallesa' (H),'Changlot Real' (M), 'Morisca' (M), 'Gordal Sevillana' (M), 'Lechín de Sevilla' (M), 'Oblonga' (M), Picual' (M), 'Ascolana Tenera' (M), 'Royal de Cazorla' (M), 'Mollar de Cieza' (M), 'Koroneiki' (M), 'Azapa' (L), 'Cerezuola' (L), 'Chemlali '(L), 'Dulzal de Carmona' (L), 'Frantoio' (L), 'FS-17' (L), 'Gordal de Alchidona' (L), 'Gordal de Hellín' (L), 'Lechín de Granada' (L), 'Manzanilla Cacereña' (L), 'Manzanilla de Sevilla' (L), 'Nevadillo Negro' (L), 'Villalonga' (L) (Penyalver *et al.*, 2006)
'Leccino' (M), 'Picholine' (H) (Godena *et al.*, 2012)
'Canino' (H), 'Bosana' (H), 'Casaliva' (H), 'Cerasuola' (H), 'Frantoio' (H), 'Maiatica di Ferrandina' (H), 'Tonda di Strongoli' (H), 'Coratina' (M), 'Ascolana Tenera' (M), 'Giarraffa' (M), 'Leccino' (M), 'Moresca' (M), 'Nocellara del Belice' (M), 'Ogliarola Barese' (M), 'Ogliarola Campana' (M), 'Ogliarola salentina' (M), 'Ortice' (M), 'Pizz'e Carroga' (M), 'Rotondella' (M), Sant'Agostino' (M), 'Taggiasca' (M), 'Zaituna' (M), 'Tonda Iblea' (M), 'Buscionetto' (L), 'Carolea' (L), 'Cellina di Nardò' (L), 'Cima di Mola' (L), 'Cucco' (L), 'Dritta' (L), 'Gentile di Chieti' (L), 'Grossa di Gerace' (L), 'Ogliarola Messinese' (L), 'Ortolana' (L), 'Pisciottana' (L), 'Santagatese'(L), 'Tonda di Cagliari' (L), 'Tondina' (L) (Lombardo, 2009)
'Tragolea' (H), 'Barnea' (H), 'Arbosana' (H), 880336 (H), 'Toffahi of Egypt' (H), 'Bouteillon' (H), 'Koroneiki' (H), 'Zoragi' (H), ' Bouquetier' (H), 'Balady' (H), 'Galega Vulgar' (H), 'Azapa' (M), 'Cordovil' (M), 'Verdale' (M), 'Redondilla de Logrono' (M), 'Chemlali di Sfax' (M), 'Chalkidiki' (M), 'Meski' (M), 'Uovo di Piccione' (M), 'Grossane' (M), 'Salome' (M), 'Giaraffa' (M), 'Grosse Aberkan' (M), 'Merhavia' (M), 'No. 65A' (M), 'Chetoui' (M), 'Syrogylolia' (M), 'Liguria de Catamarca' (M), '355-117' (M), 'Maurino' (M), 'Kalamata' (M),'Thrombolea' (M), 'Ascolana Tenera' (M), 'Karydolia' (M), 'Piconia' (M), 'Late Blanquette' (M), 'Vassilica' (M), 'Laccio' (L), 'Manzanilla de Sivilla' (L), 'Cucca'(l), 'Frantoio'(L), 'Lechin de Sevilla'(L), 'Oblonga'(L), 'Grappolo'(L), 'Oblonga Seedling'(L), 'Nabali' (L), 'Columello' (L), 'Verdal' (L), 'Black Italian' (L), 'Barouni' (L), 'Dwarf D' (L), 'Redding Picholine' (L), 'Souri' (L), 'Adrouppa' (L), 'Jlot' (L), 'Prunara' (L), 'Konservolia' (L), 'Toffahi of Syria' (L), 'Rubra' (L), 'Nikitskaya Krupnolodnaya' (L), 'No. 63' (L), 'Mastoidis' (L), 'Gordal Sevilliana' (L), '880383' (L), 'No.31 Sevillano' (L), 'Cayet Roux' (L), 'Rasie' (L), 'Ogliarola Messinese' (L), 'Memeli' (L), 'Arbequina' (L), 'No.1 Sevillano' (L), 'Menara' (L), 'San Francesco' (L), 'Cypress 31' (L), 'Campanil' (L), 'Nab Tamri' (L), 'No.12 Sevillano' (L), 'Obliza' (L), 'Gaidourelia' (L), 'Megaron' (L), 'UC 49-14' (L), 'Sigoise' (L), 'Mission' (L), 'Mixanil' (L), 'Cerignola' (L), 'Picual' (L), 'Dolce del Marocco' (L), 'Aggezi Shami' (VL), 'DOLE 116' (VL), 'Massabi' (VL),'DOLE 90' (VL), 'Leccino' (VL), '#6' (VL), 'Thoumbolia' (VL), 'Aglandau' (VL), 'Rigali' (VL), 'Empeltre' (VL), 'UC52.24.1 hybrid' (VL), 'Nikitskaya #1' (VL), 'Picholine Marochaine' (VL), 'Ascolana dura' (VL), 'Karolia' (VL), 'Coratina' (VL), 'Cailletier' (VL) (Salman *et al.*, 2020)
'Piantone di Mogliano'(H), 'FS-17'(H), 'Frantoio'(H), 'Carboncella' (M), 'Maurino' (M), 'Arbequina' (M), 'Ascolana Tenera' (L), 'Leccino' (L), 'Piantone di Falerone' (L), 'Rosciola Colli Esini' (L) (Valverde *et al.*, 2020)

[a]The list includes results of both artificial inoculations and in field evaluations; it is not intended to be exhaustive.
[b]The susceptibility level of cultivars was rated phenotypically as high (H), medium (M), low (L) and very low (VL), to facilitate the comparison.
[c]The original names of olive cultivars and genotypes are not always reported correctly by the authors.

to the eye spots of the male peacock's tail. OLS was described for the first time as *Cycloconium oleaginum* in 1845 near Marseille, France. Until 1893, the disease was quite rare and considered a botanical curiosity rather than a pathogen. The spread of OLS at the end of the 19th century in the Mediterranean region was related to the very favorable environmental conditions that characterized those years. Initially, the disease was reported with a strong impact in the central-northern areas and was very dangerous in Tuscany (central Italy), where, starting from 1894, pioneering trials with Bordeaux mixture were scheduled. Subsequently, very destructive attacks were recorded in southern regions characterized by high humidity and mild temperatures throughout the year. In 1964 and 1966, in the Calabria region, cupric sprays with airplanes were applied on about 30,000 hectares (Agosteo and Schena, 2011).

Fig. 7.3.4. Round spots incited by *Venturia oleaginea* on infected leaves. On the left, two leaves with blackish sporulating spots.

Symptoms and damage

The disease causes symptoms on leaves, shoots, leaf stalks, fruit pedicles, inflorescences and fruits. The typical 'peacock's eye' spots appear on the upper leaf surface. They can vary in size (2–10 mm in diameter) and in number. There is a strong correlation between disease incidence (percentage of infected leaves) and severity (number of spots per leaf). The first symptom of the disease appears as small, roundish spots of a blackish color; subsequently, the spots are arranged in concentric circles with an olive-green to dark-olivaceous center and peripheral rings with reddish-brown to yellow or green halos. As a consequence of the presence of cytokinins, the infected areas can also appear as green to dark-green circular spots on a chlorotic leaf blade (Fig. 7.3.4). The colors of spots and concentric halos can depend on the stage of the disease, the season and the variety. The black color of the initial spot on the leaf surface is due to the presence of the brown conidia and to the sooty color of the walls and other parts of the epidermal cells. The presence of the brown conidia on the upper leaf surface gives to the sporulated spots the typical black and velvety appearance (Fig. 7.3.4). Intense attacks occur on the leaves of susceptible cultivars in humid areas (Fig. 7.3.5).

Fig. 7.3.5. Severe foliar attacks of *Venturia oleaginea* on the canopy of an olive tree of the susceptible cv. 'Carolea'.

Leaf susceptibility largely depends on cultivar and age. Young leaves are more susceptible and show a shorter incubation period than older ones (López-Doncel *et al.*, 2000). Disease severity on the youngest leaves was about three to five times higher than on the oldest leaves (Obanor *et al.*, 2011). Recent analyses with a quantitative PCR detection method have revealed the presence of the pathogen in the asymptomatic new leaves, in very early stages of growth, when the new leaves open (leaves with a mean length of 5.4 mm) (Scibetta *et al.*, 2020).

On the lower leaf surface, symptoms are evident on the central vein, as dark-brown elongated streaks. An atypical symptom, known as 'white spots', appears on the upper blade of leaves as round, white spots when unusual weather conditions occur, e.g. when dry weather follows a rainy period. Another rather rare type of white spots is 'summer spots', which occurs on old lesions that turn whitish after the release of conidia in late spring or summer (Fig. 7.3.6).

Attacks on tender shoots result in small brown sunken spots. OLS causes the appearance of small brown spots on the petioles of the leaves, often at the base, near the shoot, which are quite unnoticeable but cause leaf fall, sometimes without showing symptoms on the upper leaf blade. Some reports of defoliation without typical foliar symptoms might refer to this type of infection, although an accurate diagnosis is quite difficult. Dark-brown to black spots, which can enlarge and merge, can be also observed on fruit pedicles and lead to early fruit drop or make the drupes wither. The incidence of this symptom may be related to a specific varietal susceptibility which may not be associated with the leaf susceptibility. Infections on olive fruits are quite rare, but can have a high incidence in the very rainy periods of late summer and autumn. Attacks may involve green or, more frequently, drupes near maturity. On green drupes the disease causes small, sunken brown lesions. The drupes shrivel and deform due to the blocked growth of infected tissues. The brown spot may be surrounded by a reddish halo in relation to the cultivar and/or environmental conditions. Near maturity, the infected areas are irregular in shape, sunken, and remain green with a greyish to brown efflorescence, due to conidial production.

The most serious damage caused by OLS to olive plants is the early fall of symptomatic leaves. Defoliation occurs intensively after recurrent leaf spot infections, above all on the lower branches, producing reduction of leaf coverage, losses of vegetative vigor and death of shoots. In addition, the weakened plants are easily subject to other adversities, mainly insect attacks, which aggravate their vegetative state and

Fig. 7.3.6. 'White spots' (*left*) and 'summer spots' (*right*) incited by *Venturia oleaginea* on olive leaves.

can lead to death, especially in young plants. Undefended trees may decline rapidly in productivity. The fall of infected leaves occurs after a variable time and is closely related to the attainment of extreme temperature conditions, as a result of the thermal shock suffered by the leaf tissues exposed to the atmospheric factors by the lesion of the cuticle caused by the fungal eruption. Thus, in the olive-growing areas of the southern Mediterranean region, characterized by mild winters, the fall of the new annual, symptomatic, leaves occurs in three to four months, from March to June, coinciding with the return of warm weather conditions (Agosteo and Zappia, 2007). Similarly, in olive-growing regions characterized by cold winters, the thermal shock of low temperatures causes the leaf falls in this season.

Premature leaf losses cause the non-conversion of axillary buds, with the reduction of flowering shoots in favor of macroblasts. Infections can reduce the lengths of olive inflorescences and fruit set, with severe reductions in olive fruit yields (Issa *et al.*, 2019). When infections affect the pedicles of the fruits they can cause the premature fall of the olives, while direct attacks on the drupes produce a delay in ripening and a decrease in oil yield, without affecting the organoleptic quality of the oil.

Life cycle

V. oleaginea is an obligate biotroph pathogen. Infection takes place starting from conidia. The inoculum for new infections originates from sporulating spots on infected leaves of the tree canopy. The infected leaves fallen on the ground play no role for infections, since the conidia cannot be transported on the trees.

After active penetration in the host, *V. oleaginea* remains restricted to the outer cuticular layer of the epidermal cell wall, where the mycelium expands to form round, flat, mono-layered, immersed colonies. At the end of the vegetative growth and of the incubation period, starting from the sub-cuticular hyphae, short bulbous single-celled brown-olivaceous conidiophores (10–17 x 10–12 μm) erupt outside by piercing the leaf cuticle. Conidiophores produce, in succession and one at a time, brown-olivaceous conidia with oval-pyriform shape, truncate at the base and elongated at the top, single-celled when young and two-celled at maturity, measuring 15–30 x 9–15 μm. A collar at the apex of the conidiophores marks the production of each conidium and up to 10–11 collars can be observed on each conidiophore.

About 24–48 h of leaf wetness were necessary within the optimal temperature range (16–24°C) to detach conidia from infected leaves (Saad and Masri, 1978). Conidia are usually washed down by rainwater and are not effectively disseminated by air currents. Infections are most severe in the lower parts of the canopy, where the inoculum comes from the upper parts of the canopy and humidity is also higher. Limited lateral spread is ensured by winds carrying conidial water droplets or, in the absence of rain, by humid winds (RH above 70%) during autumn and winter (Lops *et al.*, 1993). Conidia in the air have been detected up to 40 m away from infected trees (Viruega and Trapero, 1999). Dissemination by vectors such as insects has also been demonstrated.

The germination of conidia occurs within a wide range of temperatures (from 2–3°C to 28–34°C) with the optimum between 16°C and 21°C. Most conidia germinate within 15–18 hours within the optimal temperature range. Over a wide temperature range (5–25°C) the percentage of germination is directly related to the duration of leaf wetness (Obanor *et al.*, 2008, 2011) and greatly influences the severity of the disease (Viruega and Trapero, 2002). At the optimal temperature for infections (20°C) and with 100% RH, six hours of leaf wetness are required for the germination of the conidia. The newly formed conidia show a latency period, during which germination does not occur. It can persist for 8–10 days, but the action of low temperatures can reduce it. The viability of the conidia is long-lasting on hanging leaves and gradually decreases becoming very low after about 6–7 months from differentiation. On detached leaves they lose viability within seven days.

Once the conidia germinate, the establishment of infections need free moisture or saturated air for at least 14 hours. Differentiation of appressoria occurs six hours after the first signs of germination and it is more rapid at 15°C but does not occur at 25°C, after 48 h of incubation (Obanor *et al.*, 2008). As a consequence, approximately 1–2 days of leaf wetness and saturated atmosphere conditions are required for infections.

The duration of the incubation period varies widely, especially as a function of the climatic conditions after infection. Field investigations, with implementation of early diagnosis, have revealed that incubation periods can vary from a minimum of about 10–14 days, on new leaves in spring, with optimum temperature and RH condition, up to a few months (about 90–140 days), when infections are blocked by extreme temperature conditions and/or dryness. Generally, the longest incubation periods occur for late spring infections that enter into latency, blocked in their evolution by the onset of high temperatures and summer dryness. Even the spots already formed in spring can stop their growth until autumn when, coinciding with the first rains, they start to grow again, form new rings and sporulate.

Traditionally two distinct periods of annual infections are reported, in the autumn and spring, with a long stasis of pathogen activity in winter and summer. This pattern corresponds to the climatic conditions of the northern olive-growing areas of the Mediterranean basin, such as central and northern Italy, characterized by colder winters. In olive-growing regions characterized by mild winters such as southern Italy, southern Spain, north Africa and the Middle East, the two above-mentioned periods of infection can be merged into one, without interruption, only with an extended stasis in summer (Viruega and Trapero, 1999; Agosteo and Zappia, 2007). At most, low temperatures in winter can affect how quickly infection occurs and symptoms develop. Sometimes, dry fall conditions can further delay the onset of leaf spot symptoms until winter, favoring longer incubation periods (latency) of late-spring infections. In these conditions the maximum rate of infection is frequently recorded in late winter or early spring.

In all growing areas, when rainy weather conditions occur, spring is the most critical period for infection since they involve the new and more susceptible vegetation. In the last two decades or so, due to climatic change and global warming, the infectious cycle of the disease has changed in the southern Mediterranean regions, progressively shifting from autumn, often dry, to winter, rainy and milder. In specific field surveys, the incidence of the disease on the new annual vegetation began to increase in December and peaked in March, with over 80% of the leaves of the previous year's new vegetation infected (Agosteo and Zappia, 2007). In these conditions a very high inoculum amount is available for early infection to the new spring vegetation and the evolutions of spots may take place rapidly, in the same season. In the Calabria region, southern Italy, in the last decades, record levels of infection have been recorded in late winter, with very severe defoliation and total loss of production on susceptible varieties.

Detection methods

Biological diagnosis of OLS is not feasible as it is extremely difficult to isolate the pathogen on nutrient media *in vitro*. The diagnosis is based on the observation of symptoms, which is rather easy when leaf spots are visible. However, since the incubation periods are often very long, the early diagnosis is very problematic. This has led in the past to misinterpretation of the resistance or immunity of young leaves to infections. An important contribution to the study of the epidemiology of the pathogen was given by Loprieno and Tenerini (1959) with the development of a simple method for the early diagnosis of infections in symptomless leaves. The leaves are immersed in a warm (50–60°C) 5% NaOH or KOH solution for two to three minutes, until small, roundish black spots appear where the fungal colonies are localized. For young leaves the solution should not be heated (about 20°C). Black spots caused by *V. oleaginea* in the treated leaves can be discriminated from other types of similar stains since they are opaque to light when the leaves are observed in backlight. A more simple application that enabled the detection of 95% of spots, followed by farmers and farmers' associations, for field diagnosis, is to carry out the immersion of the leaves in NaOH solution at room temperature (10–30°C) for about 20 minutes. The early diagnosis method has been widely applied to study the life cycle of *V. oleaginea* and to determine the duration of the latent phase. However, important epidemiological aspects remain controversial because this detection method is ineffective during the early stages of infection and enables the identification of infected leaves only if the pathogen is at an advanced stage of colonization (maximum detection was achieved 28 days after inoculation). It also does not provide accurate quantification and cannot be done on very young, tender leaves that are damaged by sodium hydroxide during the immersion process.

Recently, a quantitative PCR (qPCR) assay has been developed, which proved very specific and sensitive to detect and quantify the pathogen (Scibetta *et al.*, 2020). Analyses conducted with this molecular method revealed the presence of the pathogen in the asymptomatic leaves from very early stages of growth (leaves with a mean length of 5.4 mm). These findings are particularly important for improving OLS management strategies.

Control

GENETIC RESISTANCE. The existence of different degrees of susceptibility to OLS among olive cultivars and clones is well known (Table 7.3.4). In general, the resistance to OLS is one of the most important characters to define varietal adaptability in a specific area. The diffusion of the very resistant cv. 'Leccino' in Tuscany was due to the appearance of OLS in this region. Similarly, attempts to introduce the cv. 'Carolea' into humid olive-growing areas of southern Italy failed because of high susceptibility of this cv. to OLS. The use of resistant cultivars is of paramount importance in the case of super-intensive orchards because of the excessive canopy thickness and high leaf coverage.

The resistance or susceptibility to the disease has been related to different characters. A positive correlation was found between resistance and thickness of the palisade cells in leaves. Resistant cultivars contain higher contents of oleuropein aglycone, a strong inhibitor of the cell-wall-degrading enzymes produced by the pathogen. The measurement of the content of oleuropein aglycone was suggested as a biochemical marker for the selection of resistant cultivars. The presence of pectin substances in the cutin of the cuticolar layer is a condition that leads to plant susceptibility. In the cuticle of wild olive (oleaster) and in rustic and resistant varieties, such as 'Leccino', there is an almost complete absence of pectin. This different chemical composition was also related to the thickness of the cuticular layer since susceptible varieties were characterized by a thicker one. Some of these traits do not always represent constitutional properties since they are often greatly influenced by specific plant nutrition conditions. Consequently, disease resistance can partially or totally be overcome in stressing environmental conditions, and varietal response to disease depends on the environmental conditions. The response to *V. oleaginea* attacks may be also influenced by clonal heterogeneity within olive cultivars. These factors, together with the varying degree of susceptibility of the different organs (i.e. leaves, fruit pedicles, fruits), may in part explain the numerous conflicting reports on varietal resistance from different growing areas. For example, the cv. 'Manzanilla' is reported as highly susceptible in Spain and Italy, highly resistant in Egypt and of low susceptibility in California. Similarly, the varieties 'Taggiasca' and 'Frantoio' are considered very susceptible in their area of origin (Italy) but are reported to be resistant in Montenegro and New Zealand, respectively (Table 7.3.4).

AGRONOMIC CONTROL METHODS. Cultural practices are a valuable tool for the control of the disease. Low-density plantations and accurate pruning are very important to

Table 7.3.4. Response of olive varieties to infections by *Venturia oleaginea*[a].

Variety	R/S*	Variety	R/S*	Variety	R/S*
Abou-Satl	R	Favolosa (FS-17)	MS, MR	Olieduterza	S
Achouppa	S	Farga	S	Oliva di Cerignola	S
Aglandau	MR	Favarol	S	Olivella	MR
Agrapalo	R	Frantoio	HS-R	Orbetana	R
Alameno	S	Gemlik	HS	Ottobratica	R, HR
Alfafara	S, MS	Gentile di Chieti	S, R	Palomar	HS
Aloreña	S	Gerboui	HR	Pendolino	HS-R
Amellau	S	Giarraffa	S	Pendolo	S
Arbequina	MS, HS	Gordal de Granada	S	Piangente	R
Arbosana	R, HR	Gordal sevillana	MS	Picholine	R
Ascolana	MR,R	Gordal de Archidona	R	Picholine	S
Ayvalik	HS	Grignan	R	Picholine Langudoc	MR, R
Aurina	S	Groia	R	Picual	HS, S
Bardhi Tirana	R	Grossaio	HS	Picudo	HS, S
Becarut	R	Grossane	MR	Pignola	MR
Bouitellan	MR	Grossa di Gerace	MR	Pisciottana	R
Barnea	S, MR	Grozdaka Lastovka	R	Pizz'e Carroga	S
Barouni	S-R	Haouzia	HR	Punteruolo	R
Bella di Spagna	HR	Hojiblanca	HS, S	Quercetano	R
Bianchera	MR	Itrana	HS, R	Racimal	R
Biancolilla	MS, MR, R	Kalamon	S-R	Rama pendula	R
Blanqueta	HS	Kaliniot	HS, R	Ravece	HR
Blanquette	S	Koroneiki	R	Razzaio	MS
Bosana	HS	Ladoelia	S	Razzo	HS, S
Caiazzana	R	Languedoc	R	Redondil	S
Calatina	R	Lastovka	R	Roggianella	R, HR
Callosina	S	Lazzero di Prata	HS	Royal de Cazorla	S, MS
Canino	S, R	Leccese	R	Rosciola	S
Changlot real	S	Leccino	R, HR	Rotondella	R
Carboncella	S, HS, R	Leccio del corno	R	Rotondello	MR
Carolea	HS, S	Lechin de Sevilla	HR	Rougette	R
Carpellese	MR	Lechin de Granada	HS	Rougot	S
Carrasqueño	S	Leucocarpa	R	Rubra	MR
Casaliva	S	Levantinka	S	Saint Agnes	R
Cassanese	R, HR	Limki	S	Salonenque	MR
Castellano	MR	Limli	MR	San Felice	MS
Cellina	R, HR	Limoncillo	MS	San Francesco	R
Cellina di Nardò	R	Loaime	S	San Pardo	HS
Cerasuola	MS	Locques Amellau	S	Santa Caterina	S-R
Chaibi	MR	Lucio	S	Sant'Agostino	S, MS, MR, HR
Chemlal de Kab.	MR	Madonna dell'Im.	R	Sevillano	MR, R
Chètoui	S	Maiatica di Ferr.	S	Sevillenca	S
Chondrolia Chalk.	S	Makharia	S	Shemlali	R
Cima di Mola	HR	Manzanilla blanca	S	Silvestre	R
Cirmignola	MR	Manzanilla de Sevilla	HS	Sigoise	S
Cipressino	HR	Manzanilla de Jàen	HS	Sinopolese	S, MS
Columbello	MR	Manzanilla de Zahara	S	Syrian	S
Coratina	HS, S, R	Manzanilla cacerena	MS	Sorani	R
Corbella	HR	Manzanilla de Montefrio	HR	Souri	S
Cornezuelo	HS	Morisca	HS	Spanish	R
Cornia	MR	Nevadillo de S. Martos	MR	Susi	R

Continued

Table 7.3.4. Continued.

Variety	R/S*	Variety	R/S*	Variety	R/S*
Cornicabra	HS, S	Nevadillo blanco	HS	Tanche	S
Corniola	S	Nevadillo negro	HS	Taggiasca	S, R
Correggiolo	S, MS	Nevado azul	HS	Temprano	R
Crognolina	S	Nocellara del Belice	S, MS, MR	Tonda iblea	MR
Cucco	MR, R	Nocellara etnea	S, MR	Uljarica	S
Cucco Teshi	S	Nocellara messinese	R	Uovo di piccione	S
Curatora	S	Nociara	MS	Vera	S
Dda de Marocco	R	Nostrale di Rigali	R	Verdale	HS
Doebli	R	Oblica	S	Verdial	HS, S
Dolce Agogia	R	Oblizia	S	Villalonga	HS
Dolce di Rossano	HS, S	Ocal	HS	Volon	S
Drobnica	S	Ocro-Violacea	R	Zaity	R
Dûzica	S	Ogil Blanco	MR	Zaituna	R
Elcina	R	Ogliarola barese	R	Zorzaleno	R
Empeltre	HS, S	Ogliarola messinese	S, HS	Žutica	HS, MS
Escarabajillo	R	Ogliarolo	S		

[a]From Agosteo and Schena (2011), partially modified and updated
*S, Susceptible; MS, Moderately susceptible; HS, Highly susceptible; R, Resistant; MR, Moderately resistant; HR, Highly resistant

reduce foliage, allow air circulation, limit moisture on the leaves, favor rapid drying of sensitive organs and enable the coverage of protective fungicides. Similarly, accurate soil drainage and other practices aimed at reducing soil humidity can contribute towards reducing the disease incidence. An important role is also played by a balanced nutritional condition of the plant. Excessive nitrogen intake and calcium deficiency in the soil predispose the plant to disease. Since the beginning of the 19th century, the importance of lime in favoring the resistance of plants has been highlighted, pointing out the usefulness of amendments with limestone in soils poor in calcium carbonate.

CHEMICAL CONTROL. Chemical control is always indispensable in areas with high or medium risk of infection on susceptible cultivars. Treatments with preventive fungicides have to be scheduled to cover and protect plant canopy during the periods of infection and before fungal sporulation and dissemination. They are traditionally performed using cupric fungicides, such as Bordeaux mixture and copper oxychlorides, which are also authorized to be used in organic farming. Other preventive fungicides, such as the ditiocarbamates, ziram and mancozeb and the guanidines, dodine, which can also exert a penetranting action, are useful applied in place of cupric salts to avoid their phytotoxic effect, that reduces the growth of young tender shoots, and the pouring of copper ions in the soil. In humid environments, due to the lack of persistence, it is necessary to perform many treatments. Recently, in the European Community, more persistent, penetrant, active ingredient such as the systemic demethylation inhibitor (DMI), difenoconazole, fenbuconazole, tebuconazole and the mesostemic strobilurin (QoI fungicides), kresoxim-methyl, pyraclostrobin and trifloxystrobin, were authorized to use against OLS, sometimes in the form of mixture and with specific prescriptions.

To schedule the timing of fungicide applications the risk of infections may be assessed by monitoring climatic factors together with the inoculum of the pathogen in the field. The pathogen may be monitored in infected leaves during the latent phases, especially in summer, with the early diagnosis method. A basic treatment in spring, before flowering, when flower clusters are expanded and new shoots have three to four knots (about half or two-thirds of the vegetative growth) is generally recommended to prevent infections on the young, more susceptible, leaves and to reduce the number of sporulating spots in autumn. The quantitative PCR method (Scibetta *et al.*, 2020), revealed the presence of the pathogen in a very early growth stage, when external new leaves open, suggesting the need to anticipate spring treatments and redefining the traditional scheduling, which seem to be too late to prevent new infections.

One treatment, with cupric salt, before vegetative recovery in late winter, is recommended to obtain an eradicating action of the inoculum from the trees, thus reducing the risk of infection on the new spring vegetation. Copper ions, in fact, penetrate into the cuticle through the lesions opened by the pathogen and exert a phytotoxic action on infected leaves, causing their premature fall to the ground, where the conidia become unavailable for further infections. The convenience of this treatment must be carefully evaluated since, in the presence of a high incidence of the disease, it might cause an excessive early fall of the leaves and a consequent reduction of flowering induction and production.

In the last years some studies were conducted for the construction of multiple regression models correlating the disease cycle to the environmental factors (Obanor *et al.*, 2011). New management approaches to optimize disease control and reduce the impact of fungicides, based on the availability of suitable leaf wetness sensors, include forecasting systems of disease risk in the field (Roubala *et al.*, 2013). An updated review addressing OLS has been very recently published (Bubici *et al.*, 2023).

7.3.3.2 Cercosporiose

Introduction

The causal agent of cercosporiose (also named cercosporiosis) is the mitosporic fungus *Pseudocercospora cladosporioides* (synonyms, *Mycocentrospora cladosporioides* and *Cercospora cladosporioides*), a Dothideomycetes belonging to the family of Mycosphaerellaceae, whose teleomorph is not known, although it is commonly accepted that *Pseudocercospora* species have teleomorphs in the genus *Mycosphaerella*.

Olive cercosporiose is one of the most widespread diseases of the olive tree in many cultivation areas. Its presence was often ignored by farmers, both for the non-specificity of its symptoms on the upper leaf surface and for the frequent coexistence with 'olive leaf spot', with which it was confused. Its importance as a cause of defoliation has often been underestimated because it is believed to affect only leaves older than one year, therefore already physiologically old and in any case destined for fall. The increased diffusion and severity of the disease in many olive-growing areas of the Mediterranean basin can be ascribed to various causes such as high density in new plants, poor pruning in traditional extensive cultivations and global warming with increasing winter temperatures.

Symptoms and damages

Symptoms concern, more intensely and widely, the mature leaves (on average, one year old) but can occur on young leaves (< one year), starting from no earlier than 5–6 months of age, mainly when rainy and mild autumn occurred (Nigro *et al.*, 2002; Agusti Bisiach *et al.*, 2016).

At the beginning, the disease occurs only on the underside of the leaves while the upper surface is asymptomatic. These symptoms, which consist of mycelial fruiting structures of the pathogen, have diagnostic value and show themselves as gray-lead, velvety spots, arranged in circles or irregularly shaped, with diameters ranging from 3 to 4 mm (Fig. 7.3.7). As the disease progresses, the spots coalesce and can affect the entire lower lamina or remain confined by the midrib only on half of it. In severely infected leaves, and above all during the winter, the intense sporulation and the presence of the dark sclerotial bodies gives a more intense black color to the abaxial surface of the leaf, which has caused the disease to be confused with 'sooty mold' (Fig. 7.3.8). As the lesion become larger and older, disease symptoms appear on the adaxial surface, initially as chlorotic areas, pale green or yellowish, rounded or irregular in shape, located along the margin and/or

Fig. 7.3.7. Cercosporose symptoms on the abaxial leaf blade.

Fig. 7.3.8. Intense black sporulation of *P. cladosporioides* on the lower leaf blade.

at the apex of the leaf, which progressively necrotize. The symptoms on the upper surface of the leaf are non-specific and can be confused with the symptoms caused by abiotic factors or anthracnose (*Colletotrichum* spp.). Usually, *P. cladosporioides* infect old, mature leaves. For this reason, in the recent past, it has been considered less harmful than olive leaf spot as a cause of defoliation. More recently, extensive surveys in Puglia, southern Italy, demonstrated that younger leaves, 4–5 months old, can also be infected (Nigro *et al.*, 2002). In Calabria, the first infections of cercosporiose on the leaves of the new spring vegetation began to be observed starting from October. The infected leaves fall prematurely, producing intense defoliation, which can lead to twig dieback, reduced vegetative vigor and plant production.

Infections to petioles, fruit peduncles and young shoots appear as blackened spots of various shape and extension, similar but less dark in color than those caused by olive leaf spot. Infections of the peduncle may cause abundant fruit falls before the harvest.

Symptoms to the drupes are less frequent and occur only in years very favorable to infections. In years of strong attacks symptoms were found on nearly 100% of affected olives (Del Moral and Medina, 1985), which leads to a poor quality of the oil, increasing notably acidity. They may vary considerably according to variety, stage of maturity and weather conditions. On green olives are observed red-brown, circular or irregular-shaped, more or less depressed lesions, ranging between 3 mm and 10 mm in diameter. Ripe drupes, according to the cultivars, show 2–7 mm large craters surrounded by a pale yellow halo or slightly sunken, wrinkled and often soft areas, 5–10 mm large, with darker and suberose subepidermic spots in the pulp (Nigro and Ferrara, 2011).

Life cycle and epidemiology

The life cycle of *P. cladosporioides* has been little studied but largely depends on the conditions of humidity and temperature, location, soil characteristics and density of the foliage. It was generally considered similar to olive leaf spot cycle since both diseases often occur simultaneously in the same tree, although the infections of *P. cladosporioides* take place later than those caused by *V. oleaginea*.

On the abaxial side of infected leaves the pathogen differentiates stromata on which emerge, gathered in cespitula, flexuous, long, straight or curved conidiophores, mostly unbranched, measuring 110–125 × 3–4.5 μm. Conidiogenous cells are terminal and bear one or more scars. They produce single, subcylindrical, guttulate, very pale to medium olivaceous-brown conidia, with obtuse apex and obconically truncate or subtruncate base. Usually, conidia are straight or often slightly curved, with rounded edges, 3–8 septate, measuring 27–55 × 3.5–4.5 μm. They contain many vacuoles and are easily confused with mycelial fragments. *P. cladosporioides* is characterized by the production of dark sclerotial bodies, constituted by diplodiform elements, which appear to be structures of resistance against adverse conditions (Nigro and Ferrara, 2011).

The fungus grows very slowly in culture media and forms dense olivaceous mycelia colonies. There is conflicting information about the ability of the fungus to form conidia on cultural media. In many cases, formation of conidia in culture was very rare and attempts to find a suitable medium were not satisfactory (Ávila *et al.*, 2004, 2020). This has made it very difficult to carry out studies on the epidemiology of the fungus. Conversely, abundant conidium formation was reported after incubation of the fungus on potato dextrose agar (PDA) at 25°C for 15 days (Triki and Rhouma, 2008) or four weeks at 20°C (Henricot *et al.*, 2009). This would suggest the existence of a genetic variability within the population of this fungus. As a matter of fact, it was demonstrated that two phylogenetic groups of *P. cladosporioides* co-exist in Spain; the isolates from Catalonia were clearly different from those collected in Andalusia. Similarly, significant differences were observed in *in vitro* growth rates between *P. cladosporioides* isolates in Spain (Ávila *et al.*, 2004, 2005).

P. cladosporioides is active throughout the year and the infection is particularly severe in dense plantations with little aeration, being limited in many cases to the lowest part of the tree canopy. Mild temperatures, high rainfall and humidity before the harvest in autumn, favor infection on the drupes.

The minimum temperature for *P. cladosporioides* growth was slightly lower than 5°C, the optimum range 21–22.5°C and the maximum near 30°C (Ávila de La Calle *et al.*, 2004). Conidium production occurs mainly between the months of October and March, during rainy periods or relative humidity above 80% and average temperatures between 10 and 20°C. The percentage of germination of the conidia is maximal between 15 and 25°C (Ávila *et al.*, 2005).

The infection process begins when a conidium comes into contact with the host and initiates the inoculation process. Penetration of the germination tube or epyphitic mycelium into the host occurs through stomata and wounds. The pathogen spreads in the spongy mesophyll, colonizing the tissue intercellularly, obtaining the nutrients directly from the apoplast and maintaining a biotrophic relationship in the early stages of its development. When climatic conditions are favorable, with rain or high relative humidity, the pathogen evades from colonized tissues, either through the stomata or epidermal lesions, producing conidiophores gathered in cespitula. This process begins about 10–15 days after the penetration and the production of conidia requires temperatures ranging from 12°C to 28°C. As the disease progresses, the chlorotic areas on the adaxial surface of the leaves appear. Under unfavorable environmental conditions the fungus produces the dark sclerotial bodies on the abaxial surface of the leaves.

The dispersion of the pathogen occurs at a short distance, from leaf to leaf, by conidia or fragments of mycelium, favored by the wind and due to rain splashes. The pathogen survives during unfavorable periods, mainly dry and hot weather, in the infected leaves that remain on the tree, as latent infection or through resistance structures. The fallen leaves play an important role in the survival of the pathogen but the conidia produced pose a low risk for new infections, since they lose their viability if the optimal conditions for their dispersion do not occur within few days.

The length of the incubation period is highly variable and depends on both environmental conditions and on some characteristics of the plants (cultivar, type and age of infected organ, etc.). In general, it can be quite long and can exceed 9–10 months (Agusti Bisiach *et al.*, 2016).

Some significant differences were detectable in the scientific literature. In pathogenicity tests on detached leaves was detected an incubation period of at least 30 days in Spain (Ávila *et al.*, 2020) and 10 days in Tunisia (Triki and Rhouma, 2008). The length of the latency period was 11 months in the pathogenicity tests on the olive plants (Ávila *et al.*, 2020) and 13–15 days in Tunisia and the UK (Triki and Rhouma, 2008; Henricot *et al.*, 2009). On detached fruits it was at least seven days (Ávila *et al.*, 2020).

Control strategies

IPM strategies against cercosporiose involve genetic (varietal) resistance, cultural practices and fungicides application. When planting in humid areas it would be advisable to choose resistant varieties.

The information on varietal susceptibility is based on field observations, therefore they do not always appear univocal and may show some contradictions.

In Italy the cultivars 'Biancolilla', 'Calatina', 'Carolea', 'Coratina', 'Cellina di Nardò', 'Cima di Melfi', 'Dolce di Rossano', 'Frantoio', 'Grossa di Gerace', 'Moraiola', 'Moresca', 'Nociara', 'Ogliarola', 'Ottobratica', 'Pendolino', 'Quercetana', 'Picholine', 'Roggianella', 'Rosciola' and 'Sinopolese' are considered susceptible or highly suscebtible; as opposed to the cultivars 'Carboncella', 'Cucco', 'Duzica', 'Elcina', 'Giaraffa', 'Maiatica', 'Nocellara del Belice', 'Nocellara Etnea', 'Oblica', 'Ogliarola Messinese' and 'Tonda Iblea', which are considered less susceptible or more resistant. The cultivar 'Leccino' is considered either resistant or susceptible by different authors.

In Spain, the following are considered highly susceptible: 'Arbosana', 'Blanqueta de Elvas', 'Cipressino', 'Frantoio', 'Gordal de Vélez', 'Rubio', 'Lechín de Sevilla', 'Maraiolo', 'Picudo', 'Quercetana', 'Rosciola', 'Verdial de Huevar'. Susceptible: 'Blanqueta de Elvas', 'Carolea', 'Conservolia', 'Coratina', 'Cornicabra', 'Dolce Agogia', 'Empeltre', 'Farga', 'Gordal Sevillana', 'Hojiblanca', 'Leccino', 'Lechín de Granada', 'Morrut', 'Negrinha', 'Perillo de Jaén', 'Picholine', 'Picudo', 'Sevillenca', 'Verdial de Badajoz', 'Villalonga', 'Zorzaleña'. Resistant: 'Alameño de Marchena', 'Arbequina', 'Blanqueta', 'Canetera', 'Carbonella', 'Chemlali', 'Cobrançosa', 'Cornicabra de Mérida', 'Elcina', 'Empeltre', 'Farga', 'Lemeño', 'Manzanilla de Abla', 'Manzanilla de Agua', 'Manzanilla de Almería', 'Manzanilla de Guadix', 'Manzanilla de Hellín', 'Manzanilla de Jaén', 'Manzanilla de Sevilla', 'Manzanilla de Tortosa', 'Manzanilla del Piquito', Marroquí, 'Megaritiki', 'Meski', 'Nabali', 'Picholine', 'Picual', 'Racimal de Jaén', 'Rechino', 'Temprano', 'Varudo', 'Verdial de Vélez', 'Malaga', 'Wardan'. In Montenegro, the local cultivars showed to be practically immune. In Greece, the cultivars 'Amfissis' and 'Konservolia' are very susceptible, while 'Kalamon' is highly tolerant to *P. cladosporioides* (Ávila and Trapero, 2010; Nigro and Ferrara, 2011; Agosteo, 2012).

Cultural measures, such as plant spacing and pruning, are advisable in order to favor the ventilation of the trees and reduce the time of foliar wetting.

Traditionally, cupric treatments carried out against olive leaf spot have also been considered effective against olive cercosporiose and specific treatments for this disease have rarely been scheduled if not on cultivars resistant to olive leaf spot and susceptible to cercosporiose.

Results of recent control trials in Italy indicate that the common cupric treatments against olive leaf spot are not as effective against cercosporiose and should be reconsidered, especially as regards to the timing of application. Spring treatments are too early and the better results are obtained by applying the cupric products from July to November. Four treatments proved to be much more effective than two (Guario *et al.*, 2000). Doubts on the effectiveness of copper compounds in controlling olive cercosporiose are due to their fungistatic rather than fungicidal activity towards *P. cladosporioides* (Nigro and Ferrara, 2011). A defense strategy based on a single application in spring of strobilurin (QoI fungicides) or a mixture of copper compounds with strobilurin and demethylation inhibitor (DMI) fungicide or SDHI (Succinate dehydrogenase inhibitors fungicide), such as trifloxystrobin + tebuconazole and pyraclostrobin + boscalid, even without the application of supplementary copper, is effective under field conditions in years with unfavorable climatic conditions for infections, and allows to reduce the number of treatments with copper compounds. Another treatment in the autumn may be needed for infections above 1%. Treatments in October or March were more effective than treatments in May. Fungicide applications at the end of October would be useful for protecting the leaves at the beginning of conidial production, whereas treatments in February–March would have to act at the end of epidemic development, when temperature rises up and rainy events may occur (Agusti Bisiach *et al.*, 2016; Romero *et al.*, 2020). According to the European legislation, specific prescriptions exist regarding strobilurin treatments on olive trees.

7.3.3.3 Anthracnose

Olive anthracnose (OA) may be regarded as the most damaging disease of olive fruit worldwide. It has been reported from all continents where olive is grown, including Europe, Africa, Asia, Americas and Oceania. However, it attains high economic relevance, causing heavy yield losses (up to more than 90% of fruit yield) and compromising the oil quality (high acidity, off-flavor, reddish color and reduction of polyphenols, α-tocopherol, and β-sitosterol) even with a low (around 5%) fruit-rot incidence, in olive-growing areas where disease-conducive agroecological conditions occur, including high humidity and persistent rainfall in autumn, the widespread use of susceptible olive varieties, and a great amount of inoculum reservoirs. The disease is endemic, with frequent severe epidemic outbreaks favored by heavy rains in autumn, in the regions of Alentejo in southern Portugal, central provinces of Andalusia in southern Spain, in the Salento Peninsula in Apulia and in the provinces of Reggio Calabria and Vibo Valentia in Calabria, southern Italy. In certain areas of Portugal and southern Italy, OA may be as damaging as olive fruit fly (*Bactrocera oleae*). It was also reported to cause severe damages in Australia, South Africa, Argentina, Uruguay, Greece and Montenegro, is emerging in Tunisia, and might be a potential threat in new areas where olive crop is expanding such as some provinces of China and India, the Arabic peninsula and the states of Florida and Hawaii, USA. The first published scientific report of anthracnose was from Portugal and dates back to 1899. However, the symptoms of the disease were recorded by Arabs in the Iberian Peninsula since the 11th century. Vernacular names of the disease are *gaffa* in Portugal, *aceituna jabonosa* (soapy fruit) or *momificado* (mummification) in Spain and *lebbra* (leprosys) in Italy. In Spain, the term *repilos* is commonly used to indicate, collectively, anthracnose, peacock spot and cercosporiosis.

The most typical symptoms of OA are fruit rot and mummification. In moist conditions fruits show a soft brown to dark-brown rot with an abundant production of an orange sticky, gelatinous matrix embedding conidia, while in a dry environment the fruits dehydrate, mummify and lose weight. Affected fruits fall prematurely to the ground and only a few mummies remain attached to the tree canopy. Mature drupes are mostly affected, but in favorable environmental conditions green olives of susceptible cultivars may also be affected. In cultivars with oblong drupes, rot often starts from the apical end (Fig.7.3.9) while in cultivars with large, round drupes, symptoms are subcircular, dark, depressed, sunken lesions with the typical, orange, gelatinous conidium masses (Fig. 7.3.10). Leaf-yellowing, defoliation and dieback of twigs and branches can also occur during severe epidemics in late spring, causing long-lasting effects on vigor and yield of the trees.

Several (so far 18) *Colletotrichum* species in the *C. acutatum*, *C. gloeosporioides* and *C. boninense* species complex were reported to be associated with OA; none of them is host-specific and only some of them have been proved to be aggressive on olive drupes. Moreover, some of these *Colletotrichum* species are common while others occur sporadically. More than one of these species coexist but one or two are more virulent and prevail over the others in each olive-growing region. *C. godetiae* (syn. *C. clavatum*) and *C. acutatum sensu stricto* (*s. str.*), e.g. are dominant in olive orchards of southern Italy; while C. *acutatum s. str.* and *C. gloeosporioides s. str.* are dominant

Fig. 7.3.9. Anthracnose on green olive drupes of cv. 'Ottobratica'.

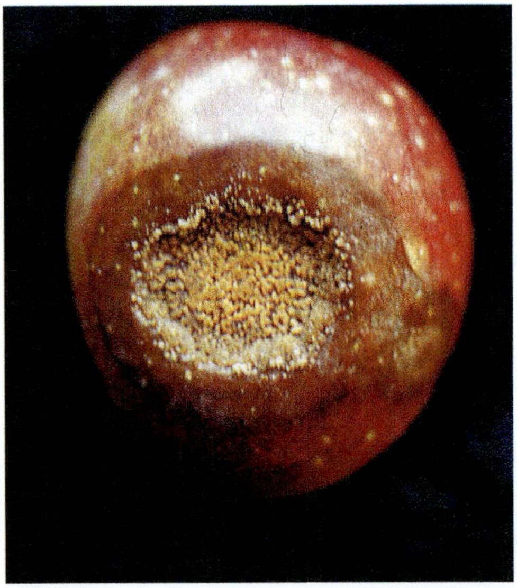

Fig. 7.3.10. Anthracnose on a mature olive drupe of cv. 'Itrana'.

and secondary species, respectively, in Tunisia; *C. godetiae* prevails in Andalusia; *C. nymphaeae* in southern Portugal; both *C. nymphaeae* and *C. godetiae* in central Portugal; *C. acutatum s. str.* in olive-growing areas of southern hemisphere, such as Australia and South Africa. Evidence has been provided that the structure of the *Colletotrichum* population associated to OA in a given region is dynamic. So, for example, presently *C. acutatum s. str.* is the prevalent species in southern Italy and Greece while *C. godetiae* was, until recently, the main species responsible for OA in both countries. Moreover *C. godetiae* is the dominant species in Andalusia (Spain), the first olive producing area in the world, *C. nymphaeae* is the prevalent species associated with OA in south western Portugal and has been recently reported as the second most common species, after *C. acutatum*, in Uruguay. It is noteworthy that wherever severe epidemic outbreaks of the disease occur, species of the *C. acutatum* complex are dominant.

Conidia of the different species of *Colletotrichum* responsible for OA are the infective propagules; they are produced in acervuli on infected drupes and are spread by rain splashes and windblown rain droplets, causing secondary infection cycles. Mummified drupes and infected twigs are regarded as the main inoculum reservoirs. Since 2005, it has also been proved that latent infections of green fruitlets may occur in spring during bloom. They may be the primary source of inoculum for infections in autumn, leading to a change in the approach to chemical control. Olive fruit infection occurs at all stages of its development from flower bud emergence to ripening. The incidence and severity of the disease vary considerably depending on the weather and environmental conditions, the amount of inoculum, the susceptibility of the olive cultivar, the virulence of the pathogen population and the time of year the fruit is harvested. OA can be favored by the activity of olive fruit flies, which may increase the fruit susceptibility by wounding and acting as conidium carriers. Even olives of resistant cultivars may be infected if they are interplanted with very susceptible cultivars. In southern Portugal, where the weather in autumn/winter is conducive for the disease, severe outbreaks occur in super-intensive orchards of the moderately susceptible cultivar 'Arbequina'. In Andalusia (Spain), where the cultivars 'Hojiblanca' and 'Picudo' (both highly susceptible) are grown, and in Calabria (southern Italy), where the cultivars 'Ottobratica' and 'Sinopolese' (highly susceptible and susceptible, respectively) prevail, severe epidemic outbreaks occur in autumn if the weather conditions are conducive despite the routine application of 2–3 copper-based fungicide treatments. The susceptibility of the drupes increases with maturity and, in general, the later the harvesting, the higher is the incidence of OA. Consequently, early harvesting or the use of late-maturing cultivars have been suggested as eco-friendly means to escape the disease and can be regarded as part of an integrated strategy to manage OA, including also chemical treatments and tree pruning to reduce the amount of inoculum and enhance the aeration of the canopy. However these solutions have some drawbacks: immature fruits have a lower oil content, are more difficult to detach and harvest mechanically and are more sensitive to winter frosts. Therefore, the use of olive cultivars resistant to OA is the most effective management method and can be combined with other control measures such as chemical treatments. The susceptibility/resistance of olive cultivars to OA was evaluated in several countries but in some cases results are contradictory due to the difficulty of comparing cultivars with different ripening periods and standardizing the conditions in both artificial inoculation-tests and field-experiments. The results of a recent screening of a large set of olive cultivars of worldwide origin, the World Olive Germplasm Bank of Córdoba, Andalusia, Spain, for the susceptibility to resistance to OA are summarized in Table 7.3.5. There is evidence that olive resistance to OA is under multi-gene control and crosses between resistant cultivars produce a high proportion of resistant hybrids.

Satisfactory results in the control of OA have been obtained with fungicide treatments. Copper-based compounds (Bordeaux mixture or copper oxychloride) are routinely used to control OA in most olive-growing regions and are preferred by growers because of the low cost, long persistence and wide spectrum of activity including other major bacterial and fungal diseases, such as olive knot, scab and cercosporiosis. Three treatments from late September to the end of December were demonstrated to be effective in southern Italy. Also, triazole (tebuconazole and hexaconazole) and QoI (strobilurin) (azoxystrobin and pyraclostobin) fungicides were proved to be effective. However, not all these fungicides are registered for treatments on olive or against OA. Theoretically, spray treatments in spring can prevent blossom blight and early fruit infections thus reducing the primary inoculum for fruit infections in autumn. In Australia, two treatments with azoxystrobin, before flowering and fruit set, respectively, were suggested. However, in southern Italy, spring and summer treatments proved to be scarcely effective if in autumn weather conditions were conducive to the disease, confirming that OA, being a typical polycyclic disease, is highly weather-dependent. Not surprisingly, fungicide treatments were proved to be more effective on moderately susceptible than on highly susceptible olive cultivars. A substantial limit to the use of fungicides to control OA is the short time interval useful for their optimal application and the prohibition of carrying out fungicidal treatments during the pre-harvest interval, which coincides with the period of greatest susceptibility of the drupes.

For more details on OA, the reader is referred to exhaustive reviews of authors from four major olive-producing countries of the Mediterranean region: Spain, Italy, Portugal and Greece, addressing this disease (Cacciola *et al.*, 2012; Talhinhas *et al.*, 2018; Kolainis *et al.*, 2020; Moral *et al.*, 2021).

7.3.3.4 Verticillium wilt

Verticillium wilt of olive (VWO) is a vascular disease caused by the polyphagous and hemibiotrophic, soilborne fungus *Verticillium dahliae* Kleb. It is currently considered one of the most serious diseases affecting olive worldwide due to its lethal effects on trees (Jiménez-Diaz *et al.*, 2012). Two VWO syndromes have traditionally been distinguished: an acute form named 'apoplexy' (Fig. 7.3.11) and a chronic form named 'slow decline'. The 'apoplexy' syndrome is characterized by sudden outbreaks and severe wilting of main and secondary branches, occurring mainly in late winter to early spring in the Mediterranean basin. Leaves first become chlorotic, then turn light-brown and undergo progressive rolling inward, remaining attached to the branches. The final result is a rapid and extensive dieback of twigs and branches. Apoplexy occurs more frequently

Table 7.3.5. Susceptibility of olive cultivars from a large collection (World Olive Germplasm Bank of Córdoba, Spain) to anthracnose caused by *Colletotrichum godetiae* (syn. *C. clavatum*, in the *C. acutatum* species complex), as evaluated phenotypically in terms of fruit rot incidence (after Moral *et al.*, 2017)[a]. (CC BY 4.0)

Cultivars[b] sorted by susceptibility level[cd]
Highly Susceptible (HS)
Abbadi, Acebuchera, Adkam, Agouromanakolia, Azapa, Bolvino, Borriolenca, Bouteillan, Cañivano Negro-55, Chalchali, Chorruo de Castro del Río, Corbella, Cornicabra de Mérida, Dulzal de Carmona, Farga, Forastera de Tortosa, Fulla de Salze, Gatuno, Gerboui, Gordal Sevillana, Grosal de Cieza, Habichuelero de Baena, Hojiblanca, Imperial, Imperial de Jaén, Jabaluna, Jaropo, Limoncillo, Llorón de Atarfe, Loaime, Lucio, Lucques, Machorrón, Mahati-1010, Manzanilla de Almería, Manzanillera de Huércal, Manzanillo de Cabra, Merhavia, Meski, Morisca, Morisca de Mancor, Negrillo de Iznalloz, Negrillo de la Carlota, Negrillo Redondo, Nevadillo de Santisteban del Puerto, Nevado Azul, **Ocal**, Ocal-25, Ojo de Liebre, Palomar, Pavo, Picudo, Picudo Blanco de Estepa, Rachati, Rechino, Safrawi, Salgar Redondo, Salonenque, Sant Agostino, Sayfi, Sevillana de Abla, Sevillenca, Temprano, Uovo di Piccione, Uslu, and Verdial de Cádiz.
Susceptible (S)
Abbadi Abou Gabra-84, Abbadi Shalal, Abou Kanani, Adramitini, Alameño Blanco, Alameño de Cabra, Alameño de Montilla, Amargoso, Amygdalolia Nana, Argudell, Ascolana Tenera, Asnal, Ayvalik, Barri, Blanqueta, Carolea, Castellana, Chorreao de Montefrío, Cipressino, Cirujal, Cordobés de Arroyo de la luz, Cordovil de Castelo Branco, Cornezuelo de Jaén, Cornicabra, Doebli, Dulzal, Escarabajillo, Escarabajuelo de Atarfe, Escarabajuelo de Posadas, Figueretes, Galega Vulgar, Gemlik, Gordal de Granada, Gordal de Hellín, Gordal de Vélez Rubio, Grosal Vimbodí, Izmir Sofralik, Jlot-841, Kaesi, Kalamon, Kolybada, **Lechín de Sevilla**, Lentisca, Levantinka, Macho de Jaén, Mahati-846, Manzanilla de Piedra Buena, Manzanilla de Agua, Manzanilla de Sevilla, Manzanilla Prieta, Mastoidis, Mawi, Mollar Basto, Mollar de Cieza, Morona Negral de Sabiñan, Negrillo de Estepa, Negro del Carpio, Nevadillo Blanco de Jaén, Nevado Basto, Nevado Rizado, Olivo de Mancha Real, Pajarero, Perafort, Pico Limón de Grazalema, Picual de Almería, Polinizador, Racimal, Shami, Sorani, Tempranillo de Yeste, Torcio de Huelma, Toruno, Varudo, Vera, Verdal de Manresa, Verdala, Verdale, Verde Verdelho, Verdial de Badajoz, Verdial de Huévar, Verdiell, and Zarza.
Moderately Susceptible (MS)
Abou Satl Mohazam, **Arbequina**, Arroniz, Azulejo, Barnea, Beyaz Yaglik, Bodoquera, Caballo, Cañivano Negro, Carrasquenho de Elvas, Carrasqueño de Alcaudete, Carrasqueño de la Sierra, Carrasquillo, Cerezuela, Changlot Real, Chorruo, Çobrancosa, Cordovil de Serpa, Cornicabra de Jerez, Cornicabra-1, Domat, Dwarf D, Erbek Yaglik, Escarabajuelo de Úbeda, Gaydoyrelia, Genovesa, Habichuelero de Grazalema, Itrana, Kelb et Ter, Khashabi, Kiraz, Klon-14-1081-1, Klon-14-1081-2, Konservolia, Lastovka, Lechín de Granada, Maarri, Majhol-1059, Manzanilla Cacereña, Manzanilla de Abla, Manzanilla del Piquito, Mari, Masabi, Mission, Moojeski, Morrut, Myrtolia, Negrillo de Arjona, Oblica, Olea ferruginea, Olivo de Maura, Picholine, Plementa Bjelica, Pulazeqin, Rapasayo, Redondilla de Logroño, Reixonenca, Royal de Calatayud, Royal de Cazorla, Royal de Sabiñan, Sabatera, Sandalio, Tanche, Toffahi, Valanolia, Villalonga, and Vinyols.
Resistant (R)
Abbadi Abou Gabra-10, Acebuche de Caravaca, Alfafara, Aloreña de Iznalloz, Arbosana, Athalassa, Beladi, Belluti, Bent al Kadi, Biancolilla, Bical, Buga, Buidiego, Canetera, Carrasqueño de Jumilla, Carrasqueño de Porcuna, Chalkidikis, Chemlal de Kabilye, Chemlali, Chetoui, Coratina, Corralones de Andújar, Curivell, Datilero, Dokkar, Dolce, Elmacik, Enagua de Arenas, Fishomi, Grit Eytini, Grossanne-67, Joanenca, Kalokerida, Kotruvsi, Leccino, Llumeta, Majhol-1013, Majhol-152, Manzanilla de Hellín, Manzanilla de Montefrío, Manzanilla de San Vicente, Maurino, Memecik, Mixani, Moraiolo, Patronet, Pecoso, Picholine Marocaine, Pico Limón, **Picual**, Real Sevillana, Redondilla de Grazalema, Sevillano de Jumilla, Shengue, Sinop, Sollana, Vallesa, Vaneta, Verdial de Vélez-Málaga, Zaity, and Zard.
Highly Resistant (HR)
Ayrouni, Azul, Bosana, Callosina, Caninese, Crnica, Dolce Agogia, Empeltre, **Frantoio**, Frantoio A. Corsini, Grappolo, Istarska Bjelica, Kan Çelebi, Kato Drys, Koroneiki, Mavreya, Megaritiki, Menya, Ouslati, Pendolino, Pequeña de Casas Ibañez, Perillo de Jaén, Piñonera-76, Racimal de Jaén, Rosciola, Royal de Calatayud-4, Selvatico, Toffahi-1000, Ulliri i Bardhe i Ti, Ulliri i Kuq, Wardan, and Zalmati.

[a]Cultivars were rated phenotypically during epidemic outbreaks which occurred three times over a period of 10 years.
[b]Cultivar susceptibility was evaluated on different dates depending on the ripening period of each cultivar, as the susceptibility of drupes to anthracnose increases when they are mature.
[c]Cultivars in **bold** were suggested by the AA as representative of each susceptibility class.
[d]Cultivars were grouped into five susceptibility classes: highly susceptible (HS); Susceptible (S); Moderately susceptible (MS); Resistant (R); Highly resistant (HR).

Fig. 7.3.11. Four-year-old olive tree cv. 'Coratina' severely affected by the apoplexy syndrome of Verticillium wilt, a rapid and extensive dieback of twigs and branches.

in young plants and commonly is fatal for the tree. The 'slow decline' syndrome is characterized by heavy defoliation of branches, leaf chlorosis and mummification and death of inflorescences. It begins in spring and progresses slowly to early summer. Defoliation usually affects only part of the tree and still-green or dull-green leaves may remain attached to the distal end of the shoots. The xylem of affected branches, or even the main trunk of the tree, often show a purple hue or a brown to dark-brown discoloration (Fig. 7.3.12). The severity of VWO infections depends on the virulence of the *V. dahliae* pathotype that infects the tree. Traditionally, *V. dahliae* isolates infecting olive are classified into defoliating (D) and non-defoliating (ND) pathotypes, according to the type of symptoms they induce on olive trees. Moreover, different pathogenic races of the pathogen have been identified (Jiménez-Diaz *et al.*, 2012). The D pathotype is highly virulent and causes severe symptoms including wilt, chlorosis, defoliation, stunting and, eventually, death of the tree. The ND pathotype causes milder symptoms than the D pathotype, usually only a moderate wilting. The D pathotype is widespread across areas where cotton has been previously grown for many years, while the ND pathotype occurs commonly in fields where potato has been cultivated prior to olive planting. Surveys in olive orchards of Andalucía (southern Spain), the largest olive

Fig. 7.3.12. Brown discoloration of the vascular tissue of an olive twig caused by *Verticillium ahlia*.

cultivation area in the world, showed that D isolates prevailed over ND isolates (67.7% versus 32.2%, respectively) (López-Escudero and Mercado-Blanco, 2011). Similarly, an overwhelming prevalence of D pathotype isolates was reported in all major olive-growing regions of Turkey (Dervis *et al.*, 2010). *V. dahliae* survives in soil by means of microsclerotia, which are also the primary inoculum of the pathogen in natural conditions and its main dispersal propagules. Microsclerotia survive for a long time in the soil and germinate, giving rise to infective hyphae that penetrate the plant roots and grow within the xylem vessels, producing mycelium and conidia. Optimum temperature for *V. dahliae* ranges from 22 to 25°C, maximum 30–35°C. In the Mediterranean region during summer, both xylem colonization by the pathogen and symptom expression slow down or stop temporarily.

During the last decades, in many olive-growing areas the incidence of VWO has increased markedly, mainly in young, newly established orchards (5- to 10-year-old trees). However, there are reports of severe attacks of VWO even in traditional orchards and old trees. Commonly, symptoms observed in old trees are less severe and transient (Cirulli and Montemurro, 1976; Thanassoulopoulus *et al.*, 1979). There are several, often concomitant causes favoring the expansion, incidence and severity of the disease: (i) the use of pathogen-infected plant material to establish new olive orchards; (ii) the widespread use of propagation by rooted cuttings; (iii) the planting of new olive orchards in infested soils; (iv) the introduction of irrigation, the high planting density, either intensive (> 200 tree/ha) or super-intensive (2000 tree/ha, arranged in hedgerows), the widespread use of more productive yet susceptible olive cultivars, such as 'Arbequina' and 'Picual', which are better suited than other cultivars to super-intensive plantings, and more generally the shift from traditional to modern and more intensive cultivation systems; (v) the increasingly widespread distribution of highly virulent isolates of the pathogen such as the D pathotype (López-Escudero and Mercado-Blanco, 2011). In particular, the responsibility of irrigation in favoring VWO has been repeatedly demonstrated while the effects of fertilization and soil-management system have been less explored. Irrigation water provides a long-distance dispersal of *V. dahliae* propagules and fosters the increase of the pathogen inoculum in soil (López-Escudero and Mercado-Blanco, 2011). Soil and water salinity may exacerbate the disease severity and cause breakdown of cultivar resistance. These effects are enhanced by drought stress. A correlation was found between soil tillage and the disease incidence. Tillage was reported as a means for *V. dahliae* inoculum dispersal and was postulated to be responsible for wounds of superficial roots that may favor penetration by the pathogen. On the other hand, many broadleaf weeds may be a reservoir and a potential source of inoculum. The use of cover crops could be a sustainable soil-management option to reduce tilling or avoid the application of herbicides.

Factors that make it difficult to control VWO include its vascular nature, the wide host range of *V. dahliae*, which comprises industrial, staple and horticultural crops such as cotton (*Gossypium hirsutum*), potato (*Solanum tuberosum*) and pistachio (*Pistacia vera*), as well as the long survival of its propagules in soil. As a consequence, an integrated management strategy, based on exclusion or sanitation of infested soils for new plantings, the use of genetic resistance and pathogen-free propagation material, a proper soil and irrigation management, inoculum reduction by soil solarization and biological control, is the way to cope with this disease. Biological control, in a broad sense including the use of bioproducts, soil organic amendants and natural substances, is promising and feasible (Mulero-Aparicio *et al.*, 2020); however, it has not yet been used on a large scale. As for other vascular plant diseases caused by soilborne pathogens, the use of resistant olive varieties or rootstocks is the most effective, economically efficient and ecologically sustainable means to cope with VWO and should be the core of integrated disease-management strategies (Serrano *et al.*, 2021). To this aim, olive breeding and selection programs have been developed in several countries, including Greece, Italy, Spain, Turkey and the United States (California). Results of phenotypic screenings of olive germplasm for resistance/susceptibility to VWO are summarized in Table 7.3.6. Not all screening tests reported in the table are comparable and contradictory results may depend on different experimental conditions, inoculation method, type of inoculum, intra-varietal variability, nomenclatural discrepancies and synonymy between cultivars in different countries. Concerning the inoculation method, root dipping and stem injection were reported to be equally effective in resistance evaluation of olive to Verticillium wilt and more reliable than transplanting into infested soil (Jiménez-Diaz *et al.* 2012). However, there is not general consensus on this (Cirulli *et al.*, 2008). An uneven distribution of the pathogen in soil may affect field tests and, similarly, lack of information on the nature of the testing pathotype can skew comparisons. López-Escudero *et al.* (2004) showed that cultivars highly resistant to the ND pathotype such as 'Coratina', 'Fragivento', 'Manzanilla de Sevilla' and 'Verdial de Alcaudete' were highly susceptible to the D pathotype. 'Changlot Real', 'Empeltre', 'Frantoio', 'Oblonga' and 'Koroneiki' were consistently rated resistant to the ND pathotype across different Mediterranean countries.

Cultivars designated with the same name may actually be different and, by contrast, clones with different names may be the same cultivar; very probably, e.g. 'Oblonga' is a clone of 'Frantoio' (Jiménez-Diaz *et al.* 2012). Efforts have been made to select olive and wild olive clones that would be used as rootstocks, including, e.g., 'Oblonga', clone Berkeley 117, 'Yarmouk' lines and wild olives OffVert, OutVert and StopVert (Colella *et al.*, 2008; Jiménez-Diaz *et al.* 2012). The effectiveness of grafting in the control of Verticillium wilt of olive has been demonstrated by several authors. Interestingly, the grafting of 'Coratina' and 'Leccino' (both susceptible) onto 'Frantoio' (resistant) as a rootstock was proved to be resistant while, *vice versa*, the grafting of 'Frantoio' onto 'Coratina' and 'Leccino', respectively, was susceptible, suggesting the resistance of 'Frantoio' is expressed in the plant roots (Bubici and Cirulli, 2012).

For more details on VWO, the reader is referred to exhaustive reviews on the disease (Jiménez-Diaz *et al.*, 2012; López-Escudero and Mercado-Blanco, 2011; Mercado-Blanco and López-Escudero, 2012; Montes-Osuna and Mercado-Blanco, 2020).

7.3.3.5 Armillaria and Rosellinia root rot

Introduction

Armillaria mellea and *Rosellinia necatrix*, the causal agents of fibrous root rot (FRR) and white root rot (WRR), respectively, are both very polyphagous soilborne fungal pathogens.

A. mellea belong to the class Basidiomycetes, ord. Agaricales, and is commonly

Table 7.3.6. Resistance of olive genotypes/cultivars to Verticillium wilt[a].

Cultivar or genotype and resistance/susceptibility class[b]
'Bella di Spagna' (S), 'Nocellara' (S), 'Ogliarola' (S) (Ruggieri, 1948)
'Manzanillo' (S), 'Redding Picholine' (S), several olive dwarf rootstocks (S), *Olea ferruginea* (S), *O. chrysophylla* (S), *O. verrucosa*®, olive seedlings from open pollination of 10 olive cultivars and wild olive from Russia and North Africa (some progenies from 'Frantoio' or a descendant from 'Arbequina' called 'Allegra' were R; progenies from 'Manzanillo','Chemlali', 'Mission', 'Redding Picholine', 'Chetoui', 'Nevadillo' and 'Liguria' were S). (Wilhelm and Taylor, 1965)
'Konservolia' (S), 'Kalamon' (S), 'Manaki' (S), 'Koutsourelia' (S), 'Matolia' (S), 'Megaritiki' (S) (Thanassoulopoulos *et al.*, 1979)
'Cornicabra' (S), 'Kilis Yaglik' (S), 'Saloneque' (S), 'Verdale' (S), 'Verdial de Huevar' (S) (Caballero *et al.*, 1980)
'Manzanillo' (S), 'Mission' (S), 'Obliza' (S), 'Sevillano' (S) (Wilhelm, 1981)
'Hojiblanca' (S), 'Manzanilla' (S), 'Picual' (S) (Blanco-López *et al.*, 1984)
'Carolea' (MS), 'Cipressino' (MS), 'Coratina (MS) (S), 'Cassanese'(HS), 'Nocellara del Belice' (HS), 'Nocellara Etnea' (HS), 'Tonda Iblea' (HS), 'Uovo di Piccione' (HS) (Pennisi *et al.*, 1993)
'Hojiblanca' (S), 'Picual' (S) (López-Escudero *et al.*, 2004)
'Frantoio' (R), 'Coratina' (R), 'Frangivo' (R), 'Oblonga' (R), 'Kalamon' (R), 'Ascolana Tenera' (S), 'Cellina' (S), 'Leccino' (S), 'Manzanillo' (S), 'Chemlalie'(S), 'Konservolia' (S), 'Mission' (S), 'Picual' (S) (Cirulli *et al.*, 2008)
'Changlot Real' (R), 'Dolze Agogia' (R), 'Empeltre' (R), 'Frantoio' (R), 'Grosal de Albocafe' (R), 'Kato Drys' (R), 'Manzanilla Picúa' (R), Oblonga (R), 'Ag'ezi Shami-1' (MR), 'Aloreña-1' (MR), 'Bodoquera' (MR), 'Carrasqueño de Porcuna' (R), 'Chemlali' (MR), 'Cipressino' (MR), 'Corneja' (MR), 'De Sal' (MR), 'íot' (MR), 'Kiti' (MR), 'Koroneiki' (MR), 'M'elia' (MR), 'Negrillo de Andujar' (MR), 'Pajar'ra' (MR), 'Sevillenca', 'Verdial de Vélez-Málaga-1' (MR), 'Amigdaloia' (S), 'Arbosana' (S), 'Ascolana Tenera' (S), 'Athalassai' (S), 'Azapa' (S), 'Blanqueta' (S), 'Carrasqueño de Alcaudetei' (S), 'Coratina' (S), 'Dokkar' (S), 'Escarabajuelo de Úb'da' (S), 'Farga' (S), 'Figueretes' (S), 'Hasou' (S), 'Gordal Sevillana' (S), 'Ka'amon' (S), 'Kalogerida' (S), 'Lechín de Sevilla' (S), 'Llumeta' (S)' 'Lucques' (S), 'Maurino' (S), 'Mission Nyeland' (S), 'Moraiolo' (S), 'Moraiolo T'. (S), 'Corsini' (S), 'Negrillo de Arjona' (S), 'Nevadillo Negro' (S), 'Pico Limón de Grazalema' (S), 'Vallesa' (S), 'Zarza' (S), 'Arbequina' (HS), 'Ayvalik' (HS), 'Azul' (HS), 'Borriolenca' (HS), 'Bout'illan' (HS), 'Callosina' (HS), 'Cañivano Negro-1' (HS), 'Carolea' (HS), 'Carrasqueño 'e Lucena' (HS), 'Cipressino' (HS), 'Chalchali-1' (HS), 'Chemlal de Kabilye' (HS), 'Chorrúo de Castro del Río' (HS), 'Cobrancosa' (S), 'Corbella' (HS), 'Cordobés Arroyo de la Luz' (HS), 'Cornicabra' (HS), 'Dulzal' (HS), 'Dulzal de Carmona' (HS), 'Fulla de Zalce' (HS), 'Gordal de Hellín' (HS), 'Hendeño' (HS), 'Hojiblanca' (HS), 'Imperial de Jaen' (HS), 'Leccino' (HS), 'Lechí de Granada' (HS), 'Lucio' (HS), 'Macho de Jaén' (HS), 'Manzanilla' (HS), 'Manzanilla de Jaén' (HS), 'Manzanilla Cacereña' (HS), 'Manzanilla del Piquito' (HS), 'Manzanilla de Sevilla' (HS), 'Megantiki' (HS), 'Meski' (HS), 'Mission Moojeski '(HS), 'Mollar de Cieza' (HS), 'Morisca' (HS), 'Morona' (HS), 'Nabali' (HS), 'Negral' (HS), 'Negro del Carpio' (HS), 'Nevadillo Blanco de Jaen' (HS), 'Nevado Azul' (HS), 'Ocal' (HS), 'Pajarero' (HS), 'Pendolino' (HS), 'Picholine Marrocaine' (HS), 'Picual' (HS), 'Picudo' (HS), 'Rapasayo' (HS), 'Redondilla de Logroño' (HS), 'Royal de Cazorla' (HS), 'Valanolia' (HS), 'Verdial de Alcaudete' (HS), 'Villalonga' (HS) (Trapero *et al.*, 2010)
'Blanqueta' (HS), 'Bosana' (HS), 'Lucques' (HS), 'Belluti' (HS), 'Picholine' (S), 'Picual' (S), 'Doebli' (S), 'Safrawi' (S), 'Pavo' (S), 'Uovo di Piccione' (S), 'Zard' (MS), 'Zaity' (MS), 'Lechin de Sivilla' (MS), 'Uslu' (MS), 'Sinop' (MS), 'Sevillana' (MS), 'Cirujal' (MS), 'Caninense' (R), 'Gemlik' (R), 'Moitul-607' (MS), 'Barnea' (R), 'Selvatico' (R), 'Majohl-1059' (R), 'Redonda del Loño' (R), *Olea ferruginea* (R), 'Aglandau-170' (R), 'Kalmett' (R), 'Manzanilla de Saicente' (R), 'Manzanilla de Lo-1816' (R), 'Verdial Budajoz' (R), 'Manzanillo Cobés de Cabra' (R), 'Lastovka' (R), 'Suca' (R), 'Chorrou de Castro del Rio' (R) (García-Ruiz *et al.*, 2014)

[a]From Lòpez-Escudero and Mercado-Blanco (2011), partially modified and updated.
[b]Cultivars were rated by their resistance/susceptibility level as Resistant (R), Moderately Resistant (MR), Moderately Susceptible (MS), Susceptible (S) and Highly Susceptible (HS).

known as honey fungus. The reproductive stage of this fungus is represented by edible basidiomata (mushrooms) typically grouped in clusters around the base of the infected trees or stumps. Five inter-sterile groups, considered as true species, are known in

Europe for the genus *Armillaria*. Among them, *A. mellea* plays the most significant role as plant pathogen and, to date, it is the only species reported on olive as a pathogen of primary importance and the most serious threat to olive growing in some areas of the Mediterranean basin.

R. necatrix, now named *Dematophora necatrix* after a recent taxonomic revision and the resurrection of the genus *Dematophora* (Wittstein *et al.*, 2020), belong to the class Ascomycetes, order Xylariales. Infections of *R. necatrix* on olive trees have been reported by several authors but it is considered of secondary importance on olive in comparison to Armillaria root rot, although the two diseases may be frequently confused, being indistinguishable from the symptoms on the canopy and rather similar symptoms on below-ground organs. Recent field surveys have revealed an increasing trend of its diffusion especially in nurseries and young olive plantations (Schena and Agosteo, 2011).

Both pathogens are very polyphagous causing severe root rots on woody species belonging to angiosperms and gymnosperms, forest trees and fruit crops, shrub and herbaceous species, although *R. necatrix* host range is much wider, including weeds and algae.

Symptoms

A. mellea and *R. necatrix* are soilborne pathogens responsible for root and collar rots. As a consequence of root damage, affected trees show general non-specific declining symptoms on the canopy. Depending on environmental conditions and tree age, the two diseases can lead to a rapid decline of plants, which quickly die (apoplexy), or to a progressive weakening of trees, which can remain alive for several years.

R. necatrix, in contrast to *A. mellea*, frequently attacks and kills rapidly very young orchard plants (often during the first year after planting) (Fig. 7.3.13). Apoplexy is common for seedlings in the nurseries and for new plantations in heavy soils and irrigated orchards but sometimes may involve also adult plants. Infected trees in the field can die in a few weeks after the appearance of the first symptoms; leaves wilt, dry up and remain attached to the tree for months (Fig. 7.3.14). These symptoms usually occur after a period of water or physiological stress.

In the case of progressive weakening, very common for *A. mellea*, the trees develop a generally unhealthy appearance. Leaves show a change of color, becoming progressively chlorotic to yellowish, reduced in size and fall prematurely, with extensive dieback of the branches (Fig. 7.3.15). Asymmetric infection of the root system can result in the death of only a single main branch. Infected young trees die in a short time, some months or a few years, depending on tree age. Adult trees can remain alive and be quite productive for several years and they can manage the disease by continuously generating new roots in the superficial soil layers. Generally, on these plants, death occurs only when the infection affects a large part of the root branches or reaches the crown. In advanced infection phases, diseased trees are easily uprooted since all small and medium roots are destroyed and the root system is very limited. Serious symptoms on the canopy are detected only when the infection is advanced and the pathogen is widely spread on the root system and in the soil, often too late to save the plant. The infection has probably already affected neighboring plants that are still asymptomatic, to which phytoiatric measures must be addressed.

A. mellea develops its mycelium below the bark at the level of the cambium, between the sapwood and the bark. The bark rises and appears cracked and fissured and darker in color, due to its progressive necrosis. It can be easily removed from the wood and will expose the characteristic white mycelial mats and/or the rhizomorphs (Fig. 7.3.16). The white mycelial mats are marked by irregular, fan-like striations (mycelial 'fans'). Rhizomorphs are highly differentiated organs with apical growth, black to reddish brown, with a peripheral cortex and up to 5 mm wide. Rhizomorphs growing beneath the bark are flat, while those that develop in the soil are cylindrical and about half as wide. *A. mellea* can reach a certain height along the trunk and can behave like a

Fig. 7.3.13. Young declined olive trees cv. 'Carolea' infected by *R. necatrix*.

wood-destroying fungus, causing white rot of wood and forming in dead wood dark lines (zone lines), which divide the colonized wood into different, more or less independent, volumes.

Specific symptoms for *Rosellinia* and *Armillaria* infections can be observed and compared. *R. necatrix* covers the roots with numerous little white mycelial or grayish strands, finger-like fans, sometimes having the appearance of spiders' webs, which develop on and into the bark of the crown and of the large roots of dying trees (Fig. 7.3.17). The fans can be observed at different levels in the invaded bark; they are not particularly localized at the cambium level as in *A. mellea*, whose mycelial fans are larger and continuous. *R. necatrix* occurs also in the soil surrounding the roots with an increasing density close to them. In high-humidity conditions, the abundant white cottony mycelium is visible with the naked eye at ground level. In dry conditions, pieces of infected roots transferred to damp chambers rapidly develop the characteristic white mycelium and, eventually, sheets of microsclerotia. Rotted bark appears sunken and dark-brown and a distinct margin is usually visible between the infected and healthy bark. *R. necatrix* does not rot the wood, the sapwood becomes dry and the presence of the fungus is marked by black halos, which are very different from the distinct zone lines characteristic of *A. mellea*.

Fig. 7.3.14. Apoplexy of an adult olive plant caused by *A. mellea*.

Disease cycle and epidemiology

Both *R. necatrix* and *A. mellea* can survive on their numerous woody and herbaceous hosts and there are a number of weeds that can host the pathogens and promote their survival. *A. mellea* can survive for a considerable time as mycelial masses in the dead host tissues buried in the soil. The mycelium is often compartmentalized and protected in heart wood by the zone lines. The period of conservation of the fungus depends mainly on the size of the wood fragments. On entire stumps, fungal survival can be estimated at several decades. The fungus can also survive as mycelium in latent cortical necrosis and on living roots of symptomless trees.

R. necatrix does not colonize woody tissues and, therefore, can survive for shorter periods compared to *A. mellea*. This period can vary from a few months to a few years. The role of microsclerotia, produced in high quantities both in nature and *in vitro*, seems to be of lesser importance.

Local spread of the pathogens in the soil can occur starting from infected roots or

Fig. 7.3.15. Young olive tree with extensive dieback of the branches, showing a cluster of basidiomata of *A. mellea* at the base of the trunk.

saprophytic mycelia on dead wood in the soil, through direct root contact between host plants, or via mycelial strands (*R. necatrix*) or rhizomorphs (*A. mellea*), which grow through soil from infected plants to healthy ones (Pérez-Jiménez, 2006). Recently, it was speculated that climatic changes and global warming have indirectly enhanced the activity of rhizomorphs increasing the susceptibility of trees (Kubiak *et al.*, 2017).

Infections of *A. mellea* concern only the large or medium roots which are lignified and suberized. No wound is required for such penetration. Rhizomorphs attach receptive organs by means of mucilage and produce short branches, the 'infection cones', which directly enter the root bark. After penetration, the infection cone separates into individual hyphae which colonize the bark. A second aggregation phase then occurs in the cambium, giving rise to mycelial fans and flat subcortical rhizomorphs. In addition to the mechanical action, a number of enzymatic activities have been involved in this process. The infection process in *R. necatrix* is similar to that described for *A. mellea*.

As far as the spread pattern is concerned, both pathogens are characterized by their occurrence in patches, which extend

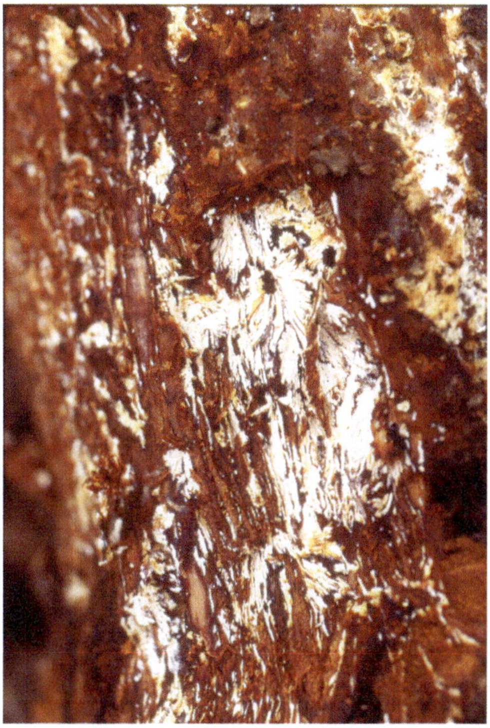

Fig. 7.3.16. Mycelial mats of *A. mellea* under the bark.

Fig. 7.3.17. Mycelium of *R. necatrix* at the base of the trunk and on the roots of a young olive plant cv. 'Carolea'.

progressively in a circular pattern. Long-distance dispersal of *R. necatrix* and *A. mellea* mainly occurs with infected propagating materials, which are frequently asymptomatic during the early phases of the infection process. The wide spread of the disease in some ancient olive-growing areas was due to the method of propagation by 'ovuli', that were removed from the base of infected unstable plants uprooted by the wind (Agosteo, 2010). Furthermore, infested soils and infected plant debris can be distributed via cultural practices or water. Another risk to spread the pathogens is due to the application on the crops, as mulch, of coniferous barks directly obtained from the forest, which are common substrate for the development of *A. mellea*. In another way, this type of sterilized mulch, added from antagonistic species, is proposed as a new method of control.

In *A. mellea*, quite long-range dissemination also occurs via basidiospores, which can germinate, generally on fresh-cut stumps, initiating new outbreaks. Conversely, *R. necatrix* ascospores are very rare and conidia produced in synnemata at the base of dead plants and on roots seem to play a minor role in the diffusion of the disease.

Control

Control of FRR and WRR must be carried out using preventive criteria, avoiding the introduction of causative agents in olive orchards. A healthy area can be isolated from a close infected one by digging a trench, at least 80 cm deep, that prevents the passage of the rhizomorphs of *A. mellea*.

The first important step is to evaluate the health status of propagation materials and soil before new plantations are set up. Some reports indicate that the pathogens are frequent in nurseries and contamination of orchards from nurseries very probably occurs frequently.

The health status of soil, apart from the development of molecular tools to directly detect pathogens (Schena and Ippolito, 2003), can be assessed more easily and without the need for specialist contributions, by the Armillaria trap method. Trapping involves placing a nutrient source, such as logs or bark, in the soil for four to six months. Such food bases become colonized by *Armillaria* if it is present in the soil. Traps are then examined for evidence of colonization, such as mycelial fans and rhizomorphs (Ip and Yongqing, 1992). 'Trap-bag' technique, based on the burial in the ground of conifer bark placed into nylon mesh sacks, showed to be more sensitive than 'trap-log' that utilizes freshly cut aspen stakes.

Pathogens exclusion can be carefully managed in nurseries, using a number of costly chemical and physical applications, such as fumigation and soil disinfection with steam, and verifying, through specific tests, the status of health of the plant organs that are introduced. Eradication from field-infected soil, especially for *A. mellea*, also after a great numbers of years with repeated cultivation of cereals and other non-host species, is extremely difficult because of its ubiquitarian presence as saprophyte in plant residues, deep penetration into the soil, resistance to drought, and tolerance to a wide range of soil pH.

CULTURAL CONTROL METHODS. Preventive agronomic practices are the main way to manage *A. mellea* and *R. necatrix* root rot.

In infected fields, control strategies should be aimed at managing the disease rather than eradicating it. In a context of integrated control strategies, cultural control methods play a major role (Fox, 2000).

It is widely reported that *A. mellea* is a more successful pathogen when the host is under stress conditions such as drought, waterlogging, mechanical damage or lack of pruning. Some plant species are naturally resistant to the fungus and, though the fungus may infect them, little damage occurs. However, if they are weakened in any way (usually from over-watering), they may become susceptible and killed by the fungus. Therefore, appropriate pruning, balanced manuring, rational irrigation and proper soil-management practices should be focussed to avoid stresses to the trees and to favor the generation of new roots.

Particular attention should be given when a new olive orchard is planted in potentially infested soils. All infected plants, together with contaminated roots and other plant debris, need to be uprooted and removed from the field as completely as possible, for instance by subsoiling. The debris can be left on the soil surface because they dry out quickly and are unsuitable for the fungus. Infested nude soils should be cultivated with non-host species (cereals are commonly suggested) or left uncultivated for a certain number of years. In the case of *R. necatrix*, a period of two to four years before planting a new olive orchard seems to be enough, whereas a longer, not well-defined period is necessary for *A. mellea*.

Certain soil types are more conducive to root rot pathogens. Heavy clay soils should be avoided in new olive groves, as they favor water stagnation and create lack of oxygen and conditions of radical asphyxia, which promote infections. To prevent waterlogging, soil drainage must be carried out and the tree should be planted in a raised bed. Other bad cultivation practices on the plantation can lead to radical suffering, such as positioning the plants too deep in the planting hole and filling mounds of soil at the base of the trunk to give the plant more stability in windy areas. Much attention must be paid to properly burning the tips of the chestnut support posts, which otherwise act as a trap for *A. mellea*, transferring the infection to the plants. In any case they must be removed as soon as possible. Drip irrigation is preferable to other irrigation methods, avoiding directly wetting the collar of plants.

Symptomatic infected young plants should be carefully eradicated. Replanting a tree in place of an infected one often results in the death of young trees within a few years. The planting hole after eradication should be left empty for a few years and disinfected with lime or replanted with a non-susceptible host. More attention should be paid to neighboring plants that are still asymptomatic but could probably be infected. These plants, as on adult plants with chronic symptoms, found to be infected in only a few roots or a small part of the root collar, can be saved for a time by carefully removing the soil from around the base of the tree, to a depth of 20–30 cm, to expose the root collar and buttress roots to aeration and drying, from mid-spring to autumn. This operation prevents the pathogen from spreading on the roots by passing or girdling the root collar. It can be carried out with great precision, without causing any damage and clearing also the fine roots, with the air-spading technique, which uses high-volume compressed air to excavate trees roots. The exposed roots and the surrounding soil are traditionally treated with lime or concentrated solutions of copper salts. Alternatively, they can be treated repeatedly with suspensions of antagonistic species. The soil must be replaced with *Armillaria* free porous filling material, such as sandy soil and gravel, eventually together with bark mulch pre-treated with *Trichoderma* spp. before the first heavy frost. Infected bark and wood on large roots, buttress or trunk, should be excised back to healthy tissue.

CHEMICAL CONTROL. In the past, the eradication of *A. mellea* and *R. necatrix*, as well as other soilborne pathogens, from infested soils, especially in the nurseries, has been based on fumigations with methyl bromide, whose use has been drastically limited according to the Montreal Protocol. Other soil fumigants such as methyl-isothiocyanate and its generators (metham sodium and dazomet) do not always provide complete disease control. The application to the soil of water-soluble chemicals provided inconsistent levels of protection in field conditions. Most recently, the systemic sterol biosynthesis inhibitor (DMI) fungicides, cyproconazole, hexaconazole, propiconazole and tetraconazole have been reported to control Armillaria root rot in vineyards and peach orchards. In field trials it has been demonstrated that cyproconazole is very effective to control *A. mellea* on apple, walnut and kiwi fruit, even on trees with several symptoms of the disease, applied as soil drench at least in autumn and in spring at the concentration of 30 mg L^{-1} (Thomidis and Exadaktylou, 2012).

In experimental conditions *in vitro*, olive oil wastewater significantly reduced the mycelial growth of both *A. mellea* and *R. necatrix*. However, in the presence of olive cuttings, wastewater significantly stimulated the production of *Armillaria* rhizomorphs and the colonization of the host tissues (Agosteo and Macrì, 2008). These conclusions are consistent with the experimental observation that phenolic chemicals produce an extensive growth of rhizomorphs. Therefore, phenolic fungicides are counterproductive for the control of *A. mellea* (West and Fox, 2005).

PHYSICAL CONTROL. Steam disinfection of the soil is a valid control strategy for the eradication of soilborne pathogens but its applications in the field have high costs, and for *A. mellea* it is necessary to treat, as for fumigants, to at least 60 cm depth of soil, without the certainty of completely eradicating the pathogen. Soil solarization is a natural soil-heating technique that can be accomplished in warm climates during summer. This technique was found to be very effective in controlling WRR disease in established apple and avocado orchards. The increment in soil temperatures achieved by soil solarization can be high enough to reduce viability of *Armillaria* spp. only in the upper soil layer. The incorporation of additional suppressive factors is often necessary for improving the efficacy of soil solarization, which must be assumed in an integrated disease-management strategy. Suppression of solarized soil has been also attributed to the activation of antagonistic microorganisms such as *Trichoderma* spp. Soil solarization has improved the introduction and spread in the soil of antagonistic strains of *Trichoderma* spp. with better activity against the pathogens.

BIOLOGICAL CONTROL. Introducing bio-control agents and promoting their establishment could represent a complementary strategy for the control of this disease. Quite a large number of bio-control agents have been proposed as effective antagonists of *R. necatrix* and/or *A. mellea*. Several species and strains of *Trichoderma* can be used as bio-control agents against these two pathogens. *Trichoderma* spp. are common soil and decaying wood colonizers; they are the most studied fungal bio-control agents and some of them have been commercialized as bio-fungicides. The ability of this antagonist to replace *Armillaria* from buried wood was demonstrated. Maintaining the viability of introduced *Trichoderma* spp. in soil for a long time is often difficult, especially in the rhizosphere. More expensive and complicated practices must be employed by growers to apply *Trichoderma* spp. directly in the soil, near the roots. This type of inoculation method requires specific devices and several applications over time. A method to add *Trichoderma* spp. in the soil through carrier media was recently developed to prevent *Armillaria gallica* infections on strawberry. With one application of bark mulch (a mixture containing equal proportions (v/v) of pine, fir and larch bark), pre-treated directly by growers with *T. atroviride* SC1, the health of plants was preserved over an entire production cycle (Pellegrini *et al.*, 2014). The addition of barks pre-treated with *Trichoderma* is also applicable to cover exposed roots after removal of the soil around the base of the infected trees. Natural soil microorganisms can be particularly effective after pathogens have already been weakened by a previous application of other stressors such as chemical treatments, heating (soil solarization) or drying. In this context, all cultural practices favoring the increase of natural microbial populations such as organic manuring and/or the application of organic amendments can be beneficial.

7.3.3.6 Phytophthora root and fruit rot

As a typical Mediterranean plant, olive has been traditionally cultivated in arid lands. However, during the last decades in many olive-growing countries, the olive cultivation has been extended to different types of soil, and irrigation has become a common practice in olive orchards. An emerging phytopathological problem of olive trees growing in wet or waterlogged soils is root rot caused by *Phytophthora* spp. The genus *Phytophthora* (Pythiaceae, Peronosporales, Oomycota, Chromista) presently comprises more than 200 described taxa. The number of accepted species during the last years has increased dramatically as a consequence of the advent of DNA sequencing and the introduction of molecular phylogenetic systematics. Moreover, there has been a substantial advancement of the methods and techniques to detect in and recover from soil these typical soilborne pathogens. Several *Phytophthora* species have been reported as causative agents of tree decline, leaf chlorosis, wilting, defoliation, and twig dieback, as a consequence of root rot and basal stem cankers on olive worldwide, including *Phytophthora acerina, P. bilorbang, P. cactorum, P. cinnamomi, P. citricola sensu lato, P. cryptogea, P. drechsleri, P. heterospora, P. inundata, P. megasperma, P. nicotianae, P. oleae, P. palmivora, P. pini* and *P. plurivora*. These species differ in aggressiveness, temperature requirements for growth, geographical distribution, and ecology. *P. palmivora*, alone or in association with *Verticillium dahliae*, was reported as a causal agent of rot of fine roots and wilt of young olive trees in nurseries and new plantings. *P. oleae* was reported to be responsible of rot of olives in commercial groves in southern Italy. Moreover it was recovered from soil and roots of wild olive trees in protected natural areas in Spain and Sicily (southern Italy). *P. bilorbang* and *P. inundata*, alone or in association with other *Phytophthora* species, are responsible for root rot and wilt of olive trees in clay soils after flooding, acting as opportunistic, albeit aggressive, root pathogens. *P. heterospora*, has been found associated with stem lesions and root and collar rot of young olive trees in nurseries and recently established commercial plantings in southern Italy. In moist environments, some typically soil-inhabiting species, like *P. nicotianae* and *P. oleae*, have occasionally adapted to an aerial lifestyle and may infect above-ground parts of olive trees such as drupes, leaves, and twigs causing fruit rot, leaf drying, filloptosis, and twig dieback.

In general, preventive measures are adopted to manage root rot of olive trees caused by *Phytophthora* species, including cautious use of irrigation, avoiding clay soils for new plantings and soil waterlogging in mature olive orchards as well as prophylaxis in nursery plant production. Only *P. cinnamomi*, which actually is not a common pathogen in olive nurseries of the Mediterranean region, is in the list of pathogens affecting the quality of olive nursery plants. For more details on Phytophthora root and fruit rot of olive the reader is referred to a review on this subject (Cacciola *et al.*, 2011), a description of a new *Phytophthora* species, *P. oleae* (Fig. 7.3.18), named after olive, which presently is the only known host (Ruano-Rosa *et al.*, 2018), a very recent description of *P. heterospora*, a new species closely related

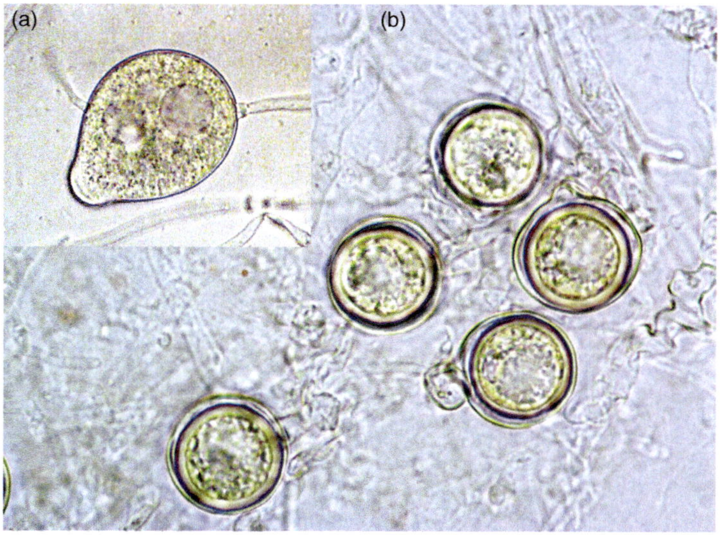

Fig. 7.3.18. (a) persistent semipapillate sporangium of *Phytophthora oleae*; (b) gametangia and oospores.

to *P. palmivora*, comprising olive among its host plants (Scanu *et al.*, 2021) and the report of olive as a new host of *P. bilorbang* (Santilli *et al.*, 2020).

References

Agosteo, G.E. (2010) Avversità dell'olivo ed evoluzione dei relativi sistemi di difesa in Calabria fra XVIII e XX secolo. *Petria* 20 (suppl.1), 41–70.

Agosteo, G.E. (2012) Principali malattie e difesa integrata dell'olivo. *Protezione delle colture* 3, 10–26.

Agosteo, G.E. and Macrì, C. (2008) Attività delle acque reflue olearie su funghi ad *habitus* terricolo patogeni dell'olivo. In: Zimbone, S.M. (ed.) *Valorizzazione delle acque reflue e dei sottoprodotti dell'industria olearia e agrumaria*. Aracne Editore, Rome, pp. 109–113.

Agosteo, G.E. and Schena, L. (2011) Olive leaf spot. In: Schena, L., Agosteo, G.E. and Cacciola, S.O. (eds) *Olive Diseases and Disorders*. Transworld Research Network, Trivandrum, Kerala, India, pp. 143–176.

Agosteo, G.E. and Zappia, R. (2007) Epidemiological study of Olive scab in Calabria. In: *Integrated Protection of Olive Crops*. *IOBC/WPRS Bulletin* 30, 233–244.

Agusti Bisiach, C., Ávila, A., Romero, J., Raya, M.C., Roca, L.F. and Trapero Casas, A. (2016) *Bases para la gestiòn integrada del emplomado del olivo*. *Vida rural*, October, 40–48.

Almeida, R.P.P. (2016) *Xylella fastidiosa* vector transmission biology. In: Brown, J.K. (ed.) *Vector-mediated Transmission of Plant Pathogens*. The American Phytopathological Association, St Paul, Minnesota, pp. 165–173.

Ávila, A. and Trapero, A. (2010) Grupo de Patología Agroforestal de la Universidad de Córdoba. El Emplomado del olivo y del acebuche. *Junta de Andalucía. Consejería de Medio Ambiente*. Available at: https://www.juntadeandalucia.es/medioambiente/portal_web/web/temas_ambientales/montes/plagas/fichas_plagas_enfermedades/emplomado_olivo.pdf (accessed 20 November 2020).

Ávila, de la Calle, A., Benali, A. and Trapero Casas, A. (2004) Morphological and cultural variability of *Pseudocercospora cladosporioides*, causal agent of olive cercosporiose. *Boletín de Sanidad Vegetal. Plagas (España)* 30, 369–384.

Ávila, A., Groenewald, J.Z., Trapero, A. and Crous, P.W. (2005) Characterisation and epitypification of *Pseudocercospora cladosporioides*, the causal organism of Cercospora leaf spot of olives. *Mycological Research* 109, 881–888.

Ávila, A., Romero, J., Agusti Bisiach, C., Benali, A., Roca, L.F. *et al.* (2020) Phenotypic and pathogenic characterization of *Pseudocercospora cladosporioides*, causal agent of cercospora leaf spot of olives. *European Journal of Plant Pathology* 156, 45–65.

Benjama, A., Walali, L., Jananti, L. and Moukhli, A. (1987) *Field reaction of different varieties of olive (Olea europaea L.) to olive knot disease caused by* Pseudomonas savastanoi *pv.* savastanoi *(E.F. Smith) Stevens. Third International Working Group on* Pseudomonas syringae *pathovars*, Lisbon, 1–4 September, pp. 41–52.

Blanco-Lopez, M.A., Jimenez-Dıaz, R.M. and Caballero, J.M. (1984) Symptomatology, incidence and distribution of Verticillium wilt of olive trees in Andalucıa. *Phytopathologia Mediterranea* 23, 1–8.

Boscia, D., Altamura, G., Ciniero, A., Di Carolo, M., Dongiovanni, C. *et al.* (2017) Resistance to *Xylella fastidiosa* in different olive cultivars. *L'informatore Agrario*, 11, 59–63.

Bubici, G. and Cirulli, M. (2012) Control of Verticillium wilt of olive by resistant rootstocks. *Plant and Soil* 352, 363–376.

Buonaurio, R., Moretti, C., Passos da Silva, C., Cortese, C., Ramos, C. *et al.* (2015) The olive knot disease as a model to study the role of interspecies bacterial community in plant disease. *Frontiers in Plant Science* 6, 434.

Buonaurio, R., Almadi, L., Famiani, F., Moretti, C., Agosteo, G.E. *et al.* (2023) Olive leaf spot caused by *Venturia oleaginea*: An updated review. *Fronters in Plant Science* 13, 1061136.

Caballero, J.M., Pérez-Hernández, J., Blanco-López, M.A. and Jiménez-Díaz, R.M. (1980) Olive, a new host of *Verticillium dahliae* in Spain. In: *Proceedings of the 5th Congress Mediterranean Phytopathological Union, Patras, Greece*. Hellenic Phytopathological Society, Athens. .

Cacciola, S.O., Faedda, R., Pane, A. and Scarito, G. (2011) Root and crown rot of olive caused by *Phytophthora* spp. In: Schena, L., Agosteo, G.E. and Cacciola, S.O. (eds) *Olive Diseases and Disorders*. Transworld Research Network: Trivandrum, Kerala, India, pp. 305–327.

Cacciola, S.O., Faedda, R., Sinatra, F., Agosteo, G.E., Schena, L. *et al.* (2012) Olive anthracnose. *Journal of Plant Pathology* 94, 29–44.

Çağlayan, K., Faggioli, F. and Barba, M. (2011) Viruses, phytoplasmas and diseases of unknown etiology of olive trees. In: Hadidi, A., Barba, M., Candresse, T. and Jelkmann W. (eds) *Virus and virus-like diseases of pome and stone fruits*. The American Phytopatological Society, St Paul, Minnesota, pp. 289–297.

Cavalieri, V., Dongiovanni, C., Tauro, D., Altamura, G., Di Carolo, M. *et al.* (2018) Transmission of the codiro strain of *Xylella fastidiosa* by different insect species. *Proceedings of the XI European Congress of Entomology, Naples, Italy*, pp. 144–145.

Cirulli, M. and Montemurro, G. (1976) A comparison of pathogenic isolates of *Verticillium dahliae* and sources of resistance in olive. *Agriculturae Conspectus Scientificus* 39, 469–476.

Cirulli, M., Colella, C., D'Amico, M., Amenduni, M. and Bubici, G. (2008) Comparison of screening methods for evaluation of olive resistance to *Verticillium dahliae* Kleb. *Journal of Plant Pathology*, 90, 7–14.

Colella, C., Miacola, C., Amenduni, M., D'Amico, M., Bubici, G. *et al.* (2008) Sources of Verticillium wilt resistance in wild olive germplasm from the Mediterranean region. *Plant Pathology* 57, 533–539.

Coletta-Filho, H.D., Francisco, C.S., Lopes, J.R.S. and Almeida, R.P.P. (2017) Homologous recombination and *Xylella fastidiosa* host–pathogen associations in South America. *Phytopathology* 104, 120–125.

Cornara, D., Bosco, D. and Fereres, A. (2018) *Philaenus spumarius*: when an old acquaintance becomes a new threat to European agriculture. *Journal of Pest Science* 91, 957–972.

Cornara, D., Saponari, M., Zeilinger, A.R., de Stradis, A., Boscia, D. *et al.* (2017) Spittlebugs as vectors of *Xylella fastidiosa* in olive orchards in Italy. *Journal of Pest Science* 107, 305–312.

Del Moral, J. and Medina, D. (1985) El 'repilo plomizo' del olivo causado por *Cercospora cladosporioides* Sacc., enfermedad presente en España. *Boletín de Sanidad Vegetal. Plagas (España)* 12, 31–35.

Dervis, S., Mercado-Blanco, J., Erten, L., Valverde-Corredor, A. and Pérez-Artés, E. (2010) Verticillium wilt of olive in Turkey: a survey on disease importance, pathogen diversity and susceptibility of relevant olive cultivars. *European Journal of Plant Pathology* 127, 287–301.

Di Serio, F., Bodino, N., Cavalieri, V., Demichelis, S., Di Carolo, M. *et al.* (2019) Collection of data and information on biology and control of vectors of *Xylella fastidiosa*. *EFSA Supporting Publications.* DOI: 10.2903/sp.efsa.2019.EN-1628

Dongiovanni, C., Cavalieri, V., Bodino, N., Tauro, D., Di Carolo, M. *et al.* (2018a) Plant selection and population trend of spittlebug immatures (Hemiptera: Aphrophoridae) in olive groves of the Apulia region of Italy. *Journal of Economic Entomology* 112, 67–74.

Dongiovanni, C., Fumarola, G., Di Carolo, M., Tauro, D., Ciniero, A. *et al.* (2018b) Olive spittlebug, insecticides in comparison. *L'informatore Agrario* 24–25, 51–56.

EFSA (2018) Update of the *Xylella* spp. host plant database. *EFSA Journal* 16, 5408.

Fox, R.T.V. (2000) *Armillaria Root Rot: Biology and Control of Honey Fungus*. Intercept Press, Andover, UK.

García-Ruiz, G.M., Trapero, C. and López-Escudero, F.J. (2014) Evaluación the resistencia a *Verticillium dahliae* de variedades del Banco de Germoplasma Mundial de Olivo de Córdoba. *Phytoma España* 260, 53–56.

Giampetruzzi, A., Morelli, M., Saponari, M., Loconsole, G., Chiumenti, M. *et al.* (2016) Transcriptome profiling of two olive cultivars in response to infection by the CoDiRO strain of *Xylella fastidiosa* subsp. *pauca*. *BMC Genomics* 17, 475.

Godena, S., Dminić, I. and Đermić, E. (2012) Differential susceptibility of olive varieties to olive knot disease in Istria. *Journal of Central European Agriculture* 13, 85–94.

Goodwin, P.H. and Zhang, S. (1997) Distribution of *Xylella fastidiosa* in southern Ontario as determined by the polymerase chain reaction. *Canadian Journal of Plant Pathology* 19, 13–18.

Guario, A., Parisi, V., Mazzotta, E., Murrone, N. and Pitotti, V. (2000) Criteri di difesa della cercosporiosi dell'olivo (*Mycocentrospora cladosporioides* Cast.) in ambienti pugliesi. Proceedings *Giornate Fitopatologiche* 2000, *Perugia, Italy*, Vol. 2, pp. 87–92.

Henricot, B., Gorton, C., Denton, J. and Denton, G. (2009) *Pseudocercospora cladosporioides*, the cause of leaf spot on olive, a pathogen new to the United Kingdom. *Plant Pathology* 58, 803.

Hernandez-Martinez, R., Costa, H.S., Dumenyo, C.K. and Cooksey, D.A. (2006) Differentiation of strains of *Xylella fastidiosa* infecting grape,almonds, and oleander using a multiprimer PCR assay. *Plant Disease* 90, 1382–1388.

Hosni, T., Moretti, C., Devescovi, G., Suarez-Moreno, Z.R., Fatmi, M.B. *et al.* (2011) Sharing of quorum-sensing signals and role of interspecies communities in a bacterial plant disease. *The ISME Journal* 5, 1857–1870. (Available at: https://www.juntadeandalucia.es/medioambiente/portal_web/web/temas_ambientales/montes/plagas/fichas_plagas_enfermedades/emplomado_olivo.pdf (accessed 16 September 2020).

Iacobellis, N.S. (2001) Olive knot. In: Maloy, O.C. and Murray, T.D. (eds) *Encyclopedia of Plant Pathology*. John Wiley and Sons, New York, pp. 713–715.

Ip, D.W. and Yongqing, H. (1992) *Armillaria* trapping in a *Pinus koraiensis* plantation in northeast China. *Journal of Northeast Forestry University* 3, 1–12.

Issa, T., Almadi, L., Jarrar, S., Tucci, M., Buonaurio, R. *et al.* (2019) Factors affecting *Venturia oleaginea* infections on olive and effects of the disease on floral biology. *Phytopathologia Mediterranea* 58(2), 221–229.

Jiménez-Diaz, R.M., Cirulli, M., Bubici, G., Jiménez-Gasco, M.d.M., Antoniou, P.P. *et al.* (2012) Verticillium wilt, a major threat to olive production: current status and future prospects for its management. *Plant Disease* 56, 304–329.

Kolainis, S., Koletti, A., Lykogiannis, M., Karamanou, D., Gkizi, D. *et al.* (2020) An integrated approach to improve plant protection against olive anthracnose caused by the *Colletotrichum acutatum* species complex. *PLOS ONE* 15, e0233916.

Krugner, R. (2010) *Evaluation of Pathogenicity and Insect Transmission of Xylella fastidiosa Strains to Olive Plants*. Final Report 2010. California Olive Committee, 3–11.

Krugner, R., Sisterson, M.S., Chen, J., Stenger, G.C. and Jhonson, M.W. (2014) Evaluation of olive as a host of *Xylella fastidiosa* and associated sharpshooter vectors. *Plant Diseases* 98, 1186–1196.

Kubiak, K., Zólciak, A., Damszel, M., Lech, P. and Sierota, Z. (2017) *Armillaria* pathogenesis under climate changes. *Forests* 8(4), 100. Available at https://www.mdpi.com/1999-4907/8/4/100/htm (accessed 16 September 2020).

Lelliot, R.A. (1988) *Pseudomonas syringae* subsp. *savastanoi* (Smith) Janse. In: Smith, J.M., Dunez, J., Phillips, D.H., Lelliott, R.A. and Archer, S.A. (eds) *European Handbook of Plant Diseases. Blackwell Scientific Publications*, Oxford, UK.

Lombardo, N. (2009) Descrizione delle principali cultivar di olivo da olio e da tavola italiane. In: Fiorino, P. (ed.). *Olea – Trattato di olivicoltura. Edizioni Agricole de Il Sole 24* ORE, Edagricole, Bologna, Italy, pp. 169–193.

Lopes, J.R.S., Daugherty, M.P. and Almeida, R.P.P. (2010) Strain origin drives virulence and persistence of *Xylella fastidiosa* in alfalfa. *Plant Pathology* 59, 963–971.

López-Doncel, L.M., Viruega-Puente, J.R. and Trapero-Casas, A. (2000) Respuesta del olivo a la inoculacion con *Spilocaea oleagina*, agente del repilo. *Boletín de Sanidad Vegetal. Plagas (España)* 26, 349–363.

López-Escudero, F.J. and Mercado-Blanco, J. (2011) Verticillium wilt of olive: a case study to implement an integrated strategy to control a soil-borne pathogen. *Plant and Soil* 344, 1–50.

López-Escudero, F.J., del Río, C., Caballero, J.M. and Blanco-López, M.A. (2004) Evaluation of olive cultivars for resistance to *Verticillium dahliae*. *European Journal of Plant Pathology* 110, 79–85.

Loprieno, N. and Tenerini, I. (1959) Metodo per la diagnosi dell''Occhio di pavone' dell'olivo (*Cycloconium oleaginum* Cast.). *Phytopathologische Zeitschrift* 34, 385–392.

Lops, F., Frisullo, S. and Rossi, V. (1993) Studies on the spread of the olive scab pathogen, *Spilocaea oleagina*. *Bull. OEPP/EPPO* 23, 385–388.

Marcelo, A., Fernandes, M., Potes, M.F. and Serrano, J.F. (1999) Reaction of some cultivars of *Olea europaea* L. to experimental inoculation with *Pseudomonas syringae* pv. *syringae*. In: Metzidakis, I.T. and Voyiatzis, D.G. (eds) Proccedings of the Third International Symposium on Olive Growing, Chania, Greece, 22–26 September 1997, Vol. 2. *Acta Horticulturae* 474, 567–571.

Marchi, G., Mori, B., Pollacci, P., Mencuccini, M. and Surico, G. (2009) Systemic spread of *Pseudomonas savastanoi* pv. *savastanoi* in olive explants. *Plant Pathology* 58, 152–158.

Martelli, G.P. (2011) Infectious diseases of olive. In: Schena, L., Agosteo, G.E., and Cacciola, S.O. (eds.) *Olive Diseases and Disorders*. Transworld Research Network, Trivandrum, Kerala, India, pp. 71–88.

Martelli, G.P. (2016) The point on *Xylella fastidiosa* and rapid olive tree desiccation. *L'informatore Agrario* 24, 40–48.

Mercado-Blanco, J. and López-Escudero, F.J. (2012) Verticillium wilt of olive and its control: the heat is on. *Plant and Soil* 355, 17–21.

Mina, D., Pereira, J.A., Lino-Neto, T. and Baptista, P. (2020) Impact of plant genotype and plant habitat in shaping bacterial patobiome: a comparative study in olive tree. *Scientific Reports* 10, 3475.

Montes-Osuna, N. and Mercado-Blanco, J. (2020) Verticillium wilt of olive and its control: what did we learn during the last decade? *Plants* 9, 735.

Moral, J., Xaviér, C.J., Viruega, J.R., Roca, L.F., Caballero, J. *et al.* (2017) Variability in susceptibility to anthracnose in the world collection of olive cultivars of Cordoba (Spain). *Frontiers in Plant Science* 8, 1892.

Moral, J., Agustí-Brisach, C., Raya, M.C., Jurado-Bello, J., López-Moral, A., *et al.* (2021) Diversity of *Colletotrichum* species associated with olive anthracnose worldwide. *Journal of Fungi* 7, 741.

Mulero-Aparicio, A., Varo, A., Agustí-Brisach, C., López-Escudero, F.J. and Trapero, A. (2020) Biological control of Verticillium wilt of olive in the field. *Crop Protection* 128, 104993.

Nascimento, R., Gouran, H., Chakraborty, S., Gillespie, H.W., Almeida-Souza, H.O. *et al.* (2016) The type II secreted lipase/esterase LesA is a key virulence factor required for *Xylella fastidiosa* pathogenesis in grapevines. *Scientific Reports* 6, 18598.

Nigro, F. and Ferrara, M. (2011) Olive cercosporiosis. In: Schena, L., Agosteo, G.E. and Cacciola, S.O. (eds) *Olive Diseases and Disorders*. Transworld Research Network, Trivandrum, Kerala, India, pp. 247–258.

Nigro, F., Ippolito, A., Gallone, P., Carmignano, P. and Romanazzi, G. (2002) Cercosporiosis of olive in Apulia and attemps to control the disease. Proceedings, 4th International Symposium on Olive Growing. *Acta Horticulturae* 586, 773–776.

Nigro, F., Boscia, D., Antelmi, I. and Ippolito, A. (2013) Fungal species associated with a severe decline of olive in southern Italy. *Journal of Plant Pathology* 95, 668.

Obanor, F.O., Walter, M., Jones, E.E. and Jaspers, M.V. (2008) Effect of temperature, relative humidity, leaf wetness and leaf age on *Spilocaea oleagina* conidium germination on olive leaves. *European Journal of Plant Pathology* 120, 211–222.

Obanor, F.O., Walter, M., Jones, E.E. and Jaspers, M.V. (2011) Effects of temperature, inoculum concentration, leaf age, and continuous and interrupted wetness on infection of olive plants by *Spilocaea oleagina*. *Plant Pathology* 60, 190–199.

Pasquini, G., Marzachì, C., Poggi Pollini, C., Faggioli, F., Ragozzino, A. *et al.* (2000) Molecular identification of Phytoplasmas from olive trees in Italy. *Journal of Plant Pathology* 82(3), 213–219.

Pellegrini, A., Prodorutti, D. and Pertot, I. (2014) Use of bark mulch pre-inoculated with *Trichoderma atroviride* to control Armillaria root rot. *Crop Protection* 64, 104–109.

Pennisi, A.M., Cacciola, S.O., Magnano di San Lio, G. and Perrotta, G. (1993) Evaluation of the susceptibility of olive cultivars to Verticillium wilt. *Bulletin OEPP/EPPO Bulletin* 23, 537–541.

Penyalver, R., García, A., Ferrer, A., Bertolini, E., Quesada, J.M. *et al.* (2006) Factors affecting *Pseudomonas savastanoi* pv. *savastanoi* plant inoculations and their use for the evaluation of olive cultivar susceptibility. *Phytopathology* 96, 313–319.

Pérez-Jiménez, R.M. (2006) A review of the biology and pathogenicity of *Rosellinia necatrix*, the cause of white root rot disease of fruit trees and other plants. *Journal of Phytopathology* 154(5), 257–266.

Purcell, A.H. (1989) Homopteran transmission of xylem-inhabiting bacteria. In: Harris, K.F. (ed.) *Advances in Disease Vector Research*, Vol. 6. Springer, New York, pp. 243–266.

Redak, R.A., Purcell, A.H., Lopes, J.R.S., Blua, M.J., Mizell, R.F. 3rd *et al.* (2004) The biology of xylem fluid-feeding insect vectors of *Xylella fastidiosa* and their relation to disease epidemiology. *Annual Reviews in Entomology* 49, 243–270.

Romero, J., Ávila, A., Agustí-Brisach, C., Roca, L.F. and Trapero, A. (2020) Evaluation of fungicides and management strategies against cercospora leaf spot of olive caused by *Pseudocercospora cladosporioides*. *Agronomy* 10, 271.

Roubala, C., Regisa, S. and Nicot, P.C. (2013) Field models for the prediction of leaf infection and latent period of *Fusicladium oleagineum* on olive based on rain, temperature and relative humidity. *Plant Pathology* 62, 657–666.

Ruano-Rosa, D., Schena, L., Agosteo, G.E., Magnano di San Lio, G. and Cacciola, S.O. (2018) *Phytophthora oleae* sp. nov. causing fruit rot of olive in southern Italy. *Plant Pathology* 67, 1362–1373.

Ruggieri, G. (1948) Ricerche ed esperienze su una tracheoverticilliosi dell'olivo. *Olivicoltora* 3, 6–9.

Saad, A.T. and Masri, S. (1978) Epidemiological studies on olive spot incited by *Spilocaea oleaginea*. *Phytopathologia Mediterranea* 17, 170–173.

Salman, M., Greenhut, R., Preece, J., Ferguson, L. and Kluepfel, D. (2020) Field evaluation of olive (*Olea europaea*) genotypes for resistance to *Pseudomonas savastanoi* pv. *savastanoi*. *Journal of Plant Pathology* 102, 663–670.

Santilli, E., Riolo, M., La Spada, F., Pane, A. and Cacciola, S.O. (2020) First report of root rot caused by *Phytophthora bilorbang* on *Olea europaea* in Italy. *Plants* 9, 826.

Saponari, M., Boscia, D., Nigro F. and Martelli G.P. (2013) Identification of DNA sequences related to *Xylella fastidiosa* in oleander, almond and olive trees exhibiting leaf scorch symptoms in Apulia (southern Italy). *Journal of Plant Pathology* 95, 668.

Saponari, M., Loconsole, G., Cornara, D., Yokomi, R.K., De Stradis, A. *et al.* (2014) Infectivity and transmission of *Xylella fastidiosa* by *Philaenus spumarius* (Hemiptera: Aphrophoridae) in Apulia, Italy. *Journal of Economic Entomology* 107, 1316–1319.

Scanu, B., Jung, T., Masigol, H., Linaldeddu, B.T., Jung, M.H. *et al.* (2021) *Phytophthora heterospora* sp. nov., a new pseudoconidia-producing sister species of *P. palmivora*. *Journal of Fungi* 7, 870.

Schena, L. and Agosteo, G.E. (2011) White and fibrous root rot of olive. In: Schena, L., Agosteo, G.E. and Cacciola, S.O. (eds) *Olive Diseases and Disorders*. Transworld Research Network, Trivandrum, Kerala, India, pp. 259–279.

Schena, L. and Ippolito, A. (2003) Rapid and sensitive detection of *Rosellinia necatrix* in roots and soils by Real time Scorpion-PCR. *Journal of Plant Pathology* 85, 15–25.

Scibetta, S., Agosteo, G.E., Abdelfattah, A., Li Destri Nicosia, M.G., Cacciola, S.O. *et al.* (2020) Development and application of a qPCR detection method to quantify *Venturia oleaginea* (Castagne) Rossman & Crous in asymptomatic olive (*Olea europaea* L.) leaves. *Phytopathology* 110, 547–555.

Scortichini, M., Chen, J., De Caroli, M., Dalessandro, G., Pucci, N. *et al.* (2018) A zinc, copper and citric acid biocomplex shows promise for control of *Xylella fastidiosa* subsp. *pauca* in olive trees in Apulia region (southern Italy). *Phytopathologia Mediterranea* 57, 48–72.

Serrano, A., Rodríguez Jurado, M.D., Román, B., Bejarano-Alcázar, J., de la Rosa, R. *et al.* (2021) Verticillium wilt evaluation of olive breeding selections under semicontrolled conditions. *Plant Disease* 105, 1781–1790.

Sisto, A., Cantore, P. and Iacobellis, N.S. (2001) *Preliminary results on the response of olive cultivars to artificial inoculation with Pseudomonas syringae pv. savastanoi*. Proceedings of the 11th Congress of the Mediterranean Phytopathological Union and 3rd Congress of the Sociedade Potuguesa de Fitopatologia, Evora, Portugal, pp. 240–242.

Su, C.C., Deng, W.L., Jan, F.J., Chang, C.J., Huang, H. *et al.* (2016) *Xylella taiwanensis* sp. nov., causing pear leaf scorch disease. *International Journal of Systematic and Evolutionary Microbiology* 66, 4766–4771.

Talhinhas, P, Loureiro, A. and Oliveira, H. (2018) Olive anthracnose: a yield- and oil quality-degrading disease caused by several species of *Colletotrichum* that differ in virulence, host preference and geographical distribution. *Molecular Plant Pathology* 19, 1797–1807.

Thanassoulopoulos, C.C., Biris, D.A. and Tjamos, E.C. (1979) Survey of Verticillium wilt of olive trees in Greece. *Plant Disease Reporter* 63, 936–940.

Thomidis, T. and Exadaktylou, E. (2012) Effectiveness of cyproconazole to control Armillaria root rot of apple, walnut and kiwifruit. *Crop Protection* 36, 49–51.

Trapero, C., Birem, F., Abho-Sheker, F.M., Caballero, J., Mercado, J. *et al.* (2010) Avances en la resistencia del olivo a la verticilosis causada por *Verticillium dahliae*. *Vida Rural* 316, 36–38, 40–44.

Triki, M.A. and Rhouma, A. (2008) First report of *Pseudocercospora cladosporioides*, the causal agent of Cercospora leaf spot of olive trees, in Tunisia. *Phytopathologia Mediterranea* 47, 262–265.

Valverde, P., Zucchini, M., Polverigiani, S., Lodolini, E.M., Lopez-Escudero, F.J. *et al.* (2020) Olive knot damages in ten olive cultivars after late-winter frost in central Italy. *Scientia Horticulturae* 266, 109274.

Viruega, J.R. and Trapero, A. (1999) Epidemiology of leaf spot of olive tree caused by *Spilocaea oleagina* in Southern Spain. *Acta Horticulturae* 474, 531–534.

Viruega, J.R. and Trapero, A. (2002) Effects of temperature, wetness duration and leaf age on infection and development of OLS. *Acta Horticulturae* 586, 797–800.

Wells, J.M., Raju, B.C., Hung, H.Y., Weisburg, W.G., Mandelco-Paul, L. *et al.* (1987) *Xylella fastidiosa* gen. nov., sp. nov.: Gram-negative, xylem-limited fastidious plant bacteria related to *Xanthomonas* spp. *International Journal of Systematic Bacteriology* 37, 136–143.
West, J.S. and Fox, R.T.V. (2005) Stimulation of *Armillaria mellea* by phenolic fungicides. *Annals of Applied Biology* 140, 291–295.
Wilhelm, S. (1981) Sources and genetics of host resistance in field and fruit crops. In: Mace, M.E., Bell, A.A. and Beckman, C.H. (eds) *Fungal Wilt Diseases of plants*. Academic Press, New York, pp. 299–376.
Wilhelm, S. and Taylor, J.B. (1965) Control of Verticillium wilt of olive through natural recovery and resistance. *Phytopathology* 55, 310–316.
Wittstein, K., Cordsmeier, A., Lambert, C., Wendt, L., Sir, E.B. *et al.* (2020) Identification of *Rosellinia* species as producers of cyclodepsipeptide PF1022 and resurrection of the genus *Dematophora* as inferred from polythetic taxonomy. *Studies in Mycology* 96, 1–16.
Young, J.M. (2004) Olive knot and its pathogens. *Australasian Plant Pathology* 33, 33–39.

8.1 Olive Leaf Extract

T. Ganino and A. Fabbri
Department of Food and Drug, Parma University, Italy

Olive leaves are one of the most abundant byproducts of olive cultivation. The term 'olive leaves' refers to the mixture of leaves, herbaceous shoots, small twigs and one- to two-year-old shoots and branches obtained after pruning and harvesting olive trees (Souilem *et al.*, 2017). Olive byproducts can derive from agronomical practices (pruning), from harvesting and from fruit cleaning before oil extraction. In Italy and Spain it is estimated that waste of olive orchards (fresh matter) ranges from 1 to 11 t/ha (Spinelli and Picchi, 2010), thus representing a great amount of vegetal remains. Usually, olive leaves are burned together with cultivation and pruning residues, turning out to be waste without any value, but, on the contrary, becoming an environmental hazard (Talhaoui *et al.*, 2015). Nonetheless, olive leaves are byproducts with an intrinsic high added value, as they are an important source of bioactive compounds.

8.1.1 Olive Leaves and Olive Leaf Extract (OLE) Composition

In recent years, beneficial health properties, *in vivo* and *in vitro*, have been attributed to regular consumption of olive leaves and their extracts (OLE). Among these properties are antioxidant effect, antihypertensive activity, reduction of body mass and fat storage, and hypoglycemic effect (Walker, 1997; Medina *et al.*, 2019).

In Table 8.1.1 is indicated the composition of olive leaves; the high standard deviation is due to the fact that these data are a mean of different studies and authors (Molina-Alcaide and Yáñez-Ruiz, 2008). Moreover, as for olive fruits for oil production, for leaves it is very important to know how to prevent degradations, especially the enzymatic one, in order to preserve all the molecules of interest (Andreoni, 2011).

The main secondary metabolites found in olive leaves are phenolic compounds, present as flavonoids, most of which are flavones. Their presence is related to UV protection and the greatest amount of these components is found in young leaves (Karabourniotis *et al.*, 1995). The presence of polyphenols in leaves is also the consequence of reactivity to pathogen attack and response to insect injuries (Vogel *et al.*, 2015). Olive leaves contain significant concentrations of bioactive compounds, such as phenolic derivatives, alditols and pentacyclic triterpenes, all of them with very interesting pharmacological properties (Erbay and Icier, 2010). The main triterpene of the olive leaf is oleanolic acid, followed by significant concentrations of maslinic acid and minor levels of ursolic acid, erythrodiol and uvaol. The literature on

DOI: 10.1079/9781789247350.0024

Table 8.1.1. Average chemical composition (g/kg dry matter) of olive leaves. The standard deviation is calculated taking into consideration several authors' studies. (From Molina-Alcaide and Yáñez-Ruiz, 2008)

Paremeters	Contents (g/kg DM)	S.D.
Organic matter	880	32
Ether extract	56.4	19.8
Gross energy (MJ/kg DM)	19.7	2.6
Protein bound condensed tannins	100	33
Amino acid N (g/kg N)	887	4.0
Acid detergent insoluble N	8.16	6.9
Neutral detergent fiber	406	24
Acid detergent fiber	302	34
Acid detergent lignin	199	69
Total extractable polyphenols	25.3	2.0
Total extractable tannins	1.0	–
Total extractable condensed tannins	2.28	2.16
Total condensed tannins	9.49	1.65
Free condensed tannins	2.98	0.78
Fiber bound condensed tannins	2.30	0.70
Protein bound condensed tannins	3.65	0.72

DM, dry matter; S.D., standard deviation.

the biological activity of oleanolic acid is extensive, and the pharmacological importance of this compound has been stressed (Tsai and Yin, 2008; Martin *et al.*, 2010; Castellano *et al.*, 2013).

There are five groups of identified phenolic compounds in *Olea europaea* L.: oleuropeosides (oleuropein and verbascoside); flavones (luteolin-7-glucoside, apigenin-7-glucoside, diosmetin-7-glucoside, luteolin and diosmetin); flavonols (rutin); flavan-3-ols (catechin); and substituted phenols (tyrosol, hydroxytyrosol, vanillin, vanillic acid and caffeic acids) (Vogel *et al.*, 2015). Most of these compounds are known to be biologically active as possessing antioxidant, anticarcinogenic, antiinflammatory, antimicrobial, antihypertensive, antidyslipidemic, cardiotonic, laxative, and antiplatelet functions. Hydroxytyrosol and its derivatives (e.g. oleuropein complex and tyrosol) may be successfully exploited as functional compounds; the most abundant phenolic compound in olive leaves is oleuropein, followed by hydroxytyrosol. Oleuropein and the products of its hydrolysis are also compounds of great biological interest. Oleuropein content in dry leaves is up to 60–90 mg/g and several pharmacological properties are attributed to this compound, including the antioxidant, antiinflammatory, antiatherogenic, anticancerogenic, antimicrobial, and antiviral activities (Carluccio *et al.*, 2003; Tripoli *et al.*, 2005; Visioli *et al.*, 2002). Moreover, oleuropein has been shown to be cardioprotective and exhibits anti-ischemic and hypolipidemic activities (Andreadou *et al.*, 2006; Singh *et al.*, 2008). In the olive leaves, other molecules, such as hydroxytyrosol, tyrosol, tocopherol, elenolic acid derivatives, hydroxycinnamic acids and flavonoids, are present for a total amount ranging between 1% and 14% of dry matter. All these compounds are known to positively act with a synergistic effect. Following the EC Regulation No. 1924/2006, according to which only compounds approved by the European Food Safety Authority (EFSA) may justify 'health claims' of the product, the EFSA has stated that there is a cause–effect relationship between the consumption of olive polyphenols and protection of LDL particles from oxidative damage, considering not only olive fruit, oil and olive mill wastewaters and pomace, but also vegetation residues from pruning (European Food Safety Authority, 2011).

In this context, the utilization of olive leaves is a promising option to increase the profitability of olive groves. On the other hand, the use of olive leaves in ethnomedicine has a long tradition in the whole Mediterranean area, playing a prominent cultural role for many populations. As a dietary component, olive leaves have been consumed in the form of extracts, whole, or as herbal powder, mainly used for herbal infusions (González-Ortega *et al.*, 2020). Leaves are considered a cheap and rich source of valuable compounds that can be obtained with relatively cost-effective processes (Xynos *et al.*, 2014; Difonzo *et al.*, 2017). In the last decades, several formulations based on olive leaves have been produced industrially in the US, and, more recently, also in northern Italy (Fabbri *et al.*, 2004; Romero-Garcia *et al.*, 2014). To improve the recovery of leaf polyphenols, many studies have evaluated methods of sample storage and preparation, and optimized the extraction techniques, since endogenous enzymes may cause oleuropein degradation (Stamatopoulos *et al.*, 2012; De Leonardis *et al.*, 2015; Ahmad-Qasem *et al.*, 2016).

The determination of phenolic compounds in olive leaves has been developed using different extraction systems, but the most common has been the solid-liquid extraction; this system involves the maceration of the olive leaves in a solvent, like methanol, ethanol or acetone (Talhaoui *et al.*, 2015). New extraction systems – microwave-assisted extraction, pressurized liquid extraction, supercritical fluid extraction, and ultrasound assisted extraction – have recently been used to reduce extraction costs and accelerate the process (Talhaoui *et al.*, 2015).

Microwave-assisted extraction is a technique characterized by short extraction time, low energy requirements and high quality and quantity of OLE (Taamalli *et al.*, 2012).

Pressurized liquid extraction is a technique which uses a solid-liquid extraction, but with controlled and constant pressure, and variation of the extraction parameters, like temperature, time etc. (Xynos *et al.* 2014).

Supercritical fluid extraction is a technique that avoids the use of toxic solvents and allows an easy and rapid extraction. This technique also has the positive characteristic of keeping high the quality of the extract, because the extraction is carried out in the absence of light and air (reduction of oxidations) (Lamuela-Raventós *et al.*, 2014).

Ultrasound-assisted extraction is very fast, and a very efficient method for quantity and quality of extraction (Cifà *et al.*, 2018). Taamalli *et al.* (2012) reported that each extraction technique was adequate depending on the class of compounds to be extracted. The authors report that microwave-assisted extraction and solid-liquid extraction are more efficient for extracting polar compounds (oleuropein derivates, apigenin rutinoside and luteolin glucoside), while supercritical fluid extraction resulted in the best technique for apigenin and diosmetin extraction.

8.1.2 Factors Affecting Leaf Phenolic Compounds

Different factors can alter OLE components such as leaf age, geographical origin, cultivar, phenological stage during sampling, proportion of branches on the tree, moisture content and leaves storage conditions.

Orak *et al.* (2019) assert that the profile of phenolic compounds of OLE had reference values for cultivar classification. Indeed, the authors affirm that olive genotypes are characterized by different levels of antioxidants and these differences can determine the potential use of leaves of appropriate olive genotypes as a source of bioactive compounds for different purposes (i.e. cosmetics, food and pharmaceutics). Bilgin and Şahin (2013) found that the climatic differences could make significant differences in leaf polyphenols. According to the same authors, the accumulation of polyphenols tends to decrease in trees cultivated in windy and humid-air environments.

Seasonal variation in chemical composition is well known in plants, and is associated with the biosynthesis, stability and degradation of secondary metabolites. In Liangshan, Wang *et al.* (2019) have shown that the amount of OLE decreased in summer and autumn but increased in spring and winter, then harvesting time seems to influence the concentration of OLE. Other authors have worked on this aspect, but their conclusions are not always in agreement. Some authors (Pasković *et al.*, 2020; Medina *et al.*, 2019; Talhaoui *et al.*, 2015) indicate that the maximum phenolic content is in the cold season, while in Turkey, Yazici *et al.* (2012) indicate that, when the rainfed regime was applied, the highest oleuropein content was observed in the leaves collected in the summer season. Probably the discordance is due to the environmental condition and to the olive cultivars. Further studies will need to be made to clarify this aspect.

Another factor that plays an important role in the accumulation of antioxidant compounds in the leaves is the geographic location (Blasi *et al.*, 2016; Brahmi *et al.*, 2013). This could explain why, in some studies, the period of maximum concentration of these substances in the leaves is different. A study conducted in Turkey found that leaves from the same olive cultivar contained higher amounts of antioxidant compounds when grown in the north of the country.

Other abiotic factors can contribute to change the OLE content. Salinity can modify the content of antioxidant molecules in different plant tissues. According to Petridis *et al.* (2012), salinity induces the synthesis of phenolic compounds, especially oleuropein, both in the leaves and in the roots. The authors assert that there is no indication that phenolic compounds are translocated from leaves to roots and therefore that each organ has the ability to synthesize the amount of polyphenols essential in order to respond to salt stress.

The accumulation of OLE in the leaves is related to agronomical practices: irrigation is one of the practices that can affect the concentration of these substances in the leaves. In this sense, recent studies have demonstrated that the concentration of oleuropein in leaves was highest when the input of water was low and decreased noticeably under the 100% ET_c irrigation treatment (Višnjevec *et al.*, 2018).

Fertilization also seems to play a role in the accumulation of phenolic compounds in the olive tree. Studies carried out with the application of compounds containing boron have shown that boron deficiency causes a greater synthesis of some phenolic compounds, such as quercetin 3-O-rutinoside, p-coumaric acid, luteolin 7-O-glucoside, apigenin 7-O-rutinoside, and decreased the concentration of quercetin aglycone (Talhaoui *et al.*, 2015).

8.1.3 OLE and the Health Potential

The olive tree has been present in the history of Mediterranean medicine and herbal medicine for several millennia. In the areas where the plant was present, all civilizations have used the products of this plant as food, fuel for lighting, and as medicine. In the 16th century, many authors reported the use of OLE to cure eye ailments as a healing agent, or to relieve skin ulcers and more (Mattioli, 1557). Historically, the OLE was used against hypertension (Hashmi *et al.*, 2015). This seems to be due to the purifying effect caused by hepato-renal activity. Other activities of the compounds contained in the leaves are: antioxidant effect; antimicrobial action; and hypoglycemic, diuretic and cholesterol effects (Hashmi *et al.*, 2015; Acar-Tek and Ağagündüz, 2020).

In recent years there has been a revival of interest in olive leaves and their use in traditional and modern medicine. This is due to a greater attention of scientific research towards these aspects and to the abundant antioxidant content in the olive leaf (Fabbri and Pedrazzini, 2001; Ghanbari *et al.*, 2012).

Talhaoui *et al.* (2015) report, in their review, a complete list of the health potential of olive leaf phenolic compounds. In Table 8.2.2 is a synthesis of the review.

8.1.4 OLE in Food

In recent years, the byproducts derived from the olive oil chain have been used in different sectors. In the case of the leaves, the extracts obtained from them are used in different application fields due to their proven biological activity. Olive leaves are abundant and at an extremely low cost, but, at the same time, have a high potential in food applications. Djenane *et al.* (2019) use extracts of oleaster leaves for improved microbiological safety and shelf life. In this study, the authors conclude that the addition of OLE resulted in an improvement in microbiological quality and shelf life. The authors also point out that adding OLE to the product did not lead to changes in taste and consumer acceptance. Testa *et al.* (2019) studied the efficacy of OLE in marinated anchovies processing in order to extend their shelf life and preserve their quality. In this case, in the anchovy-marination process, OLE has delayed the oxidative deterioration with positive effect on texture, appearance and organoleptic characteristics. The authors suggest the use of OLE in the food industry as a natural preservative and antimicrobial additive. A similar study was conducted on yogurt (Cho *et al.*, 2020). In this case, the authors investigated the OLE effect on the physical, sensory and chemical properties. The results obtained show that, after the addition of OLE, the product exhibits significant changes in the titratable acidity, antioxidant capacity and syneresis. Furthermore, also in this case, it was found that the addition of these extracts does not change the sensory properties of the product.

Recently, the oleuropein extract was used in emulsified food matrices; indeed, some authors tested the influences of this compound to the emulsification process (Souilem *et al.*, 2014) and in complex formulations, like mayonnaise (Giacintucci *et al.*, 2016).

Table 8.2.2. Some health potential of olive leaf phenolic compounds with the description of the effects and their mechanisms of action. (From Talhaoui *et al.*, 2015)

		Compounds	Effects	Mechanisms	References
In vivo	Animal studies	Oleuropein 24.54%, hydroxytyrosol 1.46%, the flavone-7-glucosides of luteolin and apigenin (1.38% and 1.37%, respectively) and verbascoside (1.11%), diosmetin-7-glucoside luteolin, diosmetin, rutin, catechin, vanillin, vanillic acid, caffeic acid	Antioxidant and radioprotective effects	Flavonols, flavan-3-ols and flavones with catechol structure are the most efficient quenchers for ABTS•+ radical cation. Hydroxyl radical (OH) scavenging capacity	(Castillo *et al.*, 2010)
		Oleuropein	Antioxidant effects on intestine mucosal damage induced by absolute ethanol	Significant increases in the levels of antioxidant enzymes such as GPx and CAT for oleuropein and oleuropein plus ethanol groups in comparison with ethanol-treated rats, whereas oleuropein treatment abolished the increase in TBARS concentration	(Alirezaei *et al.*, 2014)
		Oleuropein and its hydrolysates-(hydroxytyrosol) rich extracts	Antidiabetic and antioxidant effects	Increase of antioxidant enzyme expressions and/or activities	(Jemai *et al.*, 2009)

Continued

Table 8.2.2. Continued.

		Compounds	Effects	Mechanisms	References
		Hydroxytyrosol and its triacetylated derivative	Lipid-lowering and antioxidant effects	Inhibition of the absorption of dietary cholesterol in the intestine or its production by the liver or stimulation of the biliary secretion of cholesterol and cholesterol excretion in the feces	(Bouaziz *et al.*, 2008)
		OLE consisting of oleuropein (19.8%), luteolin-7-O-glucoside (0.04%), apigenin-7-O-glucoside (0.07%), quercetin (0.04%) and caffeic acid (0.02%)	OLE modulates cold restraint stress-induced oxidative changes in rat liver	Synchronization of antioxidant enzymes and inhibited lipid peroxidation in liver	(Dekanski *et al.*, 2011)
		Oleuropein (356 mg/g), tyrosol (3.73 mg/g), hydroxytyrosol (4.89 mg/g), and caffeic acid (49.41 mg/g)	Amelioration of gentamycin nephrotoxicity	Inhibition of lipid peroxidation, enhancing renal glutathione content, and antioxidant enzymes activity	(Tavafi *et al.*, 2012)
	Human studies	Oleuropein	Soothing effect in the treatment of UVB-induced erythema	The mechanism of action of the lenitive efficacy of oleuropein is not actually fully understood. The activity could be related to the property of oleuropein that exhibits inhibitory effects on 'reactive nitrogen species' including the radical nitric oxide (NO·).	(Perugini *et al.*, 2008)
	Animal studies	Oleuropein	Decrease of the number and size of lipid droplets in FFA-treated cells and reduced intra-cellular triglyceride accumulation	Reduction of FFA-induced extracellular signal-regulated kinase activation but had no effect on c-JunN-terminal kinase or AKT activation	(Hur *et al.*, 2012)
In vitro	**Human studies**	Oleuropein	Inhibition of platelet activation in healthy, non-smoking males	Ability to scavenge H_2O_2, which is produced during the arachidonic acid metabolism cascade, which leads to platelet aggregation	(Singh *et al.*, 2008)

Continued

Table 8.2.2. Continued.

Compounds	Effects	Mechanisms	References
Oleuropein and hydroxytyrosol	Apoptotic cell death of human breast cancer MCF-7 cells	Significant block of G1 to S phase transition manifested by the increase of cell number in G0/G1 phase	(Han *et al.*, 2009)
Oleuropein and its semi-synthetic peracetylated derivatives	Anti-proliferative and antioxidant effects. The peracetylated compounds exerted higher antiproliferative effects than oleuropein.	An arrest of cell-cycle progression, associated with a strong antioxidant activity	(Bulotta *et al.*, 2011)
Hydroxytyrosol-rich extract	A dose-dependent growth inhibition of MCF-7 cells	Cell-cycle arrest in the G1 phase Down-expression of Pin1 which in turn decreased the level of cyclin D1	(Bouallagui *et al.*, 2011)
Aromadendrine (dihydrokaempferol)	Antiinflammatory activity	Decrease markedly the expression of membrane molecules (ICAM-1) and the release of inflammatory-soluble factors (MCP-1 and IL-8) induced by IFNγ and histamine in normal keratinocytes NCTC 2544	(Venditti *et al.*, 2013)
Oleuropein (19.8%), luteolin-7-O-glucoside (0.04%), apigenine-7-O-glucoside (0.07%), quercetin (0.04%) and 0.02% of caffeic acid	Protective effect on the peripheral blood leukocytes against adrenaline induced DNA damage	Synergistic activation of several molecular mechanisms such as ROS scavenging and increasing the antioxidant capacity of cells	(Čabarkapa *et al.*, 2014)
Oleuropein, hydroxytyrosol, hydroxytyrosol acetate, luteolin, luteolin-7-O-glucoside, and luteolin-O-glucoside	Antiproliferative effect against cancer and endothelial cells	Inhibition of the cell proliferation	(Goulas *et al.*, 2009)
Apigenin-7-glucoside, oleuropein	Apigenin-7-glucoside of the OLE was mainly responsible for the HL-60 differentiation and oleuropein showed to exert an influence over this differentiation.	Growth inhibition of HL-60 cells treated with the OLE can be attributable to differentiation-mediated cell-cycle arrest and/or apoptosis.	(Abaza *et al.*, 2007)

8.1.5 Conclusions

Recent literature indicates a greater attention to OLE research. These studies highlight the potential of this byproduct in various sectors, especially in the food industry. Many studies have been conducted by relating the composition of olive leaves with biotic and abiotic variables. Orchard management, cultivar, environment and cultivation area are just some of the variables that can modify the content of these molecules in the olive leaf. There are many compounds present in this byproduct, but oleuropein is the main phenolic compound in olive leaves. More research is needed for the utilization of these compounds, not only oleuropein, in food and pharmaceutical processes. Finally, research will also have to focus on *in vivo* studies to evaluate the real bioavailability of these compounds and the biological activities of phenols derived from OLE.

References

Abaza, L., Talorete, T.P., Yamada, P., Kurita, Y., Zarrouk, M. and Isoda, H. (2007) Induction of growth inhibition and differentiation of human leukemia HL-60 cells by a Tunisian gerboui olive leaf extract. *Bioscience, Biotechnology and Biochemistry* 71(5), 1306–1312.

Acar-Tek, N. and Ağagündüz, D. (2020) Olive leaf (*Olea europaea* L. folium): potential effects on glycemia and lipidemia. *Annals of Nutrition & Metabolism* 76(1), 10–15.

Ahmad-Qasem, M.H., Ahmad-Qasem, B.H., Barrajón-Catalán, E., Micol, V., Cárcel, J.A. and García-Pérez, J.V. (2016) Drying and storage of olive leaf extracts: influence on polyphenols stability. *Industrial Crops and Products* 79, 232–239.

Alirezaei, M., Dezfoulian, O., Sookhtehzari, A., Asadian, P. and Khoshdel, Z. (2014) Antioxidant effects of oleuropein versus oxidative stress induced by ethanol in the rat intestine. *Comparative Clinical Pathology* 23(5), 1359–1365.

Andreadou, I., Iliodromitis, E.K., Mikros, E., Constantinou, M., Agalias, A. *et al.* (2006) The olive constituent oleuropein exhibits anti-ischemic, antioxidative, and hypolipidemic effects in anesthetized rabbits. *The Journal of Nutrition* 136(8), 2213–2219.

Andreoni, N. (2011) Recupero di composti fenolici da materiale di potatura dell'olivo. *I Convegno Nazionale dell'Olivo e dell'Olio* 1, 76–79.

Bilgin, M. and Şahin, S. (2013) Effects of geographical origin and extraction methods on total phenolic yield of olive tree (*Olea europaea*) leaves. *Journal of the Taiwan Institute of Chemical Engineers* 44(1), 8–12.

Blasi, F., Urbani, E., Simonetti, M.S., Chiesi, C. and Cossignani, L. (2016) Seasonal variations in antioxidant compounds of *Olea europaea* leaves collected from different Italian cultivars. *Journal of Applied Botany and Food Quality* 89, 202–207.

Bouallagui, Z., Han, J., Isoda, H. and Sayadi, S. (2011) Hydroxytyrosol rich extract from olive leaves modulates cell cycle progression in MCF-7 human breast cancer cells. *Food and Chemical Toxicology* 49(1), 179–184.

Bouaziz, M., Fki, I., Jemai, H., Ayadi, M. and Sayadi, S. (2008) Effect of storage on refined and husk olive oils composition: stabilization by addition of natural antioxidants from Chemlali olive leaves. *Food Chemistry* 108(1), 253–262.

Brahmi, F., Mechri, B., Dhibi, M. and Hammami, M. (2013) Variations in phenolic compounds and antiradical scavenging activity of *Olea europaea* leaves and fruits extracts collected in two different seasons. *Industrial Crop Production* 49, 256–264.

Bulotta, S., Corradino, R., Celano, M., D'Agostino, M., Maiuolo, J. *et al.* (2011) Antiproliferative and antioxidant effects on breast cancer cells of oleuropein and its semisynthetic peracetylated derivatives. *Food Chemistry* 127(4), 1609–1614.

Čabarkapa, A., Živković, L., Žukovec, D., Djelić, N., Bajić, V., Dekanski, D. and Spremo-Potparević, B. (2014) Protective effect of dry olive leaf extract in adrenaline induced DNA damage evaluated using in vitro comet assay with human peripheral leukocytes. *Toxicology in Vitro* 28(3), 451–456.

Carluccio, M.A., Siculella, L., Ancora, M.A., Massaro, M., Scoditti, E. *et al.* (2003) Olive oil and red wine antioxidant polyphenols inhibit endothelial activation: antiatherogenic properties of Mediterranean diet phytochemicals. *Arteriosclerosis, Thrombosis, and Vascular Biology* 23(4), 622–629.

Castellano, J.M., Guinda, A., Delgado, T., Rada, M. and Cayuela, J.A. (2013) Biochemical basis of the antidiabetic activity of oleanolic acid and related pentacyclic triterpenes. *Diabetes* 62(6), 1791–1799.

Castillo, J.J., Alcaraz, M. and Benavente-García, O. (2010) Antioxidant and radioprotective effects of olive leaf extract. In: Preedy, V.R. and Watson, R. (eds) *Olives and Olive Oil in Health and Disease Prevention*. Academic Press, pp. 951–958.

Cho, W.Y., Kim, D.H., Lee, H.J., Yeon, S.J. and Lee, C.H. (2020) Quality characteristic and antioxidant activity of yogurt containing olive leaf hot water extract. *CyTA-Journal of Food* 18(1), 43–50.

Cifá, D., Skrt, M., Pittia, P., Di Mattia, C. and Ulrih, N.P. (2018) Enhanced yield of oleuropein from olive leaves using ultrasound-assisted extraction. *Food Science and Nutrition* 6, 1128–1137.

De Leonardis, A., Macciola, V., Cuomo, F. and Lopez, F. (2015) Evidence of oleuropein degradation by olive leaf protein extract. *Food Chemistry* 175, 568–574.

Dekanski, D., Selaković, V., Piperski, V., Radulović, Ž., Korenić, A. and Radenović, L. (2011) Protective effect of olive leaf extract on hippocampal injury induced by transient global cerebral ischemia and reperfusion in Mongolian gerbils. *Phytomedicine* 18(13), 1137–1143.

Difonzo, G., Russo, A., Trani, A., Paradiso, V.M., Ranieri, M. *et al.* (2017) Green extracts from Coratina olive cultivar leaves: antioxidant characterization and biological activity. *Journal of Functional Foods* 31, 63–70.

Djenane, D., Gómez, D., Yangüela, J., Roncalés, P. and Ariño, A. (2019) Olive leaves extract from algerian oleaster (*Olea europaea* var. *Sylvestris*) on microbiological safety and shelf-life stability of raw halal minced beef during display. *Foods* 8(1), 10.

Erbay, Z. and Icier, F. (2010) The importance and potential uses of olive leaves. *Food Reviews International* 26(4), 319–334.

European Food Safety Authority (2011) Use of the EFSA comprehensive European food consumption database in exposure assessment. *EFSA Journal* 9(3), 2097.

Fabbri A. and Pedrazzini, M. (2001) *L'olivo come pianta erboristica. Erboristeria Domani* 4, 58–71.

Fabbri, A., Galaverna, G. and Ganino, T. (2004) Polyphenol composition of olive leaves with regard to cultivar, time of collection and shoot type. In: *V International Symposium on Olive Growing* 791, pp. 459–464.

Ghanbari, R., Anwar, F., Alkharfy, K.M., Gilani, A.H. and Saari, N. (2012) Valuable nutrients and functional bioactives in different parts of olive (*Olea europaea* L.): a review. *International Journal of Molecular Sciences* 13(3), 3291–3340.

Giacintucci, V., Di Mattia, C., Sacchetti, G., Neri, L. and Pittia, P. (2016) Role of olive oil phenolics in physical properties and stability of mayonnaise-like emulsions. *Food Chemistry* 15(213), 369–377.

González-Ortega, R., Faieta, M., Di Mattia, C.D., Valbonetti, L. and Pittia, P. (2020) Microencapsulation of olive leaf extract by freeze-drying: effect of carrier composition on process efficiency and technological properties of the powders. *Journal of Food Engineering* 285, 110089.

Goulas, V., Exarchou, V., Troganis, A.N., Psomiadou, E., Fotsis, T., Briasoulis, E. and Gerothanassis, I.P. (2009) Phytochemicals in olive-leaf extracts and their antiproliferative activity against cancer and endothelial cells. *Molecular Nutrition & Food Research* 53(5), 600–608.

Han, J., Talorete, T.P., Yamada, P. and Isoda, H. (2009) Anti-proliferative and apoptotic effects of oleuropein and hydroxytyrosol on human breast cancer MCF-7 cells. *Cytotechnology* 59(1), 45–53.

Hashmi, M.A., Khan, A., Hanif, M., Farooq, U. and Perveen, S. (2015) Traditional uses, phytochemistry, and pharmacology of *Olea europaea* (olive). *Evidence Based Complementary and Alternative Medicine* 2015, 1–29.

Hur, W., Kim, S.W., Lee, Y.K., Choi, J.E., Hong, S.W. *et al.* (2012) Oleuropein reduces free fatty acid-induced lipogenesis via lowered extracellular signal-regulated kinase activation in hepatocytes. *Nutrition Research* 32(10), 778–786.

Jemai, H., El Feki, A. and Sayadi, S. (2009) Antidiabetic and antioxidant effects of hydroxytyrosol and oleuropein from olive leaves in alloxan-diabetic rats. *Journal of Agricultural and Food Chemistry* 57(19), 8798–8804.

Karabourniotis, G., Kotsabassidis, D. and Manetas, Y. (1995) Trichome density and its protective potential against ultraviolet-B radiation damage during leaf development. *Canadian Journal of Botany* 73(3), 376–383.

Lamuela-Raventós, R.M., Vallverdú-Queralt, A., Jáuregui, O., Martínez-Huélamo, M. and Quifer Rada, P. (2014) Improved characterization of polyphenols using liquid chromatography. In: Watson, R. (ed) *Polyphenols in Plants*. Academic Press, pp. 261–292.

Martín, R., Carvalho-Tavares, J., Hernández, M., Arnes, M., Ruiz-Gutierrez, V. and Nieto, M.L. (2010) Beneficial actions of oleanolic acid in an experimental model of multiple sclerosis: a potential therapeutic role. *Biochemical Pharmacology* 79(2), 198–208.

Mattioli, P.M. (1557) *I discorsi nei sei libri della materia medicinale di Pedacio Dioscoide Anasarbeo*. Venezia.
Medina, E., Romero, C., García, P. and Brenes, M. (2019) Characterization of bioactive compounds in commercial olive leaf extracts, and olive leaves and their infusions. *Food & Function* 10(8), 4716–4724.
Molina-Alcaide, E. and Yáñez-Ruiz, D.R. (2008) Potential use of olive by-products in ruminant feeding: a review. *Animal Feed Science and Technology* 147(1–3), 247–264.
Orak, H.H., Karamać, M., Amarowicz, R., Orak, A. and Penkacik, K. (2019) Genotype-related differences in the phenolic compound profile and antioxidant activity of extracts from olive (*Olea europaea* L.) leaves. *Molecules* 24(6), 1130.
Pasković, I., Lukić, I., Žurga, P., Majetić Germek, V., Brkljača, M. *et al.* (2020) Temporal variation of phenolic and mineral composition in olive leaves is cultivar dependent. *Plants* 9, 1099. https://doi.org/10.3390/plants9091099
Perugini, P., Vettor, M., Rona, C., Troisi, L., Villanova, L. *et al.* (2008) Efficacy of oleuropein against UVB irradiation: preliminary evaluation. *International Journal of Cosmetic Science* 30(2), 113–120.
Petridis, A., Therios, I., Samouris, G. and Tananaki, C. (2012) Salinity-induced changes in phenolic compounds in leaves and roots off our olive cultivars (*Olea europaea* L.) and their relationship to antioxidant activity. *Environmental and Experimental Botany* 79, 37–43.
Romero-García, J.M., Niño, L., Martínez-Patiño, C., Álvarez, C., Castro, E. and Negro, M.J. (2014) Biorefinery based on olive biomass: state of the art and future trends. *Bioresource Technology* 159, 421–432.
Singh, I., Mok, M., Christensen, A.M., Turner, A.H. and Hawley, J.A. (2008) The effects of polyphenols in olive leaves on platelet function. *Nutrition, Metabolism and Cardiovascular Diseases* 18(2), 127–132.
Souilem, S., Kobayashi, I., Neves, M.A. Sayadi, S., Ichikawa, S. and Nakajima, M. (2014) Preparation of monodisperse food-grade oleuropein-loaded W/O/W emulsions using microchannel emulsification and evaluation of their storage stability. *Food Bioprocess Technology* 7, 2014–2027. https://doi.org/10.1007/s11947-013-1182-9
Souilem, S., Fki, I., Kobayashi, I., Khalid, N., Neves, M.A. *et al.* (2017) Emerging technologies for recovery of value-added components from olive leaves and their applications in food/feed industries. *Food Bioprocess Technology* 10, 229–248. https://doi.org/10.1007/s11947-016-1834-7
Spinelli, R. and Picchi, G. (2010) Industrial harvesting of olive tree pruning residue for energy biomass. *Bioresource Technology* 101, 730–735.
Stamatopoulos, K., Katsoyannos, E., Chatzilazarou, A. and Konteles, S.J. (2012) Improvement of oleuropein extractability by optimising steam blanching process as pre-treatment of olive leaf extraction via response surface methodology. *Food Chemistry* 133(2), 344–351.
Taamalli, A., Arráez-Román, D., Ibañez, E., Zarrouk, M., Segura-Carretero, A. and Fernández-Gutiérrez, A. (2012) Optimization of microwave-assisted extraction for the characterization of olive leaf phenolic compounds by using HPLC-ESI-TOF-MS/IT-MS(2). *Journal of Agricultural and Food Chemistry* 60(3), 791–798.
Talhaoui, N., Taamalli, A., Gómez-Caravaca, A.M., Fernández-Gutiérrez, A. and Segura-Carretero, A. (2015) Phenolic compounds in olive leaves: analytical determination, biotic and abiotic influence, and health benefits. *Food Research International* 77, 92–108.
Tavafi, M., Ahmadvand, H. and Toolabi, P. (2012) Inhibitory effect of olive leaf extract on gentamicin-induced nephrotoxicity in rats. *Iranian Journal of Kidney Diseases* 6(1), 25–32.
Testa, B., Lombardi, S.J., Macciola, E., Succi, M., Tremonte, P. and Iorizzo, M. (2019) Efficacy of olive leaf extract (*Olea europaea* L. Cv. Gentile di Larino) in marinated anchovies (*Engraulis encrasicolus*, L.) *process. Heliyon* 5(5), e01727.
Tripoli, E., Giammanco, M., Tabacchi, G., Di Majo, D., Giammanco, S. and La Guardia, M. (2005) The phenolic compounds of olive oil: structure, biological activity and beneficial effects on human health. *Nutrition Research Reviews* 18(1), 98–112.
Tsai, S.J. and Yin, M.C. (2008) Antioxidative and anti-inflammatory protection of oleanolic acid and ursolic acid in PC12 cells. *Journal of Food Science* 73, 174–178.
Venditti, A., Serrilli, A.M., Rizza, L., Frasca, G., Cardile, V., Bonina, F.P. and Bianco, A. (2013) Aromadendrine, a new component of the flavonoid pattern of *Olea europaea* L. and its anti-inflammatory activity. *Natural Product Research* 27(4–5), 340–349.
Visioli, F., Poli, A. and Gall, C. (2002) Antioxidant and other biological activities of phenols from olives and olive oil. *Medicinal Research Reviews* 22(1), 65–75.
Višnjevec, A.M., Valenčič, V., Hladnik, T., Podgornik, M., Bandelj, D. *et al.* (2018) Impact of weather conditions and drought stress on primary and secondary metabolites of olives from Slovenian Istra. *Acta Horticulturae* 1199, 69–74.

Vogel, P., Machado, I.K., Garavaglia, J., Zani, V.T., de Souza, D. and Dal Bosco, S.M. (2015) Polyphenols benefits of olive leaf (*Olea europaea* L) to human health. *Nutrición hospitalaria* 31(3), 1427–1433.
Walker M. (1997) *Olive Leaf Extract*. Kensington, New York.
Wang, B., Qu, J., Feng, S., Chen, T., Yuan, M. *et al.* (2019) Seasonal variations in the chemical composition of Liangshan olive leaves and their antioxidant and anticancer activities. *Foods* 8(12), 657.
Xynos, N., Papaefstathiou, G., Gikas, E., Argyropoulou, A., Aligiannis, N. and Skaltsounis, A.L. (2014) Design optimization study of the extraction of olive leaves performed with pressurized liquid extraction using response surface methodology. *Separation and Purification Technology* 122, 323–330.
Yazici, I., Karabey, F., Akay, Ü., Kaya, Ü. and Demiray, H. (2012) The effect of different irrigation levels on the oleuropein contents of olive tree (*Olea europaea* L. cv. Memecik) in the western coastal region of Turkey. *African Journal of Biotechnology* 11(90), 15664–15677.

8.2 Valorization of the Olive-oil Chain Byproducts as Soil Amendment

L. Regni, F. Famiani and P. Proietti
Dipartimento di Scienze Agrarie, Alimentari e Ambientali, Università degli Studi di Perugia, Italy

8.2.1 The Byproducts of the Olive Oil Industry

The environmental impact of olive oil production is important as oil extraction can require a considerable amount of water and generates huge quantities of liquid and solid wastes (olive mill waste – OMW) in a limited time (2–3 months of olive harvesting). Moreover, OMW contains natural components with potential bacteriostatic and phytotoxic activity, which can limit the possibility of microbial digestion or spread on cultivated soil. The OMW derived from the olives' processing are a liquid effluent, called vegetation water or oil mill wastewater (OMWW), and a solid effluent called pomace. By applying pressure (traditional method) or a 'three-phase' centrifugation oil-extraction system, effluents are virgin pomace and OMWW. By applying a 'two-phase' centrifugation oil-extraction system, the effluent is a wet pomace. Among the possible valorization and/or recycling methods proposed for these wastes (drying and solvent extraction of residual oil, biogas production, recovery of useful chemicals, composting etc.), the direct land-spreading on agricultural surfaces appears operationally simple and economically feasible.

Indeed, in many olive-growing countries, for example Italy, the agronomic use of OMWW and pomace by directly spreading in the field is a widespread method of legal disposal. Such use has taken great interest not only because it is relatively inexpensive and feasible, but also because it allows providing organic matter to the soil, enhancing, therefore, its C sequestration. Progressive C depletion in soils subjected to intensive farming is a source of great concern, due to worsening of physical, chemical and biological soil properties, leading to degenerative phenomena, of which erosion and loss of fertility are the most obvious and worrying aspects. About this aspect, the techniques of agricultural land management that maximize the accumulation of C in the soil are assuming increasing importance and interest, also for the provisions of article 3.4 of the Kyoto Protocol, according to which countries that joined it have committed to reduce not only the levels of CO_2 emissions but also to increase C sequestration. However, the spreading in the field of OMWW and pomace should be implemented with rationality since the little-stabilized organic materials may inhibit or reduce the development of crops, due to the presence of tannins, fatty acids and phenols, to the competition for nitrogen among soil microorganisms and roots, due to a high C:N ratio and to roots anoxia caused by microorganism oxygen consumption.

 The Olive: Botany and Production (eds A. Fabbri *et al.*)
DOI: 10.1079/9781789247350.0025

8.2.2 Agronomic Use of OMWW

8.2.2.1 Main physico-chemical characteristics of OMWW

It is estimated that in the Mediterranean countries, more than 30×10^6 m^3 of OMWW are produced during the harvest and processing season (Barbera *et al.*, 2013). The OMWW consists of vegetation water, about 40–50% of the weight of the processed olives, a low oil content (about 3%) and water used for the washing of olives and extraction plants (10–15% of the weight of the processed olives) and by any water dilution of the paste (40–60% of the weight of processed olives), used in continuous oil-extraction systems to facilitate the process (Suzzi and Tofalo, 2009). Therefore, the total amount of OMWW produced for every 100 kg of processed olives is 50–60 kg using the pressure oil-extraction system and 10–20, 90–130 and 30–45 kg, respectively, using oil-extraction two-phase, three-phase, and three-phase with water recycling, systems (Peri and Proietti, 2014; Amirante, 2003). The use of different volumes of water in the different oil-extraction systems determines a different level of dilution of the compounds present in solution and in suspension that represent 5–12% of the OMWW (Amirante, 2003).

The physico-chemical characteristics of the OMWW depend on the climatic conditions of the cultivation area, the cultivar, the cultural practices, the state of ripeness of the olives, and the type of extraction process adopted (Ilarioni and Proietti, 2014; Nasini and Proietti, 2014; Aviani *et al.*, 2012). The OMWW contain many organic compounds and their color is due to the presence of polymeric phenolic compounds (corresponding to 10% of the organic matter) (Jarboui *et al.*, 2008; Justino *et al.*, 2012; Zbakh and El Abbassi, 2012). The biodegradability of OMWW depends on phenol content and is expressed as biochemical oxygen demand (BOD5), which has a range of 35– 110 g/l, while chemical oxygen demand (COD) ranges from 40 g/l to 196 g/l. The OMWW do not contain significant amounts of synthetic chemicals or additives, mainly because they derive from an exclusively mechanical extraction process. Furthermore, several experimentations have shown the absence in OMWW of toxic substances or pathogenic microorganisms and viruses, as can occur for urban sludges, but they can contain many bacteria, aerobic and facultative anaerobes, yeasts and molds (Bedbabis *et al.*, 2015; Zbakh and El Abbassi 2012; Vivaldi *et al.*, 2013). However, the OMWW have a high pollution potential due to the high salt, organic matter and phenolic compounds (catechol, hydroxytyrosol, tyrosol and oleuropein) contents and the high acidity. In addition, the mucilage present in suspension (10,000–15,000 mg l^{-1}) increases the viscosity and the adhesiveness of the OMWW. Besides the high values of organic matter, the OMWW have a low biodegradability (Amirante, 2003). For the above-mentioned reasons, national laws regarding the protection of waters against pollution (e.g., in Italy, the law n. 319/76 and subsequent modifications) have banned the discharge of OMWW in streams or urban sewage, if not after proper treatment. Indeed, the seasonality of OMWW production and the characteristics of biological toxicity make the management of their treatment in the treatment plants of the urban wastewaters difficult (Toscano and Montemurro, 2012). To reduce the polluting potential of OMWW, various wastewater treatment systems have been proposed (incineration, ultrafiltration, concentration etc.), which, however, are not able to reduce the pollutant rate to the levels set by law at affordable costs for most of the mills of small size (Toscano and Montemurro, 2012; Yay *et al.*, 2012; Hanafi *et al.*, 2011; Moraetis *et al.*, 2011). Furthermore, these treatment systems produce sludges that are difficult to dispose of and their high energy consumption constitutes a cause of pollution (Yay *et al.*, 2012). The lagooning treatment, even if it does not require large investment, is difficult to practice since it has a very slow activity and can produce bad smells that spread on large surfaces (McNamara *et al.*, 2008; Roig *et al.*, 2006).

8.2.2.2 Main agronomic properties of OMWW

Generally, the OMWW do not cause significant damage to soil and crops; on the other hand, they often produce positive effects. Consequently, the spreading of the OMWW on agricultural soil is not considered only as a simple disposal system but also as an agronomic technique aimed at improving the physical and chemical properties of soil and plant nutrition (Chartzoulakis *et al.*, 2010; Mekki *et al.*, 2006).

The most important properties of the OMWW from the agronomic point of view are the low pH, the presence of nitrogen, almost exclusively in organic form, the richness, although with considerable fluctuations, in potassium, phosphorus and phenols, and the presence of yeasts, fungi and especially bacteria, cellulolytic in particular, while nitrifying bacteria are not present. Phenolic substances are characterized by antimicrobial and phytotoxic activity and by limited biodegradability (Mekki *et al.*, 2006, 2013). About the potential drawbacks linked to spreading, mainly related to the high content in salts and organic substances, in particular phenolic substances and fatty acids, and the acidity, the soil retains as a filter the suspended solids, while clay, humus and newly formed organic colloids fix the mineral salts (carbonates, sulfates, calcium humate, hydroxides of iron and aluminium etc. are formed) (Di Bene *et al.*, 2013). The bad smell following OMWW spreading is very low and less disturbing than normal manuring (Proietti *et al.*, 1995). From the available literature, it emerges that, in general, OMWW do not cause damage to soil and crops and, indeed, frequently involve positive effects. In olive groves, for example, phenomena of toxicity or effects on product quality (with the exception, in some cases, of an increase in the phenol content of the oil), were not detected, while intensification of vegetative-productive activity was present (Magdich *et al.*, 2020; Chebab *et al.*, 2019).

The spreading of OMWW modifies, after a few weeks, some physical characteristics of the soil such as porosity, stability of aggregates, retention and movement of water (Chatzistathis and Koutsos, 2017). Indeed, the administration of OMWW, due to the water-repellent action of the fatty particles adsorbed in the superficial layers of the soil, can initially decrease water infiltration (because of the temporary reduction of porosity in the superficial layers). This action attenuates in a short time with the decomposition of these substances and then the porosity improves compared to the original one, with consequent improvement of air and water dynamics in the soil and reduction of the formation of the superficial crust, which, reducing the infiltration of water, in sloping land is one of the causes of erosion phenomena, while in those on the plain determines stagnation problems.

Moreover, the stability of soil aggregates on which depend the soil aeration and drainage of water, increases significantly a few weeks after the administration of OMWW. This is due to the cementing action of the substances originated by the decomposition of the organic fraction of OMWW.

Soil water retention increases a few weeks after the administration of OMWW. This is attributable to the enrichment in organic matter and the increase in soil porosity. About the chemical characteristics of the soil, the spreading of OMWW, which have a pH of around 5.5 and 5, causes a slight and temporary decrease in soil pH, which, however, a few months after administration, returns to the previous levels or reaches slightly higher values.

Regarding the effect of OMWW on the chemical fertility of agricultural soil, the richness of OMWW in potassium, nitrogen, phosphorus, magnesium and organic matter can improve soil fertility. Indeed, results from various experiments indicate that assimilable phosphorus and exchangeable potassium increase in treated soils (Chatzistathis and Koutsos, 2017). A nitrogen immobilization effect is also evident. According to Moraetis *et al.* (2011), the application of weathered OMWW used in their study (1744 m^3 over the irrigation period) added the following concentrations of nutrients to the field: 11800 kg ha^{-1} yr^{-1}C, 1033 kg ha^{-1} yr^{-1} total N, 23.8 kg ha^{-1} yr^{-1} P and 4161 kg ha^{-1} yr^{-1} K. Consequently, the use of

OMWW could eliminate phospho-potassium fertilization and cut nitrogen fertilization in half. OMWW also contain heavy metals, mostly cadmium, zinc and manganese, but since these are present in very low concentrations they are not of major concern. In some contexts, among the beneficial effects resulting from the disposal of OMWW on agricultural land, should also be included the contribution of water, although the quantities are relatively modest and the distribution of wastewater occurs in periods usually not subject to water shortages.

In conclusion, the studies on the influence of the OMWW on crop productions and the chemical, physical and biological properties of agricultural soil showed no real risks of irreversible degradation in soil or toxic effects on crops, with appropriate dosage and rational distributions. OMWW can therefore be considered as a slow-effect fertilizer (Barbera *et al.*, 2013; Amirante, 2003). Consequently, the controlled spreading of OMWW on agricultural soil (fertigation) can be considered as a technique of great interest, since it provides fertilizing substances to the soil, allowing reduction of the use of chemical fertilizers and avoiding OMWW treatments, which, in addition to wasting useful substances, cause energy consumption.

8.2.2.3 Storage and transport

When the olive oil mill is near the farm, the OMWW can be collected from the decanters in mobile tanks and spread directly on the fields; otherwise, it can be collected in storage basins (IOOC, 2004). The basin is the simplest and most economic system of storage; however, it is exposed to risks of soil and subsoil water contamination. For this reason, the storage of the OMWW generally requires prior notice to the competent bodies and should be implemented in tanks or basins suitably waterproofed according to specific rules (size, shape, materials, coverage etc.). The storage allows avoiding spreading until rains persist and the soil is saturated with water, to protect groundwater and surface water and to avoid soil compaction. Storage tanks or basins should be large enough to contain the OMWW during periods when spreading is to be prevented, due to agronomic or climatic causes or regulations (Roig *et al.*, 2006; McNamara *et al.*, 2008). Legislation generally prohibits mixing OMWW (or even pomace) with other effluents and waste.

During storage, the concentration of pollutants in OMWW decreases by about 15% because the concentration of some fermentable organic compounds (sugars) decreases due to the metabolic activity of the microorganisms present; such as those belonging to the genus *Pseudomonas*, which can grow at the expense of readily degradable compounds, such as phenols (Amirante, 2003). During storage, the development of offensive odors from the OMWW is not significant (Proietti *et al.*, 1995). Transport of OMWW (as well as wet pomace) must be made through suitable means of transport to prevent spills and hygienic-sanitary incidents.

8.2.2.4 Spreading methods

The spreading must be done ensuring a homogeneous distribution of OMWW on the whole surface of the soil, for example using a sewage spreader suitably modified that gives the possibility to distribute the OMWW by sprinkler or furrow on the surface of the cultivated land or incorporated in the soil surface layer using small spades or other tools installed on the same spreader (Bianchi *et al.*, 2000). For olive trees, the OMWW can be spread between the rows at a distance of 0.5–1 m from the tree trunks depending on the soil texture (IOOC, 2004). The OMWW incorporation in the soil accelerates oxidative degradation of phenols and facilitates interaction between organic material and soil, thus preventing it from being dragged off by run-off or percolation. So, when possible (soil without permanent green cover), it is preferable to incorporate OMWW in the surface layer of soil (e.g. with a vibrating tine cultivator) after spreading, with benefits also to soil porosity and also to avoid potential visual nuisances and/or unpleasant smells (IOOC, 2004). Considering the load capacity of a sewage spreader (about 5–6 m^3), the spreading of 80 m^3 of OMWW on a hectare of soil requires a working day. This, of course, represents a non-negligible cost, albeit offset by the fertilizer value of OMWW. To simplify the distribution of OMWW on cropland, an automated irrigation system was also successfully tested, connecting directly the oil mill to the area of spreading. It consists of (i) a storage tank of wastewater; (ii) an electric pump automatically activated depending on the level of the OMWW in the tank; and (iii) a self-propelled machine to irrigate (Proietti *et al.*, 1995).

The OMWW spreading must be conducted following national regulations. In Italy, for example, it cannot be applied in soils that are not used for agricultural purposes or that are less than 10 m away from watercourses or the beginning of the sandy shore, or that are located less than 200 m from towns or 300 m from protected areas, where there is uptake of water for human consumption, or, for crops, where groundwater is located less than 10 m deep or on frozen soil, snow-covered, waterlogged or flooded, or with a slope more than 15% (this limit, according to some regional laws, can be increased to 25% in the presence of soils covered with grass or if there are drainage systems or where systems simultaneously distribute and incorporate the OMWW in the soil). Also, forests, garden areas for public use and quarries are excluded from spreading.

The spreading of the OMWW should not be carried out in soils with high levels of salinity since the high salt content of the OMWW can, albeit temporarily, further increase the salinity of the soil and consequently cause a decrease in the stability of aggregates. In normal soils, respecting the doses envisaged by law, the OMWW does not cause harmful increases in soil salinity (IOOC, 2004).

From a practical standpoint, the application of OMWW does not affect the chemical and biological quality of the soil when it is rich in clay and calcareous (avoid acid and sandy and stony soils). Similarly, OMWW can be spread on shallow (20–30 cm) finer-textured soils, which have better water- and fertilizer-storage capacity. Chemical analysis of the soil is strongly recommended before OMWW spreading to determine two parameters: pH and electrical conductivity (salinity) (IOCC, 2004).

Clay-loam soils, with a high percentage of clay and high cation-exchange capacity (CEC), seem to be the best soil types for OMWW application. Also, the carbonate content of the soil is important to exert efficient buffer power, so avoiding drastic changes in the soil pH could cause immobilization of different macro- and micronutrients with resulting changes in the microbiological activity of the soil (IOOC, 2004).

8.2.2.5 Spreading load

In some countries, the maximum spreading load allowed is established by national or local laws. In Italy, the agronomic use of OMWW (as well as pomace) is allowed in compliance with the limit of acceptability of 50 m^3 ha^{-1} $year^{-1}$ for the OMWW derived from pressure-extraction system (traditional old method) and 80 m^3 ha^{-1} for those derived from the centrifugation extraction system.

In addition, to determine the spreading load, it is necessary to take account of water infiltration of the soil: if this is low (< 5 mm h^{-1}), it is necessary to use lower spreading load than those above-mentioned to prevent the occurrence of surface run-off (which initially are also favored by the temporary decrease in infiltration caused by the water repellency of the OMWW fat component) with pollution risks for surface waters in sloping soils or waterlogging phenomena in the plain land (Chiesura *et al.*, 2005). The spreading load has to be reduced even for soils with high water infiltration (> 150 mm h^{-1}) since excessive percolation of OMWW organic fraction can cause pollution of groundwater.

8.2.2.6 Time of spreading

Regarding the OMWW time of spreading, it is necessary to comply with regional laws if any. Based on the results of numerous experiments, it can be stated that for the spring-summer crops (especially maize and sunflower), spreading should be done two to three months before sowing, since pre-emergence effect was detected when less than 40 days occur between spreading and sowing. On the autumn-winter cereals (wheat and barley), the spreading could also be done with crops, preferably without exceeding the dose of 40 m^3 per ha/year (Barbera *et al.*, 2013). On the contrary, on tree crops (especially olive), there are no particular time limitations and so the major limit is represented by the slope of treated soils that in rainy periods can cause run-off or accessibility problems. To prevent leaching, OMWW must not be applied on rainy days or when soil is frozen (IOOC, 2004).

8.2.2.7 Composting of OMWW together with solid matrices

An interesting perspective to enhance the benefits of OMWW is composting, which is represented by the natural process of biological transformation of the organic matrix to produce a stabilized material, mineralized, rich in minerals, sanitized and not phytotoxic. Indeed, the organic component of the OMWW is essentially non-humified, while from the agronomic point of view the most efficient form of organic matter is the humified one. The OMWW to be composted should be added to a solid substrate rich in cellulosic and with an adequate C:N ratio (25:30) (Plaza *et al.*, 2007; Paredes *et al.*, 2005). Experiments have demonstrated that the imbibition with the OMWW of solid matrices having high adsorbing power and suitable porosity (e.g. olive pomace, straw, wood chips etc., with the addition of urea, if necessary, to ensure an adequate C:N ratio) and size of particles allows mineralization and humidification of the substrate, resulting in the formation of a final product that is stable and without phytotoxic effects.

8.2.3 Agronomic Use of Pomace

8.2.3.1 Main physico-chemical characteristics of pomace

The pomace is the solid byproduct originated from the processing of olives, consisting of the fibrous part of the fruit, pit, about 5% of residual oil and an amount of water ranging from 25% to 30% using a pressure oil-extraction system (traditional old method), to 50–55%, 55–70% and 50% using three-phase, two-phase and water-saving centrifugation systems, respectively (Toscano and Montemurro, 2012; Suzzi and Tofalo, 2009; Amirante, 2003). The two-phase system reduces the consumption of water and does not produce OMWW but produces only a humid pomace with a slightly acidic pH, a high concentration of organic matter (especially fibers), potassium, oil and phenols. This pomace has a wet, soft consistency, and this makes it difficult to transport it, and watertight caissons for the transport are necessary (Suzzi and Tofalo, 2009). The total amount of pomace produced per 100 kg of olives is around 87 kg and 62 kg in the two-phase and three-phase centrifugation systems, respectively (Peri and Proietti, 2014). Except for the water content amount, the pomace has similar characteristics in the pressure and centrifugation oil-extraction systems. The chemical composition of pomace varies within very wide limits and it depends upon the cultivar of the olive trees, fruit-ripening stage at harvesting, and, above all, the processing method, which, in particular, strongly influences the water content. The most important characteristics of pomace are high values of chemical oxygen demand (COD) and biochemical oxygen demand (BOD), low pH (around 5.2), and high phenols and potassium content (Mechri *et al.*, 2011; Alburquerque *et al.*, 2004).

8.2.3.2 Main agronomic properties of the pomace

The pomace does not contain heavy metals, toxic pollutants or pathogens: it is constituted by unfermented organic vegetal substance. Therefore, it can be agronomically used, by spreading both raw or after composting on agricultural soil, as an amendment or, following composting, to prepare nursery substrates (Regni *et al.*, 2017a, 2020).

Being a solid byproduct, compared to the OMWW, for the spreading of the pomace, more attention is needed to obtain a good uniformity of distribution. The problem is greater with pomace deriving from a two-phase continuous oil-extraction system since this pomace tends to form lumps unless it is made in advance to dry.

As for the OMWW, the use of pomace as an organic amendment could determine some problems due to its high organic load and mineral salt content, low pH and the presence of phytotoxic compounds (Proietti *et al.*, 2015; Gigliotti *et al.*, 2012; Del Buono *et al.*, 2011).

The use of pomace as an amendment on agricultural soils, with the spreading on the soil surface in grass-covered soils or incorporated at a depth of 10–25 cm in tilled ones, is a widespread practice, mostly in olive groves, since, in addition to being economical and easily practicable, it has positive effects on the characteristics of the soil and trees' vegetative and productive activity (Lozano-Garcia *et al.*, 2011).

In a case study conducted in central Italy, for example, the distribution in an olive grove of large quantities of pomace (100 t/ha) from a three-phase extraction system for four consecutive years caused in the soil a slight decrease in pH, an increase in the content of organic matter, total nitrogen, exchangeable potassium, magnesium and assimilable phosphorus. The amendment has also increased vegetative and productive activity without modifying the oil content percentage of the fruits (Nasini *et al.*, 2013). In the same olive grove, the amendment with the pomace produced by an innovative two-phase extraction system, distributed composted and not composted for three consecutive years, increased the soil available P, while the exchangeable K increased only in the plots treated with non-composted pomace. In both soils amended with fresh or composted pomace, no increase in organic C in the soil was found, indicating a probable priming effect (Proietti *et al.*, 2015). Most studies report that potential changes in soil chemistry and microbiology are likely to occur during the first weeks of organic amendments application (Blagodatskaya and Kuzyakov, 2008). Regarding the microbiological changes, it seems that the amendment with pomace may represent a temporary enrichment with a readily available C source for microbial growth (Di Serio *et al.*, 2008; Mekki *et al.*, 2006; Saadi *et al.*, 2007).

In general, from the results of various experiments, it emerges that the contribution to the soil of pomace or its derived compost promotes, over time, an increase in fertility because it increases, although often to a very variable extent, the content of organic matter and nutrients (total nitrogen, assimilable phosphorus and exchangeable potassium), without significantly modifying the pH and salinity; and improves the soil structure and its stability, and increases the water capacity with positive repercussions on available water.

Regarding the risk of groundwater pollution in pomace as well as in OMWW, it should be noted that the N is contained mainly as ammonium. Therefore, when applied to the soil, this ion, positively charged, is adsorbed by soil colloids and then leaching is avoided. After that, the ammonium is oxidized to nitrate, negatively charged, and not adsorbed by soil colloids resulting in leaching. Nitrate leaching occurs when its quantity in soil solution exceeds the plant adsorption.

Various countries have legislated the use of pomace. The regulations must be taken into consideration. For example, in Italy, the spreading of pomace on agricultural soils can be made following modalities set out in the law n. 574 of 1996.

Therefore, the agronomic use of the pomace is a valid option for its disposal and valorization.

8.2.3.3 Storage and transport

The storage and transport of the olive pomace must be done in compliance with current legislation, as well as in compliance with safety requirements (e.g. the containers for the storage of wet pomace must be properly waterproofed and covered with tarpaulins to avoid percolation and infiltration).

8.2.3.4 Spreading methods

The spread of pomace is done with the same machine used for spreading manure. When possible (not applicable if a permanent green cover exists on the soil), it is preferable to lightly plough the soil after pomace spreading.

8.2.3.5 Spreading load

Italian law (n. 574 of 1996 and subsequent modification and integration) states that the volume of pomace allowed to be spread is 50 m^3 when it is produced by pressure (traditional old method) olive mills and 80 m^3 when produced by centrifugation olive mills. Nasini *et al.* (2013) and Proietti *et al.* (2015) did not observe negative effects on soil and trees of olive groves when spreading 50 t/ha of pomace. A lower dosage was used for the crops. For example, Brunetti *et al.* (2005) applied 20 t/ha on *T. turgidum* and Tejada and Gonzalez (2004) applied 40 t/ha on maize.

8.2.3.6 Time of spreading

The best period for pomace spreading is autumn/winter (López-Piñeiro *et al.*, 2008). The spreading has to be avoided before vegetative growth resumes. Rainy days should also be avoided. Moreover, spreading should be avoided on water-saturated or frozen soils.

8.2.3.7 Pomace composting

For agronomic use, pomace may also be previously composted, mixing them with other byproducts of food chains such as stalks, orchard pruning residues, straw etc. These matrices have the function of increasing the porosity of the mass and so improving its suitability for composting. To a greater degree than pomace, compost improves the physico-chemical characteristics of the soil, adding organic matter and plant nutrients, especially carbon and potassium, but also nitrogen and phosphorus, in a consistent proportion (Proietti *et al.*, 2015; Gigliotti *et al.*, 2012; Del Buono *et al.*, 2011).

The results of some experimentations have shown a good effect of the pomace when composted with olive tree prunings, grape stalks and straw. For the pomace resulting from the three-phase extraction system, good compost has been obtained by mixing it with olive pruning residues shredded in the ratio 1:1 (Regni *et al.*, 2017b). The mixture of wet pomace deriving from a two-phase plant with shredded olive tree prunings (22%) and cereal straw (8%) (w/w) also produced composts with good agronomic characteristics (Regni *et al.*, 2020).

Composts obtained in these experimentations show a high content of organic matter, especially humic-like substances, a balanced C:N ratio and high concentrations of plant nutrients. About the latter, an aspect of particular agronomic and environmental interest is that almost all the nitrogen of the compost is in organic form, as this allows provision to the soil slow-release nitrogen, not susceptible to leaching. Moreover, pomace composting allows decrease in water content and shows a beneficial effect in the reduction of potential phytotoxicity of pomace.

The compost derived from pomace can also represent a material to be used as an alternative, partially or totally, to the peat in the production of nursery substrates for species with medium salt tolerance such as olive, without negative effects on production and with economic and environmental benefits (Regni *et al.*, 2020). Great interest also assumes the possibility that the compost can develop in the substrate repressive effects against some plant pathogens (Bonanomi *et al.*, 2010; Suzzi and Tofalo 2009).

8.2.4 General Conclusion: Strengths, Weaknesses, Threats and Opportunities on the Agronomic Use of OMW

Controlled spreading of OMW or derived compost is an economically and environmentally viable way to exploit them. Indeed, the reuse of OMW leads to a decrease in pollutant wastes and reduces the use of chemical fertilizers, improves the chemical, biological and physical soil properties and enhances crop yields. The application of OMW, especially if composted, is an answer to limit the loss of soil fertility and enhance organic matter content and thus increase C sequestration of the agroecosystems. This is very important where the environmental conditions (hot and arid) make it particularly hard to maintain soil fertility with the dangerous consequences of desertification.

References

Alburquerque, J.A., Gonzalvez, J., Garcıa, D. and Cegarra, J. (2004) Agrochemical characterization of 'alperujo', a solid by-product of the two phase centrifugation. *Bioresource Technology* 91(2), 195–200.

Amirante, P. (2003) I sottoprodotti della filiera olivicola-olearia. In: *Olea Trattato di Olivcoltura*. Edagricole, Bologna, Italy, pp. 291–303.

Aviani, I., Raviv, M., Hadar, Y., Saadi, I., Dag, A. *et al.* (2012) Effects of harvest date, irrigation level, cultivar type and fruit water content on olive mill wastewater generated by a laboratory scale 'Abencor' milling system. *Bioresource Technology* 107, 87–96.

Barbera, A.C., Maucieri, C., Cavallaro, V., Ioppolo, A. and Spagna, G. (2013) Effects of spreading olive mill wastewater on soil properties and crops, a review. *Agricultural Water Management* 119, 43–53.

Bedbabis, S., Trigui, D., Ben Ahmed, C., Clodoveo, M.L., Camposeo, S., Vivaldi, G.A. and Ben Rouina B. (2015) Long-term effects of irrigation with treated municipal wastewater on soil, yield and olive oil quality. *Agricultural Water Management* 160, 14–21.

Bianchi, B., Castellano, S.and Montel, G.L. (2000) Spargimento in campo delle acque di vegetazione con sistemi di controllo della portata e modelli territoriali di riferimento. In: *Atti del convegno dell'AIIA 'La ricerca multidisciplinare ed integrata per la valutazione e la gestione dei sistemi agricoli'*. Campobasso, 27–28 giugno.

Blagodatskaya, E. and Kuzyakov Y. (2008) Mechanism of real and apparent priming effect and their dependence on soil microbial biomass and community structure: critical review. *Biology and Fertility of Soil* 45, 115–131.

Bonanomi, G., Antignani, V., Capodilupo, M. and Scala, F. (2010) Identifying the characteristics of organic soil amendments that suppress soilborne plant diseases. *Soil Biology and Biochemistry* 42, 136–144.

Brunetti, G., Plaza, C. and Senesi, N. (2005) Olive pomace amendment in Mediterranean conditions: effect on soil and humic acid properties and wheat (*Triticum turgidum* L.). *Journal of Agricultural Food Chemistry* 53, 6730–6737.

Chartzoulakis, K., Psarras, G., Moutsopoulou, M. and Stefanoudaki, E. (2010) Application of olive mill wastewater to a Cretan olive orchard: effects on soil properties, plant performance and the environment. *Agriculture, Ecosystems & Environment* 138(3–4), 293–298.

Chatzistathis, T. and Koutsos T. (2017) Olive mill wastewater as a source of organic matter, water and nutrients for restoration of degraded soils and for crops managed with sustainable systems. *Agricultural Water Management* 190, 55–64.

Chehab, H., Tekaya, M., Ouhibi, M., Gouiaa, M., Zakhama, H. *et al.* (2019) Effects of compost, olive mill wastewater and legume cover crops on soil characteristics, tree performance and oil quality of olive trees cv. Chemlali grown under organic farming system. *Scientia Horticulturae* 253, 163–171.

Chiesura, A., Marano, V., De Francesco, P. and Maraglino, A. (2005) Verso la sostenibilità della filiera olivicola: trattamento, recupero e valorizzazione dei sottoprodotti oleari. UNASCO, Rome.

Del Buono, D., Said-Pullicino, D., Proietti, P., Nasini, L. and Gigliotti, G. (2011) Utilization of olive husks as plant growing substrates: phytotoxicity and plant biochemical responses. *Compost Science & Utilization* 19, 52–60.

Di Bene, C., Pellegrino, E., Debolini, M., Silvestri, N. and Bonari, E. (2013) Short- and long-term effects of olive mill wastewater land spreading on soil chemical and biological properties. *Soil Biology and Biochemistry* 56, 21–30.

Di Serio, M.G., Lanza, B., Mucciarella, M.R., Russi, F., Iannucci, E., Marfisi, P. and Madeo, A. (2008) Effects of olive mill wastewater spreading on the physico-chemical and microbiological characteristics of soil. *International Biodeterioration & Biodegradation* 62(4), 403–407.

Gigliotti, G., Proietti, P., Said-Pullicino, D., Nasini, L., Pezzolla, D., Rosati, L. and Porceddu, P.R. (2012) Co-composting of olive husks with high moisture contents: organic matter dynamics and compost quality. *International Biodeterioration & Biodegradation* 67, 8–14.

Hanafi, F., Belaoufi, A., Mountadar, M. and Assobhei O. (2011) Augmentation of biodegradability of olive mill wastewater by electrochemical pre-treatment: effect on phytotoxicity and operating cost. *Journal of Hazardous Materials* 190(1–3), 94–99.

Ilarioni, L. and Proietti, P. (2014) Olive tree cultivars. In: Peri, C. (ed.) *The Extra-Virgin Olive Oil Handbook*. John Wiley & Sons, Ltd., Chichester, UK, pp. 59–67.

IOOC (2004) Recycling of vegetable water & olive pomace on agicultural land (CFC/IOOC/04).

Jarboui, R., Sellami, F., Kharroubi, A., Gharsallah, N. and Ammar, E. (2008) Olive mill wastewater stabilization in open-air ponds: impact on clay-sandy soil. *Bioresource Technology* 99(16), 7699–7708.

Justino, C.I.L., Pereira, R., Freitas, A.C., Rocha-Santos, T.A.P., Panteleitchouk, T.S.L. and Duarte, A.C. (2012) Olive oil mill wastewaters before and after treatment: a critical review from the ecotoxicological point of view. *Ecotoxicology* 21(2), 615–629.

López-Piñeiro, A., Albarràn, A., Rato Nunes, J.M. and Barreto, C. (2008) Short and medium term effects of two-phase olive mill waste application on olive grove production and soil properties under semiarid Mediterranean conditions. *Bioresource Technology* 99, 7982–7987.

Lozano-García, B., Parras-Alcántara, L. and del Toro Carrillo de Albornoz, M. (2011) Effects of oil mill wastes on surface soil properties, runoff and soil losses in traditional olive groves in southern Spain. *CATENA* 85(3), 187–193.

Magdich, S., Rouina, B.B. and Ammar, E. (2020) *Waste Biomass Valorization* 11, 1359–1372.

McNamara, C.J., Anastasiou, C.C., O'Flaherty, V. and Mitchell, R. (2008) Bioremediation of olive mill wastewater. *International Biodeterioration and Biodegradation* 61(2), 127–134.

Mechri, B., Cheheb, H., Boussadia, O., Attia, F., Ben Mariem, F., Braham, M. and Hammami, M. (2011) Effects of agronomic application of olive mill wastewater in a field of olive trees on carbohydrate profiles, chlorophyll a fluorescence and mineral nutrient content. *Environmental and Experimental Botany* 71, 184–191.

Mekki, A., Dhouib, A. and Sayadi, S. (2006) Changes in microbial and soil properties following amendment with treated and untreated olive mill wastewater. *Microbiological Research* 161(2), 93–101.

Mekki, A., Dhouib, A. and Sayadi, S. (2013) Review: effects of olive mill wastewater application on soil properties and plants growth. *International Journal of Recycling of Organic Waste in Agriculture* 2, 15–21.

Moraetis, D., Stamati, F.E., Nikolaidis, N.P. and Kalogerakis, N. (2011) Olive mill wastewater irrigation of maize: impacts on soil and groundwater. *Agricultural Water Management* 98, 1125–1132.

Nasini, L. and Proietti, P. (2014) Olive harvesting. In: Peri, C. (ed.) *The Extra-Virgin Olive Oil Handbook*. John Wiley & Sons, Ltd., Chichester, UK, pp. 89–105.

Nasini, L., Gigliotti, G., Balduccini, M.A., Federici, E., Cenci, G. and Proietti, P. (2013) Effect of solid olive-mill waste amendment on soil fertility and olive (*Olea europaea* L.) tree activity. *Agriculture, Ecosystems & Environment* 164, 292–297.

Paredes, C., Cegarra, J., Bernal, M.P. and Roig, A. (2005) Influence of olive mill wastewater in composting and impact of the compost on a Swiss chard crop and soil properties. *Environment International* 31(2), 305–312.

Peri, C. and Proietti, P. (2014) Olive mill waste and by-products. In: Peri, C. (ed.) *The Extra-Virgin Olive Oil Handbook*. John Wiley & Sons, Ltd., Chichester, UK, pp. 283–302.

Plaza, C., Senesi, N., Brunetti, G. and Mondelli, D. (2007) Evolution of the fulvic acid fractions during co-composting of olive oil mill wastewater sludge and tree cuttings. *Bioresource Technology* 98, 1964–1971.

Proietti, P., Federici, E., Fidati, L., Scargetta, S., Massaccesi, L. *et al.* (2015) Effects of amendment with oil mill waste and its derived compost on soil chemical and microbiological characteristics and olive (*Olea europaea* L.) productivity. *Agriculture, Ecosystems & Environment* 207, 51–60.

Proietti, P., Palliotti, A., Tombesi, A. and Cenci, G. (1995) Chemical and microbiological modifications of two different cultivated soils induced by olive oil waste water administration. *International Journal of Agricultural Science* 125, 160–171.

Regni, L., Gigliotti, G., Nasini, L., Agrafioti, E., Galanakis, C.M. and Proietti P. (2017a) Reuse of olive mill waste as soil amendment. In: Galanakis, C.M. (ed.) *Olive Mill Waste: Recent Advances for Sustainable Management*. Academic Press, Cambridge, Massachusetts, 97–116.

Regni, L., Nasini, L., Ilarioni, L., Brunori, A., Massaccesi, L., Agnelli, A. and Proietti, P. (2017b) Long term amendment with fresh and composted solid olive mill waste on olive grove affects carbon sequestration by prunings, fruits, and soil. *Frontiers in Plant Science* 7, 2042.

Regni, L., Pezzolla, D., Gigliotti, G. and Proietti, P. (2020) The sustainable reuse of compost from a new type of olive mill pomace in replacing peat for potted olive tree. *Agronomy Research* 18(2), 1444–1454.

Roig, A., Cayuela, M.L. and Sánchez-Monedero, M.A. (2006) An overview on olive mill wastes and their valorisation methods. *Waste Management* 26(9), 960–969.
Saadi, I., Laor, Y., Raviv, M. and Medina, S. (2007) Land spreading of olive mill wastewater: effects on soil microbial activity and potential phytotoxicity. *Chemosphere* 66(1), 75–83.
Suzzi, G. and Tofalo, R. (2009) Trattamento dei reflui. In: Pisante, M., Inglese, P. and Lercker, G. (eds) *Collana Coltura & Cultura – L'ulivo e l'olio*. Bayer CropScience, Bologna, Italy, pp. 690–695.
Tejada, M. and Gonzalez, J.L..(2004) Effects of foliar application of a byproduct of the two-step olive oil mill process on rice yield. *European Journal of Agronomy* 21(1), 31–40.
Toscano, P. and Montemurro, F. (2012) Olive mill by-products management. In: Muzzalupo, I. (ed.) *Olive Germplasm – The Olive Cultivation, Table Olive and Olive Oil Industry in Italy*. InTech Open Access, London, pp. 173–200.
Vivaldi, G.A., Camposeo, S., Rubino, P. and Lonigro A. (2013) Microbial impact of different types of municipal wastewaters used to irrigate nectarines in Southern Italy. *Agriculture, Ecosystems and Environment* 181, 50–57.
Yay, A.S.E., Oral, H.V., Onay, T.T. and Yenigün, O. (2012) A study on olive oil mill wastewater management in Turkey: a questionnaire and experimental approach. *Resources, Conservation and Recycling* 60, 64–71.
Zbakh, H. and El Abbassi, A. (2012) Potential use of olive mill wastewater in the preparation of functional beverages: a review. *Journal of Functional Foods* 4(1), 53–65.

9.1 Multifunctionality of Olive Groves

L. Regni[1], A. Brunori[2], H. Mairech[3], L. Testi[4] and P. Proietti[1]
[1]*University of Perugia, Department of Agricultural, Food and Environmental Sciences, Perugia, Italy;* [2]*PEFC Italy;* [3]*National Institute of Meteorology (INM), Research and Development Department, Unit of Climate and Climate Change Studies, Tunis, Tunisia;* [4]*Department of Agronomy, Institute for Sustainable Agriculture, Spanish National Research Council (IAS-CSIC), Cordoba, Spain*

9.1.1 Ecosystem Services in Agriculture

Ecosystem services (ES) are defined as the benefits derived directly or indirectly from ecosystems (MEA, 2005). The focus on ES can be traced back to the well-known study 'The value of the world's ecosystem services and natural capital', published in *Nature* (Costanza *et al.*, 1997), in which, for the first time, the relationship between natural capital and the services produced was defined. More recently, the report *The Economics of Ecosystems and Biodiversity* (TEEB, 2008), which emerged after the G8+5 meeting of environment ministers held in Potsdam in 2007, highlighted the importance of evaluating ES to optimize biodiversity management. Agriculture and, more generally, agroforestry play a complex role in relation to ES. Indeed, if on one hand, agricultural production processes use ES generated by the surrounding territory, on the other hand, agriculture can provide ES to society. The biophysical quantification and monetary evaluation of services allow, besides assessing the environmental costs related to changes in the state of agro-ecosystems (changes in land use, technological innovations), to correctly define and plan public intervention. Ecosystem services are essential to support human activities, but their importance is currently partly ignored because many of these services are not traded on the market and do not have a price that is indicative of their social value. In this context, a useful tool for the effective management of ES is the so-called ES payments (PES). Wunder (2005) defines the latter as a 'voluntary transaction in which a well-defined ES is purchased by a purchaser who receives it from an ES supplier, provided that the supplier ensures the supply of the ES'. This implies that in order to have a PES, some conditions are necessary: the presence of a well-defined service that is the object of a voluntary transaction, and that this takes place between a 'producer' and a 'beneficiary' with temporal continuity. However, given that, from the point of view of the type of goods, ES often correspond to public goods, therefore an economic institution that allows the existence of the market is often necessary. Consequently, in this case it is not entirely correct to speak of PES but frequently, a public operator creates programs based on incentive, a compensation for the loss of income (TEEB, 2008). They aim at remunerating the farmers (or public entities) who protect or supply the ES, and for the higher attributable

DOI: 10.1079/9781789247350.0026

costs to adopt techniques and processes compatible with the supply of the ES (Rasul, 2009). In order to implement effective policies for the management of the ES, it is necessary to proceed through two steps: the quantification in biophysical terms of the services and their monetary valuation. The quantification of services is done in biophysical units (e.g. amount of CO_2 absorbed by plants, amount of water absorbed by forest areas, amount of soil preserved annually from erosion). Both the biophysical and monetary quantification of ES are subject to difficulties linked to the lack of knowledge of ecosystem processes. Moreover, from an economic point of view, many of the ES lack an appropriate market (e.g. support services including resilience, nutrient or carbon cycle and primary production whose economic evaluation represents a particularly complex issue). Therefore, in order to quantify ES economically it is necessary to use evaluation methods that allow to assign a value to ES based on: (i) the willingness of farmers to accept compensation to use up part of their income by changing agricultural practices to generate more ES; and (ii) the willingness of ES beneficiaries to pay to receive more services (Swinton *et al.*, 2007). The evaluation of ES is certainly a complex process. The evaluation of each service relies on the integration of various scientific expertise from different fields as well as the availability of adequate data. The latter may show quality problems related to accuracy, scale, updating etc. Nevertheless, the methodology for the assessment of some ES is quite established, such as the estimation of the improvement of the carbon balance. Discussion for this ES focusses rather on deepening the data sources and testing other methodologies of assessment.

In this framework, it is easy to understand which ES olive growing is able to provide, starting from primary production and supply of materials and fuels, to the role of climate regulation and stabilization of the hydrogeological regime, up to important and intangible cultural and landscape values. The main functions carried out by olive groves, both from the productive as well as the ecosystem services point of view, are those of hydrogeological defense, landscape value, biodiversity protection and climate change mitigation.

9.1.2 Olive Growing and Ecosystem Services

In many areas, due to the soil, climate and structural conditions of the traditional farms, olive growing has become economically unsustainable. As a result, traditional olive groves are often abandoned due to low productivity, difficult mechanization and high labor costs. This tendency is dangerous not only from the food supply point of view but, more importantly, for the loss of the numerous ecosystem services sustained by these olive groves.

It is therefore necessary to identify and implement actions to restore economic sustainability of olive growing, especially in these areas where it is easier to generate ES and where their value can have a bigger incidence compared to areas with a high revenue from olive growing. Among these actions, an important opportunity would be the expression in monetary units of the ecosystem services carried out and then their payment to the growers to maintain the persistence of olive groves on the territory and therefore preserve the ES they supply. To achieve this goal, it is necessary to identify appropriate indicators to quantify the extent of these services in the real-life farms, rewarding those who implement low-impact production methods aimed at preserving or enhancing ES.

Furthermore, there are important ecosystem services related to olive growing that could be the subject of payments such as preservation of rural landscape, recreation, cultural heritage, biodiversity conservation and soil conservation. Unfortunately, no quantification and monetization methods have been developed so far for these ES.

In olive growing, and as already experimented in forestry, as stated before, an ES that could already be monetized is the improvement of the carbon balance. The potential of olive-tree groves for sequestering carbon has been confirmed under some soil conservation practices (Sofo *et al.*, 2005; Proietti *et al.*, 2014, 2017; Lopez-Bellido *et al.*, 2016; Proietti *et al.*, 2016; Brunori *et al.*, 2017; Mairech *et al.*, 2020). Indeed, carbon balance improvement can be achieved by reducing carbon emissions in the air (lower energy consumption for soil management, no burning of pruning, reduction in the use of pesticide and synthetic fertilizers, lower energy consumption for byproducts management, reuse of byproducts for energy purposes, etc.) and/or by increasing the carbon sequestered in the soil (on-the-spot chopping of pruning, grassing, use of organic fertilizers, soil amendment with byproducts, etc.), and in the trees' permanent structure (roots, trunk, main branches).

The Kyoto Protocol recognizes the importance of the agro-forestry sector in mitigating climate change. In fact, through photosynthesis, plants are able to sequestrate CO_2 in above- and below-ground biomass, countering therefore the increasing atmospheric CO_2 emissions. To strengthen this role, the Kyoto Protocol encourages the introduction of eco-sustainable agricultural practices and the improvement of agricultural systems efficiency in order to enlarge the organic carbon pools and reduce atmospheric CO_2 emissions.

Several studies have confirmed the long-term capacity of olive groves to store carbon in the soil and their improvement under specific soil management system (e.g. on-the-spot chopping of pruning, grassing, use of organic fertilizers, soil amendment with byproducts, etc.) (Nieto *et al.*, 2010; Regni *et al.*, 2017; Massaccesi *et al.*, 2018; Mairech *et al.*, 2020). It has also been proved that some agricultural management, such as irrigation and intensification, helps reinforce the carbon sequestration potential of olive groves acting as carbon sinks (Mairech *et al.*, 2020). Furthermore, it has to be considered that trees store large amounts of carbon in their structures for many years. When trees are felled, if the wood is burned to produce energy, no new carbon is released into the atmosphere, as in the case of fossil fuels. If, on the other hand, wood is used to produce long-lasting products, such as in construction or furniture etc., carbon is removed from the atmosphere for decades, sometimes for centuries.

Among the different tree species, the carbon storage in standing trees is, of course, more durable and abundant in more long-lived ones that have a high wood mass, such as the olive tree. Moreover, cultivation practices oriented to increase the organic matter in the soil (e.g. soil management through the use of grass cover or the incorporation of pruning residues or pomace to the soil etc.), which often characterize the management of olive groves, enhance the accumulation of carbon in the soil and reduce CO_2 emissions for its management.

Despite the important capacity of olive groves to sequester atmospheric CO_2 (also in view of the widespread use of this crop), there is still little knowledge about the potential of these practices, and therefore their role in mitigating climate change. Furthermore, work is needed to define precisely the optimal cultivation strategies to be adopted specifically in different olive grove systems, on the one hand to increase carbon sequestration potential and on the other to reduce emissions from the entire olive-growing chain in order to maximize the ratio between CO_2 absorption and emissions.

9.1.3 Improved Carbon Balance and Positive Impacts on Companies

In relation to the containment of impacts related to CO_2 emissions and other climate-altering greenhouse gases, also referred to as GHG, an important tool is the determination of the carbon footprint (CF), an environmental indicator that measures the impact of human activities on the global climate and expresses, in CO_2 equivalent (tCO_{2eq}), the total direct or indirect emissions of greenhouse gases in all phases of a product's life. CF is a codified and standardized method (ISO 14040:2006, ISO 14044:2006, ISO/TS 14067:2018). CF, by accounting for CO_{2eq} emissions, highlights the most critical environmental phases (carbon assessment) of a production cycle, thus enabling action to be taken to reduce the company's

Table 9.1.1. Ecosystem services provided by olive groves and indicators that can be used for their quantification.

Ecosystem service	Indicators
Preservation of biodiversity	✓ Olive tree groves area as a percentage of the total area ✓ Density of forests and natural areas around the olive tree groves ✓ Abundance of animal and plant species/varieties
Carbon sequestration	✓ Amount of C sequestered by tree's non-permanent structures ✓ Amount of C sequestered by tree's permanent structures ✓ Amount of soil organic carbon
Soil conservaton	✓ Abundance of soil macrofauna ✓ Area affected by wind and/or water erosion ✓ Area affected by soil compaction ✓ Organic matter content ✓ Macronutrients soil content
Recreational and landscape services	✓ Number of visitors and visitor satisfaction ✓ Number of sightings of rare species ✓ Value/price assessment by visitors ✓ Preservation of high-value old trees ✓ Maintenance of special infrastructure

environmental impact (carbon management) and, therefore, improve the company's environmental performance.

Considering the balance between sequestration and emission of carbon for the agronomic phase of the olive oil production chain, in a case study of conventional olive grove in Umbria, already in the fifth year after planting the balance became positive, that is, the olive grove has begun to sequester carbon (Proietti *et al.*, 2014). In this study, the determination of CF has shown that the practices more impacting in terms of CO_2 emissions were chemical fertilization, pesticide treatments (emissions related to the production of fertilizers/pesticide and to the relative transport in the company), and the transport of olives from the field to the mill. To improve the carbon balance, it is possible to implement sustainable strategies also in the transformation phase. For example, the use of olive stone as fuel allows recovery of more energy than that spent for the olive-pitting machine. The consequence is a 5% reduction of the CF relative to the exaction process (Proietti *et al.*, 2017).

In another case study related to the production of organic extra-virgin olive oil in Umbria (Proietti *et al.*, 2017) the CF related to the production of 1 L of oil was 1.01 kg-CO_{2eq}/L; this value is lower than that found in other studies for other regions of Italy and especially for other countries. In particular, the lowest value of CF in the cultivation phase of the case study compared to other regions of Italy (0.51 vs. 1.1 kgCO2e-q/L oil) is mainly due to the proximity of the mill to the olive grove (about 1.5 km). The amount of CO_2 sequestered, including carbon sequestration in the soil was almost double the quantity emitted (1.0 kgCO_{2eq}/L oil vs. 1.8 kgCO_{2eq}/L oil). So, it can be assessed that the production of extra-virgin olive oil with sustainable techniques can be a tool to counteract climate change.

A low CF can add value to a process or product. Indeed, the carbon label is the eco-label that certifies greenhouse gas emissions related to the life cycle of a product and can also indicate the reduction in emissions that the product itself allows compared to alternative products. Reducing the CF for companies can lead to a substantial competitive advantage as the carbon label is increasingly perceived by the most advanced consumers as an index of quality and sustainability of companies. It is also directly linked to the current environmental emergency connected to climate change. In recent decades, in fact, there has been a growing awareness among a wide range of consumers of their ability to influence production strategies through 'positive selective pressure', i.e. choosing products with lower environmental impact as a form of consumption compatible with environmental protection for current and future generations.

It should also be considered that the assessment of a CF of a farm and its products as an improvement of a company's performance may be accounted as a mitigation option to reduce greenhouse gas emissions through the implementation of beneficial agricultural practices for olive groves compared to a 'business as usual' (BAU) scenario (i.e. the starting situation). This is the approach that unifies five different categories of certification and labeling schemes about CF in the agricultural sector, namely: (i) carbon footprint certification schemes; (ii) carbon-neutral product schemes; (iii) companies' carbon reporting schemes; (iv) criteria-based low-carbon farming certification schemes; and (v) other certification schemes relevant to low-carbon farming. Farms can be certified upon a declaration of a third-party-verified carbon footprint, upon proven emission reductions, or upon lower emissions compared to the average in the product group. A certification scheme with an innovative approach was adopted for the olive sector with a LIFE project called OLIVE-4CLIMATE (https://olive4climate.eu [accessed 26/5/23]) that encouraged the development of a voluntary market for carbon deriving from the sustainable management of olive groves.

This type of carbon credit is called 'Sustainability Credits' (SC) and represents an exchange between olive growers able to reduce their emissions (and enhance C sequestration through the application of good practices) and buyers who wish to reduce their emissions within their own company. It opens therefore the possibility for olive growers to sell these credits to potential buyers in a voluntary market. The SC represents the result of human actions focussed on generating and offering an added value to the several ecosystem services originated from agricultural practices. These practices represent actions implemented by olive growers for the reduction of climate-altering gas emissions to the atmosphere and/or for the increase in the C sequestration. The SC (1 credit = 1 t CO_{2eq}) is a quantitative indicator adopted in the carbon market operations.

Two types of carbon-offset market exist: the regulatory-compliance and the voluntary markets. The SCs are considered in the voluntary carbon-offset market, where the trade of carbon credits is on a voluntarily basis. The compliance market is used by companies and governments that by law have to account for their GHG emissions (and it is regulated by mandatory national, regional or international carbon reduction regimes). Voluntary carbon markets and trading have always encountered problems because of transaction costs related to carbon trading and market participation, and because local offset projects generate a small quantity of credits compared to high-volume, least-cost projects. The SC system should be seen as an initiative to stimulate local virtuous action that does not provide for the transfer of credit ownership to the private individual but is aimed at enhancing the ecosystem. The objective is to increase the interventions through bilateral transactions in order to recognize the positive effects generated by this activity. The type of ecosystem services provided are therefore necessarily linked to the additional management activities and to the type of agricultural practices that are applied. Agricultural or forestry management activities, capable of generating ecosystem services quantifiable in SCs, must always guarantee an increase in the carbon stock or a decrease in GHG emissions resulting in an improvement of agro-ecosystem resilience. These management activities will therefore be able to help in the adaptation as well as the mitigation of climate change.

From 2016 to today, this methodological approach has been applied in tree plantations certified for their sustainable management following an internationally recognized certification standard created by a forestry-recognized normative organization. The SC is measured in tCO_{2eq} deriving both from the annual growth of the certified managed plantation and from the cultivation practices imposed by technical standards. Certification bodies verify the implementation of the selected cultivation techniques (i.e. reduction of fossil fertilizers and their replacement with organic amendments, reduction of pesticides, reduction of cultivation operations such as ploughing, mowing etc.) The lower impact on the environment (e.g. in terms of the amount of toxic residues and nitrogen compounds in the water) is communicated together with the quantity of SC. Since 2017, big retailers bought these SCs, declaring them in their CSR (corporate social responsibility) reports.

The OLIVE4CLIMATE project was able to test and verify approaches aimed at the implementation of climate change mitigation strategies in order to improve the carbon balance, which was demonstrated in olive farms in Italy, Greece and Israel. The project's actions, besides giving results for increasing the CO_2 sinks versus sources (therefore reducing the climate impact of olive production), was able to demonstrate that generating and selling sustainability credits would constitute an income supplement, also in terms of company image, that could contribute to restoring economic sustainability to the crop, especially in marginal areas at risk of abandonment. In these areas, in fact, thanks to the presence of low-intensity olive growing, it is easier to contain emissions and increase C sequestration. The certification standard tries to reduce the costs that are part of the field audits and, following calculation of the SC, control the adopted best practices by professional agronomists with the supervision of researchers.

The focus of the project lies in the replicability of the developed techniques and activities, which assured their uptake by olive farmers and oil producers resulting in an even larger reduction of CO_2 emissions. To make this possible, a standard for sustainability credits from sustainable olive grove management was created (downloadable at https://olive4climate.eu). These SCs were obtained through the implementation of a series of best practices by a sustainable olive grove management (https://olive4climate.eu) that reduced climate-altering emissions and/or increased carbon storage. The activities that could be eligible were: (i) reduction of the use of fertilizers; (ii) management of pruning material; (iii) green cover; (iv) conservation tillage; (v) creation of a new orchard. Produced SCs were then traded in a Voluntary Sustainability Credits market where the sellers were the olive growers and the buyers were those who want to reduce their CF. In 2020, the price for a sustainability credit coming from olive farmers has been set at 16 euros, a value higher than that of the forestry sector (around 8–10 euros), probably because of the small amount available or because of the novelty of the origin of these SCs, generated in important and well-known territories from a historical and touristic point of view.

The possibility to sell the SCs generated by the ecosystem services of the olive grove and the implementation of suitable practices is offering the possibility to preserve olive groves with low profitability, considering their important environmental and landscape functions.

9.1.4 Conclusions

We can say that carbon sequestration is an important ES provided by olive growing and can be an important driver for the development of the sector. In fact, by adopting cultivation models with high environmental sustainability, the carbon balance can be improved and the oil can even have negative CF. Consequently, it can be said that olive growing can mitigate climate change because the C sequestration by the olive grove system is greater than the emissions of the entire supply chain. This is a result of great environmental interest and can also have a significant positive impact on consumers, to whom is transmitted the concept that consuming extra-virgin olive oil, produced according to low-environmental-impact production systems (aimed at the improvement of the carbon balance), has beneficial effects not only on health and the environment but also for climate change mitigation. This can result in an important added value (and consequent price premium) for the oil, which, added to the possibility of selling SCs, might help avoid the abandonment of so-called marginal olive groves.

References

Brunori, A.M.E., Sdringola, P., Dini, F., Ilarioni, L., Nasini, L. *et al.* (2017) Carbon balance and Life Cycle Assessment in an oak plantation for mined area reclamation. *Journal of Cleaner Production* 144, 69–78. DOI: 10.1016/j.jclepro.2016.12.116

Costanza, R., d'Arge, R., de Groot, R., Farber, S., Grasso, M. *et al.* (1997) The value of the world's ecosystem services and natural capital. *Nature* 387, 253–260. DOI: https://doi.org/10.1038/387253a0

Lopez-Bellido, P.J., Lopez-Bellido, L., Fernandez-Garcia, P., Muñoz-Romero, V. and Lopez-Bellido, F.J. (2016) Assessment of carbon sequestration and the carbon footprint in olive groves in southern Spain. *Carbon Management* 7(3–4), 161–170.

Mairech, H., López-Bernal, Á., Moriondo, M., Dibari, C., Regni, L., Proietti, P., Villalobos, F.J. and Testi, L. (2020) Is new olive farming sustainable? A spatial comparison of productive and environmental performances between traditional and new olive orchards with the model OliveCan. *Agricultural Systems* 181, art. no. 102816, DOI: 10.1016/j.agsy.2020.102816

Massaccesi, L., De Feudis, M., Agnelli, A.E., Nasini, L., Regni, L. *et al.* (2018) Organic carbon pools and storage in the soil of olive groves of different age. *European Journal of Soil Science* 69(5), 843–855. DOI: 10.1111/ejss.12677

MEA (Millennium Ecosystem Assessment) (2005) *Ecosystems and Human Well-being: Synthesis.* Island Press, Washington, DC.

Nieto, O.M., Castro, J., Fernández, E. and Smith, P. (2010) Simulation of soil organic carbon stocks in a Mediterranean olive grove under different soil-management systems using the RothC model. *Soil Use & Management* 26(2), 118–125. DOI: https://doi.org/10.1111/j.1475-2743.2010.00265.x

Proietti, P., Sdringola, P., Brunori, A., Ilarioni, L., Nasini, L. *et al.* (2016) Assessment of carbon balance in intensive and extensive tree cultivation systems for oak, olive, poplar and walnut plantation. *Journal of Cleaner Production* 112, 2613–2624. DOI: 10.1016/j.jclepro.2015.10.009

Proietti, S., Sdringola, P., Desideri, U., Zepparelli, F., Brunori, A. *et al.* (2014) Carbon footprint of an olive tree grove. *Applied Energy* 127, 115–124. DOI: 10.1016/j.apenergy.2014.04.019

Proietti, S., Sdringola, P., Regni, L., Evangelisti, N., Brunori, A. *et al.* (2017) Extra virgin olive oil as carbon negative product: experimental analysis and validation of results. *Journal of Cleaner Production* 166, 550–562. DOI: 10.1016/j.jclepro.2017.07.230

Rasul, G. (2009) Ecosystem services and agricultural land use practices: a case study of the Chittagong Hill Tracts of Bangladesh. *Sustainability: Science, Practice & Policy* 5(2), 15–27.

Regni, L., Nasini, L., Ilarioni, L., Brunori, A., Massaccesi, L., Agnelli, A. and Proietti, P. (2017) Long term amendment with fresh and composted solid olive mill waste on olive grove affects carbon sequestration by prunings, fruits, and soil. *Frontiers in Plant Science* 7, 2042. DOI: 10.3389/fpls.2016.02042

Sofo, A., Nuzzo, V., Palese, A.M., Xiloyannis, C., Celano, G., Zukowskyj, P. and Dichio, B. (2005) Net CO2 storage in Mediterranean olive and peach orchards. *Scientia Horticulturae* 107(1), 17–24. DOI: https://doi.org/10.1016/j.scienta.2005.06.001

Swinton, S.M., Lupi, F., Robertson, G.P. and Hamilton, S.K. (2007) Ecosystem services and agriculture: cultivating agricultural ecosystems for diverse benefits. *Ecological Economics* 64, 245–252.

TEEB (2008) *The Economics of Ecosystems and Biodiversity: An Interim Report.* European Commission, Brussels.

Wunder, S. (2005) Payments for environmental services: some nuts and bolts. Cifor Occasional Paper no. 42. CIFOR, Jakarta, Indonesia.

9.2 Towards Modern Olive Polycultures

A. Rosati[1], K. Barkaoui[2,3] and P.É. Lauri[3]
[1]*Consiglio per la ricerca in agricoltura e l'analisi dell'economia agraria (CREA), Centro di ricerca Olivicoltura, Frutticoltura e Agrumicoltura;* [2]*CIRAD, UMR ABSys, F-34398 Montpellier, France;* [3]*ABSys, Univ Montpellier, CIHEAM-IAMM, CIRAD, INRAE, Institut Agro, Montpellier, France*

9.2.1 Olive Polycultures: A Historical Excursus

Since the beginning of its ancient cultivation, the olive tree was typically grown together with other arable and tree crops and grazing animals (Sestini, 1963; Lelle and Gold, 1994; Vannucci, 2009; Eichhorn *et al.*, 2006). Columella (1st century BC), in his *De re rustica*, advised sowing wheat every two years in combination with olive trees to improve olive productivity. These polycultures, also known as agroforestry systems, made it possible to satisfy more of the farmer's needs by producing diverse goods and greater overall yield on the same land. In the last century, olive polycultures, like other fruit-tree-based polycultures, were drastically reduced (Eichhorn *et al.*, 2006; Mosquera-Losada *et al.*, 2012; Dupraz *et al.*, 2018), giving way to specialized cultivation, which allowed mechanization and economy of scale, thus increasing the efficiency of the increasingly scarce and expensive labor. Today, while olive is the most extensively cultivated tree crop in agroforestry systems (Paris *et al.*, 2019; Lauri *et al.*, 2019), most olive agrosystems are managed only for olive (table or oil) production. Olive agroforestry is practiced only on about 8% of the olive area in Europe (data from the LUCAS database, further elaborated from Lauri *et al.*, 2019) and probably to a larger extent in the southern Mediterranean. Additionally, in Europe, the majority of olive agroforestry (about 66%) is silvopastoral systems where livestock graze the olive agrosystem. The rest of the area is woody polycultures (i.e. olive tree-other tree; 25%), arable polycultures (olive tree-arable crop; 6%) and vegetable crop polycultures (olive tree-vegetable crop; 2.6%). Further statistics for individual EU olive-producing countries are reported in Table 9.2.1.

The specialization of olive cultivation brought about an oversimplification of the system, often reducing it to a monoculture with continuous tillage and weed control. This has led to drawbacks such as soil structural degradation and erosion (Abid and Lal, 2009), especially in olive agrosystems cultivated on slopes (Martínez *et al.*, 2006; Fleskens and Stroosnijder, 2007; Romero *et al.*, 2007), which are the majority, and a drastic reduction in the biodiversity of the agrosystem. In these intensive systems, soil erosion reduces soil fertility, taking away the soil's upper and more fertile layer. Fertility is further reduced by a decrease in soil organic matter due to the continuous tillage and the manure replacement with chemical fertilizers (Palese *et al.*, 2014 and references therein). Consequently, fertilizers and other

DOI: 10.1079/9781789247350.0027

chemicals distributed in the olive agrosystem are easily lost with precipitations both by leaching and run-off and by soil erosion, to the detriment of both the olive agrosystem and the environment, with water pollution and eutrophication. Greenhouse gases are emitted from the production, transportation, and distribution of fertilizers and other chemicals, and from soil tillage with machinery (fuel consumption). Decreasing soil organic matter further contributes to the release of carbon into the atmosphere (Vignozzi *et al.*, 2019 and references therein) and therefore to global warming.

To stem or, better, to reverse this trend, living covers (i.e. green mulching), whether temporary or permanent, are increasingly grown in olive agrosystems (Lauri *et al.*, 2019). Living covers protect the soil structure, enhance water infiltration and reduce erosion, increase (or reduce the decline of) soil organic matter and therefore enhance soil fertility, biological activity and carbon sequestration in olive agrosystems (Hernández *et al.*, 2005, Zuazo *et al.*, 2009; Gómez *et al.*, 2009; Ramos *et al.* 2011; Gucci *et al.*, 2012; Palese *et al.*, 2014; Rodrigues *et al.*, 2015; Correia *et al.*, 2015; Repullo-Ruibérriz

Table 9.2.1. Total olive area and proportions of different olive polycultural (agroforestry) systems, per olive-producing country in the EU. Data from LUCAS 2015 (2012 survey). Total olive area from Faostat 2012 (i.e. the same year as the LUCAS survey).

Country	Total olive area (1000 ha)	Olive monoculture %	Sylvopastoral systems %	Woody polycultures %	Silvoarable (row crops) %	Silvoarable (vegetab. crops) %
EU-28	4838	91.90	5.36	2.02	0.50	0.21
ESP	2504	94.53	3.88	1.49	0.1	0
ITA	1125	89.17	4.95	3.78	1.43	0.67
GRC	808	93.69	4.97	1.34	0	0
PRT	347	77.54	18.68	1.65	1.42	0.71
CYP	18.8	93.62	0	2.13	4.25	0
HRV	18.1	92.60	3.70	3.7	0	0
FRA	17.1	91.18	5.88	2.94	0	0
SVN	0.89	100	0	0	0	0
MLT	0.03	100	0	0	0	0

CYP, Cyprus; ESP, Spain; FRA, France; GRC, Greece; HRV, Croatia; ITA, Italy; MLT, Malta; PRT, Portugal; SVN, Slovenia

Fig. 9.2.1. Oversimplified olive agrosystems in traditional (a) (Photo: Pierre-Eric Lauri) and super-high-density or hedgerow (b) (Photo: Adolfo Rosati) systems. Continuous tillage and weed control lead to soil erosion (expecially on slopes), loss of soil organic matter and fertility, nutrient leaching and loss of biodiversity.

de Torres *et al.*, 2018; Vignozzi *et al.*, 2019). This also improves water and nutrient retention and pollutant degradation. Living covers protect the soil surface from aggregate disruption due to precipitation, preventing crusting and erosion, and reduce soil compaction due to machinery (Pardini *et al.*, 2002; Gucci *et al.*, 2012). Living covers also improve field accessibility and increase agrobiodiversity, with potential additional advantages in terms of pest control (Pardini *et al.*, 2002; Pantera *et al.*, 2018), but further studies are needed to better quantify these aspects. Using legume species is particularly useful for fixing atmospheric nitrogen and, therefore, fertilizing the olive agrosystem. Indeed, this is an ancient practice that was gradually abandoned after the introduction of synthetic fertilizers in the last century (Smil, 2001), but which becomes attractive again in light of the increasing economic and environmental costs of these fertilizers. In the short term, living covers can compete with trees for water and nutrients, especially with young trees, but in the long run, this is usually not the case (Gómez, 2005; Gucci *et al.*, 2012; Vignozzi *et al.*, 2019). Instead, residues from living covers and other organic materials (e.g. pruning materials) left on the ground as mulch can increase water storage (Palese *et al.*, 2014).

While the beneficial effects of living covers, in terms of soil protection and environmental benefits, are well established, the economic benefits for the farmer are less obvious (Gomez *et al.*, 2018; Kavvadias and Koubouris, 2019). They might be substantial in the long term (i.e. enhancing soil capital), but in the short term, soil covers might entail greater costs than economic benefits. This might explain why soil covers are still not always adopted; about 50% of olive growers still use regular tillage (Sastre *et al.*, 2016; Gómez *et al.*, 2018). However, living covers are increasingly adopted, especially when rainfall is sufficiently high and/or irrigation is available. If the olive agrosystem is managed with living covers, why not exploit them as additional crops with an economic perspective, instead of interpreting them only as a necessary cost? Ideal living cover species combine environmental services with economic profitability, producing additional income (Rosati, 2011; Rosati *et al.*, 2009).

9.2.2 Olive Agrosystems Need a Boost to Their Profitability

Olive growing, like most agriculture in developed countries, is rarely profitable without subsidies. About 70% of Mediterranean olive agrosystems are traditional and low-density (Taguas *et al.*, 2010; Kavvadias and Koubouris, 2019). These agrosystems are increasingly abandoned because of their low productivity (Stroosnijder *et al.*, 2008; Duarte *et al.*, 2008). For instance, in Italy, with the decoupling of European subsidies from the production in the current Common Agricultural Policy (CAP), many olive growers find it convenient to pocket the subsidies and leave the olive trees uncultivated (albeit complying with the conditionality rule in the CAP) (Rosati *et al.*, 2009). This leads to increased risk of abandonment, especially of marginal and less productive olive agrosystems, with environmental and socio-economic impacts, given the extensive olive cultivation area in Italy. Nor can the farmer remove the olive trees and plant a more lucrative crop, as olive trees are protected in Italy.

Olive trees are an iconic element of the landscape, and their landscape value often exceeds the productive one. An example of this value is the recognition, by the Italian Ministry for Agriculture (MiPAAF), of the Assisi Spoleto olive belt as a national historical landscape, now included also in the FAO GIAHS (Globally Important Agricultural Heritage Systems) program and nominated for UNESCO as a candidate World Heritage Site. These awards demonstrate the historical-cultural importance of olive landscapes, representing a valuable opportunity for the economic and social development of the area, if well managed.

However, most often, the beneficiaries of this development are not the farmers who maintained the olive landscape but now tend to abandon the unprofitable cultivation.

Subsidies, whether for agriculture in general or specific for landscape maintenance, can contribute to preserving these landscapes and territories, but not entirely, and in any case at a great cost to society. Finding new opportunities for additional income or new ways to reduce costs would be a win-win solution to maintain these olive agrosystems' economic viability for the owners. The current CAP aims to increase the sustainability of agriculture, rewarding (with subsidies) those who reduce the environmental impact of crops. The main goal of the CAP, but also of agriculture worldwide, is to produce more with fewer resources, therefore in a more sustainable way. In the olive agrosystem, this can be achieved with intercrops and grazing, when there is complementarity in the use of natural resources. Not all tree/crop/livestock combinations are possible or convenient, including some of the polycultures used in the past. For instance, it is probably not feasible to go back to growing wheat in specialized olive systems: the competition between the two crops, especially for light, is likely to be high with the planting layouts of most modern olive systems and there would be not enough space for the combine harvester. However, as mentioned above, most Mediterranean olive agrosystems are still traditional and low-density, in many cases with sufficient distance to consider growing wheat (Panozzo *et al.*, 2019, 2020), as well as other crops such as barley, maize, faba bean and chickpea (Amassaghrou *et al.*, 2021; Pantera *et al.*, 2018; Mantzanas *et al.*, 2015). Therefore, both traditional and innovative polycultures must be carefully studied and adapted to each situation, considering the current social, environmental and market contexts.

9.2.3 Principles for Modern Olive Polycultures

Modern olive polycultures should be resource-efficient, climate-smart, and multifunctional, i.e. they should deliver multiple ecosystem services simultaneously. In Europe and around the Mediterranean, there is an increasing societal demand for higher quality-driven, environmentally friendly and biodiversity-based agricultural production, especially in high-value products like olive oil. Multiple ecosystem services, like habitat creation for insects and rare plant species, soil protection against erosion, water-quality preservation, carbon sequestration and climate regulation are increasingly demanded and strongly depend on how biodiversity is managed and preserved within agrosystems. Additionally, multifunctionality is also a new way of considering the production process within the agrosystem. Several ecosystem services like mineral recycling, soil-health preservation, biological regulation of pests and weeds etc. support agricultural yields and may even help reduce/suppress the use of fertilizers and pesticides. Contrary to specialized olive systems that are economically viable but ecologically fragile, modern olive polycultures should take advantage of a higher level of biodiversity and ecosystem services to ensure both their environmental sustainability and profitability.

Increasing biodiversity is considered a crucial lever towards improved ecological functioning within agrosystems. Based on ecological theory and evidence, 'biodiversifying' olive agrosystems into polycultures should:

(i) improve the resource-use efficiency through the right association of plants that are complementary in resource-use strategies in space and time.

For instance, growing herbaceous crops under trees promotes deeper tree roots, favoring root segregation in the soil and better exploitation of water and nutrients (Cardinael *et al.*, 2015). Another example is the cultivation of winter crops under deciduous trees, favoring complementarity in light use during the year (Dufour *et al.*, 2013). However, the burst of deciduous-tree vegetation in spring usually overlaps with the end of the understory crop cycles, causing a decrease in crop performance due to light limitations (Friday and Fownes, 2002; Jose *et al.*, 2004; Zamora *et al.*, 2009). Olive trees are evergreen; thus, the shade is significant year-round, potentially limiting understory crops more than deciduous trees. However, Villalobos *et al.* (2006) showed that the highest oil yield is obtained when olive trees intercept about 55% of the incident photosynthetically active radiation (PAR). With higher tree densities, oil yield decreases, despite increased light interception, because both fruit set and oil content decrease in the most shaded parts of the canopy and the tree invests more in upper-canopy vegetative growth in search of light, reducing the harvest index. Therefore, about 45% of the PAR is not used and is available for understory crops, suitable for partial shade. In traditional (i.e. low-density) olive agrosystems, the available light is much more significant, and even non-shade adapted crops can be considered.

(ii) promote positive interactions among plants, especially under severe environmental conditions.

Under a global warming scenario, moderate shade, such as that under most olive agrosystems, could reduce crop yield losses by buffering the negative impact of the increasingly frequent early droughts and high spring temperatures that make excessive light detrimental to the stressed crop (Arenas-Corraliza *et al.*, 2018; Pitchers *et al.*, 2021). Indeed, while light reduction due to tree shade is the most frequent limiting factor for understory crops, tree shade can also improve crop microclimate (Lott *et al.*, 2009; Mugunga *et al.*, 2017), buffering temperature extremes (Peng *et al.*, 2015), increasing air humidity (Campi *et al.*, 2009), and reducing evapotranspiration (Coussement *et al.*, 2018) under more arid conditions. These concepts are formalized in the Stress Gradient Hypothesis (SGH), which predicts that negative interactions (competition) between species shift towards positive interactions (facilitation) with increasing abiotic stress (He *et al.*, 2013).

However, Temani *et al.* (2021) found that olive shade always reduced understory wheat yield, independent of the water deficit level. This might be because unlike deciduous trees, the evergreen olive trees provide significant shade also in winter before drought limits the performance of the understory crop. Wheat yield is strongly dependent on the biomass accumulated before flowering. A strong limitation in biomass accumulation before the onset of drought (i.e. before flowering) will reduce biomass at flowering, thus reducing yield potential independent of possible benefits of shade during the subsequent drought period. This interpretation is supported by the fact that growing wheat under olive trees reduces tillering, spike number, and grains per spike but increases individual grain weight, compared to open-field control (Temani *et al.* 2021; Panozzo *et al.* 2020). This suggests that shade reduces wheat yield potential proportionally more during the early stages before bloom (when tillering, spike number and grains per spike are affected) than after bloom (grain filling stage) when the grain size is affected. This might be because in winter/early spring, light is low and thus always limiting, while drought and high temperature are usually not limiting factors. Under these conditions, any shade reduces yield potential, while in late spring/summer, light is abundant and potentially excessive under drought, making partial shade less limiting and potentially beneficial. Perhaps, therefore, the SGH could prove more effective than found for the olive-wheat polyculture with summer understory crops that spend their whole cycle under drought and high light levels. However, unless irrigation is available, summer crops are less suitable as understory crops, due to competition for water. Further studies are needed to clarify these aspects.

Decreased yields under tree shade can be only partially compensated by increased crop or forage quality, such as increased nitrogen and protein content, due to both a dilution effect (greater nutrient availability per unit of production, Li *et al.*, 2010; Artru *et al.*, 2017; Arenas-Corraliza *et al.*, 2018) or to an extended period of crop growth under the shade, for instance, extended grain filling in cereals (Inurreta-Aguirre *et al.*, 2018; Temani *et al.* 2021).

(iii) strengthen the resilience through the 'portfolio effect' created by different plant sensitivities to adverse climate and environmental fluctuations.

In Mediterranean countries, the interannual variability in climatic conditions is increasing with an increased risk of early or severe drought (Tramblay *et al.*, 2020). Diversified agrosystems are more resilient to climate change or any other kind of environmental change, stabilize overall yields and may act as a portfolio effect (Renard and Tilman, 2019). Globalized markets may fluctuate in unpredictable manners. Different crops in polycultures have different sensitivities to environmental factors and produce a range of products on the same piece of land. Therefore, olive polycultures should promote a more stable production of goods and ecosystem services and a more stable income for the growers.

However, modern olive polycultures require that the crops/livestock cultivated/reared in the olive agrosystem are compatible with modern olive cultivation techniques (Rosati *et al.*, 2009; Rosati, 2011). This compatibility must be both from a technical point of view (possibility of mechanization, absence of mutual obstruction between crops and prevention of animal damage to plants etc.) and from an agroecological perspective. To maximize complementarity in resource use and the efficacy of positive interactions, the agrosystem should be carefully designed. The spatial design, density, and olive tree training may be adapted to leave enough light to understory crops and enable safe and comfortable maneuvers of agricultural machinery, especially in annual arable crops that require the passage of a seeder, weeder, harvester etc. Additionally, the distance between sown inter-rows and tree rows should be adjusted to minimize negative below-ground interactions between crops and trees, reduce the negative impacts of machinery used for crops on tree roots and branches, and keep the trees accessible for technical operations during the crop cycle. Furthermore increasing distance between tree rows improves ventilation and illumination, and is likely to improve tree health, particularly reducing the incidence of olive leaf spot or peacock eye disease (caused by *Spilocaea oleagina*). The possible loss in olive yield can be compensated by the increased yield of the intercrop.

When possible, the calendar and olive harvest/pruning should be carefully determined according to the other crops, or *vice versa* crops should be chosen which do not impede necessary operations for the trees, to avoid work conflicts in the field and avoid critical damage to plants (e.g. the sowing or emergence of winter crops usually takes place in the same period as olive harvest/pruning). In some instances, the traditional olive planting design and training system should be revisited to optimize the polyculture. Optimizing criteria should be defined according to the objectives targeted in terms of multifunctionality.

Understory crops and livestock must have a market that makes their cultivation and sale economically viable, or they must lead to economically viable savings in the olive agrosystem management. It should also be considered that olive farmers often have very small farms. In Italy, for instance, there are over one million olive farmers with an average farm size of about one hectare, implying that many farms are much smaller than one hectare. Therefore, a good deal of the olive cultivation is for self-consumption. Under these circumstances, increasing the number of crops and products obtained from the olive agrosystem can be desirable since direct consumption does not require the bureaucracy necessary for selling. Additionally, with self-consumption, the products are effectively valued at retail prices (replacing what the farmers would have had to buy as a consumer), while when products are sold on the market, the farmer usually gets wholesale price. Additionally, in Europe there is an increasing number of agritourism farms where products are offered directly to visitors, often after processing or cooking, offering the possibility to add value. Increasing the number of products obtained from the olive agrosystem can attract more customers interested in local production from the agritourism farm and improve the farm's economic performance and resiliency.

9.2.4 Olive Polycultures in Practice

9.2.4.1 Combining olive trees with herbaceous intercrops

In olive agrosystems managed with soil tillage, annual crops can be intercropped. N-fixing legumes could be preferred for the biological nitrogen fixation which ultimately results in nitrogen fertilization to the trees. Winter crops (autumn-spring cycle) such as broad beans, chickpeas, peas, vetches etc. are preferable to avoid water and nutrient competition in the summer. Even in these cases, however, mechanical harvesting can be a problem and competition, especially for water, can be significant as the crop cycle might end too late in the spring to avoid it: the later the crop cycle, the stronger the competition (Gómez, 2005). However, some crops may consume no more water or nutrients than natural living covers. Competition can be reduced with shorter-cycle winter crops, such as many vegetables, including turnips, broccoli raab, cabbage, cauliflower, broccoli, lettuce, chard, radishes, spinach, and fababeans and peas for fresh consumption (which are harvested earlier than dry grains). Perennial crops with low water consumption might also be considered, having the advantage of needing no soil tilling and sowing annually, thus reducing both soil erosion (Vallebona *et al.*, 2016) and environmental impact (reduced fuel consumption for tillage). A case recently studied is a wild asparagus species (*Asparagus acutifolius*), a xerophitic plant producing tender spears which have been consumed and sold from time immemorial in the Mediterranean (Venezia *et al.*, 1993; Aliotta *et al.*, 2004). This plant is currently not cultivated, but recent work shows that it can be successfully cultivated (Benincasa

Fig. 9.2.2. Wild asparagus (*Asparagus acutifolius*) cultivated along the tree row in a super-high-density (hedgerow) olive system. (Photo: Adolfo Rosati)

Fig. 9.2.3. Daffodils (*Narcissus* spp.) cultivated in the alley of a super-high-density (hedgerow) olive system. (Photo: Adolfo Rosati)

et al., 2007; Rosati and Falavigna, 2000; Rosati *et al.*, 2005; Rosati, 2001; Mantovani *et al.*, 2019). It grows easily, and can be cultivated, as an understory crop in olive agrosystems (Rosati *et al.*, 2009, 2012a, b, c; Rosati, 2011, 2014; Mantovani *et al.*, 2016).

Another example is artichoke, a crop with perennial roots but with fall-spring vegetation which dies back during the dry season, thus not competing for water during summer. Bulb crops such as daffodils (*Narcissus* spp.), tulips (*Tulipa* spp.) and others may also be good understory crops, because of their short, winter-spring cycle, which does not compete for water in the summer and does not impair olive harvest in the autumn. Their short cycle and the fact that they are not food crops, reduce problems related to the drift of chemicals sprayed on the trees.

A simpler alternative, not involving sowing or voluntary cultivation, is the harvest of the various edible species that grow spontaneously in the olive agrosystem, especially when the soil is tilled. Many of these species were traditionally consumed by the farmer's family and even marketed and could now be sold in niche markets as traditional products, for instance to restaurants, or consumed directly in agritourism farms (of which there are over 23,000 in Italy alone). Informing consumers on the beneficial health properties of these wild vegetables (Salvatore *et al.*, 2005; Simopoulos, 2004; Vardavas *et al.*, 2005) would help promote the product as is the case for the olive oil (Roselli *et al.*, 2017). Many edible species grow spontaneously in olive agrosystems (Rosati, 2011), including the spicy rocket (*Diplotaxis* spp.), the corn rocket (*Bunias erucago*), the sweet sow thistle (*Sonchus* spp.), the bitter chicory (*Cichorium intybus*), the precious rampion (*Campanula rapunculus*), the healthy purslane (*Portulaca oleracea*), the crunchy 'strigoli' (*Silene vulgaris*) and many others. Medicinal species, naturally occurring or sown, also have the potential for crop diversification and additional income.

9.2.4.2 Feed potential of the olive agrosystem

An even simpler and immediate alternative is the use of understory vegetation, natural or cultivated, as forage. This reduces the competition with trees due to earlier harvest, as forage plants can be harvested at or before flowering instead of at seed maturation. In fact, forage harvesting date can be flexible and calibrated according to the seasonal weather trend, anticipating it in case of early drought, to reduce the competition for water with olive trees. Similarly, in a year of low olive productivity, more intense olive pruning and later harvest of forage can increase forage production, providing means for increased resilience. Pardini *et al.* (2002) reviewed the work that has been carried out on choices of species and cultivars of cultivated living covers most suitable for olive agrosystems (and vineyards) and their best management, concluding that much work remains to be done. A particular case of forage potential is melliferous plants, which could entail an additional ecosystem service for pollination and an important economic outcome if exploited for honey production. Melliferous plants include many culinary herbs, which have an additional value in the herb market.

In addition to the forage potential of the natural or cultivated living cover, herbivores (particularly sheep, goats, cattle, rabbits) also crave olive leaves and twigs obtained from pruning the trees, or by grazing on suckers and lower-canopy vegetation. Some studies report the nutritive value of dried olive leaves (Delgado-Pertíñez *et al.*, 2000 and references therein), and the composition of fresh leaves (Cavalheiro *et al.*, 2015), but *in vivo* studies on the nutritive and digestibility value of fresh leaves with different livestock species and breeds would be desirable. Olive leaves contain high levels of polyphenols (particularly oleuropein), which do not appear to be toxic to rumen microflora (Delgado-Pertíñez *et al.*, 1998). Instead, they can have health-promoting properties, as demonstrated for broilers (Varmaghany *et al.*, 2013), and enhance meat and milk quality (Molina-Alcaide and Yáñez-Ruiz, 2008). They might even reduce rumen methanogenesis as it has been found for tannic acids (Yang *et al.*, 2017). Copper (or other elements or chemicals) residues from pest control contaminating olive leaves can be a concern (Molina-Alcaide and Yáñez-Ruiz, 2008), and appropriate management to avoid feeding contaminated leaves must be considered.

The exhausted pomace residue from olive processing can also be used as feed. This use has a higher value than bioenergy, composting, or direct field application and, although it has a lower value than extracting high-value molecules or products for food or pharmaceutical use (Berbel and Posadillo, 2018), it is more widely applicable. Using olive pomace as feed appears to bring no detriment to the animals and their performance as found, for example on sheep (e.g. Taheri *et al.*, 2012) and cattle (Estaún *et al.*, 2014) and instead benefits their health (Molina-Alcaide and Yáñez-Ruiz, 2008). It also benefits the consumer's health due to increased unsaturation in the fatty acid profile of the livestock products (Habeeb *et al.*, 2017). Some negative effects have been found in non-ruminant herbivores such as rabbits (Dal Bosco *et al.*, 2012), although other authors reported that negative effects occurred only when olive pomace replaced > 20–25% of the rabbit diet (Mousa, 2012). Similarly, no negative effects were observed in broilers with olive pulp up to 7.5% of the diet, while 10% reduced intake and growth (Rabayaa *et al.*, 2001). The feed use of olive pomace is facilitated by the increasingly common practice of removing the olive pits (woody endocarps), which are utilized as a valuable fuel (Mediavilla *et al.*, 2020). Olive agrosystems, therefore, considering the living cover, the pruning material and the pomace, produce several tons of fresh forage, which is currently mostly unused (Lauri *et al.*, 2019) and often represents a residue to be disposed of at high cost, i.e. by mowing the cover, chipping the pruning residues and disposing of the cake.

The different forage productions of the olive agrosystem are distributed over time, creating valuable opportunities for the distribution of forage throughout the year. Olive pruning material is, potentially, available over a long period from autumn to spring, providing the opportunity to feed livestock with fresh forage for a long time in a period when green forage is scarce. This practice is also advantageous considering that olive leaves appear to decrease rapidly in digestibility and forage quality with drying and storage (Delgado-Pertíñez *et al.*, 2000 and references therein), aside from the fact that drying and storing would probably not be economically viable. Sucker and watersprout removal can provide some forage during summer, when again there is little green forage. The de-pitted pomace is available in the autumn but can be preserved fresh in silos or dehydrated. The living cover can be available for grazing almost all year round, with peaks in spring and autumn, or whenever it rains sufficiently.

The forages and feeds produced in the olive agrosystem can be collected and brought to the animals, avoiding livestock damage to the olive trees. Alternatively, animals can graze directly in the olive agrosystem, saving the costs of haymaking, transport and storage and solving the problem of hindered mechanization in the olive agrosystem, which may not allow hay to be made easily.

9.2.4.3 Grazing in the olive agrosystems

Grazing animals must be compatible with olive cultivation: they must not damage trees. Sheep can be compatible if the trees are pruned high or if there is enough ground forage, which is usually preferred to olive leaves. Some sheep breeds are more suitable due to the tendency to prefer ground forage to olive trees. Breeds with smaller/shorter body size are preferable to taller ones. There are also experiences of livestock 'training' where animals are induced not to consume the crop to be protected (Meadows, 2008). Feeding the animals with olive pruning materials before introducing

Fig. 9.2.4. Sheep can be used to weed and fertilize the olive agrosystem and, as herbivores, they make good use of weeds, reducing feed cost. (Photo: Adolfo Rosati)

Fig. 9.2.5. A bale of fresh pruning materials removed from the field and ready to be used fresh or dried after storage, for different possible uses. (Photo: Adolfo Rosati)

them into the olive field also helps prevent damage, as once the animals have consumed abundant olive vegetation, they prefer other forages for the rest of the day.

Goats and donkeys must be avoided as they prefer the olive tree to ground forage and may debark and otherwise damage the trees causing devastating damage, especially on young trees. Cows are too tall and devour olive vegetation. Horses can be compatible if there is enough grass, which they prefer over olive vegetation, but large animals in general tend to compact the soil and must be allowed to graze for short intervals when the soil is dry, and removed as soon as the ground vegetation has been sufficiently consumed. In fact, this is a general principle that applies to all grazing animals. Smaller animals are generally safer in terms of potential damage to the trees and soil compaction. Geese are excellent grazers (Geiger and Biellier, 2006) and do no harm to olive trees because of their small size and because they do not eat olive leaves. Their palmate feet reduce soil compaction, and the fact that they do not scratch the soil means that

Fig. 9.2.6. Chickens are effective weeders in the olive agrosystem. However, excessive grazing intensity and duration (a) may lead to bare and compacted soil, exposing it to erosion, excessive fertilization and nutrient leaching. A more appropriate grazing level (b), Photo: Alessandro Dal Bosco, University of Perugia), compared to the previous figure, prevents soil damage and provides more feed opportunities for the birds. The animals benefit from tree shade (notice all the chickens under the tree shade in (a)).

weeds are mowed but not destroyed. This can be a benefit or a drawback depending on the farmer's need to leave the living cover or destroy it in the summer. With time, however, weed species not consumed by geese take over, if not destroyed by other animals and by an occasional tillage. Chickens and other birds (e.g. turkeys, guinea hens) are very useful natural herbicides: even though they do not entirely consume the forage (as they are mainly granivorous), they destroy it by scratching. For this reason, however, their action can lead to soil compaction and complete disappearance of any vegetation, especially near the housing and feeding areas. Again, rotational grazing is essential (Dal Bosco *et al.*, 2014).

All grazing animals fertilize while weeding, spreading manure at the same time, without the need for machines. Fertilization and weeding are the two practices with the greatest environmental impact on olive cultivation (Paolotti *et al.*, 2016; Rocchi *et al.*, 2019). In previous articles, we have reported free-range chicken grazing experiences in the olive agrosystem (Rosati *et al.*, 2009, 2012a, b, c, 2014), showing how chickens, by fertilizing and weeding naturally, dramatically reduce the environmental impact of olive cultivation. Similarly, using olive agrosystems for grazing reduces the environmental impact of livestock rearing (Paolotti *et al.*, 2016; Rocchi *et al.*, 2019), in addition to providing other benefits in terms of protection from predators and meteorological events (rain, wind, sun, extreme temperatures), improving animal welfare and meat quality (Dal Bosco *et al.*, 2014, 2016; Rosati *et al.*, 2015).

Livestock can also feed on fallen fruits before (natural fruit drop) or after harvest (leftover fruits), thus also eliminating olive fly larvae that might infest fallen fruits (Castro *et al.*, 2019), in addition to turning the wasted fruits into valuable feed. Chickens can also contribute additional pest control, for instance against the olive weevil (*Otiorhynchus cribricollis*) and olive fly (*Bactrocera oleae*). Perhaps chickens could also contribute to controlling spittlebugs, including *Philaenus spumarius*, the main vector of *Xylella fastidiosa*, the bacterium causing the olive quick decline syndrome (OQDS), but no data are currently available on this potentially relevant topic. Finally, chickens or other birds can fertilize and weed not only the olive trees, but also understory crops, for example the wild asparagus or artichoke. In fact, as the mature wild asparagus plant is prickly and rigid, and chickens do not damage it, while they destroy the weeds, fertilizing and weeding both the olive trees and the asparagus crop. Similarly, the mature artichoke plant is too tall and bitter, and chickens graze underneath it, weeding it effectively without significant damage. In addition to positive benefits in controlling olive pests, chickens, like ducks, are known to consume and thus control the asparagus beetles, *Crioceris* spp. (Chittenden, 1917), which could otherwise cause damage to the wild asparagus.

Further unexplored possibilities to increase the overall productivity of olive agrosystems include intercropping with grain/legume crops that are earlier-maturing and with low vegetative development. Both these characteristics would reduce the competition for water and nutrients with the olive trees. The grain could be 'harvested' directly by grazing chickens (or other livestock species). This would reduce the costs of growing grains for the livestock rearing, leaving crop residues to act as fertilizer and mulch in the olive agrosystem. This would be especially beneficial if the crop was a N-fixing legume because of the N fixation and because legumes are richer than cereals in proteins, the costliest part of feeds, often imported from distant countries (soybean).

Rethinking olive agrosystems (and the olive oil industry) to consider their feed potential is probably one of the most promising avenues to sustainably increase and exploit their potential productivity, as already recommended by the FAO (Sansoucy *et al.*, 1985). Combining olive growing with livestock rearing, whether on the same farm or in separate farms, would significantly increase overall productivity while reducing costs and environmental impact in both the olive and the livestock sectors, promoting a circular economy (Berbel and Posadillo, 2018).

9.2.4.4 Use of olive byproducts to improve soil fertility

When livestock integration in the olive industry is not possible, olive byproducts (i.e. living cover, pruning material and pomace) can at least be returned to the soil, decreasing erosion and increasing soil fertility, organic matter and carbon sequestration (Kay and VandenBygaart, 2002; Xiloyannis *et al.*, 2008; Calatrava and Franco 2011; Montanaro *et al.*, 2010; Repullo *et al.*, 2012; Palese *et al.*, 2014). Particularly, lignin-rich pruning materials favor the humification of otherwise quickly decomposable materials such as olive leaves and living mulch clippings. The mulch provided by these materials increases earthworm activity in undisturbed soils (Vršič, 2011; Pérès *et al.*, 2010) and thus improves soil structure and infiltration (Lavelle, 1988). However, using these same materials as forage, before returning them to the soil in the form of manure, allows the same benefits, while also extracting more value from them, increasing their cascading value. Currently, the feed use of these materials is very limited; for instance, in Andalusia, the main olive-producing region worldwide, 80% of the olive byproducts are used for energy production while only 5% are used as feed (Berbel and Posadillo, 2018), and these numbers do not account for the mostly unexploited feed potential of the olive living covers.

Using these materials as forages, biofuels, or any other productive use, before returning them to the soil, increases productivity per unit of land, with fewer resources, while decreasing the management costs of the olive agrosystem.

9.2.5 Future Perspectives

For millennia, intercropping and grazing in the olive agrosystem, as well as in other tree-based systems, was a common (and sustainable) practice due to the many environmental and productive advantages that it entails. Some disadvantages of the old polycultures, including hindered mechanization, low labor efficiency, and incompatibility with modern techniques such as chemical pest and disease control (chemicals needed/registered for olive might not be needed/registered for the intercrops) have almost led to the disappearance of olive-based polycultures, especially in more developed countries. However, the need for more sustainable and profitable agricultural models, and the changed socio-economic conditions favoring more sustainable products and shorter supply chains, including direct sales, are creating the conditions to reconsider polycultures, albeit in a modern way. Climate change is also forcing us to reconsider polycultures. For instance, in the Mediterranean, the population is growing rapidly, and food demand (Bodirsky *et al.*, 2015) makes food security a concern (Smith and Glauber, 2019). Climate change is predicted to increase extreme drought and high temperatures (Ahmed *et al.*, 2013), adding further uncertainty to food security, especially in the Mediterranean (Iglesias *et al.*, 2011). The olive is the most cultivated tree crop in the Mediterranean, with about 11 million hectares (FAO, 2020). Unlike in past times, most olive agrosystems currently have an unexplored and unexploited productive potential in terms of forage and other intercrops/livestock, as well as in buffering the impact of climate change/climatic extremes on intercrop performance. Given the extensive olive cultivation in the Mediterranean and worldwide, exploiting this potential is an important opportunity to produce more with fewer resources (therefore more sustainably), thus contributing to farm resilience and food security. However, the most appropriate polyculture design and techniques must be developed to find species combinations suitable to modern needs and socio-economic contexts from both an environmental and an economic perspective. Although still scarce, modern research is starting to address these issues, providing useful indications. However, many research gaps remain open. For instance, intercrops performance under trees can be improved by selecting suitable varieties adapted to partial shade (Arenas-Corraliza *et al.*, 2018; Panozzo *et al.*, 2020), while current varieties have been selected under full sun and may not exploit the shaded environment at its best. For instance, it has been shown that cultivars of gamagrass (*Tripsacum dactyloides*) with lower day respiration rates perform better as understory crops (Rosati *et al.*, 2018). Similarly, crop choice should prioritize shade-tolerant species able to outyield crops currently employed because of their greater productivity in full sun.

Providing additional income in the olive agrosystem becomes essential to maintain these valuable landscapes, otherwise not economically sustainable. However, contrary to popular belief, integrating intercrops and/or livestock in olive agrosystems is not the prerogative of small family-run farms or marginal olive systems with a landscape value but otherwise non-productive. Sometimes, as with livestock rearing, it is necessary to reach a certain farm size to make the economies of scale necessary for a profitable business. It is precisely the largest farms that have the best conditions to achieve sufficient economy of scale for each of the various crops (or livestock species) of the polyculture and that have the necessary skills and investment energies to develop innovative techniques, products, and markets.

The European Commission, intent on making future agriculture more sustainable, is investing heavily in this direction, with subsidies to both farmers and research. The olive tree, an emblematic plant of the Mediterranean, has been at the center of numerous research projects on these issues, for instance with the AGFORWARD project (2014–2017, VII Framework Program of the EU) and the BIODIVERSIFY project, funded by the Partnership for Research and Innovation in the Mediterranean Area (PRIMA 2019). Research results, including those reported here and future ones, will stimulate the adoption of both old and innovative polycultures, including olive polycultures.

References

Abid, M. and Lal, R. (2009) Tillage and drainage impact on soil quality. II. Tensile strength of aggregates, moisture retention and water infiltration. *Soil and Tillage Research* 103(2), 364–372.

Ahmed, M., Asif, M., Sajad, M., Khattak, J.Z.K., Ijaz, W. *et al.* (2013) Could agricultural system be adapted to climate change? A review. *Australian Journal of Crop Science* 7, 1642–1653.

Aliotta, G., Aceto, S., Farina, A., Gaudio, L., Rosati, A., Sica, M. and Parente, A. (2004) Natural history, cultivation and biodiversity assessment of Asparagus. *Research Advances in Agriculture and Food Chemistry* 5, 1–12.

Amassaghrou, A., Bouaziz, A., Daoui, K., Belhouchette, H., Ezzahouani, A. and Barkaoui, K. (2021) Productivity and efficiency of olive-based agroforestry systems in Morocco: case of Moulay Driss Zerhoun. *Cahiers Agricultures* 30(2). https://doi.org/10.1051/cagri/2020041

Arenas-Corraliza, M.G., López-Díaz, M.L. and Moreno, G. (2018) Winter cereal production in a Mediterranean silvoarable walnut system in the face of climate change. *Agriculture Ecosystems and Environment* 264, 111–118. https://doi.org/10.1016/j.agee.2018.05.024

Artru, S., Garré, S., Dupraz, C., Hiel, M.P., Blitz-Frayret, C. and Lassois, L. (2017) Impact of spatio-temporal shade dynamics on wheat growth and yield, perspectives for temperate agroforestry. *European Journal of Agronomy* 82, 60–70. https://doi.org/10.1016/j.eja.2016.10.004

Benincasa, P., Tei, F. and Rosati, A. (2007) Plant density and genotype effects on wild asparagus (*Asparagus acutifolius* L.) spear yield and quality. *HortScience* 42(5), 1163–1166.

Berbel, J. and Posadillo, A. (2018) Review and analysis of alternatives for the valorisation of agro-industrial olive oil by-products. *Sustainability* 10(1), 237.

Bodirsky, B.L., Rolinski, S., Biewald, A., Weindl, I., Popp, A. and Lotze-Campen, H. (2015) Global food demand scenarios for the 21st century. *PLOS ONE* 10, 1–27. https://doi.org/10.1371/journal.pone.0139201

Calatrava, J., and Franco, J.A. (2011) Using pruning residues as mulch: analysis of its adoption and process of diffusion in southern Spain olive orchards. *Journal of Environmental Management* 92, 620–629. DOI: 10.1016/j.jenvman.2010.09.023

Campi, P., Palumbo, A.D. and Mastrorilli, M. (2009) Effects of tree windbreak on microclimate and wheat productivity in a Mediterranean environment. *European Journal of Agronomy* 30, 220–227. https://doi.org/10.1016/j.eja.2008.10.004

Cardinael, R., Mao, Z., Prieto, I., Stokes, A., Dupraz, C., Kim, J.H. and Jourdan, C. (2015) Competition with winter crops induces deeper rooting of walnut trees in a Mediterranean alley cropping agroforestry system. *Plant and Soil* 391(1), 219–235.

Castro, M., Rosati, A., Pantera, A., Moreno, G. and Mosquera-Losada, M.R. (2019) Sistemas agroflorestais Mediterrânicos: situação atual e perspectivas futuras. In: Bungenstab, D.J., Almeida, R.G. de, Laura, V.A., Balbino, L.C. and Ferreira, A.D. (eds) *ILPF: inovação com integração de lavoura, pecuária e floresta*. Embrapa, Brasilia, pp. 787–807.

Cavalheiro, C.V., Picoloto, R.S., Cichoski, A.J., Wagner, R., de Menezes, C.R. *et al.* (2015) Olive leaves offer more than phenolic compounds: fatty acids and mineral composition of varieties from southern Brazil. *Industrial Crops and Products* 71, 122–127.

Chittenden, F.H. (1917) The asparagus beetles and their control (No. 837). US Department of Agriculture.

Correia, C.M., Brito, C., Sampaio, A., Dias, A.A., Bacelar, E. *et al.* (2015) Leguminous cover crops improve the profitability and the sustainability of rainfed olive (*Olea europaea* L.) orchards: from soil biology to physiology of yield determination. *Procedia Environmental Sciences* 29, 282–283.

Coussement, T., Maloteau, S., Pardon, P., Artru, S., Ridley, S., Javaux, M. and Garré, S. (2018) A tree-bordered field as a surrogate for agroforestry in temperate regions: Where does the water go? *Agricultural Water Management* 210, 198–207.

Dal Bosco, A., Mourvaki, E., Cardinali, R., Servili, M., Sebastiani, B. *et al.* (2012) Effect of dietary supplementation with olive pomaces on the performance and meat quality of growing rabbits. *Meat Science* 92(4), 783–788.

Dal Bosco, A., Mugnai, C., Rosati, A., Paoletti, A., Caporali, S. and Castellini, C. (2014) Effect of range enrichment on performance, behaviour and forage intake of freerange chickens. *Journal of Applied Poultry Research* 23, 137–145.

Dal Bosco, A., Mugnai, C., Mattioli, S., Rosati, A., Ruggeri, S., Ranucci, D. and Castellini, C. (2016) Transfer of bioactive compounds from pasture to meat in organic free-range chickens. *Poultry Science* 95, 2464–2471.

Delgado-Pertíñez, M., Gómez-Cabrera, A. and Garrido, A. (2000) Predicting the nutritive value of the olive leaf (*Olea europaea*): digestibility and chemical composition and in vitro studies. *Animal Feed Science and Technology* 87(3–4), 187–201.

Delgado- Pertíñez, M., Chesson, A., Provan, G.J., Garrido, A. and Gómez-Cabrera, A. (1998) Effect of different drying systems for the storage of olive leaves on their nutritive value for ruminants. *Annals of Zootechnology* 47, 141–150.

Duarte, F., Jones, N. and Fleskens, L. (2008) Traditional olive orchards on sloping land: sustainability or abandonment? *Journal of Environmental Management* 89, 86–98. https://doi.org/10.1016/j.jenvman. 2007.05.024

Dufour, L., Metay, A., Talbot, G. and Dupraz, C. (2013) Assessing light competition for cereal production in temperate agroforestry systems using experimentation and crop modelling. *Journal of Agronomy and Crop Science* 199, 217–227.

Dupraz, C., Lawson, G.J., Lamersdorf, N., Papanastasis, V.P., Rosati, A. and Ruiz-Mirazo, J. (2018) Temperate agroforestry: the European way. In: Gordon, A., Newman, S. and Coleman, B. (eds) *Temperate Agroforestry Systems*. CAB International, Wallingford, UK.

Eichhorn, M.P., Paris, P., Herzog, F., Incoll, L.D. *et al.* (2006) Silvoarable systems in Europe – past, present and future prospects. *Agroforestry Systems* 67, 29–50.

Estaún, J., Dosil, J., Al Alami, A., Gimeno, A. and De Vega, A. (2014) Effects of including olive cake in the diet on performance and rumen function of beef cattle. *Animal Production Science* 54(10), 1817–1821.

FAO (2020) Faostat, FAO Statistics Division. Available at: http://faostat3.fao.org (accessed 18 September 2020).

Fleskens, L. and Stroosnijder, L. (2007) Is soil erosion in olive groves as bad as often claimed? *Geoderma* 141(3–4), 260–271.

Friday, J.B. and Fownes, J.H. (2002) Competition for light between hedgerows and maize in an alley cropping system in Hawaii, USA. *Agroforestry Systems* 55, 125–137. https://doi.org/10.1023/A:1020598110484

Geiger, G. and Biellier, H. (2006) Weeding with geese. Extension, Department of Animal Sciences, University of Missouri. Available at: https://extension2.missouri.edu/g8922 (accessed 15 December 2017).

Gómez, J.A. (2005) Effects of soil management on soil physical properties and infiltration in olive orchards: implications for yield. *FAO Land and Water Bulletin* 10, 65–70.

Gómez, J.A., Sobrinho, T.A., Giráldez, J.V. and Fereres, E. (2009) Soil management effects on runoff, erosion and soil properties in an olive grove of Southern Spain. *Soil and Tillage Research* 102(1), 5–13.

Gómez, J.A., Campos, M., Guzmán, G., Castillo-Llanque, F., Vanwalleghem, T., Lora, Á. and Giráldez, J.V. (2018) Soil erosion control, plant diversity, and arthropod communities under heterogeneous cover crops in an olive orchard. *Environmental Science and Pollution Research* 25(2), 977–989.

Gucci, R., Caruso, G., Bertolla, C., Urbani, S., Taticchi, A. (2012) Changes of soil properties and tree performance induced by soil management in a high-density olive orchard. *European Journal of Agronomy* 41, 18–27.

Habeeb, A.A.M., Gad, A.E., El-Tarabany, A.A., Mustafa, M.M. and Atta, M.A.A. (2017) Using olive oil by-products in farm animals feeding. *Cellulose* 3(6).

He, Q., Bertness, M.D. and Altieri, A.H. (2013) Global shifts towards positive species interactions with increasing environmental stress. *Ecology Letters* 16, 695–706. https://doi.org/10.1111/ele.12080

Hernández, A.J., Lacasta, C. and Pastor, J. (2005) Effects of different management practices on soil conservation and soil water in a rainfed olive orchard. *Agricultural Water Management* 77(1–3), 232–248.

Iglesias, A., Mougou, R., Moneo, M. and Quiroga, S. (2011) Towards adaptation of agriculture to climate change in the Mediterranean. *Regional Environmental Change* 11, 159–166. https://doi.org/10.1007/s10113-010-0187-4

Inurreta-Aguirre, H.D., Lauri, P.É., Dupraz, C. and Gosme, M. (2018) Yield components and phenology of durum wheat in a Mediterranean alley-cropping system. *Agroforestry Systems* 92, 961–974. https://doi.org/10.1007/s10457-018-0201-2

Jose, S., Gillespie, A.R. and Pallardy, S.G. (2004) Interspecific interactions in temperate agroforestry. *Agroforestry Systems* 61, 237–255.

Kavvadias, V. and Koubouris, G. (2019) Sustainable soil management practices in olive groves. In: Panpatte, D.G. and Jhala, Y.K. (eds) *Soil Fertility Management for Sustainable Development*. Springer, Singapore, pp. 167–188. https://doi.org/10.1007/978-981-13-5904-0_8

Kay, B.D. and VandenBygaart, A.J. (2002) Conservation tillage and depth stratification of porosity and soil organic matter. *Soil and Tillage Research* 66(2), 107–118.

Lauri, P.É., Barkaoui, K., Ater, M. and Rosati, A. (2019) Agroforestry for fruit trees in the temperate Europe and dry Mediterranean. In: Mosquera-Losada, M.R. and Prabhu, R. (eds) *Agroforestry for Sustainable Agriculture*. Burleigh Dodds Science Publishing, Cambridge, UK, pp. 385–418.

Lavelle, P. (1988) Earthworm activities and the soil system. *Biology and Fertility of Soils* 6(3), 237–251.

Lelle, M.A. and Gold, M.A. (1994) Agroforestry systems for temperate climates: lessons from Roman Italy. *Forest & Conservation History* 38, 118–126.

Li, H., Jiang, D., Wollenweber, B., Dai, T. and Cao, W. (2010) Effects of shading on morphology, physiology and grain yield of winter wheat. *European Journal of Agronomy* 33, 267–275. https://doi.org/10.1016/j.eja.2010.07.002

Lott, J.E., Ong, C.K. and Black, C.R. (2009) Agricultural and forest meteorology understorey microclimate and crop performance in a Grevillea robusta-based agroforestry system in semi-arid Kenya. *Agricultural and Forest Meteorology* 149, 1140–1151. https://doi.org/10.1016/j.agrformet.2009.02.002
Mantovani, D., Benincasa, P. and Rosati, A. (2016) Olive (*Olea europaea* L.) and wild asparagus (*Asparagus acutifolius* L.) agroforestry system: asparagus performance and its best positioning in the olive orchard. In: Gosme, M. (ed.) *Proceedings of the 3rd European Agroforestry Conference*. Montpellier, France, pp. 229–231.
Mantovani, D., Rosati, A. and Perrone, D. (2019) Photosynthetic characterization and response to drought and temperature in wild asparagus (*Asparagus acutifolius* L.). *HortScience* 54(6), 1039–1043.
Mantzanas, K., Papanastasis, V., Pantera, A., Papadopoulos, A. and Burgess, A.P. (2015) System report: olive agroforestry in Kassandra, Chalkidiki, Greece. Agforward.
Martínez, J.R.F., Zuazo, V.H.D. and Raya, A.M. (2006) Environmental impact from mountainous olive orchards under different soil-management systems (SE Spain). *Science of the Total Environment* 358(1–3), 46–60.
Meadows, R. (2008) Trained ovines chomp on weeds, avoid vines. *California Agriculture* 62(1).
Mediavilla, I., Barro, R., Borjabad, E., Peña, D. and Fernández, M.J. (2020) Quality of olive stone as a fuel: influence of oil content on combustion process. *Renewable Energy* 160, 374–384.
Molina-Alcaide, E. and Yáñez-Ruiz, D.R. (2008) Potential use of olive by-products in ruminant feeding: a review. *Animal Feed Science and Technology* 147(1–3), 247–264.
Montanaro, G., Celano, G., Dichio, B. and Xiloyannis, C. (2010) Effects of soil-protecting agricultural practices on soil organic carbon and productivity in fruit tree orchards. *Land Degradation and Development* 21, 132–138. doi:10.1002/ldr.v21:2
Mosquera-Losada, M.R., Moreno, G., Pardini, A., McAdam, J.H., Papanastasis, V. (2012) Past, present, and future of agroforestry systems in Europe. In: Nair, P.K.R. and Garrity, D. (eds) *Agroforestry: The Future of Global Land Use*. Springer, Berlin, pp. 286–312.
Mousa, M. (2012) Utilization of olive pulp in feeding New Zealand white doe rabbits and their offspring under the conditions of north Sinai. *Egyptian Journal of Rabbit Science* 22(1), 1–21.
Mugunga, C.P., Giller, K.E. and Mohren, G.M.J. (2017) Tree-crop interactions in maize-eucalypt woodlot systems in southern Rwanda. *European Journal of Agronomy* 86, 78–86. https://doi.org/10.1016/j.eja.2017.03.004
Palese, A.M., Vignozzi, N., Celano, G., Agnelli, A.E., Pagliai, M. and Xiloyannis, C. (2014) Influence of soil management on soil physical characteristics and water storage in a mature rainfed olive orchard. *Soil and Tillage Research* 144, 96–109.
Panozzo, A., Bernazeau, B. and Desclaux, D. (2019) Durum wheat in organic olive orchard: good deal for the farmers? *Agroforestry Systems* 94, 1707–1717.
Panozzo, A., Huang, H., Bernazeau, B., Vamerali, T., Samson, M.F. and Desclaux, D. (2020) Morphology, phenology, yield, and quality of durum wheat cultivated within organic olive orchards of the Mediterranean area. *Agronomy* 10(11), 1789. https://doi.org/10.3390/agronomy10111789
Pantera, A., Burgess, P.J., Mosquera Losada, R., Moreno, G., Lopez-Díaz, M.L. *et al.* (2018) Agroforestry for high value tree systems in Europe. *Agroforestry Systems* 92, 945–959. DOI: 10.1007/s10457-017-0181-7
Paolotti, L., Boggia, A., Castellini, C., Rocchi, L. and Rosati, A. (2016) Combining livestock and tree crops to improve sustainability in agriculture: a case study using the Life Cycle Assessment (LCA) approach. *Journal of Cleaner Production* 131, 351–363.
Pardini, A., Faiello, C., Longhi, F., Mancuso, S. and Snowball, R. (2002) Cover crop species and their management in vineyards and olive groves. *Advances in Horticultural Science* 16(3–4), 225–234.
Paris, P., Camilli, F., Rosati, A., Mantino, A., Mezzalira, G. *et al.* (2019) What is the future for agroforestry in Italy? *Agroforestry Systems* 93(6), 2243–2256.
Peng, X., Thevathasan, N.V., Gordon, A.M., Mohammed, I. and Gao, P. (2015) Photosynthetic response of soybean to microclimate in 26-year-old tree-based intercropping systems in southern Ontario, Canada. *PLOS ONE* 10, 1–10. https://doi.org/10.1371/journal.pone.0129467
Pérès, G., Bellido, A., Curmi, P., Marmonier, P. and Cluzeau, D. (2010) Relationships between earthworm communities and burrow numbers under different land use systems. *Pedobiologia* 54(1), 37–44.
Pitchers, B., Do, F.C., Pradal, C., Dufour, L. and Lauri, P.É. (2021) Apple tree adaptation to shade in agroforestry: an architectural approach. *American Journal of Botany* 108, 732–743.
Rabayaa, E., Abo Omar, J.M. and Othman, R.A. (2001) Utilization of olive pulp in broiler rations. *An-Najah University Journal of Research* 15, 133–144.

Ramos, M.E., Robles, A.B., Sanchez-Navarro, A. and Gonzalez-Rebollar, J.L. (2011) Soil responses to different management practices in rainfed orchards in semiarid environments. *Soil and Tillage Research* 112(1), 85–91.

Renard, D. and Tilman, D. (2019) National food production stabilized by crop diversity. *Nature* 571(7764), 257–260.

Repullo, M.A., Carbonell, R., Hidalgo, J., Rodríguez-Lizana, A. and Ordóñez, R. (2012) Using olive pruning residues to cover soil and improve fertility. *Soil and Tillage Research* 124, 36–46.

Repullo-Ruibérriz de Torres, M.A., Ordóñez-Fernández, R., Giráldez, J.V., Márquez-García, J., Laguna, A. and Carbonell-Bojollo, R. (2018) Efficiency of four different seeded plants and native vegetation as cover crops in the control of soil and carbon losses by water erosion in olive orchards. *Land Degradation & Development* 29(8), 2278–2290.

Rocchi, L., Paolotti, L., Rosati, A., Boggia, A. and Castellini, C. (2019) Assessing the sustainability of different poultry production systems: a multicriteria approach. *Journal of Cleaner Production* 211, 103–114.

Rodrigues, M.Â., Dimande, P., Pereira, E.L., Ferreira, I.Q., Freitas, S. *et al.* (2015) Early-maturing annual legumes: an option for cover cropping in rainfed olive orchards. *Nutrient Cycling in Agroecosystems* 103(2), 153–166.

Romero, P., Castro, G., Gómez, J.A. and Fereres, E. (2007) Curve number values for olive orchards under different soil management. *Soil Science Society of America Journal* 71(6), 1758–1769.

Rosati, A. (2001) Un possibile futuro per l'asparago selvatico. *L'Informatore Agrario* 7, 89–92.

Rosati, A. (2011) Asparagi nell'oliveto e la produzione raddoppia. *Olivo e Olio* 7(8), 20–24.

Rosati, A. (2014) Coltivare asparagi selvatici e allevare polli in un piccolo oliveto. *Vita in Campagna* 12, 44–48.

Rosati, A. and Falavigna, A. (2000) Germinazione dei semi di asparago selvatico. *L'Informatore Agrario* 56(46), 53–55.

Rosati, A., Pepe, R., Senatore, A., Perrone, D. and Falavigna, A. (2005) Produttività dell'asparago selvatico. *L'Informatore Agrario* 8, 75–77.

Rosati, A., Caporali, S. and Paoletti, A. (2009) Olive, asparagus and animals: an agroforestry model for temperate climate in developed countries. *Proceedings of the III OLIVEBIOTEQ – For a Renovated, Profitable and Competitive Mediterranean Olive Growing Sector.* Sfax, Tunisia, 15–19 December, pp. 229–233.

Rosati, A., Castellini, C., Dal Bosco, A., Mugnai, C. and Paoletti, A. (2012a) *Manuale per la coltivazione consociata olivo, asparago selvatico, pollo rustico.* Edizioni 3A-PTA.

Rosati, A., Castellini, C., Dal Bosco, A., Mugnai, C., Paoletti, A. and Caporali, S. (2012b) Olive agroforestry: an inverse approach to agroforestry. In: Mosquera-Losada, M.R., Pantera, A., Rosati, A., Amaral, J., Smith, J. and Dupraz, C. (eds), *What Priorities for European Agroforestry?* Book of abstracts of the 1st European agroforestry conference, 9–10 October, Brussels.

Rosati, A., Concezzi, L., Dal Bosco, A., Mugnai, C. and Paoletti A. (2012c) *Olivo Asparago selvatico, Pollo rustico. Video on the olive, wild asparagus and free-range chicken system.* Available at: https://www.youtube.com/watch?v=ALw73WwUr2o (accessed on 22 December 2017).

Rosati, A., Piottoli, L., Cartoni, A., Dal Bosco, A. and Castellini, C. (2014) Polli al pascolo nell'oliveto Risparmio a tutto campo. *Olivo e Olio* 6, 4–7.

Rosati, A., Boggia, A., Castellini, C., Paolotti, L. and Rocchi, L. (2015) Polli nell'oliveto, anche l'ambiente ci guadagna. *Olivo e olio* 6, 4–8.

Rosati, A, Pang, K., Van Sambeek, J., Gold, M. and Jose, S. (2018) *Improving crop productivity in agroforestry systems: low leaf respiration is a key trait.* In: Proceedings of the 4rd European Agroforestry Conference – *Agroforestry as Sustainable Land Use*. Nijmegen, The Netherlands, 28–30 May, pp. 442–446.

Roselli, L., Clodoveo, M.L., Corbo, F. and De Gennaro, B. (2017) Are health claims a useful tool to segment the category of extra-virgin olive oil? Threats and opportunities for the Italian olive oil supply chain. *Trends in Food Science & Technology* 68, 176–181.

Salvatore, S., Pellegrini, N., Brenna, O.V., Del Rio, D., Frasca, G., Brighenti, F. and Tumino, R. (2005) Antioxidant characterization of some Sicilian edible wild greens. *Journal of Agricultural and Food Chemistry* 53(24), 9465–9471.

Sansoucy, R., Alibes, X., Berge, P.H., Martilotti, F., Nefzaoui, A. and Zoïopoulos, P. (1985) *Olive By-products for Animal Feed.* Food and Agriculture Organization of the United Nations, Rome.

Sastre, B., Pérez-Jiménez, M., Bienes, R., García-Díaz, A. and de Lorenzo, C. (2016) The effect of soil management on olive yield and VOO quality in a rainfed olive grove of Central Spain. *Journal of Chemistry* 2016. https://doi.org/10.1155/2016/4974609.

Sestini, A. (1963) *Il Paesaggio*. Touring Club Italiano, Italy.

Simopoulos A.P. (2004) Omega-3 fatty acids and antioxidants in edible wild plants. *Biological Research* 37, 263–277.

Smil, V. (2001) *Enriching the Earth: Fritz Haber. Carl Bosch and the Transformation of World Agriculture. Cambridge, Massachusetts.*

Smith, V.H. and Glauber, J.W. (2019) Trade, policy, and food security. *Agricultral Economics* 51, 159–171. https://doi.org/10.1111/agec.12547

Stroosnijder, L., Mansinho, M.I. and Palese, A.M. (2008) OLIVERO: the project analysing the future of olive production systems on sloping land in the Mediterranean basin. *Journal of Environmental Management* 89, 75–85. https://doi.org/10.1016/j.jenvman.2007.05.025

Taguas, E.V., Peña, A., Ayuso, J.L., Pérez, R., Yuan, Y. and Giráldez, J.V. (2010) Rainfall variability and hydrological and erosive response of an olive tree microcatchment under no-tillage with a spontaneous grass cover in Spain. *Earth Surface Processes and Landforms* 35, 750–760. https://doi.org/10.1002/esp.1893

Taheri, M.R., Zamiri, M.J., Rowghani, E. and Akhlaghi, A. (2012) Effect of feeding olive-pulp ensiled with additives on feedlot performance and carcass attributes of fat-tailed lambs. *Tropical Animal Health and Production* 45(1), 345–350.

Temani, F., Bouaziz, A., Daoui, K., Wery, J. and Barkaoui, K. (2021) Olive agroforestry can improve land productivity even under low water availability in the South Mediterranean. *Agriculture, Ecosystems & Environment* 307, 107234.

Tramblay, Y., Koutroulis, T., Samaniego, A., Vicente-Serrano, L., Volaire, S.M. (2020) Challenges for drought assessment in the Mediterranean region under future climate scenarios. *Earth-Science Reviews*, p.103348. https://doi.org/10.1016/j.earscirev.2020.103348

Vallebona, C., Mantino, A. and Bonari, E. (2016) Exploring the potential of perennial crops in reducing soil erosion: a GIS-based scenario analysis in southern Tuscany, Italy. *Applied Geography* 66, 119–131. https://doi.org/10.1016/j.apgeog.2015.11.015

Vannucci, S. (2009) *L'ulivo e l'olio, Storia dell'olivo, Diritti di sfruttamento economico*: Bayer Crop Science S.r.l.

Vardavas, C.I., Majchrzak, D., Wagner, K.H., Elmadfa, I. and Kafatos, A. (2005) Lipid concentrations of wild edible greens in Crete. *Food Chemistry* 99(4), 822–834.

Varmaghany, S., Rahimi, S., Torshizi, M.K., Lotfollahian, H. and Hassanzadeh, M. (2013) Effect of olive leaves on ascites incidence, hematological parameters and growth performance in broilers reared under standard and cold temperature conditions. *Animal Feed Science and Technology* 185(1–2), 60–69.

Venezia, A., Falavigna, A. and Soressi, G.P. (1993) Aspects related to the valorization of spontaneous species of Asparagus. *Agricoltura Ricerca (Italy).*

Vignozzi, N., Agnelli, A.E., Brandi, G., Gagnarli, E., Goggioli, D. *et al.* (2019) Soil ecosystem functions in a high-density olive orchard managed by different soil conservation practices. *Applied Soil Ecology* 134, 64–76.

Villalobos, F.J., Testi, L., Hidalgo, J., Pastor, M. and Orgaz, F. (2006) Modelling potential growth and yield of olive (*Olea europaea* L.) canopies. *European Journal of Agronomy* 24(4), 296–303.

Vršič, S. (2011) Soil erosion and earthworm population responses to soil management systems in steep-slope vineyards. *Plant, Soil and Environment* 57(6), 258–263.

Xiloyannis, C., Raya, A.M., Kosmas, C. and Favia, M. (2008) Semi-intensive olive orchards on sloping land: requiring good land husbandry for future development. *Journal of Environmental Management* 89, 110–119. DOI: 10.1016/j.jenvman.2007.04.023

Yang, K., Wei, C., Zhao, G.Y., Xu, Z.W. and Lin, S.X. (2017) Effects of dietary supplementing tannic acid in the ration of beef cattle on rumen fermentation, methane emission, microbial flora and nutrient digestibility. *Journal of Animal Physiology and Animal Nutrition* 101(2), 302–310.

Zamora, D.S., Jose, S., Jones, J.W. and Cropper, W.P. (2009) Modeling cotton production response to shading in a pecan alley cropping system using cropgro. *Agroforestry Systems* 76, 423–435. https://doi.org/10.1007/s10457-008-9166-x

Zuazo, V.D., Pleguezuelo, C.R., Panadero, L.A., Raya, A.M., Martínez, J.F. and Rodriguez, B.C. (2009) Soil conservation measures in rainfed olive orchards in south-eastern Spain: impacts of plant strips on soil water dynamics. *Pedosphere* 19(4), 453–464.

9.3 The Landscape of the Olive Tree: Old and New Visions

G. Barbera[1] and R. Biasi[2]
[1]*Department of Agricultural, Food and Forest Sciences, University of Palermo, Italy;*
[2]*Department for Innovation in Biological, Agro-Food and Forest Systems, Tuscia University, Viterbo, Italy*

9.3.1 Introduction

Landscape is one ecosystem service to whose maintenance agriculture is called. In a context of accelerating environmental changes, as well as profound social and cultural transformations, agriculture must face new challenges, as well-defined by the UN Goals for Sustainable Development (SDGs, Agenda 2030). In particular, the challenge of meeting the food needs of the population, the need to combat and adapt to climate change, the necessity of preserving terrestrial ecosystems, invest agriculture with a primary role and farmers with a great public responsibility. The sustainability of production chains must be based on new knowledge, skills and awareness of the common benefits that can be derived from primary production.

Today, landscape represents an essential aspect of agricultural production. It can be a factor of recognition of farms and agri-food products, counteracting the degradation of terrestrial ecosystems and the globalization of agri-food production (Biasi, 2021; Russo *et al.*, 2020). In the agronomic disciplines, the approach of evaluating the sustainability of production chains on the basis of the effects on biodiversity and landscape, as well as on the physical environment in which they occur and on the environmental resources, is now well established. Biodiversity should be considered as the set of plant and animal biological diversity associated with the production process, i.e. as agrobiodiversity, that is the set of species and their ecosystems that characterize the agroecosystem and that coexist with it, allowing the productive result. Similarly, the diversity of the ecosystems, land use and covers of an area of cultivation (at large scale), determine the quality of the agricultural landscape, which is the result of the modification of the environment by man for the purpose of its production activities. Today the landscape is, to all intents and purposes, a dimension of agricultural production, being the expression of the multifunctionality of agriculture, which, in addition to being an activity in the interest of the individual, is an essential activity for the benefits of the community. There is a close relationship between 'ecology', defined by the level of biodiversity and the balance between biological systems, and 'aesthetics' of the resulting landscape. What is perceived as a landscape of quality and beauty is objectively of high value, because of the complexity of forms and ecosystems that are expressions of ecological and environmental value.

Owing to the biological characteristics of the species *Olea europaea* and because of the history of the evolution of cultivation techniques for this crop, which has made

DOI: 10.1079/9781789247350.0028

more problematic than for other tree systems the intensification process and the mechanization of production phases, the landscape of the olive tree, in most contexts in which it is widespread, is emblematic of the traditional multifunctionality of the tree crop systems. This chapter highlights the reasons.

9.3.2 Biocultural Geography of the Olive Tree Landscape

The Mediterranean, writes Fernand Braudel (1986), its greatest historian, is the 'sea of olive groves'. Along the coasts 'we find the same trinity, daughter of the climate and history: the wheat, the olive tree, the vine, that is the same agrarian civilization, the same victory of men over the physical environment'. In fact, from whatever point of view one looks at its landscape, one cannot fail to meet, with undeniable evidence in time and space, the olive tree, and it is its presence that traces the borders of the Mediterranean lands more than the climate, the topography or its geopolitics (Guarracino, 2007). The climate is too variable, even within the common rule of the long and arid dry season; the topographical forms (the ridges and the hydrographic basins) would have excluded essential areas; political boundaries are inherently changeable. To define the boundaries of the Mediterranean, one cannot even resort to natural vegetation, which is too altered in its distribution to be indicative. However, the exception, since the time of Pliny considered the most reliable, is that of the olive tree both in the wild and in the cultivated form. In fact, nature places wild olive (*Olea oleaster*) as a characterizing element of the most widespread plant associations of the Mediterranean bush and forest, and man has tamed it everywhere. The olive tree lives from the northern borders to the southern ones. The Mediterranean one, confirms Braudel (1986), is 'an olive civilization ... the Mediterranean extends from the first olive tree that can be reached from the north to the first palm groves that arise near the desert'. Before him, Vidal de la Blanche (in Claval, 2007) wrote that 'vines and olive trees make up the ornament of the Mediterranean coasts, the similarity of the vegetation is the scene of the theater where history was made'.

The landscape is the result of the relationship between nature and history and the presence of the olive tree is also shown in it as a synthesis of the cultural richness of the Mediterranean. Predrag Matvejevic (1998) recalls that 'oil production is not just a profession, it is also a tradition. The olive is not just a fruit: it is also a relic.' Lawrence Durrel (in Broodbank, 2015) observes: 'the sculptures, the palms, the gold necklaces, the bearded heroes, the wine, the ideas, the ships, the moonlight, the winged gorgons, the bronze men, the philosophers, all this is evoked by the sharp pungent taste of black olives between the teeth'. For Aldous Huxley (2018), the olive tree is the 'emblem of Latinity'. Such is the place of the Mediterranean in human history and such is the part that the olive tree plays in it that it is well understood why, in adherence to the role it occupies in the world of symbols and in the perception of the landscape, going beyond the correct biogeographical borders, Linnaeus baptized it *Olea europaea*, testifying a belonging to western culture that also involves lands far from those where it was and is widespread (Barbera *et al.*, 2005; Barbera and Dettori, 2006).

The species became part of the Mediterranean rural landscape at least from the fourth millennium BC and the wild olive seeds found in the Uzzo cave in Sicily, in the horizons of the pedological profile of the soil dating back to the sixth millennium, already documented an occasional use for food even then. The first historical information will arrive in the Greek era and confirm some characteristics that are already typical of the polycultural tree landscape and, in particular, of what Emilio Sereni (1961) defines as the 'Mediterranean garden', represented in the Table of Alesa. Thucydides seems to refer to this typology in *The Peloponnesian War*: near Syracuse 'a land enclosed all around by a low wall, with two roads bordering it on the sides, dense with an olive plantation' (Barbera and Inglese, 2008). Thanks to the centuries-old work of olive growers, adapting to the most diverse ecological conditions, the olive tree will be present in cultivation practically everywhere in specifically adapted and very diversified systems and landscapes (Fig. 9.3.1).

In the heart of the Mediterranean, the environmental diversity finds one of its highest expressions in the Italian peninsula. Italy is a territory with complex geomorphology and high physiographic variability, characterized by peculiar soils and climates that are equally heterogeneous (Blasi *et al.*, 2014) as the olive biodiversity and the cultivation techniques carried out to adapt the needs of this crop to the different environmental conditions. Olive groves have generated a landscape so eclectic for this tree species that it can be taken as a model of representation of the diversity of the olive landscape in the whole Mediterranean basin. Furthermore, the Italian olive landscape also becomes representative of a peculiarity compared to other agricultural land uses: resilience. In fact, as multi-temporal analysis of changes in land use and land cover occurring in the last decades have demonstrated, the olive surface has not been affected by that huge erosion of the surfaces that has affected viticulture or fruit crops (ISMEA, 2021). On the contrary, olive cultivation represents a crop that can be implemented even in those areas affected by marginality and abandonment of the agricultural practice. In fact, in spite of the transformations that in the last 50 years have accompanied the evolution of agriculture from tradition to industrialization and that have led to changes in the geography of tree systems, the landscape of the olive tree has been maintained in large territories (Fig. 9.3.2a, b), while for other cropping systems, both herbaceous and tree systems, the polarization in areas of specialized production has occurred (e.g. apple growing in northern Italy, citrus and cherry growing in the south, hazelnut and kiwi production in central Italy). In the areas most favored for environmental characteristics and suitable to host industrial agriculture, intensive monocultural and simplified systems have been established, while in the marginal areas, such as mountainous or inner territories, traditional agricultural systems have

Fig. 9.3.1. The typical polycultural landscape of olive in the heart of Sicily island. (Photo: G. Barbera)

been maintained or, far more often, the landscape of agriculture has disappeared in favor of the forest (RaF ITALIA 2017–2018, 2019). By contrast, the landscape of the olive tree is unifying the Italian territory. The landscape of the olive tree continues to characterize in a very unique way the landscape both of the fertile plains, where it coexists with intensive arboricultural systems, arable land and vegetable crops, but also resists in marginal areas the process of progressive abandonment of cultivation, and frequently, on the contrary, represents the only surviving agricultural use (Biasi *et al.*, 2015; Brunori *et al.*, 2020) (Fig. 9.3.3). In fact, in the conditions of greater marginality,

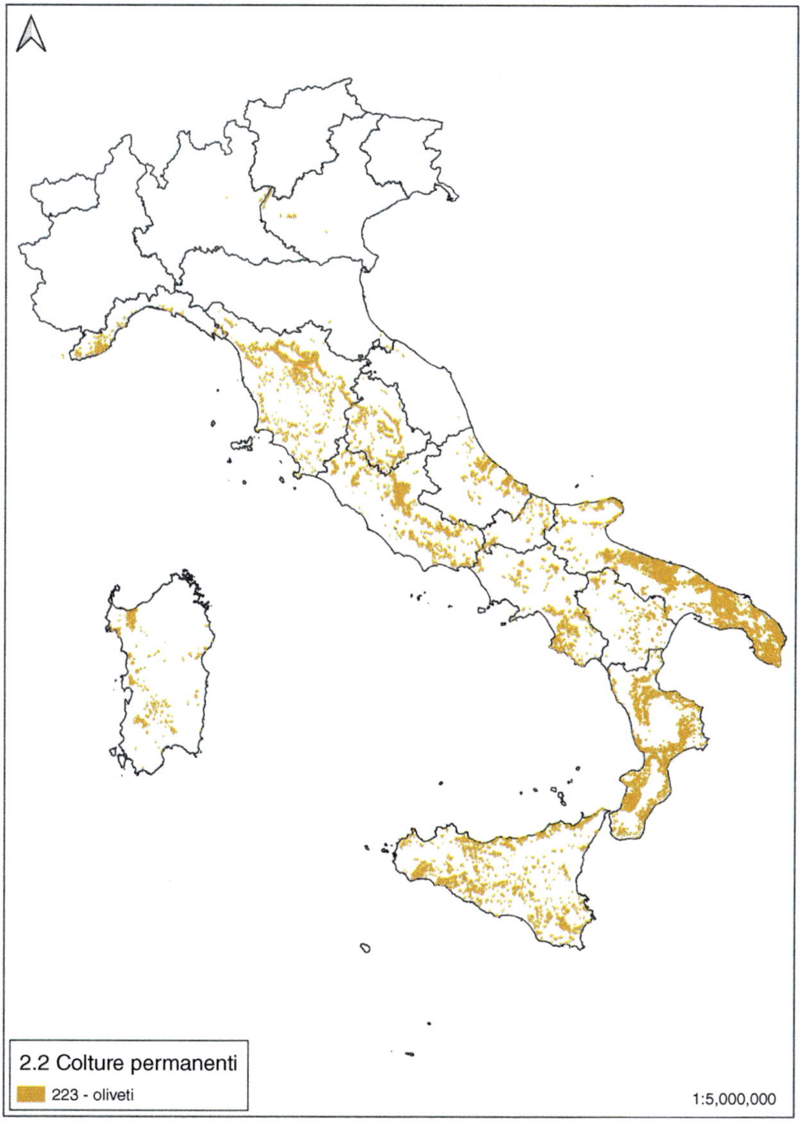

Fig. 9.3.2. (a) the geographic distribution of specialized olive groves; and (b) of heterogeneous tree-crop-based agricultural areas in the Italian peninsula, heart and model of the Mediterranean landscape. (Elaboration by C. Castaldi)

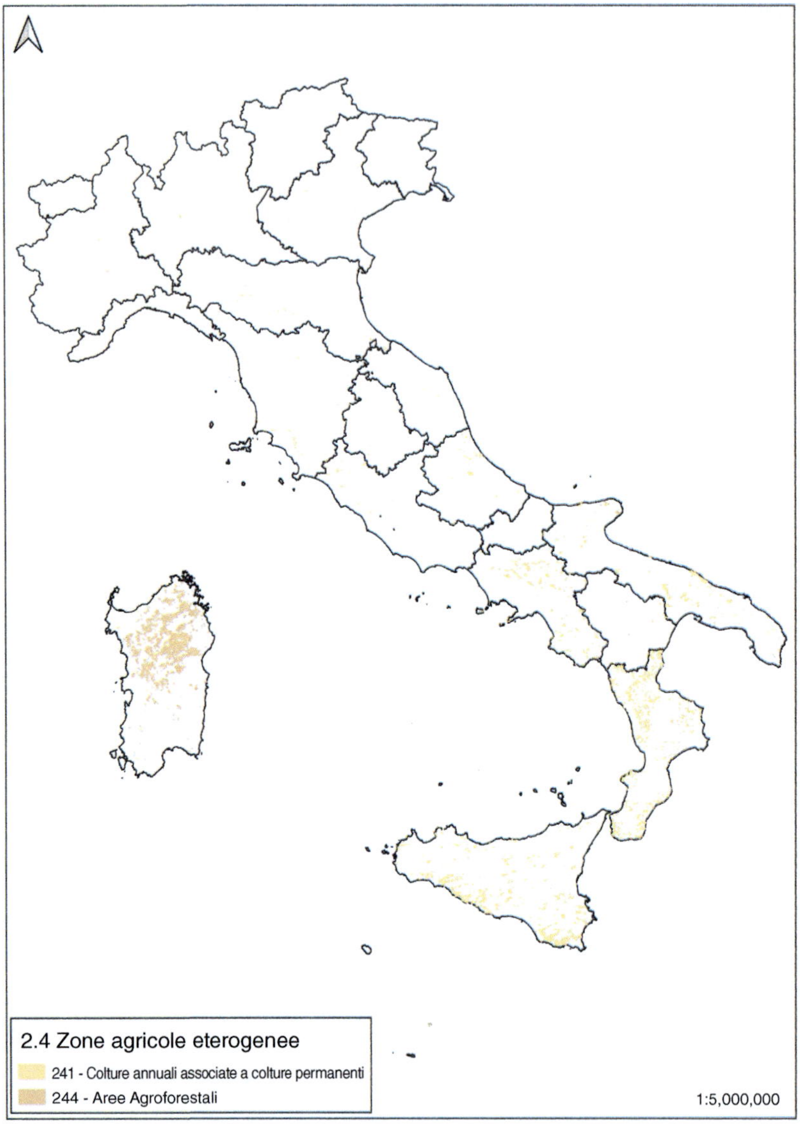

Fig. 9.3.2. Continued.

in the most sloping terrain, on the narrowest terraces, the olive tree contributes to the formation of systems and landscapes of promiscuous culture. They are polycultural landscapes that survive in the territories characterized by the exodus of farmers from the countryside, or, if abandoned, they acquire characters of semi-natural land covers, to form real forests, where the olive tree continues to thrive.

These are landscapes that can be considered the oldest in the country because they are substantially unchanged in terms of both biological, structural and territorial distribution: olive growing was largely promiscuous, associated to other crops, at least until the second postwar period (1947) when it prevailed with 1392 million hectares, compared to 835,000 hectares in specialized

Fig. 9.3.3. Resilient olive groves in marginal sloping farmlands. (Photo: R. Biasi)

cultivation. In central Italy, its privileged place was the *coltura promiscua* where it responded to the need to diversify production. The landscape of Tuscan mixed olive growing is very famous, of which Morettini (1950), one of the fathers of the science of arboriculture in Italy, emphasized the polycultural character:

> The olive tree is grown in rows; in the inter-rows the common herbaceous crops of grain, forage and vegetables are practiced in alternation. Along the row, the olive tree is ordinarily associated with the vine, more rarely fruit trees with various species. Sometimes the vines and orchards are also cultivated in rows interspersed with those of the olive tree. The distinction of the area occupied by trees and herbs is not always clear, as the cultivation of the latter is generally spread uniformly over the entire area. In the surroundings of Florence there are the most complex and intricate types of intercropping of the olive tree with other tree plants and at the same time with herbaceous plants. In fact, in addition to herbaceous plants, vines, peaches, pears, apples, mulberries, etc. are associated with the olives with promiscuity pushed to the maximum.

The landscape of mixed olive growing, despite the historical and ecological-environmental and aesthetic value inherent in the nature of the traditional agricultural system, is today residual as a consequence of a progressive transition towards monoculture and specialization promoted by the innovation of agronomic techniques starting from 1960s.

Subsequently, in the lowland areas, the monoculture of olive growing was strengthened, even if historically large plains were already characterized by the presence of the olive tree landscape: from the landscape of the specialized olive groves of the Conca d'Oro of Palermo in the mid-15th century to the '*gran selve* of olive trees', which, a century later, Leandro Alberti will see in Puglia

(Bevilacqua, 2000), to the Calabrian olive growing of Gioia Tauro in the 18th century (Inglese and Calabrò, 2002). The Calabrian olive groves are extensive monocultures, where the groves could consist of only 40–50 plants per hectare and which in any case determined a complete and uniform coverage of the soil at maturity. They were on a large scale similar to the olive groves protected from grazing and theft with 'closing' walls or hedges typical of central Italy and medieval Puglia.

The Tuscan olive groves of the early 19th century, defined as 'woodland' or 'Pisan style', with a density that reached 700 plants per hectare and those of the Ligurian tradition also responded to the rusticity characteristics of the species and the reduced cultivation requirements (Morettini, 1950). For these systems there was the possibility or the necessity of a modernization of the plantings and a rationalization of the management systems. A landscape of semi-intensive olive growing has thus established itself in which the innovations have concerned interventions for limiting the volume and height of the canopy, to favor defense practices and the mechanization of pruning and harvesting. A landscape also characterized by the introduction of nets that, today, due to their diffusion, have become a specific feature of the olive landscape during the harvesting period, and structural interventions, such as increased planting density, the introduction of irrigation systems, new soil management forms, fertilization and defense techniques for increasing the productivity of the plants; interventions which have resulted in minor changes to the cultivation model, while certainly more significant changes have taken place in the olive processing technology in terms of process and product, with the progressive and constant affirmation of extra-virgin olive oil.

However, olive groves' evolution has not only concerned the intensification of traditional low-intensity orchards (200–300 plants/ha), generally carried out in dry cultivation, but has led to new concept plants. In the 1980s and 1990s, olive groves with a density of 400–500 trees/ha spread and eventually, at the turn of the millennium and starting from Spain, super-intensive (very high density) olive groves with trees in the number of 1500–2000/ha arranged in hedges 3–4 m apart, necessarily irrigated, made their appearance. The availability of machinery for mechanical harvesting was the basis of the success of these new olive-growing systems where the technique adapted the plants to mechanization, rather than adapting the machines to the plants' growth habit, as had already happened for other tree systems that started long before the intensification of cultivation, such as apricot, cherry or plum, pioneer species of the integral mechanization of harvesting.

Today, the forms of the olive-growing landscape derive from the coexistence of different models: the promiscuous one; the semi-intensive one of traditional monoculture; and the super-intensive one whose resilience, in light of the role that is currently asked of agriculture, will depend not only on ability to ensure an income and therefore maintain a productive function but also on the ability to perform environmental and cultural functions: the so-called ecosystem services. In this respect, many olive-growing landscapes are indeed fragile and vulnerable like the large olive groves in the plains that suffer from a marginality linked to productivity and product quality, often attributable to new plant diseases, or the new-super intensive olive groves, able to provide significant production quantities and excellent quality levels, but with environmental sustainability limits typical of specialized monoculture that are based on a high demand for artificial production inputs. These latter also exhibit strong limitations in respect to the landscape, as evidenced by the metric indices of landscape ecology that attribute to an environmental ecomosaic with few, uniform, simplified ecosystems, the condition of imbalance and fragility (Tarifa *et al.*, 2021; Weissteiner *et al.*, 2016). On the contrary, the traditional olive-growing landscape constitutes a set of patches within an environmental mosaic formed by agrarian and semi-natural systems of different types with high landscape diversity, thanks to the maintenance of a high level of biodiversity even at a farm scale, which determines a complex and stable agroecosystem.

The landscapes of traditional olive growing also have an evident cultural function determined by a strong aesthetic and ethical identity. They are the result of a nature designed by the work of man and made by him harmonious and friendly. In a centuries-old olive tree, in a terrace you can find the fatigue, the work, the feelings of a community and of those who preceded us: the landscape is a representation of memory and of rural tradition. Whether or not they include farming or forests, they affect a vast area and the agricultural productive function is no longer the main one but becomes part of the multifunctionality of ecosystem services (Barbera *et al.*, 2014).

The landscape has a spatial and temporal dimension. The countryside is protean due to the action of different drivers strongly interconnected (cultural, social, technical) that act on a physical environment, which is also rapidly changing mainly owing to climate change. Consequently, some crops have 'moved' in search of new suitable pedoclimatic environments to adapt to environment changes and to meet the challenge of sustainable production.

9.3.3 A Moving Landscape

The land use based on olive groves, in contrast to the general decrease in the Utilized Agricultural Area (EUROSTAT, 2020; ISTAT, 2021), is relatively stable over time. In particular, compared to the continuous and slow erosion of other permanent crops (orchards or vineyards), the olive surfaces, and therefore the landscape of the olive tree, show to be stable in terms of total extension. But in reality it is a landscape in motion.

If, from one side, the olive landscape has been, in the past, and still is, emblematic of the Mediterranean, on the other, in the future, it may become emblematic of geographical areas much further north. The geography of the olive landscape, like that of the species with which it is often associated in the countryside, like *Vitis vinifera*, shows a tendency to move towards areas in the north of Italy and towards higher altitudes (Salvati *et al.*, 2013; Rodrigo-Comino *et al.*, 2021). The driver of this movement is climate change. The olive tree can be considered a bio-indicator of climate change and, in particular, of the impact of desertification and global warming. Although *Olea europaea* is one of the species of agricultural interest most resistant to drought – it is able to keep the stomata open and functional even for values of water in the soil very close to the wilting point – the lack of water and radiative stress highly compromises the quality of drupes and therefore of the derived oil (Fig. 9.3.4). Areas now considered suitable for olive growing could lose their environmental suitability, both as a direct result of climate change, and for the indirect consequences of a greater susceptibility to biotic stresses. As a matter of fact, climate changes may allow endemic and alien pests and pathogens to heavily attack the olive trees therefore destroying their landscape. Not only the well-known bacterium *Xylella fastidiosa,* but also an insidious pest such as *Lobesia botrana*, an alien lepidopteran in olive groves and certainly not endemic in the Mediterranean, aggressive against the grapevine, but which performs more and more frequently some growth cycles on olive trees (Sciarretta *et al.*, 2021; Semeraro *et al.*, 2021). Like climate scientists' models have predicted the expansion of viticulture to regions of northern Europe and the south of the UK, so researchers have measured an actual increase in olive grove surface in regions of northern Italy and a rise in the elevational limit of the olive crop and landscape. In the meantime, in the southern regions, especially in lowland areas, that have become increasingly inhospitable to the crop; the cultivated landscape of the olive tree disappears mainly due to the phenomenon of abandonment of agriculture (Gambella *et al.*, 2021). Where, instead, the landscape of the olive tree exhibits some resilience are the most marginal areas that are also sensitive to geomorphological features: the altitude and steep slopes. Recent studies (Biasi *et al.*, 2015; Brunori *et al.*, 2015, 2020) have shown that agroforestry

Fig. 9.3.4. The landscape of olive stress resistance in the Pantelleria (Sicily) island. Olive trees are trained to very low height to counteract wind and drought (Photo: G. Barbera)

systems based on the olive (olive-based polycultural systems) represent the only agricultural land use in areas with steep slopes and higher altitudes, where they prevent the phenomena of soil degradation, such as loss of organic matter, soil erosion and landslides, at the same time preventing wild fires. In this context the landscape of olive groves shows all its multifunctionality based on ecosystem services, in addition to supporting local economies.

The olive tree landscape also moves towards the cities and becomes part of the vegetation components that accompany the transition of the metropolis towards inclusive, sustainable and resilient contemporary habitats, in accordance with the SDG 11 of the Agenda 2030.

The city, with its peculiar microclimate and its heat islands, certainly represents a hostile environment for agricultural production, so much that only the identification of genotypes with characteristics of resistance to multiple abiotic stresses may ensure the maintenance of high production standards over time. Nevertheless, the productive areas today are part of the urban landscape, as a result of the expansion of cities that have included the countryside in the suburb; consequently, olive landscape has become a threatened landscape of the urban countryside (Gulçin, 2019). The condition of the peri-urban, the geographical space between the city and the countryside, today is so peculiar that its landscape differs from both urban and rural in configuration, functions and other characteristics (Meeus and Gulink, 2008) and increasingly appears as a heterogeneous mosaic of natural and agricultural ecosystems with their own ecosystems and dynamics of transformation. Although climate changes have increased the frequency of extreme events (heatwaves, storms, alterations in precipitation regimes), more pronounced in urban areas, cities are reinforcing their agricultural matrix (Philips, 2013; Viljoen *et al.*, 2005) because of its multifunctionality. The urban olive landscape is expanding as a consequence of the ecological design criteria for the urban open spaces. In urban and peri-urban areas, agriculture represents: (i) an economic system that supports short supply chains; (ii) a land use that provides essential ecosystem services for well-being and quality of life, like carbon storage, biodiversity maintenance, soil health, food education, social integration; (iii) a place of cultural significance; and (iv) a landscape of appreciated aesthetic value. Whether cultivated or abandoned, agricultural areas are part of the ecological network of the urban and peri-urban space. The urban countryside that olive groves strongly connote in the Mediterranean (Cecchini *et al.*, 2019; Gülçin 2019; Sini *et al.*, 2017) is configured as strategic spaces for the continuity and connection of the ecological infrastructure of the city with the natural environment, for the strengthening of city-countryside connections, for the management of a vulnerable space represented by the margins between natural and highly anthropized ecosystems. Urban and peri-urban countryside are the new productive landscapes. Furthermore, it should be considered the emergence of a new ecological sensitivity in the design of open spaces, public or private, which also involves the olive tree (Fig. 9.3.5). The contemporary design of open spaces of the city allows the agricultural use of land, for its multifunctionality and mainly because of its functions of preservation and regeneration of the soil. Increasingly in landscape design projects, the agrobiodiversity takes a prominent position in the selection criteria of the vegetational component. Designing with species of agricultural interest, designing productive spaces in the city, offers opportunities to live differently the dimension of urban areas improving their quality through the ecosystem services that urban agriculture offers (Lohrberg *et al.*, 2015; Philips, 2013).

Increasing are the actions of urban planning and urban landscape design that

Fig. 9.3.5. Olive trees utilization in urban landscape design. (Photo: R. Biasi)

valorize the pre-existing productive areas or that design *ex novo* productive spaces (Tassinari *et al.*, 2010). Urban planning is becoming 'agricultural' (De la Salle and Holland, 2010). Many farms have been incorporated into the urban fabric due to its expansion, the so-called urban sprawl, typical of many Mediterranean cities. Reconciling the pressures of landscape transformation especially due to land consumption, with the maintenance of the functions and incomes of farms and the preservation of the landscape of tree systems, means investing in products, but mainly in the multifunctionality of fruit agroecosystems and the value of interconnected ecosystem services, not least the function of carbon storage (Brunori *et al.*, 2013; Semeraro *et al.*, 2021). However, the resilience of productive urban and peri-urban landscapes must be adequately supported with targeted projects for the re-functionalization and innovation of urban agriculture (Ortega *et al.*, 2020), the agricultural landscape being a strategic production factor, element of recognition and uniqueness in the market, for new and traditional olive-growing systems, so numerous in the Mediterranean metropolitan areas.

In this context, the physiognomy of the olive landscape of tomorrow will be characterized also by the new urban olive landscapes, both at large scale in the peri-urban agricultural spaces, in the urban public open spaces, but also at detail scale in the city's tree rows or myriad balconies and terraces hosting a single olive tree.

9.3.4 Future Prospects for a Sustainable Landscape of the Olive Tree

The challenge of an ecological transition that includes a sustainable agriculture will necessarily pass through the rethinking of agricultural spaces, the redesigning of the countryside and of the agroecosystem architecture, to ensure agriculture resilience especially of permanent crops, which have been recognized as strategic biological systems for sustainable development (FAO, 2013). In fact, agriculture based on trees may allow the 'sustainable intensification' (AISSA-AAYV, 2019), the sole production strategy to achieve productivity goals, but at the same time assuring ecosystem processes and services.

Recently, different criteria of design and use of rural spaces have been proposed, aiming at the respect of crop specialization and the maintenance of high production yields together with ecosystem services. In some areas of specialized and high-density planted crops, the UAA is designed or redesigned to ensure diversity of environments and habitats, richness in biological systems, and ecological functions (Landis, 2017; Mann *et al.*, 2018). The sustainable intensification is now an essential process for the quality of the agricultural productions and includes the landscape dimension. Therefore, the certification of sustainability of agri-food productions has to include the evaluation of the impact on the landscape, which, to all intents and purposes, becomes a component of the overall quality of the agri-food products.

Tackling these issues requires a landscape approach, and it is through it that agriculture relates to other human activities, with different needs, with the other components of the biosphere (Barbera, 2019). In their recent developments, the sciences of agroecology, landscape ecology and the activity of international organizations such as UNESCO and FAO have gradually become more aware of its need. Each for the part that competes, recognizes that the landscape is the result of the interaction between the characters of nature and the history of man who has changed them to his own advantage, for food or material needs, for safety, for pleasure. It is an expression of the project of a society whose needs, material and intangible, are recognized in ecosystem services and can change in relation to the times of nature and man: the landscape, in its innate dynamism, is able to deal with the future and adapt to it, to its inevitable unpredictability, to the questions that will arise, to the needs that will arise from the past and present history of a community through participation, sharing and the meeting of different knowledge. The landscape is the result of the culture of a people, it expresses aesthetic values (beauty) through itself or the forms that represent it, and, comparing the personal and community needs with the resources and constraints of nature (living or not) with which it interacts, relates to ethics. The vision of the landscape urgently needs the passage from notional fragmentation to a multi- and transdisciplinary knowledge useful at different scales (local, global) in comparison with different human cultures, with social needs, with a dynamism that allows elasticity with respect to a future that can be foreseen but not known. The landscape should therefore not be faced as an aesthetic space but, at the same time, ecological, economic and ultimately ethical.

Regarding the recent evolution of agricultural systems, their responsibilities in determining many of the crises that have led to a serious alteration of the planetary equilibrium – to the point that it is believed to have entered a new phase of relations between man and the planet: Anthropocene, the era of the impact of human activities that have become unbearable for the equilibrium of the biosphere – it is urgent that we come to the awareness that the landscapes of contemporary agriculture are the result of a reductionist vision based on single or sectoral techniques. It should be remembered that in overcoming the 'planetary boundaries', nine parameters (climate change, ocean acidification, stratospheric ozone decrease, biogeochemical fluxes and cycles of phosphorus and nitrogen, loss of biodiversity, global use of fresh water, land use changes, increase in atmospheric aerosol, chemical pollution) that define the boundaries of the 'safe operating spaces' to be respected in order not to seriously damage the biosphere and with it the development and prosperity of human societies (Steffen *et al.*, 2015), agriculture is considered the 'major driver of terrestrial systems that go beyond planetary boundaries' (Campbell *et al.*, 2017). Both individual processes and the interaction between them contribute to the overcoming of these limits. Containment therefore requires a systemic assessment that considers the productive, environmental and cultural functions together. Moreover, the awareness that agriculture must be considered 'in relation to the multiple benefits provided to mankind' by ecosystem services is consolidated, which is how the Millennium Ecosystem Assessment (2005) lists: 'supply or procurement' of real goods such as food and raw materials; 'Regulating' the climate, air and water quality, soil formation and the mitigation of natural hazards; of 'support' in relation to habitats and biodiversity; 'Cultural' that include non-material benefits such as cultural values and identity, spiritual and intellectual enrichment and aesthetic and recreational values.

A systemic, fully landscape vision cannot be limited by points of view that reduce its value to an aesthetic or ecological one. Considering its multifunctional character, a territorial and environmental policy is needed that safeguards the traditional agricultural landscape as an asset and a resource, preventing the cancellation of historical landscapes. These should not be evaluated only for the value of heritage, a cultural asset to be safeguarded, nor as a reserve of biodiversity or as a deposit of intangible knowledge: in the spatial design they consider the whole that derives from all this in relation to the ability to organize multiple functions and the interests (tangible and intangible) of the populations. They are configurations that represent the complex needs of continuously evolving systems, ensuring the various ecosystem services also in view of the expected global changes. The model that they defend and disseminate preserves both biological and cultural diversity beyond the limits of the conception of protected areas where this leads to denying or strongly reducing human activities. In Italy, the idea is present in the initiative of the Ministry for Agricultural, Food and Forestry Policies, which established in 2012 the 'National Observatory of rural landscape and agricultural practices and traditional knowledge', currently active in cataloging through the 'National register of rural landscapes of historical interest, agricultural practices and traditional knowledge'. Very recent (May 2019) is also the dissemination by AISSA (Italian Association of Agricultural Scientific Societies) in the 'Manifesto for the Italian agricultural and forest landscape'.

References

AISSA-AAVV (2019) *Manifesto per il paesaggio agrario e forestale italiano.* Available at: https://www.soihs.it/content.aspx?id=168 (accessed 27 May 2023).

Barbera, G. (2019) *Antropocene, Agricoltura e Paesaggio.* Aboca, Sansepolcro, Italy.

Barbera, G. and Dettori, S. (2006) Traditional olive groves in the Mediterranean cultural landscapes: history, functions, future. In: *Recent Advances in Olive Groves. Special Seminars and Invited Lectures*, pp. 287–294.

Barbera, G. and Inglese P. (2008) Sistemi e paesaggi. In: Caruso, T. and Magnano, G. di San Lio (eds) *La Sicilia dell'olio.* T. Caruso e G. Magnano di San Lio Giuseppe Maimone Editore,Catania, Italy.

Barbera, G., Inglese, P. and La Mantia, T. (2005) Paesaggio colturale dei sistemi tradizionali: l'olivo in Italia. *Estimo e Territorio, Anno LXVIII*, 2, 22–31.

Barbera, G., Biasi, R. and Marino, D. (2014) *I paesaggi agrari tradizionali.* Franco Angeli, Milan, Italy.

Bevilacqua, P. (2000) Il linguaggio degli alberi nel paesaggio agrario meridionale. *Italus Hortus* 7(3–4), 7–9.

Blasi, C., Capotorti, G., Copiz, R., Guida, D., Mollo, B., Smiraglia, D., Zavattero, L. (2014). Classification and mapping of the ecoregions of Italy. *Plant Biosystems* 148 (6): 1255-1345. DOI: 10.1080/11263504.2014.985756

Biasi, R. (2021) Paesaggio agrario al centro: opportunità per le necessità future dell'agricoltura italiana. In: *Collana Quaderni della Società Italiana di Scienze del Turismo (3)*, Aracne, Rome, pp. 47–70.

Biasi, R., Brunori, E., Smiraglia, D. and Salvati, L. (2015) Linking traditional tree-crop landscapes and agro-biodiversity in central Italy using a database of typical and traditional products. *Biodiversity and Conservation* 24, 3009–3031.

Braudel, F. (1986) *Civiltà e imperi del Mediterraneo nell'età di Filippo II.* Einaudi, Turin, Italy.

Broodbank, C. (2015) *Il Mediterraneo. Dalla preistoria alla nascita del mondo classico*, Einaudi,Turin, Italy.

Brunori, E., Maesano, M., Moresi, F.V., Matteucci, G., Biasi, R. and Scarascia Mugnozza, G. (2020) The hidden land conservation benefits of olive-based (*Olea europaea* L.) landscapes: an agroforestry investigation in the southern Mediterranean (Calabria region, Italy). *Land Degradation and Development* 31(7), 801–815. https://doi.org/10.1002/ldr.3484

Brunori, E., Salvati ,L., Antogiovanni, A., Biasi, R. (2018) Worring about vertical landscape: terraced olive groves and ecosystem services in marginal land in central Italy. *Sustainability* 2018, 10, 1164. doi:10.3390/su10041164

Brunori, E., Farina, R. and Biasi, R. (2013) Vineyards as alternative carbon sinks in peri-urban areas. *Ciencia e Tecnica Vitivinicola*, vol. 28: 1091–1121.

Campbell, B.M., Beare, D.J., Bennett, E.M., Hall-Spencer, J.M., Ingram, J.S.I. *et al.* (2017) Agriculture production as a major driver of the earth system exceeding planetary boundaries. *Ecology and Society* 22, 4.

Cecchini, M., Zambon, I., Pontrandolfi, A., Turco, R., Colantoni, A., Mavrakis, A. and Salvati, L. (2019) Urban sprawl and the 'olive' landscape: sustainable land management for 'crisis' cities. *GeoJournal* 84, 237–255.

Claval, P. (2007) About Rural Landscapes: The Invention of the Mediterranean and the French School of Geography. European Landscapes and Lifestyles. In: Roca, Z., Spek, Th., Terkenli, Th., Plieninger, T. and Höchtl, F. (eds) *Mediterranean and Beyond.* Lisbon: Edicoes Universitarias Lusofonas.

De la Salle, J. and Holland, M. (2010) *Agricultural Urbanism: Handbook for Building Sustainable Food & Agriculture System in 21st Century Cities.* GFB. HB Lanarc, Edmonton, Canada.

FAO (2013) *Perennial crops for food security. Proceedings of the FAO Expert Workshop.*

Gambella, F., Bianchini, L., Cecchini, M., Egidi, G., Ferrara, A., Salvati, L., Colantoni, A. and Morea, D. (2021) Moving toward the north? The spatial shift of olive groves in Italy. *Agricultural Economics* 67(4), 129–135.

Guarracino, S. (2007) *Mediterraneo. Immagini, storie e teorie da Omero a Braudel.* Bruno Mondadori, Milan, Italy.

Gulçin, D. (2019) The influence of urban growth on surrounding Mediterranean landscapes with particular reference to land degradation of olive orchards. *Fresenius Environmental Bulletin* 28(5), 3854–3864.

Huxley, A. (2018) *L'albero di olivo.* Edizioni Henry Beyle, Milan, Italy.

Inglese, P. and Calabrò, T. (2002) *Olivicoltura e paesaggio nella piana di Gioia Tauro.* Laruffa Editore, Reggio Calabria, Italy.

ISMEA (2021) Scheda di settore: olio d'oliva. Available at: www.reterurale.it (accessed 27 May 2023).

Landis, D.A. (2017) Designing agricultural landscapes for biodiversity-based ecosystem services. *Basic and Applied Ecology* 18, 1–12.

Lohrberg, F., Lička, L., Scazzosi, L. and Timpe, A. (eds) (2015) *Urban Agriculture Europe*. Available at: chrome-extension://oemmndcbldboiebfnladdacbdfmadadm/https://www.ideabooks.it/wp-content/uploads/2016/12/Urban-Agriculture-Europe.pdf (accessed 27 May 2023).

Mann, C., Garcia-Martin, M., Raymond, C.M. and Shaw, B.J. (2018) The potential for integrated landscape management to fulfil Europe's commitments to the Sustainable Development Goals. *Landscape and Urban Planning* 177, 75–82.

Matvejević, P. (1998) *Mediterraneo: Un nuovo breviario*. Garzanti, Milan, Italy.

Meeus, S.J. and Gulink, H. (2008) Semi-urban areas in landscape research: a review. *Living Reviews in Landscape Research* 2, 1–45.

MEA (Millennium Ecosystem Assessment) (2005) *Ecosystems and Human Well-being: Synthesis*. Island Press, Washington, DC.

Morettini, A. (1950) *Olivicoltura*. Ramo Editoriale degli Agricoltori, Rome.

Ortega, M., Pascual, S., Rossell, R.E. and Rescia, A.J. (2020) Land-use and spatial resilience changes in the Spanish olive socio-ecological landscape. *Applied Geography* 117, 102171. https://doi.org/10.1016/j.apgeog.2020.102171

Philips, A. (2013) *Designing Urban Agriculture: A Complete Guide to the Planning, Design, Construction, Maintenance, and Management of Edible Landscapes*. John Wiley & Sons Inc.

RaF ITALIA 2017-2018 (2019) *Rapporto sullo stato delle foreste e del settore forestale in Italia*. Compagnia delle Foreste S.r.l. Editore, Arezzo, Italy.

Rodrigo-Comino, J., Salvia, R., Quaranta, G., Cudlín, P., Salvati, L. and Gimenez-Morera, A. (2021) Climate aridity and the geographical shift of olive trees in a Mediterranean northern region. *Climate* 9, 64.

Russo, G., Beritognolo, I., Bufacchi, M., Stanzione, V., Pisanelli, A. *et al.* (2020) Advances in biocultural geography of olive tree (*Olea europaea* L.) landscapes by merging biological and historical assays. *Scientific Reports* 10, 7673. https://doi.org/10.1038/s41598-020-64063-8

Salvati, L., Sateriano, A., Colantoni, A., Di Bartolomei, R., Perini, L. and Zitti, M. (2013) The northern shift by the geographical distribution of the olive tree – a bioclimatic indicator? *International Journal of Ecology & Development* 24(1), 1–11.

Sciarretta, A., Calabrese, P. and Trematerra, P. (2021) Evaluating the presence of larvae of *Lobesia botrana* (Denis & Schiffermüller) developing on olive trees in some central-southern Italian regions. *Phytoparasitica* 49(4), 13–20.

Semeraro, T., Gatto, E., Buccolieri, R., Catanzaro, V., De Bellis, L. *et al.* (2021) How ecosystem services can strengthen the regeneration policies for monumental olive groves destroyed by Xylella fastidiosa bacterium in a peri-urban area. *Sustainability* 13, 8778.

Sereni, E. (1961) *Storia del paesaggio agrario italiano*. Laterza, Bari, Italy.

Sini, L., Di Bartolomei, R., Biasi, R. and Salvati, L. (2017) Urban sprawl and the 'olive landscape' in a crisis city. In: Clemente, M., Biasi, R. and Salvati, L. (eds) *Crisis Landscapes: Opportunities and Weaknesses for a Sustainable Development*. Franco Angeli Editore, Milan, Italy, pp. 41–49.

Steffen, W., Richardson, K., Rockström, J., Cornell, S.E., Fetzer, I. *et al.* (2015) Planetary boundaries: guiding human development on a changing planet. *Science* 347.

Tarifa, R., Núñez, C.M., Valera, F., González-Varo, J.P., Salido, T. and Rey, P.J. (2021) Agricultural intensification erodes taxonomic and functional diversity in Mediterranean olive groves by filtering out rare species. *Journal of Applied Ecology* 58, 2267–2276. DOI: 10.1111/1365-2664.13970

Tassinari, P., Torreggiani, D. and Benni, S. (2010) Dealing with agriculture, environment and landscape in spatial planning: a discussion about the Italian case study. *Land Use Policy* 30, 739–747.

Viljoen, A., Bohn, K. and Howe, J. (2005) *Continuous Productive Urban Landscapes: Designing Urban Agriculture for Sustainable Cities*. Elsevier, Architectural Press.

Weissteiner, C., Feced, C.G. and Paracchini, M.L. (2016) A new view on EU agricultural landscapes: quantifying patchiness to assess farmland heterogeneity. *Ecological Indicators* 61, 317–327.

Index

Page numbers in *italics* refer to Figures.

Printed and bound by CPI Group (UK) Ltd, Croydon, CR0 4YY
07/07/2025
14700214-0001